The Milky Way

The Harvard Books
on Astronomy

Atoms, Stars, and Nebulae
Lawrence H. Aller

Earth, Moon, and Planets
Fred L. Whipple

Galaxies
Harlow Shapley / Revised by Paul W. Hodge

Our Sun
Donald H. Menzel

The Great Nebula near Eta Carinae. A photograph in red (H-alpha) light made with the 40-inch Boller and Chivens reflector of Siding Spring Observatory in Australia. North is toward the bottom. The estimated distance of the nebula is 2,700 parsecs. (Courtesy of Australian National University.)

The Milky Way

Bart J. Bok
and Priscilla F. Bok

Fourth Edition
Revised and Enlarged

Harvard University Press
Cambridge, Massachusetts, and London, England

Contents

Fourth edition
Third printing, 1976
Library of Congress Catalog Card Number 73-83418
ISBN 0-674-57501-6 (cloth)
ISBN 0-674-57502-4 (paper)
Printed in the United States of America

The Milky Way

1 Presenting the Milky Way

There is a way on high, conspicuous in the clear heavens, called the Milky Way, brilliant with its own brightness. By it the gods go to the dwelling of the great Thunderer and his royal abode. Right and left of it the halls of the illustrious gods are thronged through open doors; the humbler deities dwell further away, but here the famous and mighty inhabitants of heaven have their homes. This is the region which I might make bold to call the Palatine of the Great Sky.

Ovid, *Metamorphoses*, Book 1, lines 168–176.

In this book we invite you to join us on a brief tour along the road to the heaven of the Greeks. Modern science is providing the transportation facilities and, without its being necessary for you to leave your comfortable chair, we would like to show you the sights. Our plan is briefly as follows: We shall start off with a quiet evening at home, during which we shall get out maps and photographs of the territory that we are about to explore. We shall introduce you to some of the intricacies of our celestial vehicles and then we shall get under way. First we shall pay some casual visits to the Sun's nearest neighbors, but soon we shall move on to sound the real depths of our universe. We shall visit big stars and little stars and clusters of stars within the larger Milky Way system. Between the stars we shall encounter clouds of cosmic dust and gas, some of the dust clouds so dense that they hide from our view the sights beyond. We shall, of course, linger a while on our visits to the palaces of the illustrious gods on the main road, but we shall also ask you to join us on side excursions to the places of the common people away from the well-traveled main highways.

In spite of our desire to show you all of the Milky Way system, we shall have to limit our celestial itinerary. Not infrequently along the road we shall see markers such as "Unexplored Territory," "Caution, Heavy Fog," or more encouraging signs, "Men at Work; Pass at Your Own Risk." For the Milky Way is by no means sufficiently well explored to render all of it open to celestial tourists. If you are so inclined, you may stop here and there along the road, get out your celestial Geiger counters, and do a little prospecting on your own. We hope that upon your return you will not regret having taken time for the long trip.

So, let us look at our maps and photographs and lay out the plan for the journey through the Milky Way.

The Milky Way

In most of the United States and Europe the best general view of the Milky Way can be had in the late summer on a moonless night an hour or so after sunset. The Northern Cross of the constellation Cygnus is then directly overhead, Arcturus is on its way down in the west, and in the northeast the W-shaped constellation of Cassiopeia is rising into view. If you are far from the glare of city lights and neon signs, you will have no difficulty in locating the shimmering band of the Milky Way, which can be traced through Cassiopeia and Cepheus to Cygnus and then down toward the horizon through the constellations of Aquila, Sagittarius, and Scorpius.

The Milky Way from Cassiopeia to Cygnus has the appearance of a single silvery band of varying width; but between Cygnus and Sagittarius we can distinguish two bands separated by a dark space called the Great Rift (Fig. 1).

1. The Milky Way in Cassiopeia, Cepheus, and Cygnus, a composite photograph prepared from the Palomar Sky Survey. The Andromeda Nebula, Messier 31, is shown in the lower left and the bright star Vega is near the upper right. The North America Nebula and the bright star Deneb (to the right and above the nebula) are near the center of the photograph. (From the *Atlas* prepared by Hans Vehrenberg, Treugesell Verlag, Düsseldorf.)

3 Presenting the Milky Way

The western branch is quite bright in Cygnus and still discernible in Aquila, but it is lost in the wastes of Ophiuchus. The head of the Great Rift is often referred to as the Northern Coalsack. It is shown in Fig. 1 to the right of the North America Nebula. Visually, it is observed stretching to the south and west of Deneb, the brightest star in the constellation Cygnus.

There are some very conspicuous bright spots along the summer Milky Way. The star clouds of Cygnus are right overhead (Figs. 1 and 9). Though they put on a fine show, they lose out in comparison with the Cloud in Scutum—which Barnard called "the Gem of the Milky Way" (Fig. 2)—and with several bright clouds in Sagittarius. The Milky Way is still conspicuous in the constellation Cepheus, but even a cursory inspection will show that north of Cygnus it does not shine nearly so brightly as does the branch to the east of the Great Rift and south of Cygnus.

What lies beyond the horizon? Our summer night progresses. Sagittarius, Aquila, and Cygnus gradually set. As Cassiopeia rises toward the meridian other parts of the Milky Way come into view and we can follow the band through Perseus, Auriga, and Taurus;

2. The Milky Way in Aquila, Scutum, Ophiuchus, Sagittarius, and Scorpius, a composite photograph prepared from the Palomar Sky Survey. The bright star Altair is shown just below the Milky Way near the left-hand edge. The arc of Scorpius is near the right-hand edge. The Scutum Cloud is the bright star cloud to the left of the center and the Great Star Cloud of Sagittarius is to the right and slightly below the center. The connected groupings of obscuring clouds are spread in a widening band (the Great Rift) beginning as a rather narrow feature about Altair. Figure 2 is a direct continuation of Fig. 1. (From the *Atlas* prepared by Hans Vehrenberg, Treugesell Verlag, Düsseldorf.)

5 Presenting the Milky Way

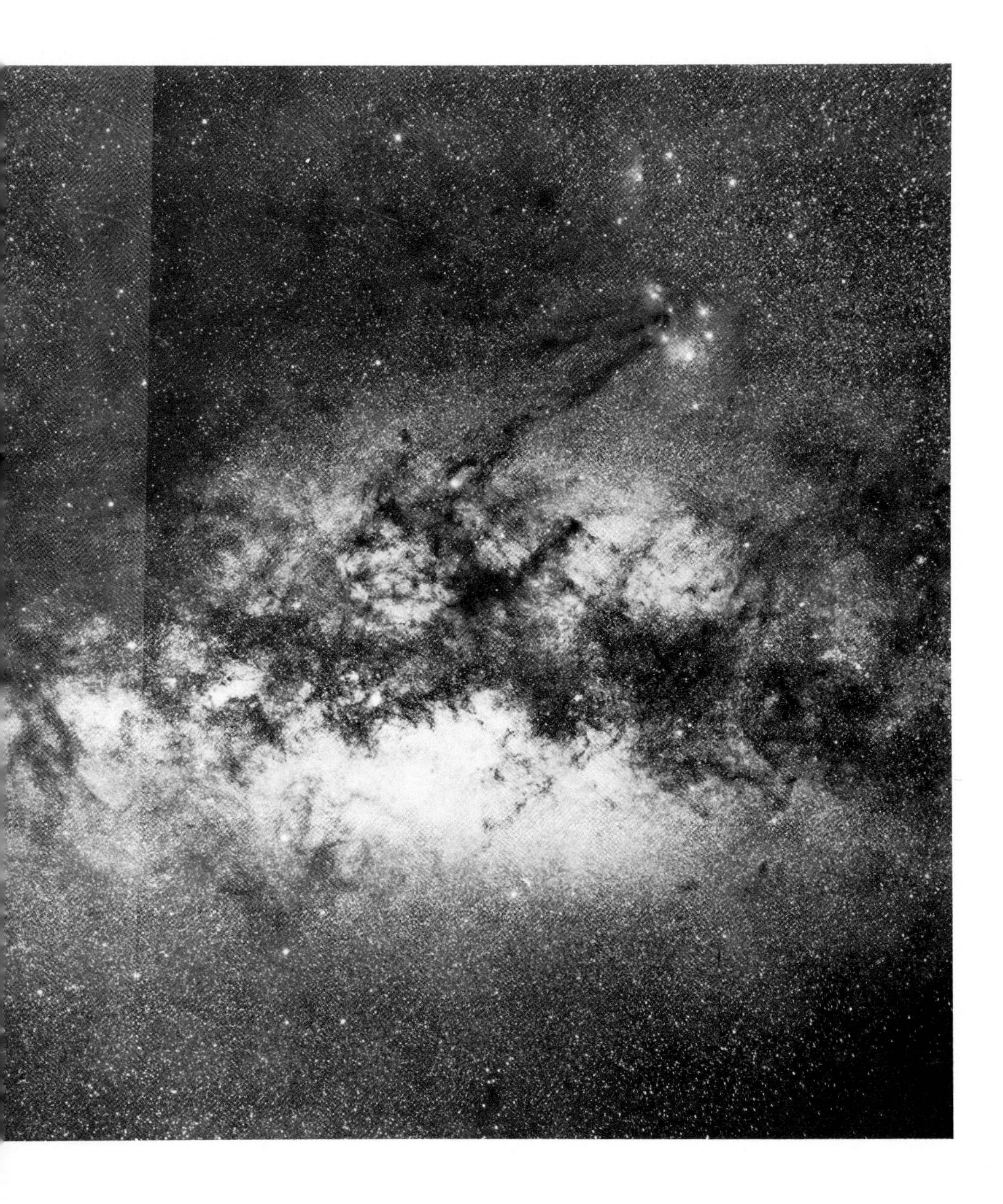

east of Taurus and Auriga it is lost in the summer dawn. But if we wait until early fall we can follow it southward through Gemini, Orion, Monoceros, and Canis Major. The Milky Way from Cygnus through Cassiopeia to Canis Major is, however, much weaker than the branches on either side of the Great Rift. In Auriga and Taurus it narrows down to a trickling little stream that is quite insignificant in comparison with the brighter sections of the summer Milky Way.

What happens to the Milky Way south of Canis Major? It is invisible from the latitudes of New York and Paris and we shall have to travel southward if we wish to see those parts. The whole Milky Way passes in review for a year-round observer in the southern tip of Florida, but for a good view we must go down to the equator or, preferably, farther south to Chile or Peru, to South Africa, or to Australia.

The section of the Milky Way from Sagittarius through Scorpius, Norma, Circinus, Centaurus, Crux (the Southern Cross), and Carina has great brilliance. In general appearance it resembles to some extent our summer Milky Way between Cygnus and Sagittarius. The star cloud in Norma is not unlike the Scutum Cloud, and the Carina Cloud appears rather similar to the Cygnus Cloud. The southern Milky Way does not show a Great Rift, such as we find from Cygnus to Sagittarius. However, it has a remarkable "dark constellation" in the connected configuration of dark nebulae stretching from the Southern Cross to Scorpius and Ophiuchus (Fig. 3). Australian aborigines referred to it as the Emu, and its ostrich-like appearance is well known to southern observers. It is best observed in the early hours of evening in July, when the Southern Cross is high in the sky and near the meridian. The Southern Coalsack represents the Emu's head, with a sharp

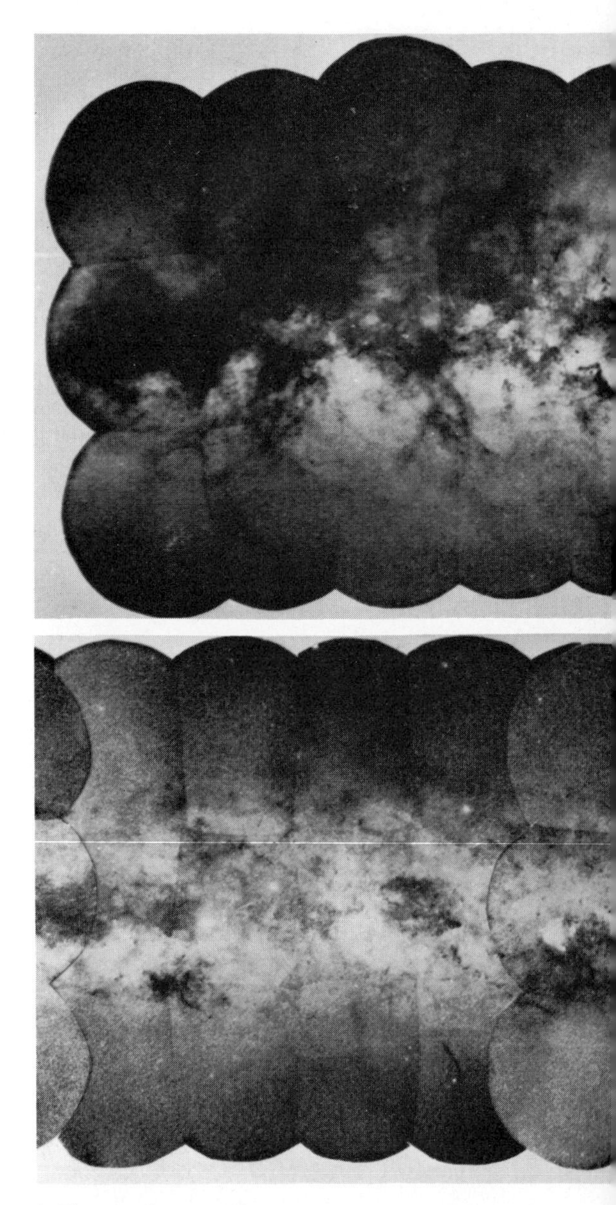

3. The southern Milky Way, a composite photograph prepared from the photographs in H-alpha (red) light reproduced in the Mount Stromlo Observatory *Atlas* by A. W. Rodgers, J. B. Whiteoak, *et al.* (Courtesy of Australian National University.)

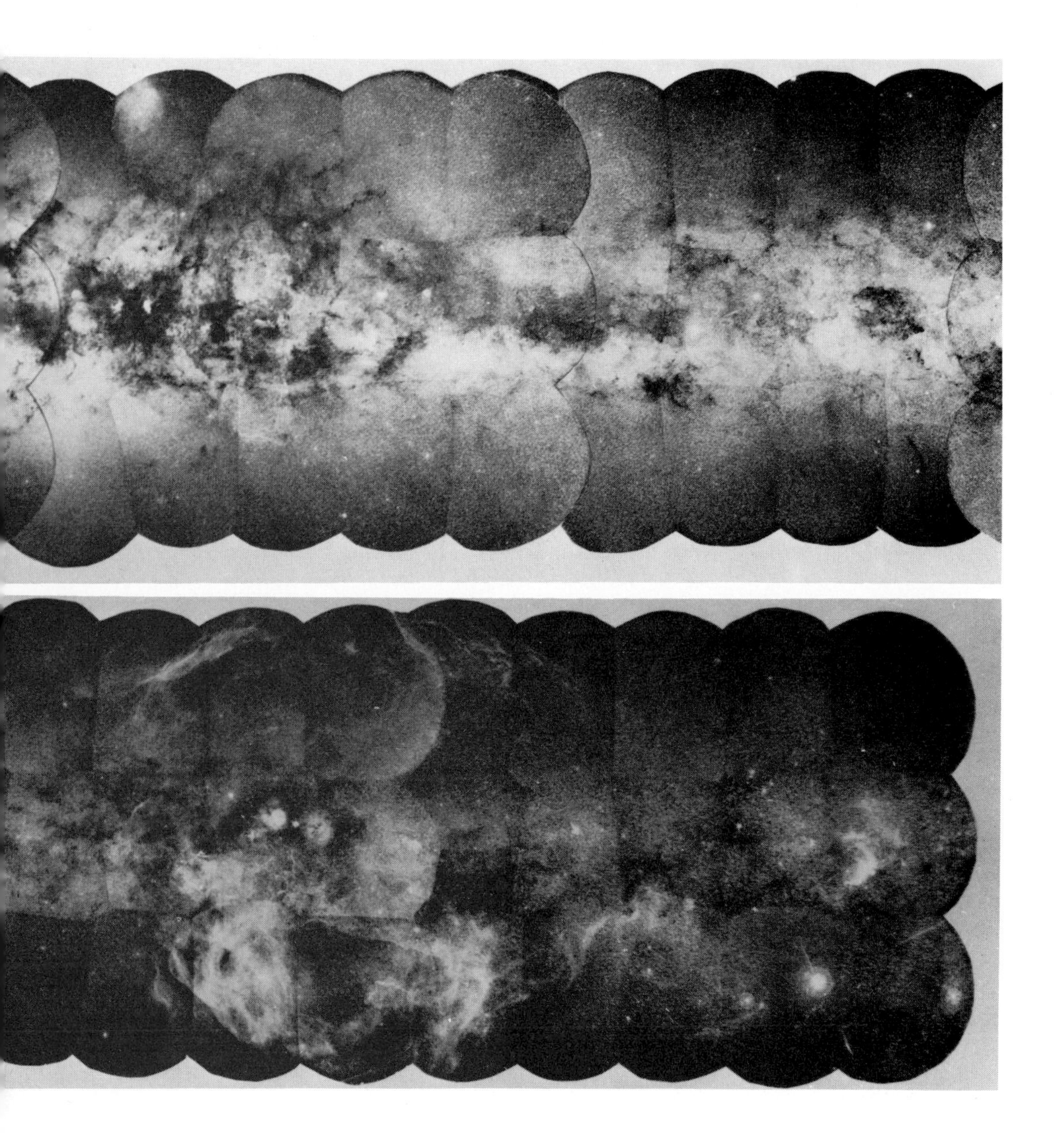

beak; the long thin neck is the narrow dark band through Centaurus to Norma, and the main body can be seen as the dark clouds in Scorpius and Sagittarius, with the thin dark legs shown by the dark lanes in Ophiuchus. The Australian aborigines can boast of having outlined the only recognized "dark constellation." It is bigger than any bright constellation, for our Emu stretches over 60° along the band of the Milky Way!

The remaining section of the southern Milky Way, which runs from Canis Major through Puppis and Vela to Carina, is in general not unlike the northern Milky Way in Cepheus and Cassiopeia. There are no marked irregularities, and a smooth band made up of thousands upon thousands of stars is clearly visible along the entire course. The whole of the band of the Milky Way forms very nearly a complete circle around the sky. We often refer to this great circle (that is, a circle that cuts the sky in half) as the *galactic circle* or *galactic equator.*

Ours are the days of large and powerful telescopes and you might well ask if there is much point to a careful naked-eye study of the appearance of the Milky Way. We earnestly believe that there is much to be learned from a survey without the use of a telescope or photographic camera. Our eyes happen to be the finest pair of wide-angle binoculars that has yet been made. A telescope is useful for the study of fine details for comparatively small sections of the sky, but no instrument is capable of revealing as well as the human eye the grand sweep of the entire Milky Way. On a good night we can directly intercompare portions of the Milky Way that are as far apart as Sagittarius and Cassiopeia. Such direct intercomparisons reveal one of the most important properties of the Milky Way, namely, that the width and the brightness of the band differ greatly from one section to another. The Milky Way attains its greatest width as well as its maximum brightness in Sagittarius. The half of the Milky Way from Cygnus through Sagittarius to Carina is generally very much brighter than the half that runs from Orion to Carina.

Telescopic Views

A good powerful pair of binoculars, or a small visual telescope, will reveal that the Milky Way is a composite effect produced by thousands upon thousands of faint stars. As we sweep across the sky with our telescope, the total number of stars in the field of view increases markedly as we approach the Milky Way. Almost two centuries ago, Sir William Herschel spent many years sweeping the sky —or, as he called it, "gauging the heavens"— with his giant reflectors. His son John later carried out the same plan for the southern sky. Their studies gave accurate data on the rates at which the numbers of stars increase toward the Milky Way. They showed that the rate of increase is very much larger for the fainter than for the brighter stars. If we use a 3-inch telescope and compare two fields, one in the Milky Way and the other in a direction at right angles to it, near one of the so-called galactic poles, then we count three to four times as many stars in the Milky Way field as near the pole. If we repeat the experiment with a 12-inch telescope the ratio is nearly ten to one.

Among the celestial objects that delight the amateur with a modest telescope of his own are the clusters of stars, the open clusters as well as the globular clusters (Figs. 11 and 12), and the beautiful nebulae. How are these prize celestial objects distributed relative to the Milky Way? Star clusters generally show a preference for the Milky Way. The open

4. The Milky Way in Sagittarius. (From a photograph taken by Ross with a 5-inch camera at the Lowell Observatory.)

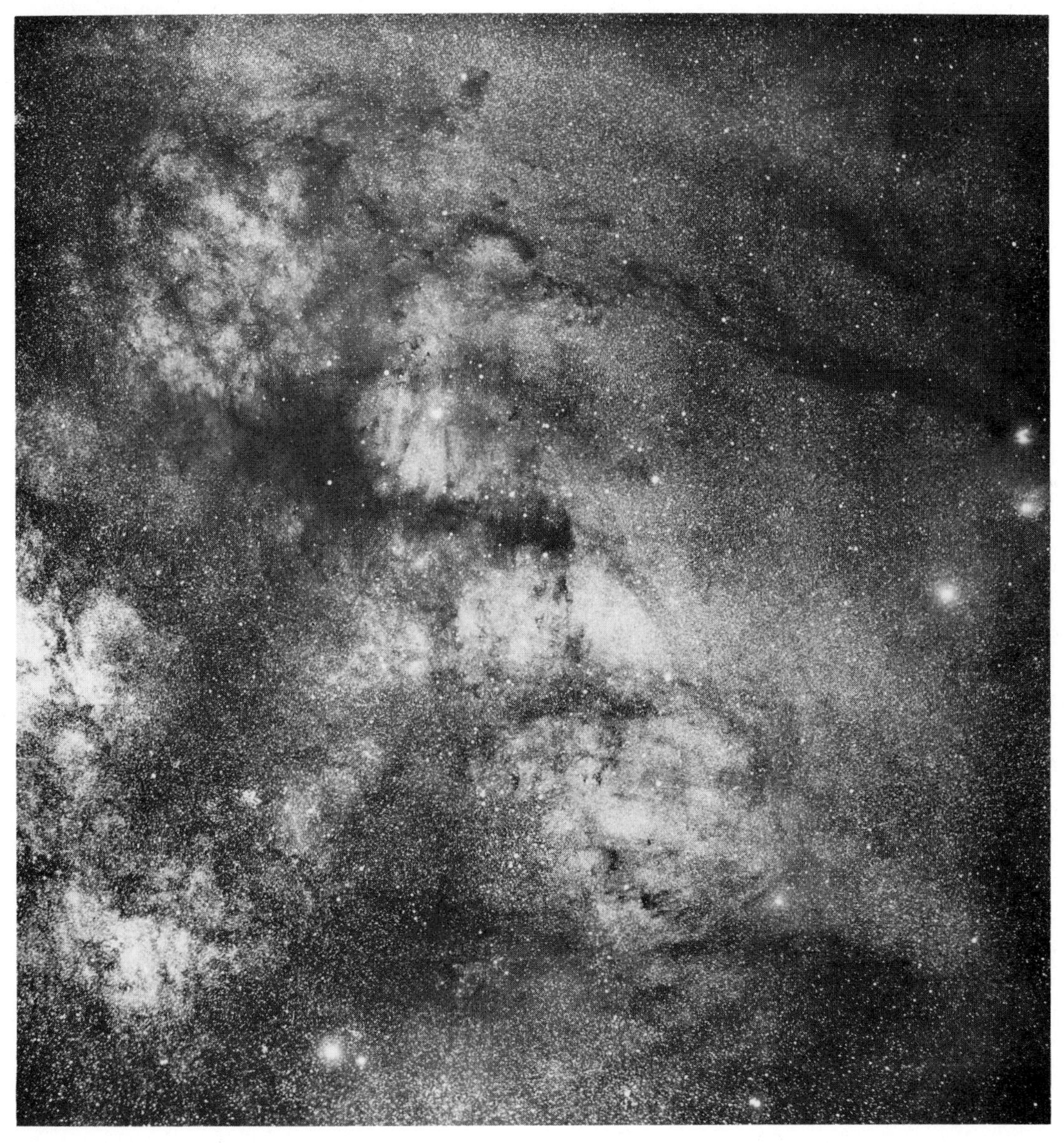

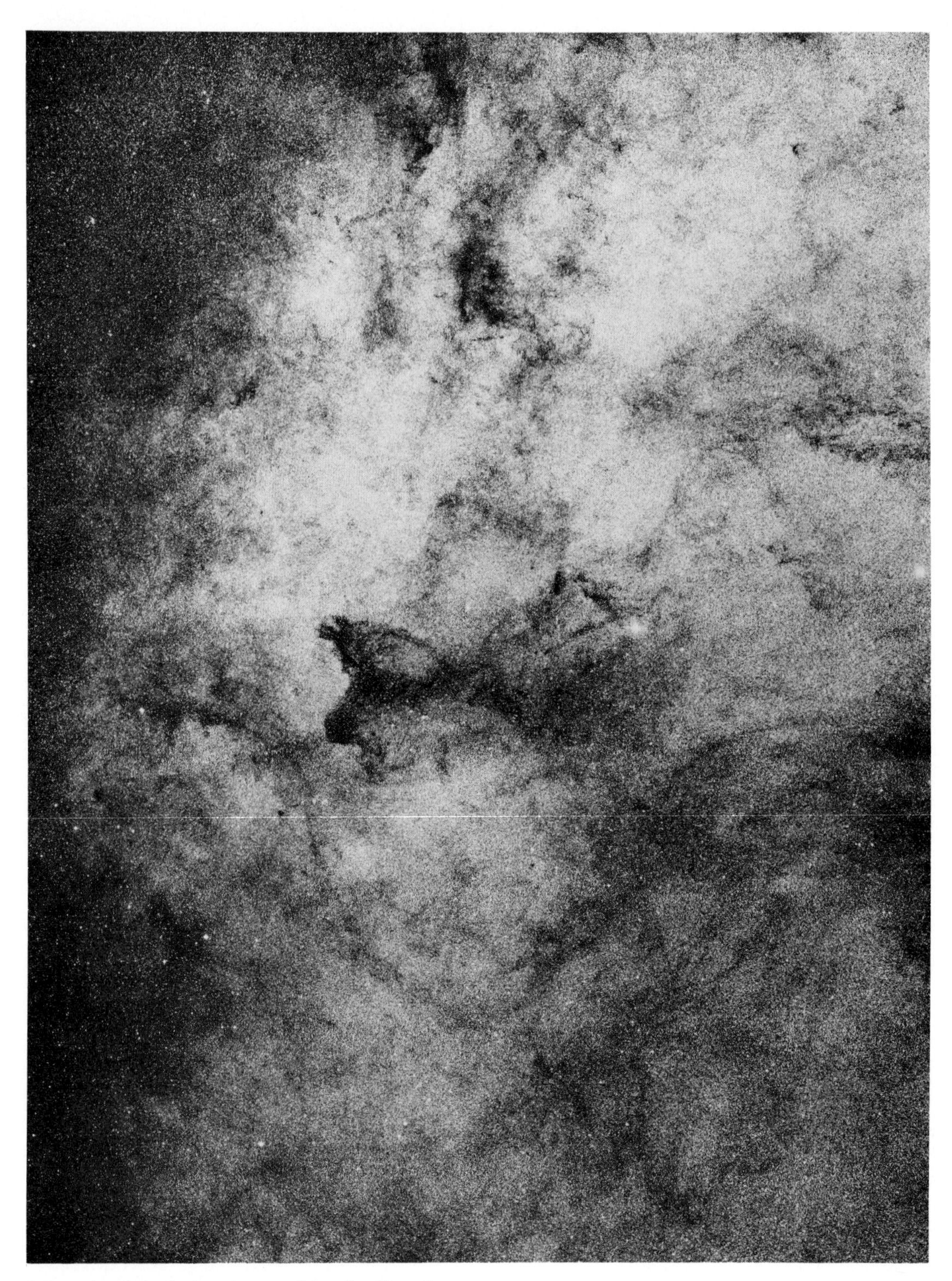

5. Star clouds in Sagittarius, marking the direction toward one edge of the nucleus of our Milky Way system. North is at the top; west is to the left. (48-inch Palomar Schmidt photograph.)

6. The Great Rift near Altair. The photograph by Ross shows the two star-rich branches of the Milky Way near the bright star Altair (lower left-hand corner) bordering the dark nebulae that mark a section of the Great Rift in the Milky Way.

7. The Great Star Cloud in Scutum, Barnard's "Gem of the Milky Way." (From a photograph by Barnard made at Mount Wilson Observatory.)

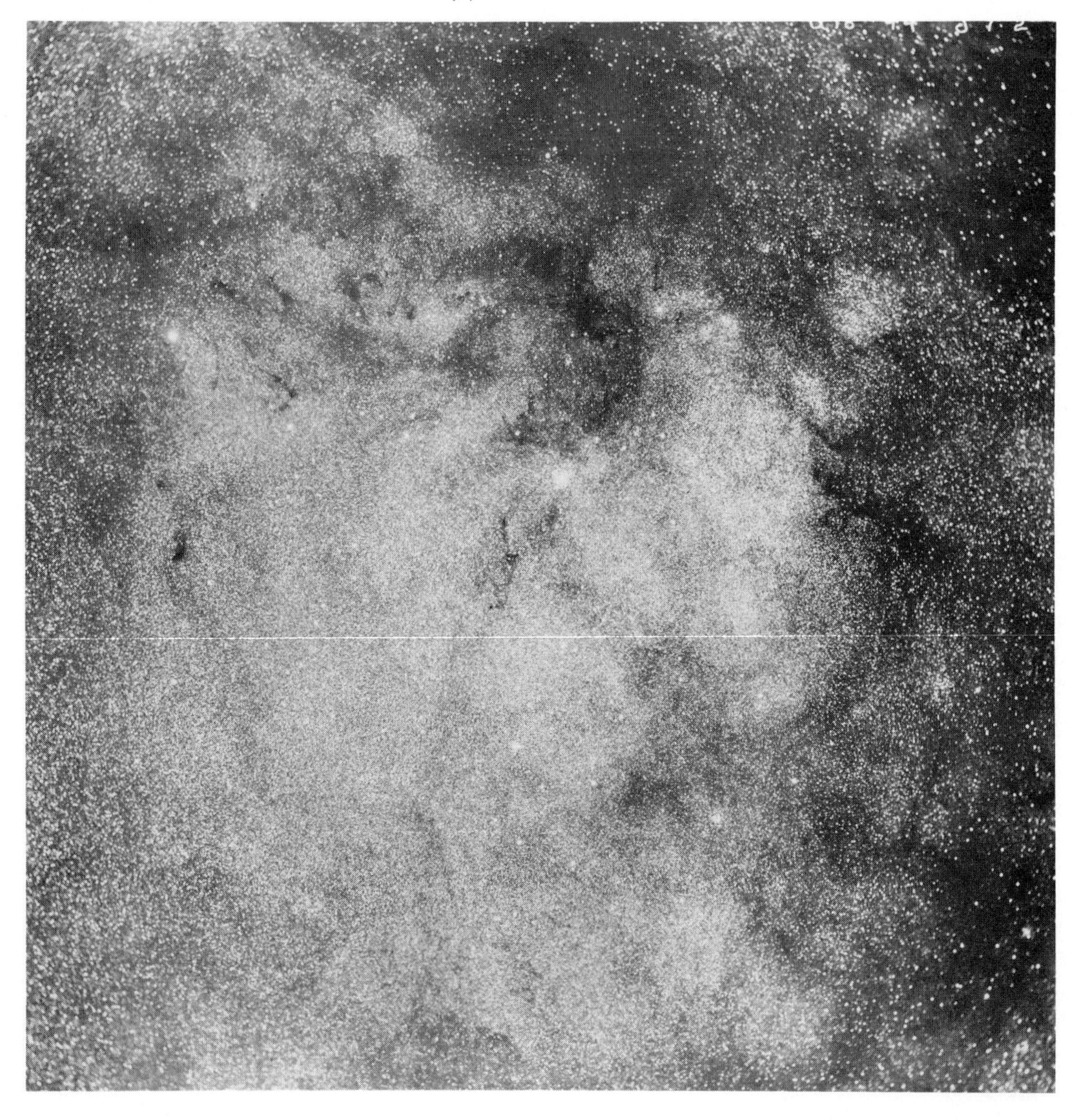

8. A region north of Theta Ophiuchi. A network of dark nebulosity overlies this rich star field. Note the remarkable snakelike formation in the lower part of the photograph and the small roundish dark nebulae near it. (From a photograph by Barnard.)

9. Bright and dark nebulosities near Gamma Cygni. (48-inch Palomar Schmidt photograph.)

10. Concentration of stars toward the Milky Way. The Milky Way in Cygnus, very rich in stars, and a small section of the Great Rift are shown in the lower left-hand corner. The photograph shows the manner in which the faint stars gradually thin out as we pass in the sky farther from the central band of the Milky Way. (From a photograph by Ross with a 5-inch camera.)

11. The galactic cluster Kappa Crucis, the "Jewel Box" of the southern Milky Way. (60-inch Rockefeller reflector, Boyden Observatory.)

clusters, of which the Pleiades, the Hyades, Praesepe, and the double cluster in Perseus are the prototypes, are mostly found close to the band of the Milky Way; faint open clusters lie almost without exception within a few degrees of the Milky Way circle. The globular clusters, such as the well-known one in Hercules, appear to prefer the regions between 5° and 20° from the Milky Way circle.

Later in this book we shall see that there are striking differences in the numbers of stars in open and in globular clusters. For the present, we shall say that the open clusters are those with somewhere between 20 and 2,000 members, whereas globular clusters may have anywhere from 10,000 to possibly 1,000,000 members. Most of the relatively nearby open clusters are readily resolved in a fairly small telescope, but it takes a large instrument to see individual stars in globular clusters. The name "open cluster" comes directly from the appearance of the cluster in the telescope.

The behavior of the nebulae may at first seem puzzling. Some of the finest nebulae, the Carina Nebula (Fig. 13) and the Orion Nebula (Fig. 70), for instance, are either in the band of the Milky Way or close to it; but there are others—especially those with spiral structure—that seem to avoid the Milky Way and occur mostly at some distance from the central band. The puzzle is easily explained. There are really two completely different varieties of objects called "nebulae"—the diffuse or gaseous nebulae, which are part and parcel of our Milky Way, and the spiral nebulae. The latter are not nebulae at all but are systems of stars in their own right, called *galaxies,* all of them far beyond the distances with which we shall concern ourselves in the pres-

12. The globular cluster Omega Centauri. (An enlargement from a photograph in red light made with the Armagh-Dunsink-Harvard telescope of the Boyden Observatory.)

ent volume. Here we limit ourselves to the study of the Milky Way system, our Home Galaxy, so to speak.

There is one further property of the globular clusters that will certainly be noted by a thorough observer with a visual telescope. He finds many globular clusters when our northern "summer" Milky Way is around, but he cannot observe many during the winter when Capella is high in the sky. This observer soon comes to the conclusion that the globular clusters in their own peculiar way favor one half of the Milky Way. They appear to be particularly fond of the region around Sagittarius, where one third of all known globular clusters are found in an area covering scarcely 2 percent of the entire sky. Open clusters and diffuse nebulae are spread more evenly along the band of the Milky Way.

Photographic Appearance

Purely visual inspection and measurement play at present a very minor role in Milky Way research. Photographic techniques came into general use about the turn of the century, and to these we have added the more recent techniques of photoelectric and radio research. When it comes to introducing beginners to the intricacies and beauties of the Milky Way, there is no substitute for photographs made with modern photographic telescopes.

Let us start with the map of the southern Milky Way that is shown in Fig. 3. It has been made by matching, cutting, and pasting together a series of black-and-white prints from photographic negatives made with an 8-inch, $f/1$, Schmidt camera at Mount Stromlo

13. The Carina Nebula. The emission nebulosity in the region of the Carina Nebula covers a large area of the sky, about 25 times the area subtended by the full Moon. The photograph was made with the 24/36-inch Curtis-Schmidt telescope of Cerro Tololo Inter-American Observatory. The frontispiece shows the section with densest nebulosity.

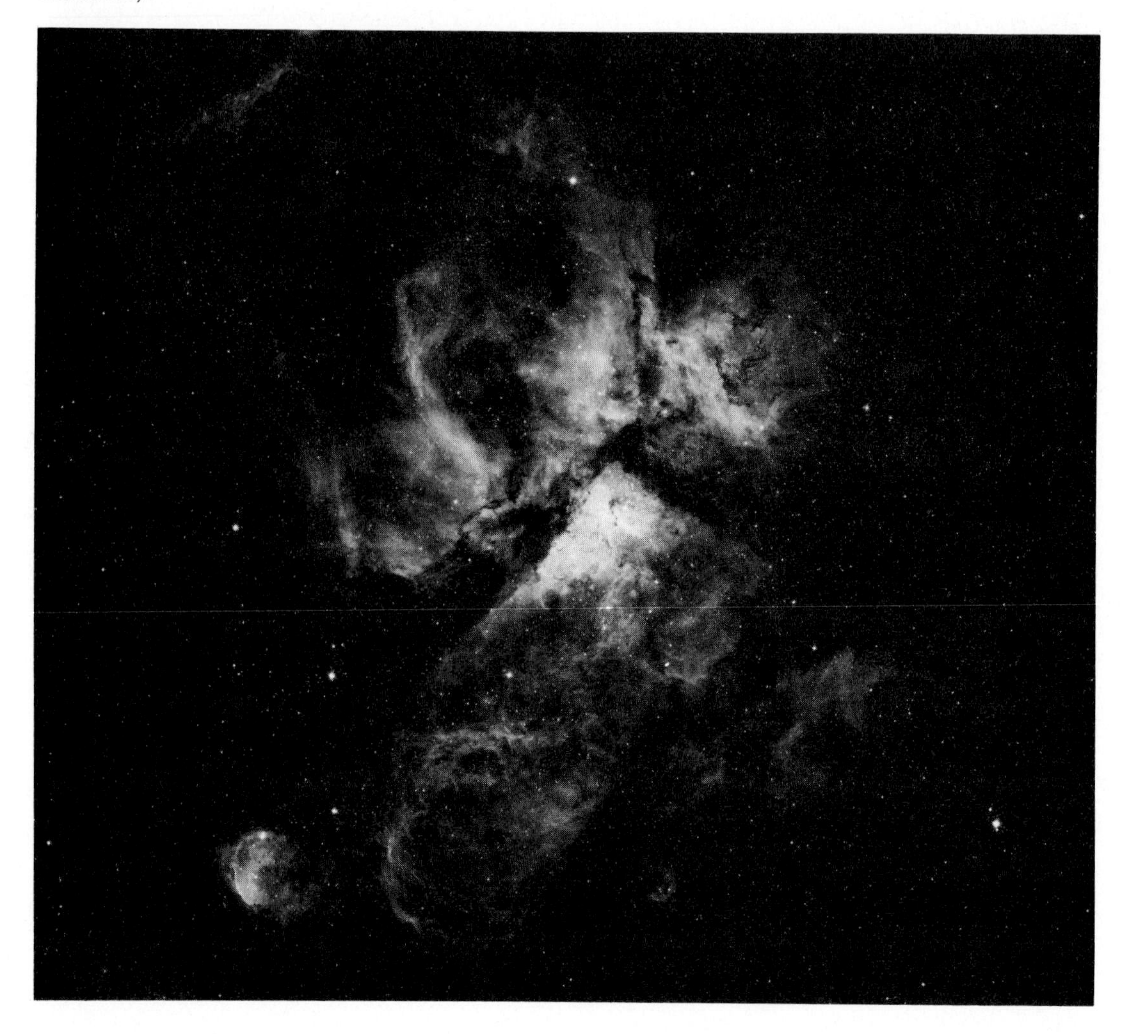

Observatory, Australia. The upper photograph covers the half circle from the star clouds in Scutum and Sagittarius, on the left, to the Coalsack and the Southern Cross. The lower photograph carries us past the Coalsack and the Carina Nebula to Canis Major and Monoceros on the right; the bright star in the lower right-hand corner is Sirius.

In spite of its small scale, the composite map shows clearly some of the important features to which we have already referred. In some places the resolution is not enough to show the individual stars, but in general we find here direct confirmation of the stellar nature of the Milky Way. The changing appearance that is noted as we proceed from one photograph to the next is not caused by atmospheric difficulties or differences in exposure time, but is mostly the result of true variations in brightness along the Milky Way. The Milky Way is very much more spectacular in Sagittarius than near Sirius.

The small-scale photographs used in the preparation of the composite map were made through a special red filter and on a red-sensitive photographic emulsion. This was done to show not only the stars of the southern Milky Way but also the gaseous nebulae that are present. These nebulae radiate profusely in the red light of the hydrogen line, H-alpha, and, by the use of the proper combination of color filter and photographic emulsion, they become very apparent. One of the finest large nebular shells is shown in Fig. 3 in the lower panel, to the left of the image of Sirius. It is named the Gum Nebula after the late Colin S. Gum, who, as a young astronomer, pioneered in filter photography of the southern Milky Way with wide-angle cameras. He was the first astronomer to draw attention to this magnificent nebula.

When we turn to detailed studies of the Milky Way, we need both a more open scale and greater penetrating power than that given by our small-aperture cameras of short focal ratio. Here is where the Schmidt-type telescopes with apertures of 24 inches or more and the large reflectors enter the picture. In the frontispiece and in Fig. 13 we show two aspects of the famous nebula near Eta Carinae, truly the most beautiful diffuse nebula of the Milky Way; because of its far southern position it is not visible from northern latitudes. Figure 13 shows the field of the nebula and its surroundings as photographed with the Curtis-Schmidt Telescope of the University of Michigan at the Cerro Tololo Inter-American Observatory in Chile. The frontispiece is an enlargement of a small portion of the nebula; here we are impressed with the grand sweep of the swirling gases and the intricate dark patterns overlying them.

The photographs in our book show principally the spectacular star-rich sections directly along the band of the Milky Way. The views are less striking when we consider photographs of fields at some distance from the galactic circle. By comparing photographs of different star fields made with the same telescope and identical exposure times, we find that the average numbers of stars per unit area of the sky are greater for fields close to or in the band of the Milky Way than for fields at some distance from the galactic circle. The larger the telescope and the longer the exposure times, the more striking becomes this contrast between counted numbers of stars for fields close to and far from the galactic circle. This observation alone suggests that the visible phenomenon of the Milky Way has great depth and that our Sun is located near the central plane of a vast star system that is highly flattened.

Our Milky Way System: A Model

If this were to be a detective story we might wish to present first all available evidence and then hide the solution in some uncut pages toward the end of the book. Our story is not so simple. The evidence is so incomplete in spots that we are nowhere near the final solution of the Milky Way mystery. Under these circumstances we might as well give away our "secrets" at the start. We shall make reading easier by providing a model of the Milky Way system. The descriptive material of the preceding pages provides a basis for such a model.

Visual as well as photographic counts show that the faintest stars are relatively most concentrated toward the band of the Milky Way. Since, on the average, the fainter stars are the more distant ones, we have thus direct proof that our Milky Way outlines a flattened system of stars. The Milky Way has great depth. Some of the stars that contribute to the Milky Way phenomenon may be only a few hundred parsecs away, but others are at distances of several thousands of parsecs from our Sun (1 parsec is equivalent to 3.26 light-years, about 20 trillion miles or 30 trillion kilometers). Since the Milky Way appears as a great band encircling the sky, cutting it into nearly equal parts, the Sun must be located close to the central plane of the system. Is it located anywhere near the center of the system? For many centuries astronomers believed this to be so, but if you have read the preceding pages carefully you will have found some indications that the Sun is located far from the center of our Galaxy. One of the most striking visual features of the Milky Way is that the half centered upon the stars of Sagittarius is wider and more brilliant than the part in Orion, Taurus, and Auriga. This means that the center of the Galaxy probably lies in the direction of Sagittarius.

There is much evidence to show that the galactic center lies in the direction of Sagittarius. The globular clusters exhibit a very pronounced concentration toward the Sagittarius region. Harlow Shapley showed that the globular clusters are among the most distant observable galactic objects; their irregular distribution is strong evidence of the existence of a distant center in Sagittarius. The concentration toward the Sagittarius region is shared by many types of objects that can be observed at great distances from the Sun, such as new stars, or novae, distant variable stars, and planetary nebulae. There is the evidence from galactic rotation, which postulates the existence of a distant center in the general direction of the Sagittarius clouds, and, finally, the indication that the strongest radio radiations come from precisely the same part of the sky.

So far we have given no indication with regard to the approximate dimensions of our Milky Way system. The evidence on this point will be presented in due course in later chapters, but we might as well complete the description of our model now. There is fairly general agreement among astronomers that the center of our Milky Way system lies somewhere between 8,000 and 11,000 parsecs from the Sun, with 10,000 parsecs representing the best compromise value.

The model is shown diagramatically in Fig. 14. There is no such thing as a sharply defined outer boundary to the Milky Way system. The solid line in our diagram is drawn through points where the space density of the stars will hardly exceed a few percent of the value near the Sun. In later chapters we shall read occasionally about varieties of stars beyond these boundaries, stars that still belong quite definitely to our Milky Way system.

One of the main problems of Milky Way astronomy is to obtain information regarding

14. A schematic model of our Galaxy. The boundaries shown outline the parts of our Milky Way system inside which the majority of the stars are located. The over-all diameter of the Galaxy in the central plane is of the order of 30,000 parsecs. Our Sun occupies a position close to the central plane at a distance of 10,000 parsecs from the center.

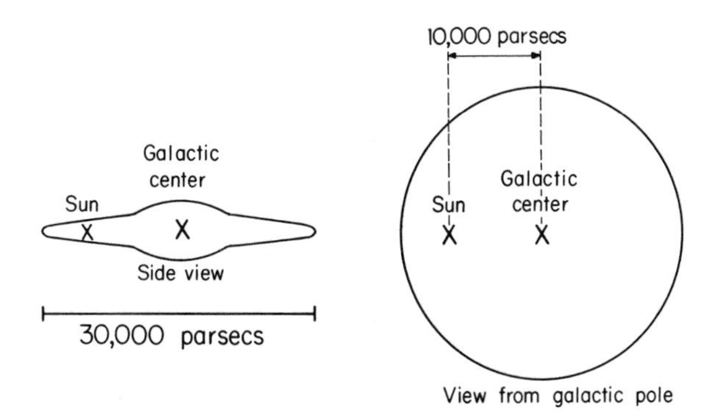

the arrangement of stars, gas, and dust near the central plane. Astronomers 50 years ago guessed that our system is probably a spiral system, and this is now well established. We shall find that, superposed on a fairly smooth conglomerate distribution of rather unspectacular stars, there exists a system of spiral arms. The spiral features can be traced rather well through the plotting of the positions of the blue-white supergiant stars and of the regions of greatest density of the interstellar gas and dust.

What about the motions in the Milky Way system? The system is apparently revolving at a terrific pace around the distant center. This is hardly surprising, for the system could not possibly stay as flat as it appears to be without a rapid rotation in the general plane of symmetry. The rate of rotation is fast; our Sun is whirled around at a speed of approximately 250 kilometers per second. That ought to be just about fast enough to suit our readers and should provide enough momentum to propel them through the basic fact-finding chapters that follow, onward into the realms of fancy.

Terms and Concepts

Each science develops its own special language, which is useful and necessary for communication but which often seems baffling to the uninitiated. To clear the atmosphere, we shall first write briefly in this section about some of the terms and concepts that all prospective students of the Milky Way should know before they delve into the field.

Milky Way and Galaxy. The physical phenomenon that we see in the sky is generally referred to either as the *Band of the Milky Way* or simply as the *Milky Way.* It stretches around the sky as very nearly a great circle

15. The great spiral in Andromeda. The spiral galaxy in Andromeda bears in all likelihood considerable resemblance to our own Milky Way system. Two companion ellipsoidal galaxies are also shown in the photograph. (48-inch Palomar Schmidt photograph.)

and it marks the central plane of our Milky Way system, shown diagrammatically in Fig. 14. Our Milky Way system, with its some 100 billion stars and with its content of interstellar gas and dust, is the stellar system to which our Sun—just another star—belongs.

There are millions of other systems not unlike ours in the observable universe, and we call these *galaxies.* In other words, our own Milky Way is just one of many galaxies and we can think of it as our Home Galaxy. Many galaxies, including our own, show spiral structure and these are spoken of as *spiral galaxies,* in contrast to the more amorphous galaxies such as the *ellipsoidal* and the *irregular galaxies.* To all but the few visual observers with very large telescopes, the spiral galaxies look quite nebulous, and visually their spiral patterns hardly reveal themselves. For that reason spiral galaxies are often called *spiral nebulae* (a name that is really out of date and wrong), and the other varieties are similarily called ellipsoidal and irregular nebulae; owners of small telescopes like to show their friends the Andromeda Nebula, which is really a spiral galaxy (Fig. 15). The only true nebulae are the gaseous nebulae, like those in Orion and Carina, or the dust nebulae like the one associated with the Pleiades cluster, or the Southern Coalsack.

The first list of clusters, nebulae, and galaxies was prepared a century and a half ago by a French astronomer, Charles Messier, and most of the brighter objects are still usually known by their *Messier numbers.* For instance, the Andromeda Nebula, a spiral galaxy, is Messier 31 and the Orion Nebula is Messier 42. Astronomers often have occasion to refer to extensive lists of clusters, nebulae, and galaxies prepared by Dreyer, the *New General Catalogue* (NGC) and its successor, the *Index Catalogue* (IC); Messier 31 then becomes NGC 221. The Messier numbers suffice as a rule for our purposes.

Magellanic Clouds. Our Milky Way system is accompanied in space by two smaller systems, the *Large* and the *Small Magellanic Clouds* (Fig. 17). Another of the Harvard Books on Astronomy, Shapley's *Galaxies,* deals with these neighboring galaxies, which tag along with our Galaxy very much in the manner in which a pair of moons accompanies a planet.

Galactic Latitude and Longitude. In order to fix the positions of stars and other celestial objects in the sky, we imagine a celestial sphere of very great radius and concentric with the Earth (Fig. 18). By extending the Earth's axis of rotation skyward, we pierce the imaginary sphere at two points called the *north* and *south celestial poles.* The north celestial pole will be directly overhead for an observer at the North Pole on the Earth. Next, we define the great circle halfway between the two poles and call it the *celestial equator.* The *vernal equinox* is then defined as the point on the celestial equator where the Sun crosses it at the beginning of our northern spring. The position of any star or other celestial object is then given by its *right ascension* and *declination.* The declination measures how far in degrees on the celestial sphere the object is north or south of the celestial equator; the right ascension measures (along the celestial equator, in hours, minutes, and seconds—$1^h = 15°$) how far it is east of the vernal equinox. Since the positions of the celestial poles, the equator, and the vernal equinox are subject to slow progressive changes, the right ascension and declination of a given object will not remain precisely constant with time and we refer all right ascensions and declinations to a given *epoch,* say 1900, 1950, or 2000.

16. The spiral galaxy Messier 101 in Ursa Major. A photograph made with the 200-inch Hale reflector. (Hale Observatories photograph.)

17. The Large Magellanic Cloud. A companion galaxy to our own Milky Way system. (ADH photograph, Boyden Observatory.)

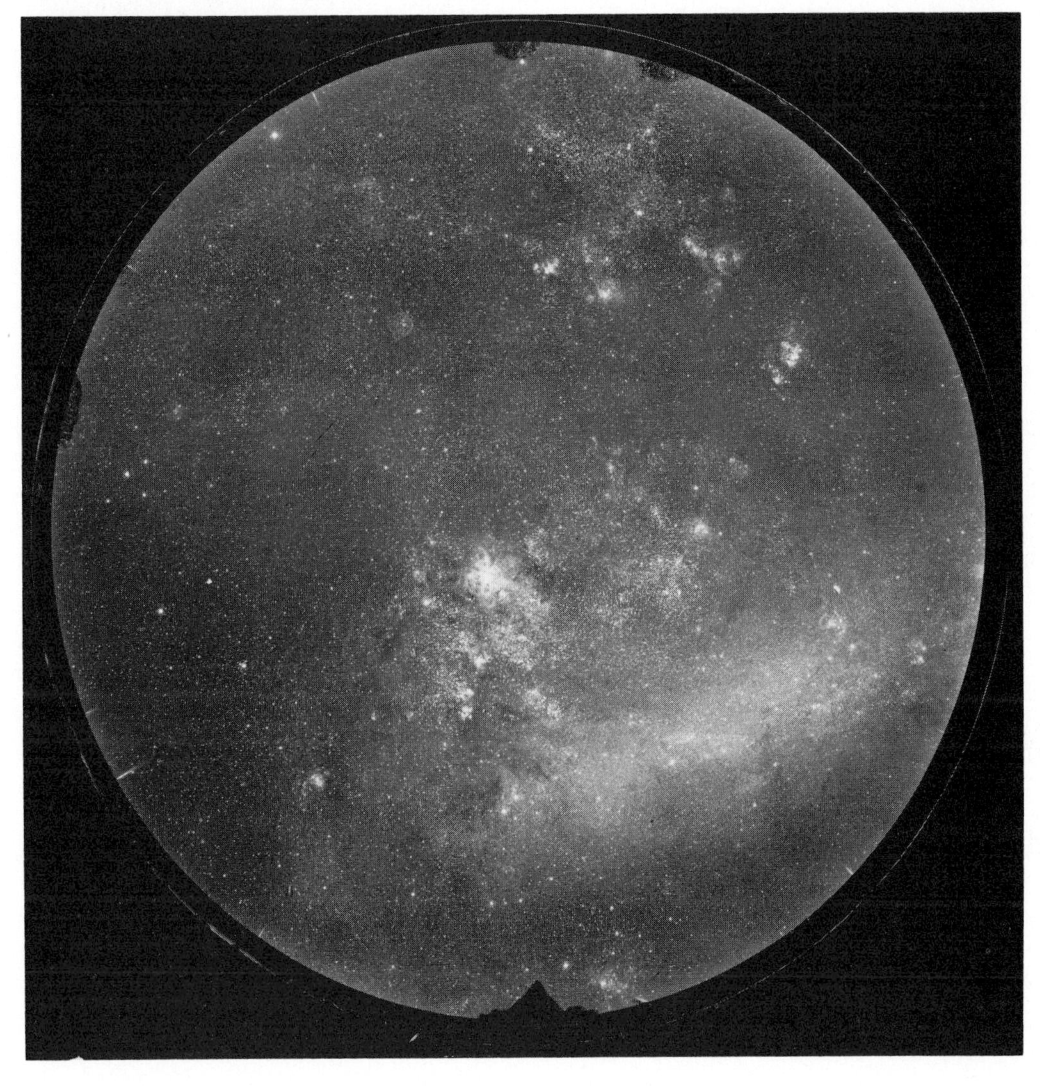

18. The celestial sphere. The diagram helps to visualize right ascension and declination. The approximate right ascension of the star shown is 75° (5 hours) and the declination is 40°N.

19. Galactic latitude and longitude. The star shown has approximately a galactic longitude $l = 100°$ and a galactic latitude $b = +20°$; its right ascension is about 20^h and its declination is approximately 70°N.

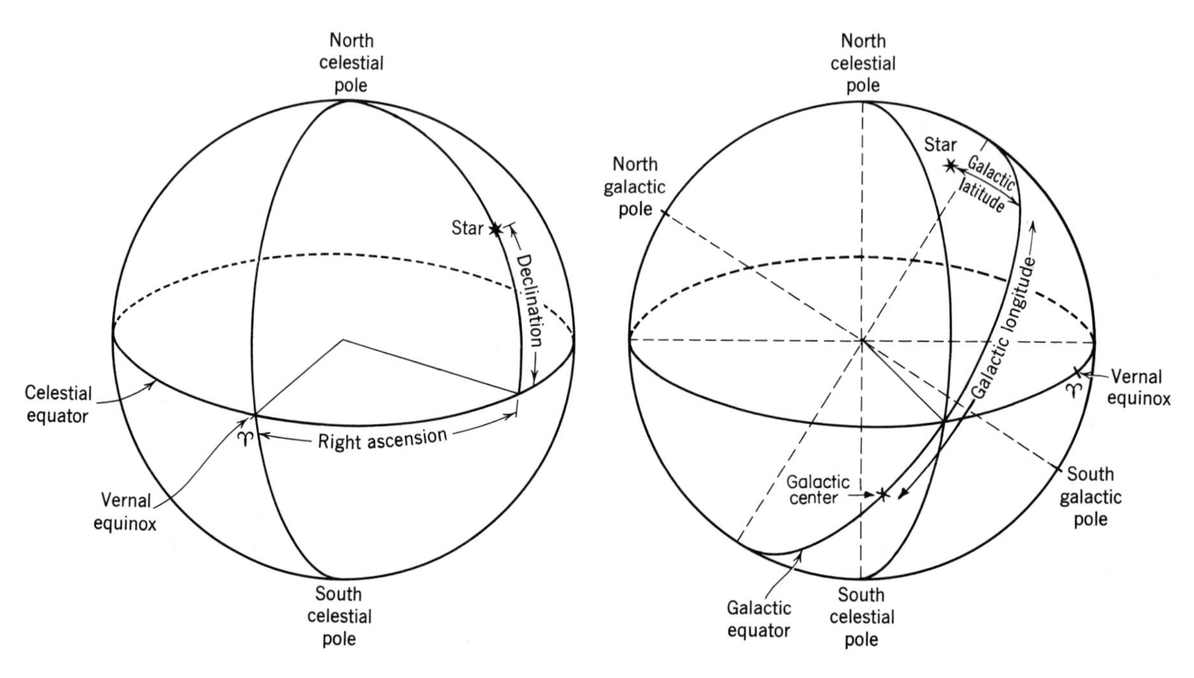

The equatorial system of defining positions in the sky is the basic one, but for studies of the Milky Way we often prefer the use of a special system, better adapted to our work. We noted that the band of the Milky Way follows roughly a great circle in the sky. For purposes of research we draw a great circle through the band of the Milky Way as it appears to us and this circle we call the *galactic equator* (Fig. 19). The points 90° away from the galactic equator on either side of it are called the *north* and *south galactic poles.* The position of any object in the sky can then be described by its *galactic latitude,* measured in degrees north or south from the galactic equator, and its *galactic longitude,* measured eastward along the galactic equator. The starting point, or zero, of galactic longitude has been established by international convention as the point on the galactic equator that marks the direction toward the galactic center as seen from the Sun. The galactic equator is inclined roughly 62° with respect to the celestial equator. The position of the galactic center (1950 epoch) is $17^h42^m37^s$ (right ascension), $-28°57'$ (declination). The north galactic pole is located (1950 epoch) at $12^h49^m02^s$ (right ascension), $+27°23'$ (declination).

20. The Steward Observatory station on Kitt Peak in Arizona. The dome of the 90-inch reflector is shown on the right, that of the 36-inch reflector on the left. The first optical pulsar was discovered with specialized electronic equipment attached to the 36-inch reflector. (University of Arizona photograph.)

Telescopes and Auxiliary Equipment

Telescopes of many varieties are the basic tools of the astronomer. For Milky Way research we need the best and most powerful telescopes available. It may be helpful to remind the reader that there is a first broad distinction between *refractors* (telescopes with lenses) and *reflectors* (telescopes with paraboloidal mirrors). For Milky Way photography, multiple-lens refractors have proved useful, especially those that permit photography with excellent definition of a large area of the sky on one single photograph; these photographic refractors are often called cameras, even though their aperture (the diameter of the primary lens) may be 15 to 20 inches. The largest telescopes in use now are the reflectors, in which the light is gathered by a carefully ground and polished paraboloidal mirror covered with a thin aluminum coating. These instruments are our most powerful light collectors, but they suffer from having relatively small areas of perfect focus.

The development of the *Schmidt-type telescope* (Fig. 24) proved a great boon to Milky

21. The Steward Observatory 90-inch reflector on Kitt Peak in Arizona. (University of Arizona photograph.)

22. The domes of the 150-inch reflector and of the 24/36-inch Curtis-Schmidt telescope of the University of Michigan at Cerro Tololo Inter-American Observatory in Chile. (Photograph by the authors.)

23. The 158-inch Mayall reflector of Kitt Peak National Observatory in Arizona. (KPNO photograph.)

24. The Armagh-Dunsink-Harvard (ADH) telescope at the Boyden Observatory, which is of Baker-Schmidt design. (Photograph by *The Friend*, Bloemfontein, South Africa.)

Way research. This telescope consists of a primary spherical mirror, with a correcting lens placed near the center of curvature of the mirror. This combination has excellent focus over a large area of the sky, together with the great light-gathering power of the large primary mirror. Modifications, such as the *Baker-Schmidt,* are extensions of the basic principle enunciated in 1929 by Schmidt of Hamburg. The most famous of all is the 48-inch Palomar Schmidt (Fig. 25), which has been used for the photography in blue and in red light of the entire sky within reach from Palomar Mountain in southern California. The *National Geographic–Palomar Sky Survey* contains 1,000 prints of the blue-sensitive photographs and 1,000 prints of the photographs in red light. Each print covers an area of the sky equal to $6° \times 6°$. No observatory library is considered complete without a set of the *Sky Survey* prints or, preferably, a set of prints on glass.

25. The 48-inch Schmidt telescope at Mount Palomar, which has completed the National Geographic–Palomar Survey.

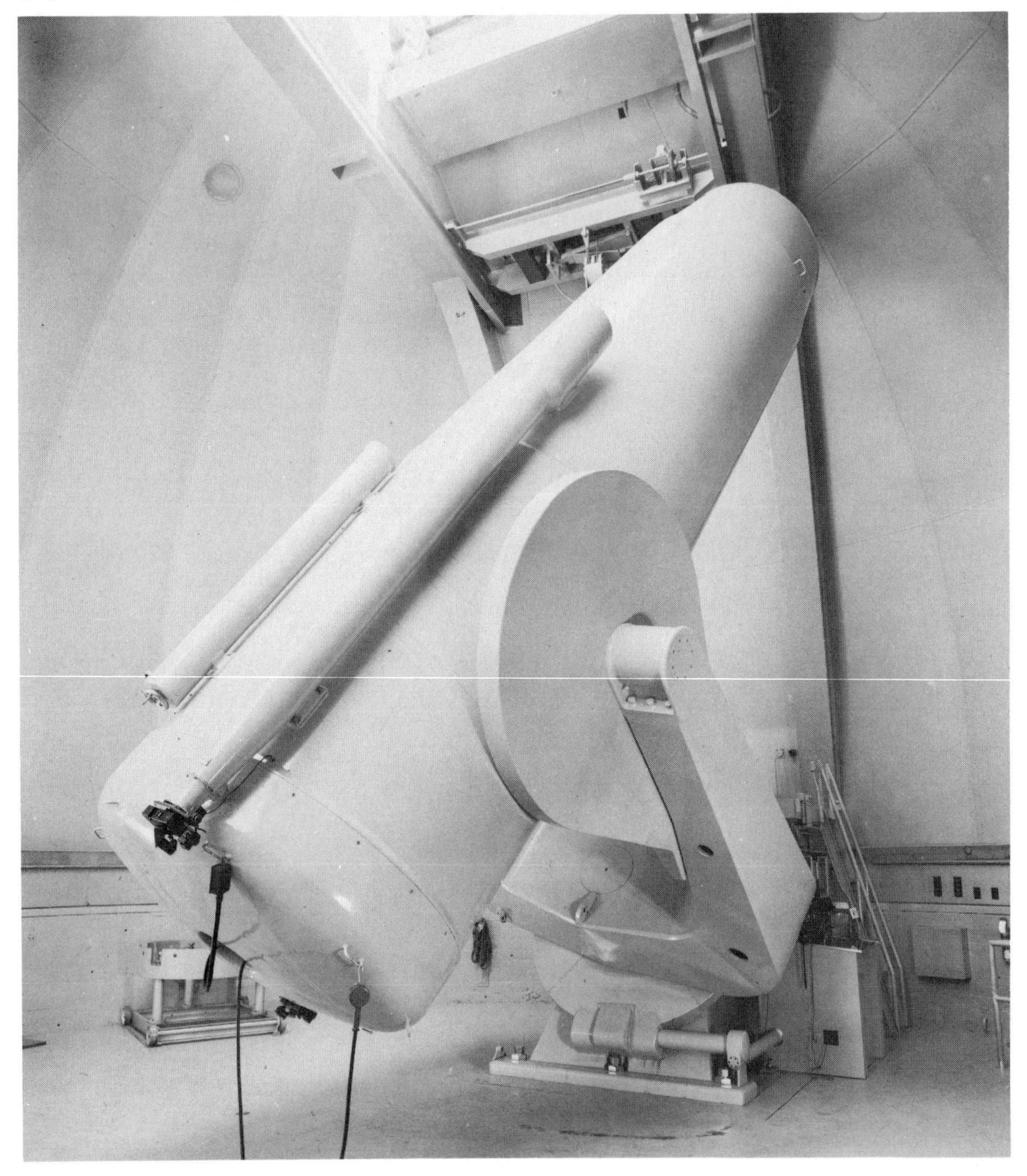

A variety of telescope that is very important to Milky Way research is the *radio telescope*. In its simplest form it is the paraboloidal reflector of optical astronomy scaled up to the dimension needed for radio studies of the Milky Way. The 100-meter steerable paraboloid radio telescope near Bonn in Germany and the precision 140-foot and the 300-foot radio telescopes at the National Radio Astronomy Observatory in Green Bank, West Virginia (Figs. 26 and 27) are among the largest Northern Hemisphere instruments used for galactic research. The 210-foot precision radio telescope at Parkes in New South Wales, Australia, has long dominated research from southern latitudes.

We shall describe in the chapters that follow several of the more useful auxiliary tools of the Milky Way astronomer, but at this early stage a few deserve to be mentioned briefly. Success in direct Milky Way photography depends wholly on the quality (speed, graininess, color sensitivity) of the *photographic emulsions* provided by the manufacturer and on the characteristics of the special *color filters* employed for each particular project. For studies involving spectra of single stars, or of star fields, we require auxiliary *spectrographic equipment*. The basic instrument is either a relatively small spectrograph used at the Newtonian or Cassegrainian focus of a large reflector, or—for highest attainable spectral dispersion—a coudé spectrograph. Photographic recording of spectra is being supplemented by automatic photoelectric recording. Photography is being assisted by image-strengthening techniques, in which use is made of the image-conversion approach. For precision work on the measurement of brightnesses and colors of stars, the astronomer depends now mostly on the *photoelectric photometer*. Basic standards are established by photoelectric techniques and the photographic plate serves as a useful tool for extending the work to many stars, using the photoelectric standards as a basis. Photographic and especially photoelectric and spectrographic work is being extended beyond the traditional violet-blue-green-yellow-orange-red parts of the spectrum. Space research probes the Milky Way beyond the near ultraviolet into the regions of shorter wavelength, all the way to the x-ray region. In recent years great advances have been made in the infrared parts of the spectrum, so much so that the earlier gap between the near infrared and the radio region has practically been filled.

Instrumental advances on all fronts are assisting further developments in galactic research. Our modern large reflectors are now often fitted with primary mirrors that have a reflecting surface slightly different from the traditional paraboloidal shape. When used in conjunction with specially designed secondary mirrors, these Cassegrain combinations provide fields for photography with wide-open scales on the photographs. One photograph can cover with near-perfect definition areas of the sky with diameters of the order of 1° or more.

Milky Way research, like all astronomy—and for that matter all modern science—is profiting tremendously from the worldwide developments of automatization and data processing. Large-scale undertakings for the measurement of hundreds of thousands of stellar brightnesses, colors, or proper motions are no longer considered beyond the realm of possibility.

The Milky Way astronomer must pay careful attention to his basic tools and to the methods available for their proper and effective use. The quality of our research depends on the accuracy and proper use of our tools.

26. The radio telescopes of the National Radio Astronomy Observatory at Greenbank, West Virginia. The three telescopes on the right represent the interferometer array used for the detailed mapping of radio sources. On the left in the foreground is the 300-foot altazimuth radio telescope, steerable only in altitude, which has been used very effectively for 21-centimeter research (see Chapters 8 and 10). In the back is shown the 140-foot radio telescope (see Fig. 27). (Courtesy of the National Radio Astronomy Observatory.)

27. The 140-foot steerable radio telescope of the National Radio Astronomy in Greenbank, West Virginia. (Courtesy of the National Radio Astronomy Observatory.)

2
The Data of Observation

Our eyes are our first tools. The sky lies open before us, for astronomy is an observational rather than an experimental science. But our eyes are weak and forgetful and therefore we build telescopes that can gather more light, cameras that can store lasting impressions, prisms and spectrographs to disperse the light into spectra, and photometers that are more sensitive than our eyes and can measure stellar brightnesses and colors with precision.

Our eyes are sensitive only to the part of the spectrum from the blue-violet to the red. Figure 28 shows the total range of wavelengths that can be detected and studied with modern techniques for gathering and recording light. Visible light is one form of electromagnetic radiation that is observed. Each electromagnetic wave is characterized by its wavelength or by its frequency of vibration. The wavelengths of light in the visual, ultraviolet, and infrared rays are measured in *angstrom units,* 1 angstrom unit being equal to 1 one-hundred-millionth of a centimeter (10^{-8} cm). But for radio radiations it may be more convenient to measure wavelengths instead in millimeters, in centimeters, or even in kilometers! Astronomers working in the radio or infrared parts of the spectrum often prefer to use the frequency of radiation instead of the wavelength. Since the wavelength λ (lambda) times the frequency ν (nu) equals the velocity of light c, we have:

$$\nu = \frac{c}{\lambda}.$$

A frequency of 1 cycle per second is called 1 *hertz/Hz,* so named in honor of the great German physicist Heinrich Hertz. Radio astronomers measure their frequencies mostly in *megahertz;* 1 megahertz (MHz) = 1 million Hz.

Figure 28 shows the ranges in wavelength and in frequency for the varieties of radiation that can now be observed. Beyond the violet or short-wave length end of the blue rays are

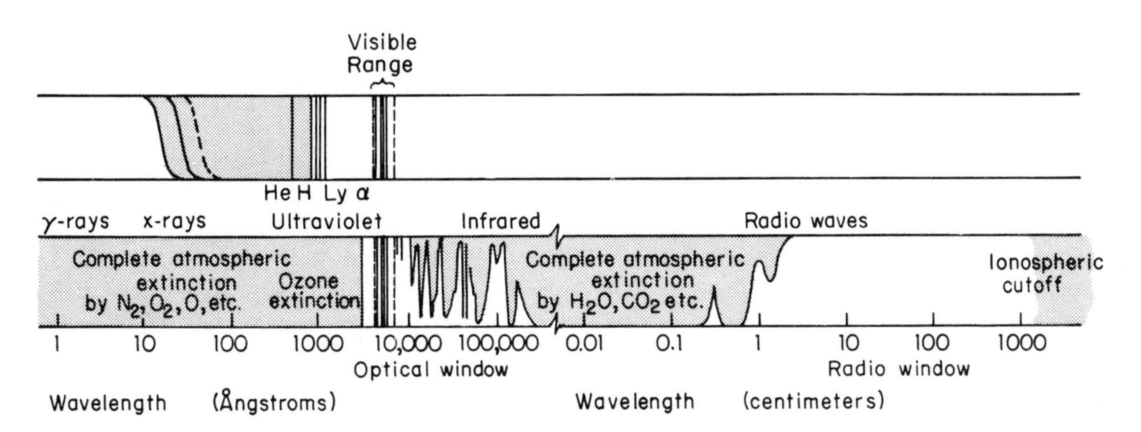

28. The spectrum of electromagnetic waves from gamma rays to long radio waves. In the lower strip, the optical and radio "windows" are indicated by white areas and the regions of atmospheric extinction by shaded areas. The upper strip shows the range of radiations accessible to detectors flown in a rocket or satellite above the Earth's atmosphere. Absorption by hydrogen, and to some extent by helium, cuts off the light of distant stars in the ultraviolet far beyond the cutoff of the Earth's atmosphere. Note, however, that gamma rays, most x-rays, and other regions of the spectrum can be observed without much interference. The narrow range of wavelengths to which the eye is sensitive is indicated. (From Lawrence H. Aller, *Atoms, Stars, and Nebulae,* revised edition, 1971.)

the invisible ultraviolet rays, the x-rays, and the gamma rays; beyond the red are the near-infrared heat rays and beyond these the far infrared and the radio microwaves. The radiation outside the visual range can be studied with the aid of the photographic plate and with special receiving devices, such as the photoelectric cell for the blue-violet. We use special infrared-sensitive cells, bolometers and thermocouples, for the infrared, and radio telescopes for the radio range. Earth satellites are helpful in carrying our instruments beyond the Earth's atmosphere. They have especially enlarged our knowledge of the far-ultraviolet radiation and of x-rays reaching us from the stars and nebulae of our Galaxy.

Stellar Brightnesses

One of the first questions that we shall want to answer is, How many stars are there? But, even if we could answer this simple question, we would immediately want to know how many there are in each successive class of brightness. At once we have met with one of the most difficult problems of the modern astronomer: How are we to measure accurately the amount of light that reaches us from individual stars?

Since the human eye was the first instrument used for observing the stars, we inquire first into the simple rules that govern visual estimates of brightness. Our appreciation of the difference in brightness of two lights is always a relative rather than an absolute matter. Two faint stars may seem to differ appreciably in brightness, but two bright stars that differ by the same absolute amount would appear almost identical, since in the latter case the difference would be a negligible fraction of the total stimulus that we receive. In other words, our eyes estimate ratios of brightness rather than absolute differences.

Hipparchus was the first to classify the naked-eye stars according to brightness. Later it was agreed that his six "magnitude" classes should be so taken that a standard first-magnitude star (the average of the 20 brightest stars in the sky) gives us 100 times as much light as a star of the sixth magnitude, which is about the limit of vision for the human eye. Since we are interested in *ratios* of brightness, we define a difference of 1 magnitude as corresponding to a ratio equal to the fifth root of 100, which is 2.512. A difference of 2 magnitudes corresponds to $2.512^2 = 6.31$, 3 magnitudes to $2.512^3 = 15.85$, 4 magnitudes to $2.512^4 = 39.82$, and 5 magnitudes to $2.512^5 = 100$.

We need perhaps to stop for a moment and consider how powerful a geometric factor of this kind can be. As viewed from the Earth, a sixth-magnitude star emits 1/100 the light of a first-magnitude star; an eleventh-magnitude star emits only 1/10,000 the light of a first-magnitude star. With the 200-inch telescope, stars of the twenty-third magnitude are recorded and measured. Such a star has $1/2.512^{22}$ or 1/630,000,000 of the light of a first-magnitude star. Please note that the faintest stars have the greatest numerical magnitudes! We start counting with magnitude zero, stars like Vega or Capella; a star like Sirius, which is brighter than Vega, is then given a negative magnitude, in this case -1.5.

Photoelectric Photometry

The instruments used for measuring with precision the brightnesses and colors of stars are called *photometers.* Among the photometers designed to surpass the eye in sensitivity and accuracy, the photoelectric photometer reigns supreme. Its accuracy is unsurpassed and it is an instrument that is relatively simple to use. The light from the star strikes a sensitive photoelectric tube, or photocell, and releases electrons, which constitute a very small current. This electric current is directly proportional to the intensity of the light falling on the photocell. The electrons released by the impinging starlight are accelerated by an electric field in the photocell and strike a second sensitive surface, where each original electron may release two or more additional electrons. This process is called *photomultiplication* and the electronic tubes that incorporate this feature are called *photomultiplier tubes;* the photo tube in standard use is referred to as the IP 21 photomultiplier cell. In most of the modern photomultiplier tubes used for astronomical work there are as many as nine successive stages of photomultiplication built into the tube so that amplication by factors of the order of 1,000,000 is achieved. The current produced by the photo tube is passed through an amplifier, and in the older instruments it is then recorded by the deflection of a pen moving across a roll of paper. Figure 29 shows schematically the type of record produced by a photoelectric photometer. In modern equipment the chart recording is dispensed with and the amplified photocurrents are numerically recorded in a form suitable for subsequent computer reduction.

Thermally excited currents are present in all photomultiplier tubes, even though no light is falling on the photosensitive surface. As a measure of the photocurrent produced by the light of a star we take the difference between the reading on our record when the star's light is striking the photosensitive surface and the reading when no light falls on the cell. Figure 29 shows how measurements

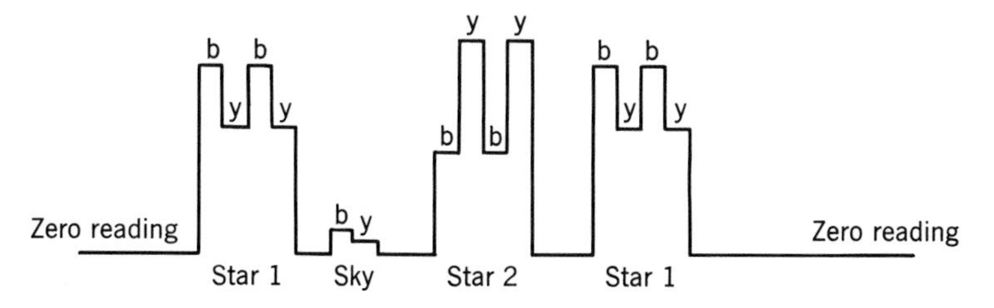

29. A schematic photoelectric tracing. The photoelectric photometer has usually one or more filters. In this schematic drawing the star is first observed in blue light, then in yellow light, and the observations are then repeated. After the first star is observed, a reading is taken on the sky where no star is visible, also in both colors. Then follow observations of star 2 and to close the sequence a repeat set on star 1. From the curves one would obtain the ratio of brightness of the two stars (equal to differences of magnitude) in blue and yellow light. The first star is brighter in blue light than in yellow light; the second is brighter in yellow light.

of the "zero" or "dark" current are made at regular intervals throughout the period of observation. The scale of brightness ratios or magnitudes is provided directly by the recording, and standard stars are observed regularly for the dual purpose of relating magnitudes to an agreed-upon zero point and of eliminating effects of changing atmospheric extinction.

Also included in the observation (Fig. 29) are so-called "sky readings." In spite of its apparent blackness, the sky background contributes a measurable brightness to the recorded deflections. Most of the radiation from the sky background is excluded from the measurements by observing the star through a small diaphragm at the focal plane of the telescope, but a little bit of sky always succeeds in peering through the diaphragm along with the star. To eliminate this sky effect, we take two readings in succession, the first of the star with the background radiation included, the second of a neighboring spot in the sky without the star. The difference between the two deflections measures the contribution from the star alone. Along the band of the Milky Way special problems arise when we attempt to measure faint stars, since it is not easy to obtain nearby sky readings free from intruding stars.

One of the advantages of the photoelectric photometer is the fact that it measures the amount of light received from the star in a direct or linear fashion: twice as much energy received by the instrument gives twice the deflection on the record. Photoelectric accuracy can be very high; it is possible to be sure of the brightness of any star to within two or three thousandths of a magnitude. Modern photoelectric methods were originally developed by Stebbins and Whitford at the University of Wisconsin and at the Mount Wilson Observatory. Photoelectric photometers are now standard equipment at practically all observatories, north and south.

Baum, at the Mount Wilson and Palomar Observatories was the first astronomer to succeed in measuring photoelectrically, with the aid of the 200-inch Hale telescope, the brightness of a twenty-third-magnitude star. To do so required special preparations and the use of techniques of photoelectric measurement that differ somewhat from those we have described, for the star in question is far too faint to be seen by the observer at the telescope and hence special ways of setting the telescope must be employed. First, a photograph of the region is taken and the difference of position between the faint star and a nearby bright star is measured. The photometer is mounted on a base that is attached to the telescope. Accurate linear scales make it possible to shift the photometer precisely and by known amounts sidewise or up or down. First the brighter of the two stars is centered on the diaphragm and then the photometer is shifted by exactly the amounts indicated by the photographic measurements. The faint star should now be centered precisely on the diaphragm and its magnitude can be determined by the customary techniques. For such very faint stars, we do not measure the instantaneous electric current produced by the very weak light of the star; instead we count the total number of light quanta, or photons, captured in the course of a few minutes to several hours and compare this number with the number of photons counted in the same interval for a somewhat brighter star of known magnitude.

Stellar Colors

The color characteristics of a system of magnitudes is determined by the range of wavelengths recorded by the receiver, be it a photocell or a photographic plate, and by the color transmissions of the filters that are used. One of the systems of magnitudes now much in use is the *UBVRI* standard system, first recommended and developed by H. L. Johnson. In this system, ultraviolet (*U*) magnitudes are obtained when a special ultraviolet filter is placed in the light path from the star to the photomultiplier cell. This filter has maximum transmission at a wavelength of 0.35 micron (1 micron = 0.001 millimeter), and it has a passband—that is, a range of wavelengths transmitted—with a width of 0.07 micron. To obtain blue (*B*) magnitudes, we use a standard filter with maximum transmission at 0.43 micron and with a passband width close to 0.10 micron; for visual (*V*) magnitudes the center of transmission by the filter is at 0.55 micron and the width of the passband is 0.08 micron. The two standard colors for the infrared have been named *R* (centered at 0.7 micron) and *I* (centered at 0.9 micron). The *UBVRI* system has been accepted by the International Astronomical Union as *the* standard system of magnitudes. Certain stars, specifically named, define the zero point from which we count magnitudes in each selected color and the measured magnitudes of all other stars are referred to these agreed-upon standards.

The color of a particular star is given by its *color index*. For example, the blue-visual color index of a star is defined as the difference between the star's blue (*B*) and visual (*V*) magnitudes; it is given as $B - V$. Similarly, one may define for each star the color indices $U - B$, $V - R$, and $R - I$.

Starlight is cut-off by the earth's atmosphere at 0.3 micron, and the cutoff is so sharp and strong that there is no chance of extending the standard system to wavelengths shorter than 0.3 micron, at least for studies from Earth-based observatories. The situation is different for observations from satellites. Extensive research is now in progress based on satellite observation. As of now, it seems that even from space observatories we may have relatively little opportunity for research at wavelengths shorter than 0.09 micron, principally because of the galactic fog that is produced by the neutral hydrogen atoms of the interstellar gas; these atoms absorb most of the far-ultraviolet radiation beyond the Lyman limit.

Photometry in the Infrared

During the past 10 years, photometry in the infrared has become a very active field. Pioneer work in the field was done by Johnson and Low at the University of Arizona and by Becklin, Neugebauer, Leighton, and others at the California Institute of Technology and at the Hale Observatories. The standard blue-sensitive photomultiplier cell can be used between the far ultraviolet (0.32 micron) and the visual (0.6 micron), but it loses sensitivity rapidly as we proceed to longer wavelengths. For work in the near infrared, photocells with cathodes of the cesium oxide–silver variety work very well; they cover nicely the range from 0.6 to 1.2 microns. These near-infrared cells are operable with the same sort of electronic equipment that is used for the IP 21 cells and cooling with carbon dioxide is all that is required to eliminate excessive background noise. In the intermediate infrared (1 to 4 microns wavelength), the most effective cells to use are the lead sulfite photoconductive cells, which must preferably be cooled to the temperature of liquid nitrogen. The Earth's atmosphere absorbs quite heavily in the 1- to 4-micron range, notably near 1.8 and 2.8 microns, but there are some clear "windows" near 1.3, 2.2, and 3.4 microns. For best results one should observe in dry areas from mountain tops 9,000 feet high or more, so as to have minimum water vapor overhead.

For observations at still longer wavelengths, one goes to a completely different type of radiation receiver. Low has been very successful with a germanium bolometer operated at liquid-helium temperatures. The region between 4 and 22 microns can thus be reached, but atmospheric absorption becomes increasingly bothersome and so does irregularly distributed radiation from the night sky. The greatest blocking in this range occurs near 6.5 microns, and the relatively transparent windows used for measurement are at 5.0 and 10.2 microns. Most of the atmospheric absorption can be eliminated if observations are made from high-flying airplanes, operating at altitudes of 40,000 to 50,000 feet. The gap between optical and radio wavelengths was really bridged when Low succeeded in making some bolometric measurements at 1 millimeter wavelength!

Standards of Brightness and Color

The first major task for the photometrist—the astronomer who measures magnitudes and color indices of stars—is to establish an extensive and comprehensive network of magnitudes and colors for standard stars, distributed all over the sky and for both the Northern and the Southern Hemispheres. The establishment of such a network involves precision measurements in well-determined filter and photocell systems of constant properties. It is also required that the observer pay very careful attention to the problems of atmospheric extinction, the effects of which must be eliminated before final values of magnitudes and color indices can be arrived at. Harold L. Johnson, now at the University of Mexico, has made such measurements for the *UBV* color system, which was first suggested by him. His lists and those of his associates give data for several hundred stars, many of them at southern declinations. Supplementary lists have been prepared to include measures in the red (*R*) and in the various infrared bands (*I* and other bands). These lists are the backbone of all modern photometry. They have almost entirely replaced the earlier photographic standards, notably the famous North Polar Sequence of the early 1900's.

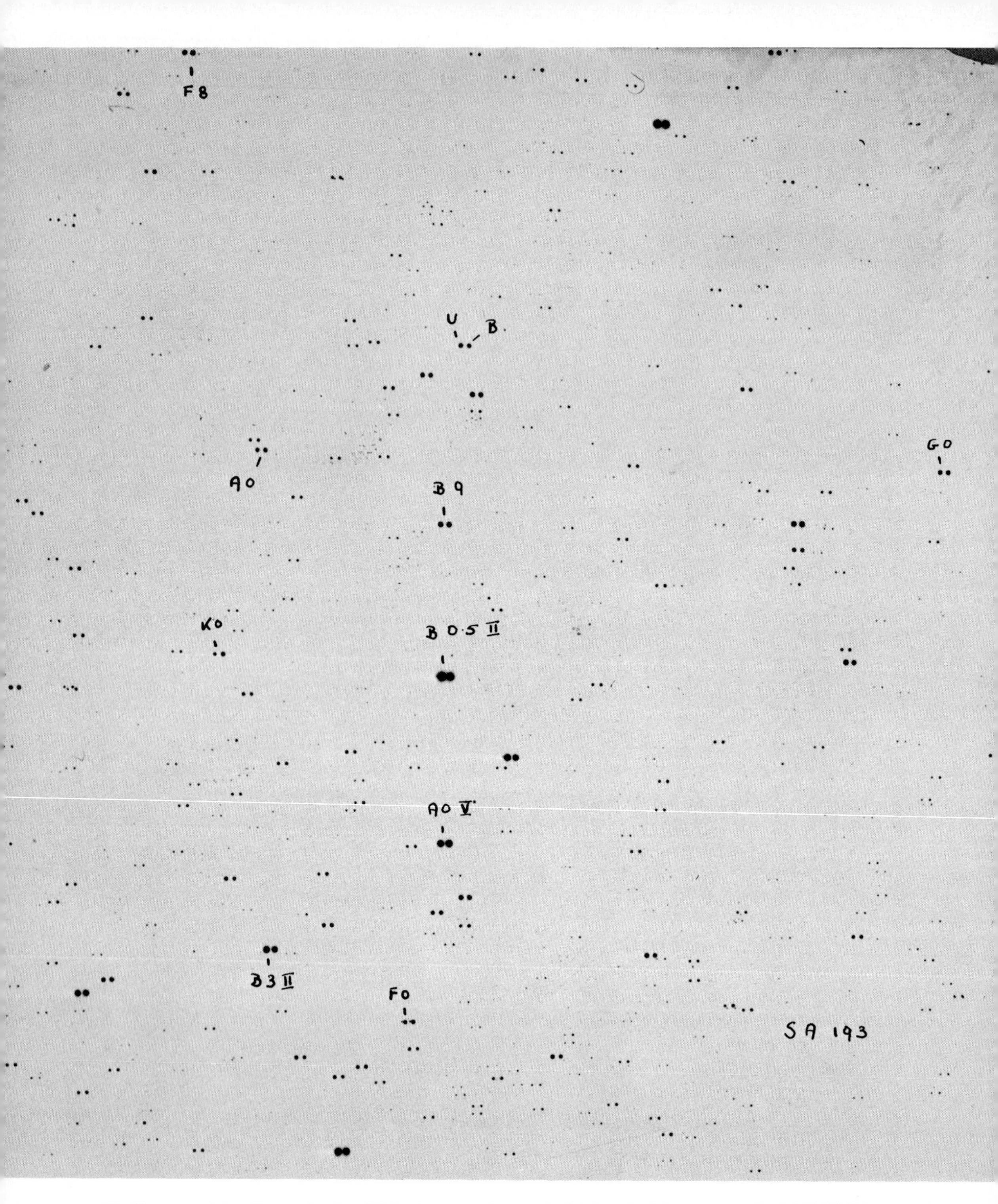

30. Photographic effects of color. This photograph shows two exposures, one, marked *U*, taken through an ultraviolet filter, the other, marked *B*, taken through the standard blue filter. To show the star images more clearly, we reproduce a negative print. The star marked K0 is a red star; its *B* image is stronger than its *U* image. The reverse holds for the blue star marked BO.5II. The remaining stars listed in order of increasing redness are marked as B3II, B9, A0, A0V, F0, and G0. A photograph made with the 40-inch reflector of Siding Spring Observatory in Australia. The southern field is that of Kapteyn Selected Area 193 (see page 65).

An astronomer who desires to obtain photoelectric substandards for any point in the sky, or who wishes to measure the UBV magnitudes for a specific star, can take as his reference star the nearest of the Johnson standards. The final result of his measurements is that, for his star, he obtains reliable values for the V magnitude and for the color indices $B - V$ and $U - B$. The next step, or steps, will depend on the nature of the research in which the astronomer is involved. When studying a variable star, he can monitor its variability by measuring at different times its V magnitude and $B - V$ and $U - B$ color indices with reference to a presumably constant standard. Or, by reference to the basic standards, he can obtain magnitudes and color indices for hundreds of stars of a special class, say white dwarfs. Since, as we noted earlier, linearity of response is built into modern photoelectric photometers, it is a simple matter to measure differences of magnitude or color index for stars differing greatly in apparent magnitude.

A different sort of problem arises when we require magnitudes and color indices for several hundred stars in a small area of the sky. This problem occurs when we study an open cluster or a globular cluster, or if we wish to measure apparent magnitudes and colors of all blue-white stars in a field along the band of the Milky Way. First we establish a *standard sequence* of magnitudes and colors involving photoelectric standards for, say, 30 stars. The sequence may well include stars in the range of visual magnitude $V = 8$ to 18, and there should be within the sequence a fair spread in colors $B - V$ and $U - B$. Next, we turn to photographic interpolation, using the photoelectric sequence as our standard reference, to find the values of V, $B - V$, and $U - B$ for all the stars in question. This is achieved by photographing the field in U, B, and V light by the selection of suitable combinations of color filters and photographic emulsions. The photoelectric U values for the sequence stars, all of which appear on our U photograph, are used as reference standards for the measurement of the U magnitudes of all other stars for which we wish to obtain this information. Similarly, we measure the B and V magnitudes for all the stars in question. The photographic measurements can all be made in the laboratory with a photographic densitometer, an instrument that measures the size and density of the photographic images of the sequence stars. Basically we use a process of direct interpolation. It is obvious that all of these processes can be executed most efficiently by proper application of modern techniques of automation and of data processing. Figure 30 shows rather nicely some of the more conspicuous photographic effects of color. The photographic plate had two exposures, one in ultraviolet light (U), the other in blue light (B). The blue stars are readily distinguished from the intrinsically red ones.

One major weakness of the *UBV* and similar systems is that the filter passbands are rather wide, often too wide for detailed research on the physical properties of the stars and of the interstellar medium. In the 1950's, Strömgren and Crawford, then at Yerkes Observatory, pioneered in the establishment of suitable systems for intermediate and narrow-band photometry. The characteristics of the *uvby* system that they developed are as follows:

	Central wavelength (microns)	*Filter passband (microns)*	
u	0.35	0.038	ultraviolet
v	.41	.020	violet
b	.47	.020	blue
y	.55	.020	yellow

The *u* filter is a normal ultraviolet glass filter, but the narrow-band *v*, *b*, and *y* are specially made interference filters. Strömgren, Crawford, and Perry have established standard *uvby* values for a large number of reference stars, very much in the manner of the *UBV* standards. It is not yet feasible to make interference filters 8 × 8 inches in area, to mimic the available much smaller, 2 × 2 inches, *vby* filters and, for the present, we cannot resort to photographic interpolation to measure large numbers of stars. However, the time may not be far off when mass determination of *uvby* magnitudes and colors by photography will be possible. Already Schreur has developed a test series of gelatine filters for photographic purposes, which mimic nicely the *uvby* photoelectric system.

One disadvantage of the intermediate-band *uvby* system compared with the broad-band *UBV* system is that with a given telescope one cannot reach as faint stars in *uvby* as in *UBV*. The limiting magnitude for *uvby* is about 2½ magnitudes brighter than for *UBV*. This lack of penetration becomes even more serious when one turns to real narrow-band photometry. Here the pass bands of the interference filters often narrow down to 0.003 to 0.005 micron, and as a result another 2½ magnitudes in limiting magnitude is lost over *UBV* photometry.

A large number of secondary sequences are distributed over the sky. At several observatories in the United States, Holland, Sweden, the Soviet Union, Great Britain, and South Africa much time and effort have been spent on standard sequences in the Kapteyn Selected Areas. About 50 years ago Kapteyn of Holland selected a network of 206 centers uniformly distributed over the whole sky to provide photographic standard magnitudes. It was Kapteyn's intention that astronomers should concentrate their efforts upon obtaining colors, proper motions, radial velocities, and spectral types for the stars in these Selected Areas, in the hope of deriving from these data the characteristic properties of the Galaxy.

Why should we want to measure magnitudes and colors of thousands of stars? For the nearby stars, the color index is a direct measure of the temperature of the star. Obviously, a blue-white star with a surface temperature of 25,000°K emits a much stronger blue radiation than a red star with a surface temperature of 3,000°K. (Temperatures of stars are measured on the Kelvin scale, which has its zero at the absolute zero of temperature, 273° below 0° centigrade.) For the more distant stars, another factor comes into play. The light of such stars is appreciably reddened by cosmic dust between the stars and our Sun. This is especially so for distant stars along the band of the Milky Way. Frequently we may predict, on the basis of the appearance of the spectrum of the star (discussed at length later in this chapter), what the value of its color index must have been before the star's light was affected by the reddening from the interstellar dust; we call this the intrinsic color index of the star. If we once know this intrinsic color index, then we can obtain from the difference between the observed and the intrinsic color indices information about the amount of reddening produced by the intervening cosmic dust. We thus derive useful data regarding the absorption of light by the cosmic dust in our Milky Way system. Color-index measurements figure prominently in all studies of interstellar absorption.

Distances and Parallaxes

When we see stars in the sky that differ greatly in brightness, we are naturally curious as to how much of this can be attributed to differences in intrinsic brightness and how much of it is due to differences in distance from us. When light travels unobstructed through space, its brightness varies inversely as the square of the distance from the source. If two stars that are in reality equally luminous appear to us to differ by 5 magnitudes (which means a light ratio of 100 to 1), the fainter one must be ten times as far away as the brighter one. How can we find the distances of the stars?

The astronomer's fundamental method of finding the distance to a star is essentially the same as that of a surveyor who wants to know the width of a river. First he measures carefully as long a base line as is practicable along the bank on his side of the river. From each end of this base line he sights on a tree or other landmark on the opposite shore and measures its angular direction. This gives him the angle-side-angle triple, well known to high-school geometers, which enables him to compute any other part of the triangle. The astronomer likewise must first carefully measure his base line and he must then measure with high precision the direction angles at which he sees the star at opposite ends of his base line. We have not enough room on the Earth to provide us with a proper base line, one long enough for the measurement of the very great distances to the stars. But the Earth moves in a very large ellipse (almost a circle) around the Sun (Fig. 31). If we but wait 6 months after one pointing, we shall have moved, without any effort on our part, 186,000,000 miles (299,000,000 kilometers), or twice the distance from the Earth to the Sun. From the two ends of this base line, the astronomer sights on a number of stars and he will find that at least the nearest ones will have shifted their positions measurably with reference to the base line, or, more precisely stated, with reference to the background of much more distant stars.

The astronomer uses as indicators of distance the shifts in the apparent positions of the stars. By definition the *astronomical unit,*

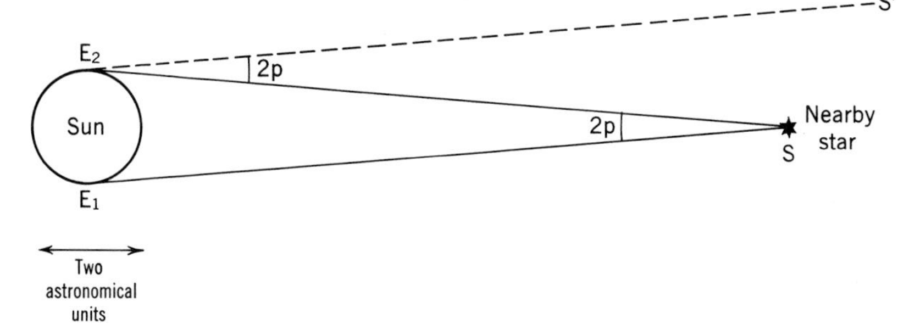

31. The parallax of a star. When the observer on the Earth moves in half a year from position E_1 on the Earth's orbit to position E_2, a distance apart equal to 2 astronomical units, the nearby star will have apparently shifted its position among the more distant stars, from direction E_2S' (parallel to E_1S) to E_2S. This angle ($2p$) is equal to the angle at the star subtended by twice the astronomical unit or is then equal to twice the star's parallax.

the mean distance from the Earth to the Sun, which is 93 million miles or 149.6 million kilometers (1 mile = 1.609 kilometers), is taken as the base line, and the angular shift that corresponds to it at the distance of the star is called the *parallax* of the star. Hence, the parallax can be defined as the angle subtended by the mean distance from the Earth to the Sun, as viewed from the star. We note that the total displacement shown by a star is equal to twice its parallax, since the distance apart of the two extreme positions of the Earth for the measurement of the parallax effect equals 2 astronomical units.

It is not difficult to compute the distance of a star from its parallax. To do so, we introduce a new unit of distance, to which has been given the hybrid name of *parsec.* A star at the distance of 1 parsec has a *par*allax of 1 *sec*ond of arc. The distance in parsecs is equal to the reciprocal of the parallax in seconds of arc: $d = 1/p$. The distance of 1 parsec is equal to 206,265 astronomical units, or 206,265 × 93,000,000 = 19,200,000,000,000 miles. Instead of the parsec, the more picturesque unit, the *light-year,* is often used. The light-year is the distance that light travels in 1 year. At the rate of 186,000 miles per second, and with about 31,600,000 seconds in a year, the light-year is equal to 5,880,000,000,000 miles. One parsec is, therefore, equal to approximately 3.26 light-years. The nearest star has a parallax of 0.75 second of arc, hence a distance of 4.30 light-years; light takes 4⅓ years to reach us from our nearest neighbor. We can also say that this star is at a distance from us of 1⅓ parsecs.

Finding the parallax of even one star is an exacting and time-consuming process. Photographs must be taken with a long-focal-length photographic telescope in order to have a large scale on the photographic plate. These photographs should be repeated at several 6-month intervals to separate the effect of the star's own motion, which is a uniform motion along a straight line on the photographic plate, from the parallax effect, which goes through its cycle once every year. If the star for which the parallax is being sought is much brighter than the stars near it, then we must develop some method of cutting down its light, since it is impossible to measure accurately relative positions of photographic images that are of very different sizes and densities.

The modern photographic method for the measurement of stellar parallaxes was developed about 70 years ago by Schlesinger, with the aid of the 40-inch Yerkes Observatory refractor. Measured parallaxes are available for some few thousands out of the many millions of stars; modern parallax catalogs list parallaxes of reasonable reliability for about 8,000 stars, and additional trigonometric parallaxes are presently being produced at a rate of only 60 to 80 per year. With the best of care, the astronomer cannot measure the parallax of a star with an error of less than 0.004 second of arc, and larger errors are not uncommon. Hence, the measured trigonometric parallax for a star that is only 50 parsecs distant may have a value anywhere in the range 0.016 to 0.024 second of arc, and the corresponding percentage uncertainty in the estimated distance is quite large. There is some hope for improvement in precision of the basic measurements. Van Altena at Yerkes Observatory is developing new techniques of measurement and reduction that he hopes will reduce the errors by 50 percent. High-quality trigonometric parallaxes are obtainable only for relatively nearby stars, and in practice parallax observers wisely concentrate their efforts upon stars within 20 parsecs of the Sun. Fortunately, there is quite a variety of stars within this distance.

In the past many observatories took part in the measurement of trigonometric parallaxes. Among the early leaders in the field, most of which are still active in this area, are the Yerkes, Allegheny, Leander McCormick, Sproul, and van Vleck Observatories in the United States, the Royal Greenwich Observatory, and the Royal Observatory at the Cape of Good Hope. All of these have traditionally concentrated their efforts on stars brighter than the tenth or, at the faintest, the twelfth magnitude. Several newcomers can now be added to the list. The most famous telescope now active in parallax work is the 61-inch reflector of the U.S. Naval Observatory at Flagstaff, Arizona. This marvelous precision astrometric reflector was constructed principally for the determination of trigonometric parallaxes of stars between the twelfth and the seventeenth magnitudes. The telescope has an annual yield of 40 precision trigonometric parallaxes. The Lick 36-inch refractor and the large refractor at Pulkova Observatory in the U.S.S.R. are also engaged upon parallax work. Luyten has attempted some studies in the field for very faint stars, mostly blue ones, with the Palomar Schmidt telescope.

An observer of trigonometric parallaxes often wishes that he might inhabit Jupiter, whose orbit has a diameter equal to 10 astronomical units, or Pluto, which would give an orbital base line of 75 astronomical units. The long-range future looks bright for the parallax observer, for it should not be many years before earth satellites will be able to bring back—or telemeter back—from outer space data on parallax shifts from stations at distances of 5 or more astronomical units from the Sun! This undertaking represents one of the most important space missions of the future.

All parallaxes determined photographically are measured relative to the average for the background stars on a photographic plate with the parallax star near its center. A small systematic correction is generally applied to correct the relative parallax to a true one. The amount of this correction is, however, uncertain. In the future, it might be well to check this correction with great care. Attempts should be made to use one or more faint galaxies as basic reference points for zero parallax. Galaxies are millions of parsecs away and hence certainly show no measurable parallax displacements of their own.

But before we get too far out into space, let us first see what results can be found from direct parallax measurements.

Absolute Magnitudes

If we know the distance of a star and also how bright it appears to us, we can find its true or intrinsic brightness. We may then compare the intrinsic brightness of each star with that of our Sun as the standard and obtain what is called the *luminosity* of the star. Or we may imagine all stars placed at the same distance and compute, from the observed apparent magnitude and the known actual distance, what the magnitude of the star would be if it were placed at this standard distance. We call this the *absolute magnitude* of the star. Long ago international amenities led us to choose 10 parsecs as the standard distance. The formula for the computation of the absolute magnitude is

$$M = m + 5 - 5 \log d,$$

where m is the apparent magnitude and d is

the distance in parsecs.* Altair, with a measured parallax of 0″.20, has an apparent magnitude $m = 0.9$. Since its distance is $1/0''.20 = 5$ parsecs, its absolute magnitude is

$$M = 0.9 + 5 - 5(0.70) = +2.4.$$

When, hypothetically, we place all the stars at the standard distance of 10 parsecs, we find that they show as wide a range in absolute magnitude as they do in apparent magnitude. The values of the observed absolute magnitudes range all the way from +18 to −10. In our imagination, we can set the Sun at the standard distance of 10 parsecs and we find that its absolute magnitude lies near the middle of the range, at +4.7; our Sun would appear as an inconspicuous star of very nearly fifth apparent magnitude if it were placed at the standard distance of 10 parsecs. If we compare any other star with the Sun as the standard—the standard candlepower, as it were—we obtain what is called the *relative luminosity* of the star. Some of the brightest stars are 14 to 15 magnitudes brighter intrinsically than our Sun; they are pouring forth $2.512^{15} = 1$ million times as much energy as the Sun. On the other hand, the faintest known star in absolute magnitude is about 14 magnitudes fainter than the Sun; the star is emitting 1/300,000 of the radiation emitted by the Sun. The ratio of intrinsic brightness between the brightest and the faintest stars is therefore at least of the order of 100 billion! This range is all the more remarkable since the corresponding ratio of masses is less than 10,000, probably not much over 1,000. There are such tremendous differences from one star to another that it is unsafe to guess whether a particular star is faint because of its low intrinsic luminosity or because it is very far away. On the average, the fainter stars will be the more distant ones, but we cannot say anything about the distance of a particular star unless we know its absolute magnitude.

* To derive this equation will be easy for those who are used to logarithms and remember how we define magnitudes. Let l be the apparent brightness of the star and L its absolute brightness. Then

$$\frac{l}{L} = \frac{(\text{standard distance})^2}{(\text{actual distance})^2} = \frac{10^2}{d^2} = 2.512^{M-m}.$$

Taking logarithms, $0.4(M - m) = 2 - 2 \log d$, and hence

$$M = m + 5 - 5 \log d.$$

Spectral Classification

Spectroscopic tools are essential to the astronomer. All large telescopes are fitted with spectrographs of various designs, and large prisms that can be placed in front of the objectives of photographic telescopes (Fig. 32) are standard equipment at several observatories. Figure 33 shows the kind of photograph that is obtained when one of these objective prisms is attached. By a slight trailing of the star's image in the direction at right angles to the dispersion, the stellar spectra, with few exceptions, are made to appear as little bands, crossed by dark lines, the so-called absorption lines. Even the most cursory inspection shows that not all spectra are alike. Certain lines of hydrogen, called the Balmer lines, are strong in some spectra, weak in others, and entirely absent from the spectra of still other stars. The lines of iron and other metals are sometimes present, and in some spectra molecular bands are the outstanding feature. These differences were a challenge to the astronomers of the late nineteenth century, when the various spectral characteristics were beginning to be studied. It was soon clear that the stars could be subdivided into a fairly small number of spectral classes, which merge gradually one into the next.

32. Mounting an objective prism. Dr. Freeman D. Miller, mounting the 6° objective prism (showing his reflection) on the corrector plate of the Curtis-Schmidt telescope.

33. Examples of spectral classification. Representative spectrum-luminosity classes are indicated for 10 stars in a southern field at right ascension 8^h24^m, declination 18°41′ south (1950). The telescope is the Curtis-Schmidt telescope of the University of Michigan (Fig. 32) now mounted at the Cerro Tololo Inter-American Observatory. The plate was selected by Dr. D. J. MacConnell, Department of Astronomy, the University of Michigan, who made the following classifications: 1, K3III; 2, F2IV; 3, A1IV; 4, A9IV; 5, K0III; 6, B2IV; 7, M2III; 8, B9IV; 9, F6V; 10, G2V. (University of Michigan Observatory photograph.)

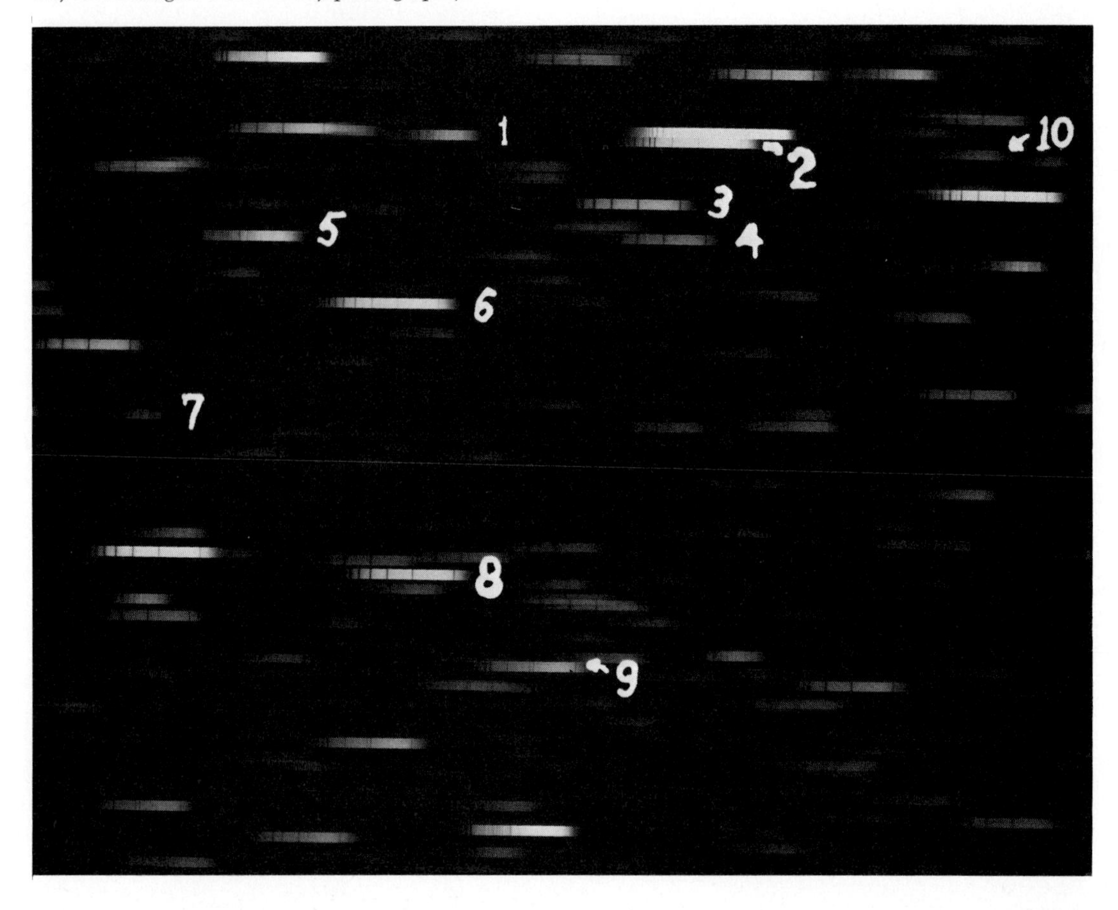

The classification that has been widely used for a long time was worked out at the Harvard Observatory under Pickering by Miss Cannon, Miss Maury, and Mrs. Fleming. The *Henry Draper Catalogue* is the work of Miss Cannon; it contains the spectral classifications of 225,320 stars of both the northern and the southern sky, and includes practically all the stars to a limit between the eighth and ninth apparent magnitudes. The *Henry Draper Extension*, which was completed after Miss Cannon's death by Mrs. Mayall, continues the classifications to the eleventh magnitude for special regions of the sky. The spectral classes were first lettered in alphabetical order, but by a "survival-of-the-fittest" development the sequence of classes has narrowed down to O, B, A, F, G, K, M. A few odd stars of classes W, R, N, C, and S, and some individuals craving for notoriety that are labeled "pec" for peculiar, supplement the main series. The O to M sequence is very strikingly a line-up in color: the O and B stars, such as Rigel and other stars in Orion, are blue; the A stars, like Sirius and Deneb, are white; Procyon and Capella, of the F and G types, are yellow; finally, the orange and red K and M stars have Arcturus, Aldebaran, Antares, and Betelgeuse as shining examples.

The surface temperature of a star, or rather the degree of ionization in its atmosphere, is the determining factor that decides where a star shall be placed in the spectrum line-up of the Henry Draper Classification. Miss Cannon found it necessary to subdivide the B stars on a decimal system, with the hottest listed as B0, B1, B2, . . . and the coolest as B8 and B9, and with B9 not very different from A0. Miss Cannon did not distinguish in her main catalogs between an intrinsically very luminous B0 star, a normal B0, and a subluminous B0 star, but the notes at the end of her catalogs show that she was aware of the presence of subtle differences indicative of absolute-magnitude effects. Miss Maury's classification, cumbersome and now wholly abandoned, was in reality a two-way classification system in which luminosity effects figure along with the temperature, or color, sequence. Chemical composition is another complicating factor. Differences in chemical composition from one stellar atmosphere to the next produce slight differences in spectral appearance. Subnormal strength of metallic lines for a given spectral class was noted in the first system of spectral classification as indicative of unusual chemical composition of a star's atmosphere. For the moment we shall leave out of consideration the effects of luminosity and chemical composition and concentrate on the main stream of spectral classes, which is a temperature, or color, sequence.

The O stars are the hottest, some with temperatures as high as 100,000°K. Their spectra can be easily recognized by certain characteristic lines emitted by ionized atoms (atoms that have lost one or more of their outer electrons) and by the far-violet extension of the spectra. Similar in spectra and temperature to the O stars, but with bright lines—emission lines—in their spectra, are the group called the Wolf-Rayet (W) stars. The emission lines are produced mostly in extended and expanding outer shells. The B stars, blue in color and with surface temperatures of the order of 15,000° to 30,000°K, show dark absorption lines of helium and hydrogen. Some B stars also show emission lines. Helium fades and hydrogen strengthens as we approach class A along the spectrum series. Lines of calcium and other metals, such as iron and magnesium, gradually increase in strength through classes F and G. Our Sun is a typical G star. In class K the calcium lines become very strong, and bands due to molecular compounds come into view. Class M stars are red, with temperatures of less than 3,000°K; their spectra show absorption bands of titanium oxide. The symbols R, N, and S refer to a series of cool stars (parallel to the K–M series) in whose spectra other compounds are present.

There is a wide range in absolute magnitude for stars of most spectral types. Capella, for example, has a spectrum very much like that of the Sun, but its absolute magnitude is +0.4, which means that Capella is 4 to 5 magnitudes (or close to 100 times) brighter intrinsically than the Sun. The surface temperatures of the two bodies are nearly the same, but their physical conditions are quite different. A *giant star* such as Capella has a much more extensive atmosphere, of much lower density, than a *dwarf star* like the Sun. These differences of density and pressure cause some spectral lines to be stronger in the giants and others to be enhanced in the dwarfs. Adams and Kohlschütter at the Mount Wilson Observatory in 1914 were the first to study the spectra of known giants and dwarfs of the same spectral class. Once the luminosity effects were noted it was possible to assign all stars of that same spectral type to either the giant or the dwarf branch, especially in the case of stars of spectral classes F to M. Adams and Kohlschütter's work was preceded by that of Miss Maury, who really discovered the first criterion of absolute magnitude. She found that there were some stars with unusually sharp spectral lines, which she designated by c; a star classified as cB3, for example, is a sharp-line B3 star. Hertzsprung showed that this characteristic is indicative of high intrinsic brightness.

In the Henry Draper system of classification, Miss Cannon introduced the practice of remarking on the sharp-line c stars, thereby indicating the most notable supergiants. For the dispersion used in the Henry Draper Classification it did not seem feasible to do more. However, Bertil Lindblad and his associates at Stockholm and Uppsala found that there were certain features in the spectra of very low dispersion for faint stars that are indicative of the star's luminosity. They developed techniques for distinguishing between giants and dwarfs among the stars of a given spectral type. In the spectral catalogs prepared after the *Henry Draper Catalogue,* notably those of the 1920's and '30's published by the Hamburg-Bergedorff, Potsdam, and McCormick Observatories, giant-dwarf separation was generally made. Yet even today the *Henry Draper Catalogue* stands first; its greatest value lies in its homogeneity and completeness.

The modern trend in spectral classification is to use an objective prism with a fairly large angle and thus produce stellar spectra in which considerable fine detail can be observed (Fig. 33). By the use of higher dispersion, we obtain greater resolution in the spectra, but it is not possible to reach very faint stars, and we are troubled by overlapping of spectra in dense regions of the Milky Way. The increased accuracy of classification, however, offsets for many problems the loss in penetrating power.

At the Yerkes Observatory, Morgan and Keenan, assisted by Kellman, developed a two-dimensional system of classification. Morgan studied the whole range of spectral types with the idea of finding "natural groups," the purpose being to use such groups in probing the galactic system. These groups are characterized by a narrow range in luminosity and by easily identified criteria in spectra of low dispersion. The latest version of the

Atlas of Stellar Spectra (by Morgan and Keenan; known to astronomers as the MK *Atlas*) is of great value for all work on accurate classification. Six luminosity classes are recognized (Fig. 36): Ia, most luminous supergiants; Ib, less luminous supergiants; II, bright giants; III, normal giants; IV, subgiants; V, dwarfs.

We shall see in later chapters that the MK *Atlas* and its system of spectrum-luminosity classification have proved invaluable for the study of our Milky Way system. They have formed the basis for the spectrum–absolute-magnitude classifications carried out by Morgan in collaboration with the Warner and Swasey Observatory. Our modern views on the spiral structure of our Milky Way system are largely based upon studies involving extensive applications of the MK system. We are not yet certain of the precise absolute magnitudes that should be assigned to each luminosity class, especially for the absolutely brightest stars, but already the success of the MK system has been so great that it supplants practically all earlier work.

In recent years astronomers have become so deeply involved with problems of spectral classification that several independent efforts have been made to improve upon the basic Henry Draper System. In these new approaches to spectrum-luminosity classification total intensities of certain spectral lines—the Balmer lines for instance—and brightness distribution in the continuous background are used as classification parameters. These can be defined numerically if we use the techniques of intermediate- and narrow-band photoelectric photometry. For example, the Strömgren, Crawford, Graham technique measures the strength of one of the Balmer lines of neutral hydrogen, H-beta, by the use of two filters, one with a 0.003-micron-wide passband, the other with a passband 0.015 micron wide. The first measures only the total intensity of the H-beta absorption line, whereas the second measures in addition quite a bit of the relatively clear continuous background. The difference of the two brightnesses appears to be dependent mostly on the absolute magnitude of the star, at least for B and A stars. The Chalonge criterion of the drop in intensity at the Balmer limit—near 3650 angstroms, where many Balmer lines crowd together—is measured in the Strömgren *uvby* system of intermediate-band photoelectric photometry. The u filter is centered close to the Balmer limit, the v filter at somewhat greater wavelengths outside the region of the crowding of the Balmer lines. The $u - v$ color index measures photoelectrically the total strength of the crowded Balmer lines.

In modern approaches to spectral classification, increasing emphasis is being placed on the inclusion of criteria in the ultraviolet. The MK *Atlas* is being supplemented by the recent (1968) *Atlas of Low-Dispersion Grating Spectra* by Abt (Kitt Peak Observatory), Meinel (University of Arizona), and Morgan and Tapscott (Yerkes Observatory). The spectra, obtained with a grating spectrograph capable of recording spectral lines in the ultraviolet almost to the Balmer limit, provide for quite a few more criteria than were attainable with the MK *Atlas*. Figures 34 and 37 show samples of spectra from the 1968 *Atlas*. Precision spectrum-luminosity classification is done by the fitting of the spectrum of a given star into the general sequence of the *Atlas*.

In the past 20 years spectral classification in the infrared has made rapid strides. Nassau, McCuskey, Blanco, and their associates at the Warner and Swasey Observatory have been the most active workers in the field in the Northern Hemisphere, whereas Henize,

34. Spectra of main-sequence stars with classes to M5. The four photographs are from Abt, Meinel, Morgan, and Tapscott, *Atlas of Low-Dispersion Grating Stellar Spectra* (Kitt Peak National Observatory, 1968). These are negative prints with the normally dark absorption lines shown white in the photograph; this makes the spectral features stand out better than on normal prints. The caption for each photograph is that shown in the *Atlas*.

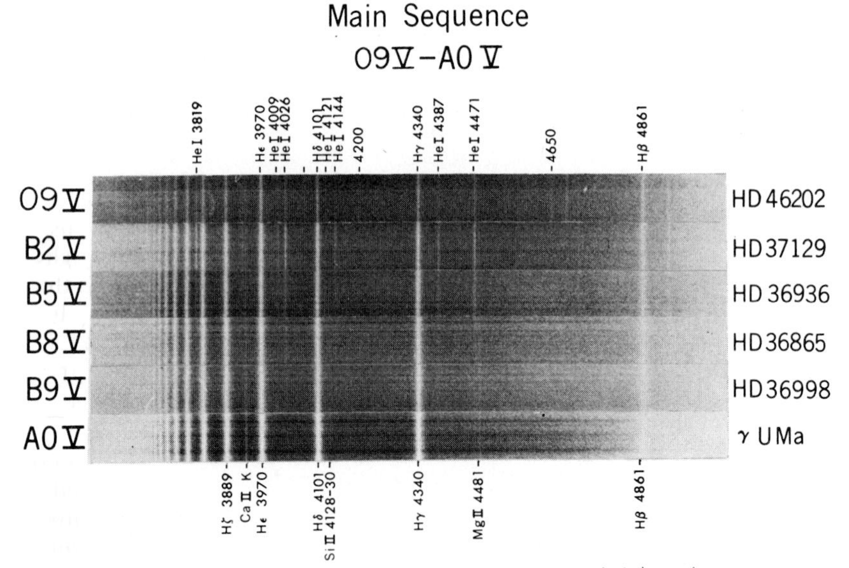

The hydrogen lines increase in intensity with spectral type and reach their maximum strength near A0 V. The lines of He I have a sharp maximum at B2 V and decrease smoothly in intensity thereafter. The stellar K-line is first used for classification near B8 and increases rapidly in intensity toward later types.

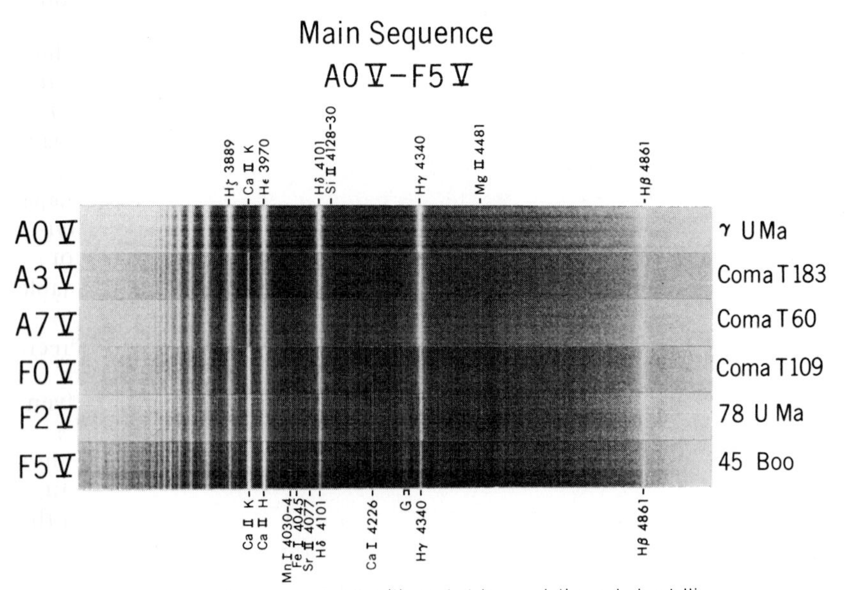

The ratio of Ca II K/Hδ changes rapidly with spectral type, and the neutral metallic lines grow stronger. The G-band first appears as a continuous feature near F2.

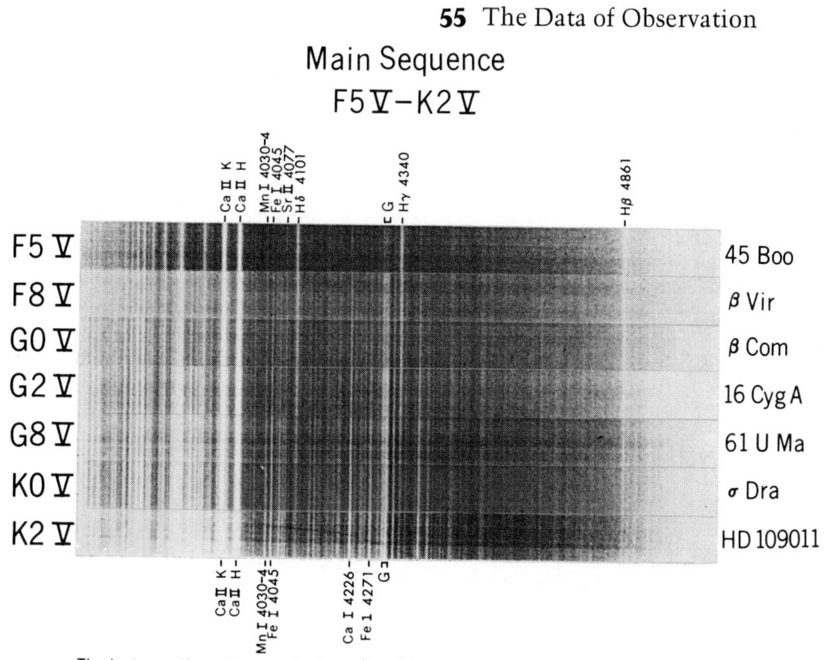

The hydrogen lines decrease in intensity, while the neutral metallic lines increase. The G-band grows in intensity to late G-types; its appearance alters thereafter.

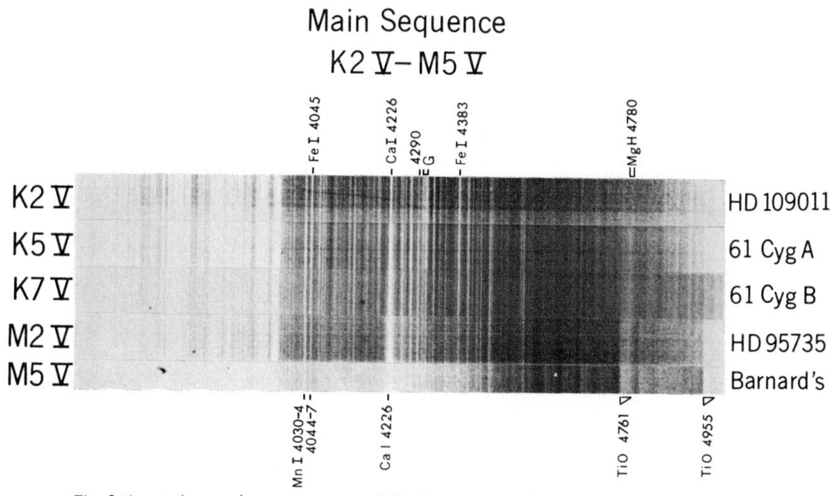

The G-band changes in appearance, and the line Ca I 4226 increases rapidly in intensity with advancing type. The bands of TiO appear near M0 and grow with decreasing temperature. An absorption band of MgH, centered near 4780 is well marked in late K-type dwarfs; it is blended with the TiO absorption near 4761 in the M dwarfs. The MgH band is an important luminosity discriminant.

35. Luminosity effects at B1. Negative prints are shown. (Reproduced from Morgan and Keenan, *Atlas of Stellar Spectra,* Yerkes Observatory, University of Chicago.)

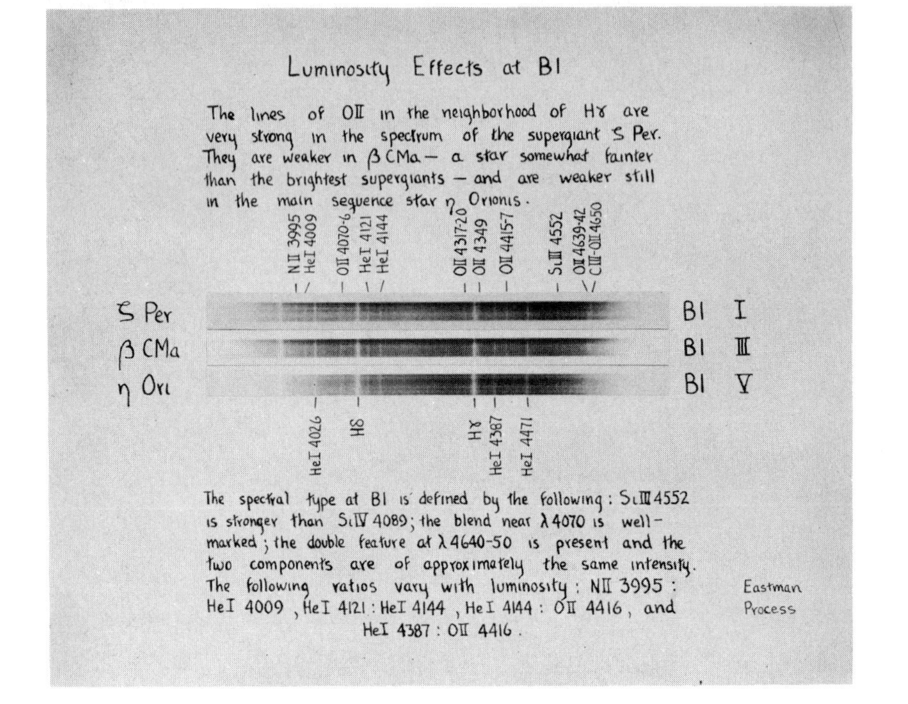

36. Luminosity classes. In the classification system first developed by Morgan, Keenan, and Kellman, one distinguishes both spectral type and luminosity class. The schematic diagram gives vertically the approximate visual absolute magnitudes for the stars of different spectral types and luminosity classes (indicated by roman numerals). (From an unpublished compilation by Matthews.)

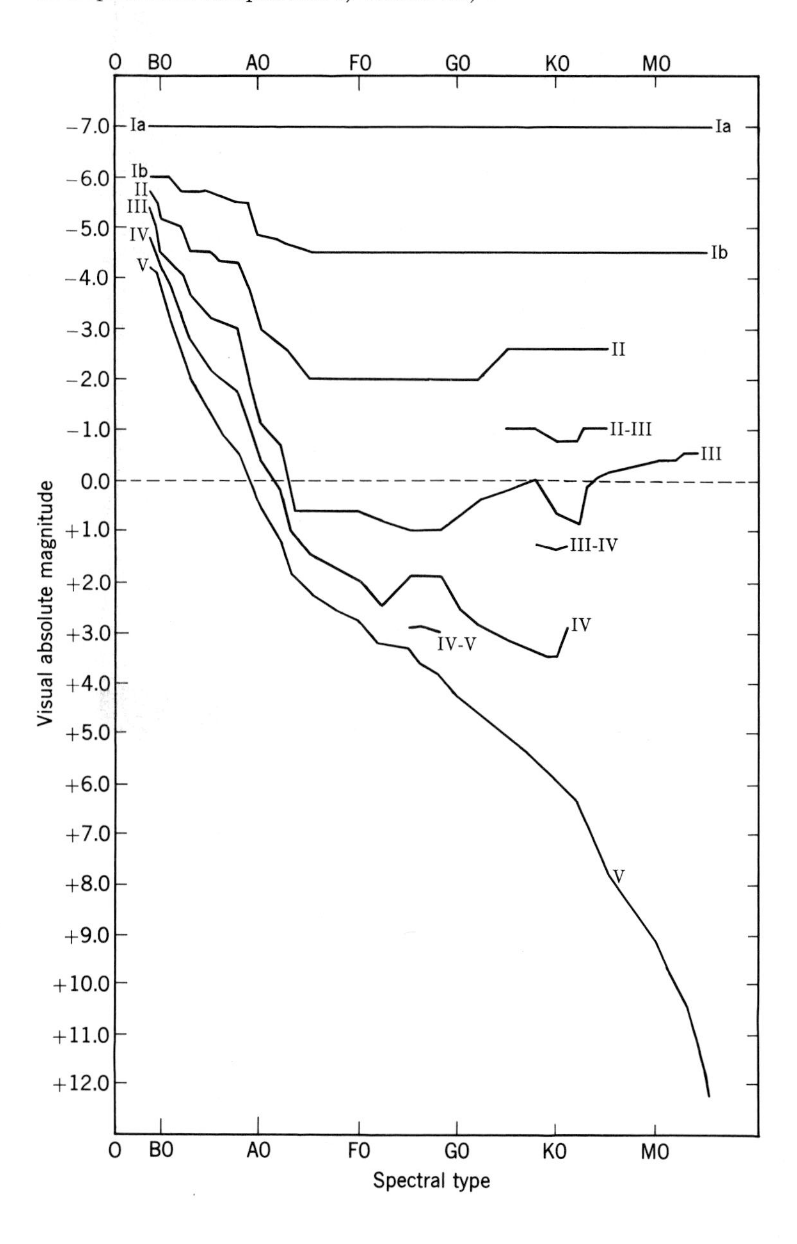

37. Spectra of hot supergiant stars. These representative spectra from the *Atlas* by Abt *et al.* show beautifully the very sharp and narrow absorption lines observed in the most luminous supergiant stars; negative prints.

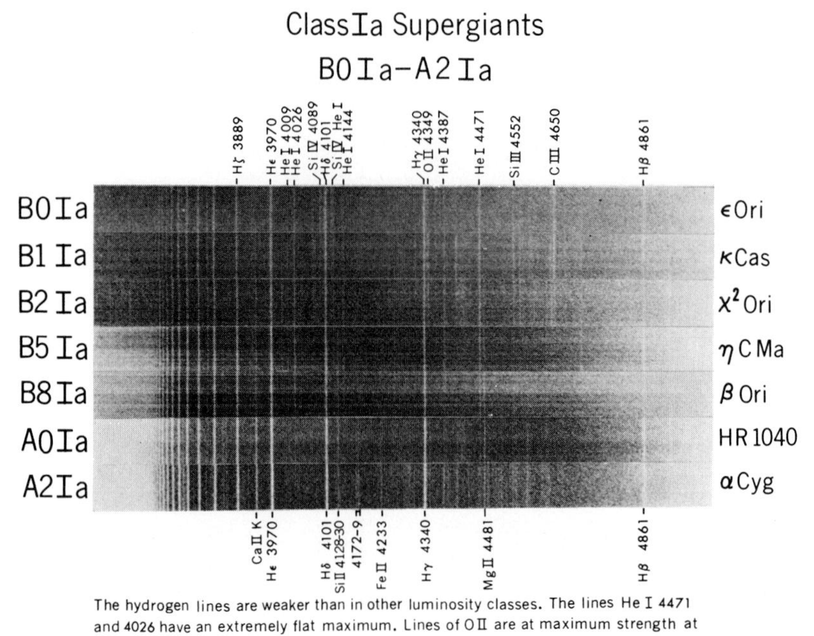

The hydrogen lines are weaker than in other luminosity classes. The lines He I 4471 and 4026 have an extremely flat maximum. Lines of O II are at maximum strength at Class B1. Lines of Fe II become outstanding in Class A.

Haro, and Westerlund have classified extensively for the Northern and Southern Hemispheres. The carbon stars and the S stars are, besides the M stars, the most common varieties of red stars. In the near infrared, Keenan distinguishes the late M stars by the presence of titanium oxide and vanadium oxide molecular bands and the S stars by zirconium oxide bands. The carbon stars—now generally referred to by the symbol C—were previously called the R and N stars. The C stars have characteristic bands of the carbon molecule and of the cyanogen (CN) molecule.

Proper Motions

In 1718, Halley compared his observations of the positions of Arcturus and Sirius with the positions as given in the catalog of Ptolemy, and he found that these stars had moved by appreciable amounts. The so-called "fixed" stars are not stationary but are constantly changing their positions in the sky. Figure 38 illustrates the motion of the star with the largest known annual displacement. This star, called Barnard's star after its discoverer, moves at the rate of 10 seconds of arc

38. A fast-moving star. Two photographs of Barnard's star, taken 11 months apart with the 24-inch refractor of the Sproul Observatory. In making the combined print, the second plate was shifted slightly with respect to the first. The arrow points to one pair of images for which the displacement differs from that for the others; we deduce that this star has changed its position perceptibly during the interval of 11 months.

per year, a Moon's diameter in less than 2 centuries. When two photographs taken a year or so apart are combined, as in Fig. 38, the runaway can be spotted.

Most stars move very slowly across the sky, so that long intervals of time are necessary for the detection of their motions. Accurate observations of position in the sky are made with the meridian circle; from the altitude at which the star crosses the meridian at a given observatory we deduce its declination, and from the precise timing of the meridian passage we obtain its right ascension. The annual *proper motions* of a star in declination and in right ascension are obtained by dividing the total displacements in declination and in right ascension by the interval in years that has elapsed between the measurements of the two positions. Proper motions are commonly measured in seconds of arc per year. The total displacement increases proportionally with time, so that, if 10 or 20 years do not suffice to show a measurable quantity, 50 or 100 years may do the trick.

The effective limit for visual observation with meridian circles lies at about the ninth apparent magnitude. To measure the proper motions of the fainter stars, we turn to photography. Suppose we have two photographs of the same part of the sky taken with the same telescope some 20 or 40 years apart. If the scale of the plate is large enough, we can measure the relative positions of the stars to within 0.01 second of arc. The differences between pairs of positions can be reduced to proper motions if there are enough stars on the plate with known proper motions determined by meridian-circle observations.

For all the bright stars and many of the faint ones the proper motions are now accurately known. For many years the catalogs prepared under the direction of Lewis and Benjamin Boss at the Dudley Observatory in Albany and by H. R. Morgan at the U.S. Naval Observatory figured prominently in researches on positions and proper motions, but these catalogs are now being replaced by others of greater precision. The Smithsonian Astrophysical Observatory has published a comprehensive *Catalog* with proper motions for nearly 260,000 stars. Photographically determined proper motions have been obtained at the Royal Observatory at the Cape of Good Hope, at the Yale and the McCormick Observatories, and at various German observatories under the auspices of the Astronomische Gesellschaft. One of the most recent accomplishments has been the undertaking by the German observatories under the leadership of Heckmann, Fricke, and Dieckvoss to determine anew the positions of the stars in the catalogs of the Astronomische Gesellschaft and we now have in the *Catalog AGK3* proper motions for 180,000 stars, mostly between eighth and twelfth magnitudes; average accidental errors amount to ±0.008 second of arc.

In the preparation of reliable catalogs of proper motions, one of the principal concerns of the astronomer is to find a reference system that is as nearly fixed in space as possible. All measured proper motions are relative to some reference frame, be it of faint and therefore presumably distant stars, or of asteroids, the motions of which are predictable from dynamical theory. There are, however, many complicating factors. The rotation of our Milky Way around the center in Sagittarius produces, for example, systematic motions for even the most distant stars of our Milky Way system.

In recent years there have been many advances toward the establishment of an improved fundamental reference system of stellar positions and of proper motions. According to Fricke of Heidelberg, who is one of the leaders in the field, the best published proper motions for eleventh-magnitude stars now have accidental errors of the order of ±0.008 second of arc and systematically they should be correct to within ±0.004 second of arc. Within a decade it should be possible to halve both errors and from then on we should be progressing rapidly on the road to perfection.

In the 1940's research was initiated toward the measurement of proper motions with reference to objects entirely outside our galactic system, namely, faint galaxies. The Lick Observatory Survey was begun by Wright in the late 1940's. The first-epoch plates were taken with the Lick 20-inch Ross astrograph in the early 1950's, mostly by Shane and Wirtanen, and the first second-epoch plates are being taken 20 to 25 years later. The measurement of the proper motions of the faint stars, mostly of fifteenth to seventeenth magnitudes, is being done by Vasilevskis and Klemola. Since the minimum distances of the reference galaxies are of the order of 100 million parsecs, the crosswise motions of these galaxies can under no circumstances amount to as much as 0.00001 second of arc per year, and the reference frame of the faint galaxies is therefore immovable and stable to a much closer tolerance than the precision of 0.001 second of arc per year that is the sort of stability we ask for in our measurements of faint stars. An independent parallel survey that is about to yield results is that undertaken by Deutsch and associates at Pulkova Observatory in the U.S.S.R. In the Pulkova Survey photographic plates are centered on bright galaxies with distances in the range 2 to 100 million parsecs. The reference points of unquestioned stability are the knots in the spiral and elliptical galaxies. As is often the case, the Southern Hemisphere has initially received much less attention than the Northern. However, Yale and Columbia Observatories have established a station in Argentina with a telescope nearly identical with the Lick 20-inch Ross Astrograph. First-epoch plates are now being taken, and in 25 years we may expect to have available many new precision proper motions for southern declinations.

In the early 1970's a new approach was developed for the measurement of fundamental star positions and proper motions. Radio astronomers perfected and applied a technique for measuring positions of radio sources with very high precision. By the use of interconnected arrays of three or more steerable radio telescopes, researchers measure the declinations of radio sources with a precision of the order of $\pm$ 0.01 second of arc and without any reference to optical standards. Differences in right ascensions between radio sources can also be measured with comparable precision, even though the radio sources may be far apart in the sky. Since they know the right ascensions of several radio sources identified with the optical equivalents of these sources, investigators can derive a reference zero-point for the right ascensions of all radio sources. There are available already radio and optical positions for about 100 radio sources, which possess optical equivalents and which are either radio stars (Algol and Beta Lyrae are radio stars!) or radio galaxies that have been identified optically. The fundamental reference system of star positions and proper motions for the future will almost certainly represent a system based in part on optical and in part on radio data.

For a star with a known distance of d parsecs, it is possible to translate the angular value of the proper motion in seconds of arc per year, which is indicated by the Greek letter μ (mu), into a linear velocity expressed in kilometers per second, V. The simple formula reads:

$$V = 4.74\mu d.$$

A star with an annual proper motion $\mu = 0.1$ second of arc per year, at a distance $d = 50$ parsecs from the Sun, will have a linear crosswise velocity $V = 23.7$ kilometers per second. The linear velocities of the stars range generally from a few to 100 kilometers per second. Higher values are rarely found and the average linear velocities are about 20 kilometers per second. Since observations show that the average crosswise linear velocities are about the same for nearby and for distant stars, the proper motions offer a fairly good distance criterion. The nearby stars have, on the average, considerably larger proper motions (in seconds of arc per year) than the more remote ones. We can be fairly certain that a star which stands out from its neighbors by its large proper motion has a good chance of being relatively nearby, and this star should be a good candidate for having its name placed on a program for the measurement of trigonometric parallaxes. The program for the U. S. Naval Observatory 61-inch reflector, for example, is almost entirely limited to stars with observed proper motions of 0.2 second of arc per year or greater.

To select the faster-moving stars for subsequent and accurate parallax measurements, the astronomer uses a device known as a "blink microscope." Two plates of the same region taken some years apart are so arranged that first one and then the other is viewed through the eyepiece of the microscope. When the alternation takes place quickly enough, the main pattern of stars appears unchanged but the stars that have a large angular motion will apparently hop back and forth in the field. We can thus distinguish those stars that are likely to be our nearest neighbors from the general run of stars. In Chapter 3 we shall see how useful it is to select these nearest neighbors and we shall discover how crowded—or rather, just how empty—is our particular part of space.

Radial Velocities

When we deal with proper motions, we are concerned only with the angular displacements of the stars projected upon the celestial sphere. Proper motions tell us nothing about the velocities of the stars in the line of sight. If the stars are moving in space, some must be getting closer to us, others farther away. To measure these line-of-sight velocities, or *radial velocities,* we turn again to the spectrum, which yields radial velocities measured in kilometers per second.

Since light is a wave motion, it is not surprising that it has some of the characteristics of sound. We have all noticed how the pitch of an automobile horn drops suddenly as an approaching car passes. If a star is moving toward us, the wavelength of the light we receive from it is shortened and the absorption lines in its spectrum shift their positions toward the violet end of the spectrum. When the star is moving away from us, the lines are shifted toward the red. The shift in wavelength is proportional to the relative speed of the star and the observer. This correlation, known as the Doppler effect, is expressed by the simple equation

$$\frac{\lambda - \lambda_0}{\lambda_0} = \frac{v}{c},$$

where λ is the observed wavelength of a line in the spectrum of the star, λ_0 the laboratory standard rest wavelength of the same line, v the line-of-sight or *radial velocity* of the star, and c the velocity of light (186,000 miles per second).

How do we measure the shift in wavelength for a given star? Figure 39, which is a reproduction of two spectra of Castor, shows the stellar spectra with a laboratory standard spectrum photographed on either side. The displacement resulting from radial velocity can readily be seen by direct inspection of the photograph. The spectrum-line shifts are affected by the motion of the Earth as well as by that of the star itself. The effects due to the Earth's rotation and revolution can, of course, be calculated and removed. The radial velocities of stars listed in catalogues have been corrected for the Earth's motion; they are listed as "reduced to the Sun."

One complication that sometimes enters into the measurement of radial velocities arises from the fact that many of the stars are found to have velocities that change in a regular cycle over a definite period of time. A companion star—often invisible—may be present and the two stars revolve about each other under their mutual gravitation. These spectroscopic double or binary stars are very interesting in themselves and from them we derive information about the masses of the stars, but they have delayed considerably the determination of the radial velocities of many fainter stars. Enough spectral photographs of a star with variable radial velocity must be taken to find out how much of the observed speed is due to the orbital motions in the system and how much to the velocity of the system as a whole. The two spectra of Castor shown in Figure 39 prove that the component of Castor for which the spectra are shown is itself a spectroscopic binary.

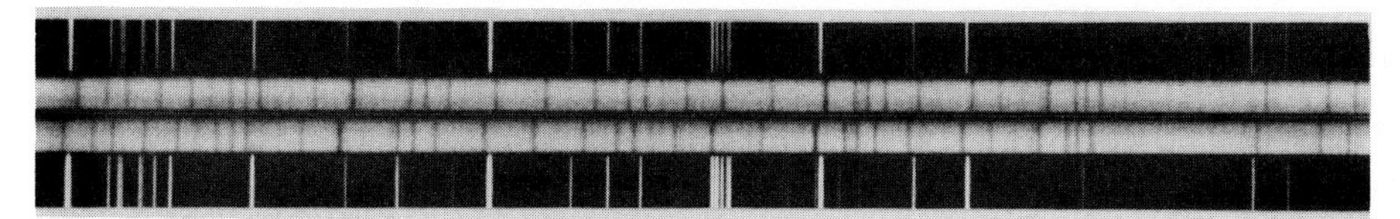

39. Radial-velocity effects in Castor. The bright bands in the center crossed by dark lines are two spectra of the same component of Castor taken at different times. The bright lines on either side are the comparison spectrum, photographed at the same time as the star's spectrum to serve as a standard of reference. Each component of the double star Castor is itself a spectroscopic binary. The upper one of the star's spectra shows a radial displacement due to a velocity of recession of 24 miles per second; the lower one shows a radial-velocity displacement of approach amounting to 32 miles per second. (Lick Observatory photograph.)

Much of the early radial-velocity work was done by Campbell, Wright, and Moore at the Lick Observatory with the 36-inch refractor and the three-prism Mills spectrograph. The Lick *Catalogue of Radial Velocities* includes almost all naked-eye stars in both hemispheres. Several observatories are now engaged upon the measurement of radial velocities of the fainter stars. We require great numbers of radial velocities for the detailed study of the Galaxy, and the task of obtaining them has been a vast one. In addition to the leading American observatories, Canadian, French, and Soviet observatories have been most active in the gathering of data for the Northern Hemisphere; the Dominion Astrophysical Observatory in British Columbia deserves honorable mention. In the Southern Hemisphere, radial-velocity measurements have been made especially in South Africa at the Cape and Radcliffe Observatories, and at Mount Stromlo Observatory in Australia. From time to time, radial-velocity catalogs have been prepared, which bring together all available data. In 1953, R. E. Wilson of Mount Wilson and Palomar Observatories (now Hale Observatories) issued such a *General Catalog*, with entries for 15,107 stars. A new catalog has been assembled under the direction of Evans of the University of Texas and another one is in preparation at Kitt Peak National Observatory by Abt.

Modern instrumental developments are helping to speed up the acquisition of data. Semiautomatic techniques for the measurement of the spectral photographs, which employ modern computer methods of data processing, are making it possible to reach faint stars and measure their radial velocities with high precision. For the brighter stars of solar type, we can now obtain radial velocities with accidental and systematic errors of the order of ± 1 kilometer per second and less. The precision goes down by a factor 2 or 3 when we go to fainter stars or measure the radial velocities of stars with few sharp lines in their spectra, such as the A and B stars, but no self-respecting observatory involved in radial-velocity work likes to publish radial velocities with probable errors in excess of ± 5 kilometers per second.

Every stellar spectrograph has four basic components. At the focus of the telescope, the image of a star is set upon a narrow *slit*. A *collimator*, of lens or mirror type, is used to focus the illuminated slit upon the *dispersive element*, now generally a precision *line grating*, and the resulting spectral band is photographed with the aid of a *camera*, now not infrequently a fast Schmidt-type camera. By placing the telescope at a mountain site with excellent seeing conditions, one obtains small star images and the slit can be closed down to a narrow width and still transmit most of the star's light. The grating spreads the light into a spectrum with minimum light loss and the high-quality camera records it all with precision. Mirrors are now generally covered with coatings of aluminum instead of silver, which assures permanent high reflectivity even in the ultraviolet; the camera photographing the spectrum has either highly reflective mirrors or lenses made of transparent optical glass, covered with a transparent antireflection coating. The photographic manufacturers, notably the Eastman Kodak Company, have traditionally provided astronomers with photographic emulsions of great sensitivity, capable of recording the spectra of the stars for each desired wavelength interval. At Kitt Peak National Observatory, the fast ultraviolet spectrograph on the 36-inch reflector is capable of recording the spectrum of a seventh-magnitude star with a *dispersion* of 63 angstrom units per millimeter inside 6 minutes; an eleventh-magnitude star would yield a spectrum of sufficiently good quality for measurement in about 3 hours.

We now have available special *image-conversion tubes,* which intensify each image by a factor 10 or greater before it is finally recorded. Such equipment is becoming standardized and is coming into general use for radial-velocity measurements. The gain in speed more than offsets the slight loss of precision in the radial velocities obtained with an image-tube attachment. With the 200-inch reflector and an image-conversion tube, spectra with a dispersion of 128 angstrom units per millimeter are now obtainable for eighteenth-magnitude stars with exposure times of 5 hours. New digitized measuring engines help to increase the precision of measurement of the radial velocities from the spectrum photographs. The task of the person who measures the plate is simplified and the speed of recording is increased.

Photographic recording has been the traditional means of obtaining a spectrum suitable for radial-velocity measurements. Photoelectric techniques show great promise for future development. One method—pioneered by Griffin in Cambridge, England—is to use a mask that carefully images the principal spectrum lines of a star and to shift the mask laterally for minimum response, which marks the position at which the absorption lines are centrally located in their mask positions. Instantaneous measurement of the star's radial velocity at the telescope is then possible. Griffin used a 234-aperture mask to mimic the absorption lines in the spectrum of Arcturus, and he finds it possible to measure radial velocities of K stars (with spectra not unlike that of Arcturus) to the ninth magnitude with errors of ± 2 kilometers per second, using the Cambridge 37-inch reflector. There are many possible variations of the Griffin technique. A related, but different, approach is rapid scanning of small portions of the spectrum.

Thus far we have dealt only with the measurement of radial velocities for one star at a time. Beside the spectrum of each star, we impress on our photograph a comparison spectrum, produced by a luminous source mounted near the slit of the spectrograph. This gives the necessary reference lines against which we can measure the shifts of the lines in the star's spectrum. It would speed up the determination of radial velocities of faint stars if we could derive from one single photograph the radial velocities of a considerable number of stars. Numerous attempts have been made to obtain radial velocities *en masse* from objective-prism plates, which record many stars simultaneously. The hardest problem is to obtain some reference point against which to measure the shift due to radial velocity. One approach, originally developed by Pickering and improved by Bok and McCuskey at Harvard Observatory, was to use a liquid absorption filter, a vessel filled with a solution of neodymium chloride and placed in front of the photographic plate. In each spectrum a molecular absorption band is produced, which closely resembles a stellar absorption line. It serves as the desired reference zero for radial-velocity measurements. To date, the absorption method has not given results of the desired accuracy, but the basic technique is capable of further development. Good results have been obtained by Fehrenbach at the Haute Provence Observatory and at the European Southern Observatory with a scheme in which the objective prism is rotated through 180° several times in the course of one exposure. The result is that each star has two images, one with the violet end of the spectrum to the left, the other with the violet to the right. The distance between the two images of the same spectrum line in a star can be used to derive the displacement due to the star's radial velocity. The Fehrenbach technique is already providing useful material relating to the radial velocities of faint stars.

Cooperation in Research

The tradition of cooperation in research is deeply rooted in the astronomical profession. There are many reasons why this is so. First, the number of research astronomers in the world is not large (hardly more than 2,500) and there is so much work to be done that some sharing of effort is necessary to achieve common research goals. Second, we are all dealing with the same physical universe, and some consultation between workers in each area of research activity is necessary if we wish to avoid duplication of effort. Third, it is difficult to combine results by different observers unless a special effort is made to have all observations on a comparable basis: for example, in magnitude and color measurements we prefer to have all observers accept common standards and use clearly specified color filters that permit reduction from one color system to another; in proper-motion work, we prefer to have all of our motions referred to a single established fundamental system of positions and motions. Fourth, the telescopic and auxiliary equipment is often so expensive that one institution or one nation (especially some of the smaller nations) cannot afford to build what it would like to have available for its own use, and thus cooperative ownership and management become essential.

Organized cooperative research relating to the structure and motions of our Milky Way system originated in the second half of the nineteenth century. The two most famous examples of the results of such research were the *Astrographic Catalogue*, an attempt to photograph the entire sky in zones of declination, with each participating observatory assuming responsibility for the measurement of the positions of the stars on the photographs in its assigned zone, and the *Astronomische Gesellschaft Catalogue*, primarily a German undertaking, to provide accurate meridian-circle positions for large numbers of stars. The most far-reaching effort to effect international cooperation was that initiated by Kapteyn in 1904 at Groningen in Holland—the so-called Plan of Selected Areas. Kapteyn's Plan captured the imagination of the astronomical world and as a result of his appeal we now possess for many stars in the Selected Areas accurate measurements of positions, proper motions, radial velocities, magnitudes, spectral types, and color indices.

In 1918 the need was felt for a broad international organization embodying the whole of astronomy, and, in consequence, the International Astronomical Union was established. The IAU—by which three initials the organization is known throughout the astronomical world—held its first formal meeting in 1922 in Rome; it has met since then once every 3 or 4 years, with only one longer interval between meetings, that caused by World War II. The principal continuing function of the IAU is to provide through its fifty permanent commissions a medium for international contact between workers in the same field the world over. The IAU has been responsible for the initiation of many cooperative projects and it serves as the clearing house for older programs like those mentioned above. Whenever the need arises, the IAU, with the support of UNESCO, sets up conferences with a limited number of specialists in attendance. These Symposia are generally initiated by one of the Commissions of the IAU; they are attended by 100 to 200 experts in the field (including a good representation of young workers in the area) and the papers are published in a Symposium Volume. IAU Commission 33, on Galactic Structure and Dynamics, has been active in sponsoring several of these Symposia. In addition, Commission 33 has been a cosponsor for several related Symposia, one on Interstellar Matter and Gas Dynamics, for example. One of the key features of these Symposia is that they are of a truly international character.

A more recent form of cooperation is one by which astronomers of different observatories share common research facilities. In the past decade several national observatories have been created. In the United States, the Kitt Peak National Observatory, with headquarters in Tucson, Arizona, and the National Radio Astronomy Observatory at Greenbank, West Virginia, with headquarters in Charlottesville, Virginia, are the two major developments, the first for optical astronomy, the second for radio astronomy.

At Kitt Peak National Observatory, 60 percent of the telescope time is assigned to visitors from other observatories, the remaining 40 percent reserved for research by Kitt Peak staff members. The Observatory has more than half a dozen telescopes, with apertures ranging from 16 to 158 inches (Fig. 23). These telescopes are fitted with the best of auxiliary equipment: spectrographs, photometers, polarimeters, infrared-recording equipment, and so on. Kitt Peak National Observatory has a major installation for solar research. Closely related to Kitt Peak National Observatory is the Cerro Tololo Inter-American Observatory in Chile (Fig. 22), with equipment not unlike that at Kitt Peak, and with its own 158-inch reflector under construction. Chilean and other Latin American astronomers have privileged access to this facility. All of these installations have been built with funds provided by the U. S. National Science Foundation, which also pays the cost of operation.

Also supported by the National Science Foundation is the National Radio Astronomy Observatory at Greenbank, West Virginia (Figs. 26 and 27). Its principal instrumentation is a beautiful 140-foot precision steerable radio telescope and a 300-foot meridian radio telescope precise enough for research to 11 centimeters wavelength. There is also a small array of movable antennas and various special-purpose instruments. Work will soon begin on a Very Large Array for high-resolution interferometry at radio wavelengths. It will be built at a high-altitude site in New Mexico.

The astronomers of Holland, Sweden, West Germany, France, and Belgium have combined their efforts toward the establishment of a European Southern Observatory. It was built and is operated jointly by the five nations and it is also located in Chile. The largest telescope under construction for this Observatory is a 140-inch reflector. Australia and Great Britain have under way a joint effort for the construction and installation of a 150-inch reflector, to be located on Siding Spring Mountain in New South Wales, Australia; a large British Schmidt telescope is already in operation at the site.

Comparable efforts of collaboration between observatories are now in effect in many parts of the world. In the United States, the University of California astronomers at Berkeley, Santa Cruz, La Jolla, and Los Angeles jointly operate Lick Observatory on Mount Hamilton, with a 120-inch reflector as their major telescope. Universities with high-quality astronomy departments located in climates that are not really suitable for astronomical observations often put their more powerful instruments at good sites, either within or outside the borders of their country. Leiden Observatory has long had a southern station at Hartebeestpoortdam in the Transvaal, South Africa, the British have maintained observatories in South Africa, and the Uppsala Observatory has had its Schmidt telescope in Australia at Mount Stromlo Observatory. In the area of radio astronomy two major national ventures are doing much toward fostering international cooperation. The 100-meter radio telescope of the Max Planck Institute near Bonn, Germany, has many astronomers from other nations on its staff, and the Westerbork Array in Holland, a mile-long string of 12 interconnected radio telescopes, is used by radio astronomers of many nationalities. There are many cooperative ventures in the Soviet Union, the best known of which is the newly established Soviet Observatory in the Caucasus, with a 250-inch reflector as its primary instrument. Small wonder that astronomers are good world travelers and fine ambassadors of international good will!

3
The Sun's Nearest Neighbors; Stellar Populations

How does the Sun rank among the stars? Is it very brilliant, average, or somewhat dull? As students of the Milky Way, we are interested in the answers to these questions not so much because we care about the Sun itself, but rather because we wish to know what varieties exist among the stars and just what proportions there are of different kinds of stars in a sample of space, and how our sample compares with other parts of the Milky Way.

The Brightest Stars and the Nearest Stars

We shall consider first two lists of stars. The first (Table 1) includes the bright stars that we know by name, together with some bright far-southern ones that cannot be seen from northern latitudes. These stars show a wide range in color, from blue Rigel and Spica, yellow Capella, and orange Arcturus to red Betelgeuse and Antares. For all these stars, values of the parallax and distance are available, but for the more distant ones the values are uncertain; their trigonometric parallaxes are of the same size as the unavoidable errors of measurement.

Figure 40 shows how the brightest stars vary in spectral class and in absolute magnitude. All spectral classes are present, but we note that 11 out of the 20 are the very hot B or A stars. All but one of these are more luminous than the Sun, with its visual absolute magnitude of +4.8. In fact, Rigel may shine as brightly as 52,000 Suns and five others are 1,000 or more times as bright as the Sun.

The star that comes closest to being a twin of our Sun in spectrum and luminosity is our nearest neighbor, Alpha Centauri, the brighter component of a visual binary star. It is only 1.33 parsecs away. Deneb, which is probably the most distant star in Table 1, is at about 450 parsecs, a distance too great for the accurate measurement of its parallax; the

Table 1. The twenty brightest stars.

No.	Name	Visual apparent magnitude, V	Spectral luminosity class	Parallax (arcsec)	Visual absolute magnitude, M_V	Distance (parsec)	Visual luminosity (Sun = 1)	Remarks
1	Sirius	−1.47	A1V	0.375	+1.4	2.7	23	(1)
2	Canopus	−0.73	F0Ib	.018	−3.0	56	130	
3	Alpha Centauri	0.33	G2V	.751	+4.7	1.33	1.1	(2)
4	Vega	.04	A0V	.123	+0.5	8.1	52	
5	Arcturus	.06	K2IIIp	.090	−0.2	11.1	100	
6	Rigel	.08	B8Ia	—	−7	250	52,000	(3)
7	Capella	.09	G8III+F	.072	−0.6	13.7	145	(4)
8	Procyon	.34	F5IV	.288	+2.6	3.5	7.6	(5)
9	Achernar	.47	B5IV	.023	−2.7	44	1,000	
10	Beta Centauri	.59	B1II	.016	−3.4	62	1,900	(6)
11	Altair	.77	A7V	.198	+2.3	5.1	10	
12	Betelgeuse	.80	M2Iab	—	−5	150	8,300	(7)
13	Aldebaran	.86	K5III	.048	−0.7	20.8	160	(8)
14	Spica	.96	B1V	.021	−2.4	48	760	(9)
15	Antares	1.08	M1Ib	.019	−2.5	53	830	(10)
16	Pollux	1.15	K0III	.093	+1.0	10.8	33	
17	Fomalhaut	1.16	A3V	.144	+2.0	6.9	13	
18	Beta Crucis	1.24	B0.5IV	—	−5	175	8,300	
19	Deneb	1.26	A2Ia	—	−7	450	52,000	(11)
20	Regulus	1.36	B7V	.039	−0.7	25.8	160	(12)

Remarks

(1) Sirius possesses a white-dwarf companion, V = 8.5.

(2) Alpha Centauri is a triple system. The brightest star is accompanied by a companion, V = 1.5, K6. Proxima Centauri, V = 10.7, M5eV, is the third member of the system.

(3) The trigonometric parallax of Rigel is not measurable; the listed values of M_V, distance, and luminosity are approximate. Rigel is a spectroscopic binary and has a binary star, V = 7.6 and V = 7.6, SpB5, nearby.

(4) Capella is a triple system, with companions, V = 10.0, M1, and V = 13.7, M5; Capella itself is a spectroscopic binary.

(5) Procyon itself is a spectroscopic binary. It is a triple system consisting of a close companion with V = 13.5, and a second companion with V = 12.2. The close companion is a white dwarf, spectral type F.

(6) Beta Centauri is a spectroscopic binary and has a faint component, V = 8.7.

(7) Betelgeuse is a semiregular variable and a spectroscopic binary. Its trigonometric parallax is not measurable; the listed values of M_V, distance, and luminosity are approximate.

(8) Aldebaran has a dM2 component, V = 13.5.

(9) Spica is variable and is a spectroscopic binary.

(10) Antares is a semiregular variable. It has a dwarf B4V companion, V = 5.

(11) The trigonometric parallax of Deneb is not measurable; the listed values of M_V, distance, and luminosity are approximate values only.

(12) Regulus has a dK1 companion, V = 7.9, and possibly two fainter companions.

Notes

1. Alpha Crucis is not listed with the twenty brightest stars, since it is a beautiful optical double, with one component, V = 1.58, B1IV, the other V = 2.09, B3n, both stars too faint for inclusion in our list.

2. The data for Table 1 are all taken from the Yale *Catalogue of Bright Stars* (third ed., 1964).

3. In the calculation of the visual luminosities, the visual absolute magnitude of the Sun has been taken to be +4.8 and visual apparent magnitude −26.8.

40. The twenty brightest stars. The diagram is based on the data of Table 1 and shows the visual absolute magnitudes plotted against spectral type. (Data from the Yale *Catalogue of Bright Stars.*)

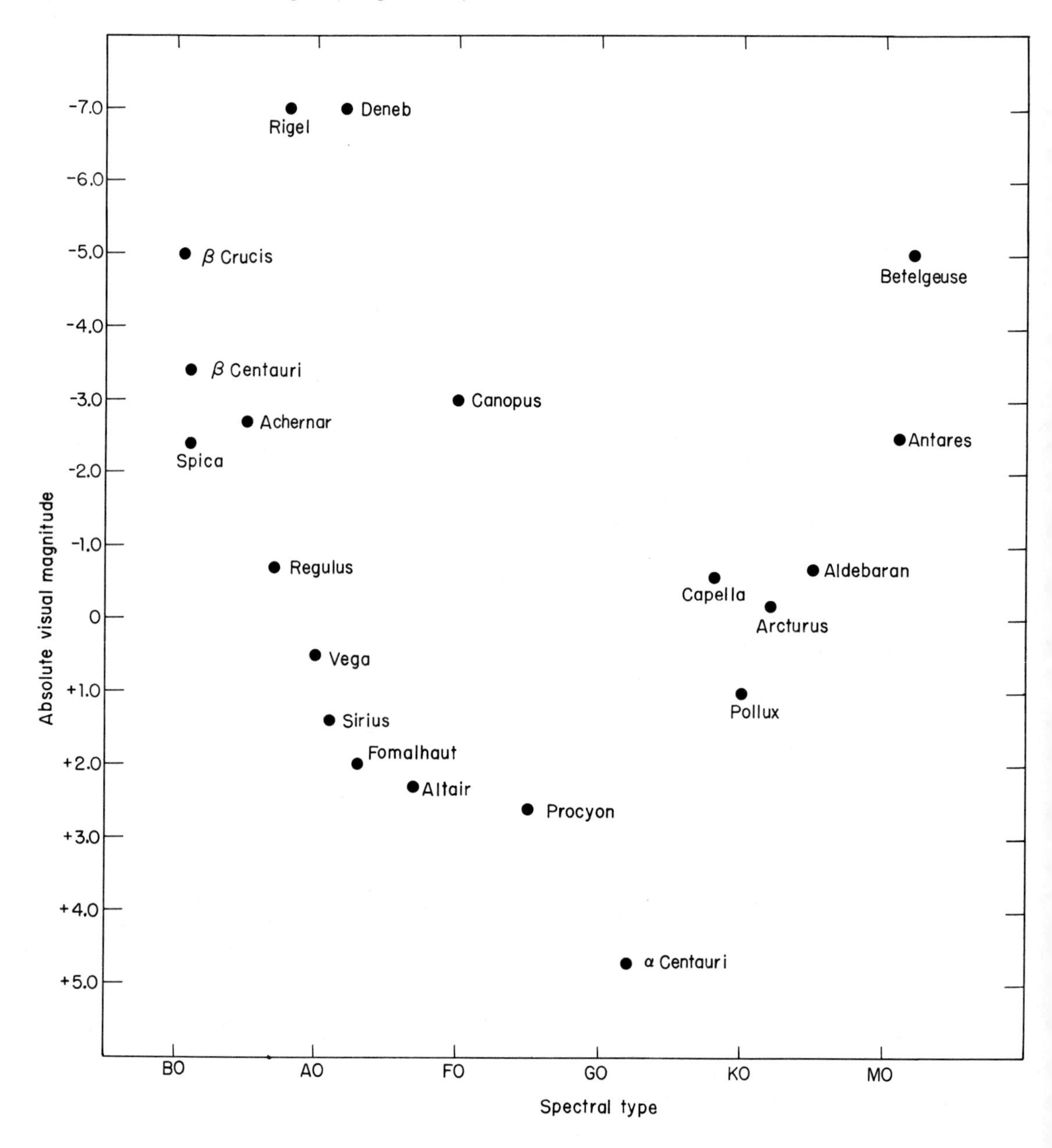

same applies to Rigel, Betelgeuse, and Beta Crucis. If we consider the volume of space that we have covered before we reach a star such as Deneb, we see that it is some $(450/1.33)^3$ or about 40,000,000 times as large as that we would have to explore to find Alpha Centauri. We may catch a minnow in our hands close to shore, but we must sail far if we wish to harpoon a whale! We begin to suspect that such distant stars as the blue supergiants Rigel and Deneb must be very rare objects in space as compared with our Sun and stars like it.

The M-type star Betelgeuse is also a very luminous star, but it is closer to the Sun than Rigel. Since its surface is comparatively cool, it must be very large to give off so much light. Betelgeuse and Antares are among the very few stars for which the diameter can be measured; the instrument used is called an interferometer. The diameter of Betelgeuse has been found to be about 600 times that of the Sun; that of Antares is only slightly less. A star so big that Mars could move in its orbit around the Sun inside that star is indeed a giant! Betelgeuse is variable in brightness and apparently also changes in size with an irregular period.

Let us next look at the list of nearby stars in Table 2. Included are all stars that are known to be within a distance from the Sun of 5 parsecs. We note that four of the brightest stars in this list—Sirius, Altair, Procyon, and Alpha Centauri—are also in Table 1. They are conspicuous stars in our sky because they are nearby rather than because of their exceptional intrinsic luminosity. The rest of the stars are much fainter, both apparently and absolutely. Figure 41 shows how these stars rate in spectral class and in absolute magnitude.

Our first list is complete, but we cannot be sure that the same is true of our second list. A few stars have been added in the past 20 years and a certain amount of reshuffling has taken place as more accurate values of the parallaxes have been determined. No doubt there will be more additions in the future. There are many faint stars with large proper motions for which parallaxes have not yet been measured, and in time we shall undoubtedly add more stars to the number of our faint neighbors. It is not believed, however, that the number will be increased very greatly. From the average speed of the stars near our Sun, the average total density of matter in our region of space can be estimated. This value, which is a little above 0.1 solar mass for a cube 1 parsec on a side, is very close to the value of the average space density that we find in the sphere with a radius of 5 parsecs, if we make reasonable assumptions as to the masses of the nearby stars and take account of the contributions from the interstellar gas and dust.

There are some very real differences between the kinds of stars on our two lists. The first list contains almost entirely what are known as giants and supergiants. They are of all spectral classes from B through M. Our second list comprises the dwarfs, or, as we prefer to call them, the "main-sequence stars." We see in Fig. 41 the very definite tendency for the stars to fall along a diagonal line, the faintest ones being the reddest; this diagonal line is called the *main sequence.* We find no stars of class K or M that are of the same absolute magnitude as our Sun. We see by an inspection of Figs. 40 and 41 that K and M stars are either intrinsically brighter than our Sun or very much fainter. We clearly see a division into giants and dwarfs. Among the nearby stars (Table 2), we find none of the brilliant B stars or G, K, or M giants. These

Table 2. Stars nearer than 5 parsecs.[a]

No.	van de Kamp no.	Name	Visual apparent magnitude	Spectral class	Parallax (arcsec)	Visual absolute magnitude	Visual luminosity	Distance (parsec)
1	1	Sun	−26.8	G2		4.8	1.0	—
2	2	Alpha Centauri A	0.1	G2	0.760	4.5	1.3	1.32
3	3	Alpha Centauri B	1.5	K6	.760	5.9	0.36	1.32
4	4	Alpha Centauri C	11.	M5e	.760	15.4	.00006	1.32
5	3	Barnard's star[b]	9.5	M5	.552	13.2	.00044[b]	1.81
6	4	Wolf 359	13.5	M8e	.431	16.7	.00002	2.32
7	5	BD+36°2147[b]	7.5	M2	.402	10.5	.0052[b]	2.49
8	6	Sirius A	−1.5	A1	.377	1.4	23.	2.65
9	6	Sirius B	8.3	DA	.377	11.2	0.0028	2.65
10	7	Luyten 726-8A	12.5	M6e	.365	15.3	.00006	2.74
11	7	Luyten 726-8B	13.0	M6e	.365	15.8	.00004	2.74
12	8	Ross 154	10.6	M5e	.345	13.3	.0004	2.90
13	9	Ross 248	12.2	M6e	.317	14.7	.00011	3.15
14	10	ϵ Eridani	3.7	K2	.305	6.1	.30	3.28
15	11	Luyten 789-6	12.2	M6	.302	14.6	.00012	3.31
16	12	Ross 128	11.1	M5	.301	13.5	.00033	3.32
17	13	61 Cygni A	5.2	K5	.292	7.5	.083[b]	3.42
18	13	61 Cygni B	6.0	K7	.292	8.3	.040	3.42
19	14	ε Indi	4.7	K3	.291	7.0	.13	3.44
20	15	Procyon A	0.3	F5	.287	2.6	7.6	3.48
21	15	Procyon B	10.8	—	.287	13.1	0.0005	3.48
22	16	Σ 2398 A	8.9	M4	.284	11.2	.0028	3.52
23	16	Σ 2398 B	9.7	M5	.284	12.0	.0013	3.52
24	17	BD+43°44 A	8.1	M1	.282	10.4	.0058	3.55
25		BD+43°44 B	11.0	M6	.282	13.13	.00040	
26	18	CD−36°15693	7.4	M2	.279	9.6	.012	3.58
27	19	τ Ceti	3.5	G8	.273	5.7	.44	3.66
28	20	BD+5°1668	9.8	M4	.266	11.9	.0014[b]	b 376
29	21	CD−39°14192	6.7	M1	.260	8.8	.025	3.85
30	22	Kapteyn's Star	8.8	M0	.256	10.8	.0040	3.91
31	23	Kruger 60 *A*	9.7	M4	.254	11.7	.0017	3.94
32		Kruger 60*B*	11.2	M6	.254	13.2	.00044	3.94
33	24	Ross 614 *A*	11.3	M5e	.249	13.3	.0004	4.02

Table 2 (continued).

No.	van de Kamp no.	Name	Visual apparent magnitude	Spectral class	Parallax (arcsec)	Visual absolute magnitude	Visual luminosity	Distance (parsec)
34		Ross 614 *B*	14.8	—	.249	16.8	.00002	4.02
35	25	BD−12°4523	10.0	M5	.249	12.0	.0013	4.02
36	26	Van Maanen's Star	12.4	DG	.234	14.2	.00017	4.27
37	27	Wolf 424 *A*	12.6	M6e	.229	14.4	.00014	4.37
38		Wolf 424 *B*	12.6	M6e	.229	14.4	.00014	4.37
39	28	G158−27	13.8	M	.226	15.5	.00005	4.42
40	29	CD−37° 15492	8.6	M3	.225	10.4	.00058	4.44
41	30	BD+50°1723	6.6	K7	.217	8.3	.040	4.61
42	31	CD−46°11540	9.4	M4	.216	11.1	.0030	4.63
43	32	CD−49°13515	8.7	M3	.214	10.4	.0058	4.67
44	33	CD−44°11909	11.2	M5	.213	12.8	.00063	4.69
45	34	Luyten 1159−16	12.3	M8	.212	13.9	.00023	4.72
46	35	BD+15°2620	8.5	M2	.208	10.1	.0076	4.81
47	36	BD+68°946	9.1	M3.5	.207	10.7	.0044[b]	4.83
48	37	L145−141	11.4	—	.206	12.6	.0008	4.85
49	38	BD−15°6290	10.2	M5	.206	11.8	.0016	4.85
50	39	40 Eridani A	4.4	K0	.205	6.0	.33	4.88
51	39	40 Eridani B	9.5	DA	.205	11.2	.0027	4.88
52	39	40 Eridani C	11.2	M4e	.205	12.8	.00063	4.88
53	40	BD+20°2465	9.4	M4.5	.202	10.9	.0036[b]	4.95
54	41	Altair	0.8	A7	.196	2.3	10.	5.10
55	42	70 Oph-iuchi A	4.2	K1	.195	5.7	0.44	5.13
56		70 Oph-iuchi B	6.0	K6	.195	7.5	.083	5.13
57	43	AC+79°3888	11.0	M4	.194	12.4	.0009	5.15
58	44	BD+43°4305	10.1	M5e	.193	11.5	.0021[b]	5.18
59	45	Stein 2051 A	11.1	M5	.192	12.5	.0008	5.21
60		Stein 2051 B	12.4	DC	.192	13.8	.0003	5.21

Note: D indicates a white dwarf; DA a white dwarf of spectral type A.
[a]Based on P. van de Kamp, *Annual Review of Astronomy and Astrophysics* 9 (1971), 104–105, Table 1.
[b]Has an invisible companion.

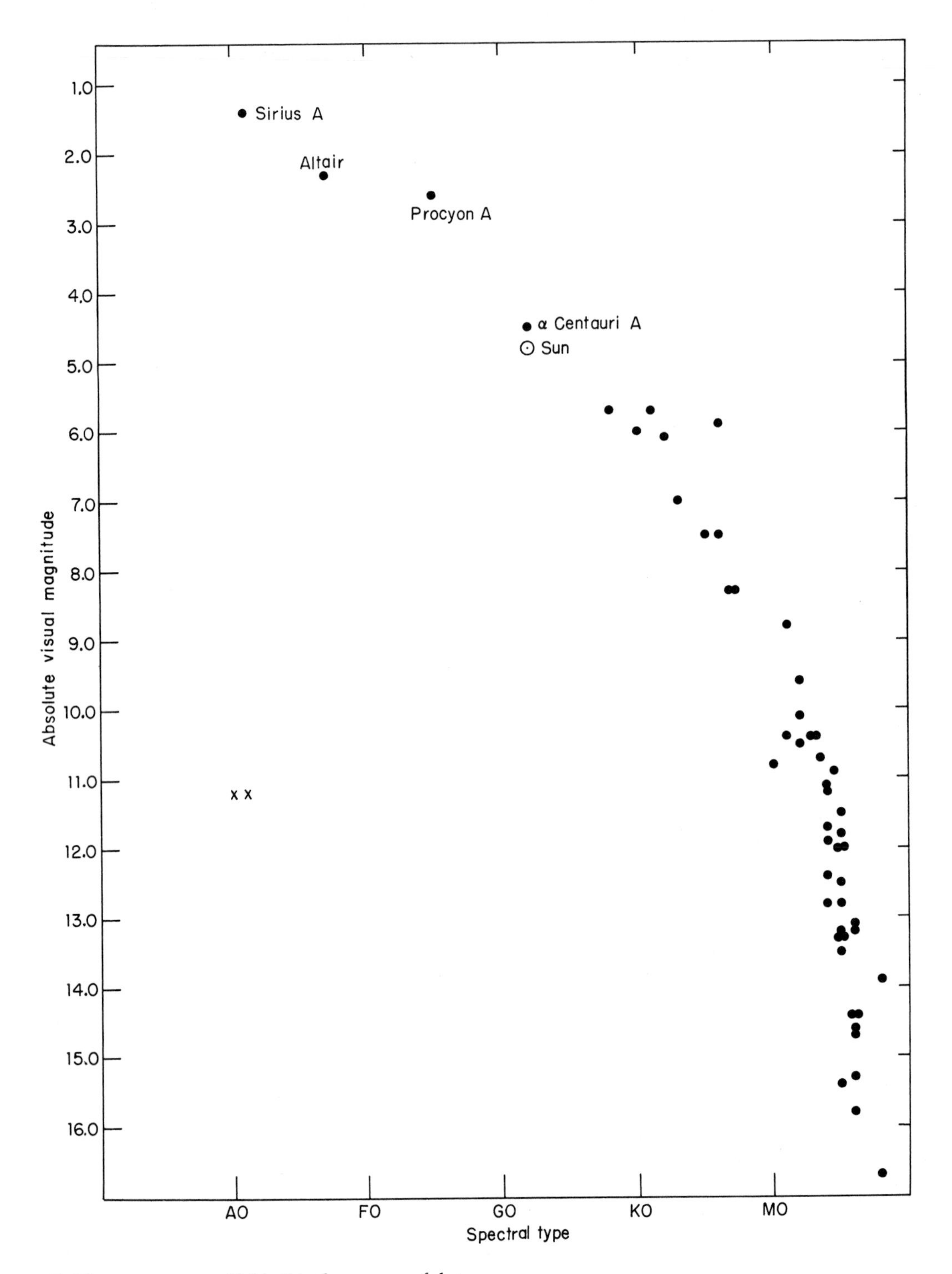

41. The nearest stars. Table 2 is the source of data for this diagram, in which visual absolute magnitude is plotted against spectral type. Only the two white dwarf A stars (DA) are plotted. (Data from van de Kamp.)

show up among the naked-eye stars because of their high intrinsic luminosities, but actually they are very rare objects in space, and those that we see are all well beyond our 5-parsec sphere.

The most common variety of star in Table 2 is the faint red dwarf of class M. These stars make up almost two-thirds of the list of our neighbors and range in luminosity from about 1/100 to 1/60,000 that of the Sun. Van de Kamp has pointed out that the faintest known star, which is less than one-tenth as bright as the faintest star in our list, lies outside the 5-parsec sphere. It is van Biesbroeck's star, BD +40°4048, with an estimated visual absolute magnitude $M_v = +18.6$, spectral class M5, visual luminosity 1/300,000 that of the Sun, and a distance from the Sun of about 6 parsecs. We may reasonably expect that there will be additions to our list at the tail end of the line.

Many of the M dwarfs are known as "flare stars." Their usual luminosities are very low, but they may on occasion brighten by 2 magnitudes or more for brief intervals, and several have repeated this flare-up more than once. At least five stars in Table 2 are flare stars. Bright lines are observed in their spectra, which are classified with a letter e added after their assigned spectral class.

We should mention that our list of 44 nearby stars (including the Sun) is really a list of 44 systems. Eleven of these 44 stars are double, and two are triple. In addition there are seven stars with as yet unseen companions. These invisible companions are indicated by the perturbation in proper motion of the visible star. Their masses begin to approach the range of planetary masses as we know them in our solar system and are of the order of a few hundredths of a solar mass; the largest planet, Jupiter, has a mass 0.001 that of the Sun. The companion of Ross 614A has been found by Miss Lippincott to be a star of very low mass, one-twelfth that of the Sun. Luyten has found a double star for which the mass of each component is even smaller, probably not more than one twenty-fifth the mass of the Sun. The gap between stars and planets seems to be closing!

Our list of the nearest stars contains six blue-white stars of very low intrinsic brightness, the representatives of the class of *white dwarfs.* These stars constitute a most interesting group among our neighbors; the most famous white dwarf is the companion of Sirius. Two others are companions to bright stars, and two are single. When the companion of Sirius was discovered, its high temperature combined with its low luminosity suggested a most unusual object and probably a very rare one. No one had previously considered the possibility of the existence of stars with masses only slightly less than that of the Sun but with radii hardly larger than that of the Earth. In a recent tabulation of stars within 20 parsecs of the Sun, Gliese lists 49 white dwarfs. The searches of Luyten and others have shown that white dwarfs are as common objects as stars like our Sun. Altogether, Luyten has identified about 3,000 "certain, probable, and possible" white dwarfs, which he calls "the easiest stars to identify and the hardest to observe." The criteria for their discovery are a large proper motion and a color index comparable to that of an unreddened B or A star.

The bulk of the known white dwarfs are of about fourteenth apparent magnitude. For more than half of those discovered, spectra and parallaxes have been determined, so that they can be fitted into a spectrum–absolute-magnitude array. They form a progression not quite parallel to the main sequence, with luminosities from $M_v = +10$ to $+15$, hence 0.01 to 0.0001 the brightness of the Sun. For the stars that are strictly white dwarfs, the range in color index $B - V$ is between -0.1 and $+0.6$. Several are found as companions to other stars, so that their masses can be determined and their sizes estimated. Most of them seem to range in mass between 0.1 and 1 solar mass. Their sizes fall between the diameter of the planet Mercury and that of the planet Uranus, hence one-third to four times the size of the Earth. Luyten has recently estimated that 5 percent of all stars are probably white dwarfs.

The Hertzsprung-Russell Diagram

The discussion of the brightest and the nearest stars has given us a good idea of the kind of stars that exist in space. Figures 40 and 41 show the relation between absolute magnitude and spectral class for the brightest stars and the nearest stars. Such arrays are known as Hertzsprung-Russell diagrams, carrying the names of two of the great astronomers of our time.

Figure 42 shows a schematic drawing of the diagram with the main sequence and the giant branches drawn in. This diagram is one that applies to the stars that are thought of as being among the Sun's neighbors, taken in the larger sense, and spoken of as Population I. These stars are known to be characteristic of the stellar population in the arms of the spiral galaxies, and most of the varieties occur also in the spaces between the arms; Population II, which has a very different spectrum-luminosity diagram, is characteristic of the outer halo and of the inner parts of a galaxy, the sections where no spiral structure is present, and is especially found in elliptical galaxies. The division into Population I and Population II seems to be a very significant distinction when it comes to showing how spectral characteristics and peculiarities of motion are related, and considerations of populations will be vital when we come to discuss the origins of stars, their probable evolution, and their ultimate fates. We note that Population I contains very bright B stars, supergiants of all classes, main-sequence stars, and the white dwarfs.

Figure 42 includes one new variety of stars, the *subdwarfs.* These are mostly Population II dwarfs of an extreme variety, whose atmospheres are almost pure hydrogen, with an unusually small percentage abundance of the heavier elements, the metals. These are the stars that presumably originated for the greater part in the earliest stages of cosmic evolution.

The diagram indicates only the mean values of the absolute magnitudes of the stars. There will be a certain spread about the mean values of these absolute magnitudes, but in general the stars conform reasonably well to the rules set up by the majority.

The Sun's Motion

There is excellent evidence to show that our Sun moves with a speed of about 18 to 20 kilometers per second with respect to the stars within a distance from it of 100 parsecs or so. The evidence for this solar motion appears clearly if we examine the radial velocities of the naked-eye stars. Figure 43 shows a projection of the sky, so drawn that equal areas on the sphere are equal areas on the paper. The sky has been divided into 94 equal areas. For each of these regions the available radial velocities of the naked-eye stars have been averaged. Altogether, the radial velocities of 2,149 stars are used, so that in each area there are on the average 20 to 25 stars.

If the Sun were at rest and if the stars were

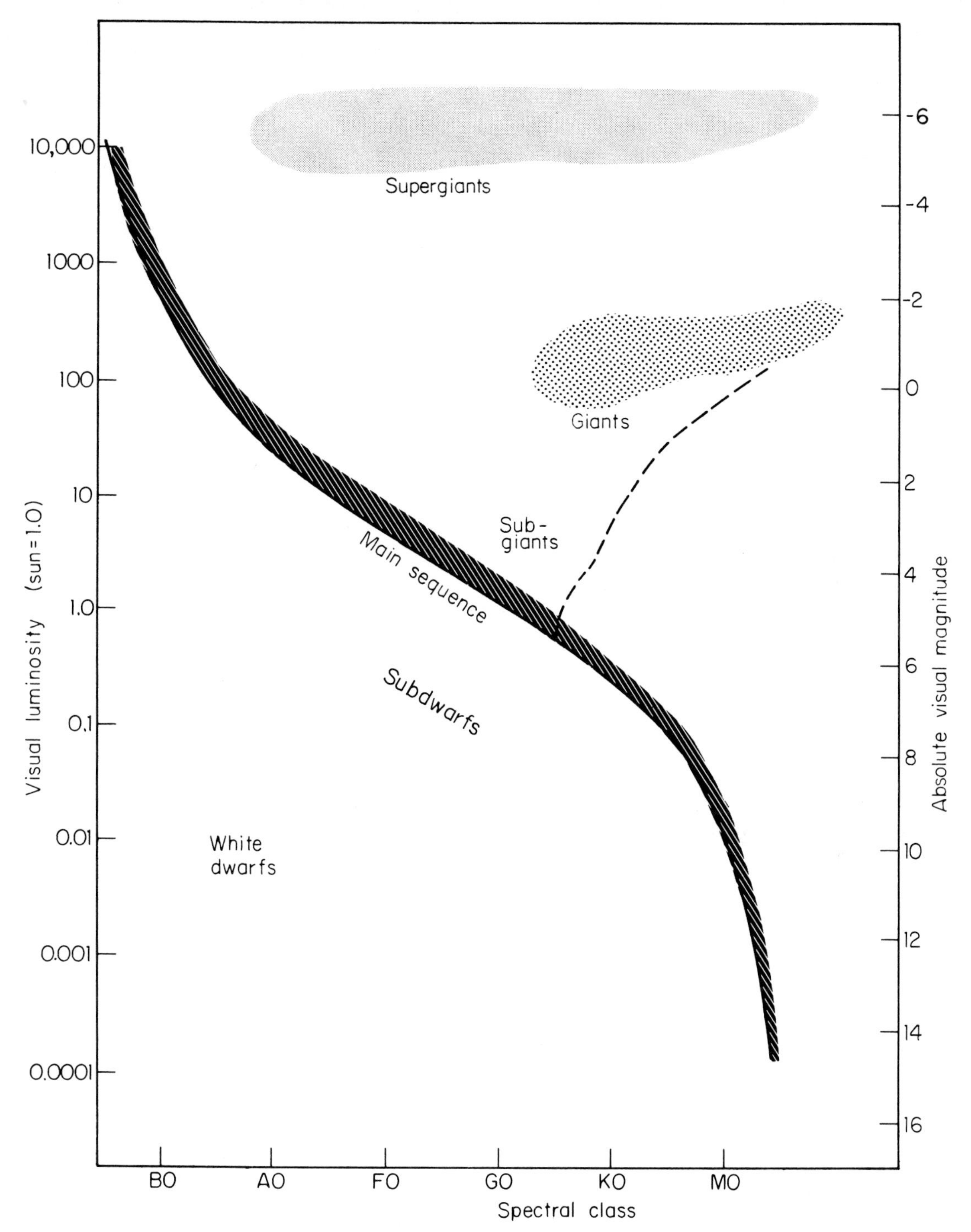

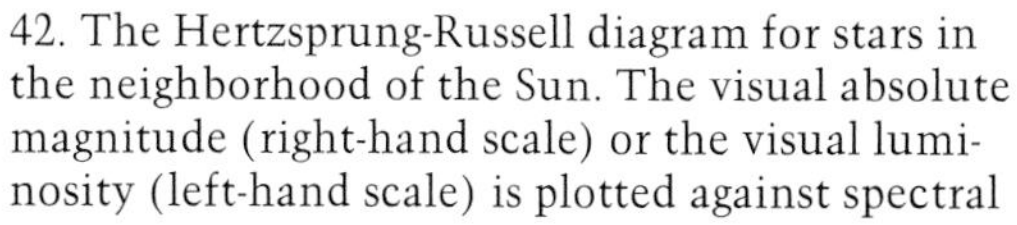
42. The Hertzsprung-Russell diagram for stars in the neighborhood of the Sun. The visual absolute magnitude (right-hand scale) or the visual luminosity (left-hand scale) is plotted against spectral class. Such a diagram was first plotted by H. N. Russell in 1913. (From Lawrence H. Aller, *Atoms, Stars, and Nebulae,* revised edition, 1971.)

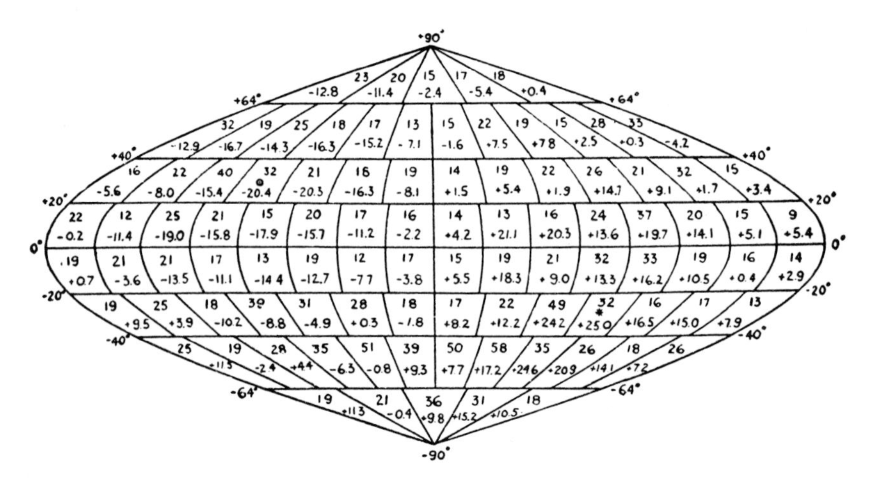

43. The Sun's motion from radial velocities. Averages of the radial velocities for 2,149 naked-eye stars measured at the Lick Observatory. The position of the apex of the Sun's motion is shown by a small circle near the position of greatest average negative radial velocity.

moving at random, there should be roughly as many positive as negative values for the radial velocities and the resultant average should be close to zero in each of the areas. Figure 43 shows that this is not so. The stars near the circle in the upper left-hand part of the figure have an average radial velocity of -20 kilometers per second; those near the asterisk in the lower right-hand part average $+20$ kilometers per second. Since a negative value indicates approach, it would seem that, as viewed from the Sun, all the stars in one part of the sky are marching toward us; in the opposite region they are moving away.

With reference to the naked-eye stars, the Sun is moving toward a point in the constellation Hercules, not far from the bright star Vega, at the rate of 20 kilometers per second. The circle in Fig. 43 is the *apex* of the sun's motion; the asterisk is the *antapex.*

At the rate of 20 kilometers per second, in the course of a year (about 31,600,000 seconds) the Sun will travel 630,000,000 kilometers, or the equivalent of 4.2 astronomical units. Our Earth moves steadily along with the Sun at the same rate. After an interval of 25 years we are more than 100 astronomical units from our starting place. We can take sights on the stars as we march along and measure their average displacements.

Let us see how the proper motions of the stars are affected by the Sun's motion. Figure 44 is the same projection of the sphere that we had in Fig. 43 but this time we have chosen it to assist us in examining the proper motions of 726 A stars of the fifth apparent magnitude. They are divided into 42 groups according to their positions in the sky, and for all groups we have determined the average proper motions, which are shown by the lengths of the arrows and the directions in which they point.

You will notice that most of the arrows seem to be directed away from the solar apex and toward the solar antapex. With respect to the Sun, the stars seem to be moving toward

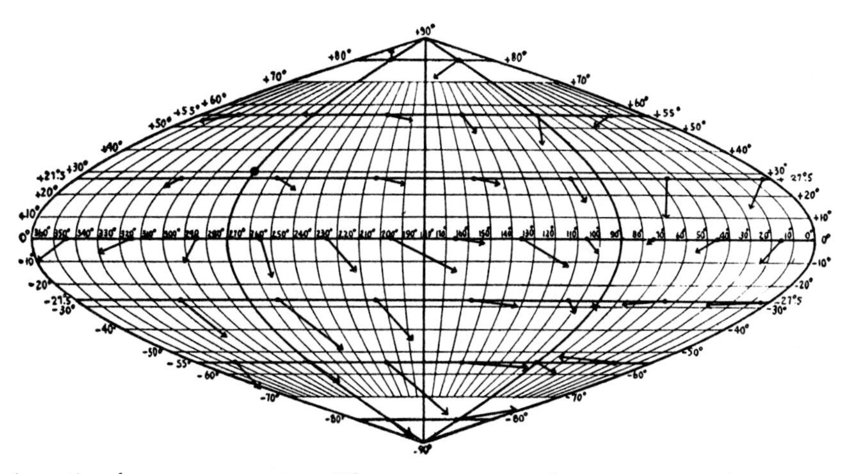

44. The Sun's motion from proper motions. The arrows represent the average directions and amount of proper motions for 726 A stars of the fifth apparent magnitude.

the antapex. What is really happening is that the Sun is moving in the opposite direction with respect to the A stars.

Since proper motions are angular displacements on the sky, they will tend to be largest along the circle at right angles to the direction in which we are traveling. In Fig. 44 they should be greatest along the projection of the circle that falls halfway between the apex and the antapex, and this is indeed observed. The average of these maximum lengths is of the order of 0.040 second of arc per year for the A stars in Fig. 44, that is, for the A stars of fifth apparent magnitude.

Mean Parallaxes

We may assume that the A stars of fifth apparent magnitude yield a solar motion from radial velocities not unlike that shown by the mass of the bright stars (Fig. 43), that is, of the order of 20 kilometers per second toward the vicinity of Vega. Hence we know that, at the average distance of the fifth-magnitude A stars, 20 kilometers per second corresponds to an average annual proper motion of 0.040 second of arc. In making a comparison between the two values, we bear in mind one outstanding difference between radial velocities and proper motions. Radial velocities do not depend on the distances of the stars. So long as a star is bright enough to appear on a spectrum plate, its radial velocity can be determined and it matters not at all whether the star is nearby or distant. Proper motion, on the other hand, varies with the distance, growing smaller for a given linear velocity as distance increases. The effect of the solar motion on the stars will therefore depend on the average distance of the group of stars under investigation. The effect on average proper motion will be largest for the nearby stars. You see now why we chose the proper motions of a group of A stars all of one apparent magnitude. For the A stars there is no division into giants and dwarfs, such as occurs in later-type stars, so that all the A stars of the fifth magnitude will be at about the same distance. What is the average distance for the A stars of fifth apparent magnitude?

By definition, the parallax of a star is the

angular displacement corresponding to one astronomical unit at the distance of the star. We have an angular displacement of 0.040 second of arc per year, which corresponds to 20 kilometers per second or 4.2 astronomical units per year for the group of stars. The mean parallax of our A stars is therefore $0.040/4.2 = 0.0095$ second of arc and their average distance is of the order of 105 parsecs.

This distance is well beyond the distance for which accurate trigonometric parallaxes can be obtained. We shall have to remember that it is only an average distance and that it may be considerably in error for an individual A star. But it is a reliable average and we can go one step further and compute from it the corresponding average absolute magnitude for our A stars. For a star of apparent magnitude $m = 5.5$ at a distance $d = 105$ parsecs, the absolute magnitude can be computed from the formula $M = m + 5 - 5 \log d$, which gives in our special case $M = +0.4$ as the mean absolute magnitude for our A stars. The average A star in our sample is intrinsically about 60 times as luminous as our Sun.

We are really solving two major problems through an analysis of the radial velocities and proper motions of the fifth-magnitude A stars. First, we show from the radial velocities of these stars that they exhibit the effects of the standard solar motion in radial velocities, and we can check on the direction of the Sun's motion with respect to the average of our A stars from an analysis of their proper motions. Second, we can derive from the material a mean parallax, or mean distance, for the A stars of fifth apparent magnitude, and follow this up by deriving the mean absolute magnitude for the stars under investigation.

Our method of measuring *mean parallaxes* can be applied to any group of stars with known proper motions, provided that the stars are evenly distributed over the sky. The method is still applicable for groups of stars that have an average proper motion as small as 0.010 second of arc per year and that are, therefore, four times as far away as the A stars in our special example. Radial velocities are readily obtainable for enough stars of any particular group to allow us to check that the solar motion agrees with the usual value found from other groups of stars. If all is well, we can immediately combine the results and compute a mean parallax, a mean distance, and a mean absolute magnitude. Our method will not, however, produce correct results if, unfortunately, we were to select a group that has peculiar systematic motions of its own.

The method of mean parallaxes has one great advantage over the basic trigonometric method: the total displacement from which the mean parallax is found increases with time. By waiting a longer interval, we can obtain increasingly more accurate values for the proper motions and so more reliable mean parallaxes. If the stars are so distant that we do not get a measurable effect in 10 years, we can wait 50 years. With the aid of trigonometric parallaxes we cannot reach beyond distances of the order of 50 parsecs, but through the use of the mean-parallax method we can gather information that is still reasonably accurate for average distances up to 400 parsecs or more from the Sun. Research on mean parallaxes continues to be of importance for extending our basic scale of distances. The most extensive and significant early researches in this area were done by A. N. Vyssotsky and Emma Williams Vyssotsky.

Luminosity Functions

We can see from the comparison of the brightest stars and the nearest stars that from these alone we cannot satisfy our desire for completeness in both numbers and kinds of stars. In our tiny bit of space within a radius of 5 parsecs we are fairly complete as to total numbers. But we are totally lacking in the bright B stars and the red giants that are so conspicuous among the brightest stars. If we go to a sufficient distance to include at least some of these, we need such a large volume of space that we are far from having complete information as to just how many other stars it contains.

Yet the astronomer wants to know how many stars of given absolute magnitude and spectral class are on the average present in a given volume of space. It lies within our power to count the stars to given limits of apparent magnitude. But counts alone are not enough. The astronomer does not want to see the sky as primitive people see it, as a curved surface on which bright lights appear. What we must add is the third dimension, so that we can see how the stars are spread out in space. We obviously wish to obtain the complete tabulation of the absolute magnitudes for a typical sample volume in the Milky Way, at least for the parts of the system in which our Sun is located. This tabulation, which lists the numbers of stars per unit volume for successive intervals of absolute magnitude, is known as the *luminosity function.*

It is clear that the problem of deriving the luminosity function will have to be tackled piecemeal. We can build up the general luminosity function by studying first the separate tabulations for different spectral classes and then putting them together in true proportions to obtain the total picture. We shall naturally be very curious to find out whether the same mixture of spectral classes holds in different parts of the Milky Way. But from stars with classified spectra, we shall principally obtain data on the stars intrinsically more luminous than our Sun and we know already that the great majority of stars are intrinsically fainter than the Sun and that mass spectral classification alone will not do the trick. We shall consider first the faint end of the luminosity function, after that the bright end, and then combine the two.

Proper motion is a powerful tool for the selection of the nearer stars, those with measurable trigonometric parallaxes, from among the stars at large. A star at a distance of 4 parsecs and moving at a rate of 20 kilometers per second of linear crosswise motion will have an observed annual proper motion close to 1 second of arc; about the same proper motion would be observed for a star at 6 parsecs distance moving at the rate of 30 kilometers per second. Since the linear crosswise speeds of stars range generally from 5 to 50 kilometers per second, with the majority between 20 and 30 kilometers per second, we may expect the majority of the stars with proper motions of the order of 1 second of arc to have measurable trigonometric parallaxes. Parallax observers naturally concentrate their efforts on varieties of stars that show promise of having measurable trigonometric parallaxes, so it is not surprising that parallax determinations are available for a sizable sample of all known stars with large proper motions. The sample is sufficiently large to permit the use of statistical techniques to correct for the absent stars and make the census figures complete and representative ones. The faint end of the general luminosity function is by now fairly well known from an analysis of available data on trigonometric parallaxes for stars with known proper motions in excess of 0.2 seconds of arc per year. The most complete study of the faint end of the luminosity function has been made in recent years by Luyten of the University of Minnesota.

Trigonometric parallaxes fail us miserably when we attempt to derive the luminosity function for stars with absolute magnitudes like that of our Sun and brighter. For these stars we turn to studies involving proper motions and radial velocities and also to work on the precise determination of spectral types and luminosities. Fortunately, we do not have to depend exclusively on the evidence from mean parallaxes for the stars beyond the reach of the trigonometric method. Spectrum-luminosity classification is now possible for all stars for which objective-prism spectra of not too low dispersion are obtainable. By careful inspection of the spectra, one may obtain an estimate of any star's absolute magnitude with an uncertainty that need not exceed one-half magnitude. In other words, we can take a fairly accurate star-to-star census and combine the results to derive the distribution function of absolute magnitudes—the luminosity function—per unit volume for all stars of all spectral types taken together. The resulting general luminosity function is shown in Fig. 45.

By a combination of the results of the various methods we can thus obtain reliable data on the distribution of the absolute magnitudes of the stars in the Sun's neighborhood. The pioneer investigations of Kapteyn were followed by extensive studies by Seares and van Rhijn. The method by which we determine at least the bright end of the general luminosity function, by combining the luminosity functions of successive relatively small intervals of spectral class, was first used in 1932 by van Rhijn and Schwassmann.

The bright end of the general luminosity function has been determined with precision through studies done at the Warner and Swasey Observatory. The work of McCuskey and his associates not only gives a very valuable check on the luminosity function for the region directly around the Sun, but provides useful information regarding the variations in the shape of the function for regions within 500 parsecs of the Sun in the galactic plane. McCuskey finds for absolute magnitudes -1 to $+1$ somewhat greater numbers than shown by the curve of Fig. 45, but on the whole there is little that can be done to improve the original results of van Rhijn and Schwassmann. In the galactic plane, there are fluctuations of the order of 30 percent in the relative frequencies of the various absolute magnitudes. The curve shown in Fig. 45 represents in a way the combination of two luminosity functions, one derived from proper motions, the other from a combination of separate functions for each spectral class. The curve shown represents the number of stars in a cube 1,000 light-years (about 300 parsecs) on a side.

Marked systematic variations are found if we consider the frequency distributions of absolute magnitudes at some distance from the galactic plane. The absolutely brightest stars—B and A, for instance—thin out far more rapidly on a proportional basis than do the G and K dwarfs, and the shape of the luminosity function at a few hundred parsecs above or below the galactic plane differs markedly from that shown in Fig. 45, the principal difference being a decided deficiency for the brighter absolute magnitudes at a height of a few hundred parsecs above or below the plane as compared with the relative numbers in the plane.

To supplement the schematic Fig. 42, we list in Table 3 the mean values of the visual absolute magnitudes and the spreads around these means (dispersions) for the stars of each spectral group. The values listed are applicable to a sample volume of space in the part of the Galaxy where our Sun is located. The values given for the main-sequence (Luminosity Class V) stars of spectral groups

45. The luminosity function. The numbers on the vertical scale are the numbers of stars in a cube 1,000 light-years on each side for the photographic magnitudes shown on the horizontal scale. (From data compiled by van Rhijn, McCuskey, Kuiper, and Luyten.)

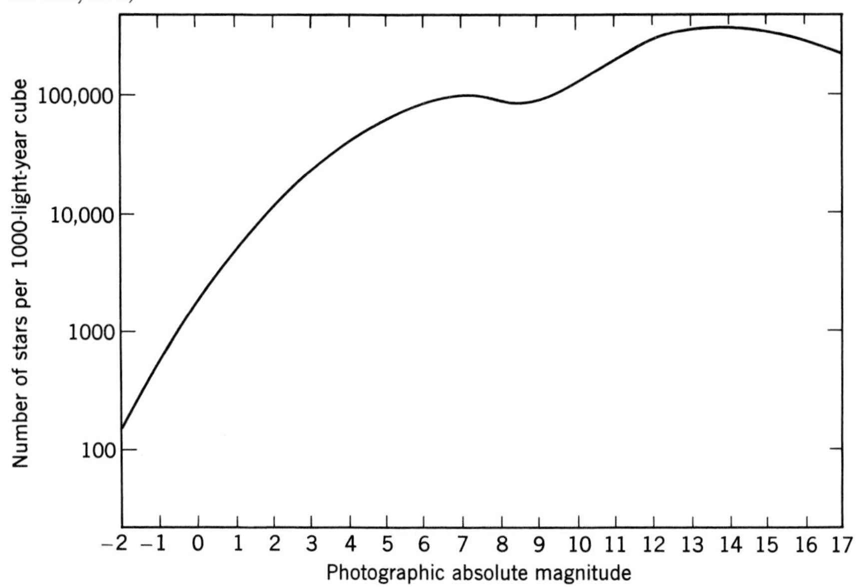

G5 and earlier and those of the subgiants (Luminosity Class IV) and of the regular giants (Luminosity Class III) are from McCuskey. We have added the standard main-sequence values for the K5V and M5V stars.

The bright end of the luminosity curve is the most important part for the study of the distribution of the stars in space. However, the dwarfs constitute the bulk of the population. Their luminosities are equal to or less than that of the Sun, so that it takes many fainter ones to shine as brightly as the giants. But their masses are not too different from the Sun's mass and these faintly shining bodies play an important role in determining the gravitational properties and the motions of the system.

Table 3. Mean visual absolute magnitudes and dispersions for different spectral groups.

Spectral group	Mean visual absolute magnitude	Dispersion around mean
B5	−1.0	±0.5
B8–A0	+0.2	.5
A1–A5	+1.6	.4
A7–F5	+2.8	.5
F8V–G2V	+4.5	.4
G5V	+5.0	.3
K5V	+7.2	.3
M5V	+12.3	.3
F8III–K3III	+0.9	.8
F8IV–K3IV	+3.2	.8
K5III–M5III	0.0	.6

Populations I and II

If we refer to the familiar distribution of absolute magnitudes and varieties of objects near the Sun as Population I, we find a very different distribution at great distances from the Sun, especially for the Sagittarius section. Following Baade, we refer to the latter as Population II. Population I, with its abundance of intrinsically very luminous objects, is apparently characteristic of the outer parts of our own and other galaxies, whereas Population II prevails in the central cores of galaxies like our own and in the thin outer halo of our Galaxy.

Historically it is interesting to note that the realization of a separation between the two populations came not as a result of studies of our own Galaxy, but rather from Baade's researches on our nearest large neighbor galaxy, Messier 31, the famous Andromeda spiral. Messier 31 is so far from us—its distance is of the order of 650,000 parsecs—that we cannot hope to observe in it stars with absolute magnitudes comparable to that of our Sun. Even with a powerful telescope like Palomar's 200-inch Hale reflector it is impossible to detect stars as faint as $M = -1$, stars that are intrinsically 200 times as bright as the Sun. Baade reasoned that it should be possible, however, to find with relative ease anywhere in Messier 31 the stars with absolute magnitudes $M = -3$ and certainly stars with $M = -5$ and brighter. Such stars are clearly shown in the outer parts of Messier 31, where spiral structure prevails, but Baade found it impossible to photograph in normal blue light any individual stars in the central regions of Messier 31. However, he succeeded with the help of red-sensitive photographic plates in resolving the nucleus of Messier 31 and the elliptical galaxy NGC 205 that accompanies it. The brightest stars proved to be red stars with visual absolute magnitudes between -3.5 and -4.0 and with color indices of the order of 1 magnitude and greater.

Baade concluded from these observations that the stars that constitute the nucleus of Messier 31 are very different (Population II) from those that are found in the outer spiral regions (Population I). The contrast between the two populations is not a minor one; the brightest stars of Population I are blue and have absolute magnitudes as high as -7 to -9, whereas the brightest stars of Population II are red and have absolute magnitudes of -3 at the most. Moreover, interstellar gas and dust abound in Population I, whereas in Population II gas and dust are either absent or obviously very minor constituents.

Baade's distinction between the two populations represents a major advance in the understanding of our own and other galaxies. We shall have to refer to it frequently, for not only does it possess tremendous significance for the study of the structure of galaxies, but the population approach has far-reaching consequences for problems of stellar birth and evolution.

In the present book we shall, for reasons of simplicity, refer mostly to the two basic populations. The reader should be warned, however, that the picture of two essentially different varieties of stars is an oversimplification. It is true that it seems to be supported by the spiral-galaxy data, but we are dealing there always with stars that are at least 5 absolute magnitudes brighter than our Sun; even our best equipment does not permit us to observe the common stars of moderate magnitude. The authors—and along with them quite a few of their colleagues—prefer to think in terms of at least three basic varieties of stars: Population I, responsible for the principal spectacular spiral features; the *Common Stars,* first so named by Oort, which inhabit the more or less amorphous regions between the spiral arms; and Population II, the nuclear population, which spills over to some extent into the outer parts.

Perhaps it is most fitting to think in terms of six major groupings, somewhat in the following manner:

Extreme Population I consists of the objects that show a very marked preference for association with structural spiral features. We include in our listing the interstellar gas and the cosmic dust, which are highly concentrated in the spiral arms and which provide the "building blocks" from which the new stars are formed. The stars associated with Extreme Population I are cosmically very young; the ages since formation from the interstellar gas and dust for the stars of Population I are 20 to 50 million years at the most, less than one-fifth of the period of one galactic revolution, hence less than one-fifth of a *cosmic year.* Some Population I stars have ages of less than 1 million years and some may have been caught in the act of formation, or directly after they had become stars. Population I is found only within a few hundred parsecs of the central galactic plane. It is confined to a wafer-thin galactic layer, a ring with an inner radius of 5,000 parsecs and an outer radius of 15,000 parsecs, about 500 parsecs thick.

Stars of Extreme Population I occur generally in regions where clouds of interstellar gas and dust are plentiful. The presence of *emission nebulae* (Chapter 8) is characteristic of regions with Population I; these nebulae can be found and studied by optical and by radio techniques. *Neutral atomic hydrogen* (H I) is recognized through its 21-centimeter radiation. *Cosmic grains* (solid particles) are found in dark nebulae (Chapter 9)—often with associated interstellar molecules present —and they can be detected by the reddening effects produced in the light of distant stars.

The varieties of stars associated with Extreme Population I are:

1. O to B2 stars, single, in clusters, or in associations;
2. Late-type supergiants, including especially those cepheid variables with periods in excess of 10 to 13 days;
3. Wolf-Rayet stars;
4. B emission stars;
5. T Tauri variable stars;
6. Certain infrared objects.

Normal Population I stars are principally the somewhat older stars, with ages since formation possibly as great as 2 or 3 cosmic years. These stars are old enough to have diffused away from their places of origin in spiral arms and from the structures to which they belonged originally. They are generally found close to the central plane of the Galaxy. The principal varieties of stars associated with Normal Population I are:

1. The B3 to B8 stars and the normal A stars. Open clusters with their brightest stars of these spectral types, but without O to B2 stars, are Normal Population I. These clusters do not fit well into observed spiral patterns and they have generally no associated interstellar gas;
2. Stars with strong metallic lines, spectral classes A to F;
3. The less brilliant supergiant red stars.

Disk Population stars, Oort's Common Stars, are next in line. They have cosmic ages in the range between 1 and 5 billion years, that is, between 5 and 25 cosmic years. Our Sun is one of the Common Stars. This population includes the great masses of the inconspicuous stars found mostly within about 1,000 parsecs of the central plane in a galactic belt with an inner radius of 5,000 parsecs and an outer radius of 15,000 parsecs, centered upon the galactic center. The following varieties of stars belong to the Disk Population:

1. G to K normal giant stars (Class III);
2. G to K main-sequence stars (Class V);
3. Long-period variable stars with periods greater than 250 days;
4. Semiregular variable stars;
5. Most disk planetary nebulae;
6. Novae (somewhat questionable);
7. Older open clusters.

Intermediate Population II stars are those that have considerable spreads in velocities, sufficiently high to be found farther above or below the galactic central plane than 1,000 parsecs. Whereas Population I stars and Common Stars move in nearly circular orbits around the galactic center, the Intermediate Population II stars move in quite elongated orbits around the same center. Many of these stars may have originated at distances of no more than 5,000 parsecs from the galactic center. Most of the really old stars (with ages of 50 to 80 cosmic years) may belong to this class. The principal varieties classed under Intermediate Population II are:

1. High-velocity stars;
2. Weak-line stars;
3. Long-period variable stars with periods less than 250 days, but greater than 50 days;
4. W Virginis cepheid variables;
5. RR Lyrae variable stars;
6. White dwarfs;
7. Oldest known open clusters.

Halo Population stars are those formed in the earliest stages of development of the Galaxy, which was then presumably much less flattened than at present. In all likelihood the building of elements heavier than hydrogen and helium had not progressed very far. The principal stars generally associated with the Halo Population are:

1. Subdwarf stars;
2. Halo globular clusters;
3. RR Lyrae stars;
4. Weak-line stars (extreme);
5. Highest-velocity stars.

Nuclear Population stars are a variety about which we know the least. The great strength of the sodium *D* lines in the spectra of the nuclei of many spiral galaxies, first noted by Spinrad, and the great strength of the cyanogen (CN) molecular bands (Morgan and Mayall) may suggest that the nuclear regions of our own and other spiral galaxies contain vast numbers of M-type dwarf stars. The stars listed below are generally considered as Nuclear Population stars, but we recognize from the start that such assignments are quite uncertain because of obvious observational limitations:

1. RR Lyrae stars;
2. Relatively metal-rich globular star clusters;
3. Planetary nebulae;
4. Vast numbers of M-type dwarf stars;
5. Giant stars (G to M) with strong cyanogen bands;
6. Infrared objects.

We note that RR Lyrae variables are listed as possibly belonging in some way to Intermediate Population II and also to the Halo and Nuclear Populations. Obviously some further sorting is in the offing!

The gradual progression from Baade's two populations to the sixfold classification we suggest is very much in line with the views advocated about 1950 by Soviet astronomers, notably Parenago and Kukarkin. They argued from the start in favor of a continuous sequence of stars grouped by their physical characteristics and motions into separate subsystems of the Galaxy. The Soviet astronomers have all along been the strongest supporters of a minimum of three major subdivisions: a highly flattened, an intermediate, and a roughly spherical component.

4 Moving Clusters and Open Clusters

On a clear night we notice, apart from the general hit-or-miss arrangement of the stars, a few places where the stars are closely clustered and seem to belong together. The Pleiades or the Seven Sisters, the Praesepe or Beehive Cluster, the Hyades in Taurus, the double cluster *h* and Chi Persei—all of these have been known since antiquity. To these naked-eye clusters, telescopic surveys have added many more. There are two varieties of star clusters: the open clusters and the globular clusters. In the present chapter, we shall be concerned with the open clusters only.

Moving Clusters

The stars of an open cluster are close together in space; they are not merely a transitory chance arrangement. If a cluster is real and has a lasting quality, then all its stars must share a common motion. They should, therefore, move through space in parallel paths and with identical speeds. If the group covers a large area of the sky—as do, for example, the Hyades—and is near enough to possess measurable proper motions, then the arrows that represent the directions of proper motion for the individual stars in the cluster will all seem to converge to one point on the celestial sphere. We generally refer to the open clusters that are close enough to us to show such measurable proper motions as *moving clusters.* The Hyades are the prototype of a moving cluster.

Lewis Boss first detected the convergent motion of the Hyades when he was preparing his *General Catalogue* (1914) of proper motions. For the stars in this cluster the proper motions are large, and it is possible to sort out the stars that belong to the cluster from the field stars (Fig. 46). Since the distances can be measured, it is possible to build up a picture of the cluster, discover the kinds of stars it contains, and determine how closely they are packed. In 1952 Van Bueren published a

46. The convergence of the proper motions of the Hyades cluster is shown by this diagram. All stars brighter than the ninth magnitude belonging to the cluster have been mapped. The sizes of the dots are a measure of the magnitudes of the stars. The arrows show the proper motion displacements that may be expected in the course of the next 18,000 years. It is apparent that the stars share a common motion in space. (This figure is by Van Bueren, who made the study at the Leiden Observatory.)

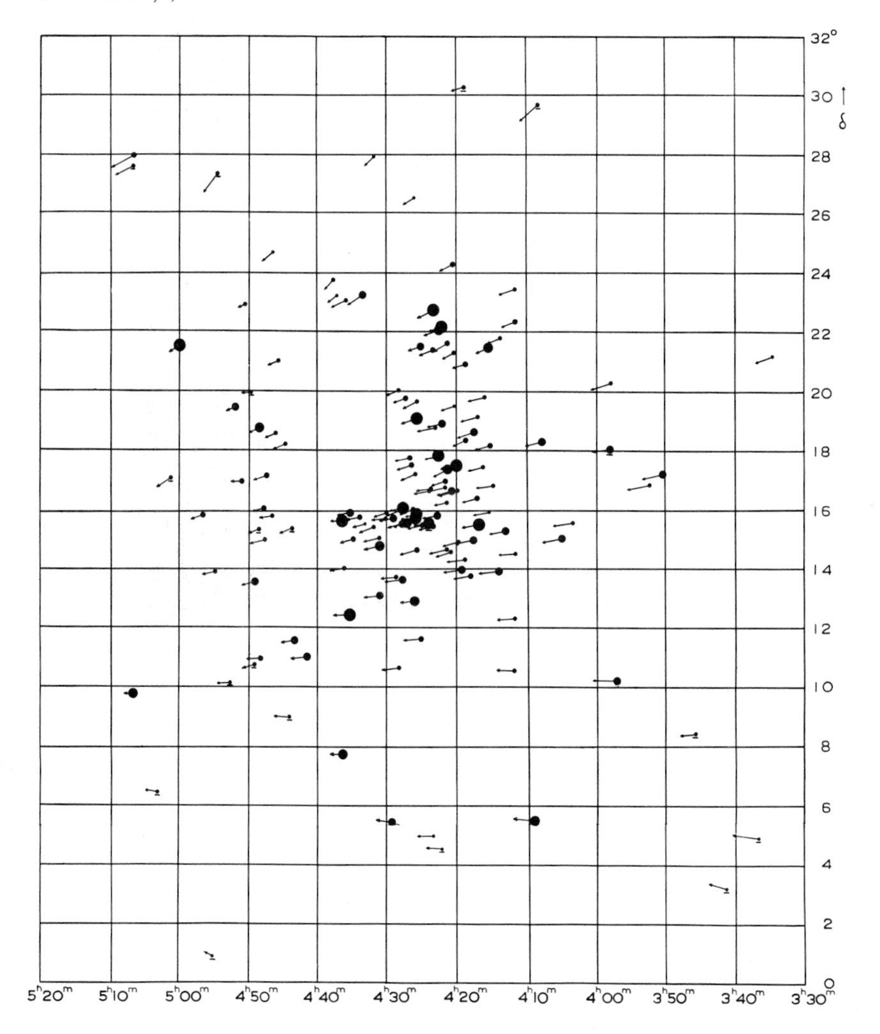

comprehensive study of the Hyades cluster, which he finds to be a flattened system with its shortest axis perpendicular to the galactic plane and about two-thirds as long as the axis in the plane. The cluster is nearby—only about 40 parsecs from the Sun—and Van Bueren lists about 350 stars as probable cluster members, from the brightest to stars intrinsically much fainter than our Sun. The Hyades members are concentrated in a somewhat irregular fashion toward the center of the cluster. The majority of the cluster members are G and K main-sequence stars. The bluest stars in the cluster are of spectral type A2 and there are a few G and K giants. In the core of the cluster the average density is at least three times that of the stars in the region around the Sun.

The Hyades moving cluster has been the subject of many investigations during recent years. Careful studies have been made of the colors and brightnesses of the recognized members, and many additions to Van Bueren's list have been published. The most recent comprehensive survey has been made by Van Altena, who has combined his new discoveries with those of Giclas and associates of Lowell Observatory and those of Luyten at the University of Minnesota. Van Altena estimates that his search is 93 percent complete for the central region of the Hyades cluster. The main sequence in the Hertzsprung-Russell diagram is very clearly marked, and there is a parallel subdwarf sequence. We now have a good list of Hyades members between visual apparent magnitudes 8 and 18, stars that have absolute visual magnitudes in the range +5 to +15. Eggen and Greenstein have found 15 white dwarf stars that are probable members of the Hyades moving cluster.

Another moving cluster that is playing an increasingly important role in galactic research is the Scorpius-Centaurus cluster. This is a moving group that is especially rich in B stars. Many of the stars that outline the magnificent arc of the Scorpion in the heavens belong to this moving group and so do most of the blue-white stars in Ophiuchus. The southern half of this great moving stream cannot be observed from northern latitudes, but it shines in all its beauty for southern observers. They see it as an isolated grouping of blue-white stars in Centaurus near galactic latitude +10°. Kapteyn first called attention to this moving stream in 1918, and the stream has since been the subject of many investigations. The checking on membership has progressed slowly. In recent years, Garrison has investigated the northern section and Glaspey has completed the analysis in the southern section. It appears that a good number of A stars, and some F stars, share the motions of the easily recognizable B-star members.

Moving clusters have a very special place in astronomical affairs, for it is from them that we can obtain precise basic information about absolute magnitudes of the recognized members. The principle by which this is done is really quite simple. When we plot on the celestial sphere the proper motions of the stars that we suspect of membership, then, as we have noted already, all of the proper motion vectors will seem to pass through one point on the sky, the *convergent* of the stream motion. This convergent point marks the direction toward which the stream is moving as viewed from the Sun and Earth. We have here a close parallel to the case of a meteor shower, in which all of the meteors seem to move across the heavens in paths intersecting at one single *radiant* point.

For each certain member of a moving cluster like the Hyades, we can determine the star's radial velocity (measured in kilometers per second) in addition to its proper motion. Knowing the distance in degrees between each star member and the convergent, we can deduce from these observed radial velocities the total stream motion in kilometers per second. Now, the proper motions toward the convergent are known in seconds of arc per year. For each star member, the linear velocity, in kilometers per second, corresponding to the proper motion can be readily predicted once we know the total stream motion and the position of the convergent. Hence we have for each star member both the measured proper motion in seconds of arc per year and the predicted linear velocity corresponding to it in kilometers per second. The distance of the star can thus be found. This result is most important, for it makes it possible to assign *individual* distances to all stars that are members of a moving cluster with a well-defined stream convergent. The Hyades cluster and the Scorpius-Centaurus cluster both fall in this category.

Moving clusters are of great importance to the astronomer in that they permit him to determine precise distances and absolute magnitudes for the member stars. We found in Chapter 3 that trigonometric parallaxes are excellent for purposes of calibration to distances of 20 parsecs from the Sun and Earth, but that they become increasingly less reliable for greater distances. The Hyades moving cluster, at an average distance of about 40 parsecs from the Sun, is excellently placed to yield distances for a number of stars beyond the reach of the trigonometric-parallax techniques, and the Scorpius-Centaurus cluster extends our range to well over 200 parsecs. It is especially important that, with the aid of the Scorpius-Centaurus moving cluster, we can obtain precise distances and absolute magnitudes for individual B and A stars. These are the varieties that are very scarce within 20 parsecs of the Sun and Earth.

The System of Open Clusters

The total number of moving clusters is small, since almost all open clusters are so far away that they do not show measurable proper motions. But in spite of the absence of observable motion we can learn much about the more distant open clusters. Some are rich in numbers of stars, others are little more than slight condensations in the sky. There are some 400 known clusters that are strictly open clusters; the Pleiades cluster (Fig. 47) is probably the best known of these. The majority of open clusters are located close to or in the band of the Milky Way. There are probably many more than 400 in our Milky Way system, but the more distant clusters are not noticed against the rich stellar background along the Milky Way. Furthermore, the great majority are hidden from our view by the intervening cosmic dust close to the galactic plane. The further cataloguing of open clusters continues to be a major task facing astronomers.

Much has been learned about the distances and physical characteristics of open clusters from the classification of the spectral types of the member stars. The most significant early research in this field was done in the late 1920's by Trumpler and by Shapley. In 1930 Trumpler summarized his researches in a paper published in a *Lick Observatory Bulletin* which today stands out as a classic of its time. Both he and Shapley stressed differences in spectral composition for some of the better-known clusters, and they intimated—as since has proved to be significant—that these differences might well have importance

47. The Pleiades cluster. The Pleiades are embedded in nebulosity. (From a photograph by Barnard.)

for problems of stellar evolution. From Trumpler's study there came the first conclusive proof of the presence of interstellar absorption at low galactic latitudes. We should note here that the first relevant observations for the law of space reddening followed also from studies by Trumpler on the intensity distribution of the continuum in the spectra of some heavily obscured distant stars.

During the 1930's, it became evident that much of value was to be learned from the Hertzsprung-Russell diagrams (generally referred to in the professional vernacular as H-R diagrams) of open clusters. There are several ways in which an H-R diagram can be plotted. The first method is to plot spectral class (horizontally) against apparent or absolute magnitude (vertically). In a second method, the measured color index of each individual cluster star replaces the spectral class. Now that it is possible by photoelectric techniques to measure color indices of faint stars quickly and precisely, the second method has become the more useful of the two.

Figure 48 shows a composite H-R diagram in which the principal branches of the color-magnitude arrays are plotted for some of the best-known clusters. In the diagram we find vertically the absolute visual magnitude M_v and horizontally the color index $B - V$, the difference between the blue magnitude B and the visual magnitude V. In this diagram our Sun would be located at the point near $B - V = +0.6$, $M_v = +5$. Upon inspection, we note that the Hyades cluster falls somewhere near the middle of the diagram and that it and Praesepe are rather similar in that they have no really blue stars and no stars as bright as absolute magnitude $M_v = 0$, that is, no stars as much as 100 times as bright as our Sun. In other words, these clusters contain a few mild giant stars, but they have no blue giants and certainly no supergiants among their members. Let us now examine some of the clusters that occupy the upper left-hand section of Fig. 48. Here we find some familiar star clusters, the Pleiades and the double cluster *h* and Chi Persei. The Pleiades cluster has a steep blue branch, with some stars reaching $M_v = -2.5$, that is, with 1,000 times the brightness of the Sun. The cluster *h* and Chi Persei outdoes all others, with blue supergiants and a few red ones at $M_v = -6$, fully 25,000 times as bright intrinsically as the Sun! But, turning to the lower half of the diagram, we find some inconspicuous clusters, like Messier 67, in which the bluest star is hardly bluer or brighter than the Sun.

There are two reasons why diagrams like Fig. 48 are of great interest to the astronomer. The first is that the information contained in them is helpful in the study of the distances of remote open clusters; the second is that from diagrams of this nature we may learn much about the ways in which the stars gradually evolve. In the present chapter, we shall concern ourselves principally with the first of these problems, saving the evolutionary implications for Chapter 11.

With the exception of not more than half a dozen nearby clusters, the distances of the remaining several hundred open clusters are too great to be measured by either the trigonometric or the moving-cluster technique. But it is possible to measure, by a combination of photographic and photoelectric techniques, the colors and magnitudes of very faint stars in open clusters. Color-magnitude arrays have now been measured for more than 100 open clusters. In every case one ends up with a diagram in which observed color (such as $B - V$ in Fig. 48) is plotted against the apparent visual magnitude V or blue

48. The Hertzsprung-Russell diagrams for the seven galactic open clusters *h* and Chi Persei, NGC 2362, the Pleiades, the Hyades, Praesepe, NGC 752, and Messier 67 have been combined in this diagram by Johnson and Sandage. The main sequences of the clusters converge toward the faint end but there are great differences in the upper regions of the main sequence and the giant and supergiant branches.

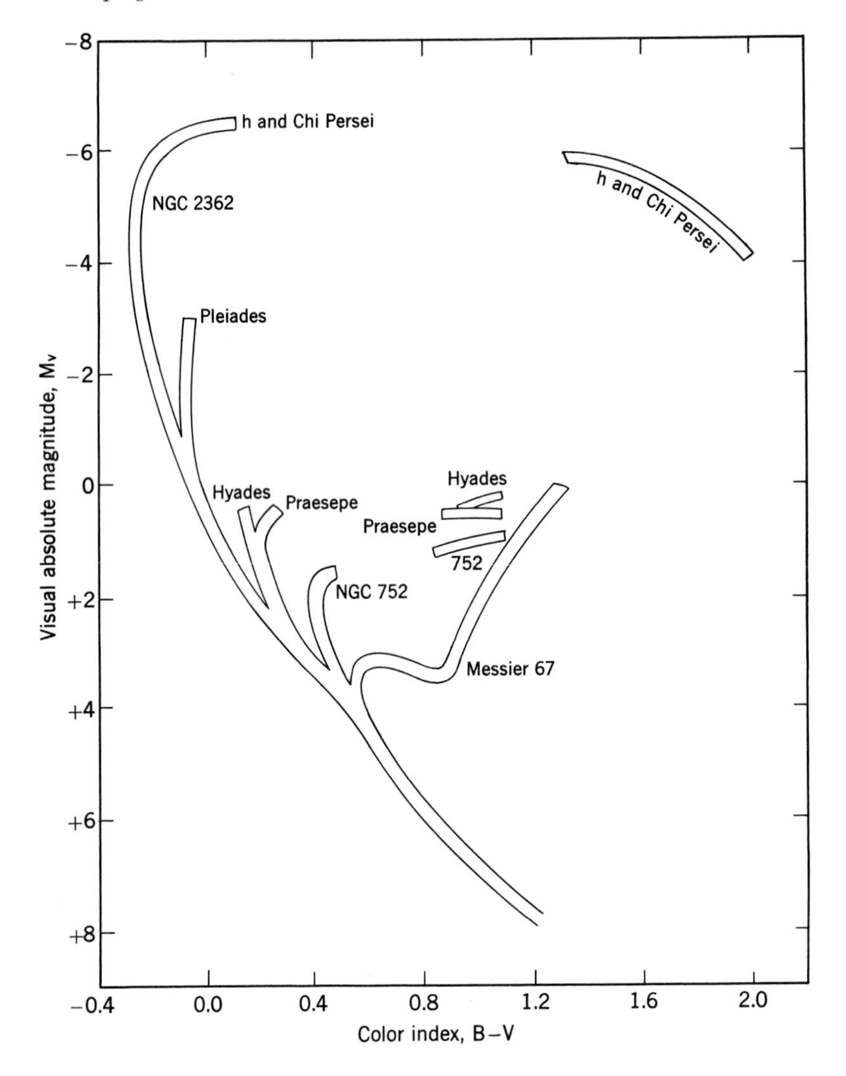

magnitude B of each cluster star. One may draw through the observed points the best-fitting curve and attempt to judge by inspection to which curve in Fig. 48 it corresponds. It is generally not difficult to decide whether the cluster one is dealing with is like the Pleiades, like h and Chi Persei, like the Hyades or Praesepe, or if perhaps it represents an extreme case, like Messier 67. In the absence of any appreciable interstellar absorption, the procedure of finding the distance to the cluster is simple and direct: one notes at various values of $B - V$ the corresponding values of M_v from Fig. 48 and the observed values of V. To check the resulting distance, one should attempt to obtain, if possible, the spectral types on the Morgan-Keenan system at least for the brightest members. From these spectral types and luminosity classes we can then readily check on the absolute magnitudes M_v of some of the individual bright stars and thus on the distance. A few well-determined spectral types and luminosity classes help immeasurably in disentangling the complex situations that often arise in practice.

Interstellar absorption and the accompanying reddening, which affects the observed colors $B - V$, complicate matters considerably, but with some care the distance problem is still amenable to relatively straightforward handling. It is simplest when the spectral types and luminosity classes are available for a few of the brighter members of the cluster. From the spectral data one can then predict for each star the intrinsic value of $B - V$ and derive the amount of reddening by noting the difference between it and the observed value of $B - V$ for the same star; we call this difference the *color excess* $E(B - V)$. From the scattering and absorption properties of the interstellar medium (see Chapter 9), we know the value of the factor by which we must multiply the color excess to obtain the corresponding total visual absorption; for $E(B - V)$ color excesses the conversion factor is close to 3. Multiplication by this factor then yields the total photovisual absorption between the star in the cluster and the Sun and Earth. If we call this absorption A_v, then $A_v = 3\,E(B - V)$ and the relation between the relevant quantities reads simply

$$M_v = V + 5 - 5 \log d - A_v.$$

In this formula we know every quantity with the exception of d, the distance of the cluster.

The great majority of the open clusters studied to date lie within 3,000 parsecs of the Sun and the remaining ones with known distances are almost all within 5,000 parsecs of the Sun. The clusters are mostly at distances less than 200 parsecs above or below the central galactic plane. In our disk-like Milky Way system open clusters are found most often in a thin layer near the central plane, the band in which—as we shall see in subsequent chapters—spiral structure prevails. Open clusters have a distribution that is totally different from that of the globular clusters (Chapter 5), which either inhabit the central regions of our Galaxy or are found to great distances from the galactic plane; globular clusters seem to avoid the outer thin layer of spiral structure. Open clusters belong for the greater part to Population I, whereas the globular clusters are of Population II.

It is obviously important that we should learn more about the distribution in space of the known open clusters with poorly determined distances. The most direct solution of this problem is through studies of colors and magnitudes. Here the photoelectric technique is the basic one but it may—and often must—be supplemented by photographic observations. The photoelectric observations need to provide only a limited number of precise standards of magnitudes and colors. We generally select the thin outer parts of a cluster for our photoelectric work. With the aid of what we may call "photographic interpolation," we can then determine colors and magnitudes for a hundred or more stars by measuring their magnitudes from photographs in two or three colors with reference to the standard stars.

There are several methods in vogue for research of this sort, one of the most effective being that developed by W. Becker of Basel, who measures his magnitudes in three carefully selected wavelength intervals and then succeeds in a straightforward manner in separating effects of space reddening from those produced by the intrinsic colors of the stars. Becker's method has yielded a detailed picture of the system of open clusters.

Wide-band photometry on the *UBV* system has been applied on a large scale to cluster studies, and intermediate and narrow-band techniques are being used increasingly in studies of moving clusters and of open clusters. It appears more and more as though the photoelectric techniques are capable of higher precision than the techniques of spectrum-luminosity classification. Not all clusters are of the same age and the chemical compositions of the atmospheres of the member stars may differ appreciably from one cluster to the next. We shall deal with the evolutionary effects in later chapters, but we should note here that astronomers, notably Harold Johnson, have obtained one basic reference line, the Zero-Age Main Sequence, which is very close to the main sequence shown in Fig. 42. It provides the basic reference line for stars of average chemical composition which, following their birth from the interstellar medium, have contracted onto the main sequence.

Once the stars, through a process of contraction from the interstellar medium, have arrived on the main sequence, they derive their energy of radiation chiefly from the continual building up of helium nuclei from hydrogen nuclei. The principal evolutionary effects that follow as these stars use up their internal hydrogen supplies of nuclear fuel is a trend toward the upper right-hand part of Fig. 48. The intrinsically brightest stars in the upper left-hand part of the figure will exhaust their hydrogen supplies most quickly and, as calculations show, they will move rapidly to the right in the diagram. Gradually the disease of hydrogen exhaustion spreads downward along the main sequence. Obviously, the clusters that still have very luminous blue-white stars (in the upper left-hand corner of Fig. 48) are the youngest, and the older clusters will show a chopped-off main sequence, with a few evolved red giants in the upper right-hand corner.

Associations and Aggregates

In addition to the tightly knit clusters of stars, there are more loosely connected distant groupings known as *stellar associations* and *aggregates.* For many years astronomers had been aware of the existence of some very extended moving clusters—the Scorpius-Centaurus cluster of B stars, for instance—recognized by their common space motions. The Russian astronomer Ambartsumian was the first to demonstrate that there exist a considerable number of loose groupings, called *associations,* most of which are too distant to show detectable parallelism of proper motions. These occur especially among the very luminous—and presumably very young—O and early B stars. The mutual gravitational attraction between the stars is too weak to hold the association permanently together, but it may have not existed long enough to have been torn asunder by the gravitational forces of the Milky Way, or to have drifted apart owing to the individual motions of the stars. These associations have been studied primarily by Ambartsumian in the Soviet Union and by Morgan at the Yerkes Observatory. The familiar Orion region, with its bright and dark nebulae and an abundance of O and B stars, has several associations, which has led Morgan to refer to it as an *aggregate.*

Every amateur astronomer is familiar with the beautiful Orion Nebula and the associated Trapezium cluster. The nebula itself is a large cloud of ionized hydrogen gas made luminous by the ultraviolet radiation emitted by the hot stars of the Trapezium cluster. In the same general region of the sky, one finds a very loose grouping of O and B stars, slightly elongated in shape, and extending over a far greater volume of space. Radio observations by Menon at the G. R. Agassiz Station of the Harvard Observatory and by others reveal that all these features are imbedded in a very large cloud of neutral hydrogen with a diameter of the order of 100 parsecs and with a total mass of the order of 50,000 to 100,000 solar masses; the famous Orion Nebula, with a total mass probably no greater than 1,000 solar masses, is just a little sore spot of ionized hydrogen in the larger complex. Many strong infrared sources are found associated with the Orion complex—probably indicative of protostar formation. Within this complex are several well-known gaseous and dust features, notably the Horsehead Nebula, a beautiful sight when photographed with the 200-inch Hale reflector (Fig. 49). One of the most striking features is a faint arc of nebulosity, first photographed by Barnard. It represents radiation from ionized hydrogen and is apparently formed at the edge of the expanding large neutral gas cloud, possibly caused by a shockwave phenomenon that occurs as the huge gas mass (expanding, according to Menon, at a rate of about 10 kilometers per second) bumps into the surrounding interstellar matter. In an extensive study of the Orion aggregate, Parenago has found evidence of a rotation of the entire system of stars and nebulosity. The radio observations of Menon confirm the presence of this rotation. We thus see the Orion aggregate as an enormous interstellar boiling pot, mostly of neutral hydrogen gas. Some of it apparently is condensed into relatively young stars, which ionize some of the gas and cause it to shine. Interstellar dust is sprinkled liberally throughout the complex.

Blaauw has made a remarkable discovery of the expansion of certain associations. For 17 stars near Zeta Persei, all within 30 parsecs of one another, he found that proper motions and radial velocities indicate a group expansion of 12 kilometers per second (Fig. 50). At this rate the group would have expanded to its present size in 1,300,000 years, a very short interval astronomically speaking; it is obvious that the age of the whole system is probably of that order. Although at first we may be startled at the shortness of the time, we cannot help but be pleased since the derived ages of the hottest O and B stars in this association are probably of the same order; this suggests some sort of explosion, a little more than 1,000,000 years ago, in which the stars were produced and then shot out into space.

Blaauw and Morgan have studied other associations in similar fashion. For a group of about 30 stars near the O star 10 Lacertae they find a rate of expansion of the order of 8 kilometers per second, suggesting a probable age of the order of 4,200,000 years. According to Blaauw, there is evidence that the Scorpius-

49. The Horsehead Nebula in Orion (south of the star Zeta Orionis) photographed in red light with the 200-inch Hale reflector.

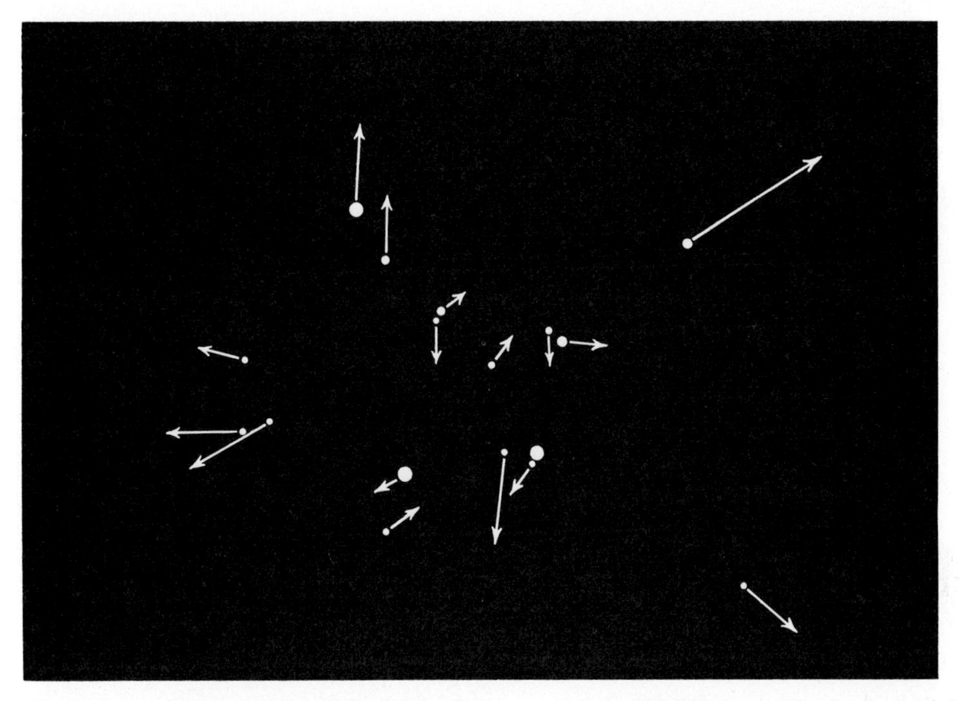

50. An expanding association. The figure shows the Zeta Persei association with the arrows indicating the directions in which the stars are traveling and the distances they will cover during the next 500,000 years. The motions of this group have been studied by Blaauw. (Courtesy of *Scientific American.*)

Centaurus moving cluster is an expanding association, with the slow expansion rate of 0.7 kilometer per second. This gives a probable age of 70,000,000 years, which seems in line with the fact that the Scorpius-Centaurus cluster does not contain excessively luminous, very hot, and hence very young O and B stars, but seems to specialize in the more sedate and older varieties of B stars.

By far the most spectacular expansion phenomenon has been observed by Blaauw and Morgan in connection with the Orion aggregate. The three stars AE Aurigae, Mu Columbae, and 53 Arietis have motions that seem to carry them away from the center of the Orion aggregate at rates in excess of 100 kilometers per second (Fig. 51). They appear to be the hottest and the fastest pellets shot into space by the big Orion explosion, which must have taken place about 2,500,000 years ago.

O and B stars are not the only variety of stars occurring in associations and aggregates. The T associations, discovered and so named by Ambartsumian, contain an abundance of T Tauri-type variable stars. These irregular variable stars are commonly found especially at the edges of very obscured regions. They are probably very young stars of low mass, and they can only be recognized either by their variability in light or by bright-line features in their spectra. They occur in groupings, some of them near O–B associations, others by themselves, but always in regions of the sky where cosmic dust is plentiful. In addition to the Soviet astronomers, Haro at the Tonanzintla Observatory in Mexico and

51. Runaway stars from Orion. The stars AE Aurigae, Mu Columbae, and 53 Arietis are fast-moving stars. When their paths are retraced they are found to intersect in the constellation Orion. It is assumed that they originated in the bright O association and have erupted from it. (Diagram by Blaauw and Morgan, courtesy of *Scientific American.*)

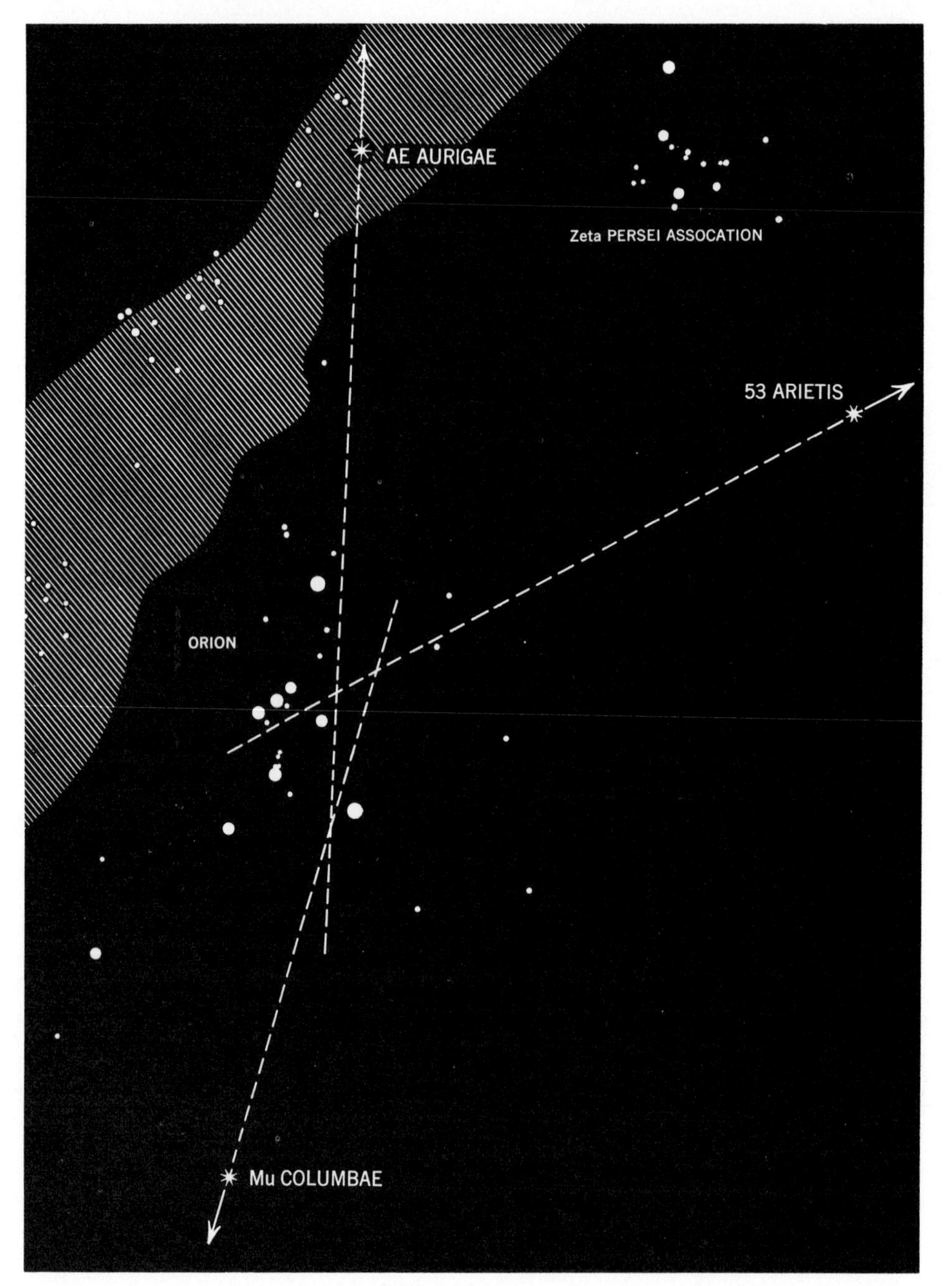

Herbig at the Lick Observatory have searched for and studied the T associations. Holopov in the Soviet Union has catalogued the principal T associations and he was one of the first to show that the T Tauri variable stars are found especially at the edges of dark nebulae.

The Orion Nebula and surroundings form a T association. Apart from the many emission objects discovered in and near the Orion nebulosity, there is an abundance of irregularly varying stars, especially in the regions of dark nebulosity. Originally, many astronomers were inclined to consider seriously the hypothesis that these variations might be caused by variable thickness in the obscuring clouds passing over the stars. Parenago has disproved this hypothesis. The fact that many of the variables have emission lines in their spectra in spite of rather faint absolute magnitudes suggests that the cause of the variability and the emission features lies either in the star's atmosphere or, possibly, in its interaction with the surrounding cosmic dust and the associated interstellar gas. Ambartsumian is of the opinion that the T Tauri variables are young stars, formed quite recently from the surrounding dust and gas clouds, but that their variability and the presence of emission lines are effects caused by internal disturbances in these youthful stars rather than by interaction phenomena between the stars and the interstellar clouds. Haro and Herbig find support for this suggestion in the presence of excessive ultraviolet radiation in the spectra of these objects. Herbig has some photographs which suggest that we may actually be observing some stars in the process of formation. The variable star FU Orionis is related to the T Tauri variety of variable stars. It came onto the cosmic scene almost unannounced when Herbig found that in 1936 it increased in brightness by 6 to 7 magnitudes. Haro has recently discovered a variable star in Cygnus which in 1969 rose in brightness by at least 6 magnitudes. The T Tauri stars must be evolving very fast. Kuhi has estimated that matter escapes from these stars at a rate of about 1 solar mass in 1.5 million years.

5
Pulsating Stars and Globular Clusters

In the preceding chapters we have gradually moved out from the immediate vicinity of the Sun into other parts of our Galaxy, reaching more remote clusters, some as far as 5,000 parsecs from the Sun and Earth. However, we have covered only a small part of the whole of our Home Galaxy. The study of spectrum-luminosity classes and of precise colors and magnitudes of faint stars has taught us much about the properties of the nearer parts of our Galaxy, but a grand picture of the outline of the Galaxy has not appeared as yet. Our established lamps of known standard brightness just cannot be observed at sufficiently great distances to give us the over-all picture. We must find new standard lamps if we wish to penetrate into the heart of our Galaxy, and beyond. The pulsating stars, especially the long-period cepheid variables and the RR Lyrae stars, provide these lamps. They are the standards that give us the distances to the globular clusters. The outlines of the system of the globular clusters will in turn yield the first definite clues regarding the over-all outline of the whole of our Galaxy.

Pulsating Stars

Among the stars that vary in brightness, some do so because they are changing in size; the star is alternately expanding and contracting. Density waves propagate outward into the lower atmosphere, which expands for a while, then contracts. The star's light varies during the sequence of expansion followed by contraction. The period of light variation

52. Light curve of Delta Cephei. The diagram illustrates the changes in apparent magnitude. The period of the light variation is 5 days 9 hours.

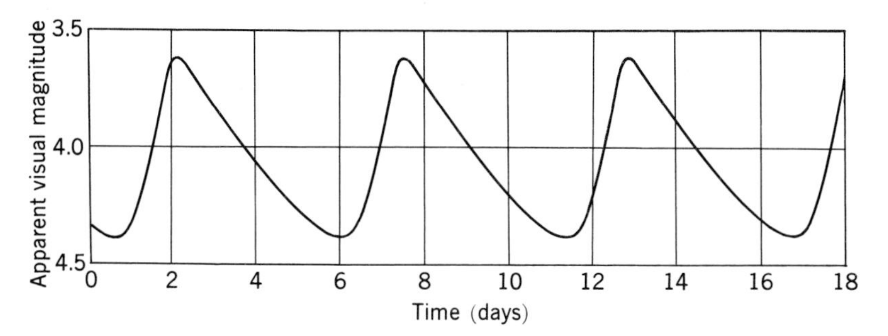

may lie anywhere between 80 minutes and 100 days, but each star has its own characteristic period, which remains constant for most stars within very narrow limits. The brightest star of this class is the star Delta of the constellation Cepheus, from which these stars received the name of *cepheids.* Delta Cephei is an easily identified star; if it is closely observed for a week or two, it is seen to change its brightness between the third and fourth apparent magnitudes in a regular pattern, repeating itself every 5 days 9 hours. Figure 52 shows the light curve; the time is plotted along the horizontal scale and the apparent magnitude along the vertical scale. The curve shows that Delta Cephei rises quickly to its greatest brilliance, fading away more slowly. Over and over again, unvaryingly, it repeats this pattern of changing brightness.

Together with the change in luminosity there occurs a change in color, so that the star becomes redder as it becomes fainter. With the aid of the spectrograph it is shown that the radial velocity of the star varies with the same period as the change in brightness. This variable radial velocity is interpreted as originating from a periodic swelling and shrinking of the star, a real *pulsation.* The time of greatest velocity of approach comes generally at or near the time of maximum light and the greatest velocity of recession comes at or near the time of minimum. The pulsation theory was first advanced by Shapley to explain the behavior of this star and the other cepheid variables. This theory was developed by Eddington, and later M. Schwarzschild showed that, in addition to the standing waves in the main body of the star, running waves appear near the surface.

A wide range of periods has been observed, but certain periods are favored. Many of these stars have periods of nearly a week or slightly longer. There are, however, fainter stars in the sky that also wink on and off but in shorter periods, of the order of half a day (Fig. 53). The short-period cepheids are frequently called "cluster variables," because Bailey at Harvard first found that they are present in abundance in globular clusters. We shall prefer to call them RR Lyrae stars, after their brightest member.

When we collect the RR Lyrae stars in one group and the pulsating stars of longer periods

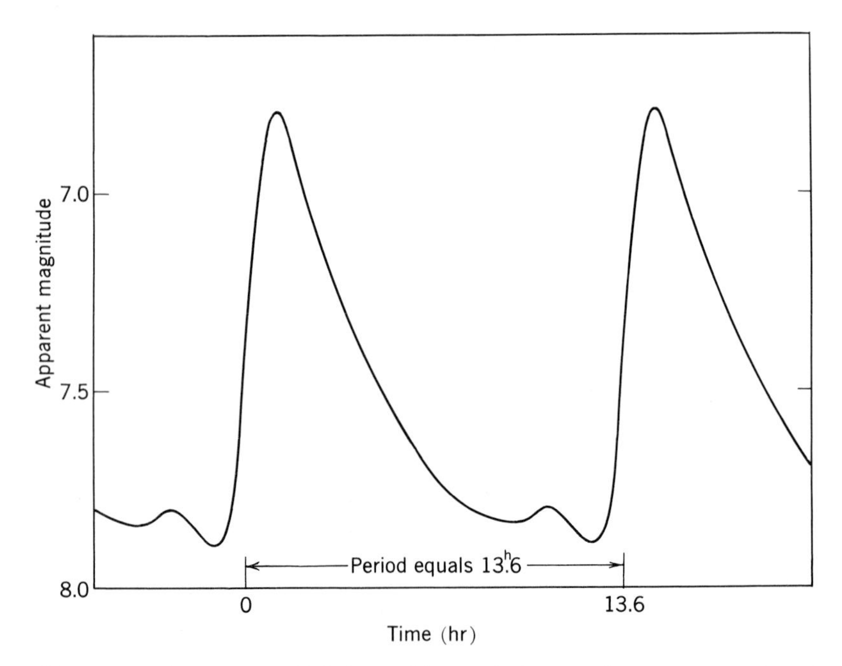

53. A sample light curve of RR Lyrae. The period of this rapidly varying star is 0.567 day = 13.6 hours; its range in brightness is 1.1 magnitudes. The star shows secondary changes in brightness, which cause small periodic changes in the shape of the light curve. (From data assembled by Walraven at the Leiden Observatory.)

in another group, we find that they may be alike in that they are pulsating stars, but they are found to differ in other important respects. The longer-period, or classical, cepheids are mostly at low galactic latitudes and are slow-moving bodies; both their proper motions and their average radial velocities are small. The RR Lyrae stars, on the other hand, move rapidly and are found scattered over the sky; they do not cling to the plane of the Milky Way. Here is a warning that the two varieties of pulsating stars are different sorts of objects, probably different in origin and in age; we must not treat them as members of one single homogeneous group.

The RR Lyrae stars are true Population II stars. They are the sort of stars that inhabit the thin outer halo of the Galaxy and the central nucleus, but they avoid the central disk and the regions of spiral structure. The long-period cepheids that we observe most readily belong to Population I, but here we must be careful. There exist also long-period variable stars with light curves that differ not too much from those of the regular cepheids, but with different mean absolute magnitudes and with characteristics of Population II objects.

During the past 50 years, the pulsating variables have played an important though rather controversial role in the drama of our unfolding knowledge of the outline and dimensions of the universe of galaxies. The first act opened in 1910, when Miss Leavitt of the Harvard Observatory studied the variable stars in the Small Magellanic Cloud. Their periods were found to vary from a few days to over

100 days. She found a very important relation between their periods and their average brightnesses: the longer the period, the brighter the star appears on the average.

Hertzsprung, and later Shapley, recognized this relation as being an intrinsic quality of the stars. Since the thickness of the Small Magellanic Cloud is small compared with its distance, all stars in this star system may be assumed to be essentially equally distant, and a relation between the period and the apparent magnitude will really be one between period and absolute magnitude. The difficulty is to find the constant, or *modulus,* that must be subtracted from all apparent magnitudes in order to change them to absolute magnitudes.

If we only knew the absolute magnitude of a single cepheid of known period, then we could use it to determine the zero point of the period-luminosity curve, and this curve could then be used to give the absolute magnitude of any cepheid for which the period can be observed. Unfortunately, the nearest galactic cepheids are too distant for accurate measurement of their trigonometric parallaxes. Hence it became necessary to determine their average parallaxes from the proper motions by the method of mean parallaxes described in Chapter 3. But the proper motions of these stars are small and their distribution over the sky is not uniform. Both factors introduce uncertainties in the parallax determinations, but it was quite clear that the brightest cepheids are very distant and highly luminous.

The RR Lyrae variables are quite a different story. Since they have high linear velocities, they are especially suited for studies of mean parallaxes. The average radial velocity of a long-period cepheid variable will generally not exceed 20 kilometers per second, but for the RR Lyrae stars velocities of the order of 100 kilometers per second are by no means uncommon. The high linear velocities lead to large proper motions for the RR Lyrae variables of the tenth magnitude and brighter. At the same apparent magnitude, the proper motions of the RR Lyrae variables are far larger than those of the longer-period cepheids. The larger sizes of the proper motions make possible fairly accurate determinations of the mean parallaxes for the RR Lyrae variables.

Originally, the mean absolute magnitudes of the RR Lyrae variables came out close to $M = 0.0$ on both the photographic and the visual scales. Over the years, it has been suggested that this figure might be too high by as much as half a magnitude and mean values for the absolute magnitude between $M_v = +0.5$ and $M_v = +0.7$ are more generally assigned to them. Most astronomers now use the value $M_v = +0.6$ as a suitable mean. The precise value is very important, for, as we shall see below, the assumed mean absolute magnitude of the RR Lyrae variables is the quantity from which the distances to the globular clusters and to the center of our Galaxy are derived.

Until the late 1940's it was assumed that a single period-luminosity curve would include both the RR Lyrae variables and the classical long-period cepheids. The zero point for calibration to absolute magnitude was found by assuming that the absolute magnitude obtained by simply extrapolating the observed curve for periods of the order of a couple of days to periods of the order of 1 day would yield the absolute magnitude of the RR Lyrae variables, that is, $M_v = +0.6$ or slightly fainter. There was some uneasiness expressed about the fact that Shapley and his associates had never been able to discover any RR Lyrae variables in the Magellanic Clouds, but few

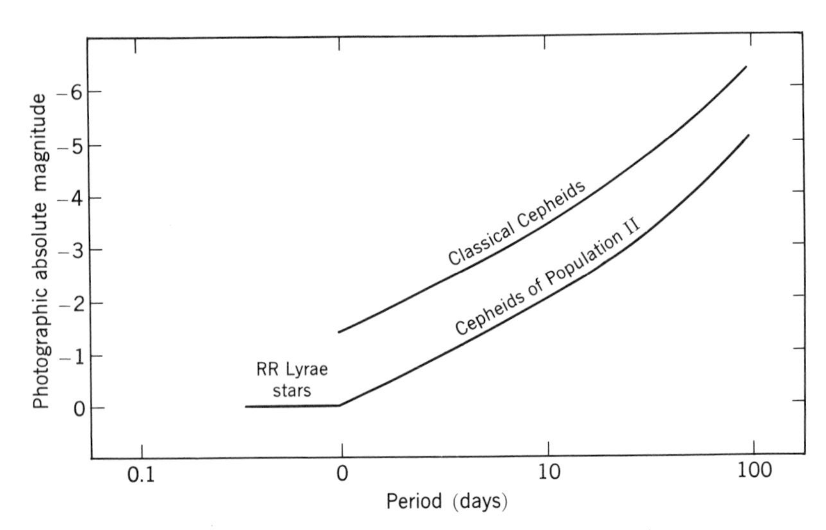

54. The period-luminosity relation. The precise form of the relation is still in doubt, but there is strong evidence to show that the curve for the classical cepheids lies approximately 1.5 magnitudes brighter than that for the Population II cepheids.

astronomers were ready to suggest from this negative observation alone that it is not permissible to extend the period-luminosity curve as indicated. It should be noted that very few astronomers read with care a paper published in 1944 by Mineur in France in which, from a study of motions, he found indications that $M_v = 0.0$ to $+0.3$ probably represents a good mean value for the RR Lyrae variables, but in addition recommended a correction of -1.5 magnitudes to the accepted zero point of the long-period cepheids.

The question became an acute one about 1950, when Baade was unable to detect the RR Lyrae variables in the Andromeda spiral galaxy. Regular long-period cepheids had been discovered in abundance by Hubble 25 years earlier and, on the basis of the accepted period-luminosity relation, the distance to the spiral galaxy had been determined. If the original figure for the distance of the Andromeda galaxy, Messier 31, had been correct, the RR Lyrae variables would have been observed in abundance at about $m = 22$, a figure close to, but still well above, the brightness limit for the 200-inch Hale reflector. Baade did not find the RR Lyrae variables as expected and he concluded rightly that the zero point of the period-luminosity relation for the long-period cepheids required a correction of just about the amount suggested by Mineur. He recognized, moreover, that the few long-period cepheids observed in globular clusters were of a brand different from the majority of those observed nearby in our Galaxy or in the Andromeda galaxy. The globular-cluster variety of long-period cepheids is Population II (as are the RR Lyrae variables), whereas the regular long-period cepheids, like Delta Cephei and the stars found in the Andromeda galaxy, are Population I. We are now prepared to deal with two period-luminosity relations, one for Population I, another for Population II; the relative shift in zero point for the two parallel curves is of the order of 1.5 magnitudes, the Population I objects being the brighter (Fig. 54).

One concluding comment is in order: a

correction of 1.5 magnitudes is no small matter. It means a doubling of the scales of distance and of diameter for all objects whose distances had been determined earlier with the aid of the older, erroneous curve. In other words, if the correction of 1.5 magnitudes is accepted, then all distances found before 1950 for galaxies outside our own had been underestimated by a factor 2. Our Galaxy escaped a comparable revision of its distance scale only because the RR Lyrae variables were used in establishing its scale, and their assumed absolute magnitudes were not changed appreciably.

Globular Clusters

Whereas there are at most two or three thousand stars in a rich galactic cluster, a globular cluster may contain as many as 100,000 stars (Figs. 55, 56, and 57). There are 120 known globular clusters in the sky and we believe that, unlike the galactic clusters, not more than another 100 lie hidden from us. The globular clusters appear in all galactic latitudes and therefore they are not so often veiled by obscuring matter as are the open clusters; their high intrinsic brightnesses render them conspicuous at great distances. They do not "melt into the landscape" as do the distant open clusters.

The distribution in galactic longitude of the globular clusters is very striking and shouts to all who will listen of the eccentric position of the Sun in our Galaxy. All but a few of the known globular clusters are in one half of the sky and one-third of these are found in a region of Sagittarius that covers only 2 percent of the sky (Fig. 58). The globular clusters form a vast system of their own, concentric with the Milky Way system, but spherical in outline.

Until Shapley's classic investigations of 1916–1919, no estimates of the distances of globular clusters were available. His studies of the RR Lyrae variables discovered by Bailey and others led to the first accurate distance estimates for these far-away objects. With the aid of the 60-inch and 100-inch reflectors of Mount Wilson Observatory, Shapley photographed great numbers of the variables frequently enough to obtain accurate light curves, and, consequently, the distances of all globular clusters with known cluster variables were found.

For the globular clusters that lacked variable stars, estimates of the distances were found from the magnitudes of the brightest stars. Shapley noticed that the brightest stars in the globular clusters—unlike the brightest stars in the neighborhood of the Sun—are not blue-white giants, but are red giants of absolute magnitude about -3, much brighter than the typical red giants found near the Sun.

Rough estimates of distance for the faintest and most distant clusters were made from the observed apparent diameters and apparent integrated magnitudes. Here Shapley was on shaky ground, since he had to make the assumption that all globular clusters are intrinsically identical objects and that apparent differences are due to one variable only—the distance.

At present, cluster-type variables have been identified in about one-third of the globular clusters. Mrs. Hogg has listed 1,100 known variables. For 20 more clusters rough distances have been found from the brightest stars. To find the distance from our Sun to a given globular cluster, we must know not only the apparent magnitude of the RR Lyrae variables in the cluster, but also the amount of intervening space absorption between the Sun and the cluster. This is determined from

55. The globular cluster Messier 92. A photograph of 85 minutes' exposure made with the 61-inch astrometric reflector of the United States Naval Observatory Station at Flagstaff, Arizona. (U. S. Naval Observatory photograph.)

56. The southern globular cluster 47 Tucanae. A photograph of 60 minutes' exposure made with the 40-inch Boller and Chivens reflector of the Siding Spring Observatory of the Australian National University. (Mount Stromlo Observatory photograph.)

57. The southern globular cluster Omega Centauri. A photograph of 60 minutes' exposure made with the Armagh-Dunsink-Harvard telescope of Baker-Schmidt design of the Boyden Observatory in South Africa. Note that this cluster is definitely flattened, suggesting the presence of rotation. (Harvard Observatory photograph.)

58. The direction of the galactic center. The circles mark the positions of globular clusters for this section of the Milky Way. One-third of all known globular clusters are in this photograph, within an area of only 2 percent of the sky. (Harvard Observatory photograph.)

the amount of space reddening. Mayall has found the integrated spectral types for a number of globular clusters; they vary between A5 and G5, with most of them between F8 and G5. Because of this narrow range of integrated spectrum, the true color index of a globular cluster can be predicted fairly well, and, from the difference between the observed and the intrinsic colors of the cluster, the reddening produced between the Sun and the cluster may be found. By multiplying this color excess by a suitable factor, the total amount of intervening absorption is obtained and the distance can be corrected for absorption effects. Stebbins and Whitford measured the color indices for 68 globular clusters. Their derived total absolute magnitudes for the globular clusters with known RR Lyrae variables range from $M = -5$ to $M = -8$, which is too wide a range to give much confidence in the distances derived from integrated apparent magnitudes alone.

The known system of globular star clusters in our own Galaxy is roughly spherical, with its center located at a distance of 10,000 parsecs from the Sun in the direction of the constellation Sagittarius. We have every reason to suppose that it is concentric with our own galactic system of stars, dust, and nebulae.

Population Characteristics of Globular Clusters

Shapley first pointed out that the stars found in globular clusters differ in several respects from those found in the neighborhood of the Sun. The brightest stars are red and they are some 3 magnitudes brighter than the brightest blue-white stars of the cluster. Shapley was unable to observe stars much fainter than about absolute magnitude $+1$; hence the limit refers to stars that are considerably brighter than our Sun.

With the 200-inch Hale telescope on Mount Palomar it is possible, for the nearest globular clusters, to make extensive studies of colors and magnitudes of stars as faint as our Sun. Thorough studies have been made of Messier 92 by Arp, Baum, and Sandage; of Messier 3 by Sandage and H. L. Johnson; and of Messier 13 by Baum and others. The basic magnitude sequences are generally established by photoelectric measurement. Some of the standard-sequence stars—those of the twenty-third magnitude—are so faint that they cannot be seen visually through the eyepiece of the telescope, though their positions are measurable on photographic plates. They are determined by offsetting for difference of position against a brighter star. For the faintest stars it may be necessary to carry on lengthy comparison measurements for star and background and thus obtain a photometer deflection for the star corrected for background brightness.

After the basic magnitude sequence is established photoelectrically, the magnitudes of other stars may be interpolated by photographic means. Some 1,100 stars were measured in both Messier 3 and Messier 92. Colors can be determined by using blue and yellow filters with the photoelectric photometer and blue- and yellow-sensitive plates with suitable filters for the photographic work.

Assuming the RR Lyrae stars to possess a mean absolute magnitude of $+0.6$, or thereabouts, the distance moduli for the nearer globular clusters are found to be about 14 magnitudes. This means that the difference $m - M$ amounts to $+14$ or $+15$ magnitudes. Hence a star with $m = 20$ will have an absolute magnitude of $+6$, not unlike our Sun; observations to the twentieth magnitude or fainter give a good segment of the main sequence to the limit of stars of the same luminosity as the Sun. Figure 59 shows a typical color-magnitude array for a globular cluster.

59. The observed color-magnitude array for the globular cluster Messier 3. A diagram prepared by Johnson and Sandage. The visual apparent magnitudes V are shown on the vertical scale; color indices $B - V$ are plotted horizontally. The features noted in the text are identified. (Courtesy of *Astrophysical Journal*.)

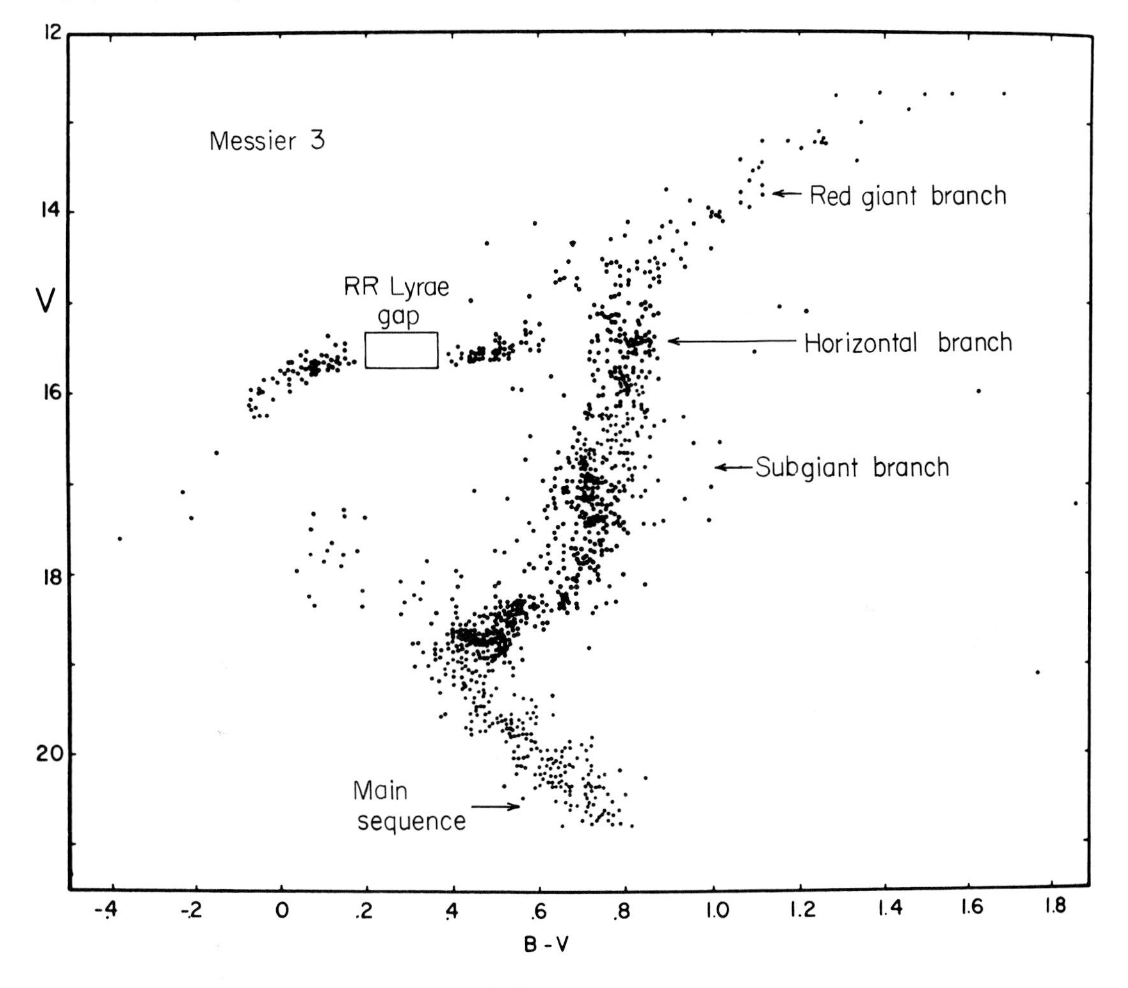

This cluster, Messier 3, also known as NGC 5272, has been studied by H. L. Johnson and Sandage. The diagram shows vertically the visual apparent magnitude V, and horizontally the color index $B - V$. It will be noted that the brightest stars, which are quite red, have apparent visual magnitudes a little brighter than $V = 13$ and that the faintest stars measured by Johnson and Sandage have values of V close to 21. Several features of this diagram should be noted. First of all, we see that between $V = 19$ and $V = 21$ the stars exhibit a well-determined *main sequence.* Whereas in most open clusters the main sequence would continue upward and toward the left, there is a turn-off point at $V = 19$ in the diagram for Messier 3. The stars in the globular cluster have clearly evolved away from the main sequence toward the right, and there is a nearly vertical stretch near $B - V = +0.7$ from $V = 18.5$ to about $V = 15$. We note also two other main features of the diagram for stars brighter than $V = 16$. The first is a scattering of bright stars that outline the sequence toward the upper right-hand corner, the brightest stars, as already noted, having values of V that are a little smaller than 13. All of these bright stars are quite red, their $B - V$ values ranging from 1.2 to 1.7; this is the *red-giant branch.* The stars in the vertical branch near $B - V = 0.7$ are called the subgiants. The second feature is a well-defined *horizontal branch* near $V = 15.7$. It runs from $B - V = -0.1$ to $B - V = +0.6$, but there is a clear gap in this sequence between $B - V = +0.2$ and $B - V = +0.4$. We should mention here that in this particular gap we find in Messier 3 nearly 200 RR Lyrae variable stars! Hence the RR Lyrae variables seem to be a feature of the horizontal branch. At the present time the horizontal-branch stars that seem to be constant in brightness, stars on either side of the RR Lyrae gap, are being studied extensively with the most powerful spectrographs available to astrophysicists. The physics of these particular stars can give us some clues to the reasons why in the middle of the horizontal branch we find a section with lots of RR Lyrae variables. We seem to have evidence of the presence of a critical stage in the nuclear evolution of the interiors of these stars. In all probability the horizontal-branch stars, and their cousins the RR Lyrae variables, are stars that have evolved from the red-giant stage after nearly complete exhaustion of their hydrogen nuclear fuel supplies.

Most globular clusters have color-magnitude arrays not unlike that of Messier 3. The principal differences are in the positions of the turn-off points from the main sequence and the thickness of the subgiant branch, which connects the horizontal branch and the giant branch. The RR Lyrae gap lies sometimes a bit toward the red and in other cases a bit toward the blue from that shown for Messier 3.

There are basically two ways of judging the distance of a globular cluster and of calibrating the vertical scale of apparent visual magnitudes V in terms of visual absolute magnitude M_v. The first is to assume that $M_v = +0.6$ for the RR Lyrae variables; the second is to match the main sequence with the zero-age main sequence as derived from nearby stars and open clusters. Both systems have their weaknesses and points of strength, but, generally speaking, the distance moduli thus arrived at are in fair agreement.

The number of RR Lyrae stars differs very greatly from one globular cluster to the next. Messier 3 has about as many as any of them and there are apparently some clusters that have none. However, wherever RR Lyrae vari-

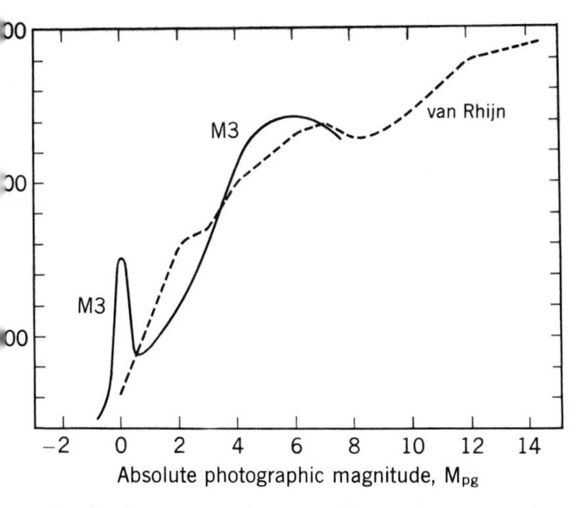

60. The luminosity function for Messier 3. Sandage has counted stars in successive intervals of photographic absolute magnitude in the globular cluster Messier 3. In the diagram his derived luminosity function is compared with that of van Rhijn in Fig. 45, which applies to the vicinity of the Sun. (Courtesy of *Astronomical Journal.*)

ables are found, there are apparently no stars of constant brightness at that particular position in the color-magnitude array. Hence it follows that the RR Lyrae gap is truly an evolutionary stage of instability in the globular-cluster stars.

Globular clusters are the oldest known objects related to our Galaxy. The turn-off point from the main sequence is indicative of the probable age of the cluster. The ages of globular clusters range from 5×10^9 years to 2×10^{10} years. The globular clusters truly all have the appearance of being old and worn-out stellar systems. Neither optically nor by radio-astronomical techniques does one find evidence of appreciable amounts of free gas or dust associated with globular clusters. In them the processes of star formation have apparently ceased to operate long ago.

Early studies of the luminosity functions in globular clusters were made by Shapley about 50 years ago. The subject remained rather neglected until Sandage made a study of the luminosity function of Messier 3. Since his photographic plates reached to apparent magnitude 22.5 and since the distance modulus of this cluster is about 15, he observed stars with absolute magnitudes from -3 to $+7$. From plates of different exposure times, he counted the number of stars of different magnitudes to within 8 minutes of arc from the center. His luminosity function for Messier 3, based on 44,000 stars, is given in Fig. 60, where it is compared with the luminosity function of van Rhijn for the neighborhood of the Sun.

The hump at zero magnitude is very evident. It includes some 200 cluster-type variables and other stars with color indices to the blue of $+0.4$. The drop after $M = +5.5$ Sandage feels is undoubtedly real, though the slope of the curve beyond that point is of course uncertain.

From the number of stars of each absolute magnitude it is possible to compute the contribution of stars of each magnitude to the total light emitted by a cluster. The value from the summation of all contributions is $M = -8.09$, which is in good agreement with the integrated absolute magnitude found by Christie. Ninety percent of the light of the cluster comes from stars brighter than the fourth absolute magnitude, that is, brighter than the Sun, although the total number of stars like the Sun and fainter is very large.

Since it is impossible at present to go below $M = +7$, Sandage uses the van Rhijn luminosity function to estimate the total number of stars of all luminosities in the cluster and from these numbers he can estimate the cluster's total mass, which he finds to be 1.4×10^5 solar masses. Wilson and Miss

61. Distribution of globular clusters. The diagram shows the distribution of globular clusters with known distances projected on a plane passing through the Sun and the galactic center perpendicular to the central galactic plane. The position of the Sun and of the galactic center are indicated. The vertical scale represents the height *z* of the cluster above (+) or below (−) the galactic plane measured in kiloparsecs (1 kpc = 1,000 parsecs). The horizontal scale x is measured in the same units. (Diagram prepared by Arp, reproduced in *Galactic Structure*, courtesy of University of Chicago Press.)

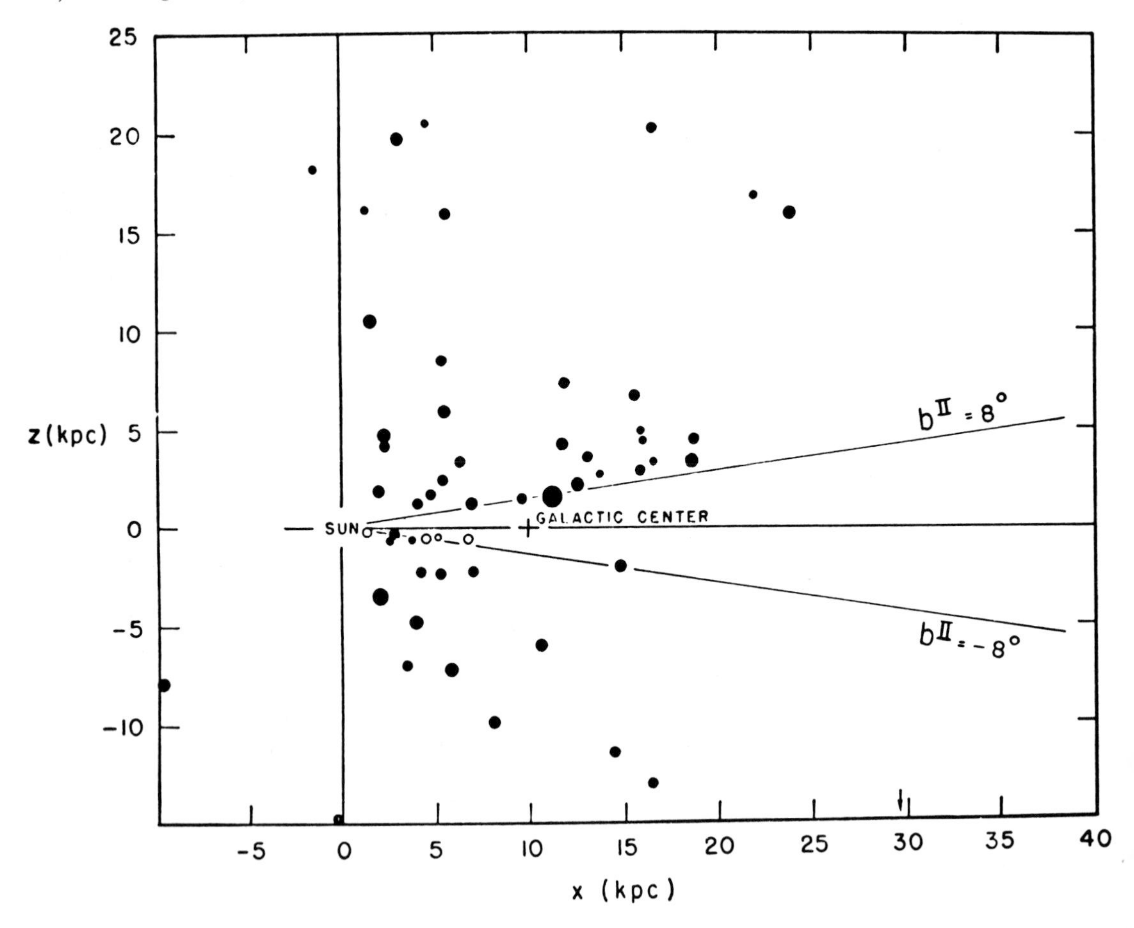

Coffeen estimated the total mass of Messier 92 from dynamical theory as 3.3×10^5. In view of the greater luminosity of Messier 3 as compared with Messier 92, its dynamical mass should be of the order of 5.6×10^5, larger than the above value by a factor 4. It is still uncertain whether this discrepancy is real. If the dynamical value for the mass is correct, then the result may indicate that the percentage of faint stars is much greater in a globular cluster than in the neighborhood of our Sun.

The System of Globular Clusters

In spite of uncertainties in the mean absolute magnitudes assigned to RR Lyrae stars, these stars still provide the best standard lamps for fixing the distances to globular clusters. We now have available good photoelectric photometry to the limit of the brightness of the horizontal branch for close to 25 globular clusters, and for these one can obtain reasonably reliable distances, which, incidentally, can be corrected for the effects of interstellar absorption. A few years ago Arp put together all of these distance estimates and prepared a list of distances for 21 globular clusters. Messier 3, the cluster to which we referred at some length in the preceding section, is among those and Arp gives it a true distance modulus of 15.13, to which corresponds a distance of 10,600 parsecs. For clusters without reliable variable-star photometry, approximate distances can be derived on the basis of the mean absolute magnitude for the 25 brightest stars. Shapley used this criterion in his investigations of half a century ago. Arp assigns to the mean absolute magnitude of the 25 brightest stars a value $M_v = -0.8$. This helps in fixing the distances for quite a few more globular clusters. Other criteria, such as integrated brightness and apparent diameter, have been used in the past, but these have been shown to be too unreliable for good work.

Figure 61 is Arp's diagram of the distribution in our Galaxy of the globular clusters of known distance. The diagram represents a cut in the plane passing through the Sun and the galactic center at right angles to the central galactic plane. The positions of the Sun and of the galactic center are marked in the diagram. The Sun is probably 10,000 parsecs distant from the galactic center. The tilted lines drawn in the diagram starting at the Sun make a section in which most globular clusters will be hidden from our view by interstellar obscuration close to the central plane. The globular clusters can be divided into two groups, one fairly well concentrated to the center of our Galaxy, the other far from the galactic plane. We note that studies of integrated spectra of globular clusters for the two groups show that there seem to be rather marked differences in chemical composition between the central group and the outlying clusters. The latter seem to be generally metal-poor in their spectra, whereas metallic lines do show up more prominently in the spectra of the clusters found close to the center of our Galaxy.

6
The Whirling Galaxy

Thus far we have dealt only with the purely structural aspects of the Milky Way problem. Now we shall turn to considerations involving the motions of the stars and the forces that control them; we shall see what may be learned about the *dynamics* of the system from studies of the regularities in the proper motions and radial velocities of the stars and of clusters of stars.

We found the Milky Way system to be a highly flattened disk, embedded in a very tenuous, more or less spherical, halo. The characteristic spiral arms, outlined by the O and B stars and by bright and dark nebulae, lie in or very near to the central plane of the Galaxy. At the center of the disk there is a very dense nucleus, apparently composed mostly of older stars. The Sun is probably about 10,000 parsecs from this dense central nucleus, but it is within 30 parsecs of the central plane.

It should have been evident all along that the flattened shape of our Galaxy is indicative of rapid symmetric rotation about an axis perpendicular to the plane, but it was not until 1926–27 that the researches of Lindblad and Oort first gave conclusive proof of this *galactic rotation.* In the first section of this chapter we shall present the evidence to show that our Sun and the great majority of the stars in our vicinity move about the center in roughly circular orbits and with speeds of the order of 250 kilometers per second. Most of the regularities observed in the motions of the stars near the Sun and those at distances up to a few thousand parsecs can be understood on the basis of this galactic rotation.

The Sun's Motion around the Galactic Center

The highly flattened shape of the system of O and B stars, and even of the system of stars like our Sun, suggests that these subsystems of our Galaxy are in rapid rotation. The situation is quite different for the globular clus-

62. Spectrum of the globular cluster Messier 15. This photograph shows the spectrum of the cluster in the middle, with a comparison spectrum (see Fig. 39) above and below the spectrum of the cluster. Rarely are the absorption lines as sharp as those shown here. (Photograph by Kinman, courtesy of Lick Observatory.)

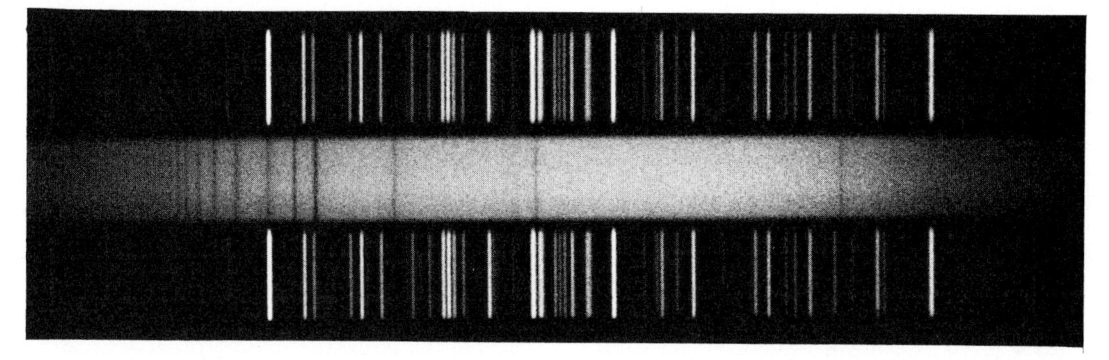

ters, which form a tenuous and more nearly spherical subsystem surrounding the flattened principal Milky Way system. Inside a system of nearly spherical shape there may be large random motions, but there is probably no rotation of the system as a whole. On the average, the globular clusters should therefore define a system more or less at rest with respect to the center of our Galaxy. From the study of the observed radial velocities of globular clusters, we should be able to obtain information about the Sun's motion relative to the system of globular clusters and hence presumably with reference to the center of our Galaxy.

Observed radial velocities are available for 70 globular clusters. The majority of these radial velocities were measured by Mayall at the Lick Observatory; his basic list contains 50 values. Radial velocities for 20 additional globular clusters, mostly at southern declinations, were obtained by Kinman at the Radcliffe Observatory in South Africa. Mayall found individual values ranging between +291 and −360 kilometers per second. Kinman's values do not cover quite so wide a range. The spectral lines are broad and fuzzy, and precise measurement of the radial velocities is far from simple (Fig. 62). If we assume that the globular cluster system as a whole has at best only a small residual circular velocity of rotation with respect to the center of our Galaxy, then we should be able to obtain from the observed radial velocities of the globular clusters the average velocity of the Sun with respect to the system of globular clusters. Such a study would presumably yield a lower limit to the circular rotational velocity of the Sun around the galactic center. For an accurate determination of the motion of the Sun relative to the system of globular clusters, we require a fair number of clusters in directions of galactic longitude $l = 90°$ and $l = 270°$; the radial velocities of these clusters would show the full effects of the Sun's motion (directed more or less toward galactic longitude $l = 90°$). Unfortunately, as we see from Fig. 61, the globular clusters, when viewed from the Sun, are concentrated to one hemisphere, and the majority of them are seen in directions from the Sun that do not differ much from the general direction to the galactic center, galactic longitude $l = 0°$.

In spite of these obstacles, Mayall and Kinman were able to derive values for the Sun's velocity relative to the system of globular clusters. The maximum value of this circular velocity, which is directed to a point in the galactic plane at galactic longitude close to 90°, is 200 kilometers per second. Mayall's values are a bit higher than Kinman's. However, as Mayall pointed out many years ago, the value of 200 kilometers per second for the rotational velocity of our Sun relative to the galactic center is a minimum value, since it is not unlikely that the system of the globular clusters—which shows some slight degree of flattening—may have a residual rotation of the order of 50 kilometers per second. In summary, the Sun's motion relative to the system of globular clusters is probably of the order of 200 kilometers per second.

Evidence to show that the circular velocity of the Sun and its neighbors is greater than 200 kilometers per second has come from radial-velocity data for neighboring galaxies determined mostly at the Hale Observatories and at Lick Observatory. Together with the Large and the Small Magellanic Clouds, the spiral galaxies in Andromeda and Triangulum, and about a dozen smaller objects, our Galaxy forms a sort of local supersystem. If we assume a velocity of the order of 250 to 270 kilometers per second toward galactic longitude 90° for the Sun relative to the center of our Galaxy, we find that the average relative speeds of all galaxies that belong to our local system of galaxies are less than 50 kilometers per second.

Following the lead of the International Astronomical Union, Milky Way astronomers have in recent years come to assume a value of 250 kilometers per second as the most likely one for the circular motion of the Sun relative to the center of our Galaxy. This value fits nicely into the over-all scheme for the constants of our Galaxy.

It may seem from the arguments presented in this chapter and the preceding one that astronomers are reasonably well in agreement on the values assumed for two basic constants of our Galaxy: 10,000 parsecs for the distance from the Sun to the galactic center and 250 kilometers per second for the circular velocity of the Sun relative to the center of the Galaxy. This is not the case. There is much discussion among the experts relating to both "constants." In recent years the generally assumed value, $M_v = +0.6$, for the mean absolute magnitude of the RR Lyrae variables has been repeatedly questioned. Clube especially has come forth with evidence to show that the mean absolute magnitude for these stars may be fainter by one-half to as much as a full magnitude. However, work by Graham on RR Lyrae variables in the Large Magellanic Cloud supports the mean value $M_v = +0.6$, and shows that the spread around the mean is small, of the order of ± 0.2 magnitude. Clube's revised value of 7,000 parsecs for the distance from the Sun to the galactic center has so far not seemed acceptable to the majority of galactic astronomers. However, a 15-percent reduction (from 10,000 to 8,500 parsecs) does not seem to be out of the question. It is well that our readers should at all times bear in mind that most quoted distances in our Milky Way system are uncertain by ± 10 percent, and there are even greater percentage uncertainties in the extragalactic distance scale. We end on a cheerful note: no one has yet questioned that our Sun moves in a very nearly circular orbit around the galactic center.

The Stars of High Velocity

Most stars near the Sun move with speeds relative to the Sun not in excess of 30 kilometers per second, but there are some that exceed this limit by a wide margin. We are accustomed to refer to the stars with velocities in excess of 60 kilometers per second relative to the Sun as stars of high velocity. Miss Roman has published a catalogue of 600 such stars. In Fig. 63 we reproduce a diagram in which the velocity arrows for the high-velocity stars within 20 parsecs of the Sun are shown projected on the plane of the Milky Way system. With practically no exception, the stars are seen to move toward the half of the Milky Way between galactic longitudes $l = 180°$ and $l = 360°$ (from Auriga, through Orion, to Carina and Sagittarius-Scorpius). Not a single star of high velocity is seen to move in a direction toward $l = 90°$, which is in the general direction of the Cygnus section of the Milky Way. If we determine the average motion of the stars shown in Fig. 63 relative to our Sun, we find it directed toward galactic longitude $l = 270°$. The apex of the Sun's motion relative to the high-velocity stars agrees with that derived from the radial velocities of the globular clusters.

The RR Lyrae variables are the best-known group of high-velocity stars; they give a value for the Sun's apex close to $l = 90°$, with an average speed relative to the Sun of 130 kilometers per second. These stars are arranged in a subsystem that is not quite spherical but has a slight flattening; the RR Lyrae stars are highly concentrated in their distribution toward the nucleus of our Galaxy. The observed flattening of the system of the RR Lyrae stars suggests some residual rotation of the system as a whole, though considerably less than for our Sun; the observed degree of flattening is consistent with a rotational speed of $250 - 130 = 120$ kilometers per second for the RR Lyrae stars as a whole relative to the center of the Galaxy. The purely random motions of the RR Lyrae stars (found after correction for systematic motions) come out quite high—of the order of 100 kilometers per second.

The amazing characteristics of the motions of the high-velocity stars were first studied thoroughly in the years 1924 to 1926 by Oort and by Strömberg. One of the most extensive surveys of this variety of stars was published in 1940 by Miczaika, who at that time was able to list 555 stars with velocities in excess of 63 kilometers per second. Certain varieties of stars contain much higher percentages of high-velocity stars than others. Practically none are found among the B stars, whereas they occur rather frequently among the M stars, where we find that 19 percent of those fainter than $V = 7$ are high-velocity stars.

Bertil Lindblad, who was the originator of the theory of galactic rotation, developed this theory especially to explain the observed asymmetry in the motions of the high-velocity stars. He thought of the Milky Way system as made up of a number of concentric subsystems, each with its own peculiar degree of flattening and rotational speed relative to the galactic center. Even today this approach retains considerable validity and it has been used extensively in the early work of the Soviet astronomers Kukarkin and Parenago. According to this concept, the Sun, and all the stars with observed low velocities (less than 20 or 30 kilometers per second) relative to it, whiz around the galactic center at a rate close to 250 kilometers per second. The RR Lyrae

63. The asymmetry in stellar motions. The diagram gives the distribution in the galactic plane of the directions of the velocities of nearby stars with speeds in excess of 60 kilometers per second. The galactic longitudes of these directions are indicated. Not a single star in the diagram is found to move in a direction between galactic longitudes $l = 30°$ and $l = 150°$; the center of the sector of avoidance is at $l = 90°$.

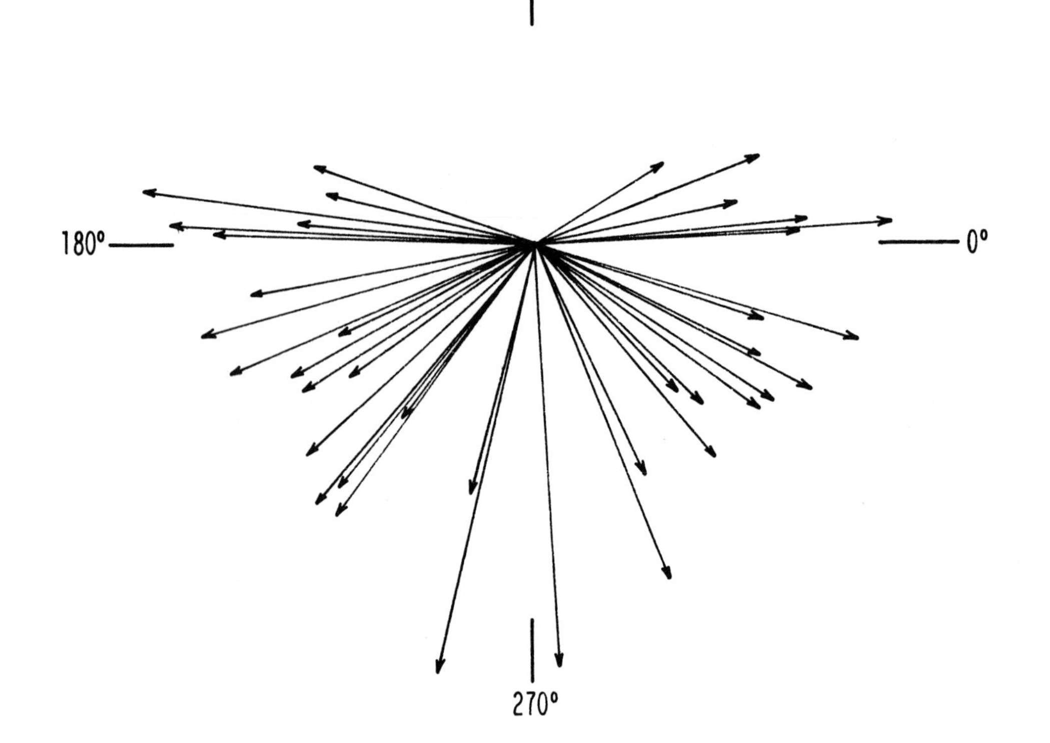

variables, moving with an average speed of 130 kilometers per second relative to the Sun in the direction opposite to where the Sun is heading, are really the laggards, and their average rotational speed is no more than $250 - 130 = 120$ kilometers per second relative to the center of the Galaxy (Fig. 64).

But, you might ask, why are there not also some stars of really high velocity moving with speeds of 100 kilometers per second relative to the Sun in the direction of galactic longitude $l = 90°$? The reason is that such stars would have velocities relative to the center of our Galaxy of the order of 350 kilometers per second, and the Galaxy apparently cannot retain such speeders. A velocity in excess of 310 to 320 kilometers per second relative to the galactic center is apparently sufficiently great to permit a star with such a velocity to escape from the Galaxy, or at least to place it in a rare outlying orbit.

Solar Motion and Star Streaming

We return now briefly to a consideration of the motions of the normal stars of all spectral classes and luminosities that are relatively near to the Sun. We saw in Chapter 3 that the Sun moves with a speed of 18 to 20 kilometers per second toward a solar apex in the constellation Hercules. This result is obtained when the motions are referred to the average for the nearby stars as a standard of rest.

The velocity of the Sun is found not to have the same value for all classes of stars. As early as 1910, Campbell at the Lick Observatory suggested that the derived value of the solar velocity varies with spectral class. He found that the B stars give a rather small solar velocity and that its value increases as we proceed from spectral class A through F to G and K but that it becomes somewhat more like the B average for the bright M stars. Later work has shown that there is a marked dependence of the value of the solar velocity upon the intrinsic luminosities of the stars, with the M dwarfs, for example, giving a much higher solar velocity than the M giants.

The researches of the Vyssotskys at the University of Virginia showed that both the position of the apex and the value of the solar velocity depend markedly on the spectral class and on the average apparent magnitude of the stars with respect to which the motion is determined. The position of the apex for the A and F stars is found at lower galactic latitude than for the other types. The differences in position of the apex and in the value of the Sun's velocity are probably due in part to effects of moving clusters, but they are caused mostly by an uneven admixture of high-velocity stars and by local streamings, different from pure galactic rotation.

We turn now to the subject of *star streaming*. Throughout the nineteenth century, astronomers were blissfully unaware of any regularities in stellar motions beyond those arising from the reflex of the Sun's motion. Then, in 1904, Kapteyn of Holland announced the discovery of two star streams. With this discovery came the first realization that the stars are not moving in a perfectly haphazard fashion, but that their motions are subject to general laws.

Kapteyn's researches dealt with the proper motions of the brighter stars in the sky. The celestial sphere was marked off into a number of sections, in each of which Kapteyn counted the number of stars moving within certain narrow limits of direction. A plot of the proper motions on a chart or on a celestial sphere showed him directly how many stars were moving within 15° of the direction of the North Pole in the sky, how many within 15° of the northwest, and so on. A diagram

64. Asymmetry and the rotation of the Galaxy. The Sun and the majority of the stars near us move around the center of our Galaxy in roughly circular orbits with velocities of the order of 250 kilometers per second. A star that would move in an elongated galactic orbit such as is shown here might have a velocity of only 120 kilometers per second at the point where the orbit reaches as far as the Sun. This star would be observed from the Sun and Earth as having a "high" velocity of 130 kilometers per second directed opposite to that of the general galactic rotation. If it were observed to have a velocity with respect to the Sun and its neighbors as high as 100 kilometers per second in the direction of galactic rotation, then its speed relative to the galactic center would be 350 kilometers per second and the star would in all likelihood not be a permanent member of our galactic system.

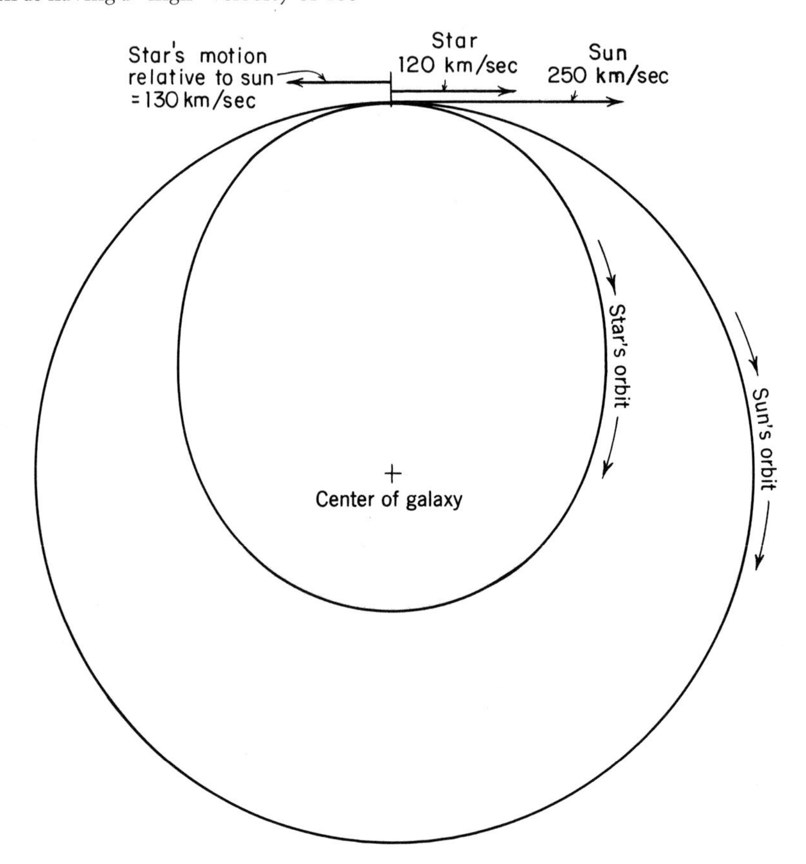

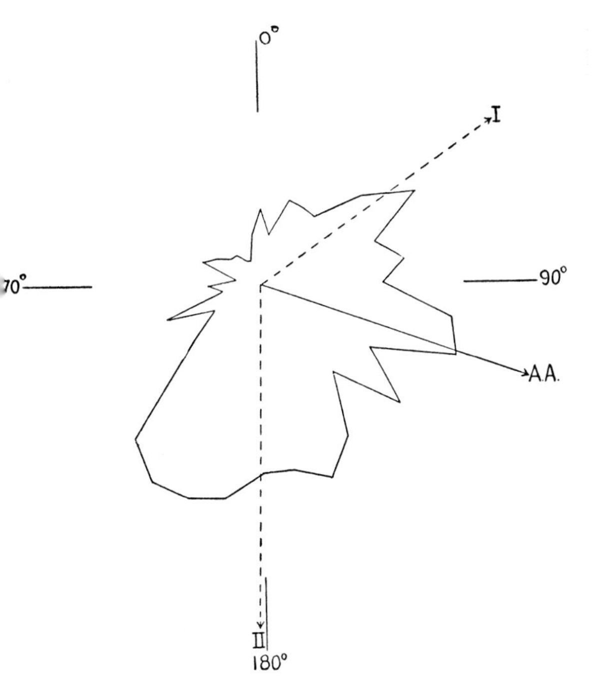

65. A typical star-streaming diagram. The diagram illustrates the distribution of the directions of proper motions, measured by Smart, for a small region of the sky; the arrow *AA* marks the direction toward the antapex. The broken arrows I and II show the directions of the two star streams for this section of the sky.

such as Fig. 65 summarizes the observed distribution in a convenient fashion. For a given section of the sky, we draw for each particular direction an arrow, the length of which is proportional to the number of stars found traveling in that direction. By connecting the ends of these arrows with straight lines, we obtain a clear representation of the distribution of the directions of proper motions for that section of the sky.

If the stars were moving perfectly at random, and if the Sun had no motion of its own, the lines connecting the points in the diagram would form a circle. The Sun's motion would have the effect of drawing this circle out into an elongated figure resembling an ellipse, the long axis of which should point away from the apex of the Sun's motion. For fields not far from the apex and antapex the figure should be almost circular, and the greatest flattening should be found along the circle on the sphere halfway between the apex and the antapex. In fact, however, we find nothing of the kind. The curves near the apex and the antapex are by no means circular, and simple figures shaped like ellipses are not observed. The characteristic figures are bilobed. Generally there are two directions that the stars in a given section seem to prefer; these are marked I and II in Fig. 65.

It was soon found that these streaming effects are not purely local ones. For each section we can draw the two preferred directions, and, if we plot these directions on a celestial sphere, the great circles defined by them converge to two points on the sphere. These points were called by Kapteyn the *apparent vertices* of his two star streams. If the observed streaming tendencies had been of a random character, we would presumably have found no regularity in the distribution of these arrows over the sphere. The fact that each stream shows a well-marked convergent point is excellent proof that the stars all over the sky show in their motions preference for either Stream I or Stream II.

The Kapteyn star streaming bears some resemblance to the convergence effect observed for the motions of stars belonging to a moving cluster, such as the Hyades cluster. In the case of the moving cluster the proper-motion vectors of all individual members are found to pass exactly through the convergent. The motions of the stars that belong to one of Kapteyn's streams show only a tendency to move along with the general stream motion rather than at right angles to it. They still insist on preserving their right to deviate considerably from the path of streaming instead of submitting to a Hyades-like regimentation.

It is not difficult to correct for the effect of the Sun's motion and derive the position of the vertices as viewed from a star supposed to be at rest with respect to the average of all of its neighbors. Kapteyn showed that the *true vertices* of star streaming, found after correcting for solar motion, fall at opposite points in the sky, one in Scutum, the other in Orion.

It was not surprising that the true vertices are 180° apart in the sky, because, by correcting for local solar motion, we have automatically balanced the motions of the two streams. But it was surprising, and highly significant, that the line of the true vertices falls exactly in the plane of the Milky Way, and we note in passing that one vertex has a direction in the sky not far from that to the galactic center. An explanation was not directly forthcoming, but at the time of Kapteyn's discovery it was realized that this represented some new major clue to the ultimate solution of the riddle of the Milky Way system.

Subsequent researches by Eddington and by Karl Schwarzschild, and more recent studies by many other astronomers, most notably the Vyssotskys at the McCormick Observatory, have confirmed and extended Kapteyn's work. Schwarzschild showed that it was unnecessary to think in terms of two specific streams of stars. He pointed out that the line of the true vertices marks the direction along which the stars prefer to move. He found methods for expressing the spreads, or dispersions, of the velocities along the direction of the true vertices and in two mutually perpendicular directions at right angles to the line of the true vertices; the spread is slightly less (0.6 to 0.7 of the first quantity) for the direction at right angles to this line in the galactic plane, and somewhat smaller still for the direction at right angles to the galactic plane.

Lindblad was the first astronomer to prove that star streaming according to the picture of Schwarzschild is a natural consequence of the rotation of our Galaxy. The majority of the stars move in orbits that are almost, but not quite, circular around the center of the Galaxy. These slightly oval orbits in the galactic plane—which are generally also slightly inclined with respect to the plane—produce observable deviations from pure circular motion, and Lindblad was able to show that, on the average, the stars can be expected to exhibit somewhat greater spreads in their motions toward or away from the center than at right angles to this direction.

According to the purest form of the theory of our rotating Galaxy, the line of the true vertices should pass precisely through the galactic center. This happy state of affairs does not exist, for the vertex for some groups of stars lies as much as 15° away from the direction to the center. The explanation of this observed deviation is probably found in the nonuniform distribution of the stars near the galactic plane. Our Galaxy has spiral arms, and we may expect a rather low density of stars and gas between them. In the arms themselves there are concentrations of many sorts, from extended star clouds and aggregates to expanding associations. We live in a part of the Galaxy that is far removed from the well-mixed state envisioned by the simplest form of the theory of a rotating galaxy. Vertex deviation may well provide the clue to the intricate relations that must exist between local variations in stellar distribution and local velocity characteristics.

Galactic Rotation

After Lindblad had presented his explanation of the observed asymmetry in stellar motions, Oort found a further proof of the general rotation of our Galaxy in the radial velocities and proper motions of the stars between 300 and 3,000 parsecs from the Sun. Oort reasoned that it is very unlikely that our Galaxy would rotate like a solid wheel. If it did, the stars would keep their same relative positions and there would be no evidence of motion except from galaxies beyond our own. Since it appears that a considerable part of the total mass of our Galaxy is concentrated near the galactic nucleus, we may expect that the motions of the stars in our Galaxy will resemble those of the planets around the Sun. Venus moves faster than the Earth and the Earth in turn outruns Mars, and in the same fashion the stars nearer the galactic center could be expected to complete a galactic circuit in shorter times than those farther from the center. We might offhand doubt whether this conclusion would hold for our Galaxy, since (see Table 4) the circular velocity of rotation around the galactic center does not vary quite in the manner of the planets of our solar system. Table 4 does show, however, that the stars closest to the galactic center complete a circuit more quickly than the more distant ones. Take, for example, the value of the circular velocity of an object at 20,000 parsecs from the galactic center. If our Galaxy were rotating like a solid wheel, the circular velocity at 20,000 parsecs would be twice as large—500 kilometers per second—as the velocity of rotation near the Sun; the value listed in Table 4 for 20,000 parsecs is only 193 kilometers per second. The more distant parts of our Galaxy obviously lag behind the nearer ones in the completion of one full revolution around the galactic center.

Table 4. The Schmidt model of circular velocities in the Galaxy.

Distance from center (parsec)	Circular velocity (km/sec)
1,000	200
2,500	190
5,000	227
7,500	250
10,000	250
12,500	235
15,000	218
17,500	205
20,000	193

Since all our observations are made from the Earth, which accompanies our Sun, we ask what observable effects in the radial velocities there might be for an observer at the Sun. Oort showed that, as viewed from the Sun, some distant stars would seem to be catching up with us, others would seem to be receding. He found that the effect in the radial velocities would go through its range of values twice as we observe completely around the galactic circle. Figure 66 shows how this comes about. In the diagram on the left, the arrows indicate schematically how the velocity of rotation decreases with increasing distance from the galactic center. In the second diagram, showing the region around the Sun, the effect of the solar motion has been removed and the arrows represent the velocities as seen from the Sun. They show the effects of the differential galactic rotation. The components of these velocities along the radii passing through the Sun will be the effects of differential galactic rotation on radial velocities.

66. The effects of galactic rotation on radial velocities. The diagram on the left illustrates a possible variation of rotational velocity with distance from the galactic center. In the diagram on the right we reproduce on a larger scale the region around the Sun. The arrows now represent the velocities as observed from the Sun. The radial components of these velocities are shown and these exhibit the variation of galactic rotation with galactic longitude, as described in the text.

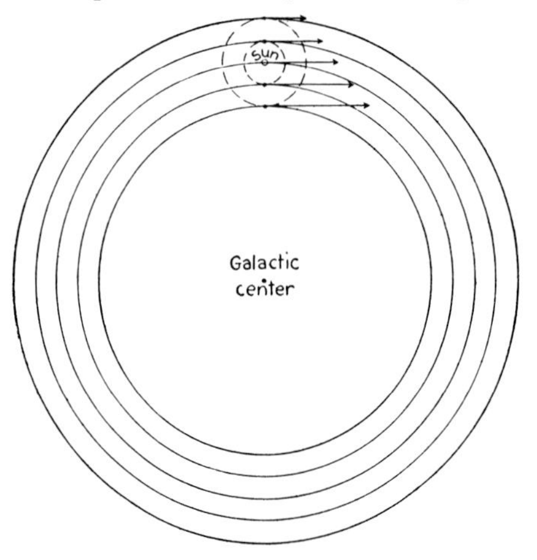

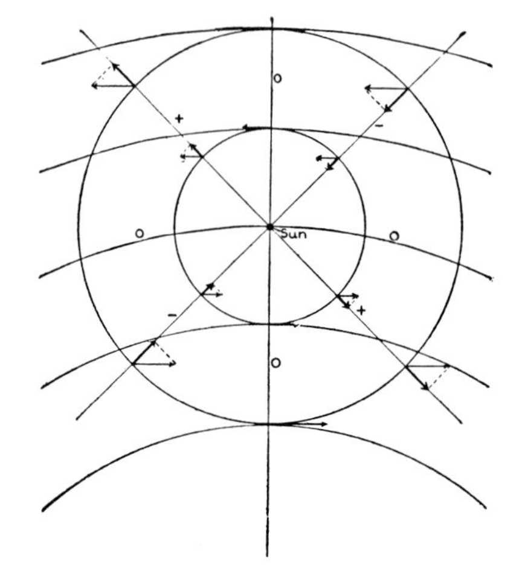

At four points along the circle there should be no approach or recession of the stars due to galactic rotation, and the observed effect in the radial velocities is there zero. These are the directions toward and away from the galactic center and at right angles to that line. Halfway between these points, one diagonal line shows the direction in which the stars will on the average be receding, and hence have positive velocities, and along the other diagonal they should have on the average negative radial velocities, corresponding to approach. As we plot the average observed radial velocities against the galactic longitude, the resulting curve should be a double sine wave, going through two maxima and two minima for the complete circle of longitudes. It is important to note that the galactic-rotation effect in the radial velocities observed at the Sun will be greatest for the most distant stars. Oort showed that for distances up to 2,000 parsecs from our Sun the effect increases very nearly proportionally to the distance.

It may be of interest to reproduce the simple mathematical expression for the galactic rotation effect in radial velocities in the form given by Oort. If V is the effect in the radial velocity, r the average distance of the stars under consideration, l the galactic longitude of the star, then $V = rA \sin 2l$. This formula may be readily checked against Fig. 66. The factor A in the formula is generally known as "Oort's constant" and it measures the maximum effect in V at the standard distance. The accepted value of Oort's constant A is 15 kilometers per second per 1,000 parsecs. The stars that have proved most useful in studies of the galactic-rotation effect are distant, intrinsically luminous objects like O and B stars, open clusters, emission nebulae, and long-period cepheids.

In the past decade a new approach to the study of galactic rotation has become possible. We can now fix with reasonable confidence an estimated distance for any single O or B star we observe. This is done by combining the absolute magnitude of the star, as found from spectrum-luminosity classification, with its observed apparent magnitude and color index. If we measure the star's radial velocity, then we have available two basic bits of information about the star: its radial velocity and its distance from the sun. A small correction is applied to the star's radial velocity (for the standard solar velocity of 20 kilometers per second) to find its value relative to the standard near the Sun moving in a circular orbit around the galactic center at the standard rate of 250 kilometers per second. Assuming that the distant O or B star moves in a nearly circular orbit around the galactic center, its observed radial velocity will yield a value for its circular velocity around the galactic center. The method that we have described had its origin in work by Camm in the 1930's, but its application has become possible only in recent years with the advent of modern techniques of spectrum-luminosity classification. Improved values of the circular velocities listed in Table 4 should become available in the years to come as distances and radial velocities are determined with precision for increasing numbers of distant O and B stars (and for cepheid variables as well).

Oort's theory of galactic rotation predicts also an effect in the proper motions of the stars. The effect in these transverse motions reaches its extreme values for directions in which the effects of radial velocity are zero. We measure proper motions not in kilometers per second but in seconds of arc per year. If, for a certain direction in the galactic plane, we go twice as far out, the linear effect will be doubled, but in the observed angular motions the effect will remain the same as that measured for the nearer stars. To the distance of 2,000 parsecs the effect in the proper motions should vary only with the galactic longitude of the stars involved and not with their distances.

Unfortunately, it is very difficult to detect and measure the galactic-rotation effects in the proper motions. The quantities to be derived from an analysis of the proper motions are small, of the order of a few thousandths of a second of arc per year. Their signs and precise values are made uncertain by possible systematic errors in the basic system of proper motions and we do not know the constants of precession and nutation sufficiently well for dependable analysis of most large bodies of proper-motion data. Clube has recently analyzed the first proper-motion results obtained from the Lick Survey, in which the motions are referred to a fixed system of faint galaxies (see Chapter 2). The Clube results seem to lend support to his low value (7,000 parsecs) for the distance from the Sun to the galactic center. In addition, Clube finds evidence that the Sun and the stars near it show a systematic motion of the order of 35 kilometers per second (or greater!) away from the galactic center. If this suggestion is supported by further evidence, the result will have far-reaching consequences for the study of the dynamics of our Galaxy.

Before we leave the subject of galactic rotation, we should inquire briefly into the shapes of the orbits in which the stars move in our Galaxy. The Sun's motion deviates only slightly from the circular galactic rotation in its neighborhood. We can think of the Sun as moving in an ellipse of small eccentricity, while at the same time it oscillates slowly back and forth perpendicularly to the galactic plane. The Sun will probably stay at all times within 200 parsecs of the galactic plane, and it should complete between two and three oscillations in the perpendicular direction in the 250,000,000 years that it takes to complete one revolution around the center of the Galaxy.

The Sun and the majority of the Population I stars have motions that differ by not more than 20 kilometers per second, that is, by less than 8 percent, from pure circular motion. Hence all of their orbits differ only slightly from pure circular orbits and these stars are in all likelihood now at more or less the same distance from the center of the Galaxy as they were at the time of their birth. The O and B stars, the cepheids, certain open clusters, and the interstellar gas clouds follow most nearly the circular orbits of pure galactic rotation. Some of these stars—notably the O and B stars—are presumably quite young and we shall see later that their probable ages are only a small fraction of the time of one galactic revolution.

The RR Lyrae stars and other fast-moving stars are presumably following elongated orbits shaped somewhat like ellipses. One wonders how close such stars may once have come to our galactic center. Martin Schwarzschild of Princeton has shown that these stars, in spite of their large peculiar motions, have probably never been much nearer to the center than half-way to the Sun. Miss Roman, then at Yerkes Observatory, has, however, discovered some weak-line F stars with very elongated galactic orbits. Some of these stars have velocities that differ so much from pure circular motion that they must have come from points within 2,000 parsecs of the center of the Galaxy. They are probably the best-known representatives of stars that were shot out by the central region of the Galaxy to our remote outpost; they must have come directly from the heart of the Galaxy. By the study of its presently observed motion, we may obviously learn much about the past history of a star.

The theory of galactic rotation represents a great advance in our understanding of the observed regularities in stellar motions. It has provided us with very reasonable explanations of star streaming and of high-velocity stars, and has further led to the discovery of the Oort effects. It is proving extremely useful in interpreting the observations of the 21-centimeter radiation from neutral hydrogen (Chapters 8 and 10).

We should, however, bear in mind that the picture of a smoothly rotating galaxy can at best be only a first and very rough approximation to the true state of affairs in our complex galactic system. The presence of spiral arms, of expanding clusters and streams, of irregularities of distribution and motions, all contribute to the intricacies of the problem. At the same time, these phenomena may hold the key to the past and to the future development of our Galaxy.

7 The Nucleus of Our Galaxy

The Nucleus

Before the work of Shapley (1916–1919), it had been generally assumed that our Sun is not far from the center of our Galaxy and that the stars thin out in all directions from the Sun. Indications were that the diameter of the whole system could hardly exceed 10,000 parsecs. To Shapley belongs the ever-lasting credit of having shown that the Sun and Earth are nowhere near the center of the Galaxy. He gave conclusive evidence of the presence of a massive galactic nucleus in the general direction of the Sagittarius star clouds. Modern estimates place this nucleus at a distance of 10,000 parsecs from the Sun. The apparent thinning out of the stars in all directions from the Sun was explained later by the presence of an interstellar absorbing medium, which dims the light of the distant stars and makes them seem farther away than they really are. Our Galaxy stretches to a distance of at least 5,000 parsecs beyond the Sun, thus yielding an over-all diameter for the central galactic disk of $2 \times 15{,}000$ parsecs $= 30{,}000$ parsecs. It is generally estimated that the total mass of the Galaxy is a little under 200 billion (2×10^{11}) solar masses, the precise value, according to the model of M. Schmidt, being 1.8×10^{11} solar masses. About 5 percent of this total mass is concentrated in the nuclear region of the Galaxy, and the remainder appears to be divided rather evenly between the flattened galactic disk interior to the Sun and the outer spheroidal shell of the Galaxy. The nucleus of the Galaxy is its gravitational center, which controls the motions of the stars in our vicinity. The high concentration of matter to the galactic center is responsible for its over-all rotational properties (see Table 4). It is clearly important that we should learn as much as possible about the nuclear regions of the Galaxy.

When the third edition of this book appeared in 1957, we devoted only a few paragraphs to the nucleus of the Galaxy. It had become evident that the center itself was hidden from view by dense overlying obscuration. Optical studies of peripheral objects, notably of RR Lyrae stars, had yielded a good value for the distance to the center of the Galaxy, a value rather nicely confirmed by studies of galactic rotation. The heavy overlying obscuration made it impossible to photograph the center in normal blue or visual light. Some guesses could be made about the population characteristics of the nuclear region, which seemed to be mostly of Population II stars, but no solid information on the subject was available. Radio-astronomical evidence had already contributed much to our knowledge of the nucleus; a strong radio-continuum source had been located at precisely the position where we would expect to find the center from optical evidence. First attempts had been made to penetrate in the infrared through the dense obscuring matter that lies between the Sun and the nucleus of the Galaxy, but nothing very definite could be said.

All this has changed in the past 15 years. The nuclear region can now be studied in a variety of ways, especially by radio-astronomical and by infrared techniques, and there is a hint that x-ray observations may contribute further in the near future. Gravity waves emanating from the nuclear region have probably been detected and, if these observations are confirmed, they would seem to suggest that the nuclear region is losing mass at a fairly rapid rate. Observationally, the nuclear region of the Galaxy can now be studied in many different ways.

Interest in the study of nuclei of other galaxies has spurred on research about the nucleus of our own Galaxy. Following the early leads of the Soviet astronomer Ambartsumian, there has come a realization that the nuclei of galaxies are the seats of major explosive events. Some of these are on most spectacular scales and involve the production of unbelievably large amounts of energy. Although the nucleus of our Galaxy seems to behave in a relatively moderate manner, it does show side effects of past and possibly of present explosions, the study of which can obviously contribute to the understanding of such phenomena in more distant galaxies. The nucleus of our Galaxy deserves a chapter of its own.

Optical Features of the Nuclear Region

The over-all picture of the surface distribution of faint stars along the band of the Milky Way directly supports the hypothesis that the star clouds in Sagittarius mark the direction toward the center of the Galaxy. The Palomar Schmidt photographs, which reach to very faint limits of apparent brightness, show that nowhere along the band of the Milky Way do the stars appear in greater numbers than in the Sagittarius section. In many ways it does seem surprising that it took until the time of Shapley's work before astronomers began to realize that the Sagittarius star clouds mark the direction to the center of our Galaxy. Observers in the Southern Hemisphere, who see the Sagittarius clouds of the Milky Way in all their glory, can hardly fail to be impressed by them. Nowhere does the Milky Way glow more brilliantly than in Sagittarius and Scorpius and nowhere else does it show such great apparent width. Telescopic views serve only to strengthen our conclusion that the Sagittarius clouds mark the direction toward the center. The evidence was there all the time, but no one took an unprejudiced look at the sky until Shapley pointed out some of the simple facts.

Unfortunately, relatively little is yet known about the detailed stellar composition of the nuclear region. The optical evidence, which does not reach to the nucleus itself (because of galactic obscuration), is very incomplete, but it suggests principally the presence of an agglomeration of older stars in the nuclear region. Radio-astronomical evidence shows that there is a thin and highly turbulent gaseous disk and a group of radio sources centered upon the nucleus itself. Strong infrared radiation is emitted by some objects in the central region.

The most direct evidence for the nuclear concentration of older stars related to Population II has come from studies of the distribution of RR Lyrae stars outside globular clusters. Walter Baade photographed a region about 4° from the direction of the galactic center that is remarkably free from local obscuration. He showed that for this particular field the total overlying photographic absorption over a distance of 10,000 parsecs is only a little more than 2 magnitudes. Baade's choice was an excellent one. A recent study by Van den Bergh shows only 1.5 magnitudes of foreground absorption overlying Baade's field. In this field, which marks the direction in which we can probably see closest to the galactic center, Baade found large numbers of RR Lyrae variables among the faint stars and, to his delight, he found a definite maximum in the numbers of these variable stars at apparent magnitude $m = 17.5$. This maximum is well above the plate limit, which is near $m = 20$. Correcting for the best available value of the total photographic absorption, and assuming a mean absolute magnitude $M_b = +0.5$, Arp finds that the maximum frequency of RR Lyrae stars occurs at a distance of 9,500 parsecs from the Sun. We note that this is very close to the distance of 10,000 parsecs between the Sun and the center of the system of globular clusters. The value of the distance is slightly larger than the original one found by Baade and by S. Gaposchkin, who assisted Baade in the measurement and analysis of these variable stars. The search for and study of RR Lyrae variables should obviously be pressed for other relatively unobscured regions in the Sagittarius-Scorpius section. Such work is now under way under the direction of Plaut at the Kapteyn Astronomical Laboratory in Holland. The first results of these studies indicate the presence of RR Lyrae concentrations at distances less than the value generally assumed for the distance to the galactic center—8,500 rather than 10,000 parsecs.

Many varieties of stars of Population II show a marked concentration toward the center of the Galaxy. Long-period and other irregular variable stars, and especially also the novae, occur in greater abundance in the Sagittarius-Scorpius section of the Milky Way than anywhere else. Near-infrared surveys, conducted at the Warner and Swasey Observatory by Nassau, Van Albada, and Blanco, and at the Mexican Observatory in Tonanzintla by Haro and associates, have revealed that the late M-type giant stars (M5 and later) show little tendency toward clustering, but that they increase in numbers uniformly toward the center of the Galaxy. Comparable results have been obtained from Southern Hemisphere surveys by Henry and Elske Smith and by Westerlund.

Another group of objects that is of interest in this connection is the planetary nebulae. These nebulae possess gaseous shells of relatively small dimensions, which are often centered upon a hot, presumably old, star that may have been a nova in the past. The studies of Minkowski and of Perek on the distribution of these planetary nebulae have revealed that the fainter objects show a great preference for the Sagittarius-Scorpius section. Generally speaking, all stars and related objects that show clear Population II characteristics appear in unusual abundance in the section of the center of the Galaxy.

There is still considerable confusion about the exact stellar population that dominates the central region of our nearest neighbor, the Andromeda galaxy, Messier 31. The spectrum of the nucleus of Messier 31 was studied in 1957 by Morgan and Mayall, who found in it strong cyanogen bands, which are normally indicative of dwarf stars. Since then, Spinrad and associates of the University of California have done some very interesting analysis of the composite spectra of the nuclear regions of our Galaxy and of Messier 31. The great intensity of the sodium *D* lines in the spectrum of the center of Messier 31 would offhand seem to suggest that most of the red M stars are not giants, but rather dwarf stars. However, a more likely explanation is that the stars at the centers of our Galaxy and of Messier 31 are a special class of stars unusually rich in metallic lines. It is perhaps not surprising that we should find near the nucleus stars that are not composed principally of hydrogen and helium, but that apparently have been made from interstellar gas that had been subjected to atomic transformations early in the history of the Galaxy. We have noted already that explosive events on a large scale seem to be characteristic of nuclei of galaxies and it does not seem out of order to suggest that atom cooking on a large scale may have gone on long ago near the center of the Galaxy. All sorts of evidence shows that there is relatively less hydrogen gas right at the center than in regions at some distance from the center.

We should state clearly that the visible great star cloud in Sagittarius marks, in all probability, only the closer edge of a rather extended dense nuclear star cloud. As early as 1952, Bok and van Wijk demonstrated that in the direction of the galactic center there is very much local obscuring matter within 2,000 or 3,000 parsecs of the Sun. Modern estimates, which are based on infrared observations of the nucleus itself, yield approximate values of 25 to 30 magnitudes for the overlying obscuration in visible light between us and the galactic center. This means that only about one part in 1,000 billion of the light emitted by the center is observable in the visible part of the spectrum! Obviously we cannot hope to study the center by the standard techniques of astronomical photography.

The Radio Center of Our Galaxy

From the very day of the discovery that radio radiation is reaching us from distant parts of our Galaxy, it has been known that the central region of the Galaxy produces strong radio emission. Jansky's discovery paper of the middle 1930's gave conclusive evidence for radio radiation coming to us from the general direction of the galactic center. The center of the Galaxy figures prominently in the first radio-continuum maps, those published in 1944 by Grote Reber, which indicate a strong maximum in the direction of the Sagittarius clouds. In 1951 Piddington and Minnet showed that there is a strong discrete radio source in Sagittarius, which obviously marks the direction to the center. Following their suggestion, we refer to this source as Sagittarius A. Over the years, many radio maps of the central regions of our Galaxy have been made. One of the most famous of these, reproduced in Fig. 67, shows the radio equal-intensity contours for radio-continuum radiation with a wavelength of 3.75 centimeters observed by Downes, Maxwell, and Meeks. The

67. A radio map of the galactic-center region. This map, by Downes, Maxwell, and Meeks, shows the distribution of radio sources at a wavelength of 3.75 centimeters, obtained with the 120-foot Haystack Antenna of the Lincoln Laboratory, antenna beam width 4.2 minutes of arc. The contours of equal brightness represent antenna temperatures, which are directly proportional to the intensity measured at each point. The source G 0.7 0.0 is at galactic longitude 0°.7, galactic latitude 0°.0. (Courtesy of *Astrophysical Journal.*)

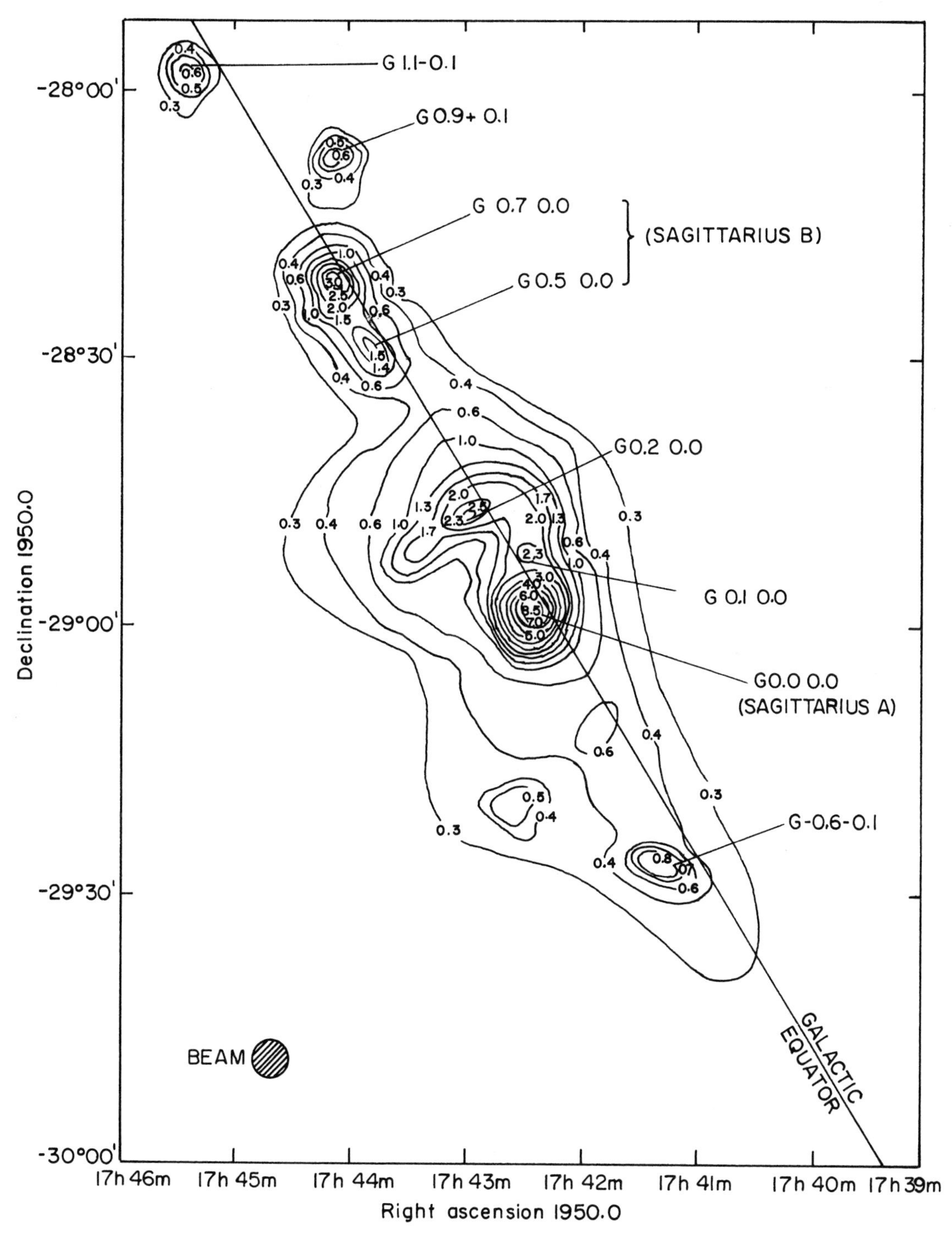

central region of the Galaxy is a conglomerate of a number of separate sources, each of which is labeled in Fig. 67 by its position in galactic longitude and latitude. The strongest of all is the source marked Sagittarius A, which is generally considered to indicate the precise position of the center of the Galaxy. The radio map contains seven named sources in addition to Sagittarius A.

There is a very interesting difference in the properties of Sagittarius A and of the seven radio sources that accompany it. When we study the relative brightness distributions of the seven other sources at different radio wavelengths, we find all seven of them to be relatively weak at longer radio wavelengths, whereas Sagittarius A has very strong radiation at these wavelengths. This alone suggests that there is quite a difference between the physics of the seven sources and the physics of Sagittarius A. An analysis of the brightness distribution with wavelength, that is, the spectra, of the seven sources shows that most of them radiate very much like the beautiful emission nebulae, to be discussed in Chapter 8, which produce their radiation by a simple process of recombination of free electrons with hydrogen nuclei. Why can we not photograph or see these radio sources? Simply because at optical wavelengths they are hidden from our view by the heavy overlying obscuration which, at visual wavelengths, amounts to between 25 and 30 magnitudes. The distribution of the continuum radiation from Sagittarius A is, however, very much like that found in distant extragalactic radio sources and in galactic supernova remnants, such as the Crab Nebula. This is the sort of radiation referred to as synchrotron radiation. It is produced only in the presence of strong magnetic fields. The radiators that produce the radiation are electrons spiraling at very high speeds around magnetic lines of force. The radio source, Sagittarius A, that marks the center of the Galaxy has all the earmarks of being an aftereffect from a mighty explosive event.

The radio fine structure of Sagittarius A has been investigated by Downes and Martin, also by Ekers and Lynden-Bell. It appears to be composed of two fairly large sources, Sagittarius A East, diameter 150 seconds of arc, and Sagittarius A West, diameter 45 seconds of arc, and a third compact source with a diameter of only 10 seconds of arc. The smallest detected unit measures less than 0.5 parsec in diameter!

Radio-continuum studies are only part of our story. For the past 20 years the central regions of the Galaxy have also been studied extensively at the radio wavelength of the 21-centimeter line emitted by neutral atomic hydrogen. This line originates from a transition in the neutral hydrogen atom from a level with very slightly more energy than the lowest level, and it is intrinsically a very sharp one; the neutral hydrogen atom at rest emits radiation at very nearly one wavelength only. If the hydrogen atom that emits the radiation is moving either toward or away from the Earth, the wavelength will be slightly shortened or lengthened according to the usual radial-velocity Doppler formula. If we observe 21-centimeter radio radiation of a given strength and at a wavelength differing by a certain amount from the normal one, then we know that this radiation originates from a cloud of neutral atomic hydrogen with a well-determined velocity of approach or recession relative to the observer. Modern radio telescopes are excellently equipped to study the radial-velocity shifts and intensities of the 21-centimeter radiation for any direction in the sky. The Dutch radio astronomers, notably Oort and Rougoor, have made most extensive studies of 21-centimeter radiation received from the section containing the galactic center. The most striking feature discovered by them at this wavelength is an *expanding spiral arm,* located at a distance of about 3,000 parsecs from the center of the Galaxy and expanding outward from it. The neutral atomic hydrogen that makes up the strong 21-centimeter radiators in the expanding arm, the section that is located between the Sun and Earth and the galactic center, comes toward us with a velocity of approach of about 50 kilometers per second. A second section of the arm can be seen beyond the center. It is moving away from us with a speed of 135 kilometers per second. It is not quite certain whether this whole feature represents a true spiral arm or a ring, but there is a very strong suggestion that it is expanding away from the center and that originally it was thrown out of the nucleus of our Galaxy. Assuming that the hydrogen in the feature was expelled by the nucleus, we can calculate that the original expulsion took place between 10 and 100 million years ago, which is a rather short time on the cosmic scale of time measurement, 0.04 to 0.4 of a cosmic year.

One of the strongest arguments to show that there is neutral atomic hydrogen between the Sun and Earth and the galactic center comes from the observation of 21-centimeter absorption lines viewed in the radio spectra of most of the sources shown in Fig. 67, including Sagittarius A. There are several other major 21-centimeter features associated with the section of the central region of the Galaxy. One of the most important of these is a rapidly rotating flattened nuclear disk of neutral atomic hydrogen with an outer diameter of about 800 parsecs, which whizzes about the center at a rate of more than 200 kilometers per second. Furthermore, there is evidence of massive ejection of neutral atomic hydrogen in directions away from the central plane of the Galaxy, hydrogen clouds that can be observed at high galactic latitudes.

Some sections near the galactic center have proved to be happy hunting grounds for radio astronomers looking for new molecular lines. Sagittarius A shows strong absorption lines from the OH radical and from formaldehyde (H_2CO), both of which, from their radial-velocity shifts, give further proof of the turbulent nature of the gases observable in the central region of the Galaxy. There appear to be no molecules to speak of inside the Sagittarius A source; the absorption features are produced presumably in clouds located between the Earth and the galactic center. The molecules seem to prefer the conditions in the source Sagittarius B, which is located about 40 minutes of arc to the north and east of Sagittarius A. In Fig. 67 Sagittarius A is the strong source marked G 0.0 0.0, whereas Sagittarius B is marked G 0.7 0.0 and G 0.5 0.0.

We should note that x-ray observations are beginning to contribute to the study of the galactic nucleus. Giacconi, Gursky, Bradt, and others have found several x-ray sources in the section of the sky that contains the galactic center. An extended region of moderately strong x-radiation, 2 degrees in diameter, has been located centered upon Sagittarius A and the infrared sources that we shall describe later in this chapter.

Infrared Studies

In 1951, three Soviet astronomers, Kaliniak, Krassovsky, and Nikonov, were the first to penetrate in infrared light through the cosmic dust and to observe features associated with the galactic nucleus. Dufay and Berthier

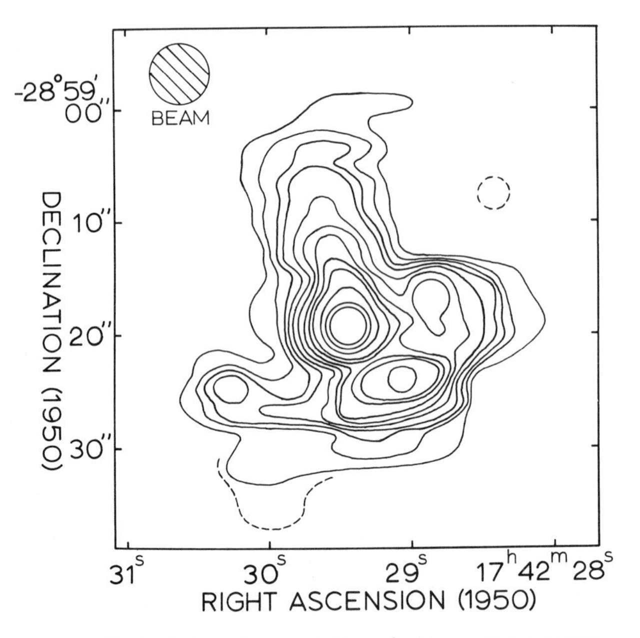

68. An infrared map of the galactic center at wavelength 10.5 microns. This map, prepared by Rieke and Low, covers only a very small region of the sky, as will be seen by comparison of the horizontal and vertical scales in this diagram with those in Fig. 67. It is restricted to the vicinity of Sagittarius A. The beam width for these observations was 5.5 seconds of arc. Contours of equal brightness are shown. There are at least four distinct sources. (Courtesy of *Astrophysical Journal.*)

in France discovered some new infrared features from their studies in 1959. A new era in infrared research began in 1968, when Becklin and Neugebauer found that infrared radiations from the galactic nucleus were observable at a wavelength of 2.2 microns (22,000 angstroms). Their strongest source has a diameter of 5 minutes of arc, which, at the distance of the galactic center, corresponds to about 15 parsecs. At its very center there is a small source with a diameter of only 15 seconds of arc, less than 1 parsec. This particular source seems to be close to the position of Sagittarius A. These strong sources are viewed against an extended infrared background, and a number of additional discrete sources were indicated by this survey.

The central infrared source has been subjected to rather extensive investigation. Studies by Low, Kleinmann, Forbes, and Auman have shown that this source has terrific intensity at somewhat longer wavelengths, between 3 and 20 microns. Its integrated luminosity is the equivalent of 1 to 10 million Suns. A very striking result is that the intensity of the source at still longer far-infrared wavelengths, especially in the range between 30 and 300 microns, is far greater than one would expect from simple interpolation between the intensities observed in the near infrared and at the shortest radio wavelengths. One surprising result of the recent infrared work by Rieke and Low has been that there seems to be a small difference between the position of the radio source Sagittarius A and the strongest source in the infrared. Sagittarius A is located about 1.5 seconds of arc to the east of the infrared source. Figure 68 shows the detailed map for the direction of the galactic center by Rieke and Low. Their beamwidth is about 5 seconds of arc and at 10.5-microns wavelength their positions are accurate with a maximum uncertainty of 2 seconds of arc!

The identifying and measuring of infrared sources observable in the wavelength region between the shortest radio waves and the normal near infrared is not proving to be an easy task. Several additional infrared sources have apparently been found. An interesting current map has been prepared by Richard Capps (Fig. 69). It shows the principal radio and infrared sources in a single diagram. One feature included in this diagram is an extended source at 100-microns wavelength discovered

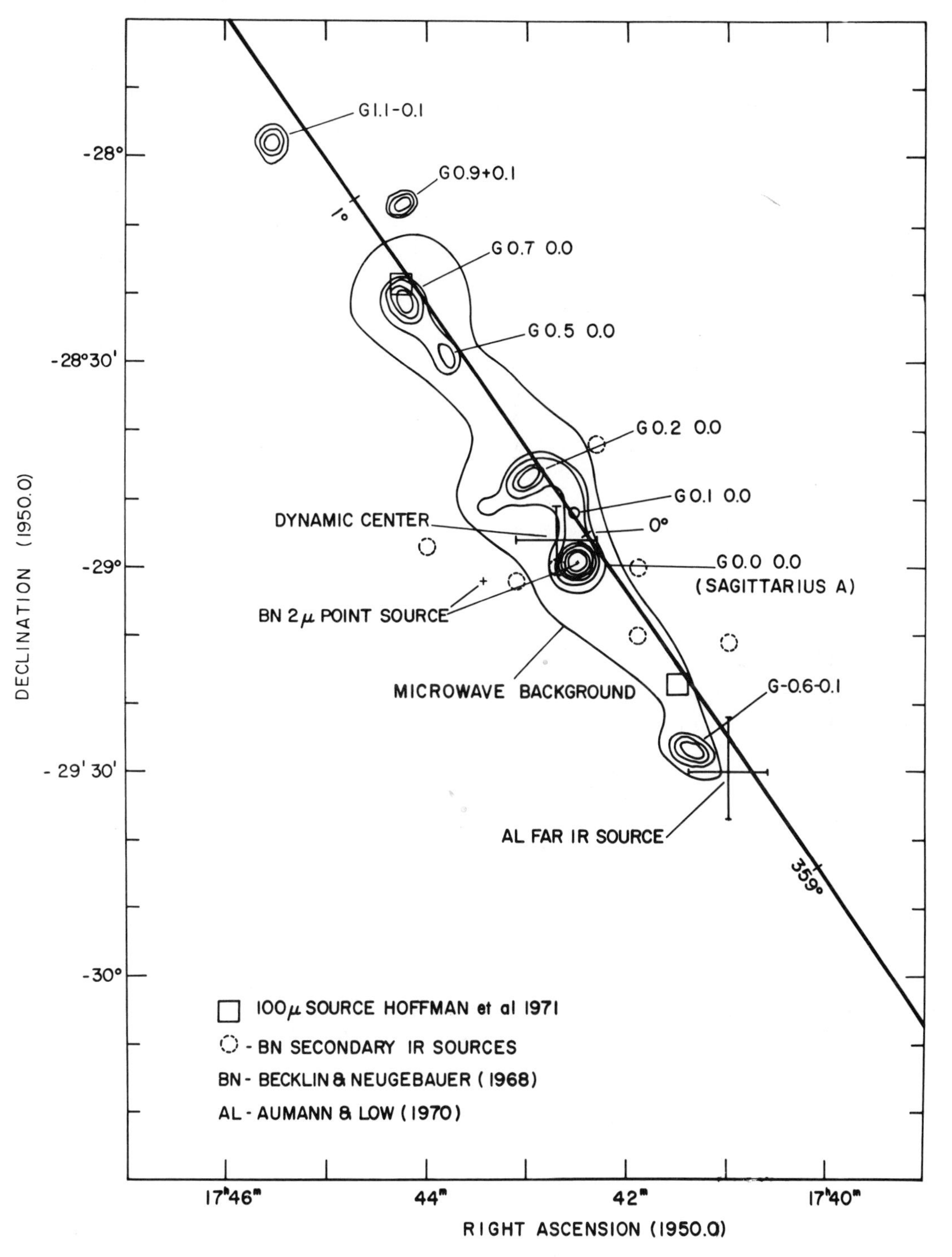

69. Infrared and radio sources near the galactic center. This map, prepared by Richard Capps, is a continuation of Fig. 67. The elongated contour is the result of observations at 100-microns wavelength by Hoffman, Frederick, and Emery. The positions of two secondary sources have been marked by squares. Broken circles give the positions of infrared sources observed at 2.2-microns wavelength by Becklin and Neugebauer. The sources detected by Auman and Low are marked by large crosses. (Courtesy of Kitt Peak National Observatory.)

and studied by Hoffmann, Frederick, and Emery; it appears as a background against which the stronger sources are observed.

There are some striking similarities between the nucleus of our Galaxy and that of the Andromeda galaxy, Messier 31. In the infrared both galaxies have a sharply defined nuclear region. The intensity distribution at 2.2 microns within 40 parsecs of the center is very similar for the two galaxies. However, Messier 31 has no equivalent to the strong radio source Sagittarius A. The most detailed comparison between the two galaxies was made in 1970 by Rubin and Ford, who showed that dynamically and physically the two galaxies, including their nuclei, are quite similar to each other. For example, the sharp nucleus of Messier 31 has a diameter of about 2 seconds of arc, 8 parsecs, a value comparable to that of the infrared nucleus of our Galaxy. Spinrad's metal-rich giants may account for the near-infrared radiation of the source at 2.2-micron wavelength found by Becklin and Neugebauer.

There has been much speculation regarding the origin of the strong infrared radiation reaching us from the nucleus of the Galaxy. The similarities between the near-infrared properties (at 2.2 microns) of the center of our Galaxy and of Messier 31, taking into account the spectral properties found by Morgan and Mayall and by Spinrad, make it seem likely that large numbers of metal-rich giant stars are an important and massive component. As of now the simplest and most reasonable explanation for the strong infrared radiation coming directly from the nucleus of our Galaxy is that advanced by Low. He suggests that we are viewing here a complex of emission nebulae and hot stars embedded in dense clouds of cosmic dust. The infrared radiation we receive would then be the infrared radiation emitted by the dense dust clouds; we never see the hot stuff deep inside.

The study of the central regions of our Galaxy has only just begun. Whereas at first the region of the galactic center seemed to be mostly out of reach for the observer, it has now become accessible to observation and study at many wavelengths. For many years, it seemed that by optical means we would be limited wholly to studies of regions peripheral to the center itself. However, with the new infrared techniques, we have succeeded in penetrating through the cosmic dust and we have found what appears to be the real kernel of the central region. Radio data provide a wealth of material. We now have reliable and detailed maps of the sources of continuum radiation; 21-centimeter and molecular research are helping to sort out what is at the center and what lies in front and beyond. High-energy gamma radiation has now been detected coming from the band of the Milky Way and a broad maximum is shown for the direction of the galactic center. Gravitational waves are only just beginning to tell their story. The center obviously bears watching!

8
The Interstellar Gas

Emission Nebulae

Many of our readers have probably viewed the Great Nebula in Orion through a telescope. The soft greenish hue of the nebulous mass, gradually dimming toward the edge of the field, its erratic though immobile shadings, smooth and mellow in spots, hard and sharp elsewhere, together with the diamond-like scintillations of the four closely packed stars of the Trapezium, present a picture of unsurpassed beauty (Fig. 70). No telescope has ever been able to resolve this glowing mass into stars, and spectroscopic evidence shows that it is truly a nebulous cloud of gas, shining in the transmitted glory of its central stars. What causes such nebulae to shine?

The Orion Nebula has a bright-line spectrum in which the lines of hydrogen, ionized oxygen, and helium predominate. Nebulae like the Orion Nebula are not self-luminous. Nearly 50 years ago Hubble showed that a very hot star is located in the immediate vicinity of every diffuse nebula that shows a spectrum similar to that of the Orion Nebula.

The physical theory that explains why and how such nebulae shine is basically quite simple. The densities and pressures in the nebulae are so low that according to earthly standards we would consider such regions to be perfect vacua. In our physical laboratories an atom is never left alone for any length of time; it is constantly bumping into one of its companions or into the walls of its container. If we wish to observe the atomic processes in their majestic simplicity, we turn to the diffuse nebulae, or to clouds of interstellar gas, which are apparently places where atoms are left alone long enough to perform without undue disturbances.

The atoms in the nebular gas are being bombarded in a leisurely fashion by the radiation from surrounding stars. The only light quanta that can produce major atomic excitation

70. The Orion Nebula. (Photograph made with the Crossley reflector of Lick Observatory.)

are those of high frequency sent out in abundance by the blue-white stars of spectral types O and B. Most quanta of lower frequencies will simply filter through the gas without bothering it, or being bothered, to any appreciable extent. If a quantum of very high frequency, hence of short wavelength, strikes a neutral atom, it may transfer enough energy to the atom to cause the expulsion of an electron. The atom is then no longer electrically neutral, but positively charged. It became an *ion* by the process known as *photoionization*. The electron is free to start off by itself on a journey through interstellar space.

What can happen to the free electron? With its negative charge, it is ready to combine with any positively charged ion that is available, but it soon discovers that there are very few such ions. Our positively charged ion is equally hampered in its search for a free electron that would return it to the neutral state. An atom once ionized may travel through the nebula for days or months before it encounters a free electron to neutralize its charge. In the interstellar laboratories physical processes operate in unhurried and leisurely ways. An atom inside a star, or in one of the physical laboratories on the earth, is constantly being bumped and jostled. The atoms in the nebulae, however, live alone and like it.

Occasionally one of the free electrons will be captured by a positively charged ion. Let us suppose that a capture is made by a hydrogen ion, that is, a *proton*. According to modern atomic theory, the neutral hydrogen atom has only a limited number of permitted orbits in which the electron can move about the proton (Fig. 71). Each such orbit has a definite energy associated with it. When an electron is captured by the proton, it can land in any one of the permitted orbits of the now neutral hydrogen atom. If the capture takes place in the tightest orbit—that of lowest energy—the whole show will be over at once; a single ultraviolet quantum will then be emitted as a result of the capture. Frequently, however, the free electron will be captured in one of the orbits of higher energy. The hydrogen atom cannot remain for more than a small fraction of a second in this excited state and the capture is followed almost immediately by a series of stepwise transitions of the captured electron from one orbit to another of lower energy. The electron cascades inward to end up in the orbit of lowest energy, where it will remain until the next ultraviolet quantum comes along to begin another sequence of disruption and ultimate recombination. During the cascading process a light quantum will be released as the result of each transition. The chances are that somewhere on the way a quantum corresponding to one of the Balmer transitions will be produced which involves a transition from a higher level to the basic level with quantum number $n = 2$. The Balmer lines in diffuse nebulae are produced during such internal adjustments in the neutral atom, which follow immediately after the capture of a free electron by a proton.

The beautiful and conspicuous emission nebulae such as the Great Nebula in Orion, or the famous nebula near the star Eta Carinae, are not isolated phenomena; they represent marked concentrations of interstellar gas, which occupy much of the space between the stars in and near the central plane of our Milky Way system, especially in the spiral arms. There are various ways in which this gaseous substratum reveals itself to the observer. Modern photographic and spectrographic techniques have enabled us to record the faint outer extensions of the larger nebulae

71. Emission of Balmer lines by diffuse nebulae. The circles represent the relative energy levels of the electron in the hydrogen atom, with 1 indicating the lowest or Lyman level, 2 the Balmer level, 3 the Paschen level, and so on. The Balmer series and the Lyman series are shown by arrows representing transitions from higher levels to the Balmer and Lyman levels, respectively. The lines *A*, *B*, and *C* show ways in which an electron may be captured by a proton to produce a neutral atom in the Balmer level. In *A* the electron is captured immediately in the Balmer level and as a result a quantum in the Balmer continuum is emitted. In *B* and *C* the capture takes place in level 5 with the emission of an infrared quantum in the continuum for level 5. In *B* this capture is followed by a direct transition to the Balmer level and emission of the H-gamma line. In *C* the first transition is from level 5 to level 3 and then follows a transition to the Balmer level with the emission of the H-alpha line. The electron in the hydrogen atom would remain in the Balmer level for 1 one-hundred millionth of a second and then settle down in the lowest or Lyman level.

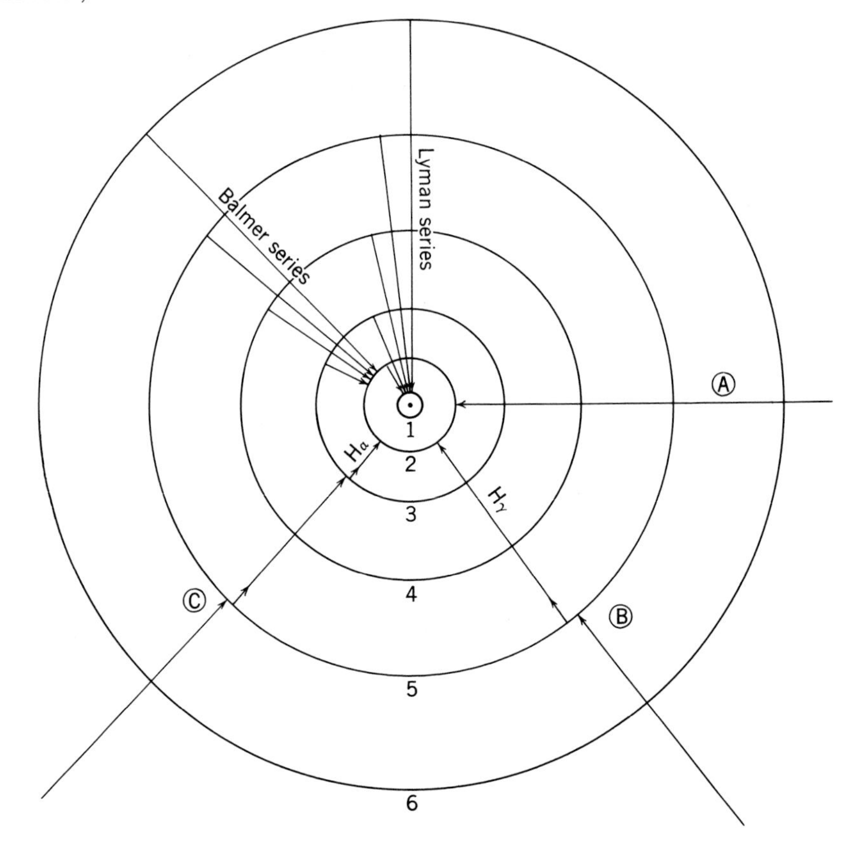

and to locate sometimes very extensive and very faint emission nebulae that had not been previously observed. The interstellar gas is also detected through the presence of certain sharp absorption lines observed in the spectra of distant stars. And, finally, fresh evidence is now coming from radio astronomy, and most recently from the Orbiting Astronomical Observatories. Added interest in studies of emission nebulae arises through the discovery by Baade that the spiral arms in remote galaxies outside our own Milky Way system are most readily traced through alignments of patches of emission nebulosity. If we can locate the emission nebulae of our own Milky Way system and determine their true distances from the Earth, then we should be able to trace the spiral arms of our Galaxy.

Abundances of the Chemical Elements in Nebulae

One might almost gain the impression from our introductory section that the only chemical element present in interstellar space and in emission nebulae is atomic hydrogen. This is not so. Helium atoms are present in considerable abundance, and the appearance of nebular spectral lines attributable to various stages of ionization of nitrogen, oxygen, neon, and some of the heavier elements shows that other elements are present as well. The first estimates of relative abundances of the chemical elements in interstellar space were made by Dunham, and improved values were published subsequently by Strömgren and by Seaton. There have been many recent determinations of relative abundances of the chemical elements, mostly by Aller and his associates, D. J. Faulkner and Helene Dickel, by Mathis, and by H. M. Johnson. Although there are minor differences from one nebula to another, one can say that, on the average, for every 10,000 hydrogen atoms, there are present 1,200 helium atoms, 1 or 2 nitrogen atoms, 3 or 4 oxygen atoms, 1 neon atom, and 1 sulfur atom, with smaller numbers of some heavier atoms, such as iron and chlorine. The casual chemical analyst in interstellar space would conclude that the interstellar gas is mostly a mixture of hydrogen and helium atoms, with a trace of impurities present!

Rather similar relative abundances of the chemical elements are found in the emission nebulae of the Large Magellanic Cloud and in the nearby spiral galaxies Messier 31 and Messier 33. It is of interest to compare the relative abundances of the elements inside stars and in emission nebulae and the interstellar gas. Early in 1973, Spitzer, Morton, and collaborators reported that heavy elements such as magnesium, phosphorus, chlorine, and manganese are less abundant in interstellar space than in the Sun by factors ranging between 4 and 10. Other heavy elements seem to be similarly depleted relative to the Sun. These results were obtained on the basis of observations made from Orbiting Astronomical Observatory No. 3, the famous Copernican Satellite. The observations suggest that heavy elements may have a tendency to stick to the cosmic grains more readily than light ones.

There is proportionately far less hydrogen and helium on Earth than there is in the Sun or in interstellar space. The reason for this can be readily understood. At the time when our Earth was formed, the light hydrogen and most of the helium atoms moved fast enough to escape from the gravitational pull of the proto-Earth. The heavier atoms did not escape. The only hydrogen that stayed with us was that already combined into molecules that were heavy enough to be retained by the crust and the atmosphere. A trace of helium remains in the upper atmosphere of the Earth.

72. The Orion Nebula in infrared light. The photograph shows clearly a marked concentration of very red stars in the region of the gaseous nebula. Comparison with Fig. 70 shows many interesting features. The sharp lines emanating from the star are diffraction patterns produced by the various supports in the tube of the telescope. The white ring around the black circle results from reflection of the light of the star against the glass back of the photographic plate. (An enlargement from a hypersensitized infrared plate by Haro made with the Tonanzintla Schmidt telescope.)

73. Objective-prism spectra for a region near the Orion Nebula. The region is that shown in the lower right-hand corner of Figs. 70 and 72 (tilted). A red-sensitive emulsion was used for this photograph by Haro at the Tonanzintla Observatory. The marks (by Haro) indicate stars with the Balmer H-alpha line in emission.

Optical Studies of Emission Nebulae

In 1937, Struve and Elvey constructed at the McDonald Observatory in Texas a mountainside spectrograph, capable of recording spectral lines of very faint and extended emission nebulosities which, before that time, had not been known to exist. More recently it has become possible to photograph these same nebulae, and other hitherto undetected nebulae, with the aid of fast red-sensitive photographic emulsions and special color filters. Direct filter photography is basically a very simple process. Earlier attempts at direct photography of very faint emission nebulae had failed because the luminous background of the sky blackened the photographic plates before the faint nebulae had registered themselves sufficiently for detection. Astronomers reasoned that, if they could only suppress the sky fog without cutting out much of the nebular light, they should be able to take very long exposures with fast cameras and so record the faint nebulae. It should be remembered that the nebulae shine in the light of specific spectral lines, but that the sky fog on our photographs is produced mostly by the aggregate light of all wavelengths from thousands of faint stars and by certain specific radiations present in the background light of the night sky. By a proper selection of color filters with narrow transmission bands, we are able to cut out all but a small fraction of the continuous light from the background stars and also to suppress most of the night-sky radiation, while permitting the light from one of the nebular emission lines to pass practically undiminished (Fig. 74). In practice this selective photography is achieved most readily through the use of the hydrogen emission line, H-alpha, with a wavelength of 6,563 angstrom units, in the red part of the spectrum. The most effective filter is a Corning red filter combined with one of the many available narrow-band interference filters. One type of filter that has thus far been successfully employed in work of this sort transmits close to 90 percent of the H-alpha light of the nebula, but almost no radiation is transmitted outside a band 50 angstroms wide centered on H-alpha. With such a filter arrangement, and with the fastest red-sensitive plates produced by Eastman Kodak, it has been found practicable to make exposures of up to 4 hours with a camera operating at a focal ratio of 1.5.

The search for faint, extended nebulosities has proved so rewarding that in recent years many observatories have participated in it. In the Soviet Union, Shajn and Miss Hase at the Simeis Observatory in the Crimea and Fessenkov at Alma Ata were the pioneers of photography showing filamentary structure overlying large sections of the northern Milky Way. In France, Courtès and Dufay worked in the same field, and in the United States some of the finest photographs have been made at the Yerkes and the McDonald Observatories, first in 1952 by W. W. Morgan, Sharpless, and Osterbrock and later by Morgan, Strömgren, and H. M. Johnson. The southern Milky Way has also received its share of attention; the first surveys for the half of the Milky Way that is richest in emission nebulosity were made by Gum in Australia and by Bok, Bester, and Wade and also by Code and Houck from the Boyden Station in South Africa. The finest available atlas of the southern Milky Way is the *Mount Stromlo H-Alpha Atlas*, a joint effort by Rodgers, Campbell, Whiteoak, Bailey, and Hunt. It covers a band 30° wide of the southern Milky Way from Sirius past the Scutum Cloud (Fig. 3). There exists, as yet, no comparable atlas for the northern Milky Way. All

74. A matched pair of photographs of the region of the North America Nebula. The left-hand photograph was made on a red-sensitive emulsion and represents the nebula almost entirely in hydrogen H-alpha light. The right-hand photograph was made on an infrared-sensitive emulsion and with a filter to exclude H-alpha light. The total absence of bright nebulosity in the right-hand photograph is to be noted. We draw attention to the fact that the shape of the North American continent is produced by dark nebulosity, especially in the "Gulf of Mexico" and in the "North Atlantic." The famous Pelican Nebula is shown in the left-hand photograph to the right of the North America Nebula. (Photographs by Dufay at the Haute Provence Observatory.)

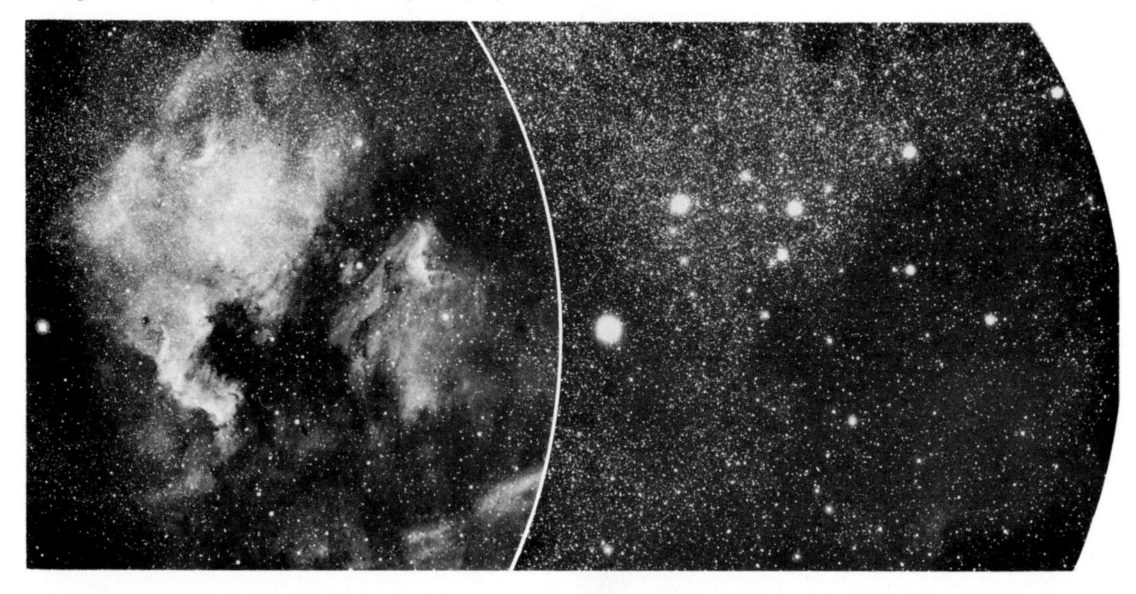

75. A bright fan-shaped nebula in Carina. The southern gaseous nebula NGC 3581, photographed in red light with the ADH telescope of the Boyden Observatory.

76(*a*). The Milky Way in Scorpius in H-alpha light. The key chart shown in Fig. 76(*b*) indicates the positions of the principal gaseous nebulae in this section of the Milky Way. These nebulae are part of an inner spiral arm of our Milky Way system. (Perkin-Zeiss camera, Boyden Observatory.)

76(b). Key chart for Fig. 76(a). The full lines indicate the principal emission nebulosities shown in Fig. 76(a); the dotted line gives the outer boundary of detectable nebulosity. The numbers refer to the positions of the features in galactic coordinates under the former system of galactic coordinates; to refer to the system now in use, 33° should be added to the given galactic longitudes. The first three figures are the longitudes, the last two the latitudes; underlining indicates negative latitudes. A section of the Sagittarius spiral arm can be traced diagonally from upper left to lower right in the diagram.

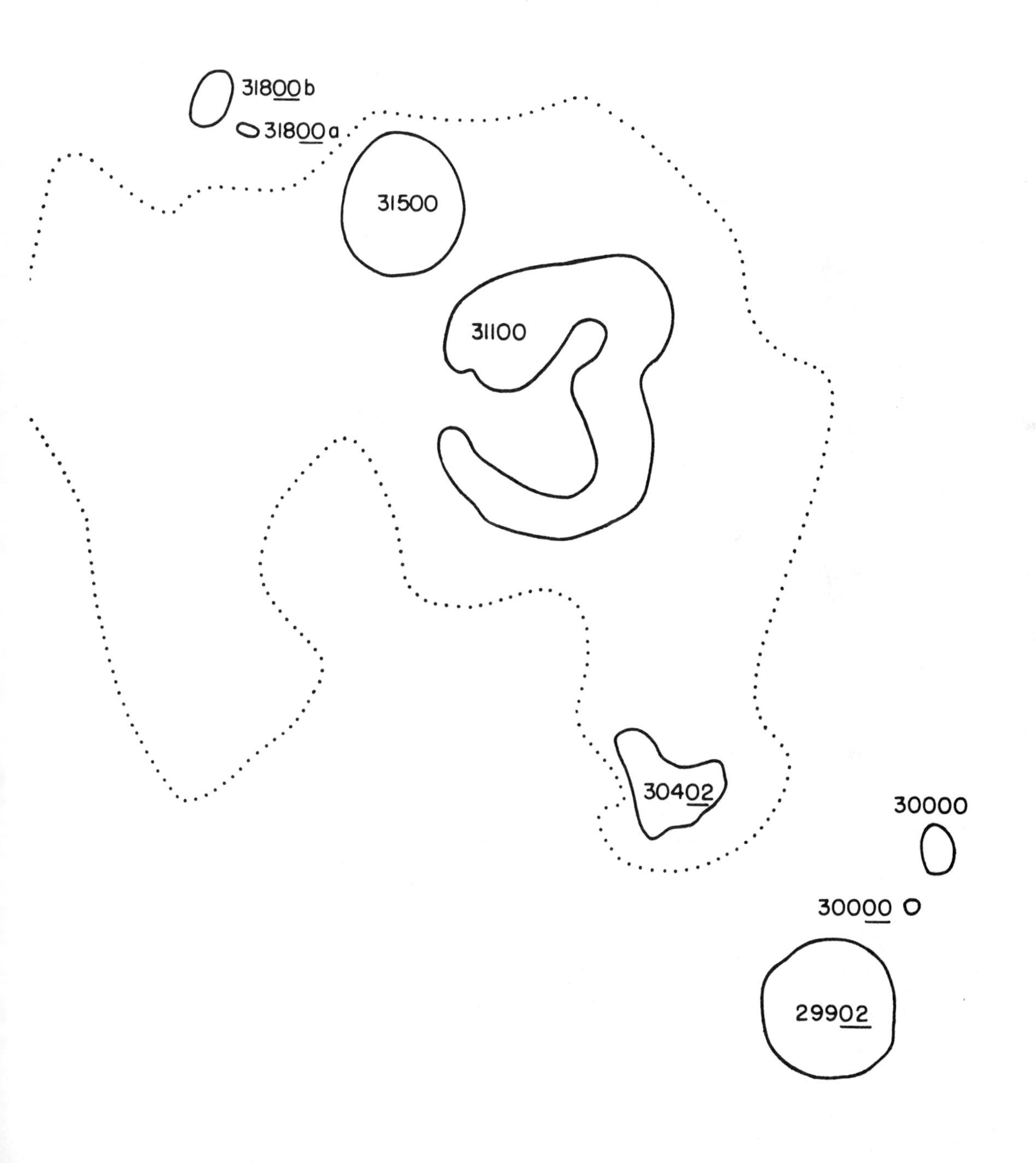

77. The Milky Way in Scorpius; H-alpha light excluded. The photograph covers precisely the same area of the Milky Way shown in Fig. 76(*a*), but here a narrow-band Baird filter was used, selected to exclude practically all H-alpha radiation. The emission nebulae are either absent or very weak. (Perkin-Zeiss camera, Boyden Observatory.)

this work with relatively small wide-angle cameras has resulted in a fairly complete mapping of the sections in which emission nebulosity, including very weak emission, occurs over rather large areas of the sky.

The linear scales of the photographs made with the small search cameras are, however, too restrictive to reveal the full detail of each nebulous structure, so that some of the finer features may remain undetected. Hence large telescopes with fairly small focal ratios (and therefore of great light-gathering power) are used to bring out the details. In this connection the 48-inch Schmidt telescope of the Mount Palomar Observatory and the 32-inch Baker-Schmidt telescope of the Boyden Observatory, known as the Armagh-Dunsink-Harvard telescope, have proved very helpful for the charting of the northern and southern Milky Way. The 24-inch Curtis-Schmidt telescope of the University of Michigan, now at Cerro Tololo Inter-American Observatory in Chile, and the 20-inch Swedish Uppsala-Schmidt telescope at Mount Stromlo Observatory in Australia, have done effective work on wide-angle photography for the Southern Hemisphere. Two additional Schmidt telescopes of large aperture and one wide-field reflector are being added to the southern arsenal: the 40-inch Schmidt for the European Southern Observatory in La Silla in Chile, the 48-inch British Schmidt for Siding Spring Observatory in Australia, and the 100-inch reflector for the Hale Observatory at Las Campanas in Chile.

There exists as yet no complete catalog of emission nebulae found from surveys made with the southern Schmidt telescopes, but considerable work has been done on the basis of the Palomar Schmidt survey. The most comprehensive catalog was published by B. T. Lynds, who was then at Steward Observatory. Her list contains close to 1,200 entries and covers all except the southernmost quadrant of the band of the Milky Way (galactic longitude 250° to 340°). A Schmidt photograph covers an area of the sky about 100 times that covered by the full Moon, and yet the scale of the photograph is sufficient to show details of structure in the nebulous filaments.

To round out the telescope picture for photography of emission nebulae, we turn finally to the large reflectors, which show the more conspicuous nebulae in all their glory. The 200-inch Hale and the 100-inch Hooker telescopes of the Mount Wilson and Palomar Observatories and the 120-inch Lick reflector are the best available telescopes for photography of nebulae from the Northern Hemisphere. The 74-inch reflectors of the Radcliffe Observatory in South Africa and of Mount Stromlo Observatory in Australia are the major instruments of the Southern Hemisphere. Four additional large reflecting telescopes are now under construction. The 158-inch Mayall reflector of Kitt Peak National Observatory is ready for Northern Hemisphere research. Three major telescopes are under construction for the Southern Hemisphere; they are the 158-inch reflector for Cerro Tololo Inter-American Observatory in Chile, the 150-inch British-Australian reflector for Siding Spring Observatory in Australia, and the 140-inch reflector for the European Southern Observatory in Chile. We are entitled to have great expectations!

Thus far the optical searches for faint, extended emission regions have concentrated largely upon nebulosities shining in the red light of the H-alpha line of hydrogen. We noted at the beginning of the present chapter that the H-alpha line originates as the result of the capture of a free electron by a single proton. Since this is the dominant process,

78. The Norma Nebula. (From a photograph by Westerlund with the Uppsala-Schmidt telescope.)

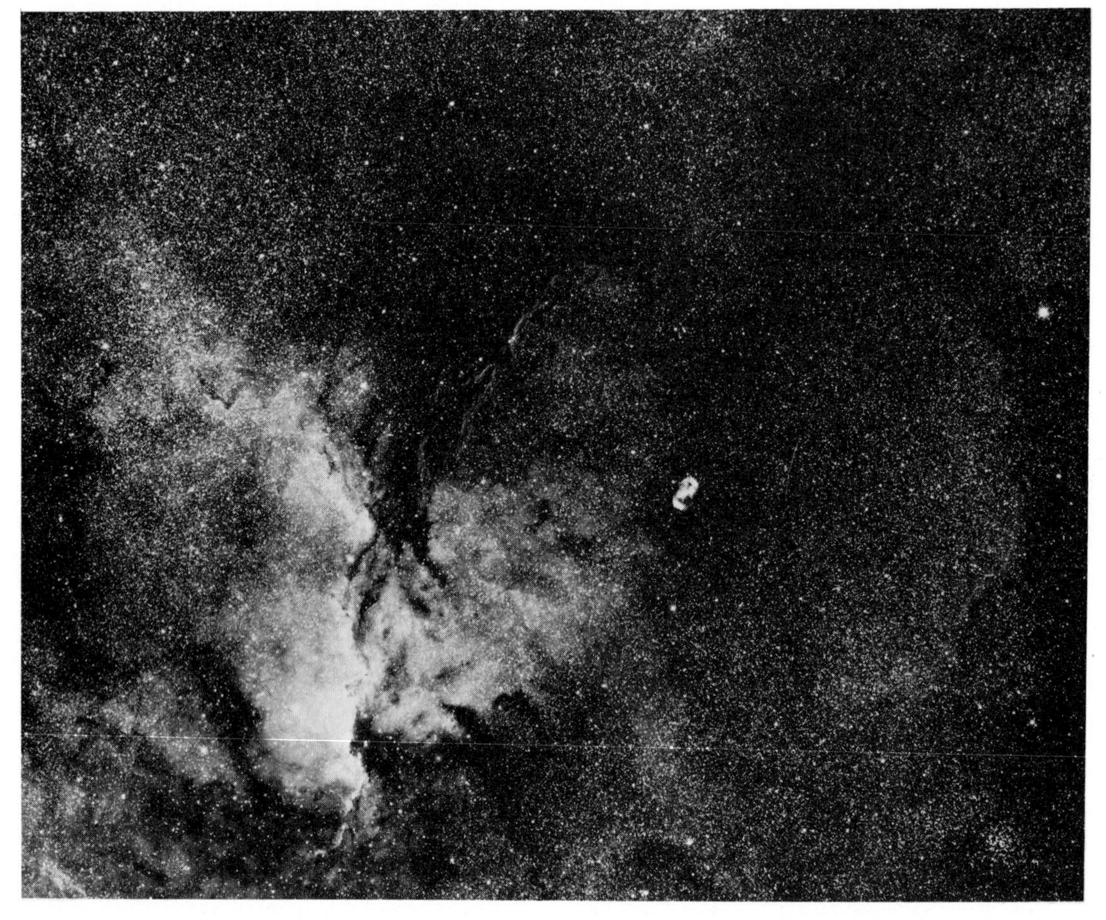

the observation of H-alpha is indicative of the presence of ionized hydrogen atoms (protons) caught in the act of recombination. In the professional jargon of the astrophysicist, the clouds of hydrogen that are largely ionized are called H II regions. The term was first introduced by Strömgren, who showed that H II regions should occur around O and B stars rich in ultraviolet radiation capable of ionizing the nearby interstellar hydrogen. Strömgren finds that, for an average density of interstellar hydrogen of the order of one hydrogen atom per cubic centimeter, a B0 star emits sufficient ultraviolet radiation to ionize all the hydrogen to a distance of 30 parsecs from the star. For the same hydrogen density, a very hot O star may produce complete ionization within a sphere of radius 200 parsecs, resulting in a hydrogen emission nebula of truly gigantic size. The supply of ultraviolet light and the resultant ionizing power decrease rapidly as we proceed to the cooler stars. An A0 star, with a still rather respectable surface temperature of the order of 11,000°K, will probably ionize the gas within a sphere with a radius of only 0.3 parsec, and the cooler stars will produce no appreciable hydrogen ionization at all. It therefore causes little surprise that near almost every H II region one finds an O or B star, or a cluster of O and B stars, that may be held responsible for exciting the nebular radiation.

We are really concerned with two varieties of regions of interstellar hydrogen, the H II regions and the H I regions. The latter contain principally neutral atomic hydrogen, whereas in the H II regions the majority of the hydrogen atoms are ionized. In the H II regions, the ionizing ultraviolet radiation from nearby stars is relatively plentiful, and the electrons are ejected at the time of ionization with sufficiently high speeds to produce temperatures in the interstellar gas of the order of 10,000°K. In the H I regions ionization does not play much of a role and there is no process by which interstellar atoms may maintain the same high speeds as for the H II regions. According to estimates by Spitzer and Savedoff, the speeds of the interstellar atoms in H I regions correspond to a temperature of 60°K, that is, 60° above the absolute zero, or, on the centigrade scale, more than 200° below the freezing point of water.

In recent years, increasing emphasis has been placed upon studies of radial velocities of emission nebulae and of spreads in radial velocities inside these nebulae. The standard approach in optical radial-velocity studies has long been to measure radial velocities from slit spectra obtained with large reflectors and refractors. The original Lick Observatory survey by Wright was carried out by these techniques. As spectrograph designs improved, it became possible to obtain radial-velocity data for increasingly less conspicuous H II regions. One of the finest studies in this area done in the middle 1960's, is that of J. S. Miller of the University of Wisconsin, who used a fast slit spectrograph attached to the 36-inch Kitt Peak reflector. He obtained data for 36 H II regions, mostly of the northern Milky Way. Such radial-velocity data figure prominently in studies of the spiral structure of our Galaxy.

Optical measurements of faint H II regions have become possible by the application of the Fabry-Perot techniques of interference-fringe photography. These techniques have been developed for studies of H II regions principally by Courtès and his associates Cruvellier, Monnet, and the Georgelins. Most of the work was done in France at the Haute Provence Observatory, some of it at an observing station in South Africa, and more recently at the European Southern Observatory in Chile. More than 4,000 individual radial velocities have been measured, and a list of radial velocities is now available for 200 H II regions. The photographic techniques developed and applied by the French astronomers are now being used by M. G. Smith at Kitt Peak and Cerro Tololo observatories for photoelectric analysis. A pressure-scanning technique enables the observer to obtain directly at the telescope the intensity distribution in the profile of an emission line, which not only yields an average radial velocity of the nebula, but also provides data on the distribution of radial velocities within the part of the nebula being scrutinized. The average radial velocity represents very useful kinematical information for the study of spiral structure of our Galaxy, whereas from the spreads of radial velocity we may learn much about the physical conditions caused by the exciting star inside the nebula.

Interstellar Absorption Lines

The interstellar gas reveals itself not only through the characteristic bright-line spectra of the emission nebulae, but also through certain sharp and narrow absorption lines found in the spectra of many distant stars. The discovery of such absorption lines goes back to 1904, when the German astronomer Hartmann showed that the absorption *K* line of ionized calcium (wavelength 3,933 angstroms) in the spectrum of the star Delta Orionis behaved in a very peculiar fashion. Delta Orionis is a blue star with B0 spectrum and was recognized to be a spectroscopic binary. Hartmann found, however, that the wavelength of the *K* line did not vary at all in the course of the binary period. The hydrogen and helium lines in the spectra of Delta Orionis were broad and fuzzy, but its *K* line was sharp and distinct. Hartmann referred to the line as the "stationary" calcium line, since it did not share in the radial-velocity shifts exhibited by the spectral lines in the binary star.

Stationary calcium lines have since been discovered in the spectra of many other early-type stars. In 1919 Miss Heger (Mrs. C. D. Shane) and Wright at the Lick Observatory found that the spectra of some early-type stars have strong stationary sodium lines in addition to their stationary calcium lines. In all cases these lines were found to be sharp and distinct.

The interstellar origin of these absorption lines was first suggested by V. M. Slipher in 1909, but his suggestion unfortunately did not receive the attention it deserved and his views were not generally accepted until more than 15 years later. It was thought at first that the stationary lines originated in the immediate vicinity of the stars in whose spectra they are found. The researches of J. S. Plaskett and of Struve proved that this explanation was incorrect. Theoretical investigations, especially those of Eddington and Rosseland, showed also that stationary lines of calcium would naturally be observed if any free gas were present in interstellar space. The total evidence left little doubt as to the interstellar origin of the stationary lines and, since 1930, they have generally been referred to as interstellar lines.

Adams and Dunham at the Mount Wilson Observatory found interstellar absorption lines attributable to neutral calcium, neutral potassium, neutral iron, and ionized titanium in addition to the lines of ionized calcium and neutral sodium to which we have already referred (Figs. 79 and 80); all told, 15 sharp interstellar absorption lines were definitely identified. Dunham also found some sharp interstellar lines in the blue-violet part of the spectrum which for some time remained unidentified. McKellar of the Dominion Astrophysical Observatory in Canada proved that these lines arose from transitions between energy levels in simple molecular compounds of carbon, nitrogen, and hydrogen. In a recent tabulation, G. Münch lists 13 lines of molecular origin as definitely identified. Finally, Merrill of the Mount Wilson Observatory has given evidence of the presence of 6 broad and as yet unidentified interstellar absorption bands, the strongest at 4,430 angstroms with an approximate width of more than 40 angstroms. It is not yet known how these broad bands are produced. The suggestion has been made that they may come from more complex molecules or from a transition stage in the formation of solid particles. But for the time being all this is frankly guesswork.

The far ultraviolet spectra of half a dozen stars observed by the Copernican Satellite

79. Multiple interstellar lines. High-dispersion spectra made by Adams at the Mount Wilson Observatory show many instances of multiple interstellar lines. This multiplicity indicates the presence of distinct interstellar clouds, each moving at its own peculiar rate.

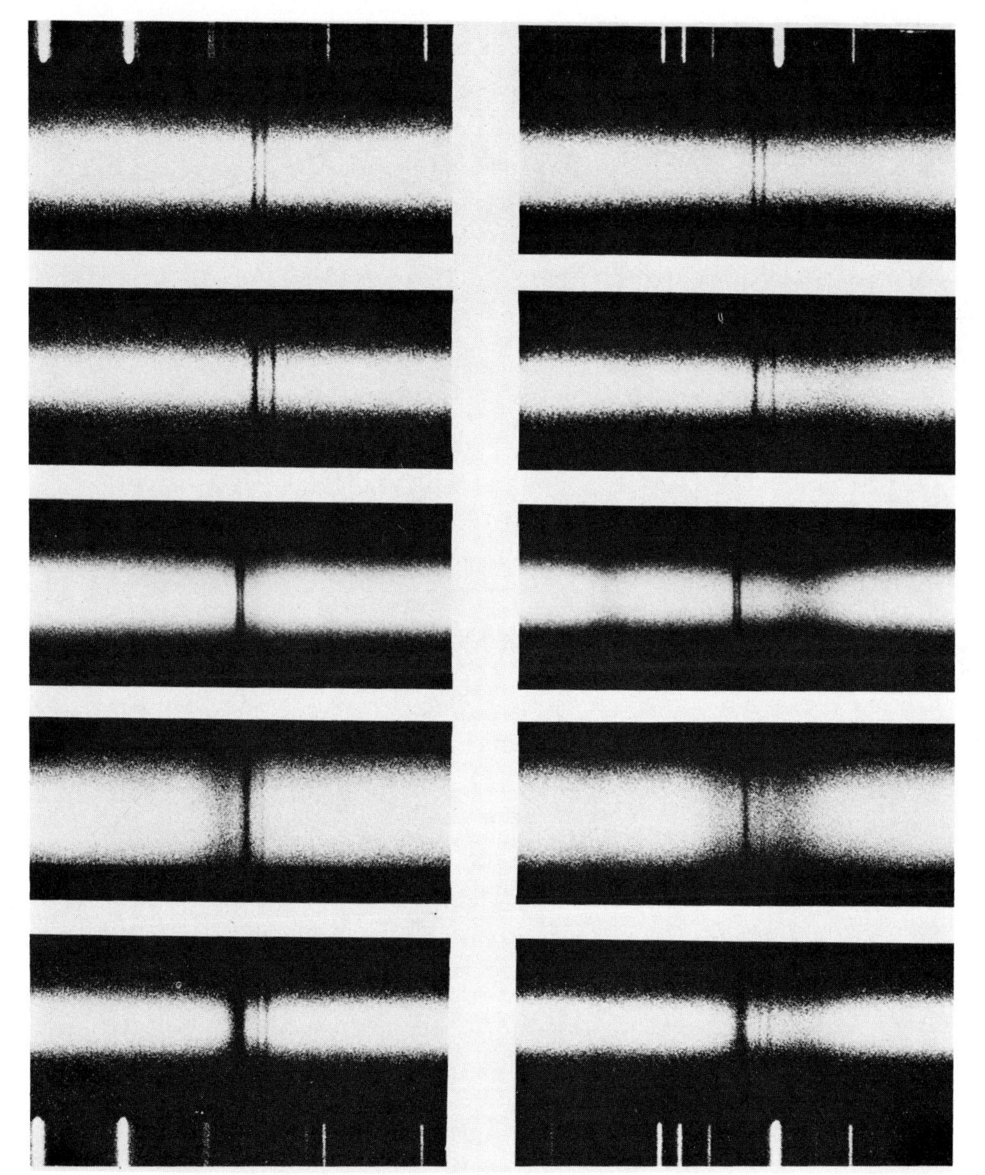

80. Interstellar lines of ionized calcium and of ionized CH. The marked sharp lines are examples of interstellar absorption lines—several of them multiple lines—discovered by Adams. (Mount Wilson photograph.)

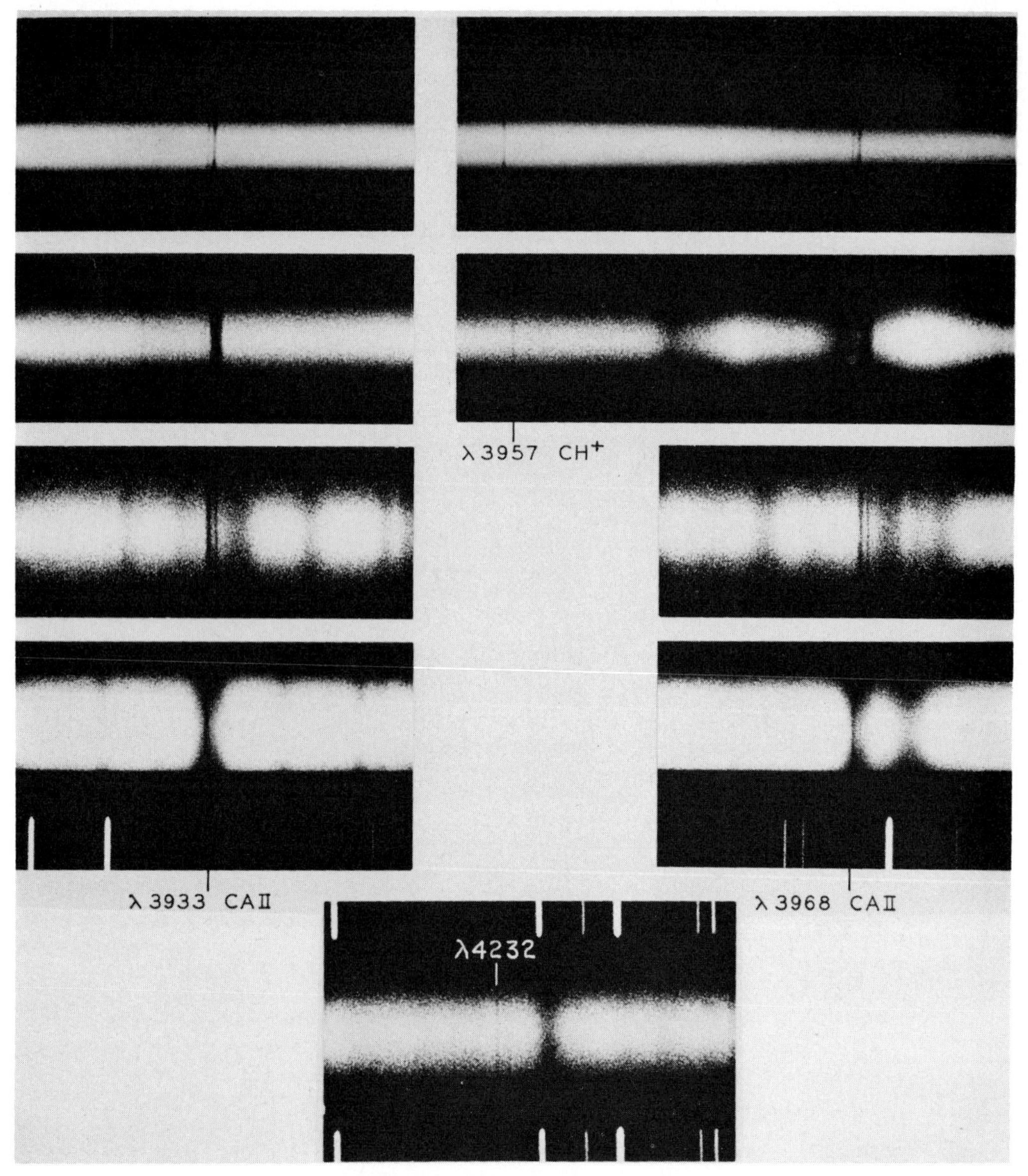

have yielded a harvest of new interstellar absorption lines. The lines produced by various ionization stages of carbon, nitrogen, oxygen, magnesium, silicon, phosphorus, sulfur (four ionization stages observed!), chlorine, argon, manganese, and iron have been identified by the Princeton group headed by Spitzer and Morton. Many absorption lines attributable to the hydrogen molecule (H_2) are present. This molecule was first detected by Carruthers in 1970 on the basis of rocket ultraviolet observations. The Copernican Satellite spectra show many H_2 absorption lines. In addition heavy hydrogen or deuterium (D) is found from telltale lines apparently produced by the HD molecule. The data indicate a surprisingly great abundance of D compared to H, the indicated ratio being that for every D nucleus there are 200 H nuclei, which represents a very much larger deuterium abundance relative to hydrogen than is found on Earth or in the Sun's atmosphere.

How do the observed interstellar absorption lines originate? Suppose we consider the interstellar *K* line with a wavelength of 3,933 angstroms. This absorption line is produced when an ionized calcium atom absorbs a light quantum of that wavelength. Since atomic phenomena take place in a rather leisurely fashion in interstellar space, we may assume initially that most ionized calcium atoms will have settled down to the lowest energy level. Then there may by chance come along a stellar quantum of wavelength 3,933 angstroms, just capable of temporarily exciting the calcium ion into a higher level of energy. Atomic physics teaches us that the calcium ion can stay in that blown-up state for only something like 1 ten-millionth of a second and that then it returns to its original lowest energy level. The energy that was absorbed is thereby released, but the new quantum of wavelength 3,933 angstroms will go off—scatter—in a direction different from that in which the previous one arrived. If we are looking at the spectrum of a certain star and there is ionized calcium between that star and us, the interstellar calcium ions will absorb and scatter many of the quanta of wavelength 3,933 angstroms that would otherwise have reached our spectrograph; hence a dark line appears in the spectrum of the star.

The traditional technique for the study of interstellar absorption lines has been through the study of spectra of very high dispersion obtained with the aid of coudé spectrographs attached to large reflectors. The most active research worker in the field has been G. Münch at the Hale Observatories. Together with Vaughan, he has developed a high-resolution photoelectric Fabry-Perot system, which shows many components in each interstellar absorption line. Another approach to the same problem has been that of C. R. Lynds and Livingston, who used the McMath Solar Telescope at Kitt Peak for high-resolution measurements of the interstellar absorption lines for some bright stars within reach of this equipment.

The interstellar gas shows considerable cloud structure. Until 1936, astrophysicists supposed that the interstellar gas was distributed smoothly through a thin layer near the central plane of the Milky Way. In that year Beals, then of the Dominion Astrophysical Observatory in Canada, found evidence of multiplicity in some of the lines, indicating that more than one cloud of gas was contributing to the formation of the interstellar absorption lines in certain stars. Subsequent studies at high dispersion of the *K* line of ionized calcium were carried out by Adams at the Mount Wilson Observatory. He showed that more than 30 percent of the stars ex-

81. Interstellar sodium lines. The spectrum of star No. 12953 in the Henry Draper *Catalogue* photographed by Münch with the coudé spectrograph of the 200-inch Hale telescope. The black central band is the negative impression of the stellar spectrum and the white lines in the dark band are two components of the D_1 line of neutral sodium and two components of the D_2 line also due to neutral sodium. There are two components to each line because between the star and our Sun there are two clouds producing neutral sodium absorption, one of which, with a radial velocity of approach of 55 kilometers per second, gives the weaker component (to the left) of each pair, the other, with a radial velocity of approach of 7 kilometers per second, yielding the stronger component (to the right) of each pair. The lines at the top and bottom are from the comparison spectrum imprinted during the exposure. They are emission lines and hence appear black in the negative print. The star is of spectral class A1 and luminosity class Ia. It is at an estimated distance of 2,000 parsecs, and it so happens that, in the limited stretch of spectrum shown, there are no stellar absorption lines present.

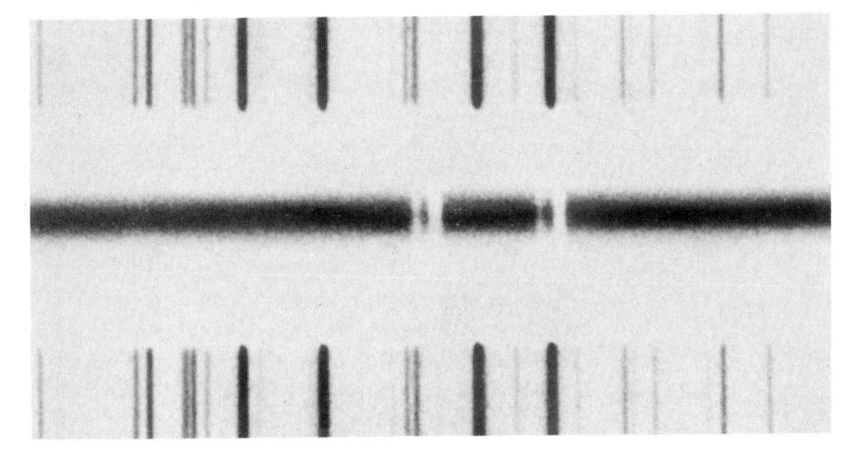

amined by him had double or triple interstellar *K* lines, and a *K* line with four components was found for four stars. It may be assumed that every observed component of the *K* line is produced by a separate and distinct interstellar cloud. The separation between the components is caused by the difference in the Doppler shifts produced by different radial velocities of the separate clouds relative to the Sun.

In recent years, G. Münch has found many stars that show multiple interstellar lines (Fig. 81); some of his spectra show as many as seven components, every one of them indicative of a separate interstellar cloud with a different radial velocity relative to the Sun. Münch's observations present most convincing evidence for the cloud structure of the interstellar gaseous medium. We note here that these multiple interstellar lines found by Münch occur even in the spectra of some stars at considerable distances from the galactic plane, thus suggesting that at least some wisps of interstellar gas have sufficient speeds at right angles to the plane of the Milky Way to reach considerable heights above or below the plane. We shall see in Chapter 11 that the great majority of the interstellar clouds discovered by Münch are near the central plane and fit nicely into the accepted spiral pattern of our Galaxy.

The interstellar gas partakes in the general

rotation of the Galaxy. This fact was established by the researches of Plaskett and Pearce at the Dominion Astrophysical Observatory. Shortly after Oort had suggested that the radial velocities of distant stars should vary in a double sine wave according to galactic longitude, Plaskett and Pearce undertook to measure the radial velocities of several hundred faint, and hence distant, early-type stars and they made a successful check of the theory of galactic rotation. The interstellar *K* line was measurable on many of the spectrograms of Plaskett and Pearce. When the radial velocities determined from the measurements of the interstellar *K* lines were plotted against the galactic longitude of the stars, the results showed very clearly the familiar effects of galactic rotation. The striking difference was that the range of the double sine wave with galactic longitude for the stars was approximately twice that shown by the radial velocities from the *K* line. The conclusion that Plaskett and Pearce drew from their curves was that the interstellar *K* line in the spectrum of a given star yields on the average a radial velocity corresponding to the halfway point between the star and the observer.

The early measurements based on interstellar lines seemed to suggest that interstellar calcium atoms are distributed rather uniformly near the galactic plane. Because of the discovery of multiple interstellar absorption lines in the spectra of many stars, this simple picture has now been abandoned. We visualize the interstellar gaseous medium as consisting primarily of a relatively flat layer of individual gas clouds, with a thin all-pervasive galactic substratum possibly connecting them. Each cloud moves around the center of the Galaxy in an approximately circular orbit. On the average the actual velocity of a cloud differs from the circular velocity of galactic rotation by ± 8 kilometers per second. With very few exceptions—which we shall note later on—the clouds that produce the observed interstellar absorption lines are located in the spiral arm that contains our Sun.

The average properties of the individual interstellar clouds can be found from a study of the components of the interstellar absorption lines shown by stars at various distances. On the average, a line of sight close to the galactic plane cuts through seven or eight of these clouds in 1 kiloparsec. Each cloud has a diameter of between 10 and 15 parsecs and a probable mass equivalent to a few hundred solar masses.

Radio Studies of the Interstellar Gas

During the past decade, radio techniques for the study of emission nebulae have become increasingly important. First of all, we learn from such studies much that is new about the physics of the nebulae, and second, we find some of the more distant emission nebulae very convenient anchor points for studies of remote spiral features in our Galaxy. These radio studies supplement in many ways the optical work on emission nebulae. One of the very useful properties of radio radiation is that it passes almost undiminished in intensity through clouds of cosmic dust. Many emission nebulae that are hidden from direct view by interstellar absorption can be detected and observed with ease by radio techniques. This is becoming increasingly important. Not only does it permit access to distant nebulae that otherwise would remain unknown, but it gives us an opportunity as well to detect and observe some relatively nearby and also many distant emission nebulae that are embedded in clouds of cosmic dust. It has become increasingly evident in recent years that protostars and other very

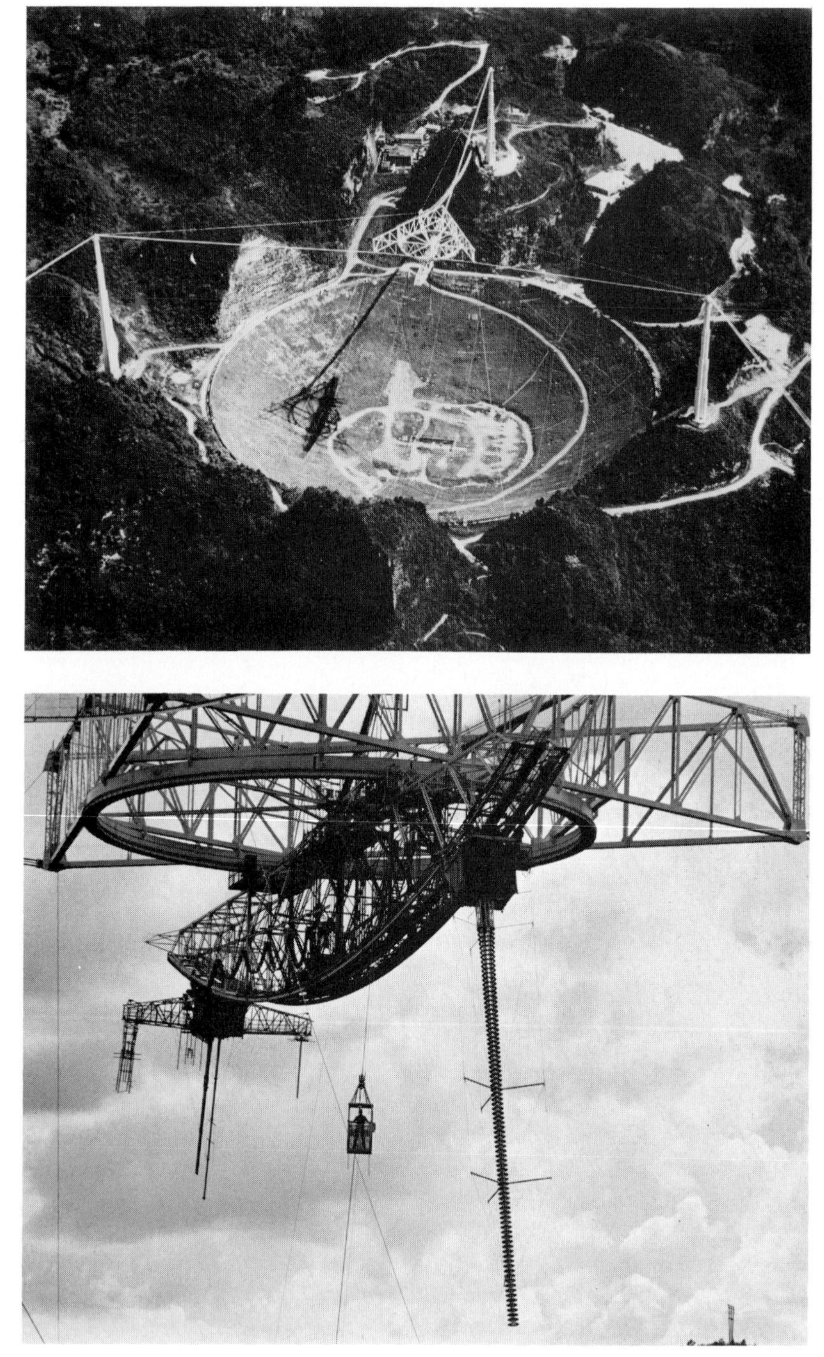

(*a*)

(*b*)

82. Four photographs of the 1,000-foot radio telescope at Arecibo, Puerto Rico. Figure 82(*a*) shows how advantage has been taken of the terrain to mount the 1,000-foot reflector firmly so that it can reflect radio beams from small areas of the sky to the receivers at the focus. The complex arrangements at the focus are shown in (*b*). Various focus probes are shown; each of these can be placed in the proper position at the precise focus for receiving and recording the radiation at the radio frequency in the range which that particular probe is capable of receiving. To improve the quality of radio reception at high radio frequencies (short wavelengths) the reflector itself was resurfaced with precision panels. Figures 82 (*c*) and (*d*) show the resurfacing work in progress during mid-1973. At the time these photographs were taken, about half the surface had been covered with the precision panels; the work was completed late in 1973. Note in (*b*) the size of the observer in the cage being transported to the focal area, and in (*c*) the size of the technician working on one of the panels. The focus probe can remain locked on one position for about an hour; the control is by computer. With its new reflector cover, the Arecibo Radio Telescope is one of the most powerful instruments for study of the interstellar gas and of pulsars and supernova remnants. (Courtesy of National Radio Astronomy and Ionosphere Center, Cornell University.)

(*c*)

(*d*)

young stars have associated with them considerable amounts of interstellar gas. This leads to the formation of emission nebulae, to which we have referred already as H II regions. However, such young star-gas configurations are often embedded in thick and dense clouds of small solid particles, which do not permit us to view the objects at normal optical wavelengths. Radio radiation is not bothered by this cosmic dust, and the young grouping, or at least the gas associated with it, can be studied nicely at radio wavelengths. We note here that some of these dust-embedded emission nebulae often also can be detected through observations in the far infrared.

The oldest technique for studying emission nebulae by radio methods makes use of the effects produced by free-free transitions. In H II regions there is of course much ionized hydrogen. The passage of a free electron close to a positively charged nucleus of the hydrogen atom, the proton, produces continuum radiation in the centimeter and decimeter range. The observed strength of the radiation depends on the number density of the protons and electrons and on the speeds with which the free electrons move about. These speeds are of course fixed by the temperature of the free-electron gas. The continuum radiation was first observed by radio astronomers in the early 1950's; by now the whole of the band of the Milky Way, north and south, has been mapped for continuum radiation, especially by Westerhout.

A second way in which ionized hydrogen can be studied by radio techniques is through the so-called alpha transitions, a method first suggested by Kardashev of Moscow. In regions with much ionized hydrogen, the capture of a free electron by a proton often will take place in one of the very high-energy levels of the neutral hydrogen atom. After capture, as we have mentioned, the electron will stay in this high level for only a minute fraction of a second, and it will then cascade down toward levels of lower energy, ultimately ending up in the level of lowest energy of the hydrogen atom, the Lyman level. The first transitions in the cascading process may well be from, say, level 158 to 157, or from 110 to 109. Happily, many such high-level transitions fall within the wavelength and frequency ranges of modern radio-astronomical receiving equipment. What we observe in each case is a quite sharp radio emission line, for which the rest wavelength is very accurately known. Any shift in frequency between the observed and rest frequencies may be attributed to radial velocity of approach or recession of the gas cloud in question. Thus not only do we have here a tool for detecting H II regions, but we are also in a position to measure the radial velocities of the gas clouds participating in the process.

It is likely that less than 10 percent of the interstellar gas in our Galaxy is in the ionized condition. The most common form of hydrogen is probably neutral atomic hydrogen. It is most fortunate that the very abundant neutral atomic hydrogen can be observed through the radiation that it emits at a wavelength of 21 centimeters. The first observational evidence for the presence of the 21-centimeter radio radiation came early in 1951, when Ewen and Purcell of Harvard detected radiation with a wavelength near 21 centimeters reaching us from some sections of the Milky Way. This discovery of the 21-centimeter radiation was made because Ewen and Purcell had followed up van de Hulst's suggestion that such radiation should be observable. Once the discovery had been announced, other radio astronomers promptly confirmed and extended it. Principally through the efforts of Dutch and Australian radio astronomers, more recently with strong support from the United States, it has been possible to map with high angular resolution the whole of the Milky Way as it appears in 21-centimeter radiation.

We noted in Chapter 7 that the 21-centimeter line results from a transition between the lowest levels of the neutral hydrogen atom. A fraction of the neutral hydrogen atoms in interstellar space are excited to the slightly higher energy level, mostly through collisions. The atom will stay for quite a while in this very slightly excited state and then return to the very lowest of the levels. The 21-centimeter line is produced as the result of this transition. It is basically a very sharp spectral line. Since we know its rest wavelength very precisely, we can use the shift between the observed wavelength and the rest wavelength as a precise measure of the radial velocity of the cloud of neutral atomic hydrogen that produces this feature. The simplest way to represent an observation is by pointing one's radio telescope toward a certain direction in the Milky Way and then slowly scanning in wavelength through the line. If the receiving apparatus is of good modern quality, it will be possible thus to obtain a 21-centimeter profile with a radial-velocity resolution of 2 kilometers per second or less.

We shall see in Chapter 10 that the 21-centimeter line of neutral atomic hydrogen provides us with a very powerful tool for the study of the spiral structure in our own and other galaxies. It now appears likely that most of the 21-centimeter radiation recorded with our radio telescopes originates in relatively cool interstellar regions. Average temperatures of the order of 120°K, 153° below zero on the customary centigrade scale, will be found in most of the regions where the 21-centimeter line originates. However, this temperature is very much of the nature of an average. We shall see later on that inside dark nebulae, which are clouds of cosmic dust, much lower temperatures may prevail.

We noted earlier in the present chapter that during the 1930's the first molecules were detected in interstellar space. High-dispersion optical spectra reveal the presence of the telltale absorption lines attributed to simple diatomic molecules, such as CH, CH^+, and CN. In the early 1960's, radio astronomy entered the picture in a big way. First, in 1963, there came the discovery of radio lines attributed to the hydroxyl radical (OH). Three major discoveries followed in 1968–69: ammonia (NH_3), water vapor (H_2O), and formaldehyde (H_2CO). In 1970 a number of additional radio lines were discovered, most notably a carbon monoxide (CO) line. It is surprising that some very complex molecules, for example methyl alcohol (CH_3OH), are found to be present in interstellar space. Our present list of interstellar molecules contains about 30 varieties. We note that most of these molecules are found in regions of space where cosmic dust prevails. The Orion Nebula, with all its associated gas and dust, the gaseous clouds near the center of our Galaxy, and the quiescent smaller and larger clouds of cosmic dust within a few hundred light-years of the Sun have proved to be the favorite hunting grounds for molecular gas. Molecules are continuing to pop up in surprising concentrations in many unexpected spots in the Galaxy. Some of these molecules are organic species with as many as seven atoms to a molecule. Carbon monoxide, formaldehyde, and the OH radical are commonly present. Molecules are often found in regions associated with strong infrared sources.

The hydrogen molecule, H_2, is probably the most prevalent type of molecule in interstellar space. Theorists have proposed mechanisms that would make it very abundant inside dark nebulae, which are composed of cosmic grains, and possibly also in dust-embedded objects that shine in the far infrared. The Princeton results obtained with the aid of the Copernican Satellite, show that molecular hydrogen is present wherever cosmic dust is found in abundance.

Galactic Cosmic Rays

Cosmic rays are charged particles that move at terrific speeds through our Galaxy. They are mostly nuclei of standard chemical elements probably produced by supernova outbursts and guided in their galactic paths by the weak magnetic fields that prevail in the Galaxy. They are an integral part of the interstellar medium, and they harbor a considerable fraction of the total energy available in the Galaxy. When we trace a galactic cosmic ray by recording its path in a specially prepared thick photographic emulsion, we really record the capture of a particle reaching us from interstellar space. The cosmic rays are for the present our only direct contact with particles that are known to have come from beyond the solar system. For this reason alone, they deserve careful study.

When cosmic rays manage to reach the Earth after penetrating our atmosphere, they have already been thoroughly disturbed in their motions by the Earth's magnetic field and by possible interplanetary fields. They have also been affected by the action of the solar wind, the stream of particles that is thrown out into space by the Sun's atmosphere. Cosmic rays were first detected about 60 years ago through the ionizing effects they produce in gaseous ionization chambers. Information regarding the directions from which they came could be obtained by tracing the continuing effects of one single charged particle on a string of ionization chambers, properly aligned. It was soon found that the Earth's atmosphere greatly affects all but the most energetic particles and that secondary cosmic-ray "showers" are observed on Earth as a consequence of highly energetic cosmic-ray particles interacting with atoms high in our atmosphere.

To learn more about cosmic-ray particles, observers traveled to the high mountains of the Andes and the Himalayas, and established cosmic-ray observatories in the American Rockies and elsewhere. Particle counters were sent up by balloons to heights of 100,000 feet and more, and observations were made in deep mines at high altitudes (like the copper mines of Peru) and from the bottoms of deep lakes, also at high altitudes. These researches made possible the study of the properties of the charged particles that compose the cosmic rays. The most abundant components were readily identified; they are the nuclei of the hydrogen atom, the proton, and of the helium atom, the alpha particle, which consists of two protons and two neutrons. But it soon became evident that nuclei of heavier elements, certainly including the iron atom with atomic number $Z = 26$, were present as well.

In the past two decades, technological improvements have assisted greatly in the further development of the observational study of cosmic rays. To the rather meager tool chests of the early cosmic-ray astrophysicists were added the thick nuclear emulsions, developed for the study of charged particles in cyclotron and synchrotron atomic research. Also, it has become possible to make use of man-made satellite vehicles to carry detection equipment beyond the atmosphere of the Earth and into interplanetary space; the Moon may provide a good base for future cosmic-ray research. The detection equipment itself is being constantly improved. Cosmic-ray counters are being made increasingly more sophisticated and new "track-etching" techniques have been applied to meteorites, with the result that elements heavier than iron have been detected. The most massive detected nucleus is now one of atomic number $Z = 106$, truly a transuranian nucleus!

Where do the cosmic rays originate—in our Sun, in our Galaxy, or beyond? The answer to these questions is that they are to some minor extent of solar origin, mostly at the lower energy levels, but that the major contributions, especially at the higher energies, come from our Galaxy—with possibly some contributions from other galaxies as well. In this book we are of course especially interested in cosmic rays originating in our Galaxy. As of now, supernova explosions seem to be the most likely source for cosmic rays originating within the Galaxy.

For the galactically interested astrophysicist, the Sun—as Peter Meyer of the University of Chicago has put it so well—is really a nuisance. Not only does the Sun produce some cosmic rays of its own, especially at times when solar flares erupt, but, most of all, the gentle but persistent solar wind of low-energy atomic particles emanating from the Sun disturbs and modifies the cosmic-ray flux through interactions between cosmic-ray particles and atomic nuclei of the solar wind. Small wonder that the cosmic-ray astrophysicist has learned that the best time for the observations of galactic and possibly of extragalactic cosmic rays is at the time of minimal solar activity, that is, near the times of sunspot minimum.

The Earth and its atmosphere further complicate the life of the cosmic-ray astrophysicist. The magnetic field of the Earth makes the cosmic rays that penetrate through it deviate from their original paths. The Earth's magnetic field affects the cosmic-ray particles to such an extent that for all except the most energetic particles it becomes very difficult to trace the original directions in which they moved before entering the Earth's magnetosphere. Moreover, interaction between cosmic-ray particles and the gases in the upper atmosphere produce secondary effects in the form of showers of ionized particles. The Earth's magnetic field and its atmosphere surely act as a shield against cosmic rays! Satellite research is assisting terrifically in the study of cosmic rays before they enter the Earth's atmosphere and are affected by its magnetic field. Ultimately it will become exceedingly important to carry satellite research beyond the inner parts of our own solar system.

Protons and alpha particles are, as we noted, the principal constituents of cosmic rays. Next in line come the elements with atomic numbers $Z = 30$ and greater, especially the iron group. The most abundant elements of low atomic weight are listed in Table 5, with their approximate cosmic-ray abundances indicated relative to silicon $= 1.0$. For comparison, we list beside these numbers the cosmic abundances of these same elements according to Cameron, who bases his values mostly on meteoritic abundances. Two basic results stand out from this tabulation. First, for most of the elements from carbon to iron the relative abundances of the atomic nuclei comprising the cosmic rays are remarkably similar to those of the elements in interstellar space and in meteoritic material. Second, cosmic rays contain excessive num-

Table 5. Abundances of elements in cosmic rays.

Chemical element	Atomic number (Z)	Cosmic ray abundance (Meyer)	Cosmic abundance (Cameron)
Lithium	3	1.2	0.000045
Beryllium	4	0.8	0.00007
Boron	5	2	0.00006
Carbon	6	7	13
Nitrogen	7	2	2.4
Oxygen	8	5	24
Neon	10	1	2.4
Sodium	11	0.4	0.06
Magnesium	12	1.2	1
Aluminum	13	0.2	0.1
Silicon	14	1	1
Argon	18	0.4	0.2
Chromium	24	0.6	0.01
Manganese	25	0.6	0.01
Iron	26	0.9	0.9

bers of atomic nuclei of the light elements lithium, beryllium, and boron as compared with the cosmic abundance of these elements.

It is interesting to note that there are electrons and protons among the cosmic-ray particles. It is not an easy matter to separate the truly cosmic electrons from those produced by the solar wind and by secondary effects in the Earth's atmosphere. Satellite observations near sunspot minimum give the best data relating to free electrons in interstellar and interplanetary space. We should add that the deuterium nucleus (heavy hydrogen, D) and the isotope of the alpha particle, He^3, have both been detected in cosmic rays by Fan and his colleagues, then at the University of Chicago. The abundance of deuterium appears to be less than 10 percent of that of He^3. These rather unstable elements are not likely to have been produced in the violent events that are probably involved in the original production of high-energy cosmic-ray particles. They must almost certainly have come about as a by-product of interaction between original cosmic-ray particles and the interstellar gas. They could not possibly have survived travel through the vast distances between galaxies. In themselves, they are strongly suggestive of a galactic origin of cosmic rays.

In recent years, there has been much controversy among the experts in the field about the origin—galactic or extragalactic—of cosmic rays. On the whole it seems that the proponents of the largely galactic origin are coming out ahead. The suggestion that has received most attention is one made by Ginzburg and Syrovetski, supported by Shklovsky, all three of the U.S.S.R., which places the origin of cosmic rays in supernova explosions within the Galaxy. In a galaxy like ours there are approximately two or three supernova explosions per century. The amount of energy produced in one of these explosions is enormous and the fact that radio-synchrotron radiation is observed as coming from the direction of known supernovae remnants, such as the Crab Nebula, indicates that large-scale magnetic fields are associated with them. The atomic nuclei that are being thrown into space as by-products of supernova explosions will be accelerated by these magnetic fields and the high energies of the cosmic-ray particles can thus be understood. Model calculations by Colgate and White and others confirm that the observed high energies are quite reasonable. There is a continuing supply of supernovae in the Galaxy. This is just as well, for the heavy nuclei and the light ones like deuterium and the isotope He^3 will not likely survive in the Galaxy for more than a few million years. One would not expect to find these varieties of cosmic-ray particles if the

83. The Crab Nebula in Taurus (Messier 1). The photograph was made in red (H-alpha) light with the 200-inch Hale reflector. It represents the remnants of a supernova observed by the Chinese in 1054. The Crab Nebula is probably a continuing source of cosmic rays. (Courtesy of Hale Observatories.)

84. A section of the Gum Nebula. The filaments shown in this photograph are probably the aftereffects of a supernova explosion that may have taken place as long as 20,000 years ago. The whole of the Gum Nebula is shown in the lower panel to the right of center in Fig. 3. (From a photograph made with the Curtis-Schmidt telescope of Cerro Tololo Inter-American Observatory.)

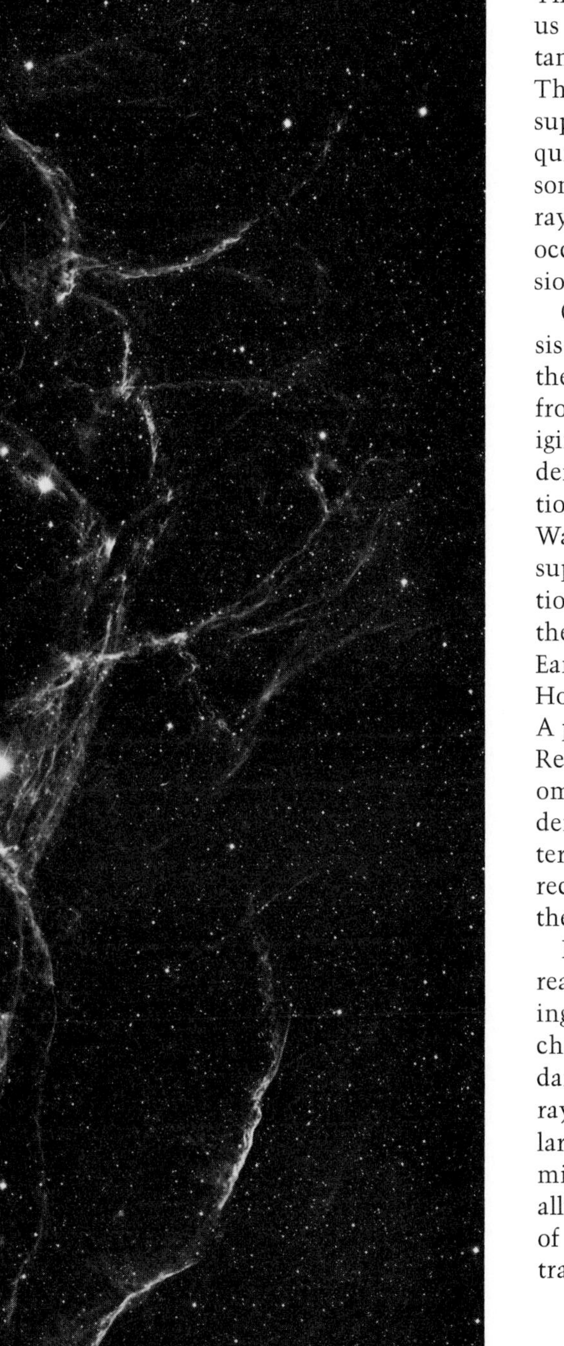

origin of most cosmic rays were extragalactic. The cosmic rays are certainly not coming to us to any great extent from galaxies at distances of the order of a few billion parsecs. The supernova hypothesis produces a steady supply of particles with just about the required amount of energy. It seems quite reasonable to look for the origin of the cosmic rays in the most spectacular events known to occur in our Galaxy, the supernova explosions.

Our principal concern about the hypothesis of the galactic origin relates to the fact that the highest-energy particles seem to come from all directions. If they were of galactic origin, then one would expect to have a preponderance of particles reaching us from directions pointing along the band of the Milky Way. To date there is no good evidence to support anything except isotropy of directions of arrival of the highest-energy particles; these are the particles most likely to arrive on Earth undiverted from their original paths. However, the story is by no means complete. A preliminary result of an analysis by Grote Reber, one of the originators of radio astronomy, indicates a considerably greater shower density coming from a band 20° wide centered upon the Milky Way than from the direction of a comparable area of the sky near the galactic poles.

From the study of cosmic rays we are already learning much that is new and interesting regarding the relative distributions of chemical elements and their cosmic abundances in galactic interstellar space. Cosmic rays give further proof for the presence of large-scale magnetic fields in the Galaxy. Cosmic rays are like tiny billiard balls that actually traverse the vast spaces between the stars of the Galaxy and, during their far-flung travels, interact with and suffer from encounters with interstellar matter. They arrive on Earth influenced by their experiences; they are ready to tell us about their adventures. As the late Arthur H. Compton, one of the early giants of cosmic-ray research, used to say, it is up to each of us to learn to read their messages from interstellar space. They are the only true galactic sidereal messengers.

During the past half century astronomers have increasingly paid more attention to the interstellar gas. In the first edition of this book (1941), we could devote a brief chapter to the interstellar gas and cover the field pretty well, but with each revision that chapter has grown in length. We have tried to present in the present chapter the most important facts about the interstellar gas and its composition, but our presentation is quite incomplete with regard to the physics of the gas. Some of these physical aspects are treated more fully in Chapters 7 and 12 of the Harvard Book on Astronomy, *Atoms, Stars, and Nebulae,* by Lawrence H. Aller (1971), but the time is obviously approaching when the interstellar gas and the cosmic dust deserve a Harvard Book of their own!

9
Dark Nebulae and Cosmic Grains

For the part of our Galaxy within 2,000 parsecs of the Sun and Earth, 80 to 85 percent of all matter is in the stars and the remainder, 15 to 20 percent, is in the gas and dust of interstellar space. According to the best current estimates, 99 percent of this interstellar matter is gaseous; tiny solid pellets account for the remaining 1 percent. In the present chapter we shall be concerned with observable effects produced by the interstellar dust particles and we shall see that much can be learned about their composition, approximate dimensions, and distribution inside and outside the cosmic clouds. Clouds of cosmic dust must reveal themselves in a variety of ways. Sometimes they shine like faintly luminous nebulae, but more often they appear as apparent star voids through which some distant stars may be seen. We can detect the presence and estimate the amount of intervening cosmic dust by the reddening and polarization effects in the light of remote stars.

Reflection Nebulae

Until 1912, it was generally supposed that all bright nebulae would show spectra similar to that of the Orion Nebula, which has a bright-line spectrum. Then V. M. Slipher announced that the nebula associated with the Pleiades has an absorption spectrum, very much like that of the brightest stars in the Pleiades cluster. Other nebulae were found to behave in a similar fashion. There are apparently two kinds of bright nebulae, one with bright-line (emission) spectra, the variety on which we wrote in the preceding chapter, the other with dark-line (absorption) spectra similar to those of the majority of the stars. Hubble showed that emission nebulae are generally near very hot stars with spectral types O, B0, or B1. He found that the nebulae with absorption spectra are associated with cooler stars. These nebulae are called *reflection nebulae,* since they shine by the reflected and scattered light from the stars that render them visible. They are truly starlit cosmic clouds. The reflection nebulae generally show no emission features because the stars that cause them to shine lack a sufficient supply of ultraviolet radiation to produce luminescence through ionization followed by recombination.

Struve, Greenstein, and Henyey made some careful early studies of the physical properties of reflection nebulae. From the observed surface brightnesses of these nebulae they deduced that the particles that produce the reflection are excellent reflectors. There is nothing gray or grimy about the dust of interstellar space. The cosmic grains resemble in their reflective power tiny hailstones much more than dust; in the language of the astronomer, the cosmic particles appear to have a high *albedo,* comparable to that of snow, though probably not quite so high. The indicated high reflectivity of these particles suggests that they are not primarily metallic. Most likely they are small icelike grains composed of simple molecular compounds of the lighter elements, such as carbon, nitrogen, oxygen, and of course hydrogen. Silicates too are identified in cosmic clouds; there is quite a bit of sandy stuff around!

Important conclusions regarding the probable dimensions of the particles may be drawn from comparative studies of the colors of these reflection nebulae and of the stars that illuminate them. We are all familiar with the observation that the light of the setting Sun is reddened considerably in its passage through the atmosphere and, in turn, that the scattered light from the Sun produces the blue aspect of our earthly sky. The contrast between the color of our Sun and the blue of our sky is explained by the phenomenon called *Rayleigh scattering.* The actual scattering of the Sun's light in our atmosphere is produced by particles with dimensions considerably smaller than the wavelength of visible light, probably mostly molecules. The reflection nebulae are somewhat bluer than the stars whose light they reflect, but the color difference is by no means so marked as in the case of our Sun and the daytime sky. From the observed differences in color between reflection nebulae and illuminating stars, we deduce that the scattering particles have average radii of the order of 0.00001 inch, one quarter of a micron.

Dark Nebulae

If a cloud of gas and dust is present in interstellar space, it will not appear as a diffuse nebula unless there is a bright star in or near it. Most clouds with cosmic dust do not shine, but such clouds will absorb and scatter the light from the stars beyond them and they will be seen as dark areas against the bright background of the Milky Way. We call them *dark nebulae.*

Dark nebulae are not conspicuous objects for visual observers. They are uninteresting regions devoid of stars, or with fewer stars than normal, and an observer will pass them by in favor of fields that are rich in stars. Sir William Herschel was the first astronomer who seriously considered the implications of the vacancies along the Milky Way, and it was also Herschel who noticed that these vacancies occur frequently in the vicinity of bright nebulae. It was not until a century after Herschel's observations that Barnard and Wolf succeeded in proving from their Milky Way photographs that many vacancies were caused by obscuring clouds rather than by real holes between the stars along the Milky Way.

We reproduce in this book several photographs of diffuse bright nebulae, emission nebulae as well as reflection nebulae. The many irregularities in the light reaching us from these bright nebulae, and the related irregularities in the distribution of the faint stars, suggest that many of the bright nebulae serve as backgrounds against which can be observed the numerous dark nebulae of various sizes that lie between these bright nebulae and our Earth. Among the finest examples are the Great Nebula near Eta Carinae (frontispiece) and the emission nebulae known as Messier 8 and Trifid (Figs. 85, 86, and 87). Lanes of obscuring material obviously overlie much of the bright structure, and we observe numerous small dark spots seen projected against the bright background. These spots are seen in the same positions from night to night and from year to year and they must surely indicate the presence of small obscuring clouds. Some of these spots have a wind-blown, turbulent appearance, whereas others—generally referred to as *globules*—have a markedly round appearance (Figs. 88, 89, and 90).

There are many known instances of association between bright and dark nebulae; the Horsehead Nebula in Orion is a good example (Fig. 49). The ectoplasmic glow around the horse's head emanates from the bright nebulosity. The horse's head is a part of the large dark nebula that covers most of the lower half of Fig. 49. If the dark nebula suggests an ominous thundercloud, then the bright nebula is the sunlit edge. The photograph of the Horsehead Nebula shows clearly the power of a dark nebula to absorb the light of the stars beyond it. If we compare the number of stars in two squares of equal size, one above and the other below the horsehead, we count at least ten times as many stars in the first square as in the second square. One can have little doubt about the power of this particular dark nebula to dim the light of the stars that lie behind it.

The Coalsack (Fig. 91), in the Southern Hemisphere, is one of the most striking dark nebulae in a region devoid of bright nebulosity. Largely because of the contrast with the brilliant Milky Way surrounding it, the Coalsack appears to the visual observer as an intensely black cloud. Telescopic observations readily reveal the presence of numerous faint stars in the apparent inky blackness of the Coalsack. Long-exposure photographs show that there are still on the average one-third as many faint stars in the Coalsack as in an adjacent clear area of the same size. It is not really so black as it appears at first. The Coalsack Nebula stood for many years as the finest example of a dark nebula free from bright nebulosity. In 1938 a minute patch of bright nebulosity in the Coalsack was found by Lindsay at the Boyden Station, and others have since been noted by Gum; some of these are probably distant gas clouds viewed through the Coalsack.

Figure 92 shows the North America Nebula, so named by Wolf of Heidelberg. The photograph gives a striking illustration of the association between bright and dark nebulosity. The "United States" is a conspicuous bright nebula and the "Gulf of Mexico" is one of the densest portions of the surrounding dark nebula. From the large numbers of faint stars that shine through the bright nebula, it is apparent that the bright nebulosity is more transparent than the dark portions.

One of the finest early photographs of a dark nebula, made by Barnard, is reproduced in Fig. 93. The dark nebula near the star Rho Ophiuchi is a part of the giant dark-nebula complex in Ophiuchus, which, according to

85. Messier 8 and the Trifid Nebula. Two of the finest emission nebulae north of the galactic center: Trifid near the top, Messier 8 (large) below. Note the overlying patterns of dark matter. (Uppsala-Schmidt photograph.)

86. The emission nebula Messier 8. A copy of one of the first photographs made with the Mayall reflector of Kitt Peak National Observatory. Messier 8 is the same emission nebula shown in the lower half of Fig. 85. Several dark markings are of roughly circular shape and these are referred to as "globules." (Photograph by D. L. Crawford.)

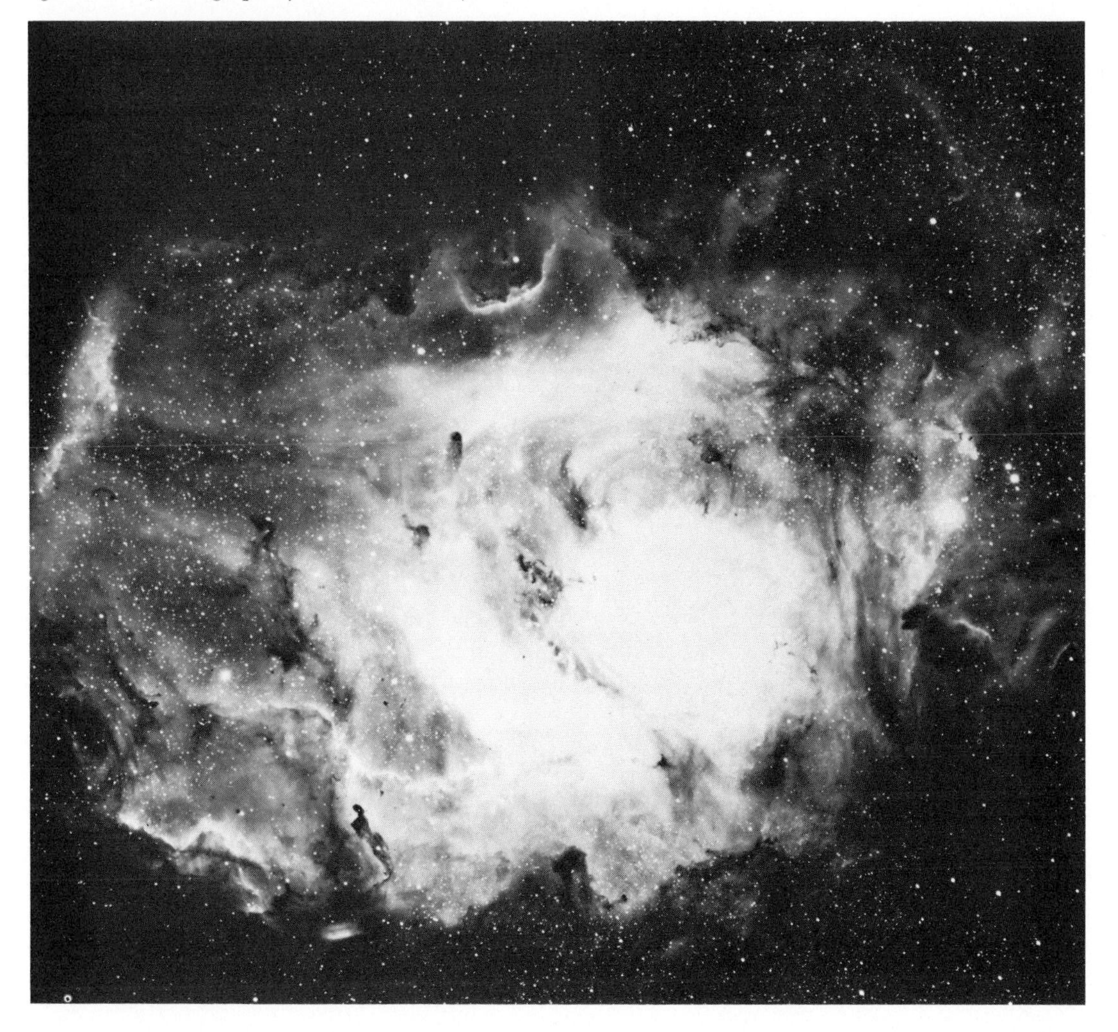

87. The Trifid Nebula in Sagittarius. (Photographed by N. U. Mayall with the Lick Observatory 120-inch reflector.)

88. The Rosette Nebula. A photograph in red light of the Rosette Nebula in Monoceros, made with the 48-inch Palomar Schmidt telescope.

89. Small globules of the Southern Milky Way. Thackeray at the Radcliffe Observatory first noted (1951) the remarkable dark markings shown here, which represent a variety of small globules. (Photograph on a red-sensitive emulsion made with the Curtis-Schmidt telescope at Cerro Tololo Inter-American Observatory.)

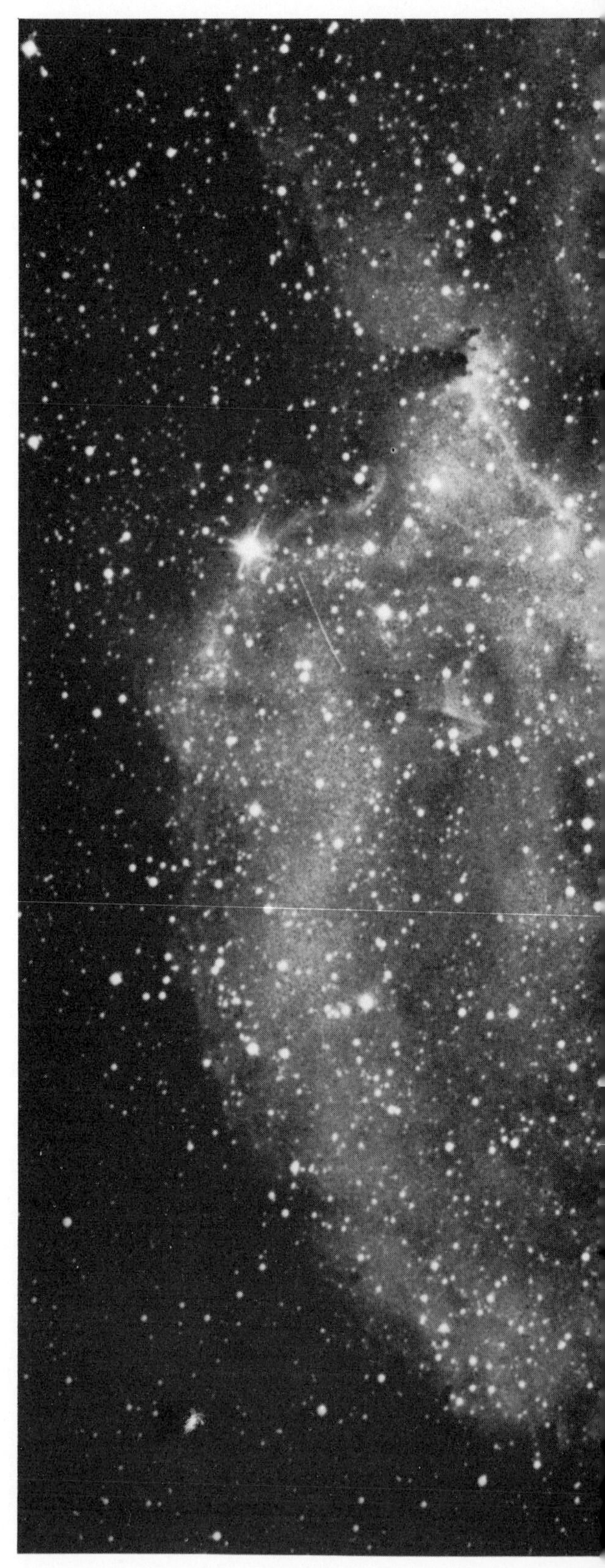

90. Globules in the Rosette Nebula. An enlarged section of Fig. 88, showing numerous globules, first noted here by Minkowski. (Hale Observatories photograph.)

91. The Coalsack and the Southern Cross. The photograph at the top is a composite of five photographs taken in H-alpha (red) light that served later for reproduction in the Mount Stromlo Observatory *Atlas* by A. W. Rodgers, J. B. Whiteoak, *et al.*; see Fig. 3. The bottom photograph was made by A. R. Hogg with a small camera. The Southern Cross is marked and also the Pointers Alpha and Beta Centauri. The Eta Carinae Nebula is shown in the lower right-hand corner.

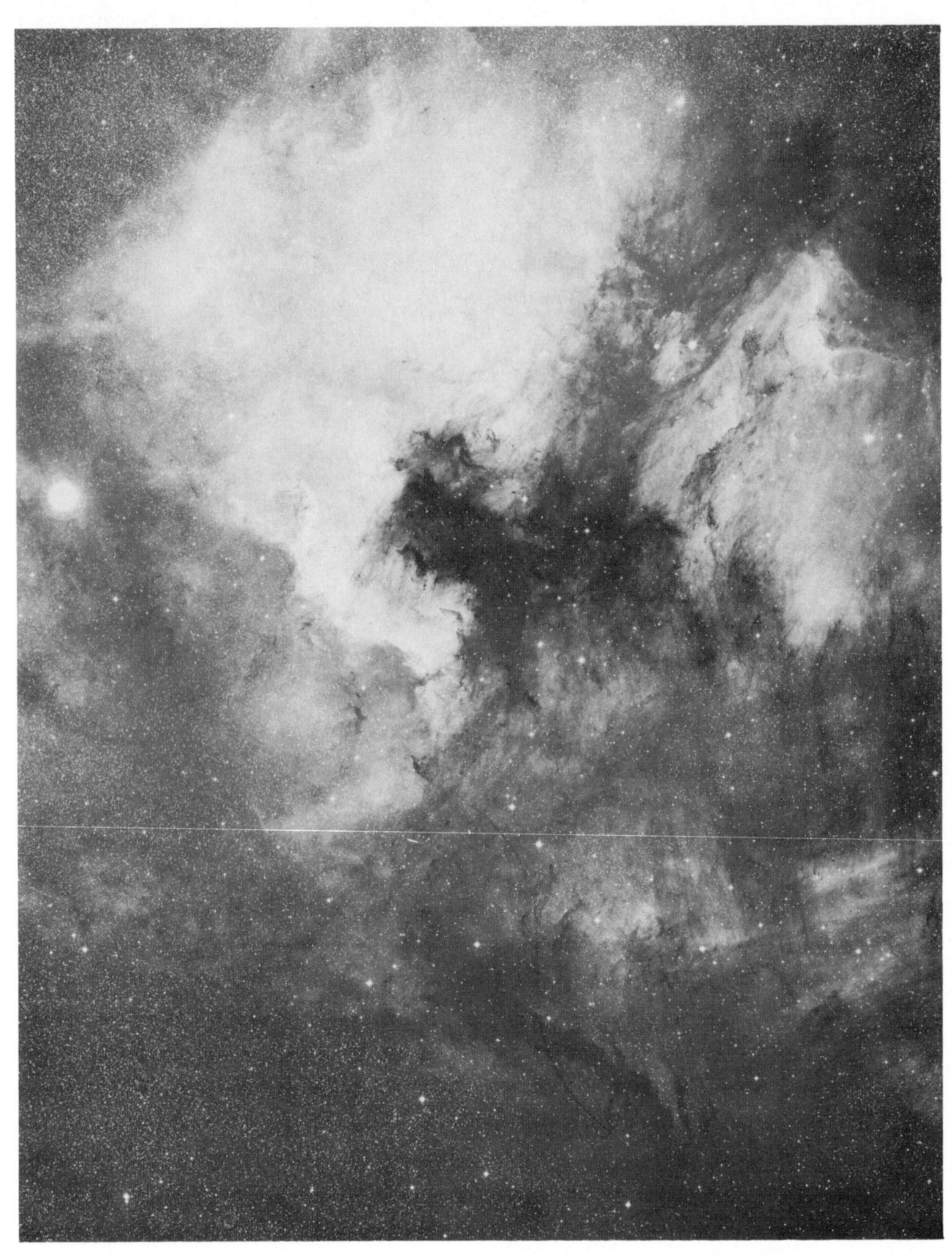

92. The North America Nebula and the Pelican Nebula. Compare with Fig. 74. (Copyright by the National Geographic Society–Palomar Observatory Sky Survey.)

93. The dark nebula near Rho Ophiuchi. The obscuration in the center of dark nebula amounts to approximately 30 magnitudes of dimming. Near the edge of the pronounced obscuration, where two distant globular clusters can be seen on the photograph—a small one below the center and a larger one to the right of it and below—the obscuration amounts to only about 2 magnitudes. Most of the bright nebulosity is of the reflection variety. The cosmic dust is strongly concentrated at a distance of 200 parsecs. (From a photograph by Barnard at the Mount Wilson Observatory made with the 10-inch Bruce camera of the Yerkes Observatory.)

the evidence from the distribution of stars and faint galaxies, covers an area of 1,000 square degrees. The great body of the nebular mass lies from 5° to 20° north of the galactic circle and lacks the background of faint stars to render it conspicuous. The low-latitude "tentacles" of the Ophiuchus dark nebula are seen projected against some of the richest portions of the Milky Way and the dark nebula near Rho Ophiuchus is one of these tentacles. Other examples of Barnard's photography are shown in Figs. 7 and 8.

Table 6 gives some indication of the types of objects that concern us. Though there is a terrific variety of dark nebulae, they can be grouped roughly according to the three categories given in the table. The largest dark clouds of cosmic dust that we observe with well-defined boundaries are listed in the first line of the table. We can estimate rather closely the distances to these clouds. Clouds of cosmic dust not only produce a dimming in the light reaching us from stars beyond them, but also have the effect of reddening this starlight. Hence it is not difficult to distinguish between foreground and background stars, and on the basis of such observations we can assign rough distances to the various clouds. From their apparent diameters in the sky, we deduce their linear diameters. The average radii given in the second column of the table are thus derived. The typical large cloud has a radius of about 4 parsecs and the total mass of its cosmic dust alone is equivalent to about 20 solar masses. This mass estimate is based upon average values of the total

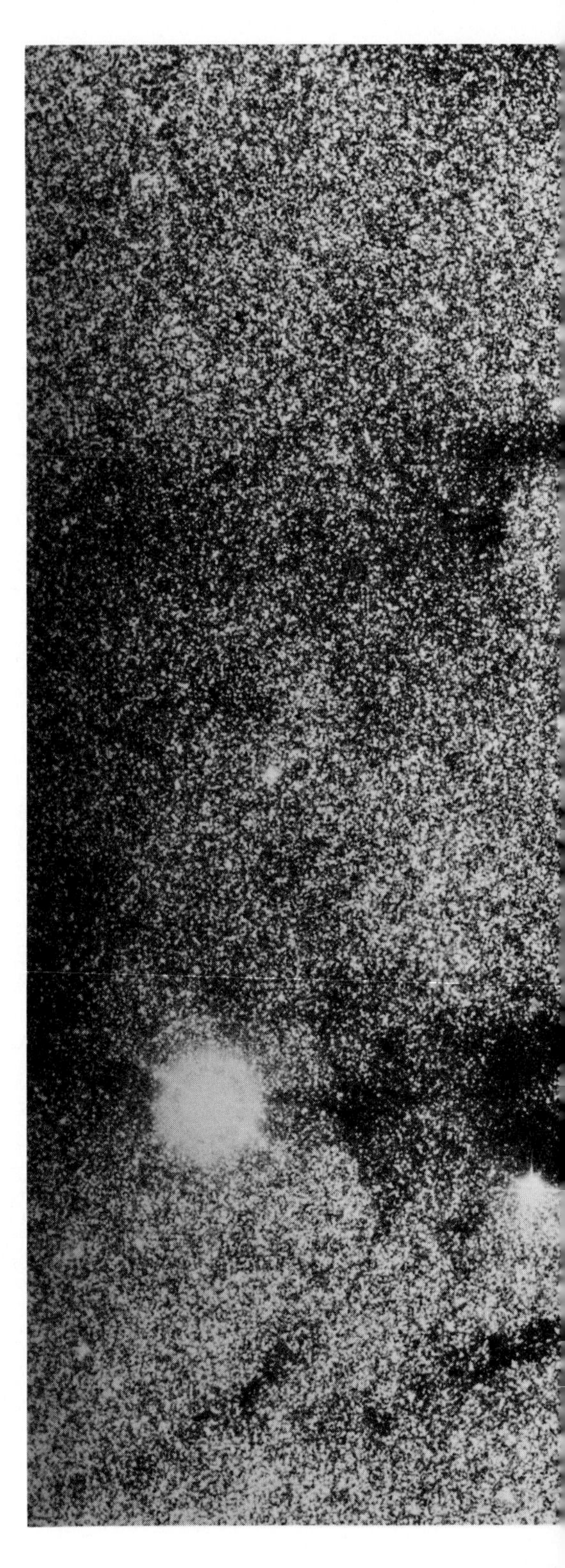

94. Varieties of dark nebulae. The section from a red plate made with the 48-inch Palomar Schmidt telescope (Sky Survey) shows the intricate dark patterns observed against the background of the star clouds in Sagittarius; see also Fig. 8. (Copyright by the National Geographic Society–Palomar Observatory Sky Survey.)

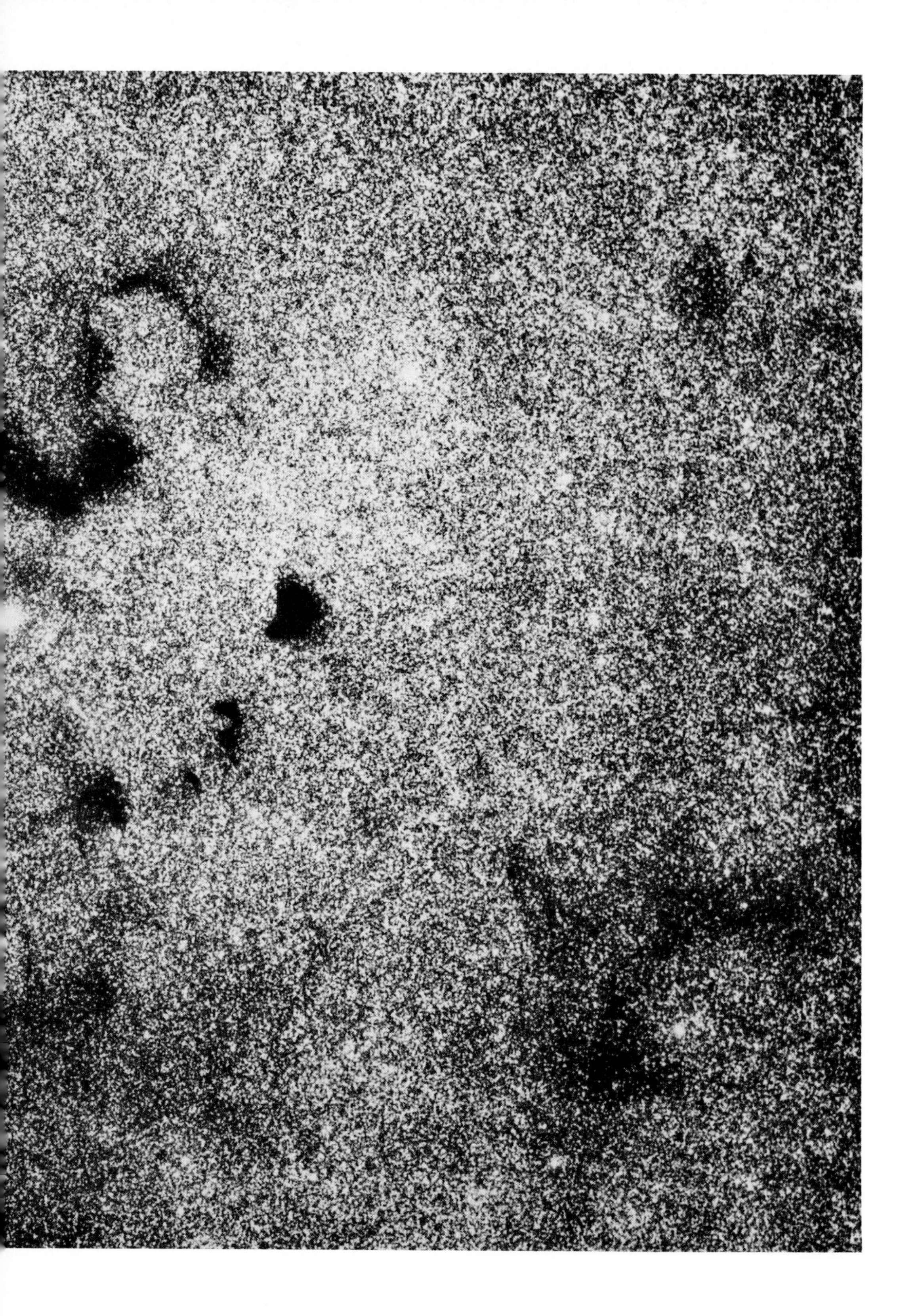

Table 6. Visible dark nebulae.

Object	Average radius (parsec)	Estimated mass (solar mass)	Accretion in 100 million years (solar mass)
Large cloud	4	2,000	1,000
Large globule	1	60	30
Small globule	0.05	0.2	0.05

absorption produced by the dark nebula and the known scattering properties of the small particles that produce the absorption in the first place. Various studies have suggested that there is much gas associated with these objects; the amount of gas is generally estimated to exceed the mass of the cosmic dust alone by a factor somewhere between 50 and 100. The estimated mass in the third column of the table includes the contribution from the associated gas. In the course of time, these clouds must sweep up considerable amounts of matter from the surrounding interstellar medium. It looks as if an amount of interstellar gas and cosmic dust equal to about half the estimated mass of the cloud is swept up in a time interval of the order of 100 million years. Roughly speaking, most of the well-delineated dark clouds should double their masses in time intervals of the order of 1 cosmic year.

In a way the globules, large and small, are the most interesting objects. The large globules often look on our photographs like "holes in heaven." In a region with a perfectly normal rich and smooth background of stellar distribution one suddenly encounters a darkened spot, which looks like an area of low sensitivity in the photographic emulsion. Longest-exposure photographs with large modern telescopes sometimes show the background stars faintly coming through the obscuring matter associated with the globule. There can be little doubt that these dark holes are roundish clouds of cosmic dust floating by themselves in interstellar space. The table shows their estimated properties.

Finally, there are the small globules. They are seen most often projected as tiny dark specks against the luminous background provided by a bright nebula. We see no background light shining through them and the assigned masses in the third line of the table are only guesses at minimum masses. It has been impossible as yet to actually measure the gas content of globules. They have associated with them such small amounts of interstellar gas, especially of interstellar molecules, that observational evidence of the presence of gas is difficult to obtain. We should mention here that radio astronomers have discovered small and very dense units that contain the hydroxyl radical, OH; these are mostly near emission nebulae, but not at the positions of the small globules.

It is really very encouraging that we observe in space two varieties of dark globules and an assortment of large dark clouds. Their numbers are considerable. Within 1,000 light-years of the Sun, a very small distance compared with the diameter of our Milky Way System, we find about a dozen large clouds and about 100 fair-sized globules. We are not at all certain how many small globules there are. Small globules can only be seen projected against the bright emission nebulae, for they would cover areas of the sky too small to be distinguishable against stellar backgrounds. We therefore cannot say at present whether the small globules are selectively associated with the peripheries of emission nebulae, or whether they occur with reasonable regularity in our part of the Galaxy. On the whole, we would favor the first suggestion, since it is quite striking that small globules are absent from some emission nebulae. The small globules are probably clouds of dust and gas literally rolled up as little dust balls by the pressure exerted by the expanding gas at the periphery of the nebula. Their formation is probably assisted by the pressure of the ultraviolet radiation emitted by the hot O and B stars that are at the heart of each emission nebula.

The smallest globules measure only 0.05 parsec across (only 10,000 astronomical units) and the expected hydrogen density is close to 100,000 times that of hydrogen in interstellar space. They seem in many ways like protostars—stars and solar systems in the making. The larger globules are more massive and of course have larger diameters than their smaller equivalents, but they still seem close to the prestellar stage.

Clouds and globules seem to be units that have no choice except to collapse gradually into protostars, or break up into clusters of protostars. Whereas the pressure waves emitted by the bright nebulae may contribute to the formation of the smallest globules, the large clouds and the large globules will probably develop rather quietly on their own by the simple process of gravitational collapse. We need not worry about the rather high masses assigned to the larger clouds. Not every dark nebula needs to collapse into a single star. Instead, star birth may happen by simultaneous collapse of many subclouds in a Coalsack, or in a large-cloud complex. Tapia has found great irregularity in the distribution of the dark matter over the area of the Coalsack, and a number of globule-like units have been discovered.

How can we learn about the extent, distance, and composition of a dark nebula? Information on the first two points can be obtained from counts of stars in the area covered by the nebula and an adjacent area. The counts for two areas of similar size, one in the obscured region, the other in a neighboring apparently unobscured region, will generally agree for the brighter stars. As we count to fainter magnitudes, however, we soon come to a point where the counts in the obscured areas fall below those for the comparison area. The percentage deficiency will generally increase as fainter stars are included in our counts, but will finally assume a constant value. The apparent magnitude for which the deficiency begins to be noticeable gives us some idea about the approximate distance from the Sun to the dark nebula; the percentage deficiency for the faintest magnitudes is a good measure of the total obscuration caused by the cloud.

Wolf was one of the first astronomers to realize the value of star counts for the study of dark nebulae (Figs. 95, 96, and 97). Statistical methods for the analysis of such star counts were developed by Pannekoek. The large spread in the absolute magnitudes of the stars of all kinds renders it difficult to compute precise distances of dark nebulae, but Pannekoek's method of analysis tells us at least whether the absorbing cloud is at 100, 200, or 600 parsecs distance from the Sun. Pannekoek showed further that counts to faint limits give accurate and useful information on the total dimming produced by a dark nebula. The Coalsack, for instance, is caused by a dark cloud at roughly 170 parsecs from the Sun. That is right next door as galactic distances go. The total absorption of the cloud averages a little more than 1 magnitude, but in some dense portions is as high as 3 magnitudes; still larger values are obtained for Tapia's globules.

Estimated distances and absorptions are

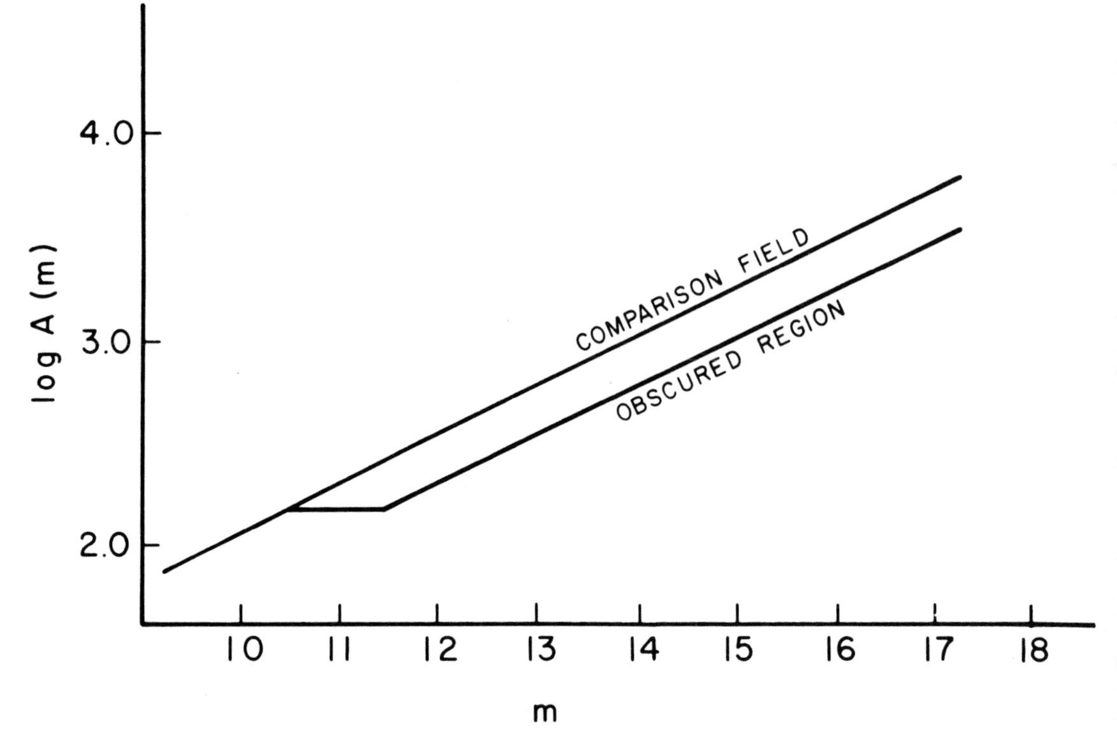

95. Schematic Wolf curves. Star counts are made for equal areas of sky in an obscured region and in a comparison field. The logarithms of the numbers of stars, $A(m)$, counted between apparent magnitude $m - \frac{1}{2}$ and $m + \frac{1}{2}$ are plotted (vertically) against the apparent magnitudes, m, shown horizontally. In the illustration the effect of the dark nebula is first noted near $m = 10.5$.

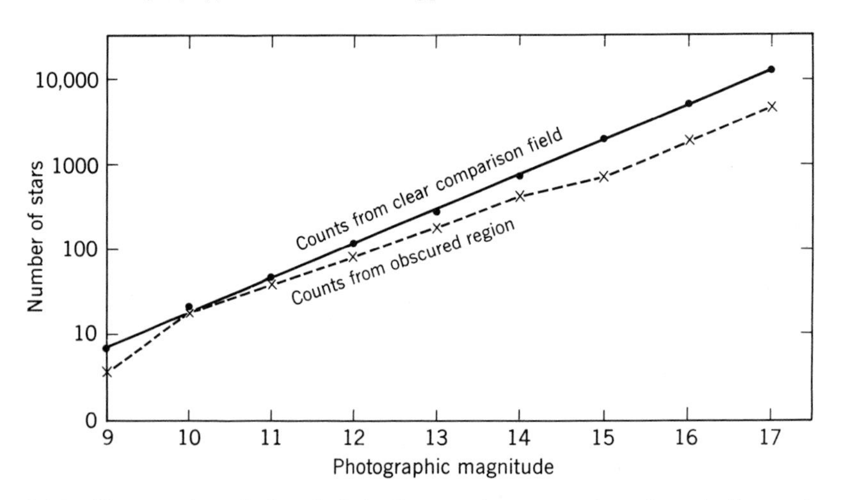

96. Wolf curves for a dark nebula in Cygnus. Star counts by Franklin at Harvard have yielded these curves for a dark nebula seen superposed on the star cloud in Cygnus. The solid curve gives the counts for the obscured field. The quantities plotted vertically are the numbers of stars between photographic apparent magnitudes $m - \frac{1}{2}$ and $m + \frac{1}{2}$, reduced to an area of 1 square degree of the sky.

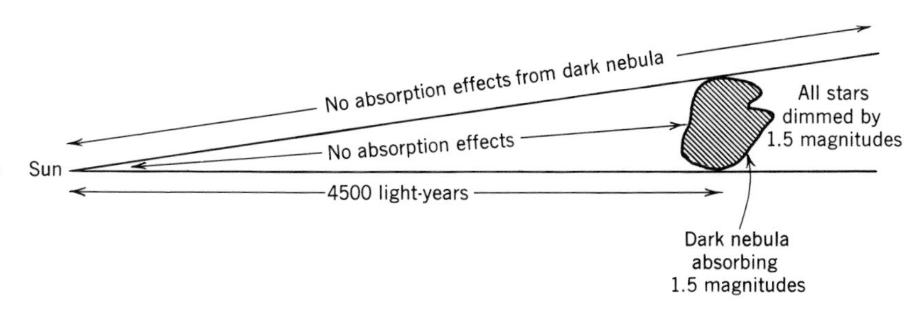

97. A model of a dark nebula. The star counts in the obscured field of Fig. 96 can be represented by assuming the presence of a dark nebula with a total photographic absorption of 1.5 magnitudes at a distance of 1,500 parsecs. (Franklin's results.)

now available for the dust clouds in most major centers of local obscuration. In the Rift of the Milky Way, the dark nebulae in Aquila and southward seem to be all relatively near the Sun, with estimated distances of the order of less than 200 parsecs; for the part of the Rift in Cygnus the distances are more of the order of 600 to 1,000 parsecs. The large dark-nebula complex in Taurus is again nearby, at 140 parsecs, but to the south we find in Orion, Monoceros, Puppis, and Vela a connected system of not-so-thick dark nebulae at 600 parsecs from the Sun.

General star counts alone tell us a good deal about the distance and absorption of a given dark nebula, but even with improved methods of analysis the results remain rather uncertain because of the great spreads in the absolute magnitudes of the stars. For definitive research on dark nebulae, we therefore turn to statistical studies based upon spectra and colors of faint stars. From the spectral statistics alone, we obtain an improved value for the distance of a dark nebula. The colors of the faint stars of known spectral type enable us to investigate in addition the extent to which the starlight has been reddened in its passage through the dark nebula.

The first extensive study of the effect of a dark nebula on the colors of faint stars observed through it was made in the early 1930's by the Swedish astronomer Schalén, who found that although the starlight is definitely reddened in its passage through the nebula, the percentage reddening is rather small. The observed degree of reddening should depend almost entirely on the average dimensions of the cosmic dust particles in the nebula. Schalén concluded that the particles that are the most effective absorbers are tiny cosmic grains with diameters of about 0.00001 inch. Unfortunately, the observed coloring effects in the spectra of stars observed through dark nebulae do not permit clear-cut decisions about the composition of the particles. Schalén made the then rather reasonable assumption that the tiny dust specks are metallic, notably compounds of iron, zinc, or copper. Much has happened since the time when Schalén did his pioneer work. It has been proved that hydrogen and the lighter elements are much more abundant in space than the heavier metals. Also the reflectivity studies of bright reflection nebulae, about which we spoke in the preceding section, suggest that many interstellar particles are probably composed of carbon, nitrogen, oxygen, and hydrogen. Most astrophysicists are now inclined to favor the hypothesis of a nonmetallic composition for the cosmic grains, though serious doubts have again arisen since the discovery of interstellar polarization, about which we shall speak later in this chapter. Infrared studies are contributing much new information about the nature and origin of particles. Some stars that shine with unusual brightness in the infrared apparently possess very extensive dusty atmospheres. It looks as if silicate particles are being formed in these atmospheres. These particles are then probably expelled into interstellar space. Silicate particles do seem to represent a common variety of interstellar dust particles.

Ratio of Gas to Dust

The coexistence of interstellar gas and dust in many sections of the Milky Way has led astrophysicists to ask whether a constant ratio is maintained between the average densities of gas and dust everywhere in space. Attempts have been made to answer this question either through purely theoretical deductions or by trying to secure critical optical or radio observations.

The purely theoretical approach has been explored most thoroughly by Spitzer and Savedoff at the Princeton Observatory. If the gas component has a total density equal to at least 10 and probably 100 times the dust component, then it can be shown that the gas will in all probability literally drag the tiny dust particles along with it. The dust becomes little more than a tracer for the gas and a fairly constant ratio between the densities of interstellar gas and dust might be maintained everywhere. One recognizes, however, that this mixing in constant proportions may extend only over fairly small volumes and that there are many factors that may produce a different mixture for two widely separated points in our Milky Way system. Theoretical studies are helpful as a guide to the interpretation of observations, but theory alone cannot give a solution to the problem.

Unfortunately, the second approach, that of observation with the aid of either photographic and spectroscopic or radio techniques, has not provided a final clear-cut answer. Observations with conventional telescopes yield no clue, for the simple reason that the neutral hydrogen atoms in the cold interstellar clouds send out no detectable radiation in the photographic, visual, or red range. The neutral hydrogen will betray its presence only by the radio radiation with a wavelength of 21 centimeters.

Early studies of the radio radiation from neutral hydrogen were made by Lilley and others with the radio telescope at Harvard's Agassiz Station. Lilley found that the large Taurus complex of dark nebulosity is a region for which the intensity of the 21-centimeter radiation is considerably greater than the average for its galactic latitude. Where there exists a very large complex of cosmic dust, there is also a greater than average concentration of interstellar hydrogen gas. Lilley has shown for the Taurus complex that the average density of the neutral hydrogen is about 100 times the average calculated mass density of the cosmic dust. Hence, there is a great deal of neutral hydrogen for each dust particle.

It seems probable that the hydrogen in the denser dust clouds is either mostly in molecular form, H_2, or frozen as an H_2 mantle onto the tiny dust particles. The latter suggestion has recently been receiving much support. The solid particles that are the mainstay of the smaller dust clouds must have very low temperatures. Wickramasinghe, Reddish, and others have suggested that solid H_2 condenses onto cosmic grains, which they consider to be composed of graphite. In dark nebulae such as the Coalsack or in the tiny globules, the temperatures of the cosmic grains may be as low as 3°K. The presence of OH—the hydroxyl radical—molecules in very small clouds at the outer rims of H II regions is further proof that molecular binding processes are at work. This suggests also that the absence of neutral atomic hydrogen does not necessarily prove that all hydrogen is absent from the object being scrutinized. The graphite cores postulated by Wickramasinghe and Reddish have radii of 0.05 micron and the radius of the hydrogen mantle may be four times as great.

We mentioned in the preceding chapter that the Princeton analysis of data obtained with the Copernican Satellite has shown that molecular hydrogen prevails in some parts of interstellar space. The telltale absorption lines produced by the molecule H_2 are strong in the ultraviolet spectra of highly reddened stars, whereas they are mostly absent from the spectra of unreddened stars. Since reddening is indicative of the presence of clouds of cosmic grains between the star and the Earth, we may conclude that molecular hydrogen prevails in cosmic dust clouds. The Princeton estimates suggest that more than one-tenth of the hydrogen associated with the cosmic dust clouds is in molecular form. The corresponding fraction is less than one part in a million for the average of the interstellar gas. The hydrogen molecule, H_2, can form and continue to exist only in the cold dense clouds of cosmic dust, where the molecule is protected from the disruptive glare of ultraviolet radiation produced elsewhere by O and B stars.

The Cosmic Haze

In our study of interstellar material we have paid much attention to the obscuring clouds. What about the regions where faint stars shine in great numbers, where there is no direct evidence for the presence of intervening dust? We have learned to expect along the Milky Way the telltale absorption lines from the interstellar gas. But the atoms of hydrogen, calcium, and sodium can only absorb light of very special wavelengths and the presence of the gas will not lead to any general scattering or reddening of the light of distant stars. If we find scattering or reddening effects for distant objects outside the obvious obscuring clouds, then we shall have to leave room on our census blanks for a general haze of cosmic dust.

If we photograph any region of the sky that is at least 25° away from the galactic belt, we generally find quite a few images of faint spiral or elliptical nebulae on our plate. We refer to these as "galaxies" because they are separate stellar systems, some of which are probably not unlike our own Milky Way system. All of these galaxies are at distances from us that place them far beyond the borders of our Milky Way system. They begin to appear on 1-hour photographs with a 10-inch refractor. A 3-hour exposure with a 16- or 24-inch telescope will often show 100 of these galaxies for some sections of the sky outside the band of the Milky Way. They appear in abundance on all regular photographs taken with large reflectors of regions outside of the Milky Way, regions with galactic latitudes 40° to 90°, north and south.

Photographs of regions in the Milky Way, even long exposures with the most powerful telescopes, may not bring out any of these galaxy images. What does this mean? Can it be that there are no faint galaxies in those directions, or are they cut off from our view by a general interstellar absorption? There can now be little doubt that general galactic haze is to blame for the absence of galaxies on our

Milky Way photographs. There are probably distant galaxies in all directions. Those along the galactic circle are hidden from our view by the interstellar material in and near the central plane of our own Galaxy.

Now that we recognize the presence of an all-pervading cosmic haze close to the central plane of the Milky Way, we should attempt to find out by how much on the average the light of a star in the Milky Way at a distance of, say, 1,000 parsecs is dimmed by the haze. To do so we study distant stars of known absolute magnitude and estimate their distances by two methods, one basically trigonometric and hence not affected by interstellar absorption, the other wholly founded on brightness measurements. The latter produces distance estimates that are too great because they contain the full effect of the extra dimming due to interstellar absorption. We find the amount of absorption in magnitudes between the Sun and Earth and the star (or cluster) by finding the amount of interstellar absorption required to bring the trigonometric and the photometric distances into agreement. Once we know the total amount of absorption and the trigonometric distance, we can obtain a value for the average amount of absorption in magnitudes per 1,000 parsecs for the direction of the star under investigation.

The O and B stars, separate, in open clusters, or in associations, are the prefered objects for the application of this method, but cepheid variable stars of long period also are of value for studies of this nature. Unfortunately, since we need stars or groups of stars with distances of 1,000 parsecs or over, trigonometric parallaxes and proper motions are of little use. Hence we turn to the galactic-rotation effects in the radial velocities of O and B stars or cepheids and use Camm's method, which we described in Chapter 6, to fix the average trigonometric distance for a group of these stars that seem to be all about equally distant from the Sun. We note that the distances thus obtained are free from effects produced by the dimming of starlight through interstellar absorption and hence represent true average distances. Next, from the known average absolute magnitudes of the stars, estimated from the spectrum-luminosity classes for O and B stars or from the period-luminosity relation for cepheid variables, we find the average distances d, uncorrected for interstellar absorption, from the formula

$$5 \log d = (m - M) + 5,$$

where m is the apparent magnitude of the star and M its absolute magnitude. We then determine how much intervening absorption we must assume to bring each pair of derived distances into agreement.

Trumpler found indications of an average dimming in the photographic range of 1 magnitude at a distance of 1,500 parsecs from the Sun for stars along the galactic equator. Rather comparable values have been derived from studies of galactic-rotation effects in radial velocities of distant stars, like the cepheid variables; the first dependable absorption estimates from cepheids were obtained by Joy. The geometric methods give at best only averages of absorption. Although such averages hold considerable interest, they give us little or no information about the absorption characteristics for a given direction in the Milky Way, for, as we have already noted, irregularity is one of the primary characteristics of the distribution of the interstellar dust that makes up the cosmic haze.

For the study of the absorption along a given line of sight, we prefer to measure and analyze space reddening suffered by individual distant stars. O and B stars, separate or in clusters, or cepheid variables are the primary targets in such studies. One measures, by photoelectric techniques or otherwise, the color index $B - V$ for the star and then predicts from the spectrum-luminosity class for the O and B stars, or from the periods of light variation for the long-period cepheids, what the value $(B - V)_0$ of the intrinsic color index of the star really is. One will generally find that the observed value $B - V$ is larger than $(B - V)_0$, the difference being attributed to the reddening effects of the intervening cosmic dust. This difference is called the *color excess* exhibited by the star, or cluster, in question.

It is not difficult to prove that space reddening is a general phenomenon. Its effects are shown dramatically in Figs. 98 and 99. A distant open cluster, which is shown clearly in Fig. 98 by Westerlund's infrared photograph (on the right), is inconspicuous in the photograph in yellow light (in the middle) and completely hidden from view in the photograph in blue light (on the left). Figure 99 shows a reproduction of two photographs of a globular cluster by Walter Baade. The photograph on the left was made on a red-sensitive emulsion, the one on the right on a blue-sensitive emulsion. The exposure times were adjusted so that stars with color indices like that of our Sun would show images of roughly equal size. It is obvious from the appearance of the globular cluster and from the field stars that interstellar absorption affects the blue photograph (right) much more than the red photograph. In some cases the effects of space reddening can even be observed visually. Observers with large reflectors who do research on the colors of B stars naturally develop the habit of checking at the eyepiece upon the identification of a given B star by noting the marked blue-white color of these stars as contrasted with the more yellow or reddish colors of other stars. But when we look at a faint B star in the section of the Milky Way that contains the galactic center, where space-reddening effects are quite marked, the faint B star may seem yellowish, sometimes even distinctly reddish.

How do we correct the distances of reddened stars for the effects of interstellar absorption? We shall learn in the section to follow that the ratio between an observed color excess and the corresponding total visual absorption is fairly constant. The ratio is determined by the size distribution and chemical composition of the interstellar particles that produce the absorption. If the color excess E_{B-V} is defined as

$$E_{B-V} = (B - V)_{\text{obs}} - (B - V)_0,$$

then the corresponding total visual absorption A_V, expressed in magnitudes, is

$$A_V = 3.2E_{B-V}.$$

The precise value of the transformation factor, 3.2, can be determined by the methods explained in the next section. This relation appears to hold for all except a few directions in the sky. The true distance of the star is then found by the relation

$$5 \log d = (m - M) + 5 - A_V.$$

Space Reddening and Cosmic Grains

It is obviously critically important for research on our Galaxy that we should learn as much as possible about the properties of the cosmic grains. They make up the many varieties of dark nebulae, large and small, and they produce the general cosmic haze. First of all, we want to know all we can learn about

98. A reddened southern open cluster. Three photographs of the same field by Westerlund show the effects of space reddening. From left to right we have a photograph on a blue-sensitive emulsion, one on a yellow-sensitive emulsion, and finally one on an emulsion sensitive to the near infrared. (Uppsala-Schmidt photographs.)

cosmic grains for their own sake, and second, we hope to derive reliable information on the value of the coefficient that relates the total absorption to the observed color excess.

The most comprehensive first attempt to derive the law of space reddening from observation was made by Whitford in 1948. By comparing the variation in brightness distribution with wavelength as observed in some highly reddened O and early B stars with the predicted brightness distribution on the basis of their spectrum-luminosity classes, he determined by how much the light of the star is reddened at each wavelength. Since the observations indicate decreasing reddening as we proceed from the blue-violet through the red toward the infrared, he assumed that there would be essentially no reddening left in the farthest accessible infrared. He then derived at each wavelength the total absorption A, expressed in conventional magnitude units, obtaining a value for each pair of wavelengths, for example those corresponding to B and V. He read from his curve the value of E_{B-V}, the color excess, which, again, can be expressed in units of magnitude. If A_V represents the total visual absorption, Whitford's basic result is that on the average—and with very small scatter—we have

$$\frac{A_V}{E_{B-V}} = 3.2.$$

Following Whitford's precepts, we can obtain the total visual absorption A_V affecting the light of a star at the standard visual wave-

length by multiplication of the observed color excess E_{B-V} by 3.2.

So far so good! But there were some thunderclouds on the horizon as early as 1948. Back in 1937, Baade and Minkowski had shown that a conversion factor 3.2 did not seem to apply to the early B stars in the Sword of Orion, and a region in Cygnus was similarly suspect. There were indications of the presence locally of conversion factors as great as 6.

The Whitford results apply to the interval in wavelength from 3,500 to about 21,000 angstroms; the Earth's atmosphere gives an automatic cut-off in the ultraviolet at 3,500 angstroms, and Whitford's equipment did not reach in the infrared beyond 21,000 angstroms. About 1961, H. L. Johnson started work further in the infrared, including observations at 50,000 and 102,000 angstroms (at wavelengths near 5 and 10 microns). He found indications that the value of the ratio given above, 3.2, is a minimum value, but that for some directions values as great as 6 or 7 may apply. At first, Johnson's results caused much consternation. Recent work in the infrared has shown that the apparent far-infrared excesses found by Johnson are mostly intrinsic to the stars observed by him and that the conversion factor 3.2 still holds for most directions in the Milky Way. Deviations from average conditions do prevail, however, in some sections of the Milky Way, notably the heart of the Orion Nebula and a region in Cygnus.

Space research is adding new information

99. Two photographs of the globular cluster NGC 6553. This heavily obscured globular cluster was photographed twice by Baade with the 100-inch Hooker Telescope on Mount Wilson. The photograph on the left is on a red-sensitive emulsion, the one on the right on a blue-sensitive emulsion. (Hale Observatories photograph.)

on the law of reddening in the ultraviolet. Boggess, Borgman, and Stecher were the first to extend the curve from 3,500 to 1,400 angstroms. They found indications of a maximum of absorbing power near a wavelength of 2,000 angstroms. More precise recent results have come from Orbiting Astronomical Observatory No. 2. Bless and Savage have published an extinction curve that has a distinct maximum near wavelength 2,175 angstroms. Their tabulation lists the ratio of the color excess $E_{\lambda-V}$ to E_{B-V}, where λ represents the wavelength in the ultraviolet. Their representative values are as follows:

λ (angstroms)	$E_{\lambda-V}/E_{B-V}$
4,400	1.0
3,300	1.9
2,500	4.1
2,175	6.8
2,000	5.6
1,500	4.3
1,250	6.0
1,100	8.4

The precise wavelength of the maximum near $\lambda = 2{,}175$ differs a bit from star to star and the observed ratios also vary slightly from star to star. The gradually increasing value of the ratio as we proceed toward shorter and shorter wavelengths is rather to be expected, but the origin of the bump near $\lambda = 2{,}175$ is still a puzzle.

Infrared researches are yielding further useful information about the interstellar medium. The spectra of cool and very red supergiants show evidence of an absorption feature near $\lambda = 97{,}000$ angstroms, the presence of which has been stressed especially by Woolf and Ney at the University of Minnesota. There is good support for the suggestion that solid silicate particles are ejected in large numbers by the atmospheres of cool supergiant stars. Evidence of such particles is found in the infrared spectrum of the Orion Nebula. There is an indication also of the occasional presence of water ice, which possibly may form a mantle around a silicate particle; however, regular ice seems quite rare. Laboratory studies by Huffman and Stapp at the University of Arizona support the hypothesis of the presence of silicate particles of rather limited ranges of diameters. The polarization data obtained by Zellner, at the University of Arizona, indicate that reflection nebulae show infrared polarizations that also suggest the presence of silicates. We should note that the silicate interstellar feature, though clearly marked, is not really a very strong one. Woolf has estimated that for every 50 magnitudes of visual absorption at 4,400 angstroms there is only 1 magnitude of infrared absorption at 97,000 angstroms!

What do all these results imply with regard to the sort of particles that we encounter in interstellar space, and their probable dimensions? We have already noted that silicate particles may well be the most abundant ones in the interstellar medium. But it seems likely that there are also present some carbon compounds and an admixture of metallic particles, notably iron particles. "Dirty-ice grains," simple frozen aggregates of carbon, nitrogen, and oxygen, combined chemically with hydrogen, have been suggested and there probably also exist some solid particles with icy mantles. The important thing to note is that more and more evidence seems to be accumulating to suggest that many interstellar particles are created first in the atmospheres of cool stars. We have already presented the case for the silicates. The case for carbon compounds being formed in atmospheres of certain cool stars is also a convincing one. The R

Corona Borealis stars are a peculiar variety of variable stars. Most of the time they are at constant brightness, but then they dim suddenly, an effect caused by changes on a grand scale in their atmospheres. Apparently very large numbers of solid carbon particles form without much warning. Upon ejection these become interstellar particles—and the star recovers, only to get ready for the next dimming! We note that this basic mechanism was suggested as early as 1938 by O'Keefe, then a Harvard graduate student.

The small particles ejected by cool stars should be fine centers for condensation once they are off by themselves in the interstellar medium. They are cold, with temperatures ranging from 100°K (173° below zero centigrade) to 3° to 5°K. We may expect that interstellar atoms which collide with the particles should frequently stick, literally freeze onto them. The prevalence of molecules in dark nebulae shows that molecule building takes place in interstellar space, and where could an atom find more hospitable surroundings to mate with another atom (or two) and form a molecule than at the surface of a cold interstellar particle inside a dark nebula? The dark nebula provides incidentally a fine shield to protect the young molecule from potentially destructive ultraviolet radiation.

Interstellar Polarization

Polaroid sunglasses take most of the glare out of the Sun's rays reflected by the surface of a roadway because they remove effectively the rays in the plane of vibration that is strongest in reflection. Effects of light polarization are observed in the scattered light from the corona of our Sun and in the atmospheres of planets, but until 1949 no one expected that the light of distant stars would become polarized as a result of its passage through the interstellar medium.

To produce interstellar polarization in the light of distant stars requires not only a preponderance of somewhat elongated particles in interstellar space, but, further, some powerful mechanism to align these particles over very great distances. This, so it was reasoned, would not likely occur and it therefore came as a great surprise when, in 1949, Hall at the U. S. Naval Observatory and Hiltner at the McDonald Observatory announced simultaneously that they had succeeded in detecting polarization effects in the light of distant stars of our Galaxy. The observations of Hall and Hiltner were made with photoelectric polarimeters. These are instruments that permit comparative measures of any star's brightness in different planes of vibration. The recording by photoelectric means guarantees a remarkably high precision of measurement, which is essential when one deals with observed effects as small as those of interstellar polarization. Even in the most extreme case, the differences in intensity for the planes of greatest and of least intensity amount to no more than 0.15 magnitude, and for the majority of the more distant reddened stars the differences are less than 0.03 magnitude.

A sufficient variety of distant stars has now been observed that we can say definitely that the polarization is produced by the solid particles of interstellar space. Polarization affects the light of all distant stars along the belt of the Milky Way, irrespective of spectral class or absolute magnitude. Strong polarization effects are observed only for stars that are highly reddened by intervening cosmic dust, even though strong reddening is not necessarily accompanied by high percentage polarization. The fact that polarization and reddening generally do go together speaks strongly in favor of the hypothesis that polarization is associated with cosmic dust.

This conclusion is strengthened by the further remarkable observation that similar polarization characteristics are often found over fairly large sections of the Milky Way. In the direction to the constellation Perseus, for example, distant stars over a large area of the sky show comparably high degrees of polarization and a close alignment of their principal planes of polarization. Such phenomena can be understood only if one assumes that the polarization is produced by interstellar clouds or cloud complexes, with dimensions of the order of several hundred parsecs, composed of elongated particles with roughly parallel alignments. In some sections of the Milky Way the parallel alignment is very marked, whereas in others the planes of polarization present a more jumbled picture. This is shown very nicely in Fig. 100.

To produce detectable polarization effects in the light of distant stars requires first of all that there be in interstellar space somewhat elongated particles and, second, that some mechanism exist to align these particles in a more or less parallel fashion over very large regions. There is now little doubt that the particles are aligned mostly by large-scale interstellar magnetic fields. To obtain the necessary degree of alignment for the explanation of the observed polarization effects requires at least egg-shaped particles, with some metals (iron?) present and, according to the best available theory, that of Davis and Greenstein, magnetic fields of the order of 0.00001 gauss. In this theory the tiny interstellar particles would be spinning rapidly. The assigned values of the magnetic fields are infinitesimal compared with those found for the Sun and the Earth. However, the effects of these extremely weak fields are felt over hundreds of parsecs.

At present, the Davis and Greenstein theory seems to meet with the most general approval. There are, as always, some remaining problems. The theory requires the presence of magnetic fields over large stretches with field strengths that are greater by a factor 10, more or less, than the strongest large-scale magnetic fields supported by radio-astronomical evidence, which indicates fields of only a few millionths of a gauss close to the galactic plane. B. T. Lynds and Wickramasinghe have suggested that graphite cores with ice mantles can most readily explain polarization effects as observed in the presence of magnetic fields with strengths in the range 0.000010 to 0.000003 gauss.

Very useful information regarding the properties of the polarizing particles can be derived from studies of the variations with wavelength of the amount of polarization shown in the highly polarized light of some stars. Gehrels and associates at the Lunar and Planetary Laboratory and Visvanathan at Mount Stromlo Observatory have especially studied these phenomena. In most stars the polarization is greatest between 5,100 and 5,800 angstroms, and the percentage polarization is found to decrease toward the red and the blue-violet. Balloon observations show different rates of dropping off of percentage polarization in the far ultraviolet. Some of these differences may have their origin in the polarization properties of stellar envelopes, rather than in the interstellar medium, a hypothesis advanced by Serkowski. However, the observed sharp decreases in percentage polarization toward the far ultraviolet seem to speak in favor of the presence of graphite grains with an ice mantle, which, according to calculations by Wickramasinghe and others, should show this sort of wavelength dependence.

One of the most remarkable phenomena suggesting the presence of large-scale magnetic fields is the alignment of wisps of dark and bright nebulosity for rather large sections

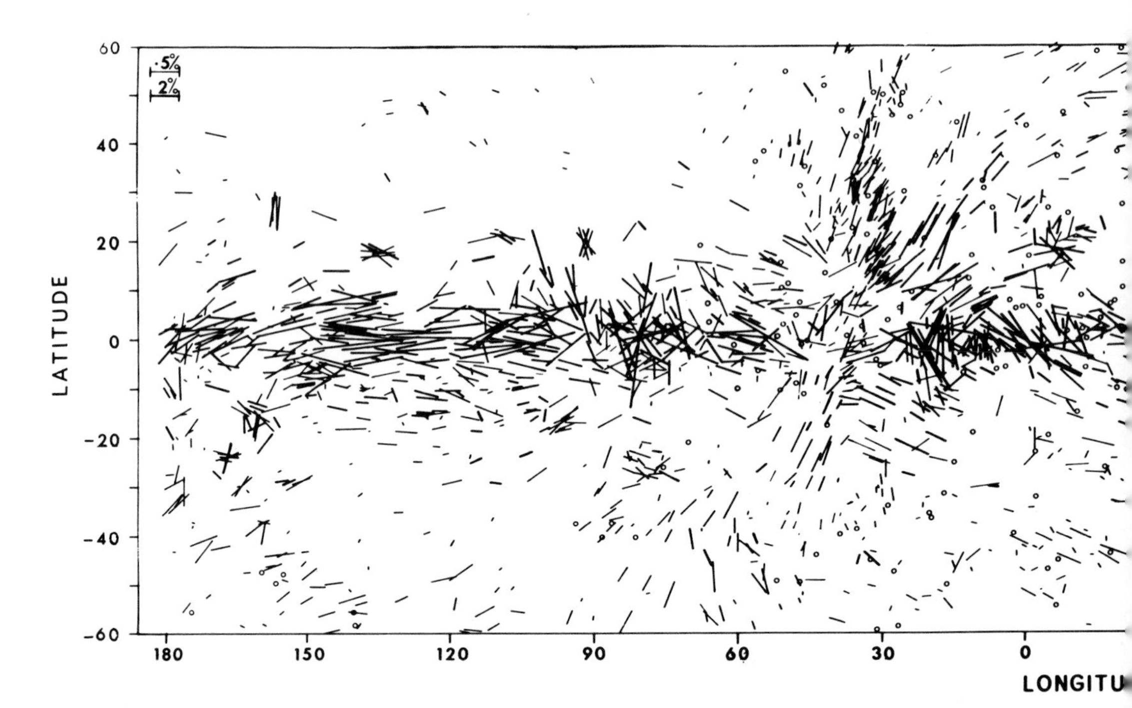

of the Milky Way; Shajn and Rouscol have especially drawn attention to this phenomenon. Shajn derived for some sections the direction of the magnetic field that he holds responsible for the observed elongations of the dark nebulosities. He found (see Fig. 101) that the orientation of this magnetic field agrees, within permissible limits of error, with that derived from the observed polarization effects for stars seen through these nebulae.

It should not be thought that polarization effects are limited to cosmic dust. Shklovsky predicted from theory that strong polarization effects might be expected from very fast-moving electrons in magnetic fields of interstellar space. Dombrovsky observed the effect in the Crab Nebula (Fig. 83); his measures were confirmed and extended by Oort and by Walraven, who found close to 100-percent polarization in certain parts of the Crab Nebula. Baade has made some very high-resolution photographs of the inner filamentary structure of the Crab Nebula and he finds filaments with diameters of the order of 1 to 3 seconds of arc that show 100-percent polarization. The polarizations produced by these so-called synchrotron-radiation effects are inherent to the object—the Crab Nebula in this case—that exhibits them. From such effects we gain no information about the properties of the interstellar medium.

Data on Space Reddening

The most complete picture of space reddening along the band of the Milky Way is obtained through measurements of color excesses of O and B stars. The first accurate photoelectric survey was completed by Stebbins, Huffer, and Whitford at the University of Wisconsin and the Mount Wilson Observatory (1939). Their basic list contains precise

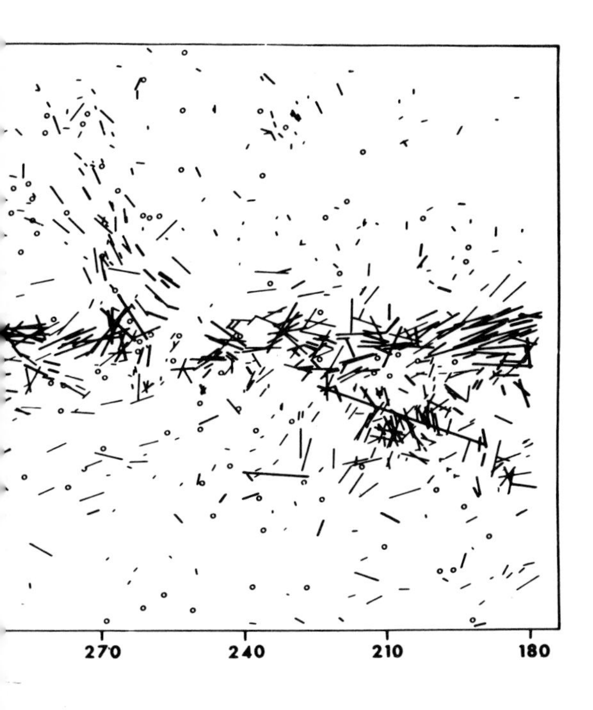

100. Optical polarization. This diagram summarizes all available data on interstellar galactic polarization. The plot shows the electric vectors of polarization and represents data for 7,000 stars. Measurements for 1,800 stars were made by Mathewson at Siding Spring Observatory; the remainder are from the catalogues by Hall, Hiltner, Behr, Lodén, Appenzeller, Visvanathan, and E. van P. Smith. Small circles: positions of stars with percentage polarization $P < 0.08$; thin lines: vectors for stars with $0.08 < P < 0.60$; thick lines: vectors for stars with $P > 0.60$. (From a paper by Don S. Mathewson in the *Memoirs of the Royal Astronomical Society.*)

101. Relation between dark nebulosities and polarization. Shajn calls attention to the tendency for long filamentary structures in emission nebulae and in dark clouds to lie parallel to the galactic equator. This figure shows a region in Perseus-Taurus with dark nebulosities. The short lines indicate the direction of the interstellar magnetic lines of force as deduced from the observations of the polarization of light for distant stars made by Hiltner and Hall.

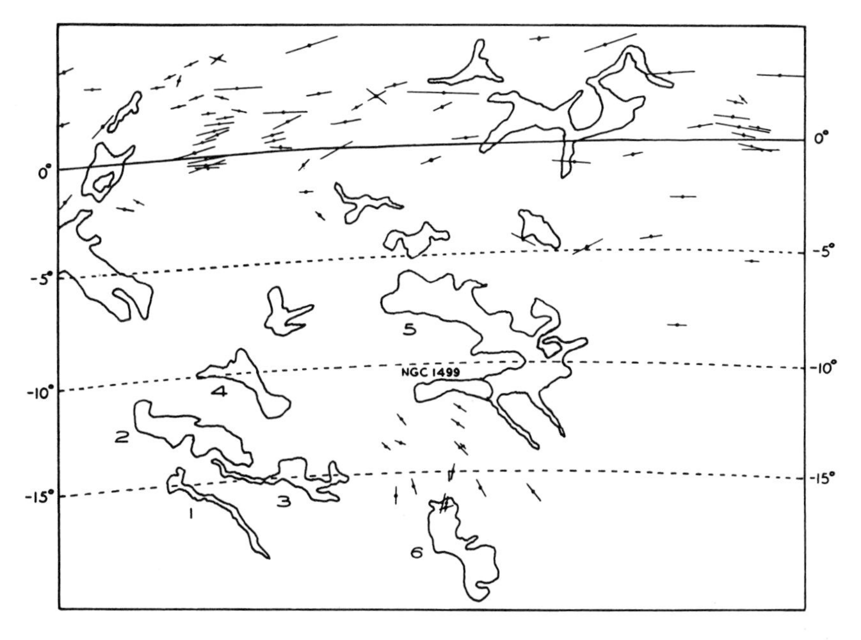

photoelectric colors for 1,332 O and B stars. Because of their great intrinsic brightnesses, even rather bright O and B stars are distant objects and they are therefore quite suitable for work on space reddening. The research has been extended and continued by many workers in the field in both Northern and Southern Hemispheres. For the majority of O and B stars we now possess precise determinations of color excess and of spectral absolute magnitudes on the Morgan-Keenan-Kellman system. Observed color indices on the *UBV* Johnson system and on the Strömgren *uvby* system are used for determinations of color excesses. A wealth of observational material on space reddening therefore is now available.

High reddening prevails in some sections of the Milky Way, in particular for the region of the galactic center. Bok and van Wijk found one faint B star in an obscured field not far from the direction of the galactic center which was so reddened that its estimated total photographic absorption is close to 6.5 magnitudes. In other words, this tenth-magnitude B star would be a fairly conspicuous star, photographically between the third and fourth magnitudes, if it were not for the obscuration between the star and our Sun. Space reddening is generally much less marked away from the direction of the galactic center, but to the best of our knowledge all faint O and B stars directly along the Milky Way show measurable effects of space reddening. This suggests that nowhere along the Milky Way is there a direction totally free from the general haze. The historic survey of Stebbins, Huffer, and Whitford showed that very highly reddened B stars are apparently absent between galactic longitudes 150° and 240°, the region away from the galactic center, but they are numerous for the region between galactic longitudes 350° and 30°, which includes the galactic center.

In recent years much attention has been given to color studies of distant cepheid variables along the Milky Way. At Mount Stromlo Observatory in Australia, Eggen and Gascoigne have studied the cepheids of our Galaxy and compared them with similar stars observed in the Magellanic Clouds. Gascoigne has made the rather startling suggestion that probably all the cepheids observed in the Galaxy are considerably reddened and that we observe unreddened cepheids only in the Magellanic Clouds. Now that we all agree, more or less, on the absolute magnitudes of the long-period cepheids, they become one of the most useful types of objects for the study of space reddening.

Faint and distant globular clusters have come in for their share of attention. Stebbins and Whitford measured photoelectric colors for all those within reach from the Mount Wilson Observatory, and Irwin has checked their survey and extended it to the Southern Hemisphere. Comparative studies of the color excesses for the very remote globular clusters in the direction of the galactic center, and for the relatively nearby B stars in the same general direction, have shown that the strong interstellar absorption is concentrated in the outer part of our Milky Way system, that is, in the region where the Sun and Earth are located. Even for the most transparent sections of the Milky Way in the direction of the galactic center, the indications are that two-thirds of the total observed obscuration occurs within 2,000 parsecs of the Sun, leaving the remaining 8,000 parsecs that separate us from the galactic center relatively free from cosmic dust. However, we do remember that the center itself is probably viewed through 27 magnitudes of visual absorption!

How regular or irregular is the distribution of the interstellar dust in the general haze? Is it smooth or does the superposition of many single dust clouds produce the total effect we observe? At present it is impossible to give a positive answer to these questions. There is apparently no region along the galactic circle where the view is perfectly clear. In a few places some galaxies shine through the haze in the galactic belt, but even there the number of faint galaxies is far below par and a total of several magnitudes of visual absorption is indicated.

Because of the irregularities in the total distribution of the faint external galaxies (they are inclined to come in bunches too!), estimates of total absorption from galaxy data for a given area may well be in error by as much as half a magnitude. In spite of such uncertainties, the observed deficiencies in the numbers of faint galaxies can tell us much about the extent of some of the largest single clouds. The large dark-nebula complexes in Orion and Taurus, in Cepheus, and in Ophiuchus are not only star-poor regions but also very deficient in faint galaxies. The strong absorption by the dark complex hides the galaxies in these directions from our view.

We should not underestimate the total effect of the isolated dark nebulae. Only the nearest of these nebulae will be discovered by an inspection of Milky Way photographs. The Southern Coalsack is conspicuous to us because it is within 200 parsecs and covers a large angular field. But we should realize that it would hardly have been discovered if it had been ten times as far away, at 2,000 parsecs. Not only would it then cover only 1 percent of its present area, but it would further lack contrast because of the many foreground stars. Greenstein has computed that the known dark nebulae alone may account for 30 percent of the total haze for distances up to 1,000 parsecs.

In the late 1930's a group of Soviet astronomers led by Ambartsumian introduced the hypothesis that the interstellar absorption may be produced entirely by chance agglomerations of small obscuring clouds, with average radii of the order of 8 parsecs, with the average absorption per cloud of the order of 0.2 magnitude, and with the average line of sight intersecting about one cloud every 20 parsecs. Several lines of evidence suggest that this is not the whole story. First, we note that the known dark nebulae are by no means distributed at random; they tend to congregate in large complexes, such as the Taurus and Ophiuchus dark nebulosities. Second, we find several large smooth stretches of the Milky Way where there is little or no indication of structure. In these sections, marked space reddening is present and nowhere is the total photographic absorption found to be less than half a magnitude at 1,500 parsecs from the Sun. Finally, we should mention that nowhere have we found any clear "holes"—patches of clear sky between the cumuli—which should occur under the pure cloud hypothesis. The current trend is then to look upon the interstellar absorbing medium as a turbulent affair. As far as we can tell today, there is probably some continuous haze, but it seems that the isolated dark dust clouds may be responsible for more than half of the observed absorption and scattering. We certainly have a cloud structure, with clouds of all sorts of dimensions contributing, and in addition there seems to be, close to the galactic plane, an all-pervading cosmic haze.

10
The Spiral Structure of the Galaxy

Our Galaxy consists basically of three parts: the nuclear region, extending to a distance of about 5,000 parsecs from our Sun; the thin disk, not more than 600 parsecs thick at the Sun, which contains the most spectacular Population I objects and in which spiral structure prevails; and the outer halo, mostly inhabited by Population II stars. In the present chapter we shall be exclusively concerned with the parts showing evidence of spiral structure. The spiral region extends in the galactic plane from about 5,000 to 15,000 parsecs from the center. We note that the Sun is placed rather centrally in this spiral belt at a distance of about 10,000 parsecs from the center.

History of Spiral Structure, 1949 to Present

In the late 1930's, when the authors wrote the first edition of this book, it seemed an almost hopeless task to undertake the tracing of the spiral structure of our Galaxy. A real breakthrough came in the late 1940's, when Baade and Mayall reported results of their studies relating to the spiral structure in the Andromeda galaxy, Messier 31. They found that the spiral arms in that galaxy were most clearly traced by emission nebulae and by cosmically young O and B stars. Clusters and associations of O and B stars were especially helpful in outlining the spiral structure. Baade called on astronomers to examine in our Galaxy the distribution of O and B associations and their related nebulosities. The challenge was accepted by W. W. Morgan of Yerkes Observatory, who, in 1951, with two of his young students, Osterbrock and Sharpless, presented the first over-all picture. They found three parallel sections of spiral arms clearly delineated. The first of these they called the *Orion Arm;* it is the arm in which they locate our Sun, near the inner edge. Next they traced a portion of the parallel outer arm, the *Perseus Arm,* about 2,000 parsecs farther away from the center of our Galaxy than the Orion Arm; third, they found evidence for a section of an inner arm, the *Sagittarius Arm,* about 2,000 parsecs closer to the center than the Orion Arm. Figure 103 shows the Morgan-Osterbrock-Sharpless diagram brought up to date; it includes data from both the Northern and the Southern Hemispheres. Based largely on researches by Wilhelm Becker and Fenkart of Basel and by Schmidt-Kaler of Bochum, the diagram shows the positions in the galactic plane of the clusters and associations with O to B2 stars and of the associated emission nebulae. The three sections of the spiral arms found by Morgan are shown rather neatly, and it is on the basis of this sort of diagram that Becker asserts that the spiral arms of our Galaxy have an average pitch angle close to 25°. The pitch angle is defined as the angle between the direction of a section of a spiral arm and the direction of circular motion.

Early in 1951 radio astronomy began to stir. Ewen and Purcell in the United States and Christiansen in Australia had followed up Van de Hulst's suggestion and discovered the 21-centimeter line of neutral atomic hydrogen. Thus they had found a way of pinpointing the hydrogen gas not only in the three local sections of spiral arms but also in very distant spiral features. Most of the latter are hidden from the optical astronomer's view by intervening thick cosmic dust clouds. The Dutch radio astronomers, led by Oort and van de Hulst, were the first to have published a radio 21-centimeter map of our Galaxy.

The basic analysis of observational material from a 21-centimeter-line survey of the distribution of neutral atomic hydrogen in the Galaxy follows a rather simple pattern. First we select the section of the Milky Way that we wish to study. For example, it may stretch from galactic longitude 275° to 305° and cover the zone in galactic latitude from −10° to +10°. We establish within this area a fine network of positions for which we desire basic information. Along the galactic equator we may place these positions 0.5° or 1° apart in galactic longitude, and we may have rows of additional positions, perhaps not quite so tightly spaced, at every degree of galactic latitude from −10° to +10°. To obtain relevant material for a network of points that is as tight as indicated, we should have access to a radio telescope with an angular resolution of about 0.5° or better. A radio telescope of 100-foot aperture with a surface precise to 1 centimeter serves very well for the purpose. We should use this instrument with receiver equipment of high frequency resolution, so that features in the line profile separated by a frequency difference corresponding to a radial velocity of 1 or 2 kilometers per second will stand out clearly as separate features. The observations at each position consist in pointing the radio telescope in the right direction, something that must be done with high precision, and then recording the profile of the 21-centimeter line in that direction. This is generally done with multichannel receiving equipment, in which each channel successively or simultaneously records the intensity of the radiation in a narrow band of frequencies. The observation for each position consists of a trace in which intensity of the hydrogen 21-centimeter radiation is plotted as a function of frequency; we call this a *21-centimeter profile.* Since for a cloud of cool hydrogen gas at rest the profile would be very narrow indeed, each frequency in the observed profile can be taken to correspond to a certain radial velocity of approach or recession of the hydrogen cloud. The rest frequency of the

21-centimeter hydrogen radiation is very precisely known and it is hence not difficult to translate each observed frequency into a radial velocity of approach or recession of the hydrogen cloud to which the observation refers. Hence the final data for an observation of each point in the network can be shown as a plot of intensity of 21-centimeter radiation versus radial velocity. Typical profiles are shown in Fig. 104.

Once the network of profiles has been obtained, the radio astronomer faces the difficult problem of assigning approximate distances to the various features in the profile that occur at certain radial velocities of approach or recession. In the original interpretation of such profiles by the radio astronomers in Leiden and in Sydney, the assumption was made that there exists a single well-defined rotation curve of circular velocities for our Galaxy. The standard curve of M. Schmidt is shown in Fig. 105. It is then possible to fix for any direction of galactic longitude precisely what would be the distance of a cloud that shows a certain radial velocity of recession or approach. This follows, of course, on the assumption that all hydrogen clouds move around the center of our Galaxy with the precise circular speeds assigned to them by the basic rotation curve.

However, the situation is not so simple. First of all, and not unexpectedly, the clouds show motions of their own, mostly of the order of ± 6 to 10 kilometers per second. What is more disturbing, it soon became evident also that our Galaxy does not possess a single average rotation curve, in which the circular velocity of galactic rotation is only a function of the distance of the cloud from the center. It was at one time thought that there were two different kinds of rotation curves, one for the northern half of our

102. The southern Milky Way. The four photographs show nicely most of the southern Milky Way from Carina through Norma to Scutum. (*a*) The Eta Carinae Nebula and the Southern Coalsack above and slightly to the right of center, photographed in red light (wavelength 6,600 to 6,900 angstroms). The brightest area below the center is the Sagittarius Cloud. The bright lines on the right are from the seeing tower and from the building of the 152-centimeter reflector at the La Silla Station of the European Southern Observatory in Chile. (*b*) A photograph taken in visual light (wavelength 4,600 to 5,800 angstroms). The Southern Coalsack is just above and to the right of the shadow of the objective and the Sagittarius Cloud is to the left of the shadow of the lower strut. The shadows shown on the right are from the dome of the Danish telescope at La Silla and from the building of the 152-centimeter reflector. (*c*) A photograph taken in blue light (wavelength 3,900 to 4,700 angstroms). This photograph, like the others, shows especially well the dark constellation of the Emu to which reference is made in Chapter 1. The Southern Coalsack marks the sharp beak of the Emu (upper right). The long thin neck of the animal and the main body are seen directly to the right of the shadow of the objective and the legs are seen below the shadow of the right strut. The bright image to the left of the Milky Way and below the center is that of the planet Jupiter; the planet is not shown quite so conspicuously in the other photographs, but with a little effort it can be readily identified in each reproduction. (*d*) An ultraviolet photograph (wavelength 3,200 to 3,800 angstroms). The Emu seems to have more than two legs! (Photographs by Schlosser and Schmidt-Kaler made with the Bochum University wide-angle camera at the Bochum Station at La Silla in Chile.)

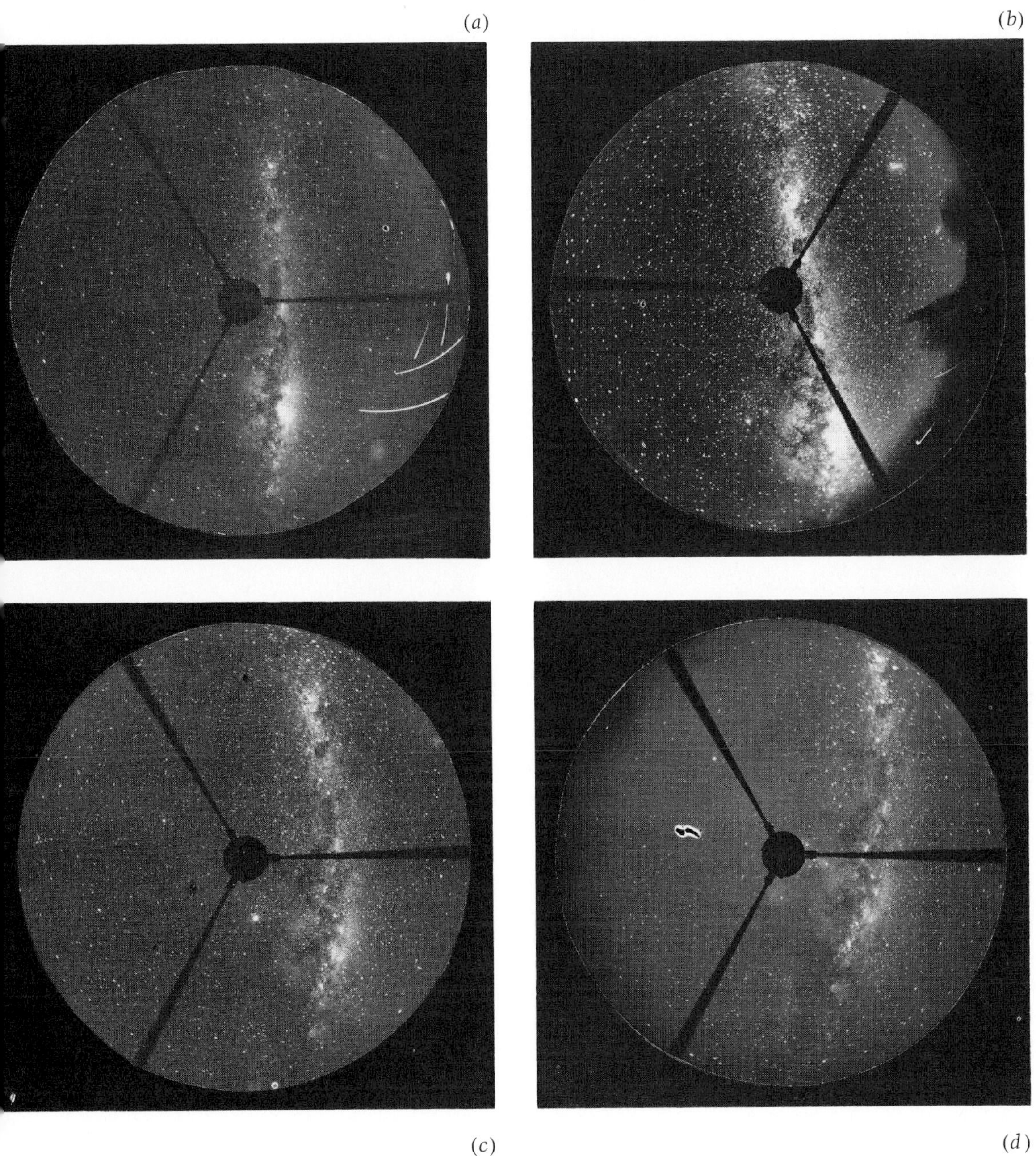
(a)
(b)
(c)
(d)

103. Optical spiral structure of our Galaxy. The Sun is at the center of the diagram. The galactic center is in the direction toward 0° galactic longitude at a distance of 10 kiloparsecs from the Sun. The principal observed sections of the Sagittarius, Orion, and Perseus Arms are shown. The directions from the Sun toward some of the key constellations along the band of the Milky Way are marked along the periphery of the diagram. (A diagram based on data from W. Becker and Th. Schmidt-Kaler.)

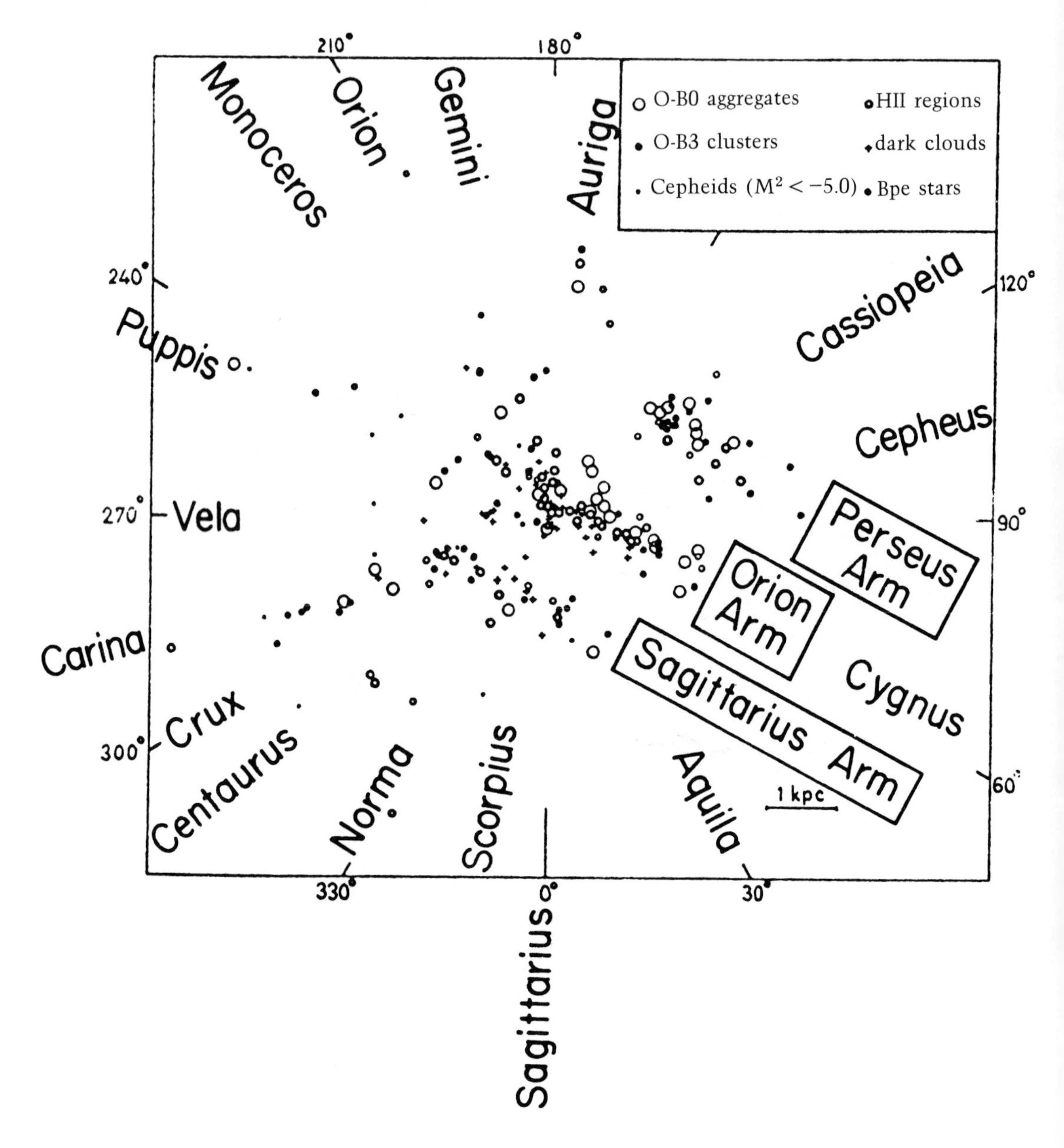

104. Typical 21-centimeter H I velocity profiles for adjacent centers in the southern Milky Way at two galactic longitudes, $l = 296^{\circ}.5$ and $l = 297^{\circ}.5$, for galactic latitude $b = 0^{\circ}.0$. The abscissae give the radial velocities of the neutral hydrogen gas. The velocities have been corrected for the effects of the Sun's local motion. The ordinate gives the radio intensities of the signal in terms of the antenna temperature T_A in degrees Kelvin. If we assume that all the peaks are due to concentrations of H I and not to velocity crowding effects, we find for the H I clouds at $l = 296^{\circ}.5$ the following radial velocities: -28, -7, $+15$ to $+20$, $+55$, and $+114$ kilometers per second.

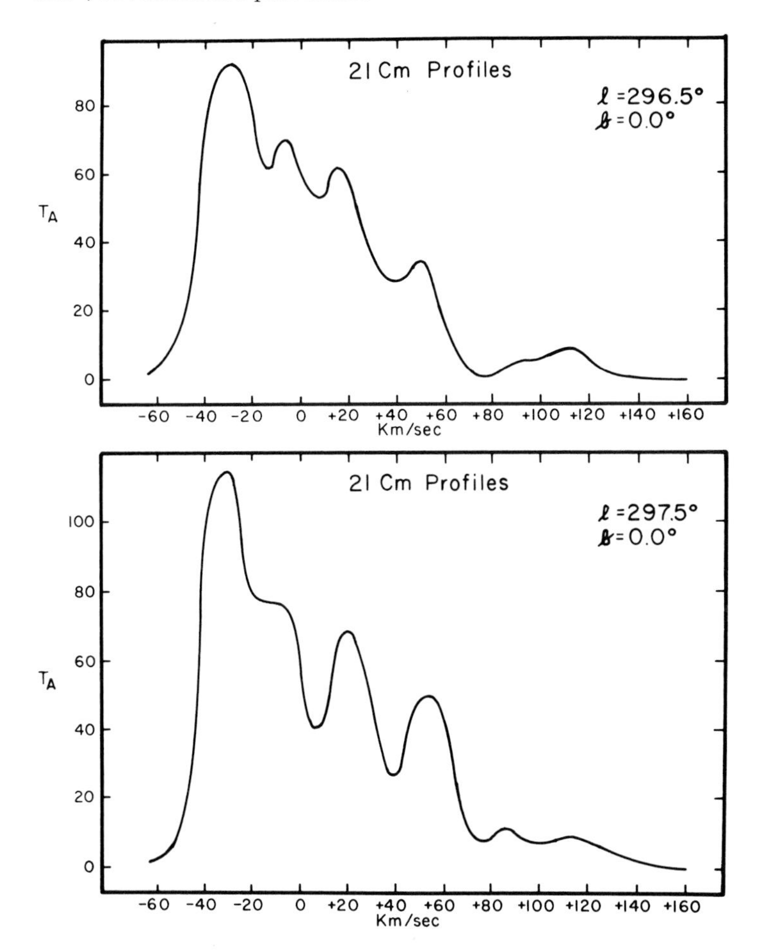

Milky Way band, and another for the southern half; Fig. 106 shows the two curves. This is probably not the case, but recent work has shown that there are large-scale velocity streamings present in the interstellar gas, which make it a difficult matter to transform the observed radial velocity of a cloud at a given galactic longitude into an estimated distance for that cloud.

The present state of interpretation of the large body of data on 21-centimeter profiles is rather confusing. A comprehensive review of the situation that we face was made at the 1969 Basel Symposium organized by the International Astronomical Union. We can do no better than to reproduce here two recent diagrams of radio spiral structure of our Galaxy—one by Kerr, the other by Weaver; these are shown in Figs. 107 and 108. The diagram presented by Kerr is based upon joint analysis by himself and Westerhout. It indicates a rather tightly wound pattern of spiral features, leading to spiral arms with average pitch angles of the order of 5°. The Weaver diagram shows a much more open spiral pattern, and he assigns pitch angles to sections of the arms in the range between 4° and 12°.

Before we attempt to reconcile Figs. 103, 107, and 108, we should have a look at some of our neighbor galaxies that obviously possess spiral structure. We obtain guidance in the analysis of our own Galaxy from the results of studies of these neighboring galaxies, for in them we can view the whole of the spiral pattern from a single photograph, or from a comprehensive analysis of 21-centimeter profiles. For our own Galaxy, we cannot hope to obtain such an over-all picture simply because the Sun and Earth are immersed in the Galaxy. To guide our attempts at analysis, we need the data from the Andromeda galaxy and from others, especially Messier 51.

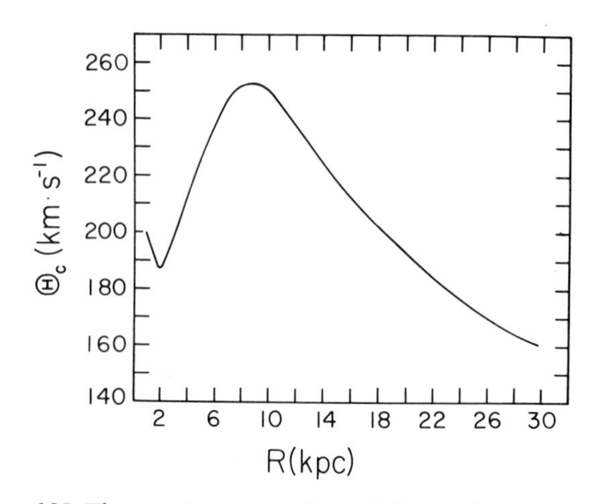

105. The rotation curve of our Galaxy. The dynamic model for our Galaxy by M. Schmidt yields the rotation curve shown in the diagram. Vertically the circular velocity Θ_c is given in kilometers per second, for distances (horizontally) from the galactic center measured in kiloparsecs.

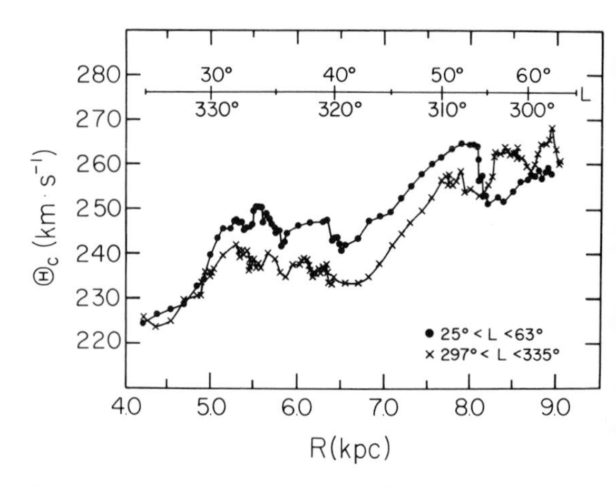

106. Kerr's two rotation curves. The galactic rotation curves for northern and southern sides of the galactic center have been derived from tangential 21-centimeter radio observations assuming circular rotation. The lack of smoothness of each curve and the differences between the curves are now being interpreted as caused by large-scale streaming associated with spiral features.

107. Radio spiral structure of our Galaxy (1967). This diagram was prepared by F. J. Kerr from 21-centimeter profiles observed in southern latitudes with the 210-foot radio telescope at Parkes, Australia (observations by Kerr and J. V. Hindman), and by A. P. Henderson from profiles observed from northern latitudes by G. Westerhout with the 300-foot radio telescope at Greenbank, West Virginia. The distribution in the inner parts is only roughly sketched in the diagram. The most noteworthy feature is the nearly circular pattern of spiral structure that emerges, with pitch angles of the order of 5°. The trough of low hydrogen density that can be traced from galactic longitude 20° to 80° is another remarkable feature. Regions of low hydrogen density are indicated by *L*. (Diagram from the University of Maryland.)

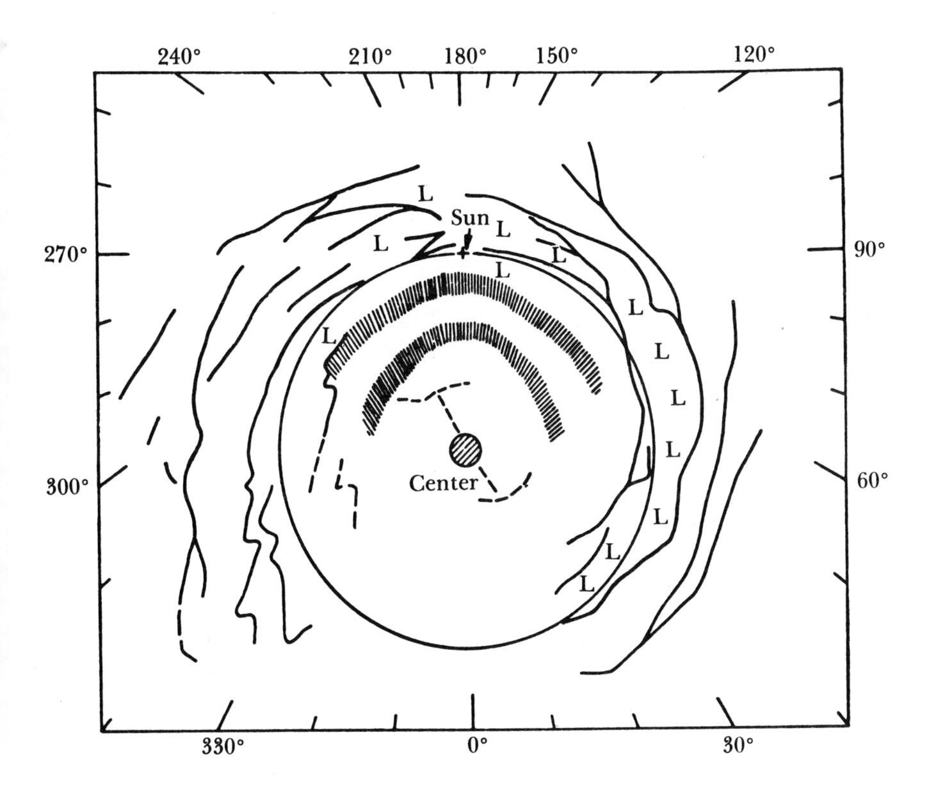

108. A preliminary map of radio spiral structure. Harold F. Weaver made this radio map on the basis of the Hat Creek 21-centimeter survey. It represents only the preliminary analysis (1972) of an extensive body of basic data. For comparison, Weaver has entered the optical data (see Fig. 103) of Becker and Schmidt-Kaler. The probable extensions of the observed features to the southern Milky Way (unobservable from Hat Creek Observatory, in California) have been sketched by Weaver.

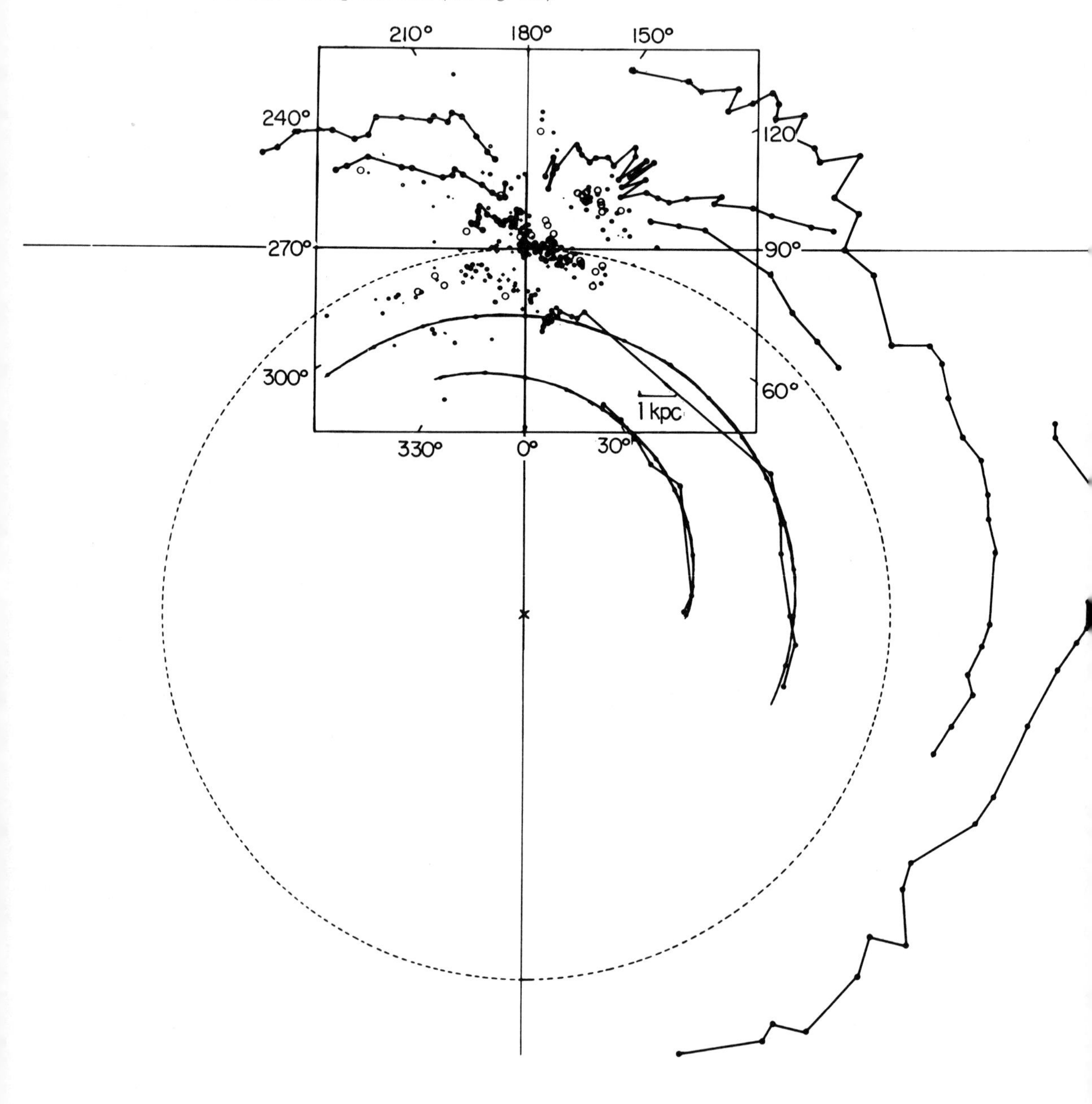

Spiral Structure in Neighboring Galaxies

The photographs of neighboring galaxies such as Messier 31 (Fig. 15), 33 (Fig. 109), 51 (Fig. 110), 81 (Fig. 111), and 101 (Fig. 16) show beautifully the sweep of the spiral structure of our nearest neighbors. There are, however, several properties known about these galaxies, which are recognized as the finest of spirals, that do not fit nicely into a straight and simple, all-inclusive spiral picture. Let us take Messier 31, the Great Spiral in Andromeda, as an example. First we reproduce the two famous photographs of it (Figs. 112 and 113) that Walter Baade made specially for the third edition (1957) of this book. In a study published in 1964, Arp reanalyzed some of Baade's material on the spiral structure of this galaxy. He rectified the photographs and drawings for a tilt of the central plane of the galaxy to the line of sight of 11°. In other words, he attempted to obtain a face-on view of Messier 31. His results are reproduced in Fig. 114. The Arp diagrams show that the emission nebulae found by Baade do give patterns that can be adjusted to logarithmic spirals, but we note from an inspection of the figure without the spirals that the basic pattern found by Arp has almost a closer resemblance to a ring structure than to true spirality. If we want to see spiral arms, we can certainly find them, and most of us would have no doubt that the basic structure in Messier 31 is of a spiral nature. And yet, the ring structure seems to have quite a bit to recommend it.

Several astronomers have attempted to trace the radio spiral structure from 21-centimeter profiles for the Andromeda spiral. The first study was made by Morton S. Roberts; more recently Rubin and Ford have again studied Messier 31 by combining the results of 21-centimeter and optical observations. There are marked H I peaks at the distances from the center where the Baade-Arp distribution charts of ionized hydrogen regions, observed optically, also show peaks. However, the highest concentration of neutral atomic hydrogen appears to be in a ring structure that mostly falls beyond the distances from the center of the Andromeda galaxy where the H II regions are found. This result bears an important similarity to one found from the analysis of hydrogen distribution in our Galaxy. It was shown some years ago by Westerhout that in our Galaxy the ionized-hydrogen distribution peaks at 5,000 parsecs from the center, whereas the neutral-hydrogen distribution peaks at distances of the order of 13,000 parsecs from the center. We find a similar situation in Messier 31.

A close examination of photographs of normal galaxies with spiral structure shows many worrisome features. In most of the beautiful spirals shown in the photographs of the *Hubble Atlas of Galaxies,* prepared by Sandage from Mount Wilson and Palomar photographs, the spiral arms are by no means smooth and continuous structures. There are often holes and bifurcations in them; loops and fringes occur in abundance. One may pity the astronomers who are placed inside some of the sections that are full of confusion and who may be attempting to unravel the details of spiral structure of the galaxies in which they find themselves. Hodge and others have traced the distribution of H II regions in several neighboring galaxies. On the whole, the spiral patterns stand out with reasonable clarity, but Hodge has found several galaxies in which there seems to be only confusion. Even with the best of optical and radio telescopes, the tasks of analysis for astronomers living and working in certain sections of

109. The spiral Messier 33 in Triangulum. The photograph with the 200-inch Hale reflector shows the central section of this spiral galaxy. (Courtesy of Hale Observatories.)

110. The spiral galaxy Messier 51 in Canes Venatici. A photograph of the main spiral (NGC 5194) and its companion (NGC 5195), made with the 200-inch Hale reflector. (Courtesy of Hale Observatories.)

111. The spiral galaxy Messier 81 in Ursa Major. A photograph of one of the most beautiful spiral galaxies, made with the 200-inch Hale reflector. (Courtesy of Hale Observatories.)

112. Outer spiral structure of the Andromeda galaxy. The negative of a Mount Wilson photograph shows a section near the upper left-hand corner of Fig. 15. Dr. Walter Baade has given the following information about this photograph: Emission nebulae along one of the outer spiral arms of Messier 31, photographed in H-alpha light. At the bottom (center) emission nebulosities belonging to the next inner arm are seen, whereas Nos. 66, 67, and 1a are scattered members of one of the outermost arms of Messier 31. (Four-hour exposure at 100-inch telescope by Baade.)

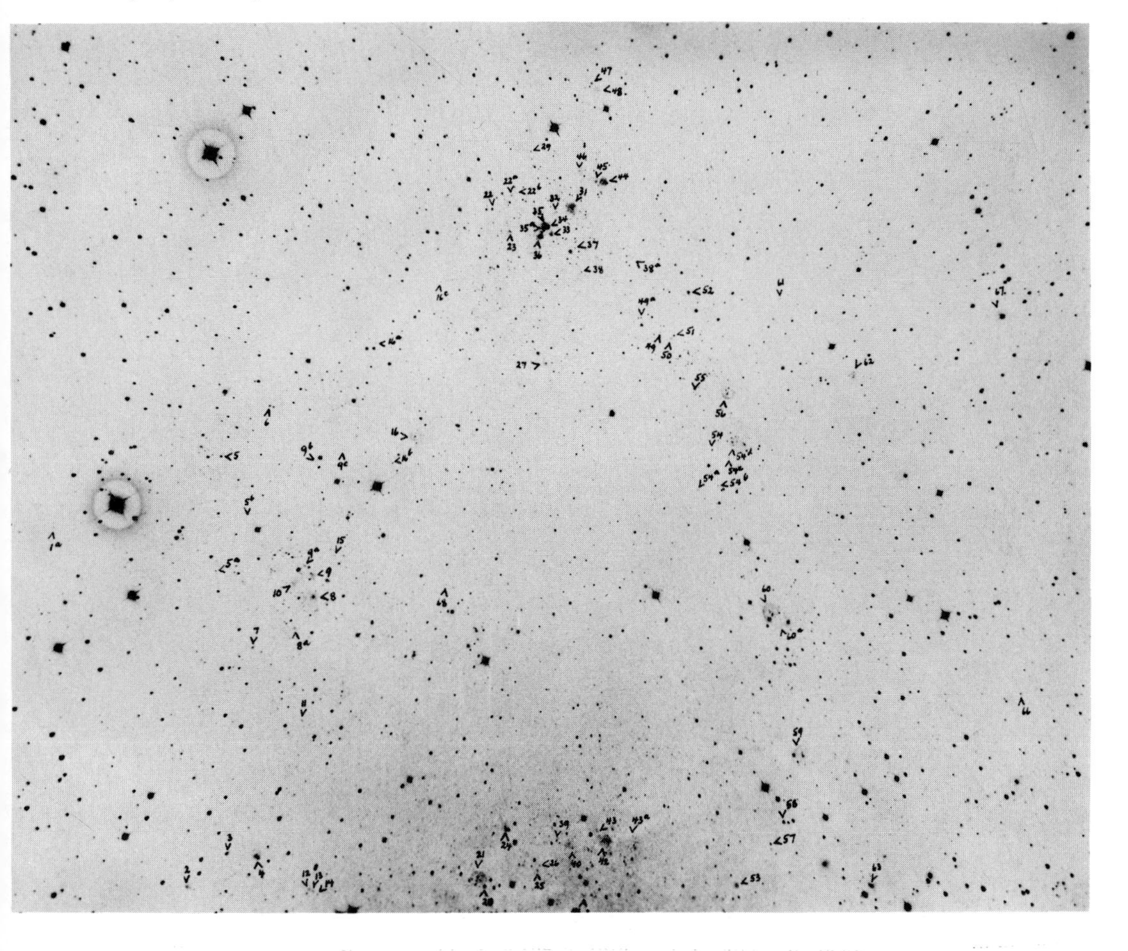

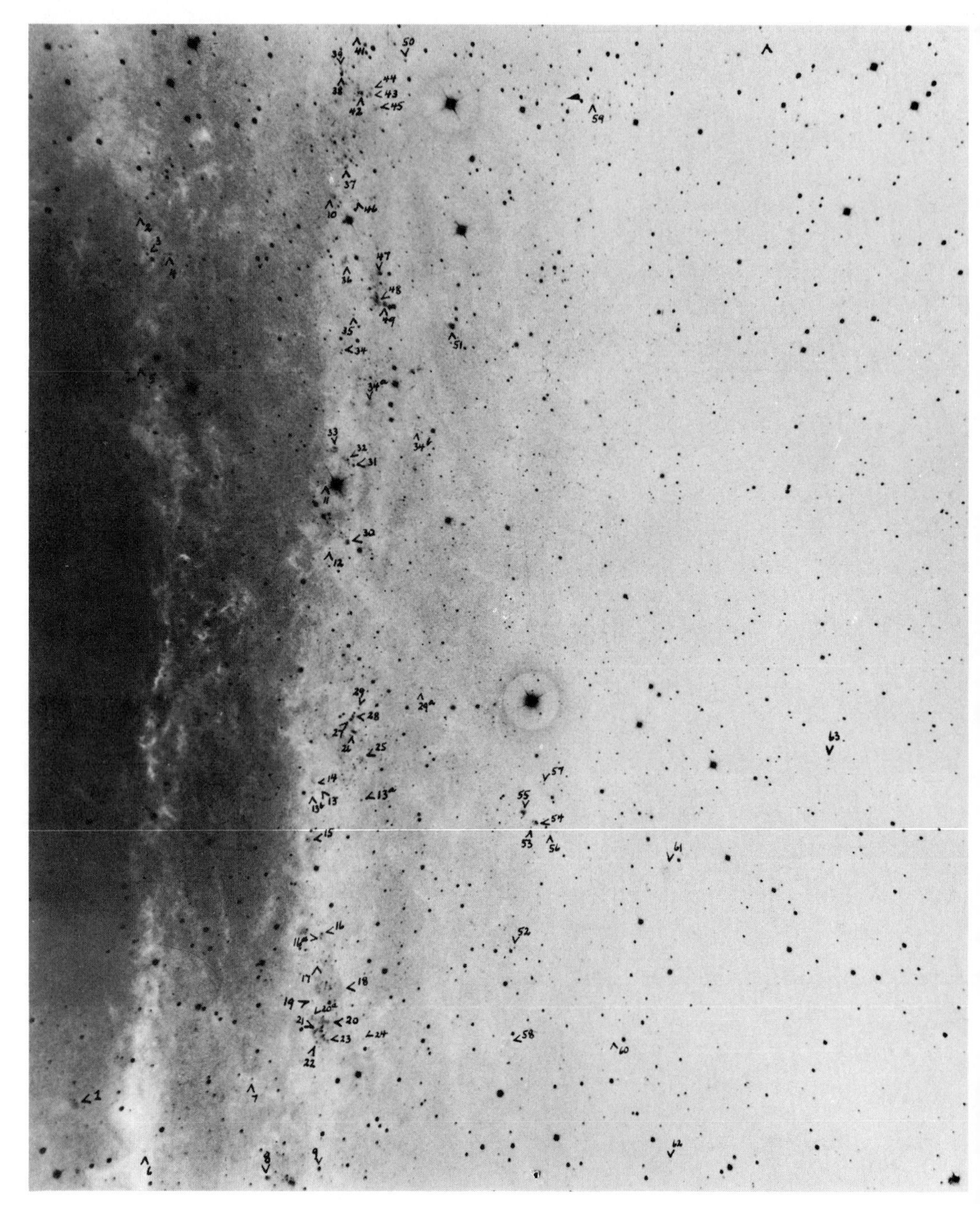

113. Inner spiral structure of the Andromeda galaxy. The negative of a Mount Wilson photograph shows a section to the right of center of the Andromeda galaxy as shown in Fig. 15. Dr. Walter Baade has given the following information about this photograph: Spiral arms of Messier 31, silhouetted against the bright nuclear region. Note that the emission nebulosities of the innermost spiral arm (Nos. 2, 3, 4, 5, and 6) lie in a dark lane; similarly, the emission nebulosities of the next spiral arm. This shows convincingly that the spiral arms are made up of large dust clouds which become conspicuous in this position because they cut off the light of the underlying Population II. (Four-hour exposure at 100-inch telescope by Baade.)

Messier 31, 33, 51, or 81 may be difficult. We would not be surprised if those astronomers in turn were to pity the poor fellows who live and work in the parts of our Galaxy not far from where we are located!

Messier 51 (Fig. 110) is one of the finest nearby spiral galaxies, seen almost face-on, with two well-developed arms that can be examined in great detail. Optically it shows very well how the most prominent spiral tracers, blue-white supergiant stars and their associated nebulosities, fix the over-all spiral pattern. Figure 115 shows side by side two photographs of Messier 51 and its companion. The photograph on the left displays beautifully the distribution of the hydrogen H-alpha concentrations. They outline very clearly the basic spiral pattern and help to define the manner in which one of the spiral arms merges into the companion. The negative picture on the right represents a normal blue photograph of the galaxy. These two photographs show that the spiral arms contain mostly gas and young stars, and that the luminous bridge connecting the main spiral and its companion has the same basic gaseous composition. Beverly Lynds has studied the distribution of obscuring matter in Messier 51 and she has concluded that the H II regions most frequently occur at the outer edges of the continuous dust lanes. Her diagram for the distribution of the dust is shown in Fig. 116. Messier 51 was one of the first to be studied with the Dutch radio interferometer, the Westerbork Array, which is a mile-long string of 12 interconnected radio telescopes. Figure 117 shows the result of the Dutch studies at a wavelength in the radio continuum near that of the 21-centimeter line, but well outside its limits. Mathewson, Van der Kruit, and Brouw find that the two strong continuous radio spiral arms coincide with the inner edges of the luminous optical arms. The radio spiral arms apparently follow very well the dust arms delineated by Beverly Lynds.

Figure 118 indicates that the spiral arms in Messier 51 are truly gaseous features. If the neutral atomic hydrogen were not actually concentrated to the spiral features in this galaxy, the 21-centimeter contours of equal intensity could be expected to wander about without reference to the optical spiral features; this obviously is not the case. One notes especially that the contours show the bridge connecting the main body (NGC 5194) and the companion (NGC 5195) as neatly as does the optical connection. To complete the picture for Messier 51, we should mention that Tully has made an optical study of the motions of the gas in the inner spiral regions. He has found some clearly defined streaming characteristics, which are in accordance with the predictions of the density-wave theory of Lin, Shu, and Yuan, to be described later in this chapter.

Though Messier 51 appears to be one of the prize spiral galaxies, it possesses one structural feature that is in a way very disturbing. This is the companion galaxy, which marks the end of one of the two major spiral arms. Arp has provided good evidence to show that the companion galaxy is truly associated with Messier 51 and he has furthermore suggested that it was probably ejected from the nucleus of Messier 51 as recently as 10 to 100 million years ago. He considers this observational evidence for a theory of Ambartsumian—which will also be described later in this chapter—according to which ejection of mass from the nucleus may be the source of all spiral structure.

The student of galactic spiral structure should always bear in mind the great variety

114. Emission nebulae in the Andromeda galaxy; two diagrams prepared by H. C. Arp. The left-hand diagram shows the positions of 688 emission nebulae (H II regions) in Messier 31, the Andromeda galaxy, as they would appear corrected for a tilt $i = 11°$ to the line of sight for the plane of M31. The diagram on the right shows the same plot with a logarithmic spiral fitted to the points. Note how difficult it is to distinguish between possible ring structures and spiral arms.

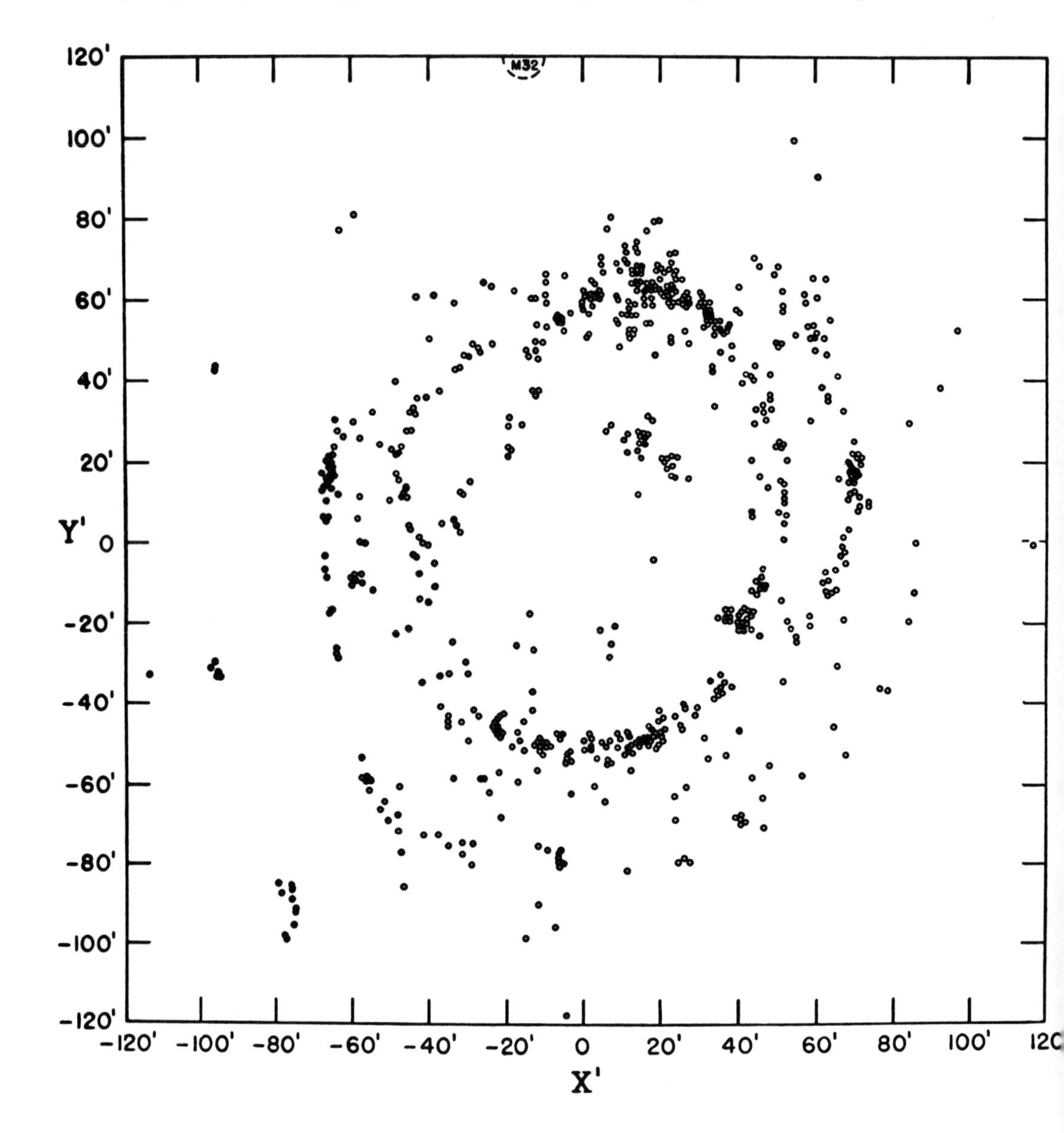

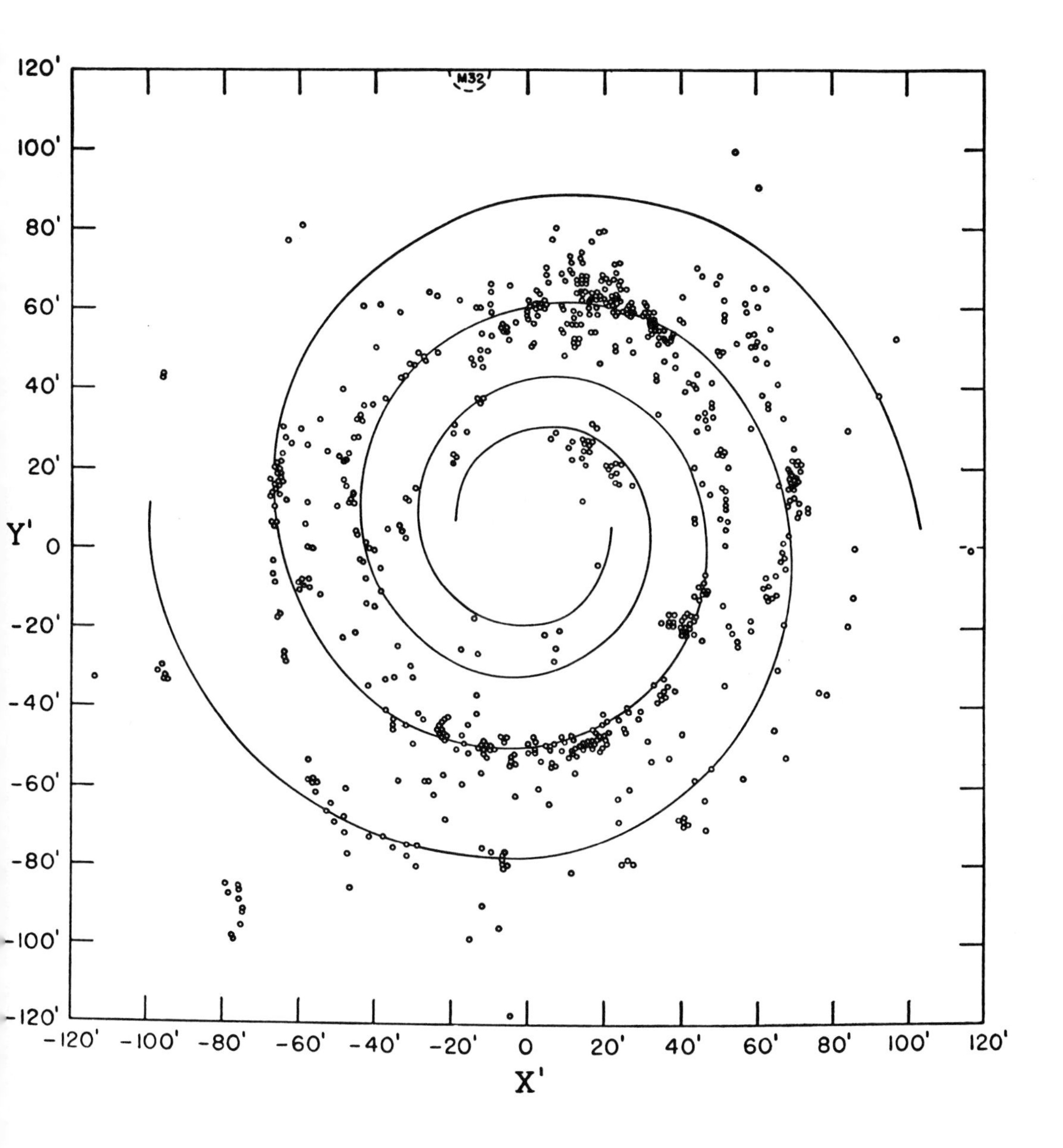
M32
120'
100'
80'
60'
40'
20'
Y' 0
-20'
-40'
-60'
-80'
-100'
-120'
-120' -100' -80' -60' -40' -20' 0 20' 40' 60' 80' 100' 120'
X'

115. A photographic composite of two prints of Messier 51, prepared by H. C. Arp of the Hale Observatories. The photograph on the left represents a superposition of a negative print of M51, taken with an H-alpha color filter with a pass band 100 angstrom units wide, and a positive print made with a comparable filter centered at a wavelength on the blue side of H-alpha. The result is a picture of M51 in the light of H-alpha only. The photograph on the right is for comparison purposes; it shows M51 in a negative print in normal blue light. The H-alpha spiral features are shown beautifully in the photograph on the left. (Courtesy of *Astronomy and Astrophysics.*)

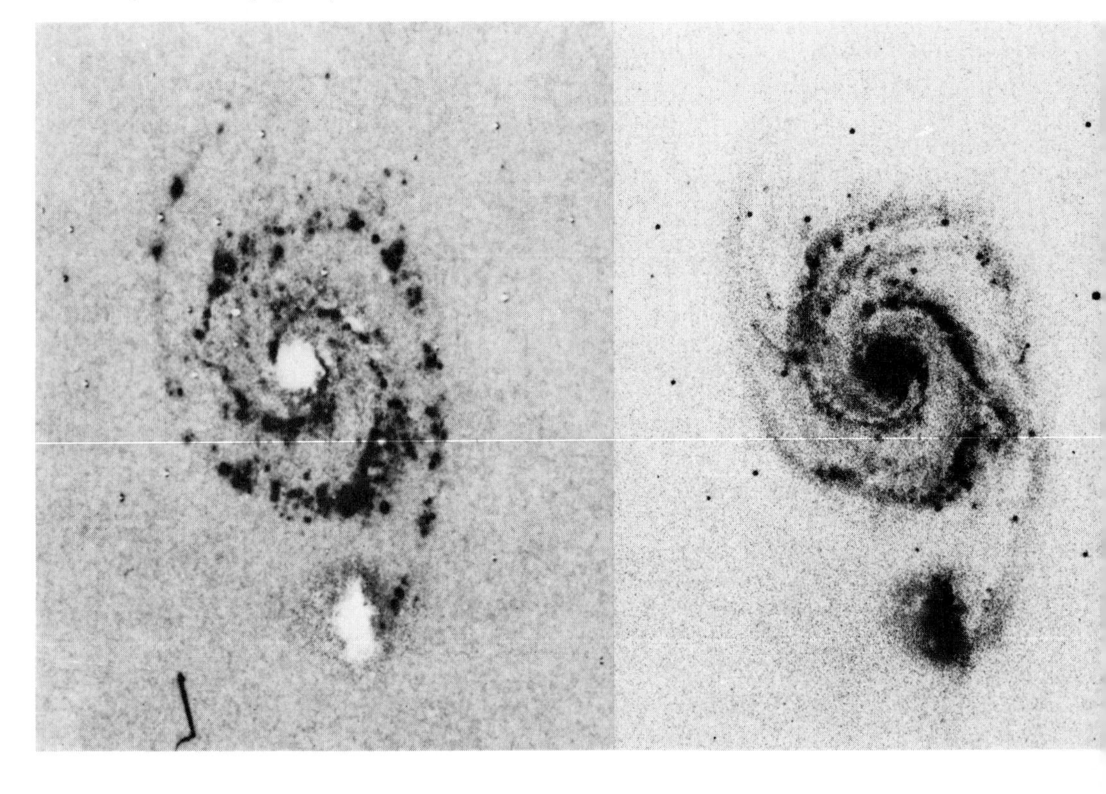

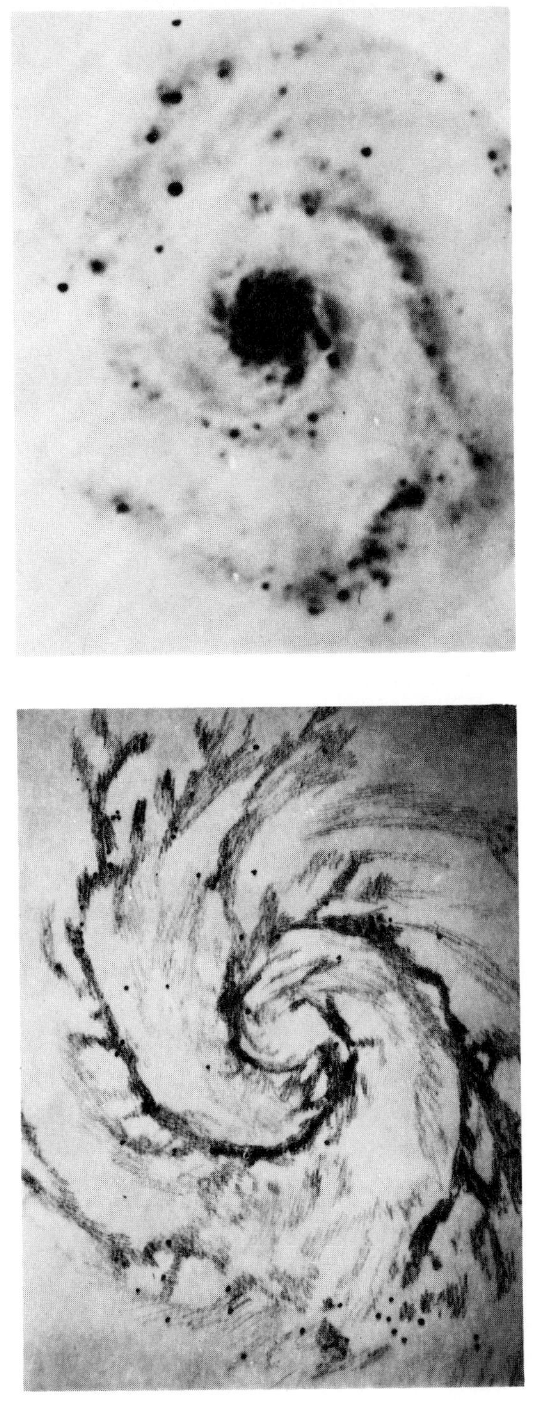

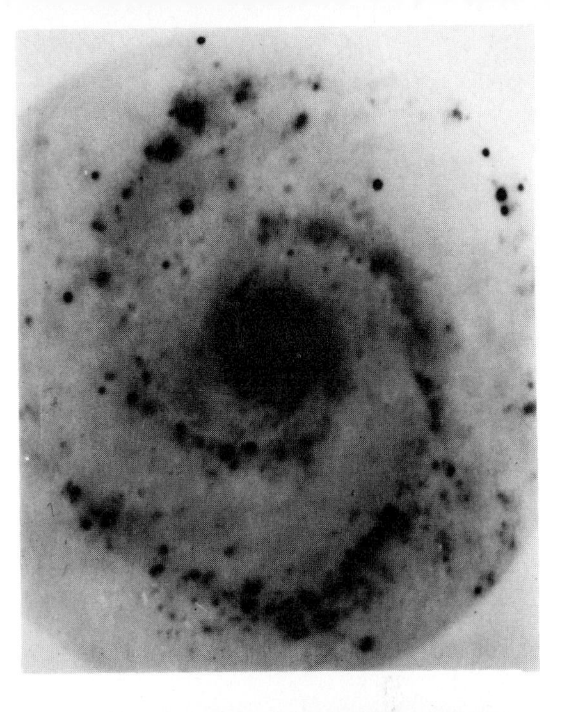

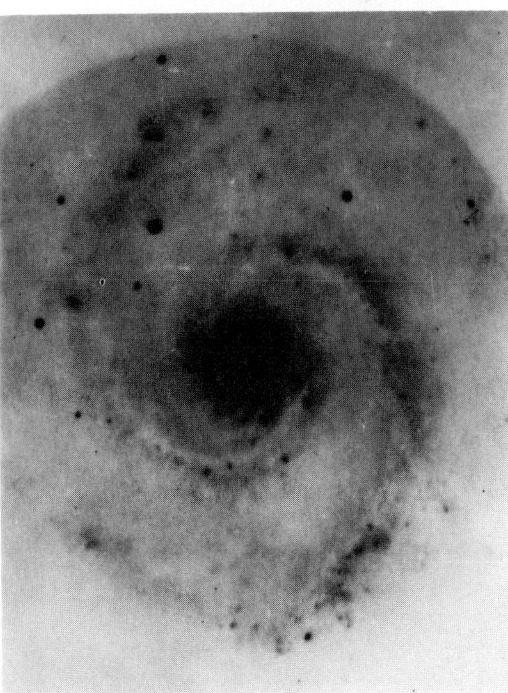

116. Obscuring matter in Messier 51. Beverly T. Lynds has prepared a series of photographs to illustrate the relation between the lanes of obscuring matter and the spiral arms in M51. The photographs at the top represent negatives of exposures through a broad-band filter in the blue (upper left) and through an H-alpha filter (upper right). The spiral arms are clearly shown. The photograph in the lower right is a negative of a plate taken with a narrow-band red filter (eliminating H-alpha) centered at 6,650 angstrom units. The drawing in the lower left is based upon one of Milton Humason's long-exposure photographs with the 200-inch Hale reflector; Dr. Lynds has drawn the dark lanes by proper shading and she shows the H II regions as black dots. (The photographs at the top were made with the 90-inch reflector of the Steward Observatory, University of Arizona.)

117. Radio emission from Messier 51. The ridges of radio-continuum radiation at 1,415 megahertz are shown superposed on Humason's photograph of M51 taken with the 200-inch Hale reflector. Dashed lines *A*, *B*, *C*, *D*, and *E* show interarm radio links, and *F* the link with the companion galaxy, NGC 5195. The ridge lines of radio emission coincide nicely with the absorption lines of Lynds shown in Fig. 116. (From a paper by D. S. Mathewson, P. C. van der Kruit, and W. N. Brouw.)

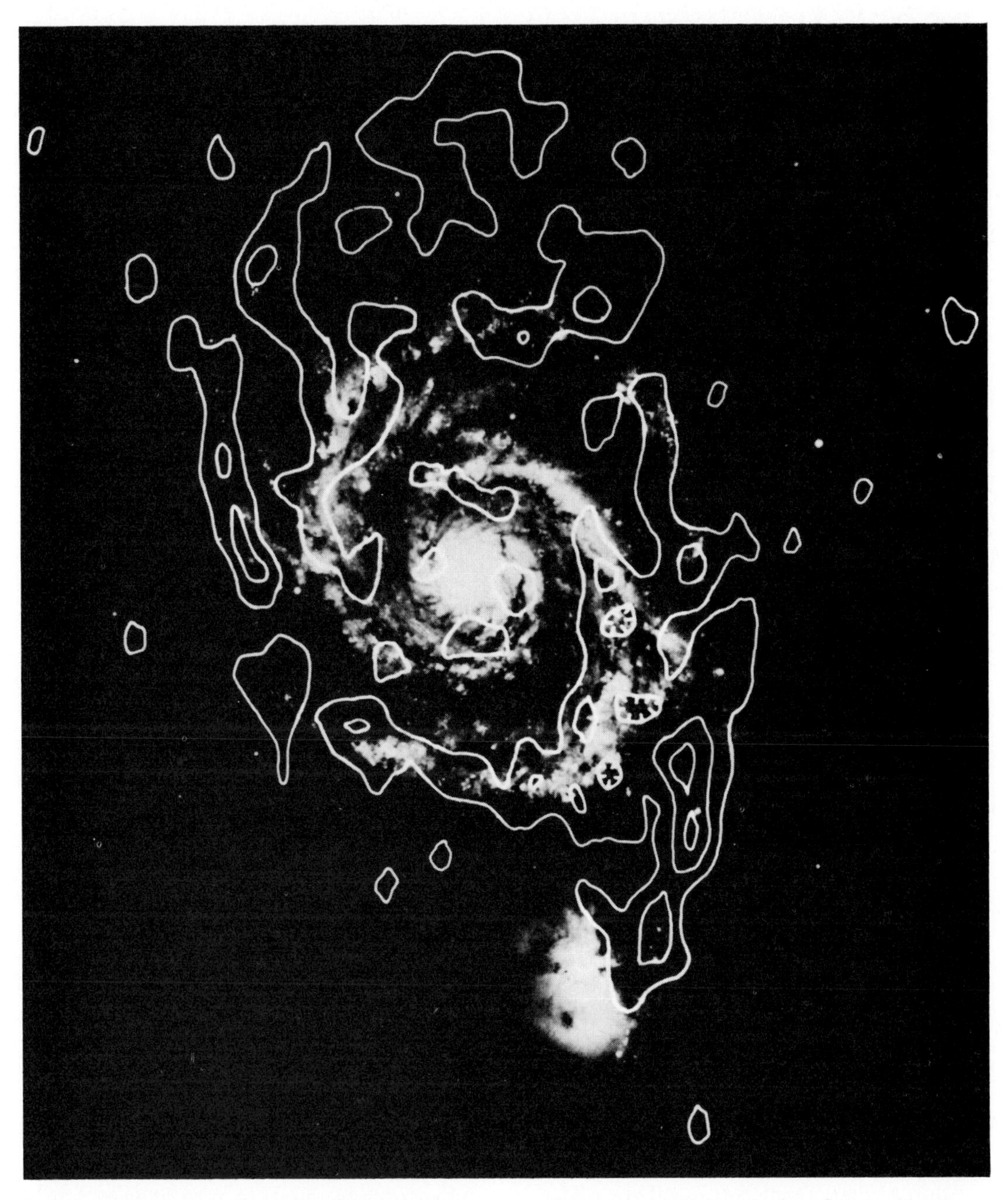

118. Equal-intensity contours for 21-centimeter radiation produced by neutral atomic hydrogen in the spiral galaxy Messier 51. The contours are shown superposed on a print of the 200-inch Hale reflector photograph reproduced in Fig. 115. Preliminary results with the Westerbork Array shown to B. J. Bok by Dr. Ernst Raimond (November 13, 1972).

of structures that occur in the spiral galaxies that have been photographed. Although it is true that the overriding property of spiral galaxies is that they show two trailing spiral arms, there are many objects that behave quite differently from this simple pattern. The nearby spiral Messier 101 (Fig. 16) is a multiple-arm spiral in its outer parts, and some of the more distant spirals show in their outer parts fireworks patterns with as many as six sections of arms or armlike features. Examples of complex patterns are shown in Figs. 119, 120, and 121. We should also bear in mind that there is a class of *barred spirals* (Fig. 122), objects with nearly circular spiral arms that seem to emanate from a central bar structure. The presence of an outer and inner ring structure appears to be quite common in spiral galaxies. Outer rings are shown beautifully in many of the barred spirals. Some astronomers have suggested that there may be evidence of a bar structure in our own Galaxy, but others—including the authors—have not accepted this suggestion.

Piddington of Australia especially has drawn attention to peculiar and interacting galaxies that do not fit into the basic two-arm spiral pattern that is shown in most of our illustrations of nearby galaxies. There are many examples of deviates in the *Hubble Atlas.* In 1959 B. A. Vorontsov-Velyaminov of Moscow published a remarkable *Atlas of Peculiar and Interacting Galaxies* and in 1966 Arp published his very useful *Atlas of Peculiar Galaxies.* Piddington has expressed the opinion that the exceptions may be the rule and that there is no basic two-arm design among spiral galaxies. He has attacked the assumption made by many, especially in the researches of one of the authors of this book, according to which spiral features are long connected streamers of hydrogen gas and young stars. Piddington doubts that there is any marked concentration of interstellar gas associated with spiral features. It seems as though the results obtained for Messier 51 with the Westerbork Array (see Fig. 118) refute Piddington's criticism and they prove conclusively that the H I spiral features appear as broad background features against which the material spiral arms are shown as narrow luminous traces. We note that very comparable conclusions may be drawn from the Westerbork 21-centimeter contours for Messier 81.

One of the most important lessons to be learned from an inspection of spiral galaxies is that the spiral phenomenon seems to possess a good deal of permanence. If it were a fleeting property of a galaxy, we would not expect to find so many of our neighbors showing spiral structure of a very similar type. If the spiral arms were in the habit of winding up, we would expect to see many more spiral galaxies with tightly wound inner spiral patterns than we seem to observe. The finest collection of galaxy photographs is contained in Sandage's *Hubble Atlas of Galaxies.* The Milky Way astronomer who is engaged upon an analysis of the spiral structure of our own Galaxy should plan to spend at least an evening each month just taking in the wonders of the *Hubble Atlas.* This is, in fact, a highly recommended activity for anyone interested in astronomy.

Spiral Tracers in Our Galaxy

O and B stars, singly, in clusters, or in associations, have traditionally proved to be about the most useful spiral tracers in our Galaxy. They are often found associated with H II regions, gaseous nebulae shining brightly in hydrogen H-alpha radiation. The stars responsible for exciting the H II regions are of

119. The spiral NGC 6946, photographed by J. B. Priser with the 61-inch astrometric reflector of the U. S. Naval Observatory Station at Flagstaff. Complex multiple structures prevail. (U. S. Naval Observatory photograph.)

120. The spiral NGC 5985, photographed by J. B. Priser with the 61-inch astrometric reflector of the U. S. Naval Observatory Station at Flagstaff. Again a very complex pattern of spiral features. (U. S. Naval Observatory photograph.)

121. The spiral NGC 5364 in Virgo. This photograph, made with the 200-inch Hale reflector, shows a spiral with remarkably wound-up continuous features. (Courtesy of Hale Observatories.)

122. The barred spiral NGC 1300 in Eridanus, from a photograph made with the 200-inch Hale reflector. (Courtesy of Hale Observatories.)

spectral class O, B0, B1, or B2, either single stars or clusters or associations of stars. A star with a spectral type between O and B2 is a blue-white giant or supergiant star. According to present evolutionary theories, such stars have been in existence for only a few, generally less than 10, million years. The O to B2 stars and the emission nebulae were considered to be most promising spiral tracers by Baade and by Morgan and his associates, and they continue to hold this place of honor today.

We described in Chapter 8 how radio-continuum radiation and radio alpha transitions contribute to our knowledge of the distribution and motions of the hydrogen gas. We noted there that the radio techniques used for these researches provide us with data on the distribution of ionized hydrogen. The alpha transitions yield, furthermore, very useful data on the radial velocities of the clouds under investigation. These two radio tracers therefore supplement very nicely the data on the distribution and motions of neutral atomic hydrogen obtained from studies of the 21-centimeter line.

Returning to the optical picture, we remind the reader that clouds of interstellar gas produce interstellar absorption lines in the spectra of distant stars. The best known of these absorption lines are the interstellar *K* line, produced by ionized calcium, and the interstellar *D* lines of neutral sodium. The most extensive work in this field so far has been done by G. Münch for stars within reach from northern latitudes; the southern Milky Way still awaits full exploration. High-dispersion spectra often show multiple interstellar absorption lines, which are presumably indicative of the presence of several gas clouds along the line of sight; for each observed component we can readily measure the radial velocity. These interstellar gas clouds are presumably concentrated in and along the spiral arms. The radial velocities found for the interstellar absorption lines make it possible to pinpoint their places of origin in very much the same way as we do with the gas concentrations found from 21-centimeter profiles and from alpha transitions. To derive the distance to individual clouds, we still require a velocity model for the Galaxy, which transforms these radial velocities for each particular direction in the galactic plane into distances from the Sun and distances from the galactic center. What we obtain basically in each case is kinematic information, that is, information about distributions of velocities. It is only by the use of a velocity model of our Galaxy that we can obtain the distances. However, any cloud that is detected through an interstellar absorption feature must lie nearer to the Sun than the star whose spectrum contains the feature.

The interplay between kinematic and structural observations promises to become increasingly important in future years. Very often we have extensive data for the spectra, magnitudes, and colors of the stars, star cluster, or association responsible for a given H II region. We can then derive the distance of the cluster or association on the basis of known absolute magnitudes and colors of embedded stars and from a knowledge of the estimated amount of intervening absorption. If we can measure as well the radial velocities of the stars and of the associated gas, we may proceed to check whether or not the observed radial velocities of the gas clouds agree with the values predicted from our model. A difference between the observed and the predicted velocities will suggest that there is something wrong with our basic velocity model of the Galaxy, or, alternatively, it will give an indication of local streaming. We await the time when we can say that our basic velocity model is reasonably well established, so that we may then use the observed differences as indicators of large-scale streamings. Recent 21-centimeter data as well as stellar and nebular optical radial velocities have shown in many places large-scale regional departures from circular motion, which suggest large-scale streaming of gas.

The increased emphasis on kinematic studies is an encouraging development. It represents an area on which we must concentrate in radio and in optical researches. We cannot stress too much the need for optical radial velocities for H II regions and their associated stars to complement radial velocities obtainable by radio-astronomical techniques. Without establishing any distance scales, we can find for certain directions in the galactic plane whether or not identical radial velocities are obtained from radio-astronomical and optical studies of the same gas clouds. In addition, we may compare radial velocities from gas clouds with those from associated stars or star clusters.

In recent years the results of many optical studies of radial velocities for H II regions have been published, most notably by Courtès, the Georgelins, Monnet, and Cruvellier of Marseille Observatory, and by J. S. Miller and M. G. Smith, associated with Kitt Peak National Observatory and Cerro Tololo Inter-American Observatory. Abundant radio hydrogen alpha radial velocities have been measured by Mezger, T. L. Wilson, Gardner, and Milne for the southern Milky Way and by Reifenstein, Burke, Altenhoff, Mezger, and Wilson for the northern Milky Way. All of this material is now being analyzed in conjunction with a rapidly growing body of stellar radial velocities. Especially useful in this respect are radial velocities for the longest-period cepheid variables and for red supergiants, which have been accumulating in recent years through the work of Feast and Humphreys.

The cepheid variables with periods greater than 10 days are fair to good spiral tracers. They are supergiant stars that presumably have already gone through the O and B stages. They have probably also been red supergiants for cosmically brief intervals before entering the phase of long-period cepheid variability. Kraft has called attention to their potential as spiral tracers. He stresses that cepheid variable stars can be detected and studied to far greater distances than are within reach for O and B stars even with the largest reflectors. As of now, it seems very likely that the longest-period cepheids are the youngest ones and hence they seem to be the best potential spiral tracers.

Dark nebulae are as yet one of the least studied groups of objects, but the time has come to use them more extensively than in the past. Good distance estimates must be made for many dark nebulae seen along the band of the Milky Way. Most dark nebulae occur in very crowded regions, where the scale of Schmidt-telescope photographs is not sufficient for careful and detailed study, and hence we must look to the large reflectors for precision work on them. Large reflectors now under construction almost always have provisions for large-field wide-scale astronomical photography, and these instruments offer beautiful opportunities for further research on galactic dark nebulae. Photography with image-conversion tubes makes it possible to reach the faintest accessible stars in reasonably short exposure times. Intermediate- and narrow-band filters, now available for photoelectric and photographic work, should prove a great help in obtaining magnitude data and precise colors.

There are many useful special groups of

stars other than O and early B stars that make reasonably good spiral tracers. The B emission stars are, according to Th. Schmidt-Kaler, among the best of these. Lindsey Smith has shown that the hot emission-line Wolf-Rayet stars are also fairly good spiral tracers.

In all work relating to O and B stars, longest-period cepheids, and dark nebulae, we are beginning to pay attention to their relative distribution within specific spiral features. We have noted already that the dark nebulae are often found concentrated along the insides of the spiral arms and there are indications that the cepheids and the O and B stars trace arms or spiral features that do not coincide precisely with each other—nor, for that matter, with the 21-centimeter features. In the years to come, we must look for possible differences in details of distribution, which may be caused by age effects coupled with basically different kinematic properties. It is clear that we cannot afford to indulge in theoretical explanations and interpretations until we have a good observational basis for our speculations.

Supergiant stars of all varieties, from the blue-white O and B supergiants to the reddest M supergiants, are among the most useful spiral tracers. The discovery of such stars from objective-prism spectral plates for both the northern and the southern Milky Way is not a difficult task. The spectrum of any star of interest can now be photographed in intimate detail, because most of our larger spectrographs are fitted with image-conversion equipment, which makes it possible to obtain high-quality spectra of fairly high dispersion in relatively short exposure times. From these spectra one can make estimates of absolute magnitudes for the stars under investigation, and their radial velocities can be determined with precision. In recent years Humphreys has been especially active in this area. Her results have demonstrated that the supergiant stars may be the key to our penetrating to great distances for many sections in the band of the Milky Way. We now possess sufficient numbers of faint photoelectric standards to make it possible to measure *UBVRI* magnitudes and colors with precision for any star of interest. If desired, the same stars can be subjected to intermediate- and narrow-band photoelectric photometry.

It seems probable that galactic magnetic fields are not the major controlling factor in the production of the observed spiral patterns. Such fields must, however, have considerable organizing influence on details of spiral structure. The observed polarization effects in the light of distant stars prove that magnetic fields are associated with spiral features. The observed polarizations are generally largest, and the alignment of the polarization vectors appears to be most regular, when we view a star across a spiral arm; small percentage polarizations, combined with haphazard orientations of the vectors, are observed when we look longitudinally along an arm. Hence, present indications are that polarization effects are helpful in the tracing of spiral features, even though we are far from understanding just what roles large-scale magnetic fields play in theories of spiral structure.

We now have available extensive and beautifully precise material on the amounts and orientations of the polarization vectors for stars along the entire Milky Way, north as well as south (Chapter 9). The original data, which had been contributed by the discoverers of interstellar polarization, Hiltner and Hall, have been supplemented by extensive new catalogues by Mathewson of Mount Stromlo Observatory and by Klare and Neckel of Hamburg. They have published catalogues and charts with polarization data for the southern Milky Way, the sections for which

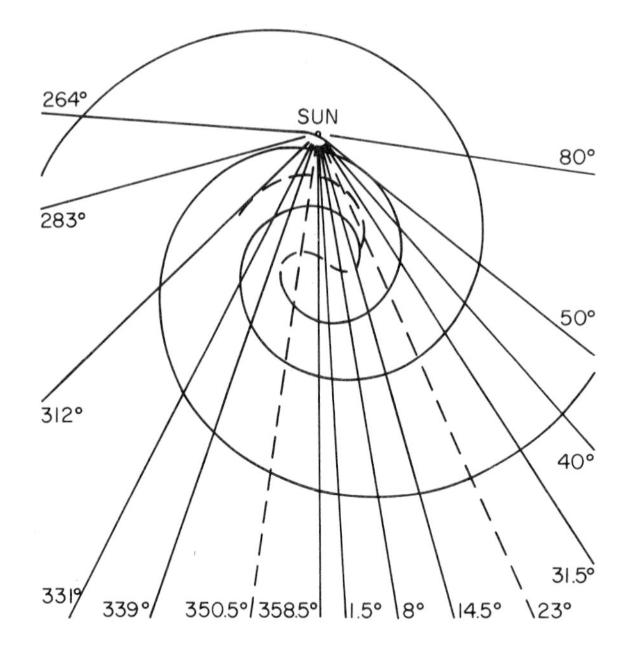

123. Spiral structure from radio edges. Anne J. Green of Sydney University has drawn this spiral diagram with a pitch angle of 12°.5 as fitting the steps in the radio-continuum observations for our Galaxy as well as several optical features. In this particular model, the Sun would be located in a spur.

such data had previously been lacking. Figure 100 shows the Mathewson polarization map.

Synchrotron radiation, which dominates at longer wavelengths in the radio range, has been helpful in defining certain edges in the longitudinal distribution of high-energy free electrons. These directions are identified as ones in which we look more or less lengthwise along a spiral arm. At Sydney University, Mills and Mrs. Green have done the most effective research along these lines. They have located some marked jumps in the distribution of the strength of the integrated synchrotron radiation plotted as a function of galactic longitude. Mrs. Green's analysis of the available data suggests that our Galaxy probably possesses a spiral pattern with average pitch angles in the range between 9° and 15°, but ring structures are not ruled out (see Fig. 123).

Pulsars and x-ray sources have been suggested as potential spiral tracers. Pulsars are probably the direct descendants of stars that have passed through the supernova stage, and supernova remnants are known to be the emitters of very powerful x-rays. However, the precise status of supernovae, pulsars, and strong x-ray sources as spiral tracers remains very much in doubt, at least for the present.

Gravitational and Other Theories

Ten to 15 years ago, the trend was to consider as the most likely hypothesis that galactic spiral structure is caused and maintained by large-scale magnetic fields, which were supposed to align the gas in spiral patterns. However, it was then discovered that the longitudinal magnetic fields average in all probability only 2 or 3 microgauss in strength, and they appeared too weak to produce the major controlling effects that are required. It is more likely that gravitational forces are responsible for the observed over-all spiral patterns and that the magnetic fields are frozen into these patterns as a by-product of the distribution of the ionized gas, mostly hydrogen (protons). Gravitational concentration of this ionized gas may locally produce magnetic fields of unusual strength. The behavioral patterns of the polarization vectors, such as those shown in Fig. 100, can thus be considered as a by-product caused by the concentration of ionized gas in spiral features.

Bertil Lindblad was for many years the lone champion of gravitational theories to explain spiral structure. In the middle 1960's C. C. Lin and his associates, Shu and Yuan, began a series of studies according to which the observed spiral features would be produced by a density wave passing through the interstellar gas. Lin starts out by assuming that the gravitational potential close to and in the central plane of a highly flattened rotating galaxy has a spiral component. He finds that gas and stars moving with near-circular velocities can, because of the effects of this spiral component, be subject to density waves, with the gas and the associated stars piling up for shorter or longer times in the gravitational spiral-shaped troughs of low potential. He investigates, with various types of spiral patterns for the gravitational potential, whether or not this piling up is a transitory phenomenon. He finds certain strict conditions that must be satisfied if the spiral potential field is to be a semipermanent one. If it is to persist, it must be sustained by the gravitational effects produced by the stars and gas that participate in the patterns of spiral behavior. Lin's theory predicts that in most highly flattened galaxies there will be a region in which spiral structure can prevail. For our own Galaxy this region lies roughly in the range of distances between 5,000 and 15,000 parsecs from the galactic center, which means that our Sun is, as it should be, right in the part of the Galaxy where spiral structure may be expected. Lin's theory predicts that the only stable and semipermanent spiral configurations will be those with two trailing spiral arms, emanating from opposite sides of the nucleus, and it also predicts quite precisely the spacing between the arms and their pitch angles. Spacings and pitch angles are determined by the nature of the underlying force field of the galaxy. The predicted pitch angles for our Galaxy are about 6°, very much in line with the radio-astronomical results of Kerr, but out of line with the Becker-Morgan results. A typical calculated spacing between arms separates them by 3,000 parsecs and that is in fair average agreement with what we find in our Galaxy. Lin's theory predicts some large-scale streamings of gas and of young stars, with average stream velocities possibly as high as 6 to 10 kilometers per second relative to pure circular motion. Such deviations have actually been observed, and of the right order. The observed deviations from pure circular motion fit quite well into the pattern of the streamings predicted for our Galaxy by Lin's theory.

In Lin's theory, the spiral arms are produced by the spiral-shaped component of the field of gravitational potential. This component produces the density waves according to which gas and young stars are piled up along the troughs of minimum potential. In other words, the gas and young stars linger relatively longer in the spiral-potential grooves than in between, while the material in the spiral arm is being constantly replenished. In the earlier picture that most of us had (with the notable exception of Bertil Lindblad), the spiral arms were considered fixed loci of objects, such as clusters, associations, and emission nebulae. It seemed inevitable that, after one or two galactic revolutions, they should wind up, since the outer objects would move at slower speeds than those closer to the nucleus. This problem ceases to exist if one accepts Lin's view that spiral arms do not contain the same stars and gas indefinitely, that they are the product of density waves, and that the only semipermanent feature is the underlying spiral component of gravitational potential. We should note that the spiral pattern of gravitational potential in Lin's theory does not rotate as fast as the Galaxy for regions near the Sun. Hence we find that the instantaneous spiral arms do not persist, and the locus of maximum spiral features shifts with time and with respect to existing spiral arms.

There is as yet no good and complete theory to explain how the spiral potential fields of force come about in the first place. There is a slow rotation of the spiral-shaped field in the central plane of the galaxy; it resembles in some ways a two-armed boomerang! In the outer parts of the galactic spiral structure, the galaxy and the field rotate at about the same speed, but in the inner parts of a galaxy the boomerang does not move as fast as do the stars, the cosmic dust, and the interstellar gas that go around the galactic center with the speed of the general circular galactic rotation.

The stars are not bothered very much by the spiral potential field. However, in troughs of low potential the gas will have a tendency to pile up, slow down temporarily, and then move on to run again with the regular speed of galactic rotation, leaving the new gas behind it to go through the same performance. A spiral feature will be observed in the direction along the curve of greatest density, and in the course of several galactic rotations the gas will pass several times through these potential troughs, thus producing a steady spiral pattern of constantly varying gas clouds. In Lin's picture the spiral potential field, or boomerang, of our Galaxy moves at the Sun's position at a rate a little more than half that of the normal galactic rotation. Agreement between rotational speed and pattern speed is found at a distance of 14,000 to 15,000 parsecs from the center of the Galaxy. It will be noted that the shape of our boomerang fixes the shape of the spiral arms that are observed at any one time. Since the boomerang does not change its shape, the spiral arms should have the same appearance for as long as the density-wave pattern persists. There is no winding-up problem in Lin's theory. It is worth noting here that the outer ring structures of neutral atomic hydrogen are found at just about the positions where the pattern moves at the normal rate of galactic rotation. Studies of the stability of the density-wave pattern have shown that a two-armed trailing spiral boomerang of potential has the best chance of surviving for several galactic revolutions.

The predictions of Lin's theory can be tested by mathematical calculations relating to the behavior of the density-wave pattern and they can also be checked by modern computer calculations for stellar systems with large numbers of gas clouds, or with hundreds of thousands of stars. Density-wave-like phenomena have been produced through computer calculations by P. O. Lindblad (Bertil's son) in Sweden and by R. H. Miller, Prendergast, Quirk, and others in the United States.

One of Lin's associates and former students, W. W. Roberts, has studied in some detail the processes that go on when the gas passes through the regions of lowest potential, that is, through the strongest part of the boomerang. He has found that sudden compression of the gas will take place and that along a wide front galactic shock waves will be produced. Narrow lanes where gas and interstellar dust are at very high density can be expected to be present. It seems that star formation might well occur under the expected conditions of high compression. The dust lanes would be at the inside of the spiral arm and the stars would migrate outward from it, with the youngest stars located closest to the shock front and the slightly older ones a bit farther removed. Star formation from interstellar gas and dust is a very complex phenomenon, and the passing of a density wave through the gas and dust cannot alone guarantee that star formation will take place on an extensive scale. However, if other conditions are right, then the passing of the density wave may well provide the required trigger action.

We should not leave the false impression that the Lin theory is the accepted gospel in spiral structure. It does represent a major breakthrough in a field where theory had made little headway. Its greatest asset to observing astronomers is that it presents them with firm predictions of the behavior, distribution, and kinematics of interstellar gas, cosmic dust, and young stars in and near spiral features.

The most severe of Lin's critics has been the Australian physicist Piddington, who, as we noted earlier in this chapter, favors a complete rejection of the density-wave theory of Lin, and especially of the shock-front hypothesis of W. W. Roberts, which is a part of the density-wave theory. Piddington has expressed the opinion that spiral arms are not truly gaseous features, but that they represent principally lines of concentration of young stars.

Piddington offers as a substitute for the approach of Lin and Lindblad a hydromagnetic theory of his own. In this theory the basic forces that are responsible for the observed spiral patterns arise through the encounter of our Galaxy (and others!) with a primordial intergalactic magnetic field. Our Galaxy's axis of rotation meets this field at an oblique angle. Piddington has shown that even a very weak intergalactic field will order the interstellar gas of our Galaxy sufficiently to produce in one galactic revolution patterns of concentrated regions of star birth resembling spiral arms. One interesting consequence of Piddington's theory is that it predicts spiral features away from the central disk of our Galaxy. Such features have apparently been detected by R. D. Davies and by Verschuur. However, Piddington's spiral features are basically more like concentrations of recently formed cosmic dust and very young stars than like gaseous concentrations. In contrast to the Lin-Shu theory, which makes firm predictions about the velocity fields associated with spiral features in the central plane of our Galaxy, the Piddington theory predicts large velocities perpendicular to the galactic disk.

All workers in the field realize that there are many theoretical and observational approaches to the study of spiral structure. The density-wave theory does not exclude the presence of important effects produced by large-scale galactic magnetic fields, and it is quite reasonable that the importance of such fields be fully explored. Nuclear ejection, of the type favored by Ambartsumian and Arp, may provide a basic motive force for initiating and maintaining spiral structure. The researches of A. and J. Toomre have shown that encounters between galaxies may produce calculable spiral-structure-like phenomena. And, in the end, as has been stressed repeatedly by Pikelner of Moscow University, the barred spirals may well provide the major key to the understanding of spiral structure. There is obviously no such thing as a simple theory of spiral structure, and neither is it practicable to describe in a general way all of the complex phenomena shown by spiral features in our own and other galaxies.

Ambartsumian is one who has urged astronomers since the late 1950's to look toward the central regions of our own and other galaxies as providing the impetus for the spiral structure we observe in the outer parts. High-energy jets of ionized gas may enter the outer parts of a galaxy after having been generated by giant explosive phenomena in and near a galactic nucleus. The region of impact may be at 3,000 to 5,000 parsecs from the center, at which interface we may expect to find clouds of very fast-moving electrons entering the more peaceful sections of the galactic disk. It is at 5,000 parsecs from the center of our Galaxy that we find the great ring of giant H II regions noted by Mezger, Westerhout, and others. Large-scale magnetic fields may play more of a role than we have assigned them in our treatment.

Spiral Structure in Our Galaxy

It would be pleasant if we could offer for the end of this chapter a good diagram of the spiral structure of the Galaxy. But an inspection of Figs. 103, 107, and 108 shows that we cannot do so at the present time. We do not yet possess for the Galaxy the "Grand Design" for which C. C. Lin began to ask a decade ago.

Why are there differences between the Kerr-Westerhout and the Weaver diagrams of radio spiral structure? One reason for the discrepancies is that different interpretations are used by the analysts for the sections of the Milky Way in which sufficient 21-centimeter data are not yet available. Other differences arise because of variations in the method of analysis of the basic 21-centimeter profiles, and because of the use of differing velocity models for the Galaxy. One of the most important sources of discrepancies appears to lie in the manner in which authors connect the major concentrations of neutral atomic hydrogen into a single all-inclusive spiral pattern. The optical astronomer is faced with equally difficult problems in the analysis of his material. For example, most of us working in the field are agreed that there is a major concentration of young stars, gas, and dust in the direction of the southern constellation of Carina, and other spirallike features have been studied in Cygnus, in Orion, and in Norma. The Perseus Arm is well established over a considerable range of distance beyond the Sun and the Sagittarius Arm is equally well delineated at distances from the galactic center 2,000 parsecs less than that of the Sun. Becker and associates tend to interconnect these features into an over-all spiral pattern in which the arms have pitch angles of about 25°, whereas to Bok and others it has seemed more reasonable to connect these same features into a pattern showing several arms with pitch angles of 6° more or less.

The game of connecting recognized radio and optical spiral features into an over-all spiral pattern for our Galaxy will undoubtedly continue in the years to come. In the end, the basic pattern will probably be revealed by analyses of radio-astronomical data, especially 21-centimeter-line data. However, the analyses will by no means be straightforward. It is relatively easy to obtain definite results relating to concentrations of neutral atomic hydrogen in various directions of galactic longitude and latitude, but, for each direction, these concentrations will be observed as coming from clouds that have specific observed radial velocities of approach or recession. We shall require an intricate model of galactic rotation and of streaming motions applicable to the gas clouds in the Galaxy before we shall be able for each direction to translate the observed velocities into distances from the Sun.

To make life even more difficult, we encounter certain disturbing purely geometric effects in 21-centimeter analysis that complicate the situation. W. B. Burton and W. W. Shane have especially called attention to such effects. For some directions in the galactic plane, that is, at some galactic longitudes, the radial velocity of approach or recession will change very slowly with distance. This has a very disturbing consequence for the 21-centimeter profile at that galactic longitude. Even if the distribution of the neutral atomic hydrogen were perfectly uniform along the line of sight, there still would be an unduly large amount of that hydrogen in the direction that possesses a radial velocity close to the critical value. At the critical radial velocity the signal will build up in intensity, which produces an abnormally strong signal for that particular direction. The consequence of the very simple geometric effect is that we get a strong signal, which at first sight might wrongly be attributed to radiation from a cloud of neutral atomic hydrogen. *Velocity crowding* is one of the most serious concerns of the radio astronomer studying the distribution of neutral atomic hydrogen in our Galaxy.

We have noted that it is quite possible to isolate and study certain definite spiral features. At Steward Observatory, Bok, Humphreys, E. W. Miller, and others have in recent years made a comprehensive study of the distribution of gas, cosmic dust, and young stars for the Carina section of the Milky Way. The optical data alone, especially studies of O and B stars by Graham, show that in Carina we are probably looking edgewise along a major spiral feature. Graham showed that the sharp outer edge of this feature is at galactic longitude 283°, and there is evidence for some sort of inner edge near galactic longitude 300°. The feature has been traced optically over a distance of 7,000 parsecs, from 1,000 to at least 8,000 parsecs from the Sun. Figure 124 shows the diagram produced by Miller. The same sort of studies have been made for a feature in Cygnus, where Dickel, Wendker, and Bieritz have studied the distribution and distances of 90 H II regions, all associated with a spiral feature stretching to at least 4,000 parsecs from the Sun in the direction of galactic longitude 75°. Sections of the Perseus Arm are equally well delineated, and the inner Sagittarius Arm is also clearly marked. The outlining and detailed studies of some of these features is exceedingly useful for an understanding of the properties of the spiral arms of the Galaxy. First of all, for a spiral feature like that observed in Carina, one may obtain the distribution of dust and gas and young stars of various ages across the feature, and thus hope to arrive at useful conclusions regarding the processes of star birth in spiral features. The Lin theory predicts certain systematic motions at the inner and outer edges of spiral arms. Observational evidence of deviations from pure circular motion associated with the Carina spiral feature has been obtained by Humphreys, who finds velocities of the order of 8 kilometers per second against galactic rotation on the inner side of the spiral feature and velocities of comparable amounts with galactic rotation on the outer side. This is of the amount and sense predicted by the Lin-Shu theory. For the Perseus Arm some very interesting phenomena have been found. Some of the gas clouds associated with the Perseus Arm appear to be rushing toward us at a rate of 20 to 30 kilometers per second, whereas the stars appear to move more or less in the circular orbits expected from basic galactic rotation.

We shall certainly continue our attempts to present diagrams showing the over-all spiral structure of the Galaxy and we shall of course try to obtain the Grand Design. However, in

124. A working diagram of the Carina spiral feature. Ellis W. Miller has prepared a diagram to show how the objects studied in the southern constellations of Vela, Carina, Crux, and Centaurus (galactic longitudes 270° to 300°) are concentrated to a curved spiral feature that can be traced between 2 and 8 kiloparsecs from the Sun. The interstellar obscuration reaches its greatest value on the inside of the feature, but there is marked obscuration on the outside as well. The center of the Galaxy is below the Sun, outside the diagram at a distance of 10 kiloparsecs from the Sun.

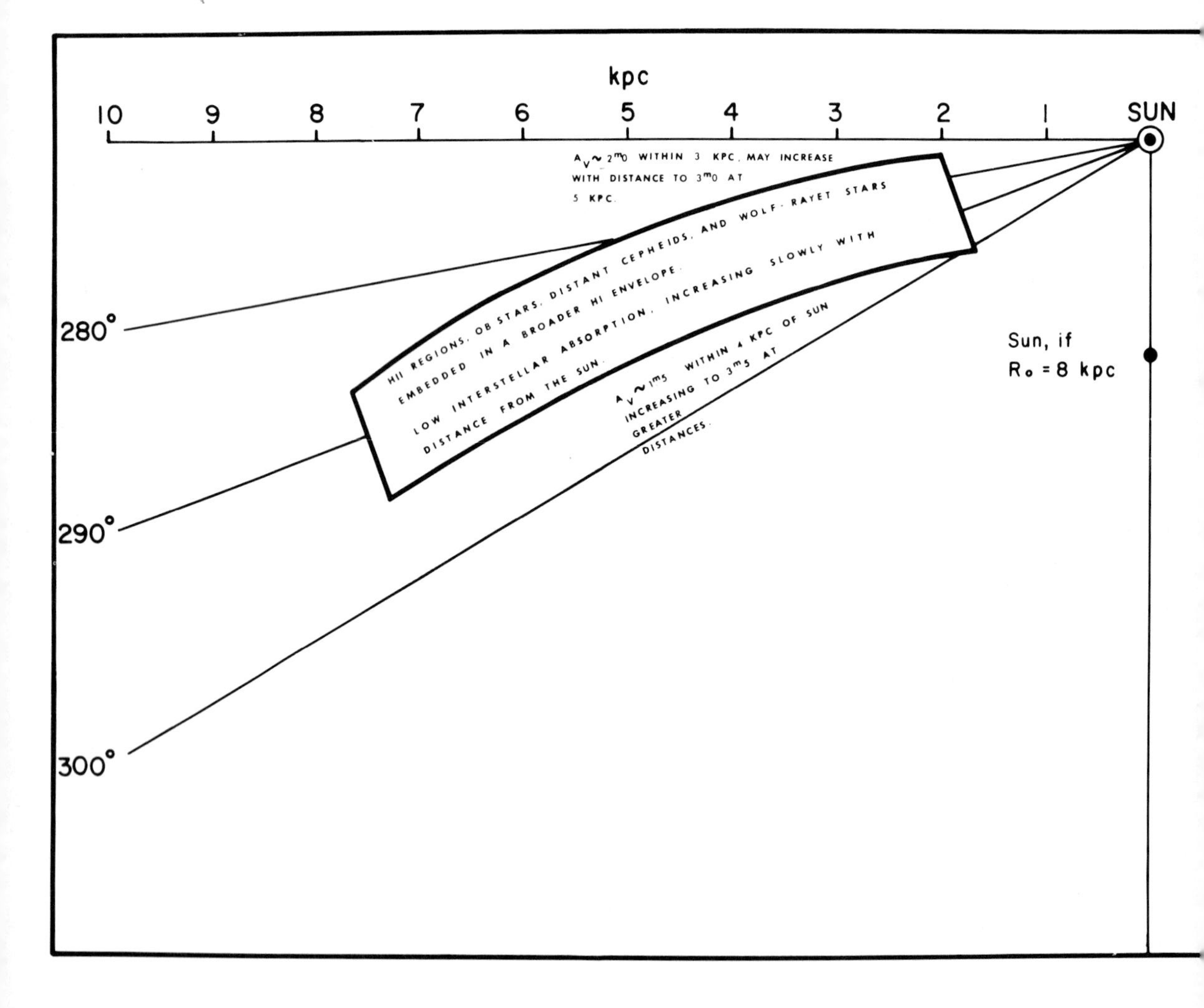

the meantime, much valuable research is being done on the detailed properties of isolated spiral features. The happiest aspect of present-day studies of galactic spiral structure is that the problems are being studied simultaneously from many angles. The theorists and users of large computers are hard at work on many aspects of the problem. Observing optical and radio astronomers consult and argue with one another uninterruptedly, and the astrophysicists are probing as best they can into the physical conditions of interstellar matter in spiral arms and in the regions between spiral arms. The full study of the spiral structure of the Galaxy is providing a stimulus for astronomical research on many fronts.

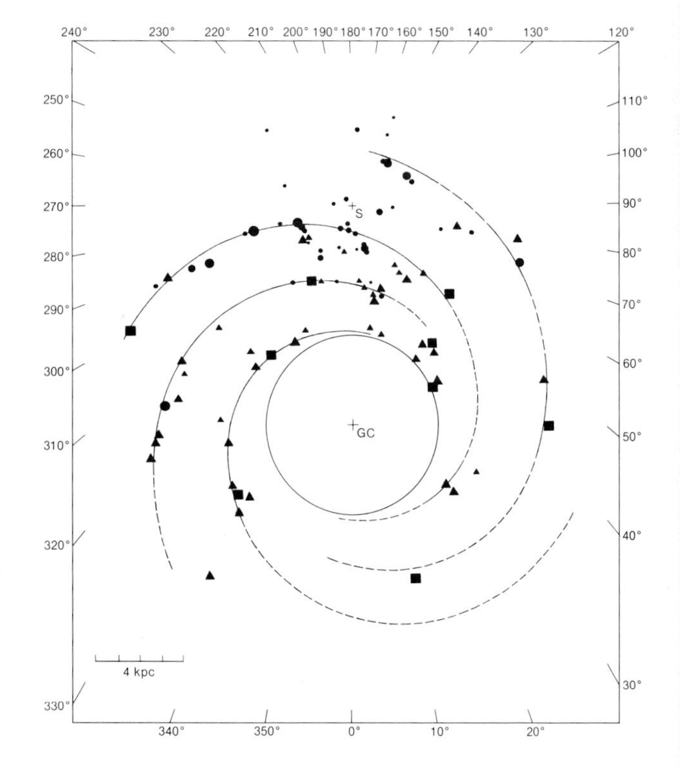

125. A 1975 diagram of galactic spiral structure prepared for the doctoral thesis (Université de Provence) of Yvonne Georgelin. The closed black circles represent the known optical HII regions, the sizes of the circles indicating the importance and strength of each emission region. The black squares mark the positions of the strong radio HII regions. The black triangles of two different sizes indicate the positions of the weaker radio HII region. The general pattern reminds one of a tightly wound system of approximately logarithmic spiral features. Galactic longitudes are shown along the outer rim. The scale in kiloparsecs is shown in the lower left-hand corner. The diagram is centered upon the galactic center (GC) and the position of the sun is shown (S). No data are given for the inner circle with a radius of 4 kiloparsecs.

11 Our Changing Galaxy

Large-Scale Changes

Our Milky Way is changing constantly. The changes are slow and gradual and not easy to detect, for on the cosmic scale our span of life represents but a fleeting moment. Whereas the year serves as a convenient unit for time measurement on the Earth and for the recording of the changes in the positions of the planets, we shall have to turn to a larger unit in considering cosmic evolution. For this purpose we choose the *cosmic year,* which is the time of one revolution of our Sun around the center of the Galaxy. One cosmic year measures 250,000,000 of our familiar solar years.

What changes take place if we think in terms of cosmic years? The stars of the Galaxy are continually being reshuffled, for as small a difference in velocity as 1 kilometer per second will serve, in the course of 1,000,000 years ($^1\!/_{250}$ cosmic year), to separate two stars by 1 parsec. In the course of 1 cosmic year, groups of stars may be dissipated, new groups may be born, and the appearance of the Galaxy may undergo profound changes.

In addition to these purely mechanical changes, the physical make-up of our Milky Way system will not remain the same over intervals of the order of 1 cosmic year. In the deep interiors of the Sun and stars the nuclear energy sources are constantly at work, but they are not inexhaustible and the rates at which they produce radiant energy may in the long run vary considerably. The O and B stars use up their large yet limited supplies of nuclear energy at such prodigious rates that their lavish displays cannot persist for more than a fraction of a cosmic year. It seems almost certain that they were formed rather recently on the cosmic scale, that is, much less than 1 cosmic year ago. The intrinsically brightest supergiant stars have probable ages since birth of the order of 10 million years or less, which is a small fraction of a cosmic year. We are thus led to consider not only stellar evolution, but also processes of star birth. What role do the interstellar gas and dust play in the formation of new stars and to what extent do the stars replenish their not inexhaustible supplies of stellar matter? These are some of the questions that will concern us in this concluding chapter. First, we shall inquire into problems related to the appropriate time scales for cosmic evolution.

The fact that the individual stars are in rapid motion does not by itself imply that the system as a whole is changing appreciably. Red and white blood corpuscles are always racing through our veins and arteries, but the amount of blood in any particular vein does not as a whole vary greatly in the course of days or months. In the same fashion, the configuration of a knot in a spiral arm might remain the same over long periods of time; while some stars move out of a particular grouping, other stars may replace them.

If the distribution of positions and motions in a galaxy suffers no large-scale changes, we say that the stellar system is in *dynamic equilibrium*. Even if we should disregard real evolutionary changes—which we can certainly not do when we are dealing with the O and B stars—the state of dynamic equilibrium will not be a permanent one. For a particular group of stars, the state of dynamic equilibrium would probably survive for as long as 100 cosmic years, but not for 1,000 or 10,000 cosmic years. Chance encounters with passing stars will begin to have their effects over such long intervals of time and the state of dynamic equilibrium will be upset, to be replaced by the more thoroughly mixed state of *statistical equilibrium*, which is dynamically semipermanent. Detailed predictions for the future are not possible, but certain trends may be demonstrated to exist and from them we may gain an insight into the time scale of dynamic evolution. We may, with considerable hesitation, attempt even to probe back into the dim past of the origin of the Galaxy.

When Stars Meet

How often, on the average, does a star come sufficiently close to our Sun to change the Sun's course appreciably? We know fairly well the average distances of separation and the velocities of the stars. We can compute without much difficulty how often a star will pass the Sun at a minimum distance less than that of Neptune. Such close approaches are rare; the Sun will, on the average, have such an encounter only once in 10,000 cosmic years, which means that it is unlikely that the Sun has suffered this sort of encounter with a passing neighbor during the 20 cosmic years of its existence. The total effect of the close passage probably would be rather minor and, although the event would be front-page news for a while, no permanent harm would be done. The orbits of the planets around the Sun would be changed, especially those of the outer planets, and Neptune or Pluto might join the interloper and leave the solar system. After the visit the Sun with its flock of remaining planets, and possibly with a few additions, would move in a direction that would deviate by about 20° from its original path.

What are the chances that the Sun might be hit by a passing neighbor? If that were to happen the results would be almost certainly catastrophic to life on Earth. But we probably have more immediate worries ahead of us, for the average interval between actual collisions of stars the size of the Sun is of the order of 1,000,000,000 (10^9) cosmic years or 250,000,000,000,000,000 (2.5×10^{17}) solar years! The only place where collisions might occur once in a while is in the central nucleus of the Galaxy, but even there the probability is not great.

The rarity of the phenomenon is best illustrated if we note that in our galactic system, with its mass equivalent to 100 billion (10^{11}) Suns, a collision between two stars should occur only once in 1 million years. If we consider that there are some regions of the Galaxy where the stars are closer together than in our neighborhood and that some stars have much larger target areas than our Sun, we might concede a somewhat higher probability for collisions. It is, however, very unlikely that even in a globular cluster, or in the dense nucleus of a galaxy, an actual collision will take place more frequently than once in 1,000 years.

From these simple computations it would seem that our chances of being put out of commission through a direct hit or close approach are very much less than the continual risk of being destroyed by comparatively small internal changes in the Sun. As small a persistent change in the Sun's total brightness as 1 magnitude would automatically result, in a fairly short interval of time, in a change of the average temperature of the earth by 75° centigrade. Such a change would not alter the planetary system as a whole, but it is improbable that life on the surface of the Earth could adjust itself to an average temperature around the boiling point of water or to conditions prevailing at 50° below zero centigrade. A mild variation in our Sun, something very much less drastic than a nova explosion, could easily put an end to all life on Earth.

But let us get back to our subject—stellar encounters. We have found already that spectacular close approaches are probably rather insignificant. We should, however, not forget that the path of our Sun is continually being changed to some extent by more distant passages. A single star passing at a distance of 0.5 parsec will change the direction of the Sun's motion by somewhat less than 1 minute of arc. In the course of time the number of encounters within a few parsecs of our Sun is, however, quite large. Computations show that the cumulative effect of all encounters at minimum distances less than 3 parsecs between our Sun and other stars will, in the course of 1 cosmic year, be about equal to the total effect produced by a single encounter of the Sun and another star at a distance of 100 or 200 astronomical units. The greater importance of the cumulative effects of many unspectacular distant passages is demonstrated by the fact that a single encounter at a distance of 100 astronomical units should on the average happen only once in 2.5×10^{13} solar years or 100,000 cosmic years. It is quite clear that we might as well forget about the single "close" encounters and the head-on collisions and remember that the distant passages, because of their far greater frequency, are more effective in producing changes in direction and speed of the Sun's motion over long intervals.

Our Galaxy has not been rotating sufficiently long for the interchange of energy between stars of different types to have become effective. From our considerations of stellar encounters it seems very unlikely that the stars would still show so much individuality in their motions if our Galaxy had existed in its present form for much longer than 100 cosmic years = 2.5×10^{10} solar years.

The Evolution of a Star Cluster

Our part of the Galaxy is characterized by the presence of many loosely connected open clusters. Consider, for example, such clusters as the Hyades and the Pleiades. From our considerations of the disruptive effects of stellar encounters, these clusters would hardly seem very stable objects. The Hyades are relatively so close to us that we have not only accurate measurements of the motions of the brighter members, but also a rather complete census of the total membership of the cluster. The densest part of the cluster lies 40 parsecs from the Sun. The moving cluster contains about 150 members within a distance of 6 parsecs from its center. The speeds of the individual stars differ by not more than 0.5 kilometer per second from the average speed of the cluster as a whole.

We can predict what will probably happen to the Hyades cluster in the course of the next 10 or 20 cosmic years. The cluster ought to stay fairly close to the plane of our Galaxy and move therefore in a region where the stars are probably spaced much as they are in the vicinity of the Sun. We can easily compute how frequently a star that does not belong to the cluster, a field star, will pass through the cluster or how frequently one of them will pass by at a given distance.

Before we trace the effects of encounters on star clusters, we should first look a little more closely at the mechanics of a cluster free from intruders and passers-by. A cluster star that wanders away from the rest of the group will generally be pulled back by the attraction of the whole mass. The cluster tries desperately to preserve its unity. But there is a villain in the piece! We know from observations on galactic rotation that the stars in the vicinity of the Sun are all subject to the pull of the galactic nucleus. The parts of the cluster that are closest to the central nucleus of the Galaxy will feel more of the nuclear pull than those that are farther away. The general galactic forces will therefore tend to shear the cluster apart. There results a regular tug of war. Can the force of attraction produced by the cluster as a whole pull hard enough to counterbalance the disrupting "tidal" forces from the galactic nucleus? If the answer is yes, the cluster will hold together, but if not, the cluster may soon be disrupted and its members strewn far and wide.

It takes some mathematical juggling to find where the dividing line will lie between clusters that can and those that cannot withstand the shearing forces of galactic rotation. A certain average critical density is computed. A cluster for which the average star density is less than the critical value cannot possibly stay together, but one with an average density above the critical value will generally have enough internal gravitational attraction to withstand the insidious disruptive forces from the galactic nucleus. A cluster for which the average density is equivalent to 10 solar masses for a cube 3 parsecs on each side should stay together if, in the course of time, it does not come much closer to the galactic nucleus than does our Sun.

I. R. King has made extensive studies of the dynamics of open and globular star clusters. He concludes that the dynamic future of a star cluster is generally established at the time of its formation from the gas and dust clouds of the Galaxy. These initial conditions determine the total number of stars of all varieties in the cluster and the degree to which the stars are initially packed; the radius of the young cluster is thus fixed. The stars move about inside the cluster with a distribution of velocities that settles down in a cosmically short time to a random Maxwellian type of distribution, and the main body of the cluster can stay that way for a long time. Matters are not so cheerful for the outer parts. Each stable star cluster has an outer boundary—the *tidal limit,* as King calls it—and the stars that lie beyond that limit will not be retained by the cluster, but will be torn from it by the differential shearing forces—the tidal forces—of the galactic nucleus. Only inside the tidal limit will the cluster exert enough total attraction to hold on to the member stars. King has developed methods for fixing the tidal limits rather precisely; they are in effect the boundaries of globular clusters. If the globular cluster moves in an elongated orbit around the galactic center (as most globular star clusters apparently do), the tidal limit is set at the position in the orbit where the maximum galactic tidal forces occur, which is at the position in the orbit of closest approach to the galactic nucleus.

What will be the probable effect of encounters between clusters and stars of the field? The dimensions of the clusters will in general be large and the passing field star will therefore not affect all cluster members in the same way. Two cluster stars that were originally pursuing strictly parallel paths will probably move after the encounter in slightly diverging orbits. The net result of each encounter will be to loosen up the cluster. This will tend to weaken the cluster's attraction on its members and the general galactic tidal force may then get a chance to do its disruptive work.

The Hyades cluster is probably dynamically safe for at least 4 more cosmic years or for 1,000,000,000 solar years. By that time the encounters with field stars will have done enough preliminary softening up to bring the cluster to the brink of disruption. About 5 cosmic years hence the galactic tidal force will become strongly effective. We may as well predict that 2,000,000,000 years from now we shall look in vain for the remains of the Hyades cluster wherever we may search. The remaining aging stars will be scattered far and wide!

We now turn to the more compact clusters, such as the Pleiades and Praesepe, whose average densities are about ten times that of the Hyades. These clusters are in a far better state than the Hyades to withstand the disruptive shearing forces of galactic rotation; the time scale for disruption from this cause alone is probably of the order of 50 cosmic years. But here a different process is at work: the denser Pleiades and Praesepe will probably suffer a gradual internal collapse as a result of the escape of many members.

The process of collapse was first suggested in 1937 by Ambartsumian and independently in 1940 by Spitzer. It operates as follows: in a fairly dense cluster encounters between two cluster members will be considerably more frequent than encounters between a cluster member and a field star. On occasion a cluster member involved in an encounter with one of its fellow members may acquire a velocity large enough to permit it to escape—to "evaporate"—from the cluster. The escaping star will carry with it more than its fair share of the available total energy of the cluster, leaving the remaining stars with a somewhat reduced average energy. When there is less energy per star, the star members will not move quite so far from the center of the cluster as formerly. Hence the cluster will shrink. In the course of time many stars will escape, each of them taking with them more than their share of the total energy and, as the depletion progresses, a gradual collapse of the cluster will ensue.

The rate of escape will be slowed down somewhat, as Chandrasekhar has shown, because a *dynamic friction* is exerted by the slower-moving stars upon the escaping stars. However, the total effect, after correction for dynamic friction, still leads to the gradual dismemberment and slow collapse of clusters like the Pleiades and Praesepe. The predicted times in which purely internal collapse should take place are of the order of a few billion solar years—20 cosmic years, more or less.

The dynamic evolution of a star cluster will be affected by two additional processes:

(1) We have so far considered only encounters between stars. However, not infrequently a star cluster must pass close to—or through—a cloud of cosmic dust and gas. The masses of the known clouds of interstellar gas and dust run into thousands of solar masses, and, when a cluster of stars passes near one of these gas and dust complexes, there must be all sorts of disturbances at work. The gas and dust complex may produce far more disturbing tidal forces than those originating from stars or from the galactic nucleus. The life of the cluster may be severely shortened by a close passage involving it and a massive interstellar cloud. The times involved in such disruptions may be cosmically brief, but the effects should be startlingly disruptive.

(2) The second process that may affect the dynamic history of a star cluster is another wholly internal physical one. More and more, we are seeing evidence that stars lose mass by escape of gases from their atmospheres, especially in the later stages of stellar evolution. A large percentage loss of mass will make a star more susceptible to encounter effects. The lighter-weight star may pick up enough energy of motion from its neighbors and from passers-by to fly away from its parent cluster.

From a purely dynamic point of view, the time scale for important changes in the open clusters of the Galaxy seems to be of the order of 5 cosmic years. At the moment the Hyades, the Pleiades, and Praesepe are among the more conspicuous local features of our Galaxy. If we were to return in 10 cosmic years, these clusters would either have evaporated or collapsed—and there seem to be no others slated to take their place. One might be tempted to think about dismembered globular clusters as possible future Pleiades-like clusters, but two considerations show how impossible this would be. In the first place, globular clusters, with their characteristic spectrum-luminosity diagrams, cannot change into clusters with Pleiades-like color-magnitude arrays. Second, calculations readily show that the rate of evaporation for globular clusters is far too slow to lead to major changes through escapes in 10 or even 100 cosmic years. Dynamically, the globular clusters are among the most stable features of our Galaxy.

Might it be possible that other clusters not unlike the Hyades and Pleiades are now being formed from field stars? From a purely mechanical point of view it seems very unlikely that a workable mechanism could be found for the building up of clusters by chance encounters of unattached stars. However, it seems probable that new open clusters are being formed steadily by the process of the breaking up of clouds of interstellar gas and dust, followed by the collapse of the separate small clouds, or globules, into the stars that are the members of the newborn clusters. Such processes are rather likely to occur. It has been suggested by McCrea and others that there may be a compensating process of formation and of evolutionary decline of open clusters and associations of O and B stars, with the number of clusters and associations presumably not varying much in the course of time. For those open clusters with stars of spectral types A and later, the rate of depletion seems faster than the rate of formation of new clusters, and these may be a vanishing species.

We have, of course, no information concerning the numbers of open clusters that there were in the sky at the time when the first cockroaches began to march over the face of the Earth, a time which, according to the paleontologists, dates 1 or 2 cosmic years back. We know even less about the make-up of our Galaxy at the time of the birth of our Earth—an event that happened presumably some 20 cosmic years ago. The rate of disappearance of the open clusters that we know today is so high that it would seem extremely unlikely that the Galaxy could have existed in its present form for much longer than 50 cosmic years.

The Expanding Universe and the Cosmic Time Scale

Before we deal further with the physical problems of star birth, ages, and evolution, we should inquire whether there is not some reasonably well-established upper limit to the age of the universe of galaxies and to the ages of the stars in these galaxies. Such a limit is indeed set by the probable age of the expanding universe of galaxies. In the present volume we are primarily concerned with our Home Galaxy, and we must refer the reader to the Shapley-Hodge volume *Galaxies* (1972) for a broad discussion of the expansion of our universe and the related cosmic time scale.

Indications are that the interval between the beginnings of the expanding universe and the present covers 10 to 15 billion years. This total age is just about of the right order for the dynamic evolution of our Galaxy to fit into the scheme of things. Sixty years ago, V. M. Slipher showed from his studies at Lowell Observatory in Flagstaff, Arizona, that some of the fainter galaxies, which he called nebulae, exhibit remarkable redshifts in their spectra. If these redshifts were interpreted as shifts caused by radial velocities of recession, then, so Slipher reasoned, several of his nebulae possess velocities of recession with respect to our Sun in excess of 1,000 kilometers per second. Around 1930, when it was recognized that Slipher's nebulae were true galaxies, Hubble and Humason of Mount Wilson Observatory extended Slipher's measurements to fainter galaxies. Since they could obtain approximate distances for these galaxies, they could study the universality of a redshift-versus-distance relation, which emerged from these studies.

Since the days in which Hubble and Humason did their basic work, the distance scale for galaxies has been adjusted considerably. Recent studies by Sandage and others, mostly based on data gathered with the 200-inch Hale reflector, show that the redshift-versus-distance relation is very nearly a linear one. If we assume that the redshifts are indicative of the presence of radial velocities of recession, which practically everyone accepts nowadays, then the redshift-versus-distance law becomes a fundamental one relating velocity of recession and distance. The whole of the universe of galaxies is apparently expanding; the rate of expansion is set by the fact that two galaxies 10 million parsecs apart seem to be receding from each other at a rate of about 550 kilometers per second, the value derived by Sandage and Tammann. Redshifts corresponding to radial velocities equal to half the velocity of light have been observed for normal galaxies, and redshifts corresponding to velocities of recession in excess of 80 percent of the velocity of light have been found for some very distant quasistellar sources, known as *quasars*. Hence the general expansion of the universe appears to be well established on the large scale. If the above expansion rate has more or less persisted in the past, then very simple calculations show that all galaxies participating in the expansion should have been very close together 17 billion years ago; this "age" suits the Milky Way astronomer very well indeed.

The foregoing calculation is of course a very rough one; for more precise results, one has to fit the observational data into a relativistic model of the universe of galaxies. We may summarize the situation by saying that the ages derived for various models of the expanding universe all come pretty close to our figure. The expansion does not need to have been a steady one. It may well be, for example, that our universe of galaxies began as a Big Bang, resulting from some gigantic explosion, and that the very rapid expansion rate that may have existed shortly after the explosion has begun to slow down. It is obvious that the time since the explosion, as derived from presently observed expansion rates, would then have to be less than the figure of 17 billion years that we have quoted. Also, it may well be that our universe is an oscillating one, which is at present in a stage of expansion, but which ultimately may collapse again upon itself.

There are many observations that support the hypothesis of the expanding universe. The quasars are almost surely galaxies that we observe now as they were 5 billion years or more ago. They show by their numbers at great distances that there was much more activity in the universe of galaxies 5 to 10 billion years ago than there is at the present time. Another line of support for the hypothesis that a terrific cosmic explosion actually took place about 10 billion years ago comes from an observation by Penzias and Wilson, as interpreted by Dicke. They observe the remains of the energy originally associated with the Big Band as a microwave background radiation with an effective temperature of 3°K, which is found over the whole sky. According to the best available observational evidence, we can now reach galaxies, distant quasars, to distances as great as 8 to 10 billion light-years, about 3 billion parsecs. Such observations imply that we can look back in time to objects now observed from the Sun and Earth as they were 8 to 10 billion years ago.

How was our Galaxy formed? An answer to this question can be given if we bear in mind that the oldest star clusters and individual stars are found at large distances from the central Milky Way plane in our Galaxy. This would seem to imply, according to a theory developed by Eggen, Lynden-Bell, and Sandage, that shortly after the Big Bang our Galaxy became a separate unit, a large near-spherical blob of gas. When condensation of some of the original gas into stars and star clusters began to take place, it occurred probably all through the blob. As time progressed, the gas began to be concentrated more and more toward the central Milky Way plane, which then acquired its present rotation. Younger stars and clusters were formed as the gas cloud became flatter and flatter, and we are now in the stage in which the central gas (and dust) cloud is remarkably thin. Star birth now seems to be confined entirely to the regions of cosmic gas and dust within a few hundred parsecs of the central Milky Way plane. According to this attractive picture, the oldest globular clusters and the oldest open clusters were formed first. Star birth and cluster birth have long ago ceased in the halo of our Galaxy. We are fortunate that these processes have not, however, ceased to operate near the central plane of the Galaxy, and we are even more fortunate that the Sun and Earth are in a position close to this central galactic plane, and in the outskirts of our Galaxy, the parts of the Galaxy where the evolutionary pots are still boiling nicely!

Energy Production inside Stars

Our estimates of the cosmic time scale were based on data from stellar motions and the arrangement of stars in systems. What about the stars themselves? They ought to have an important voice in the matter, for they are asked to supply energy for millions and millions of years. It is all well and good to talk in terms of 10 billion years, but can the stars keep shining for so long a time?

Other volumes in the series of Harvard Books on Astronomy, notably Lawrence H. Aller's *Atoms, Stars, and Nebulae* (revised edition, 1971), consider carefully the processes by which our Sun and other stars are continually supplied with energy for radiation. The familiar processes of chemical burning have no place in the scheme. It is so hot in stellar interiors that the atoms are stripped of most of their outer electrons. Transformations like those of molecular chemistry can take place in the atmospheres of some cool stars and in our laboratories on Earth, but not in stellar interiors. At the high temperatures prevailing inside the stars, physics looks for the source of stellar energy in the changes that must take place in the atomic nuclei. For our Sun, the interior temperatures rise steadily from 6,000°K at the surface to some 15 million degrees in the central region.

At these tremendously high central temperatures the atomic nuclei, freed from most of their customary neutralizing entourage of electrons, move at such speeds that violent changes occur when two of them collide. Hydrogen nuclei—the familiar protons—collide with other hydrogen nuclei. These nuclear collisions lead to the formation of heavier nuclei, notably of helium; in some stars this comes about through a transformation cycle in which the carbon nucleus figures prominently as a catalytic agent, in others through direct interaction of protons with other protons. We refer the reader to the Aller book to learn precisely how these nuclear transformations operate, but the important fact is that two protons and two neutrons with a total mass of 4.033 atomic mass units must combine to form a helium nucleus, whose mass is 4.004 on the same scale. A small fraction of the original mass—about 0.7 percent—is therefore lost in the course of the nuclear construction. This mass is transformed into radiative energy according to the famous Einstein equation,

$$E = mc^2,$$

which relates the mass loss m to the corresponding released energy E, where c represents the velocity of light. The transformation of hydrogen into helium is the most effective process of energy generation at work in stellar interiors. Stars that have exhausted their plentiful original supply of hydrogen have some secondary nuclear processes on which to call, but none are so effective in releasing free radiational energy. Annihilation of matter at very high temperatures would presumably be a still more effective process, but we do not seem to find generally in the universe temperatures high enough for this process to occur.

Let us see how well the stars can get along in their evolution if they depend entirely on the energy produced by the transformation of hydrogen into helium. Our Sun, an average dwarf G star, radiates at a rather low rate; it should use up about 1 percent of its mass of hydrogen in 1,000,000,000 years, that is, in about 4 cosmic years. Hence, with about 60 percent of its mass still hydrogen, the Sun has enough nuclear fuel left to shine as it does now for another 50 cosmic years or more. But the picture is not so cheerful when we turn to typical A stars, like Sirius, with absolute magnitudes 3 to 4 magnitudes brighter than the Sun. These stars send out between 15 and 40 times as much energy per unit time as does the Sun. The masses of A stars are no more than 2 or 3 solar masses, so that the energy that must be generated per unit mass is about 10 times as great as for the Sun. Such a star should use up 1 percent of its mass as hydrogen nuclear fuel within half a cosmic year or so and it can hardly have sent out energy at its present rate for as long as 20 cosmic years. Hence the A stars can barely have existed since the time of the birth of our Earth, and, if they were formed that long ago, they must be getting near the end of their available supplies of nuclear hydrogen fuel.

What about the O and B stars, which may shine with intrinsic brightnesses equivalent to 100,000 Suns and which have masses less than 100 times that of the Sun? Here the nuclear processes must be producing energy at rates 1,000 times as fast as in the Sun and the total supply of nuclear fuel should be exhausted in a fraction of a cosmic year. Hence the O and B stars must be of very recent origin, and we are forced to look—a not unpleasant task!—for places in our Milky Way system where the process of star birth may be observed today. Hoyle, Bondi, McCrea, and others have suggested that massive stars may replenish their exhausted energy sources by accretion, that is, by the capture of matter from interstellar space. This process is probably not a very effective one. In order to acquire any appreciable quantity of matter by accretion, a star must be practically at rest relative to the interstellar medium, and the radiation pressure it exerts on the particles of the medium must not be sufficiently great to counterbalance the tendency of matter to fall into the star's atmosphere. Neither condition appears to be satisfied by the spendthrift O and B stars.

We turn back to the OB associations and aggregates that were described in Chapter 4. We found several of these associations to be expanding. When we retraced their development in time, we found that the expansion in at least four well-established cases must have begun only 0.1 cosmic year ago. The short maximum lifetimes for O and B stars and the expansion phenomena in OB associations both suggest that the associations and their component stars are cosmically of very recent origin. Some were apparently formed a fraction of a cosmic year ago—a very short time compared with the age of our Earth, which is 15 to 20 cosmic years!

It is noteworthy that the O and B stars—which are known to be first-rate spiral tracers (see Chapter 10)—occur especially in parts of the Milky Way where interstellar gas and dust are plentiful (Figs. 125 and 126). It is tempting to hunt for relatively small condensations of interstellar gas and dust and look upon these as possible protostars. The small globules that are often seen projected against the luminous peripheral edges of emission nebulae (Figs. 89 and 90) appear to come closest to the protostars for which we are searching. They are found especially in the turbulent regions of high gas pressure near the outer boundaries of emission nebulae, the H II regions. Inside a globule contraction presumably will take place, possibly at first through outside pressure effects, and then

126. The emission nebula Messier 8. This short exposure with the 90-inch reflector of the Steward Observatory shows the core of Messier 8 and to the left of it a cluster of young hot O and B stars. Star formation has been pretty well completed in the section of the cluster, but there is still much gas left in the core region of the nebula.

127. The emission nebula Messier 16. This photograph shows the young cluster and the nebula known as Messier 16. The intricate patterns of overlying dark nebulosity are suggestive of the presence of shock fronts and turbulent conditions. From a Lick Observatory photograph with the 120-inch reflector by N. U. Mayall. (Courtesy of Lick Observatory.)

through the self-gravitation of the globule, which may grow slowly in the meantime through the accretion of gas and dust. This contraction should continue until the particles inside the globule shatter in collisions to become molecules and then atoms. Finally, the atomic nuclei move sufficiently fast that nuclear transformations result from collisions. When nuclear transformations occur, the collapse should slow down and gradually come to a halt; a star would have been born.

Globules and Dark Nebulae

The search for and study of protostars is one of the most intriguing tasks facing the Milky Way astronomer of the 1970's. A protostar will likely be the result of the collapse, through gravity or otherwise, of an interstellar gas cloud, possibly of a cloud contaminated with cosmic dust. Such a collapse may produce at first a very extended and rarefied star, with a radius of a few hundred astronomical units. A star can be said to have been born once the object sends out detectable radiation. It may well be that protostars are formed by the combined processes of collapse and fragmentation of the original interstellar cloud. Such a process would naturally produce a cluster of protostars.

The most obvious candidates for protostars are the small dark globules, which we discussed in Chapter 9 and in the previous section. Their probable masses are in the range between 0.1 and 1 solar mass. We also discussed larger globules, the variety that is not infrequently seen on our Milky Way photographs off by themselves, where they appear as roundish star voids against a background rich in stars (Fig. 94). Their radii are in the range 0.1 to 1 parsec, with estimated masses between 2 or 3 and 100 solar masses. They seem predestined for ultimate collapse into protostars. We also find some large unit dark nebulae, like the one near Rho Ophiuchi (Fig. 93), that have radii as great as 4 parsecs and most likely masses of the order of 2,000 solar masses. These may be slated for fragmented collapse into a cluster of protostars. Grasdalen and the Stroms may have discovered a young OB association embedded in dense cosmic dust right in the heart of the Rho Ophiuchi dark nebula.

It is encouraging to note that the future for the detailed study of globules, large and small, and of unit dark nebulae looks promising. Already observation has shown that molecular lines in the radio range are observed coming from the directions of some of the best-known unit dark nebulae. Heiles has studied the OH bands reaching us from the direction of the Rho Ophiuchi dark nebula, and he finds that temperatures of the order of 5° to 10°K prevail in the interior of this nebula. As our observational techniques are further developed, it may become possible to search for and study molecular radiation reaching us from the larger globules and from smaller unit dark nebulae. Studies of 21-centimeter radiation should produce data on the neutral-atomic-hydrogen content of these objects, which is probably small, and it is now possible to check on the presence of molecular hydrogen, which is plentiful. From intensity ratios of critical molecular lines, we are able to determine the velocities of turbulence associated with the cool gas. Before long, we should be able to present the theoretical astrophysicist with data on the total masses, composition, temperatures, and turbulent motions in many of these objects. There are further contributions that the observational astronomer is making. For example, in the Rho Ophiuchi cloud (Fig. 93) star counts by one of the authors have shown that there is an observable and well-established density gradient for the

128. The globule Barnard 335. The small and roundish dark nebula is seen projected against a rich star field. Its diameter measures 4 minutes of arc, which, at a not improbable distance of 300 parsecs, amounts to about one-third of a parsec, 70,000 astronomical units. Radio studies of the small dark cloud by Palmer, Rickard, Zuckerman, and Buhl show that this globule contains formaldehyde, H_2CO. No stars are observed that lie beyond the globule.

cosmic grains in the cloud. This particular cloud resembles in many ways a globular cluster of cosmic grains! Carrasco and the Stroms have made a comprehensive photometric study of the Rho Ophiuchi region. They find that the sizes of the cosmic grains in the cloud are greater than average for the interstellar medium. They ascribe this to depletion by sticking of heavy elements onto the grains. A reasonably strong magnetic field (10^{-4} gauss) appears to be associated with the cloud. They estimate that the cloud has existed as a unit for close to 1 million years.

We are now gathering increasing evidence to show that some of the larger nebulae are observed in the act of breaking up into many units. The finest example of such an event is seen in the Southern Coalsack, where Tapia has found that a fragmentation process is already in progress. The Coalsack is not a near-spherical ball of cosmic dust, acting as a single unit, and neither is it a thin sheet with a variable surface density for the grains. Instead it appears to be a conglomerate of smaller units, some of them remarkably dense already at the present stage, and presumably headed for ultimate collapse.

Star Formation and Spiral Arms

The stars commonly found associated with spiral features are without exception very young. As we have seen, typical ages since formation are in the range between 10 and 25 million years, well below 1 percent of the age of the Sun and Earth. It is therefore only natural to look for evidence of continuing star birth as a phenomenon related to spiral structure. We spoke earlier of the Lin-Shu-Yuan density-wave theory of spiral structure being the one that currently is favored by many workers in the field. W. W. Roberts has shown that a high-pressure shock wave accompanies each density wave. Therefore, if clouds of cosmic dust and gas of more than average density are present in the interstellar material, they may be compressed by the shock wave to five or ten times their original density, possibly beyond the critical stage necessary to form protostars. These shock waves may serve as a trigger mechanism that sets off the processes of star birth along a spiral arm. Roberts has shown that the conditions for the piling up of dust and gas are most favorable along the insides of the spiral arms, and here is precisely where we observe the early beginnings of star formation. One naturally expects to find protostars and newborn stars along the inner edges of a spiral arm; the spectacular O and B giant stars are the brilliant and slightly older objects that mark the central line of the spiral arm.

The distribution of interstellar gas and dust is far from uniform in our Galaxy. Hence, as the shock wave moves through the interstellar medium, it will inevitably pass through some regions of below-average density, regions where the compression of the shock wave is insufficient to produce conditions of collapse followed by the formation of a protostar. It is therefore quite understandable that along a spiral arm we do not find a smooth distribution of young clusters and associations of stars. We should expect to find voids between the obvious concentrations of protostars and young objects. This picture of star formation seems to have very strong observational backing; the various stages are as plainly visible to the eye as are plants in successive stages of development in our garden.

Supernova Explosions

Supernova explosions are among the most spectacular of celestial phenomena. In 1054, Chinese astronomers observed a gigantic supernova explosion, in the position where we now find the beautiful Crab Nebula (Fig. 83). There is a cinder left behind, which is observed as a radio and optical pulsar and which appears to be a totally collapsed neutron star, rotating on its axis in the incredibly short period of $\frac{1}{30}$ second. The Crab Nebula is at a distance of about 2,100 parsecs from our Sun and hence strictly an object belonging to our Milky Way System. During 1971 attention was drawn to another supernova pulsar, this one located about 45° south of the celestial equator in the constellation Vela. This explosion appears to have produced aftereffects detectable in the sky up to angles of 30° or more from the central pulsar. The Vela pulsar and its associated Gum Nebula (Fig. 84) are at about one-fifth the distance from the Sun to the Crab Nebula, 500 parsecs.

A supernova explosion must have two immediate effects on the surrounding interstellar medium. First, considerable amounts of gas enriched in heavy elements are added to the interstellar medium around the supernova; second, tremendous amounts of energy are transmitted to the surrounding interstellar medium in the form of explosive shock waves. The neutral atomic hydrogen that must have been present in the surrounding interstellar medium before the supernova outburst would obviously have become ionized by the outpouring of ultraviolet energy. It should be noted that the supply of fresh energy is by no means exhausted a few hundred or 1,000 years after the original outburst, for the tiny rapidly rotating neutron star continues to pour energy into the interstellar medium. We remind the reader that pulsars and supernovae are potentially most productive sources of cosmic-ray particles with very high energies, which help maintain the ionization of the surrounding medium.

One of the most striking optical features of the Gum Nebula is the presence of a remarkable filamentary structure, the sort of effect that one would expect from the passage of energetic shock waves through the surrounding interstellar medium (Fig. 84). It has been suggested that star formation may well have taken place, and may still be taking place, in the region of the Gum Nebula. Several stars of high luminosity, which are apparently young on the cosmic scale of time measurement, have been marked as possibly having originated at the same time as the star that later on produced the supernova pulsar; some of these are typical "runaway" stars. The gaseous filaments now visible on our photographs must represent highly condensed gas, and it would not be surprising if some of these filaments were to break up into strings of young stars, or protostars. Many workers in the field consider it likely that the supernova phenomenon may be conducive to the formation of protostars, but it is a bit uncomfortable that we do not see any sort of mass production of protostars taking place before our eyes and neither do we find near supernova remnants an abundance of cosmically young O and B stars.

It would be naive to look for an abundance of protostars in the regions close to recent supernovae. The collapse of a gas cloud into recognizable protostars or young massive stars, singly, in clusters, or in associations, is a process requiring at least 100,000 years, in most cases probably as long as 1 to 10 million years. The Crab supernova explosion was observed in 1054, less than 1,000 years ago, and the supernova explosion at the heart of the Gum Nebula cannot have taken place much longer than about 30,000 years ago. Even the youngest observed hot stars in the region near the Gum Nebula must have antedated the supernova explosion by hundreds of thousands of years. The condensations from which these hot stars were born must have originated from explosive events that took place long before the recent supernova outburst that produced the existing pulsar. However, there is little doubt that recent supernova outbursts must have produced conditions favorable for future star formation in their surroundings. And young stars may still be formed today through the collapse of gas concentrations caused by supernovae of the past that are no longer detectable.

The Formation of Stars

Interstellar clouds have been studied by the radiation they emit in several quite different regions of the electromagnetic spectrum: at x-ray, ultraviolet, visible, infrared, and radio wavelengths. Such studies have yielded basic information on the properties of clouds, which are presumably on their way to becoming protostars. A variety of dark clouds and infrared objects have been discovered and classified. On the basis of such information we can ask: What kinds of mechanism are responsible for the development of a protostar and its collapse into a star?

In a recent survey of theories of the formation of stars, McNally of the University of London lists several processes that may be at work; in the end he favors star formation by collapse, with gravity as the major cause. His conclusions are generally confirmed by the studies of other astrophysicists, notably Hayashi and colleagues of Japan. Larson of Yale University has drawn special attention to a process by which one of two things could happen: either a cloud will collapse into many different units and form a cluster of stars, or it will collapse much faster near the center than in its outer parts. The second alternative means that a star would be formed mostly from the material near the center of the cloud, and that the young star would be embedded in a large envelope of dust and gas. Such a star would have a truly murky atmosphere! Much attention is being given to the manner in which the extended atmosphere might collapse. It seems likely that it would rotate and that this rotation would play an important part in holding it up for some time. In one way or another the protostar must get rid of the angular momentum that is stored in the rotating cloud of gas and dust. It can do so most readily by forming dusty shells around itself and these in turn may well break up into planets. The theory of the formation of a protostar from a cloud of gas and dust seems to lead almost naturally to the formation of a planetary system.

Recent investigations of infrared objects strongly support the theory that stars are formed from collapsing clouds of gas and dust. Becklin and Neugebauer of the California Institute of Technology have discovered an infrared point source near the heart of the great nebula in Orion that is almost surely a very young star. Low and Kleinmann of the University of Arizona have found a second object close to the same region. This infrared source seems to be a compact dust nebula, probably with a newborn star or cluster of stars near its center.

Evidence of nebulae enclosed in dust shells has been forthcoming from radio-astronomical observations as well. Mezger and his colleagues have found a number of emission nebulae that emit strongly at radio wavelengths but are not detectable at visual wavelengths. The hypothesis is that these are "cocoon nebulae": brilliant nebulae embedded in clouds of interstellar grains. Their radiation in the radio region can pass through the surrounding

129. Evolutionary tracks of pre-main-sequence stars. A diagram prepared by C. Hayashi to illustrate the development of pre-main-sequence stars with masses between 0.05 and 4 solar masses contracting to the main sequence. Vertically he plots the logarithm of the intrinsic luminosity of the star in terms of the sun's luminosity and horizontally the logarithm of the surface temperature of the star. (From *Annual Review of Astronomy and Astrophysics,* 4:189, 1966. Copyright by Annual Reviews, 1966. All rights reserved.)

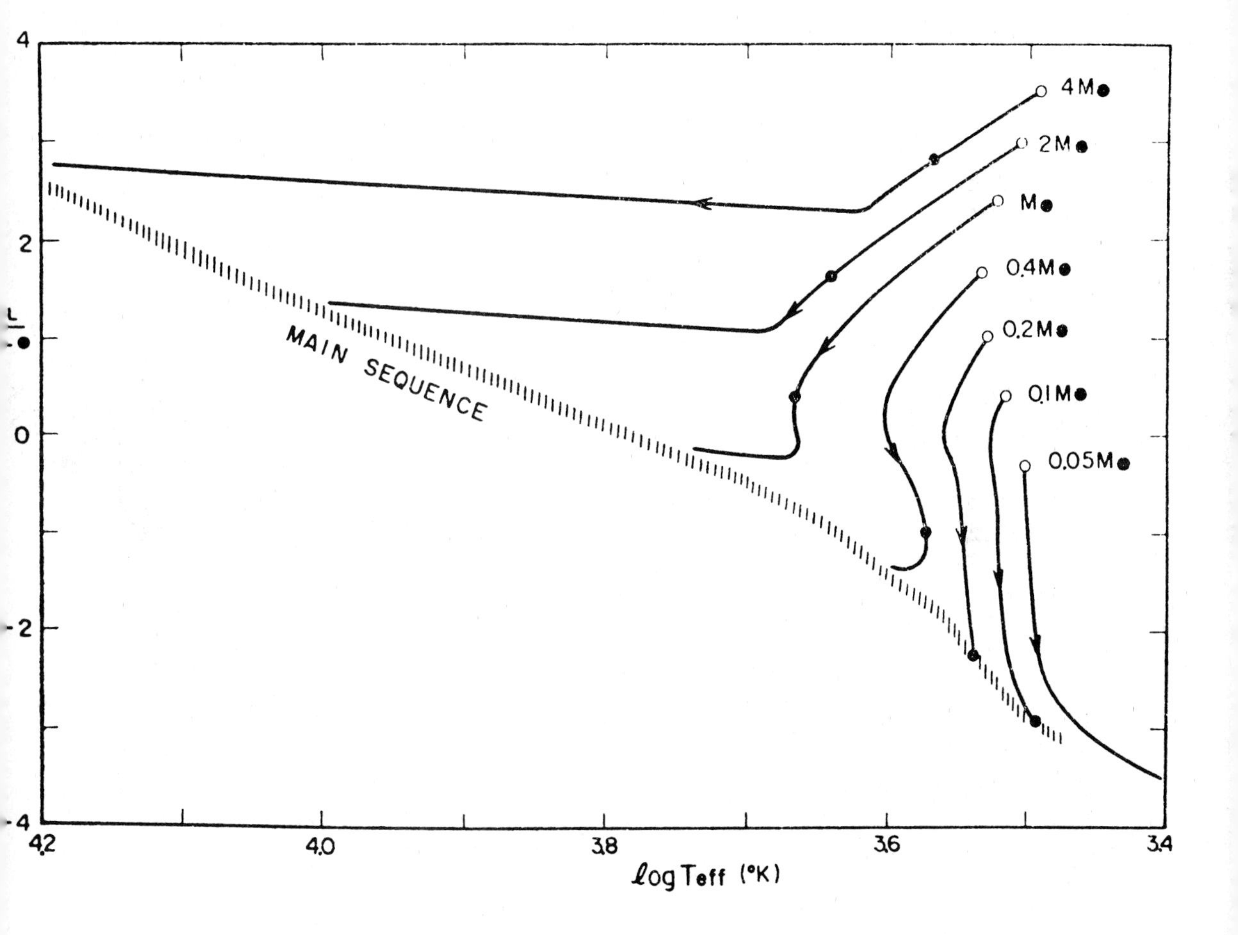

130. Nebulosity in Monoceros. This cometary nebulosity contains a dark globule near the head of the cometary structure. The globule measures about 1 minute of arc in diameter, which at the distance of the associated nebula NGC 2264 amounts to about one-quarter of a parsec, 50,000 astronomical units. There is a strong infrared source, discovered by Allen, directly above the globule. Zuckerman, Turner, Palmer, and Morris have found formaldehyde, H_2CO, in this region. From a photograph in red light with the 200-inch Hale reflector. (Courtesy of the Hale Observatories.)

dust clouds, but that in the visible region cannot. Some might be observable in the infrared.

A class of intrinsically faint stars, the T Tauri stars, almost certainly represents a very early stage of stellar evolution. T Tauri stars vary irregularly in their energy outputs, and their spectra show strong emission lines that are presumably produced in their extended outer atmospheres. The spectra of such stars also exhibit absorption lines that are formed deeper in the atmosphere. The lines are broad and fuzzy, indicating that the stars are either rotating rapidly or continually ejecting mass. There is much evidence to support the hypothesis that gases are steadily flowing out of the atmospheres of T Tauri stars. Such stars are most often found in groups, generally at the edges of or within dark nebulae. The fact that they cluster together is so marked that Ambartsumian gave them the name of *T associations.* The Mexican astronomer Mendoza has found that T Tauri stars are strong emitters of infrared radiation.

Herbig of Lick Observatory has apparently observed the formation of one truly new star, FU Orionis, which suddenly appeared on the scene in 1936. Quite recently Haro of Mexico has drawn attention to a star that behaves very much like FU Orionis: the faint variable star V 1057 in Cygnus. This star has recently flared up and is now very bright in the infrared. Haro expresses the opinion that FU Orionis and V 1057 Cygni were originally T Tauri stars and that now they have advanced to the next evolutionary stage, which is represented by a characteristic long-term flare-up. Herbig and Haro have discovered a number of small bright nebulae (referred to as Herbig-Haro objects) that are interspersed with the edges of dark nebulae, mostly in regions where T Tauri stars are abundant. T Tauri itself, the prototype for which the class is named, is embedded in such a nebula.

The general scheme we have described here is one that favors the formation of protostars through the process of gravitational collapse in clouds of interstellar gas and dust. We should point out that not every astronomer and astrophysicist favors this scheme. Layzer of Harvard has developed the following theory of the related formation of stars and galaxies. In the beginning the mass of the universe was distributed quite irregularly. Fragmentation and clustering took place on a large scale, and there were some blobs of hot, dense plasma (ionized gas) in which the gravitational field was much stronger than average and in which there were also large electric fields. Layzer considers conditions in these blobs as conducive to star formation. Another position has been taken by Ambartsumian. As we mentioned in Chapter 10, he considers it likely that violent explosions in the nuclei of galaxies (including the nucleus of our own Galaxy) may have much to do with the origin and maintenance of spiral structure. He suggests that associations of young stars along with interstellar gas and dust would be a direct result of the ejection of material by such explosions. So far neither his theory nor Layzer's has been developed to the point where it can be checked in detail by observations.

It seems quite natural that star birth should be occurring now in the spiral arms of our own and neighboring galaxies. Many dark nebulae and globules composed of interstellar gas and dust are seen almost in the act of collapsing into protostars or their close relatives. Objects that are either protostars or very young stars have also been observed. Infrared objects provide a natural link between small dark clouds and relatively normal stars. Quite a few of them may be cool, dense dust clouds with a star or a cluster of stars near the center. The cocoon nebulae may also supply new-born stars. The T Tauri stars seem to be the next stage and they help to bridge the gap between the protostars and the young stars.

Comprehensive research is continuing on the problems of change in our Galaxy and on the related questions of the birth of stars and their early evolution. It is a good thing to describe the Galaxy in all its majesty and to study the properties of its many components. The final aim, however, goes further. Deep in our hearts we want to know how the Galaxy came into being, how the stars were formed, and what is the future of the Milky Way system.

Index

advice FROM THE doggy lama

by BOB LOVKA

Photographs by Alan & Sandy Carey

Irvine, California

Nick Clemente, Special Consultant
Karla Austin, Project Manager
Ruth Strother, Editor-at-Large
Michelle Martinez, Editor
Lisa Barfield, Designer

Library of Congress Control Number: 2003102988

BowTie Press®
A Division of Bowtie, Inc.
3 Burroughs
Irvine, California 92618
Printed and Bound in Singapore
10 9 8 7 6 5 4 3 2

—INTRODUCTION—

Oh, to live your dog's life! No freeways, no office managers, no foreman, no boss—just lots of doing whatever comes naturally. Dogs seem to know what's important in life and what's okay to let slide by. They focus on the simple things, and never get stressed out over what tomorrow may bring.

Many of us are not aware that most dogs subscribe to the wisdom of a spiritual leader (the Doggy Lama) who offers advice to help the world work a little better. To the untrained eye, the signals our dogs send us may not be obvious, but by carefully studying this little book and taking the Doggy Lama's lessons to heart, you will find that your dog is telling you a lot about happiness, love, positive thinking, and relaxation. It doesn't have to be a dog-eat-dog world. Try a dose of simple advice from the Doggy Lama!

BOB LOVKA

The Dog Listener

Teach—Don't Preach

Everybody looks for a little guidance and advice, but nobody wants to be told what to do. Many times it's not what you say but how you say it that counts. Take the gentle paw approach and lead by example.

The most valuable thing you can do for yourself and others, too, is to be the best you can be!

Face It Now and Get It Over With

If you come nose-to-nose with a problem on Monday, it's best to take care of it right away because by Friday it will scare you right down to your tail! The longer you let a problem go, the tougher it becomes to solve.

Whether it's dog years or human years, there's no time to wait for the right time to show someone how much they mean to you! Actions speak louder than words. Try a little tenderness each day.

Everybody has off days, but the trick is to keep on trying. Don't get discouraged–even Lassie didn't do it all in one take!

Love is something you give; need is something you take. What it comes down to is we don't really love someone we need; we need someone we love.

A Relationship Based on Love Works Better Than One Based on Need

Be Open to Life's Little Wonders

Take time to appreciate the amazingly ordinary gifts life has to offer. When you look at life through innocent eyes you'll often find what you're looking for.

Life works in mysterious ways. There may be a reason why you can't reach that certain point in your life as fast as you want to. There could be a danger you're not ready to tackle. Sometimes you need to trust that your time will come.

Sometimes It's Best to Accept Where You Are for a While

Don't Let Your Troubles Weigh You Down

You always have a choice to either step lightly or sadly shuffle along. Look on the light side and your burdens won't feel so weighty! Whether you're a Saint Bernard or a Yorkshire terrier, only gravity can hold you down. Your spirits can always soar!

When you're down and out and just plain dog-tired, put on a happy face! Keep your focus on the beauty in every beast of a day.

Present Your Best Face to the World

Make Someone's Day!

It's important to show appreciation to others. Everyone likes to be appreciated, and you can make someone's day by showing them that they are.

Among life's gardens, you'll come across some weeds. Buck up! Every dog has his day but that doesn't mean he'll have them all!

It's Not All a Bed of Roses

Don’t Forget to Treat Yourself

Take time to appreciate yourself. A little care for the inner you can only make you better. Take pride in yourself and honor the positives!

What an amazing world it is when you live in the moment and experience all the things that are going on right now—the sounds, the colors, the scent of the breeze. Be on the lookout for the unique part of each moment.

Sometimes you'll feel lost and overwhelmed. Don't worry, it's all part of your growth experience. There will be many obstacles you can't get around, you just have to go through them.

Nobody has all the answers, but everybody has something to say. Learn to listen and you might pick up something valuable from a perspective that never occurred to you!

There's Lots to Be Learned From Other Points of View

Make the First Move!

Those who play hard to get often don't get "got." Don't beat around the rosebush with your tail between your legs—make the first move!

Just as we chase our tails, we often chase our problems around and around, never getting anywhere but dizzy. When things get to be too much, stop what you're doing, put it aside, and try again another time. To be refreshed is to be refocused.

Keep Clean; Keep Neat—What You Reflect Has an Effect

The face you put on for the world reflects how you feel about yourself and affects the way others treat you. Clean up your act, put on your best, and take on the world!

You can't force the issue in this competitive world. You can have a fit and get nowhere, or you can sit back, reassess, and let things unfold. A demand is often met with resistance while patience opens the door to wisdom.

Patience Leads to Wisdom

Find a Love to Shout About!

Give your all or nothing at all—that's what makes the best kind of love! Find the one who puts a wiggle in your tail. Don't settle for less because you don't want to be alone. When the chemistry is right, you'll feel it.

You don't need a diamond collar or a silver doggy dish to feel the joy of being alive. Find the fun and happiness in the little things, the big things will take care of themselves.

The Little Things Mean a Lot

There's No Place Like Home

You need to have a sense of belonging because the world can be a large and lonely place. You'll be at home any place you go if you care for others, keep your integrity, and have good character.

Life has its ups and downs—today the fillet, tomorrow the gristle. Don't let the highs get you too giddy or the lows pin you down. No matter what happens, know that every dog has his day!

Strive to Keep Your Cool Through It All

Realize the Importance of Friends

You can be an individual but to get the most out of life you need to share some space with others. Love, companionship, frienship, and sharing make the world a more secure and less lonely place.

Be curious and stay involved. Be on the lookout for new challenges and different ways to do things. There's an added joy when we're willing to try something new. And, be an observer—you can learn a lot by watching!

Ease up—don't take yourself too seriously! Frettin' over the past and fears of the future only waste your mind. Relax, regroup, and take advantage of your best natural medications—smiling and snoring!

You can't judge a dog by his coat. Snap judgments lead to long-term difficulties; an open mind broadens your perspective. Markings, colorings, noses, and tails are shared by everyone from beagles to bunnies—be open to discovering the individual.

Things Are Not Always As They Appear

Life's a Ball—If Only You Know It!

When you're feeling blue, nothing is fun. That's when you need to realize that you can either bounce your ball or just let it lie there. How you approach a situation determines a large part of what you'll get out of it.

What life comes down to is that we're all in it together. Everybody grows and benefits by caring, sharing, and having the idea of "we" instead of "me." Remember: We're all part of the same litter!

Bob Lovka is a Southern California-based writer whose work includes poetry, satire, humor books, and television, script, and stage show writing. Bob's connection to felines and canines keeps expanding. Homeless cats and dogs look him up constantly and have increased their appearances since Bob wrote *Purrles of Feline Wisdom, Cat Blessings, Dog Blessings, The Splendid Little Book of All Things Cat, The Splendid Little Book Of All Things Dog, Cat's Rule!, Dog's Rule!, Cat's Are Better Than Dogs,* and *Dogs Are Better Than Cats.* Bob is owned by a cat and is tormented by a jealous Lhasa apso.

Alan & Sandy Carey have been wildlife and nature photographers for more than twenty years. Their love for nature has taken them all over the world. Their work has appeared in *National Geographic, Smithsonian, Life, Readers Digest, Audubon,* and *Time* magazines. When asked what their favorite animal is, their answer is an enthusiastic, "Whatever we are photographing today!" You can also find their work on postcards, billboards, and even on some of the tail sections of Frontier airplanes.

PINELANDS FOLKLIFE

*A project of the New Jersey State Council on the Arts,
the New Jersey Historical Commission,
and the New Jersey State Museum in the Department of State*

Co-sponsored by the American Folklife Center at the Library of Congress

*Partly funded by a National Endowment for the Humanities planning grant for the exhibition
"New Jersey Pinelands: Tradition and Environment."*

PINELANDS

EDITED BY

RITA ZORN MOONSAMMY
DAVID STEVEN COHEN
LORRAINE E. WILLIAMS

FOLKLIFE

RUTGERS UNIVERSITY PRESS
New Brunswick and London

Library of Congress Cataloging-in-Publication Data

Pinelands folklife.

Includes index.
1. Pine Barrens (N.J.)—Social life and customs.
2. Pine Barrens (N.J.)—Industries. 3. Handicraft—
New Jersey—Pine Barrens. I. Moonsammy, Rita Zorn.
II. Cohen, David Steven, 1943– . III. Williams, Lorraine E.
F142.P5P56 1987 974.9 86-6713
ISBN 0-8135-1188-7
ISBN 0-8135-1189-5 (pbk.)

British Cataloging-in-Publication Information Available.

CONTENTS

ILLUSTRATIONS

Black-and-white:

MAPS

FOREWORD

This book stands at a confluence that, like Barnegat Bay, has many feeder streams of private and public initiative. The wellspring of these many initiatives in turn has been the abiding interest of local people in maintaining their way of life, and the equally abiding curiosity of others from the outside world to learn more about it.

My own relationship with the region began in 1977, when the Pinelands National Reserve was first considered by the Congress and the American Folklife Center was less than two years old. At the urging of Douglass College folklorist Angus Gillespie, my wife and I visited Waretown and the surrounding area. "You've got to come to Albert Hall," Angus said, knowing I play the fiddle and love folk music. So we attended the Saturday-night program at Albert Hall, where members of the Pinelands Cultural Society played an amiable concert of homespun music to an appreciative audience of locals and visitors. Like thousands of visitors to the Pinelands before and after us, we received our first glimpse of its culture through the Pinelands Cultural Society.

In 1977 the national impulse to save the New Jersey Pine Barrens was grounded in ecological and recreational concerns. Then, as now, many people actually thought that there was no real culture in the Pine Barrens—nothing beyond a few remnants of the region's nineteenth-century industrial heyday. The members of the Pinelands Cultural Society wanted to save the Pine Barrens, too, but their concern was for their cultural traditions. They sensed that, unless some official vehicle could be found for recognizing local culture, the cultural treasures of South Jersey would be overshadowed by Pine Barrens tree frogs and curly grass ferns in the public image of, and the planning for, the region. The Society knew that in order to protect their culture they had to start articulating it to outsiders and to each other. When they hosted my visit to Albert Hall, they were asking the government to acknowledge the worth of their traditional life and values by documenting South Jersey culture with the same care that it devoted to resources of scientific value.

Albert Hall had its roots in weekly musical gatherings that developed in the 1960s at Joe and George Albert's hunting cabin in the Forked River Mountains. At George Albert's funeral his musical friends decided to establish a permanent, formal gathering place and to name it in honor of the Albert brothers. Remembering the famed music hall in London, they chose the name Albert Hall. Thus a place was established for the formal presentation of the Pinelands' heritage and values.

At the Saturday Night Jamborees in the old warehouse on Route 9 in Waretown that constitutes the temporary Albert Hall, themes similar to those that animate this book are evident. The ancestors of the people onstage sailed as sea captains, operated water-powered sawmills, mined and logged cedar, and converted hundreds of acres of pine into charcoal. They themselves have built sneakboxes and garveys, fought forest fires, trapped snapping turtles, clammed in the bays, lumbered in the swamps, chased foxes on the plains, and harvested muskrats and salt hay in the meadows. They are the custodians of their culture and of their landscapes, of which the Pineconers sing:

> Beyond the mountain to the meadow, from the river to
> the bay—
> Beloved by folks of yesteryear and those right here today;
> The Pines along the Jersey shore seem to be the greenest
> here—
> The lakes with sparkling beauty bring us joy throughout
> the year.

This excerpt from the song "A Piney's Lament," by Janice Sherwood, is part of a large repertoire of original compositions that teach visitors to Albert Hall about local life and traditional values, that preserve and celebrate important places, events, and sentiments in the collective memory: "The Pine Barrens Song," "The Clamdigger," "A Home in the Pines," "Proud to be a Piney," "The Ballad of Julie Jane," "Forked River Mountain Blues," "Pines of Bamber," "My Double Trouble Queen," "Have You Ever Been Out on the Bay," "Jersey Moon," "Beautiful Barnegat Bay," and "Come on Down to Waretown."

Over the past decade public concern for "local lifeways" in the Pinelands has been catching up with the international interest in the region's special flora, fauna, and geology. In 1977 Linda Buki at the New Jersey State Council on the Arts inaugurated a Folk-Artists-in-Schools program in Burlington County. Building on research by folklorists Patricia Averill and Angus Gillespie, the program placed suburban students in contact with Pine Barrens traditions such as cranberry harvesting, boat building, clamming, the music of the Pineconers, and the stories of David Ridgway. From 1979 to 1983 the understanding of the region gathered force through school programs in Ocean and Cumberland counties conducted by Mary Hufford and Rita Moonsammy. These programs featured many of the traditions and tradition-bearers highlighted in this book: boatbuilding, salt haying, trapping, lumbering, clamming, oystering, fishing, cranberrying, mossing, gathering, charcoal making, fox hunting, decoy carving, quilting, cooking, singing, storytelling, and glassblowing.

Clearly, the region was rich with traditional life and well stocked with indigenous teachers; clearly, the living cultural traditions deserved a systematic survey. The foundations were already in place: the surveys and mapping of natural resources by scientists for the Pinelands Commission, the development of a human ecological model of the region by Jonathan Berger and John Sinton, and a regional network of tradition-bearers uncovered by state folk arts and folklife projects. In 1983 the American Folklife Center brought several years of discussion with the Pinelands Commission and the National Park Service to fruition by launching the Pinelands Folklife Project. The project's first task was to create an archive, a collection of documentary materials that would survey and preserve evidence of the traditional life of the region. An archive could not only document the cultural life of the region today, but could also provide a kind of map for the commission—making cultural signposts visible; illuminating the junctures of landscape, history, and traditional life and thought; and providing a cultural perspective for the decision making about the region's future.

Now we have a book to carry us a step farther. Its publication is timed to coincide with the January 1987 opening of the exhibition "New Jersey Pinelands: Tradition and Environment"; together they will synthesize and make public the great volume of work that has been generated over the past decade. The State Museum, the Historical Commission, and the Arts Council have rendered an important service by creating such fine vehicles for sharing the collaborative research of so many agencies and individuals. The book and the exhibition are partly for people who still think that there is no culture in South Jersey. But mostly they are for the people of the Pinelands, who want to keep their cultural traditions alive and who have the most at stake in our better understanding of those traditions. We are pleased to have had a part in presenting the landscape that is the magnum opus of their life and thought.

Alan Jabbour, Director
American Folklife Center
The Library of Congress

ACKNOWLEDGMENTS

Many people and organizations were involved in the publication of this book and in the preparation of the exhibition that accompanies it. We wish to express our thanks to them and acknowledge their contributions.

The staff of the American Folklife Center at the Library of Congress has been helpful in many ways throughout this project. Pinelands Folklife Project Director Mary Hufford provided assistance in formulating ideas, retrieving materials, and identifying resources.

The fieldworkers who participated in the Pinelands Folklife Project, and who also consulted with us during the preparation of the manuscript, included Thomas Carroll, the late Christine Cartwright, Carl Fleischhauer, Eugene Hunn, Jens Lund, Bonnie Blair O'Connor, Mal O'Connor, Gerald E. Parsons, Nora Rubinstein, Susan Samuelson, and Elaine Thatcher.

Photographs were taken by Joseph Czarnecki, Carl Fleischhauer, and Dennis McDonald for the American Folklife Center, by Joseph Crilley and Anthony Masso for the New Jersey State Museum, and by Michael Bergman for the New Jersey State Council on the Arts.

The staff of the Pinelands Commission assisted in locating resources for both the exhibition and the book.

Others who consulted with us about the project are William Bolger, Betsy Carpenter, Herbert Halpert, Donald Pettifer, George H. Pierson, Edward S. Rutsch, John Sinton, Gaye Taylor, Paul J. Taylor, Eugene Vivian, and Elizabeth M. Woodford.

Beryl Robichaud deserves special thanks for supplying much of the natural history material.

Librarians and archivists who retrieved information and photographs for us included Ronald Becker, Charles Cummings, Rebecca Colesar, Barbara S. Irwin, Carol Lapinsky, Robert Looney, Carl Niederer, Kathy Stavick, and Edward Skipworth.

The local historians Patricia H. Burke, John Callery, Somers Corson, Herbert Vanaman, Barbara Koedel, Everett Mickle, Polly Miller, Mary Ann Thompson, Sarah Watson, Carl West, and Joseph Wilson also helped in the project.

Staff at the New Jersey State Museum who worked on the project were Zoltan Buki, Suzanne Crilley, Wallace Conway, Karen Flinn, David Parris, and Cater Webb.

Curators at other museums and collections included Martin Decker, Alan Frazier, Thomas F. Harrington, John Hines, Ruth Hyde, Craig Mabius, and Rick Mitchell.

The following administrators of the cosponsoring agencies provided guidance and encouragement: Jeffrey A. Kesper and Barbara Russo at the New Jersey State Council on the Arts; Bernard Bush, Richard Waldron, and Howard Green at the New Jersey Historical Commission; Leah P. Sloshberg at the New Jersey State Museum; and Alan Jabbour at the American Folklife Center.

Support staff who helped with the manuscript were Elizabeth A. Crummey, Darin Oliver, Linda L. Tate, and Dolores Truchon at the New Jersey State Council on the Arts, Evelyn Taylor and Patricia Thomas at the New Jersey Historical Commission, and Gina Giambrone at the New Jersey State Museum.

Individuals who lent objects and photographs from their private collections to the exhibition include Frank Astemborski, Margaret Bakely, Tom Brown, Owen Carney, Les Christofferson, Hurley Conklin, John DuBois, Clifford Frazee, Steve Frazee, Walter Earling, Father Konstantin Federov, Lillian Rae Gerber, Albertson Huber, Jean Jones, Leo Landy, Dorothy Lilly, Ted Ramp, Michelle Rappoport, Joe Reid, Anne Salmons, Mary Ann Thompson, Virginia Durell Way, and Helen Zimmer. Many of these people deserve additional thanks for providing us with important background information, as well.

Organizations that lent objects and photographs include the Atlantic County Historical Society, Agricultural Museum of the State of New Jersey, Batsto Citizens Committee, Batsto Village State Historic Site, Burlington County Historical Society, Cape May Historical Society, Cumberland County Historical Society, Deserted Village at Allaire, Donald A. Sinclair Special Collections of the Alexander Library at Rutgers University, Gloucester County Historical Society, Hammonton Historical Society, Medford Historical Society, Monmouth County Historical Association, Newark Public Library, New Jersey Bureau of Forest Management in the Division of Parks and Forestry, New Jersey Division of Archives and Records Management, New Jersey Historical Society, New Jersey Pinelands Commission, New Jersey State Library, Ocean County Historical Society, Office of New Jersey Heritage in the Division of Parks and Forestry, and Wheaton Historical Association.

Many other people shared their knowledge of the Pinelands during interviews and the planning phase of this exhibition and book. They include Joe Albert, Fenton Anderson, Belford Blackman, George Brewer, Jr., Patricia Burke, Ken Camp, George Campbell, Larry Carpenter, Jack Cervetto, Milton Collins, Mark Darlington, Anne and Jack Davis, Frank Day, Charles De Stefano, Miguel Juan de Jesus, John Earlin, Eugene Espinosa, Lehma and Ed Gibson, Richard Gille, Lydia Gonzalez, Ed Hazelton, George Heinrichs, John A. Hillman, Nerallen Hoffman, Sam Hunt, Norman Jeffries, Malcolm Jones, Mary and Ed Lamonaca, Hazel and Leo Landy, Abbott, Stephen III, and Stephen Lee, Jr., Leonard Maglioccio, George Marquez, Haines, Henry, and Francis Mick, Herb Misner, Burtis Myers, Jr., Clara Paolino, Gladys and Harry Payne, Herbert Payne, Lou Peterson, Valia Petrenko, Charles Pomlear, Bluma Purmell, Ralph Putiri, Albert Reeves, Todd Reeves, David Ridgway, Merce Ridgway, Harry V. Shourds, Jim Stasz, Caroline and Norman Taylor, Brad Thompson, Alice Tomlinson, Orlando Torres, Dusia Tserbotarew, Ted Von Bosse, Lynwood Veach, Bill Wasiowich, Joanne Van Istendal, and Beryl Whittington.

The preparation of the manuscript of this book was partially funded by a planning grant from the National Endowment for the Humanities.

PINELANDS FOLKLIFE

PINEY
POWER

INTRODUCTION

MUCH OF THE PINELANDS REGION OF SOUTHERN NEW JERSEY seems to consist only of monotonous stretches of sand and scraggly pine, the basis for both the historic name "Pine Barrens" and the old perception of the area as infertile. Yet the region is neither uniform nor barren. It is rich in both ecological subsystems, such as cedar swamps, meadows (salt- and freshwater marshes), and forests, and natural resources, such as sphagnum moss, fur-bearing animals, and wood. And, partly because of this natural wealth, it is equally rich in folklife. In every subsystem of the Pinelands, residents have built traditional systems of resource cultivation and use that have shaped the landscape and the history of the region.

Opposite: *Orlando Torres of Vincentown. Photograph by Carl Fleischhauer. American Folklife Center, PFP201851-2-29.*

These traditional technologies, however, are merely one aspect of the folklife of the Pinelands. Like such technologies elsewhere, they are grounded in the "sense of place" of the people of the region. Sense of place is the totality of perceptions and knowledge of a place gained by residents through their long experience in it, and intensified by their feelings for it. It enables them to create indigenous systems for classifying resources, methods for cultivating them, and tools for harvesting them. But sense of place also informs their play and art. Jokes about Jersey mosquitoes, poems about the woods, and paintings of blueberry fields express sense of place and give us a native's eye for the region.

These cultural phenomena, and the people who have created them, are

Rita Moonsammy / David S. Cohen / Lorraine E. Williams

Sandy Road in the Pinelands. Photograph by Dennis McDonald. New Jersey State Council on the Arts.

an essential and fascinating, but little told, aspect of the story of the Pinelands. Ideally, the Pinelands people should provide the voice of the story, spoken in the context of the scholar's description of the natural environment and the broad historical patterns of human use of it. That conviction shapes this book, from its overview of the ecological makeup of the region and the changes in perceptions of it, through its essays on sense of place, human ecology, and traditional activities.

Ecologically, the Pinelands is remarkable for its location, its water system, and its flora and fauna. It stretches over much of the Outer Coastal Plain, that portion of New Jersey that punctuates America's Atlantic coastline like a giant comma, and encompasses one million acres, nearly 20 percent of the state's land area. The Pinelands' northern limits touch Monmouth County, and its southern edges reach down into Cape May County on Delaware Bay, so that a traveler on the Garden State Parkway may survey the varied scenery of the Pinelands National Reserve while riding south from Exit 90 at Bricktown to Exit 12 near Swainton. Not far north, west, and south of the Pinelands are several of the nation's largest cities.

Though travelers on other roads often believe that the region has a tedious sameness, it is actually a large ecosystem embracing a variety of very different, yet closely interrelated subsystems: forests, hardwood swamps, spungs (cedar bogs), streams, ponds, salt- and freshwater meadows, rivers, and bays.[1] Its most crucial and remarkable feature is the Cohansey Aquifer, a body of groundwater with a storage capacity believed to be as much as 17 trillion gallons. The aquifer links the various subsystems, from the dry, sandy uplands, where the water table is more than two feet beneath the surface, to soggy wetlands, where it rises to or above the soil surface.

The geological formation that holds the water is called the Cohansey Sand; generally, its soils are coarse, loose, and low in nutrients. Precipitation flows easily through the sandy soil and replenishes the water table, and groundwater seeps just as easily upward to feed the streams. The streams, in turn, coalesce into rivers, which form estuaries where they meet the sea.

But the very ease of water flow within the system also makes it vulnerable to pollution damage. Because this largely unpolluted aquifer is bordered by areas of rapid population growth, its protection is of serious concern to conservationists and natives.

Unusual plants and animals with interesting histories are another important feature of the Pinelands that has also prompted protection efforts. Pyxie moss and the northern pine snake are among more than 100 plants and animals that extended their ranges north thousands of years ago when the sea level was lower and nature provided a broad highway along the coast. Today, the Pines is their northernmost home. Other species—broom crowberry, for instance—extended their ranges southward as tundralike conditions were created by advancing ice sheets.

Still other species are remarkable for having adapted over the centuries to forest fires, which occur regularly in the Pinelands. The pitch pine (*Pinus rigida*), for instance, which dominates the landscape, outcompetes other tree species here because of its thick bark and its ability to put out new shoots from the base of the tree even when the top has been killed by fire.

The histories of some plants are more closely linked to humans, whose landscape artifacts—cranberry bogs, iron forges, mining pits, agricultural fields, and railroad rights-of-way—have left their mark on the Pinelands flora by providing habitats for alien species.

Though the Pinelands is often depicted as uninhabited wilderness, this area has actually been used by humans for many centuries. A thousand years ago, the region was used by the Lenape Indians (also known as the Delaware). These people are assumed to have lived in independent groups of about 100 members, and spoke dialects of the Algonkian language family similar to those of the Native Americans in what are now southern Connecticut, metropolitan New York, eastern Pennsylvania, and northern Delaware.[2] Much of what is now considered traditional culture of the Pinelands—hunting, trapping, fishing and collecting of wild plants—was already Indian tradition when the first Europeans arrived. The Lenape moved seasonally within southern New Jersey to take advantage of the availability of different foods. They fished shad from the rivers in spring, and farmed sandy riverside soils in summer. In fall, they collected nuts from the woodlands, and in fall and winter they hunted white-tailed deer in communal hunts.

Before the arrival of the Europeans, the Lenape probably used the Pinelands mainly as hunting territory. When the Brotherton Reservation was created for them in the eighteenth century at what is now Indian Mills in Burlington County, officials expected them to harvest game and wood from the forests. The reservation was too small a territory to support their seasonal hunting and gathering system, however, and by 1802 most of them either moved west or settled in New York.

By the mid-eighteenth century, the Pinelands supported a number of resource-based industries, and for 150 years the size of the basically Anglo-American population rose and fell with these various industries: lumbering, iron mining and forging, and glassmaking. In the late nineteenth century, however, railroads opened the interior and the coast to development as an agricultural and resort area, and brought European immigrants who established small towns and ethnic agricultural communities. In the twentieth century, the development of the automobile and the desire for homes outside of the cities have brought new residents to the Pinelands, many of whom work elsewhere. Today, therefore, the region has a wide variety of residents and users, including both families who continue to ply traditional occupations and visitors who explore woods and waters.

Over the years, much has been written and said about the people and the place, for the Pinelands has been perceived as unusual since the first travelers passed through. Evaluations of the region have changed, as have ways of defining its boundaries.

During the Colonial period, the region was considered a "barrens," unfit for agriculture. In 1765, in his *History of the Colony of Nova-Caesaria, or New Jersey,* Samuel Smith wrote:

> Almost the whole extent of the province adjoining on the atlantick, is barrens, or nearly approaching it; yet there are scattered settlements all

> along the coast, the people subsisting in great part by cattle in the bog, undrained meadows and marshes, and selling them to graziers, and cutting down the cedars; there were originally plenty of both the white and red sorts: The towring retreat of the former have afforded many an asylum for David's men of necessity. . . . The barrens or poor land, generally continues from the sea up into the province, thirty miles or more, and this nearly the whole extent from east to west; so that there are many thousands areas, that will never serve much the purposes of agriculture; consequently when the pines and cedars are generally gone (they are so already in many places) this will not be of much value.[3]

Although by the late nineteenth century the potential of the region for certain types of agriculture was realized and some evaluations had changed, the people of the Pinelands continued to be regarded in a stereotyped way that can be traced back to the British Isles and Europe, where the forest was considered as a refuge, and the forest dweller as one who did not fit into society.[4] The same notion was applied to both the frontiersman in eighteenth-century America and the "Piney" in nineteenth-century New Jersey. It has persisted into the twentieth century.

Legends about the origins of the people of the Pinelands were created and circulated primarily by people who lived outside the area. They variously claimed that Pineys were descended from disaffected Quakers who had rejected the tenets of the Society of Friends; Tories who had sided with the British during the Revolutionary War; mercenary Hessian soldiers who deserted the British army during the same era; pirates, smugglers, privateers, and "Pine Robbers" who took advantage of the social unrest during the Revolution to plunder and steal; and Indians from the Brotherton Reservation.

In fact, the people of the Pinelands are descended from the same families that settled other areas of South Jersey. They included Dutch, Germans, and English Quakers. The Germans, however, came by way of Pennsylvania well before the Revolutionary War, and had different surnames from those of the Hessians who fought for the British. There is some truth to the claim that smugglers and privateers worked the region, for in the years before the Revolution, smugglers operated along the coast to avoid the British Navigation Acts, and privateers operated out of these same ports during the war. However, there is no evidence that the pirates William Kidd or William Teach ("Blackbeard") ever came to New Jersey, contrary to local legends about buried treasure. Finally, the claim that the Lenape Indians left descendants (with the possible exception of Indian Ann Roberts, a basketmaker from Shamong Township) remains to be proven.[5]

Map of the boundaries of the Pine Barrens and the Pinelands National Reserve.

The stereotype of the Piney was reinforced by the notorious Kallikak study conducted in the first decades of the twentieth century by the Vineland Training School. The study purported to trace two branches of a family that descended from Martin Kallikak (a pseudonym), a Revolutionary War soldier. The "Piney" branch supposedly descended from his alliance with a barmaid and included depraved and feebleminded members. The other branch, however, supposedly intermarried with some of the "best" families in New Jersey.[6] In fact, recent research has proved that the so-called Kallikak family was from Hunterdon County, not from the Pinelands.[7] Yet people thought the study was about the "Pineys," perhaps because Elizabeth Kite, the fieldworker on the Kallikak study, also wrote a report on the people of the Pinelands.

The boundaries of the Pinelands have been defined differently at different times, and the names "Pine Barrens" and "Pinelands" illustrate those differences. In 1916, when botanist John Harshberger published *The Vegetation of the New Jersey Pine Barrens,* he identified the boundaries of the region by vegetation distribution and defined it as bounded on the east and south by bays and salt marsh, and on the north and west by farmlands and hardwood forests. Similar methods were used by other botanists.

Aspects of the region's environment have affected the study of its culture. When folklorist Herbert Halpert, then a graduate student at Indiana University, began his fieldwork in the Pinelands in the 1930s, he adopted Harshberger's concept of its boundaries, but focused on the forests north of the Mullica River rather than the agricultural area south of it. Reasoning that the culture of the farmers would be different from that of people living in the forested areas, Halpert therefore collected data mainly in Burlington and Ocean counties.

Halpert believed that the negative reputation of the people of the Pines reflected a bias of urban people toward rural people, and he set out to record their stories and songs in order to show them through their own expressions. His work took the form of an unpublished dissertation entitled "Folktales and Legends from the New Jersey Pines," and of a folk-song collection which is housed in the Indiana University Archives of Traditional Music.[8] Unfortunately, his collection never reached the general public.

In the 1960s, however, environmentalists began to focus attention on the Pinelands as a unique and sensitive region, and to emphasize the importance of the aquifer. Their view was typified in a popular book entitled *The Pine Barrens.* Its author, John McPhee, portrayed the area as a wilderness "so close to New York that on a very clear night a bright light in the Pines

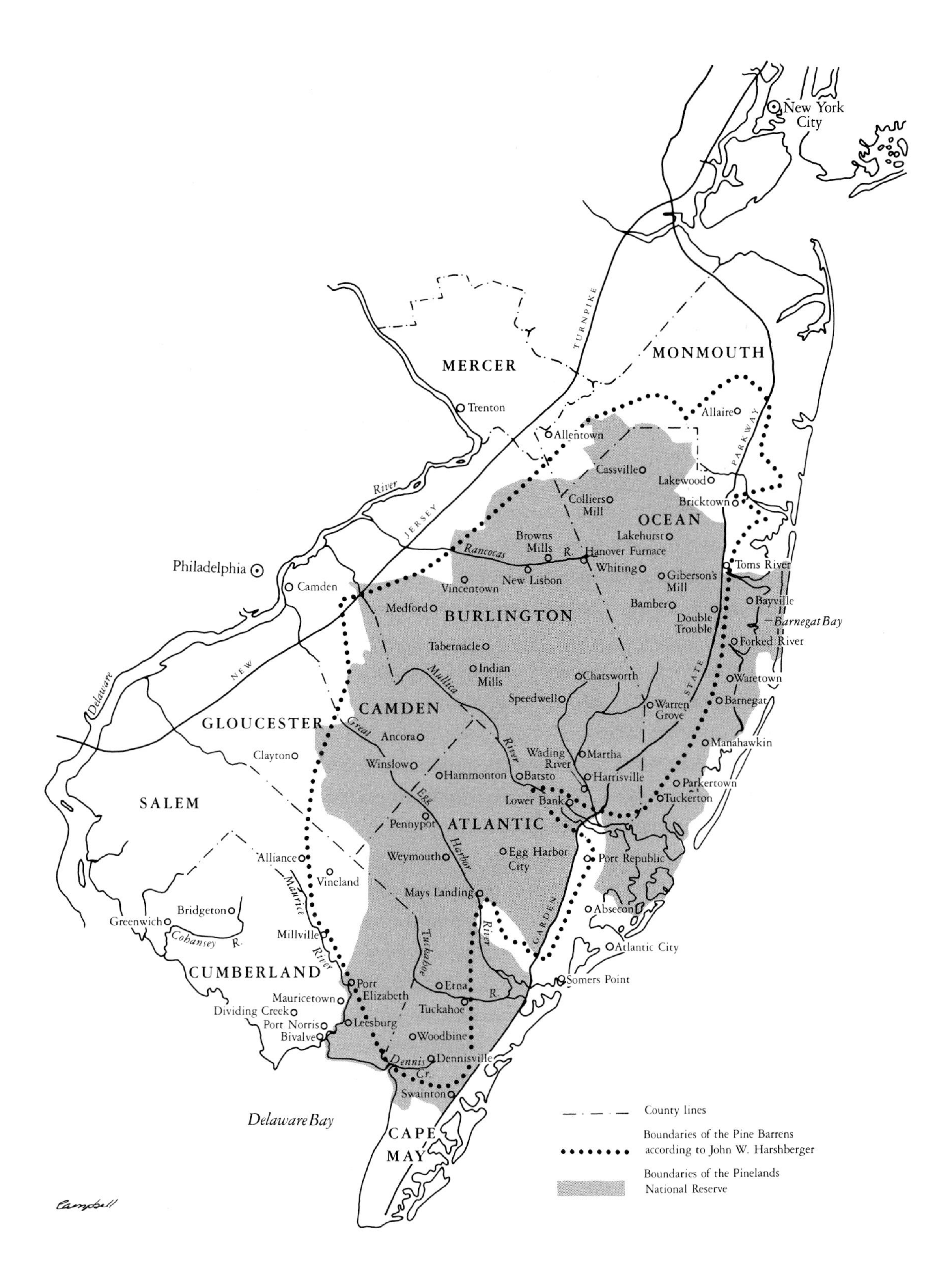

THE BOUNDARIES OF THE PINE BARRENS AND THE PINELANDS NATIONAL RESERVE

Folklorist Elaine Thatcher interviewing Clara Paolino of Ancora. Photograph by Susan Samuelson. PFP–BSS286558–14–4a. Pinelands Folklife Project Archive, American Folklife Center, Library of Congress. Hereafter, photographs from the archives will be listed with data retrieval number only.

would be visible from the Empire State Building,"[9] and stressed the importance of the Cohansey Aquifer as a supply of pure water:

> The water of the Pine Barrens is soft and pure, and there is so much of it that, like the forest above it, it is an incongruity in place and time. In the sand under the pines is a natural reservoir of pure water that, in volume, is the equivalent of a lake seventy-five feet deep with a surface of a thousand square miles. . . . The Pine Barrens rank as one of the greatest natural recharging areas of the world.[10]

Evaluations such as this gave impetus to efforts to preserve the region from rampant development. In the 1960s and 1970s, local, state, and federal agencies began to enact legislation to govern land use and to preserve the natural resources of the region. In 1978, the federal government designated the Pine Barrens as the country's first "national reserve," a protected area in which much of the land remains in private ownership although its use is subject to governmental review. In 1979, the New Jersey Legislature passed the Pinelands Protection Act and created the Pinelands Commission to regulate land use.

The legislative actions resulted in a new map of the area. In consideration of management needs, the National Parks and Recreation Act of 1978 included within the boundaries of the Reserve parts of Barnegat Bay and the barrier islands on the east, Delaware Bay to the south, and the Maurice River on the west. The northwestern boundary follows several highways. Thus, what was once called the "Pine Barrens" is different from the region today called the "Pinelands."

The environmental remapping coincided with a new approach to the region and its people by cultural researchers, one which regarded the aquifer as a unifying feature. Folklorist Mary Hufford, planner Jonathan Berger, and historian John W. Sinton, all of whom did seminal work in documenting the people of the Pinelands and their use of the land, defined the region by the interrelationships of groups and traditional activities, and emphasized the insider's point of view in descriptions of the region.

In 1983, when the American Folklife Center at the Library of Congress established the Pinelands Folklife Project under Hufford's direction, fieldworkers surveyed all the traditional activities of people throughout the Reserve, concentrating particularly on their interactions with the environment. In addition to folklorists, members of the research team included an environmental psychologist and an ethnobotanist; this underscored the importance to the project of documenting the perceptions and understandings that constitute a group's sense of place and that underlie regional folklife.

During the same period, Berger and Sinton were studying the culture of people throughout the Reserve to assist in the creation of a management plan which would be informed by the needs of traditional users. In their book, *Water, Earth, and Fire: Land Use and Environmental Planning in the New Jersey Pine Barrens,* they describe the people of the Pinelands as " . . . no

Cranberry scoop made by Dan Aker at Giberson's Mill. New Jersey State Museum Collection, 174.51. Photograph by Joseph Crilley.

more isolated, nor . . . startlingly different, from rural cultures in other mid-Atlantic sections. They share with other rural people common values, including love of place, the central role of the family, Christian morality, participation in seasonal activities (farming, hunting, trapping) and participation in voluntary organizations (civic organizations, fire companies)."[11]

Hufford, Berger, and Sinton documented the ways in which natives perceive their place and build their lives with that knowledge. Their analyses of these perceptions begin this story of the Pinelands.

Mary T. Hufford first explains how sense of place "literally begins with the senses." She describes how sensory familiarity is a basis for concepts of place that are expressed in folklife forms. Through these forms, residents both organize and tell about their knowledge of the place.

John W. Sinton's essay then traces the broad history of human use of the Pinelands and describes the regional patterns of continuity and change. He explains how human activities created the present landscape.

In the concluding essay by Rita Moonsammy, David S. Cohen, and Mary T. Hufford, traditional activities within each of five major environmental subsystems of the Pinelands are described in historical sequence. Details of folk technologies are added to Sinton's framework of resource use, and descriptions of folklife forms and the perceptions of natives illustrate Hufford's analysis of "sense of place." Wherever possible, the words of residents are used, from both historical sources such as diaries and letters and contemporary sources such as interviews conducted during the Pinelands Folklife Project.

Through these analyses and descriptions, the remarkable variety of the place and its people, and the important folklife forms that link them, will enrich the story of the Pinelands.

NOTES

1. Ecological material for this essay has been provided by Beryl Robichaud and was drawn from her previous work. See Beryl Robichaud and Emily Russell, *Protecting the New Jersey Pinelands: The First National Reserve,* forthcoming, and Beryl Robichaud, *A Conceptual Framework for Pinelands Decision-Making* (New Brunswick, NJ: Center for Coastal and Environmental

Studies, Rutgers University, 1980). Portions of the latter were directly incorporated in Chapters 2 and 3 of the Comprehensive Management Plan for the Pinelands Commission in 1980.

2. Ives Goddard, "Delaware," in *Handbook of North American Indians,* ed. William C. Sturtevant, vol. 15, *Northeast,* ed. Bruce G. Trigger (Washington, DC: Smithsonian Institution, 1978), 213–239; and Herbert C. Kraft and R. Alan Mounier, "The Late Woodland Period in New Jersey: ca. A.D. 1000–1600," in *New Jersey's Archaeological Resources from the Paleo-Indian Period to the Present,* ed. Olga Chesler (Trenton: New Jersey Department of Environmental Protection, 1982), 139–184.

3. Samuel Smith, *History of the Colony of Nova-Caesaria, or New Jersey: Containing An Account of its First Settlement, Progressive Improvements, the Original and Present Constitution, and Other Events, to the Year 1721. With Some Particulars Since; and a Short View of its Present State.* (Burlington, NJ: James Parker, 1765), 487–488.

4. David J. Fowler, "Nature Stark Naked: A Social History of the New Jersey Seacoast and Pine Barrens, 1690–1800" (Ph.D. dissertation proposal, Rutgers University, 1981), 3–6.

5. David Steven Cohen, "The Origin of the 'Pineys': Local Historians and the Legend," *Folklife Annual* I (1985): 40–59.

6. Henry Herbert Goddard, *The Kallikak Family: A Study in the Heredity of Feeble-Mindedness* (New York: The Macmillan Company, 1913).

7. J. David Smith, *Minds Made Feeble: The Myth and Legacy of the Kallikaks* (Rockville, MD: Aspen, 1985).

8. Herbert Norman Halpert, "Folktales and Legends from the New Jersey Pines: A Collection and a Study" (Ph.D. dissertation, Indiana University, 1947); "Piney Folk Singers: Interviews, Photos, and Songs," *Direction* 2 (1939): 4–6, 15; "Some Ballads and Folk Songs from New Jersey," *Journal of American Folklore* 52 (1939): 52–69.

9. John McPhee, *The Pine Barrens* (New York: Farrar, Straus, and Giroux, 1968), 5.

10. Ibid., 13–14.

11. Jonathan Berger and John W. Sinton, *Water, Earth, and Fire: Land Use and Environmental Planning in the New Jersey Pine Barrens* (Baltimore: Johns Hopkins University Press, 1985), 11, 13.

JOE'S PLACE
AFTER DARK
TRESPASSERS
WILL BE SHOT
SURVIVORS
WILL BE PROSECUTED
PLEASE
NOVISITING
AFTERDARK
JOE

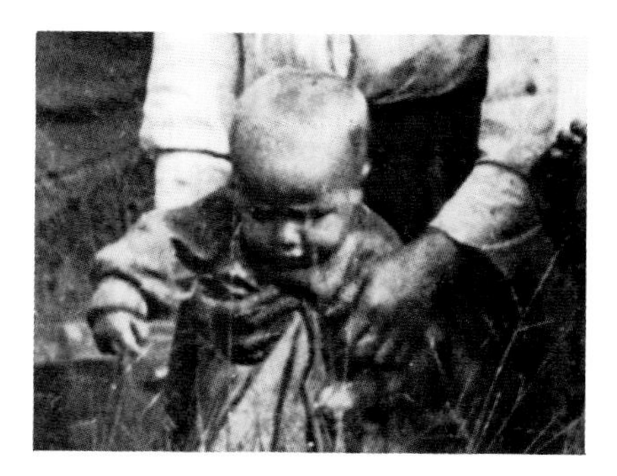

TELLING THE LANDSCAPE:

Folklife Expressions and Sense of Place

"I THINK OF TWO LANDSCAPES," WRITES BARRY LOPEZ, "ONE outside the self, the other within."

> The external landscape is the one we see—not only the line and color of the land and its shading at different times of the day, but also its plants and animals in season, its weather, its geology, the record of its climate and evolution. . . . The second landscape I think of is an interior one, a kind of projection within a person of a part of the exterior landscape. . . . The interior landscape responds to the character and subtlety of the exterior one; the shape of the individual mind is affected by land as it is by genes.[1]

Opposite: *Woodland signpost on the road to Joe Albert's cabin. Forked River Mountains, Waretown. Photograph by Mary Hufford.*

Connecting the outer and inner landscapes are landscape tellings. The word "telling" operates on several levels. In one sense, telling alludes to the way in which Pinelands residents take stock of landscape elements, assigning them values and names. They study the environment to tell what is there, perceiving differences in surroundings that at first seem indecipherable. This environmental literacy, the ability to read the environment, relies upon the accuracy of the environmental image, a collaborative vision of a given place that guides people who must constantly make their way through it, and who have for generations shaped it into what we see today.

Mary T. Hufford

In the sense that telling also applies to the act of recounting or narrating, people constantly interpret the landscapes they have mentally ordered. Their interpretations crystallize in expressive forms that fuse and transform both inner and outer landscapes—songs, names for things, paintings, recipes, rituals, and tools and technologies for working the land. Such forms and processes contain stories—telling us the elements of the landscape and then telling its story through folklife.

The living landscape is hard to pick out at first. Before it was molded, it was deciphered into categorical forms, the "pieces" discerned and ordered into backdrops for human action on the land. These distinctions are indiscernible to the motorist passing through miles of ground oak and scrub pine, a landscape that folklorist Herbert Halpert described in the 1940s as "desolate and dreary":

> In many places one can travel for miles on the through highways with no relief from the uniform bleakness save for infrequent gas stations or narrow sand roads leading off to someone's bog. Houses are few and far between, gray and weatherbeaten, in marked contrast to the well-painted structures of the surrounding farm country. The house usually rests on what seems to be pure sand, and the scrub pine comes up almost to the doorstep.[2]

For Merce Ridgway, Sr., a woodsman who made charcoal early in this century near Bamber, such a landscape was enlivened with family and occupational history. When he moved to the Shore to make his living as a bayman, he nostalgically remembered the woods in "The Pine Barrens Song":

> I left the place where I was born, many years ago.
> For times were tough and work was scarce,
> I had no choice but go.
> But I've been back there many a time, in my memory,
> Of all the places that I've been, it's there I'd rather be—
>
> Where the scrub pine, ground oak, berry bush, and sand,
> They never changed—they never will—Pine Barrens land—
> The sweet May pink and curly fern, leaves all turning green,
> And the water running red in the cedar swamp stream.[3]

Ridgway's song also records other facets of the backdrop against which woodsmen perform—sand roads through twisted pine trees, cemeteries, the smell of pine smoke from wood stoves, and ruins of towns left by failed industries. The landscape he describes is as man-made as it is natural. Humans have carved it into roads, ditches, channels, fields, settlements, and banks, and it has etched its features into the collective memory.

The song provides a short list of plants that are significant to woodsmen—evoking an impression of what the place looks and smells like. Scrub pine, ground oak, berry bushes and sand are ubiquitous, both sources of income and indicators that tell about the environment. They are scenic resources in the theatrical sense that J. B. Jackson reminds us of:

> When we speak of the "scenes of our childhood," or borrow Pope's phrase and refer to the world as "this scene of man," we are using the word *scene* in what seems a literal sense: as meaning location, the place where something happens. It rarely occurs to us that we have in fact borrowed a word from the theater to use as a metaphor. Yet originally *scene* meant stage, as it still does in French, and when it first became common in everyday speech it still suggested its origin: the world (we were implying) was a theater, and we were at once actors and audience.[4]

The actors and audiences read their cues in the natural settings. The colors of serotinous (delayed-opening) pine cones reveal how long it has been since a fire has occurred; an abundance or dearth of acorns may disclose feeding patterns of deer; vegetation indicates the kind of "bottom" (substrate) on which it grows; and in the sand woodsmen can read what animals have passed through recently (they can even identify fellow woodsmen by distinctive tire tracks). May pinks, as trailing arbutus is called locally, are a welcome sign of spring. The curly fern Ridgway mentions is not *Schizea pusilla,* a tiny, relict plant that environmentalists point to as a regional symbol, but rather the large bracken ferns (*Pteridium aquilinum*) that Koreans gather and consume in the spring. Ridgway's nostalgic portrait, painted from memory, is meaningful to people whose interior landscapes resemble his.

Outsiders to the Pinelands often marvel at the perceptive skills of natives, as Captain Lou Peterson, a Delaware Bay oysterman, observed to Jens Lund:

> I don't think that you consciously know anything. I think living around this area and growin' up around it, your senses become used to it or somethin', because, you take my father-in-law, he moved down in 1950, and he couldn't get over the fact that every native down here, when the wind changed, noticed it.[5]

Sense of place literally begins with the senses, with an ability to make sense of the environment, not only to tell what is there, but to understand the relationships between environmental elements. Outdoorsmen working in what we might call the endemic folklife habitats of the Pinelands may or may not express ecological relationships in scientific terms, but they know what the place looks, sounds, tastes, smells, and feels like at different times of the day and in different seasons. Environmental literacy is for them a survival skill that enables them to "work the cycle."

The high sensitivity to wind is important not only for watermen but also for woodsmen, who contend with the dangers of living near fire-climax forests. Awareness of the imminence of fire—the ability to read what Jack Cervetto, a Warren Grove woodsman, calls the "fire index"—is widespread among natives. "The old timers always knew to check the fire index," Cervetto observed. "Young people don't seem to check for it so much any more." The buildup of pine needles and brush on sandy soil that holds no water creates a tinder-box. Natives are tensed when a fire is overdue, like people elsewhere, who, living on the margins of disaster, are tensed for volcanic eruptions, or earthquakes, or floods. "When she's ready to burn," wrote Joanne Van Istendal, of Medford, "the wind will dry her like a bone. You gotta feel and smell the wind, be ready for a good fight. When there's a fire at the other end of the Pines, we all feel it. It's always on our minds. It binds us together."[6]

Herbert Payne, who made charcoal in Whiting, was able to "tend" his charcoal pit from his job on a construction site many miles away in Toms River. One day, noticing a sudden shift in the wind, he excused himself to the foreman and hurried home, arriving just in time to keep the charcoal from completely combusting.[7]

Woodsmen, in turn, marvel at the perceptual acuities of animals, who they may regard as teachers, rivals, saboteurs, companions, and assistants. "We're a class of animal, too," reflected Harry Payne, "a higher class of animal." He spoke admiringly of the acute senses that make foxes so hard to trap, and that enable birds to "forecast the weather":

> They know an awful lot, to keep away from the human being—because naturally we're smarter than they are to a degree, but I think they're more sensible to nature than we are. They can forecast the weather, they know when there's going to be a storm. You can tell by the way they move. In the afternoon you'll see a little cloud in the sky and you'll see the birds goin' way up high and comin' down, keep goin' way up high and flut-

terin'—you know there's gonna be a storm that afternoon—a hard thunderstorm.[8]

The trapper, as folklorist Gerald Parsons reminds us, has an extremely fine-grained image of the world:

> In the vastness of a tide marsh, he looks for the track of a mink no larger than a thumb-print. Where the fin-fisherman watches for the flash of bait fish on the surface of the ocean, or looks across the broad horizon to find a cloud of feeding sea birds, the trapper searches the ground under his boots for the glint of a few dried fish scales—otter vomit.[9]

Fur traders, botanists, herpetologists, deer hunters, and horticulturists have relied upon the senses of local woodsmen–gatherers for generations, for they have powers of discernment that seem astonishing. The historian Harold Wilson tells us that woodsmen who mined cedar near Dennisville in the early nineteenth century "progged" for the ancient fallen trees with rods in the swamps. (People who wrest their livings from the muck are, in fact, known as "proggers.") There were two kinds of trees: "windfalls," which had toppled with their roots attached, and "breakdowns," which had broken off. Wilson reports that because breakdowns were more easily mined, a log digger would first secure a chip from any log he discovered, determining by the smell of the chip whether the tree was a windfall or a breakdown.[10]

A keen sense of smell is still important to woodsmen. Tom Brown, a woodsman from Cumberland County, may avoid a place that smells like cucumbers, a sure sign that rattlesnakes are nearby: "I've been blueberrying with the wife and I said, 'Come on, let's get out of here!' She said, 'Why?' 'Let's go! Cause I *smell* 'em!' They got that odor, you know—like fresh-cut cucumbers."[11]

Elizabeth White and Frederick Coville developed the world's first culti-

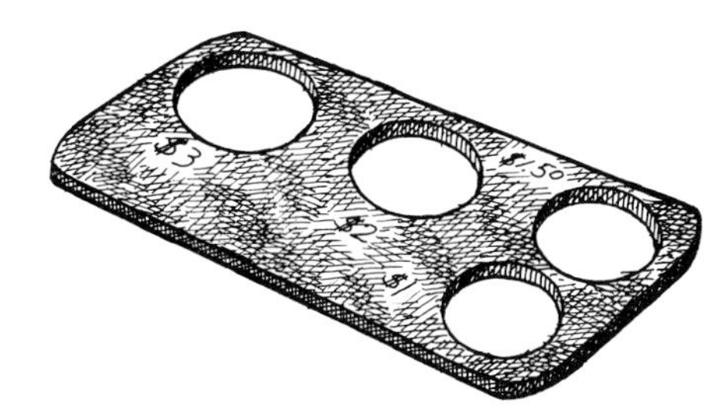

Five-eighths inch gauge to measure wild blueberries used by Elizabeth White and Frederick Coville. Drawing by Allen Carroll. Based on aluminum replica by Mark Darlington.

vated blueberries at Whitesbog in the early 1900s. They were aided by woodsmen–gatherers who, it was said, could almost navigate by the tastes of blueberries in different swamps, and by their shapes and sizes. Attempting to develop a cultivated strain from the wild blueberries, White enlisted the woodsmen in her search for bushes bearing the largest berries. She devised a pay scale for bushes bearing fruit that would not fall through the $\frac{5}{8}$-inch hole in her blueberry gauge, and she named the bushes after their finders:

> In getting the early bushes I tried to name every bush after the finder. . . . And so I had the Adams bush found by Jim Adams, the Harding bush that was found by Ralph Harding, and the Dunphy bush that was found by Theodore Dunphy. When Sam Lemmon found a bush I could not name it the Lemmon bush, so I called it the Sam. Finally, Rube Leek of Chatsworth found a bush. I did not know it was anything special at that time, and I used the full name in my notes. . . . Coville called it the Rube, which I thought was a poor name for an aristocratic bush. He finally suggested that we call it the Rubel. And the Rubel has been the keystone of blueberry breeding.[12]

At the second annual Whitesbog Blueberry Festival, held in 1985 by the Whitesbog Preservation Trust, local children went on a blueberry hunt. Each child was equipped with a $\frac{5}{8}$-inch gauge made by Elizabeth White's great-nephew, Mark Darlington. They took their gauges into the old blueberry fields around Whitesbog to find the biggest berries. The winner received a flat of cultivated blueberries. Thus, in a bit of historic re-enactment, the relationship between the horticulturists and woodsman–gatherers that produced the cultivated blueberry is commemorated.

The effort to achieve contact with nature and contact with the past is often formalized in such recreational activities. Aldo Leopold observes that a key aesthetic feature of nature contact is "the perception of evolutionary and/or ecological processes."[13] Activities such as blueberry games, deer hunting, and birding formalize the challenge of reading natural "texts," a challenge met by those who first tamed the wilderness.

"Have you ever followed a deer trail?" asked one deer hunter from the Spartan Gun Club:

> You have to be able to read signs to know what's goin' on—you find droppings, you find their beds, you find buck rubs, which are young saplings that they rub the bark off when they take the fuzz off their antlers, and

> these young saplings—about an inch and a half in diameter—these young saplings are all stripped of bark about that high off the ground where they slip the head down and get the fuzz off their antlers. They polish 'em up, get 'em ready for fightin' for a mate.[14]

Deer hunters see part of their mission as commemorative. Their annual re-enactments of the frontier days typify what Aldo Leopold termed "the split-rail fence" approach to recreation. "We have a saying," said Larry Carpenter, of the Harmony Gun Club, "'Take your boy hunting instead of hunting your boy.' It *is* a tradition. . . . It's like American heritage. We take time off of work to come up and put in the time. We keep the population under control. Everything we take we use. It's just like what happened in the eighteenth century."[15]

Jim Stasz, an avid birder from Audubon, New Jersey (where John James Audubon, "the great American woodsman," once collected a scissor-tailed flycatcher), spoke of "reading" with one's eyes closed, using only one's ears and nose: "You could close your eyes, and listen to the birds and know what plants were there. Or you could look at the plants, even just smell them at certain seasons and you could predict what birds would be there."[16]

The Christmas Bird Count is an annual event of no less significance than deer hunting, a ritual celebration that occurs at the time of year when it is most challenging to find birds. This heightens the significance of what the participants are able to report afterwards, when the birders gather to tell of their discoveries.[17]

Franz Boas wrote that "all human activities may assume forms that give them esthetic values."[18] In a similar vein, folklorist Dell Hymes commented that all communities encapsulate shared experiences in "meaningful, apposite forms."[19]

These forms include all things that are crafted out of the imagination's encounter with the land and its resources—recipes, songs, poems, paintings, crafts, tools, technological processes, rituals, festivals, recreational activities, and landscapes. They show how communities organize both natural and cultural differences, and they serve as repositories of information about both nature and society—archives for the collective memory.

Information about the landscape, its places and people, its economic and seasonal cycles, its past, and its multiple realities, is encapsulated in the names for things within it. The following informal nomenclature for Pinelands groups suggests that people, like natural resources, may be classified as ecotypes: mudwallopers, pineys, stumpjumpers, proggers, river rats, snakehunters, baymen, and woodsmen. Even terms such as "transplant," for newcomers, and "shoobee," for tourists (who wear their shoes to the beach), address the referent's degree of removal from the land.

"The very naming and distinguishing of the environment," writes Kevin Lynch, "vivifies it, and thereby adds to the depth and poetry of human existence."[20]

Consider the three categories of names for plants: botanical names—Latin binomials conferred by an international congress of botanists; common names—"specific vernacular names selected by a regional body as the preferred non-botanical name for specific kinds of plants";[21] vernacular names—the local unstandardized names that show great geographical and cultural variation. "May pinks," for example, is a local name for *Epigea repens,* which goes by the common name of "trailing arbutus." The botanical, or scientific, names provide "knowledge about" natural species, identifying them in the format and language of an international classification system, and for the use of an international community of scientists. Local names, on the other hand, encapsulate human experience with the species on this landscape, conveying that "knowledge of" the place that is born of repeated experience with it.[22]

Scientific names often memorialize people relating to the history of an international community. The roster of scientific names for plants and reptiles reads like a roll call for botanists and zoologists. Pickering's morning glory (*Breweri pickeringi*), Fowler's toad (*Bufo woodhousei fowleri*), Knieskern's beakrush (*Rhynchospora knieskernii*), and Hirst's panic grass (*Panicum hirstii*). Peter Kalm and Mark Catesby, natural historians of the eighteenth century, are commemorated in the scientific names for sheep laurel (*Kalmia angustifolia*) and bullfrogs (*Rana catesbiana*).

Local names for plants and animals, however, spring from experience with a place. Taken as a whole, they graphically catalog its sights, tastes, sounds, smells, and impressions: "sugar huck" (*Vaccinium vacillans*) is the sweetest wild huckleberry; "whompers" (*Lampropeltus gitulis gitulis*) are immense eastern king snakes that live in the swamps; "green juggers" (*Rana catesbiana*) are bullfrogs; "stinkpots" (*Sternotherus odoratus*) are turtles, so named, according to Tom Brown, because they urinate on people who pick them up; "squawks" (*Nicticorax nicticorax*) are black-crowned night-herons, so named, according to Bill Lee, for their raucous calls; "croakers" (*Micropogonias undulatus*) and "drumfish" (*Pogonias cromis*) fill the bay with their "singing," as Bill Lee put it; "puff adders" (*Heterodon platyrhinos*) inflate themselves and play dead when disturbed; "snappers" (*Chelydra serpentina*) command respect for their jaws; "rattlers" (*Crotalus horridus*) issue percussive warnings; and "woodjin's enemy" (*Comptonia asplenifolia*) is sweet-fern, laden with the sand-ticks so bothersome to "woodjins"—a portmanteau word combining "woodsman" and "Injun."

Blueberries and huckleberries continue to go by practical names. Local names for the wild berries include "upland blacks" (*Gaylussacia dumosa*), also called "grouseberries" (Marucci), "grassberries," and "dwarf hucks" (Harshberger); "black huckleberries" (*Gaylussacia resinosa*); "swamp black" (*Vaccinium attrococcum*); "sugar hucks" (*Vaccinium vacillans*); and "dangleberries" or "bilberries" (*Gaylussacia frondosa*). "There's three different types, and three different sizes of blue huckleberries," said Jack Cervetto, a Warren Grove woodsman, "Now the swamp blues (*Vaccinium corymbosum*), now they're the largest and the best ones for pie. I have cultivated berries at home, but I'll go in the swamp and get blues if my wife wants to make a pie."[23]

The environmental image, on which outdoorsmen must agree in order to communicate, is condensed in names and in other folklife expressions, many of which serve as mnemonic forms that keep the region's collective memory accessible.

"The named environment," writes Kevin Lynch, "familiar to all, furnishes material for common memories and symbols which bind the group together and allow them to communicate with one another. The landscape serves as a vast mnemonic system for the retention of group history and ideals."[24]

Local people often describe wild phenomena through domestic analogues. Tiny cedar saplings come up, said George Brewer, "like hair on a dog's back." Elizabeth Woodford compared the sound of a controlled burn to "bacon crackling in a skillet." Tom Brown likens the odor of rattlesnakes to cucumbers, and Jack Cervetto thinks of sphagnum moss as a rug. It smells, he says, "like iodine." Brad Thompson sees the Inner and Outer Coastal Plains coming together "like fingers in a line," where husbanded soil and virgin soil may be read in the alternating fields of soybeans and blueberries. Croakers, drumfish, and weakfish fill the bay with their "singing," according to Bill Lee. Hunters distinguish the two species of foxes by placing red foxes in the dog family, and gray foxes in the cat family.

Place names, linked with landscape features, encode the shared past, distinguishing members of one group from another. "Let me show you where I was born," said Richard Gille, the zoning officer for Lacey Township:

> There's a foundation there where I was born, and when I was a kid, the train used to stop here to pick up water, and there was a small little stop house—there wasn't actually a station. We called that Ostrum. But only

> somebody like me or Johnny Parker or Cliff Frazee would know if I said "Ostrum" that I was talkin' about that. That was right along the railroad. The train that ran there was steam and they had to pick up the water there. So they called that part of town Ostrum, and when I was a kid, we said we didn't live in Forked River, we lived in Ostrum.[25]

The charred remains of an event in natural history may also become a nucleus around which common memories are gathered. Lacking the touchstones to the past that ruins of settlements provide, places in the woods are more subtle, defined by the people and events that coalesce in them. "Webby was great for naming places," said Jack Davis, "Clarence Webb. He's the one come out with 'Lightnin' Point,' 'cause of this tree that was hit with lightnin'. That's known as Lightnin' Point nowadays by the older people."[26] One space may be several places at once, reflected in the names given to it by its various users. Near Woodmansie is a place that the older local people call "The Clay Pits." The lakes left there by an extractive industry are now called "Hidden Lakes" by the recreational-vehicle users who swarm there each weekend from points as far as 50 miles away.

Places and their names are sources of identity and security. Joe Albert, an octogenarian woodsman from Waretown, was frustrated in his effort to formally designate a favorite woodland crossroads:

> The main corner up here by the fire tower, I named that Star Tree Corner—it's like a star. I always keep paintin' signs and nailin' 'em up there, and guys steals 'em! . . . I put it there so it'd keep that name, you know—the younger ones don't follow things up like that. Now they tell me that sign is down by a little camp on the hill.[27]

Wilderness-seekers might be startled by signs in the woods. Yet these seekers are themselves partly defined as a community by their own shared system of place names. "Do you have names for the sections you hunt?" I asked a member of the Spartan Gun Club in Chatsworth. He was incredulous. "How can you name anything like that? It's in the woods!" Yet, as it turned out, they did have names for places:

> "You know, you might have like the Double Dirt Road or Big Hill or Sandy Ridge. . . ."
>
> "Down by the Crooked Tree. . . ."
>
> "Two or three deadfall trees together: 'Down by the Big Deadfall,' 'Sandy Ridge,' or 'Apple Pie Hill.'"[28]

Children living in Pinelands communities play at frontiering, building "forts" in the woods, as children have done for at least half a century. Their worlds are not as large, but they are as diversified as adult woodlands, which the children may regard as threatening. Frank Day, 13, of Lebanon Lakes, identified Hidden Lakes as a scary place: "The kids said all these guys go back there and chase them. . . . Probably Pineys. Hunters—wild hunters."[29] The children's landmarks are "The Old Pine Tree" or "The Rotten Out Tree"—a tree with a hole in it that squirrels live in.

Like tides with place names bobbing in them, successive generations wash over the region, and sometimes the names stick—serving, like the advice of older people, as landmarks for future generations.

It is difficult to consider the landscape apart from folklife expressions, because the landscape itself is partly the work of human hands. It has been so extensively modified that its crafted aspects are an integral part of it.

"This is a high ridge here," said Nora Rubinstein to Jack Cervetto, standing on a raised area near an old cranberry bog.

"This is a dam," Jack corrected her. "This is man-made. This is built."

What appear to be monotonous woodlands to outsiders are teeming with categorical forms in the eyes of woodsmen.

"All right," said Jack Cervetto to Nora Rubinstein, "Now see that water we just went through? That separates that island from this island. Now this is another island surrounded by water."

"How can you tell that it's a slough and not just a wet spot?" Nora inquired.

"Well, it's low all the way around it. Altogether different bottom. . . . Last week this was perfectly dry. The rain we had the other day has put some water in here."

"Does this island have a name?"

"No, no. But it's my island, and I'm pretty well attached to it."[30]

"Islands," "sloughs," and "bottom" are landscape motifs, grammatical units in the language of those who read the environment. These units may be natural entities that are mentally discerned and endowed with meaning, or they may be human constructions with historic legacies, such as channels, ditches, dams, bogs, sand roads, corduroy roads, and charcoal pits.

The old corduroy roads or crossways that criss-cross the swamps contain lessons about the old-timers' ways of harvesting and managing cedar. They are blueprints for the new roads, as Clifford Frazee observed:

> The old timers did it that way. We put the brush on the crossway, and that holds the slabs in position. We leave the slabs right there and the

> cedar won't reseed in the road, and the trees'll grow up on both sides and meet at the top. I can show you a couple of places where it was cut off like a hundred years ago. It's rotted out, but the trees don't grow up in it. The advantage is that it gives a tree five extra feet for growth. You should thin out cedar, but if you thin it too much the wind'll blow it over. We fall 'em toward the crossway, and most of 'em fall right on it. You have to notch 'em just right to have 'em fall.[31]

Puerto Rican Sacred Heart altar, home of Eugene Espinosa of Lakewood. Photograph by Dennis McDonald. New Jersey State Council on the Arts.

Walking on such a crossway conveys Frazee into the swamp, into the past, and into a reflection on the most critical problem in cedar management: striking a balance between sunlight, which makes the trees grow faster, and crowding, which keeps out the sun, but protects the trees against windthrow.

Long-time residents can often critically appraise the landscape, and appreciate subtle nuances relevant to traditions in which they have been trained. Norman Taylor, on returning to the site of his childhood home in Lower Mill after 20 years, excitedly witnessed in retrospect his father's good judgment in locating the mill at the confluence of two streams: "You can see why it's appropriate for a mill to be there, where two streams come together—look how quick the water is. . . . You can see the piling down there where the old mill used to set. That's a lot of water flowing, for as low as the water is."[32]

Farmers enjoy looking at farms that look as farms should look. For them, farms are critical scenic resources. "It's a pretty, pretty sight," said Mary Lamonaca, a peach grower in Hammonton, "when the fruit is in bloom, when the fruit is hanging, the tomato fields, and the beautiful rows of baskets."[33]

"Every orchard and every field has its own different thing, see?" said Eddie Lamonaca, her husband, showing his farming community to Mal O'Connor: "Now coming up here, you've got . . . Pine Road, a pretty nice road, you know. . . . Most of it's kept well around here. There's nothing shabby or rundown about the farms. They're meticulous."[34] The trained landscape observer knows when to applaud technical skill, and exhibitions of what Franz Boas called "feeling for form,"—whether the crafted work is a cranberry bog, a Maryland hound dog, or a Barnegat Bay sneakbox.

Folklife expressions help to keep people oriented in the face of drastic cultural change. When communities or individuals are separated from their cultural contexts, they may encapsulate that lost world in some form, using the materials at hand as touchstones that confirm their own sources. Thus, for example, Lydia Gonzalez told Bonnie Blair that Puerto Ricans in

COLECCION
ORACIONES ESCOGIDA

The Last Supper, *framed by deer antlers, and snapping turtle shell painted with a duck scene. Tom Brown's trapper cabin, Millville. Photograph by Joseph Czarnecki. PFP83–B217721–14–8.*

Woodbine "don't lose themselves," because they keep their houses "the old way, with their Sagrado Corazon [Sacred Heart] and their plastic flowers and everything, and still really cook the old way all the time." [35]

Religion and foodways—resources for spiritual and physical nourishment—are common foci for communal identity. People bind themselves to the "great traditions" of Judaism and Christianity with the "little traditions" of their own design.[36] Copies of Da Vinci's *The Last Supper,* for example, are framed differently in different cultural settings. In Valia Petrenko's Russian Orthodox home, the scene is draped with a *rushnyk* (a traditional embroidered towel), while in Tom Brown's trapper's cabin it is flanked by two deer antlers; thus, reminders of Russian culture and South Jersey nature, respectively, are linked through a painting that bridges and articulates cultural difference.

Food is an important focus of communal identity for both long-established natives and more recently arrived immigrants. "I don't lose my way of doing things," said Clara Paolino of Ancora, speaking of her ability

The Last Supper, *framed by Russian* rushnyk *(ceremonial cloth), home of Valia Petrenko, Allentown. Photograph by Dennis McDonald. New Jersey State Council on the Arts.*

to produce a wide range of Italian dishes out of the gardens of relatives and neighbors.[37] Ann Davis, of Brown's Mills, spoke of the Piney way to make snapper soup:

> I shred my snapper. I start out with cabbage, potatoes, celery, onions, carrots—cooked in stock. Then string beans, peas, whatever you want. Then I put six pounds of butter. Then . . . hard boiled eggs that I ground. That's the Piney way that we make snapper soup. In restaurants you get that brown gravy soup, and I don't like it.[38]

People "read" people by the ways they do things, just as they read the environment. Some people claim to be able to tell a New Gretna speaker from a Green Bank speaker. Decoy aficionados distinguish subregional styles among carvers—the Parkertown style, the New Gretna style, the "Head of the Bay" style, and the Tuckerton style.

Even stylistic differences in oyster shucking methods mark different re-

Peach orchard, Hammonton. Photograph by Joseph Czarnecki. PFP217721–4–4.

gional backgrounds. People from Virginia are accustomed to the long, thin oysters called "hotdogs" or "snaps" in southern Maryland.[39] They "break" their oysters open by smashing the shell. People from Maryland learn to open oysters that are shorter and rounder, and separate the two shells by "stabbing" the knife between them. When people from Maryland and Virginia converged in the same shucking houses in Port Norris they were quickly differentiated into "breakers" and "stabbers." "I'd want my back to the breaker," said Beryl Whittington, who stabs his oysters "straight out," "because the breaker throws a lot more mud. Sometimes the shell from a breaker will pop you upside the head."[40]

Certain landscape elements may reflect transplanted ethnic values. "The topography is always 'trying' to match an image," wrote Paul Shepard. "This meshing brings order out of the natural world without homogenizing it."[41] Many of the farms surrounding Hammonton, for example, are Italian, as the names on the mailboxes attest: DeMarco, Risotti, Putiri, Angelo, Bucci, Lamonaca. They also *look* Italian, according to Eddie Lamonaca: "Italian farmers, I think, like to see trees, and olive trees, or peach trees. . . . Some kind of a gnarly, European-growing-looking tree—and the ground's flat, you know, and has some kind of rolling landscape to it. . . . I think it does resemble Italy somewhat—northern Italy, and central Italy."[42]

For the same reason, Russian and Ukrainian immigrants in Cassville line their yards with colonnades of white birch trees transplanted from local bogs. Reminiscent of Eastern European forests, birch trees powerfully evoke the homeland.

The remains of some of the Pinelands National Reserve's fragile landscapes are just as symbolically powerful. For George Campbell, a fourth-generation salt-hay farmer, the man-made banks that have maintained the meadows represent a way of life that is being consumed by the ocean: "Last year I was down on my bay shore down here. My bank had broke, you know. I stood there and cried 'cause I seen a way of life leavin', you know, that I'd loved, and you can only fight Mother Nature for so long and she's gonna win. I know she's gonna win."[43]

Many technological processes in the Pinelands aspire toward the most precise balance possible between beauty and efficiency. When they achieve that balance, they are likely to be considered an art form, as Franz Boas observed:

> When the technical treatment has attained a certain standard of excellence, when the control of the processes involved is such that certain typical forms are produced, we call the process an art, and however simple the forms may be, they may be judged from the point of view of formal perfection; industrial pursuits such as cutting, carving, moulding, weaving; as well as singing, dancing and cooking are capable of attaining technical excellence and fixed forms.[44]

"Now that's an art in itself, building a corduroy road," said George Brewer, who harvests cedar in Great Cedar Swamp. The list of forms and processes accorded this status in the Pinelands is lengthy. Processes such as trapping, canning, farming, net making, boatbuilding, eeling, fox hunting, bog building, glassblowing, and lumbering all have aesthetic as well as pragmatic aspects, laden with personal and communal significance. Out of them emerge emblems of regional identity—Jersey cedar, Jersey cones, Jersey tomatoes, and Jersey garveys.

Created at the intersection of the inner and outer landscapes described by Barry Lopez are the tools for plying the exterior landscape, many of which attained fixed forms generations ago. The body of maritime tools yields a remarkably complete image of the waterman's workplace in its varied aspects at different seasons. Tools for clamming, oystering, and eeling, for example, reflect the varied nuances of "bottom"—some geological, some seasonal—below the region's diverse waters. There are, for example, eel spears for summer and winter, and at least three different kinds of heads for clam and oyster tongs: wooden, barrel, and keyport. Wooden heads are made of seasoned oak, with teeth of sawed-off steel pounded through; they are the best for soft bottom. "You can catch oysters with wooden heads when you can't catch anything but mud with the other ones," said Lou Peterson to Jens Lund. "You have to feel your tongs, and hold up on them, and you can pretty near feel the oysters goin' in."[45]

Such tools, like stories, are eloquent interpretations of the environment that may reflect the personalities of their makers and users. Rube Corlies' black-duck decoys, for example, are unmistakable, because of his effort to keep them ice-free:

> . . . Rube's ducks are high on the front, tufted breast from the bottom to the bill and the reason he made 'em that high was because in a blow, when the wind is blowing and your duck is dipping, and the temperature is falling, that bill will hit the water and it'll start to ice up, and the

Black-duck decoy made by Ruben Corlies (1882–1976) of Manahawkin. Drawing by Allen Carroll.

> first thing you know you got an icicle about that long on the duck's bill. So as long as he could keep 'em high and keep 'em out of the water, he'd keep 'em from icing up just that much longer.[46]

In recent decades, collectors have recognized the aesthetic value of such decoys, but as Ted Von Bosse, of Port Republic, pointed out, such artistry was incidental to their main function. "They made them to hunt with, not as art, but there's art in those things, and there's shape and beauty in the old ducks that's just beyond belief."[47]

The artistry in duck hunting does not begin and end with hand-carved decoys, however. Whether he uses cedar, cork, or plastic decoys, the skill of the duck hunter resides in his ability to perceive and exploit a variety of factors, including wind direction, temperature, tide, and his position relative to the ducks. His goal is to create an impression with his decoys that will attract flying ducks within shooting range. Ed Hazelton's brother-in-law was a good duck hunter, according to Ed, because he thought like his quarry:

> He was the type of fella that thought like a duck. He thought like a duck. He just knew every move they were gonna make. In other words, we'd sit there, gunning, and have the stools [decoys] out, and in would come some ducks. And they wouldn't come just the way he wanted 'em. Just exactly right. You *could* kill 'em, but he says, "They gotta do better than that." And he would go out and he would take this stool here and put it there, and this stool here and set it back there, and the next time they'd almost light in your lap. . . . He just thought like a duck all the time. He knew . . . when there would be a lid on the bay . . . and that the ducks would come in to feed in the ponds before the freeze.[48]

These are the aesthetic judgments, the critical canons that guide the development of forms. Each practitioner contributes to the form in the way that scholars contribute to their own disciplines. We see this especially in the indigenous small watercraft of the Pinelands.

"There are two kinds of garvey makers," says Ed Hazelton, "builders, and craftsmen." A builder, inattentive to details, might produce a "hairy" garvey—one that is unsanded, and comes out looking "like a crew cut" beneath its paint job. Also, a real craftsman, according to Hazelton, drills a hole for each nail to prevent cracks that eventually widen into leaks.[49]

The Barnegat Bay sneakbox, one of the most elegant forms to emerge from the region, effectively synthesizes the observations of generations of baymen of water, land, air, man, and mud. It was custom made for the marshes and estuaries of South Jersey. In its form we see every contingency neatly anticipated. Its spoon-shaped hull enables it to glide through areas marked as land on coastal maps. Only Atlantic white cedar will produce the requisite compound curves, according to builders George Heinrichs and John Chadwick, who use old family patterns. The tiny skiff—12 feet long by 4 feet amidships—is light enough for one man to haul over land between channels. It is equipped with a mast-hole, centerboard well, and detachable rudder for sailing; winter and summer sails; folding oarlocks and a removable decoy rack to suppress its profile; runners for traveling on ice; and two kinds of accessory ice hooks for breaking up slushy ("porridge") and hard ("pane") ice. Its sloping transom allows a hunter to row backward in channels that are too narrow to turn around in. It is linked in tradition with the Barnegat Bay duck stool, a decoy with a dugout (hollow-carved) body to lighten the boat's burden and keep its draft shallow enough, as the saying goes, "to follow a mule as it sweats up a dusty road."

In the sneakbox the shapes of men and meadows are fused. The planked-over deck, which keeps the gunner's legs warm, was often custom made. "They used to build a sneakbox special for a man," Sam Hunt once said. "He used to lay down on the ground and they'd draw a circle around him and build a hatch so his belly could stick out."[50]

The sneakbox incorporates many facets of the South Jersey environment—alluding in its design to independent gunners, cedar swamps, salt meadows, mud, water, ice, winter, summer, and wind.

Some tools and technologies, such as boats, decoys, corduroy roads, and orchards, are tangible creations—each instance of the form an enduring crystallization of environmental concepts. Others, such as decoy rigs, sets for catching fox and muskrat, and ways of casting nets for fish, are repeatedly produced and dismantled.

The ultimate harvest, however, is the story, captured through formalized pursuits, such as yardscaping, birding and hunting. We have seen how individuals formalize their perceptions of nature in narrative, songs, and paintings, and how Pinelands communities formalize nature contact in activities such as birding and deer, duck, and fox hunting. Such activities are wellsprings for stories that create the environmental image.

Hunting of any sort can provide a way to interact with wildlife, whether or not any game is taken. Ann Davis recalled her opportunities for witnessing wildlife dramas:

> My mother wouldn't understand why I wanted to go gunning. But I would go up into the woods and set there by myself, and listen to the squirrels cussing me out, and the blue jays, and you get in where they are, and they don't like you to interrupt what they're doing . . . and till they find you there they're just unconcerned, but when they find you, boy oh boy! You'd be surprised how they chatter, and the blue jays'll see you and they'll go up and down the woods and the road hollerin', and I said, "They're just tellin' everybody else that somebody's here." I believe that that's what it was. And a deer could come along, and the squirrels and the blue jays would tell him, "Don't go any further!" And off they would run. It's really amazing the things that you see and hear off in the woods.[51]

Fox hunting ceases in March when foxes begin raising their young, but foxes will still perform for those who know where to go, according to Norman Taylor:

> If you go out around April or May, and go where [the mother fox] has got little ones, she'll put on an act that is out of this world for you. She'll do everything in the world to get you away from that hole. She'll holler, she'll squawk, she'll let you see her, she'll run, she'll roll. She'll do anything to get you away from there.[52]

Such accounts view the landscape within a theatrical frame, a way of experiencing the landscape that differs from the scientific perspective. When we conceive of the landscape as an "ecosystem," we filter our image of it through a scientific analogue, an analogue that does not help us to fully appreciate the expressive dimension of the landscape. J. B. Jackson suggests that, in this regard, "theater" serves as a more useful metaphor, because, unlike ecosystem, it is drawn from human experience:

> All that we have so far come up with is an analogue of one sort or another, borrowed from biology or ecology or communication theory. When it is a matter of controlling or manipulating the environment, analogues can be extremely helpful; yet if we are again to learn how to respond emotionally and esthetically and morally to the landscape, we must find a metaphor—or several metaphors—drawn from our human experience.[53]

The landscapes of the Pinelands are indeed theatrical settings. Woodspeople scan them, focusing on the dramatic action, or "events," as Bill Wasiowich described them to John McPhee:

> When he is not working in the bogs, he goes roaming, as he puts it, setting out cross-country on long, looping journeys, hiking about thirty miles in a typical day, in search of what he calls "events"—surprising a buck, or a gray fox, or perhaps a poacher or a man with a still.[54]

Jack Cervetto's swamps are filled with natural dramas. Cedar trees, scraping against each other in the wind, make "music," while sphagnum moss repels mosquitoes with its iodine fumes. Against this backdrop he encounters the wildlife world, marveling at its strangeness, continually interpreting its mysteries. "The weirdest sound that I've heard in the swamp," he said, "is a snake swallowing a frog. It takes a long time for a snake to swallow a frog, and the whole time the frog is squeeching."[55]

"The green snake," he said, "They're the friendliest snake. . . . You approach that branch he'll stick his head out in front of you, but he won't bother you."

The biggest snake, the "whomper," is also called the "pilot" snake, according to Tom Brown, because tradition has it that it indicates where the rattlesnakes are.[56] "Have you seen a whomper?" Cervetto asked Eugene Hunn, "or heard 'em?"

> Oh, they're a big one. They're short, and great big, and boy, them sonofaguns, they're not afraid of anything. Now, my wife's uncle had charge of the Sim Place cranberry bog there—this oh, probably three years ago, and at night time us boys all hung around together there—Warren Grove, in the old candy store there. We used to play quoits and horseshoes and checkers and that, and he said, "Boys, I'm gonna show you somethin'." And he had one of these whompers. And this whomper had thirty-eight snakes inside of it, from twelve inches to eighteen inches long.[57]

While such natural events are not themselves works of art, they are, to borrow a concept from Robert Plant Armstrong, "affecting presences"[58] that move and amaze their witnesses. Nature is a wellspring of mysterious otherness, an "insistent and live reality," as David Wilson put it, present to its connoisseurs "the way God is to saints or the past is to humanists":

> To come face to face with a flying spider or a rattlesnake in the road unhinges habit and intensifies awareness, just as stumbling upon an ancient rune does, or encountering a burning bush. These uncommon phenomena throw the settled world into disarray. . . . Close attention must be paid and care taken if one is to make sense of it all, and if the world is to make sense still. There are some who are dead to the impact of such events, but for those who are not, every detail seems significant and its gestalt feels compelling. It is like the difference between those who encounter a poem as an "affecting presence" and those who see it only as a batch of words.[59]

Tom Brown inventories his scenic resources in verse:

> My wife has often said to me,
> "How lonesome the woods must be,"
> I answered, "No, there's too much to see"—
> I love the murmur of the trees
> As the wind softly stirs her leaves
> The bees flying to and fro
> Gathering nectar as they go
> The robin and the little wren
> Among my many feathered friends
> The cardinal with his coat of red
> The mockingbird singing overhead
> At yonder log I chance to glance
> A grouse is starting to drum and dance
> The otter from the bank its slide
> The mink that hunts a hole to hide
> The deer that drink in yonder stream
> I often see them in my dreams
> And though folks may say it's a waste of time
> I'll always have this love of mine.[60]

Painters such as Margaret Bakely, of Vincentown, and the late Win Salmons, of New Gretna, have captured such natural events in paintings. Some scenes—deer browsing in a clearing, or ducks being flushed from the water—are common, and often appear on related flat surfaces produced by those environments, such as sand dollars and snapper shells. The kind of scene that would be virtually impossible to photograph, a snapping turtle

Snapping Turtle. *Painting by Margaret Bakely of Vincentown. Photograph by Dennis McDonald. PFP235202-1-25.*

lunging at two ducks, is captured on canvas by Margaret Bakely, in a striking image of predation. Central to her painting are the turtle's giant jaws, a much ruminated topic among local people, who all know that a snapper can bite long after its head has been severed. The saying is that only two things will cause it to release its grip: thunder or the setting sun.

Such beliefs about animals, usually attributed to old-timers, may seem naive and misinformed to the scientific observer. Old-timers along the Maurice River used to say that railbirds turn into frogs. Albert Reeves, an old timer and well-known teller of landscapes there, gives them the benefit of the doubt:

> In ancient times, those old elves there, some of them thought that birds turned into different things. . . . Well, a beautiful butterfly was nothing but a worm, wasn't he at one time? Wasn't he? He was like a worm, and

> when he comes out of that thing, and he's in there—you know in the cocoon—he turns into a beautiful butterfly, doesn't he? . . . So why can't a railbird turn into a frog?[61]

The fact is that, in the experience of mudwallopers, there is a mystifying relationship between frogs and railbirds, and snapping turtles (dead or alive) are unbelievably tenacious. People offering folk explanations for such phenomena often do not profess to believe them. Like natural phenomena, such explanations are themselves food for the imagination. They are literary resources in oral tradition that vivify life in the region.

Thus far we have considered a variety of ways in which the inner and outer landscapes mutually affect each other in literal and figurative realms; how it is that people cast form and meaning upon their surroundings, which in turn sustain them both physically and aesthetically. Many of the expressions we have looked at relate to ways in which people decipher, organize, and interpret the landscapes, turning them into backdrops for dramas in which animals are often the players.

On closer inspection of some of these rituals and stories about animals, we discover the outer landscape reflecting the inner landscape in complex ways. Through behaviors like bird feeding, turtle marking, and fox hunting, for example, people construct natural texts that simultaneously explore and reflect the social and natural orders. They get animals to tell them stories about themselves.[62]

The ubiquitous bird feeder, for example, is like a small theater, often viewed between the curtains of kitchen windows. Ethnobotanist Eugene Hunn compared one birdfeeder to a morality play, having witnessed through the eyes of its keeper, Leo Landy of Nesco, a cast of "spartan" chickadees, "proletarian" sparrows, "lazy" cardinals, "mild-mannered" doves, and "aggressive" bluejays who disdain anything but sunflower seeds. "That's the biggest robbers there are, blue jays," said Elwood Watson, of Wading River. Landy's feeder is sparingly stocked with sunflower seeds, to keep the jays from becoming like "welfare types."[63] This observation of the natural world gives rise to myriad stories through which the animal world continues to mirror the human one, as it has done for thousands of years.

Sense of place is one of the gifts that old people give. In Cumberland County, box turtles are made to represent the Brown family tree. Whenever a child has been born to Tom Brown, or a grandchild or a great-grandchild, Tom has inscribed the child's name and birthdate on the shell of a box turtle, according to a custom in Cumberland County:

> Now I got a box turtle that's come back every year for twenty-seven years. Twenty-seven years ago I put "Pop-Pop" and "Dawn" on it, and put the date, and it come back again this year. It's blind. It's been blind for about five years, but it comes right to the back door and my wife'll feed it bread and we put it down in the window box for a few days and then release it. And this year, one I did for the great-granddaughter Muriel, I only just did that in '81, that one come back.[64]

Thus a living family register moves through varied landscapes on the shells of turtles.

Fox hunting is an elaborate ritual in which the quarry is more often protected than taken at the end. Any fox hunter will say that the whole purpose of chasing foxes is to hear music from the dogs. Hound dogs are for many local people what binoculars are for birders. The stories that hounds deliver are not only about the natural world, but about the hunters, for whom the hounds are extensions into the natural world, and for whom they substitute in the pack. "You stand there at night," said Milton Collins of Port Republic, "you don't see anything. That sound comes to you, and there's a beautiful *story* in it. I know that it's my dog just picked up the double and then Robly's dog took the double away from my dog, and he's runnin' it, or she's runnin' it."[65]

What *is* the story? Like other rituals, fox hunting mirrors its participants. According to fox hunters, a working dog is like a working man: it

Maryland hound dog. Drawing by Allen Carroll.

doesn't matter what he looks like, as long as he does his job. The elements of fox, dogs, landscape, and men are always the same. In each chase the themes of everyday life are recapitulated. Through their dogs the men aspire to a united effort—the production of a rich, orchestral sound. The dogs should "pack up" so tightly, as Norman Taylor said, that you could throw a blanket over them. However, there should not be so many of them that hunters cannot pick out the individual notes. The hunters also compete with one another through their dogs. They expect sportsmanlike behavior from their dogs, who should enact the social code that hard work, not laziness, will be rewarded. "I'd like to tell you a story that I just remembered," said John Earlin, of Browns Mills.

> Many years ago, Ben was runnin' a fox on the east side of [Rt.] 539, and I was trailin' a fox on the west side, and after a bit, our dogs got together. At that time I had an old Black-and-Tan dog that I . . . hadn't ever seen 'im run before. And my dogs and Ben got in, and they were really doin' the job on this gray fox. They were takin' him around and around and around. And my old Black-and-Tan, he just stood there and listened. And they run 'im maybe for an hour, an hour and a half. And out he came and he came right in the middle of the road. He run right up there, and the Black-and-Tan stood right in the middle of the road, and the old fox went right between the legs of the Black-and-Tan, and he turned around and *caught* him. And I mean to tell you, you want to hear a man that really was mad, that was Ben. He said, "The worthless son-of-a-gun," he said. "Can you imagine a dog that wouldn't run nothin', catchin' that fox!" Now he was very unhappy about the whole situation, and I couldn't hardly blame him. I think that the dog should have showed some sign of going in there.[66]

At the juncture of nature and culture, hound dogs deliver information about the unseen landscape to the listening hunters. They also enact social conflicts: the collective versus the individual, music versus din, fair versus foul play, domestic order versus the wild unknown, and merit versus chance. A pack of hounds is like a book on a shelf, and each time the hunters open it up, a new story on the old themes emerges, telling the landscapes afresh.

NOTES

1. Barry Lopez, "Story at Anaktuvuk Pass: At the Juncture of Landscape and Narrative," *Harper's* (December 1984): 49–52.

2. Herbert Halpert, "Folktales and Legends from the New Jersey Pines: A Collection and a Study" (Ph.D. dissertation, Indiana University, 1947), I:18–19.

3. Excerpt from "The Pine Barrens Song," written by Merce Ridgway, Sr., New Gretna, NJ, 1978. Permission to quote granted by Merce Ridgway, Jr.

4. J. B. Jackson, "Landscapes as Theater," *Landscape* 23 (1979): 1.

5. Interview, Jens Lund with Louis Peterson, October 22, 1983, PFP83–RJL019, Pinelands Folklife Project Archive, American Folklife Center, Library of Congress. Hereafter, references to archive materials will be listed with names, date, and data retrieval number only.

6. Correspondence, Joanne Van Istendal to Mary Hufford, May 1984.

7. Interview, Mary Hufford with Herbert Payne, May 1979.

8. Interview, Mary Hufford with Harry Payne, November 14, 1983, PFP83–RMH031.

9. Gerald Parsons, "Suggestions Concerning the Study of Fur Trapping in the New Jersey Pinelands Project," United States Government Memorandum, Library of Congress, August 13, 1983.

10. Harold F. Wilson, *The Jersey Shore: A Social and Economic History of the Counties of Atlantic, Cape May, Monmouth and Ocean* (New York: Lewis Historical Publishing Company, Inc., 1953) II:724.

11. Interview, Eugene Hunn with Tom Brown, December 1983, PFP83–AEH008.

12. Elizabeth White, personal communication, 1953, cited in William Bolger, Herbert J. Githens, and Edward S. Rutsch, "Historic Architectural Survey and Preservation Planning Project for the Village of Whitesbog, Burlington and Ocean Counties, New Jersey (Morristown, NJ: New Jersey Conservation Foundation, September 1982), 46.

13. Aldo Leopold, "The Conservation Esthetic," in *A Sand County Almanac* (New York: Oxford University Press, 1949), 173.

14. Interview, Mary Hufford with members of Spartan Gun Club, November 17, 1983, PFP83–RMH044.

15. Interview, Mary Hufford with Larry Carpenter, November 19, 1983, PFP83–RMH046.

16. Interview, Eugene Hunn with Jim Stasz, December 19, 1983, PFP83–AEH006.

17. Fieldnotes, Eugene Hunn, December 18, 1983, PFP83–FEH1218.

18. Franz Boas, *Primitive Art* (Irvington-on-Hudson, NY: Capitol Publishing Company, Inc., 1951), 9.

19. Dell Hymes, "Folklore's Nature and the Sun's Myth," *Journal of American Folklore* 88 (1975): 348.

20. Kevin Lynch, *The Image of the City* (Cambridge: MIT Press, 1960), 127.

21. Kay Young, "Ethnobotany: A Methodology for Folklorists" (M.A. thesis, Western Kentucky University, 1983), 22.

22. For a discussion of the difference between "knowledge about" and "knowledge of," see Susanne Langer, *An Introduction to Symbolic Logic* (New York: Dover Publications, Inc., 1967), 22–23. Jonathan Berger and John Sinton emphasize the need to combine scientific and experiential perspectives (which they call "cognized" and "operational," respectively, after anthropologist Roy Rappoport) in the management of the Pinelands National Reserve in *Water, Earth and Fire: Land Use and Environmental Planning in the New Jersey Pine Barrens* (Baltimore: Johns Hopkins University Press, 1985), 203–206.

23. Interview, Eugene Hunn with Jack Cervetto, June 22, 1984, PFP84–AEH014.

24. Lynch, *Image,* 127.

25. Interview, Mary Hufford with Richard Gille, September 23, 1983, PFP83–AMH008.

26. Interview, Mary Hufford with Jack and Ann Davis, September 19, 1983, PFP83–RMH010.

27. Interview, Mary Hufford with Joe Albert, August 22, 1982.

28. Hufford with members of Spartan Gun Club, PFP83–RMH044.

29. Interview, Mary Hufford with Frank Day, September 19, 1983, PFP83–RMH016.

30. Interview, Nora Rubinstein with Jack Cervetto, October 28, 1983, PFP83–ANR003.

31. Interview, Mary Hufford with Clifford Frazee, September 23, 1983, PFP83–RMH012.

32. Interview, Mary Hufford with Norman Taylor and Caroline Taylor, September 22, 1983, PFP83–AMH006.

33. Interview, Mal O'Connor with Mary Lamonaca, June 20, 1985, PFP85–AMO001.

34. Interview, Mal O'Connor with Eddie Lamonaca, June 25, 1985, PFP85–AMO005.

35. Fieldnotes, Bonnie Blair, November 11, 1983, PFP83–FBB1111.

36. The concept of "great" and "little" traditions is drawn

from Robert Redfield's formulation in *The Little Community* (Chicago: University of Chicago Press, 1956).

37. Interview, Elaine Thatcher and Sue Samuelson with Clara Paolino, October 11, 1983, PFP83–RET016.

38. Interview, Mary Hufford with Ann Davis, September 19, 1983, PFP83–RMH009.

39. Paula Johnson, "Sloppy Work For Women: Oyster Shucking in Southern Maryland" (Paper presented at the American Folklore Society 1984 Annual Meeting, San Diego, California, October 11, 1984).

40. Interview, Rita Moonsammy with Beryl Whittington, November 1982, PFP83–ARM011.

41. Paul Shepard, *Man in the Landscape: A Historic View of the Esthetics of Nature* (New York: Alfred A. Knopf, 1967), 47.

42. O'Connor with Eddie Lamonaca, PFP85–AMO005.

43. Interview, Jens Lund with George Campbell, November 7, 1983, PFP83–RJL023.

44. Boas, *Primitive Art,* 10.

45. Lund with Peterson, PFP83–RJL017.

46. Interview, Tom Carroll and Nora Rubinstein with Ed Hazelton, November 4, 1983, PFP83–RTC005.

47. Interview, Jens Lund with Ted Von Bosse, September 21, 1983, PFP83–RJL003.

48. Interview, Tom Carroll and Nora Rubinstein with Ed Hazelton, November 4, 1983, PFP83–RNR011.

49. Ibid.

50. Interview, Christopher Hoare with Sam Hunt, April 28, 1978, quoted in David Steven Cohen, *The Folklore and Folklife of New Jersey* (New Brunswick, NJ: Rutgers University Press, 1983), 122.

51. Hufford with Davis, PFP83–RMH009.

52. Interview, Mary Hufford with Norman Taylor, November 1980.

53. Jackson, "Landscapes as Theater," 7.

54. John McPhee, *The Pine Barrens* (New York: Farrar, Strauss and Giroux, 1968), 9.

55. Fieldnotes, Mary Hufford, April 8, 1984, PFP84–FMH0408.

56. Hunn with Brown, PFP83–AEH008.

57. Hunn with Cervetto, PFP84–AEH014. Hunn speculates that this species is *Lampropeltis getulis getulis,* the eastern king snake, based on local descriptions and Roger Conant's allusion to it as "swamp wamper" in *A Field Guide to Reptiles and Amphibians of Eastern and Central America,* 2nd ed. (Boston: Houghton, 1975), 202.

58. Robert Plant Armstrong, *The Affecting Presence: An Essay in Humanistic Anthropology* (Urbana, IL: University of Illinois Press, 1971).

59. David Wilson, *In the Presence of Nature* (Amherst: University of Massachusetts Press, 1978), 1.

60. Interview, Rita Moonsammy with Tom Brown, February 17, 1982.

61. Interview, Gerald Parsons with Albert Reeves, September 24, 1984, PFP84–RGP006.

62. Clifford Geertz, "Deep Play: Notes on the Balinese Cockfight," in *The Interpretation of Cultures* (New York: Basic Books, 1973) 448.

63. Fieldnotes, Eugene Hunn, December 20, 1983, PFP83–FEH1220.

64. Interview, Mary Hufford with Tom Brown, June 1984.

65. Interview, Mary Hufford with Milton Collins, December 1980.

66. Tape of fox-hunting recollections, John Earlin, Browns Mills, New Jersey.

CREATING THE LANDSCAPE:

Historic Human Ecology of the Pinelands

Opposite: *Detail, Apple harvest, Cumberland County. Courtesy of the Donald A. Sinclair New Jersey Collection. Special Collections and Archives, Rutgers University Libraries.*

NO LESS THAN IN ANY OTHER REGION, IN THE PINES THE history of settlement patterns and resource use is reflected in its various landscapes and in the contemporary activities of its residents. And, as in other regions, the array of relationships and activities is stunning. It includes the fishing, gunning, trapping, and recreational landscapes of the coastal areas, as well as the forests, bogs, fields, and military installations of the uplands.

The forces which created the human ecology and landscapes of the Pines stem from the geographic happenstance of the Pinelands' location between New York City and Philadelphia; from the indigenous cultures of the people who settled here; and from the great social, political, and economic forces, such as the Industrial Revolution, which changed the entire nation. The Pinelands is, therefore, the product of its own special circumstances as well as a reflection of regional and national trends. As J. B. Jackson wrote recently:

> No group sets out to create a landscape, of course. What it sets out to do is to create a community, and the landscape as its visible manifestation is simply the by-product of people working and living, sometimes coming together, sometimes staying apart, but always recognizing their interdependence. . . . Like a language, a landscape will have obscure and inde-

John W. Sinton

> cipherable origins, like a language it is the slow creation of all elements in society. It grows according to its own laws, rejecting or accepting neologisms as it sees fit, clinging to obsolescent forms, inventing new ones. A landscape, like a language, is the field of perpetual conflict and compromise between what is established by authority and what the vernacular insists upon preferring.[1]

The history of human ecology in the Pinelands can be divided into three periods based on major changes in the social and economic structures of the northeastern United States and the nation as a whole. The first period spans the seventeenth, eighteenth, and nineteenth centuries. Dutch, English, and New England settlers, followed by northern Europeans and blacks, came down the coast and up the rivers to harvest woods, water, and farmlands. During the second period, beginning in the 1850s, railroads brought immigrant groups from Germany, Italy, and Eastern Europe into the Pinelands. At what must have seemed like isolated spots, they established towns like those they had left: Egg Harbor City, Vineland, and Woodbine. The railroads aided the growth of truck farming, fruit farming, and tourism by linking the region with the cities. Speculators sold quarter-acre lots on what were no more than paper streets. After World War II, the third period brought suburban and second-home developments to the Pinelands.[2]

Depending on the technology of the period and the economic demand for various resources, whether fish, wood, or potable water, different sections of the Pinelands were used for different purposes. In the first half of the first historic period, from about 1680, settlement took place principally along the coasts and navigable rivers. The primary activities were maritime industries, especially shipbuilding, and subsistence agriculture. Between 1760 (the founding of the first ironworks) and 1860, the rural industries of iron manufacturing and glassmaking thrived and died. During the second period, from 1860 until 1950, berry agriculture replaced rural industry in the forest regions, while small manufacturing, commercial agriculture, and recreation increased commensurately. After World War II, residential developments grew rapidly along the coast, while small manufacturing and agriculture declined in the interior.

The first period, from 1680 to 1860, encompassed the first coastal settlements, the beginning of the Civil War, and the arrival of the railroads, and coincided with the high tide of the Industrial Revolution. This period set the broad patterns of landscapes, resource use, settlement, and cultural life that still exist in the Pinelands National Reserve—the basic elements that contribute to the contemporary sense of place.

Two overriding geographic factors laid the foundations of the human ecology that has developed over the past 300 years. First was the simple fact that the Pinelands is located near two of America's largest cities—Philadelphia and New York. The Pinelands, therefore, participated in the birth of the nation and had direct connections to the cultural and political life of what would become a megalopolis. Second, the sandy composition of most Pinelands soil limited its attractiveness to early agricultural settlers. This fact, coupled with the lack of a physical site for a major city along the coast or the Delaware estuary, dictated that the Pinelands would become what Jane Jacobs calls a "supply region," destined to supply materials, people, and services to Philadelphia and New York rather than to become a major economic generator itself. In short, the Pinelands was to remain an appendage, a situation which brought both benefits and curses. While the region could retain its rural character, it would never be able to generate enough economic activity to enrich all its residents. While Pinelands people would have the use of the region's water, woods, and wildlife, much of the land would be owned by more wealthy outsiders. While those who chose to remain in the Pines could live within the seasonal rhythms of the place, they would lose many of the young people to the excitement and riches of the cities. And, during the most recent historic period, a long series of struggles would develop between those who wanted to retain the rural character of the Pines and those who wanted to convert much of the region into metropolitan bedroom communities.

On top of the geographic factors were laid the economic activities and settlement patterns as determined by eighteenth- and nineteenth-century technology and the social and cultural composition of the settlers. The human ecology of that early period can best be appreciated by looking at a map of the Tuckahoe area, which was typical of most other parts of the Pines at that time. Tuckahoe is in the southeast section of the Pinelands region, 15 air miles from Atlantic City. The most outstanding features on the map are the nine sawmills in a 50-square mile area, all built between 1737 and 1812. The original sawmills and ponds, around which small communities of three or four households gathered, served the larger coastal and estuarine villages with their shipbuilding and commercial activities. An 1834 gazeteer described Tuckahoe as follows:

> Tuckahoe, port town on both sides of the Tuckahoe river, over which there is a bridge, ten miles above the sea, forty-six miles southeast from Woodbury (the county seat), and by post-route 192 from Washington, D.C.: contains some twenty buildings, three taverns, several stores. It is

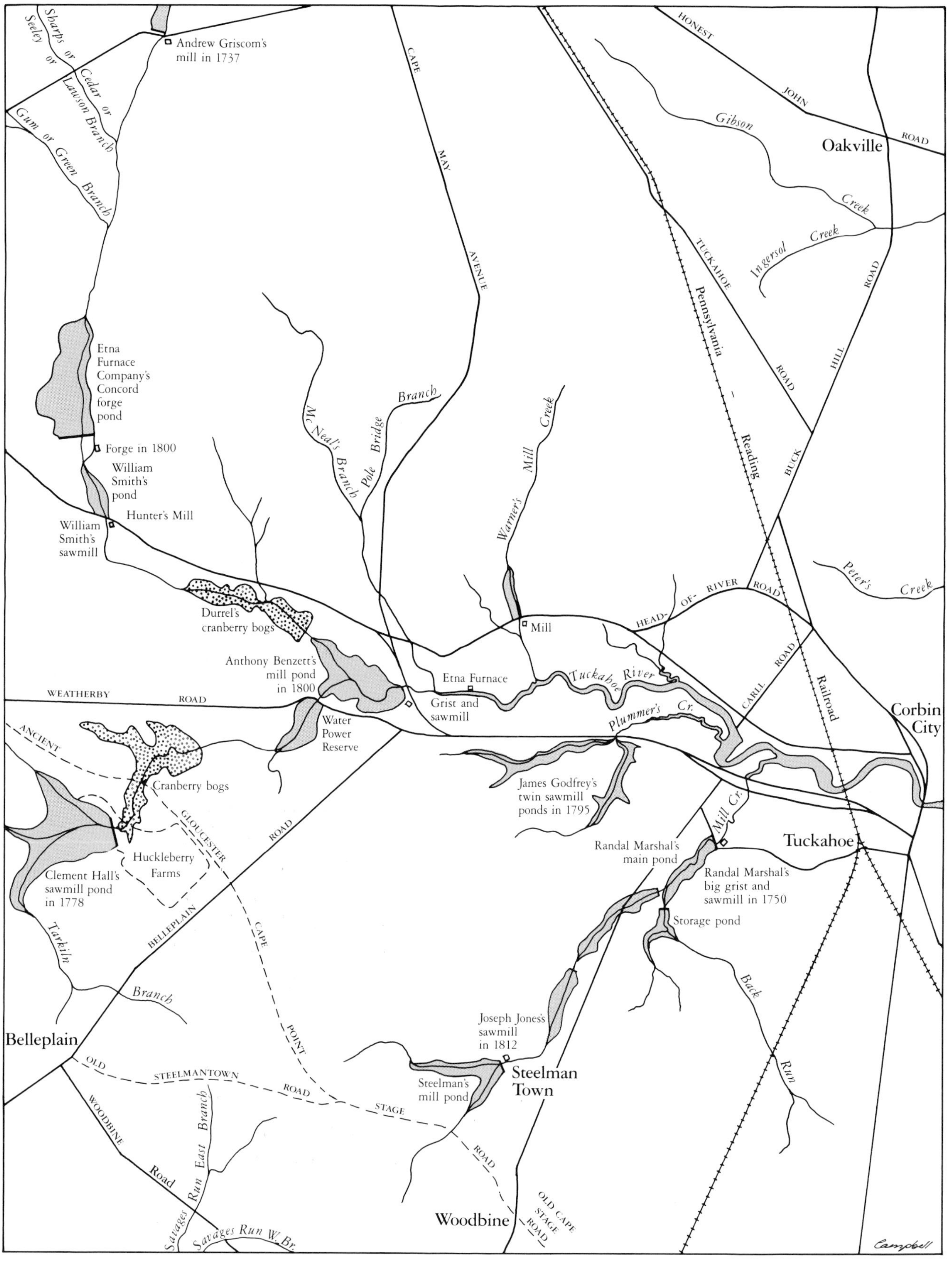

MAP OF HISTORIC SITES IN THE TUCKAHOE REGION
BASED ON A FIELD SURVEY BY CHARLES HARTMAN OF MILLVILLE
(1979)

Opposite: *Map of historic sites in the Tuckahoe region, based on a field survey by Charles Hartman of Millville (1979).*

a place of considerable trade in wood, lumber, and ship building. The land immediately on the river is good, but a short distance from it, is swampy and low.[3]

All the resource-based activities which occurred in the coastal and southern uplands region also took place in the Tuckahoe area, from fishing and shipbuilding to farming, logging, and iron making. The most important long-term activities centered on the coast. Tuckahoe was just one of many small Jersey coastal towns producing boats during the eighteenth, nineteenth, and early twentieth centuries.[4] The shipyards were the hub of economic life in the Pines for 200 years, and almost all of them were on navigable rivers, far enough upstream to be protected from northeast storms and hurricanes. Like other shipyards, the one in Tuckahoe probably produced its share of schooners over 100 tons, although most were in the 30- to 60-ton range.

In 1900 New Jersey had the largest working sailing fleet in the country, and most of the vessels were produced in South Jersey. With its demand for men and materials, this huge industry supported a wide range of auxiliary enterprises. Many kinds of wood and wood products (such as tar), as well as textiles, iron, glass, tallow, sulfur, and paint, were used in boats. The skills of carpenters, shipwrights, sawyers, joiners, carvers, caulkers, sealers, instrument makers, glaziers, blacksmiths, upholsterers, and chairmakers were required. Because the market for boats fluctuates, however, many workers had to combine several skills rather than depend solely on boatbuilding for a living.[5]

While boat- and shipbuilding provided coastal residents with most of their cash, their lives were also patterned around seasonal activities to harvest the rich resources of the environment: fishing and shellfishing, gunning, trapping, salt haying, farming, and logging. These are all activities which have continued to the present day.

It is important to understand both the richness of the coastal resources and, as can be seen from the Tuckahoe map, the relationship between the coastal areas and the forest regions, because it was the rural industries of the forest which fed the larger, more permanent communities along the coast.

The map of the Tuckahoe region shows numerous small mills scattered along all the watercourses. The earliest date for a sawmill or gristmill is 1737, for the Andrew Griscom mill in the northwest corner of the map, well inland from a coastal center and quite near Mays Landing. This location suggests that the inland sections of the Pinelands were settled before

the advent of the iron furnace towns. By extrapolation we can assume that a significant amount of activity was taking place throughout the forest regions of the Pines during the eighteenth century. We can also assume that by the middle of the nineteenth century, almost every river and stream in the Pines had been dammed for some use, whether for iron or glass, or, more commonly, lumber and grist milling. All these mill communities

Mulford's Shipyard on the Maurice River at Millville, 1875. From Atlas of Cumberland County. *Philadelphia: D. J. Stewart, 1876.*

were tied directly to the larger, more permanent coastal communities by waterways or sand roads.

We have no real documentation about these early communities other than notations in deeds and probates or archeological evidence. We do not really know what sort of people settled these hamlets, but their names suggest English or New England origins. The technologies they used for mill-

Sawmill at New Lisbon, 1894. Photograph by Nathaniel R. Ewan. Courtesy of the Burlington County Historical Society.

ing and lumbering and the kind of houses they built were similar to those found throughout the northeast United States at the time. The products they produced and the ships they built in the coastal towns were commercially valuable to merchants not only in Philadelphia and, later, New York, but to the English, who had denuded their own forests for wartime shipbuilding in the seventeenth century and were in dire need of timber from the American colonies.

From their earliest settlement, then, Pinelands communities were connected to a larger region—indeed, to world commerce. The patterns of Pinelands towns, even the architecture of the houses, were derived from general patterns found in England and the Northeast, and the dominant religion, Quakerism, stemmed from the same cultural background. Some old coastal towns, such as Barnegat and Tuckerton, still preserve part of their original character, and many of the eighteenth-century houses stand today, testaments to the original settlers.[6]

By 1760, when the first Pinelands iron furnace was founded at Batsto, the social and geographic patterns of the first historic period were already established. The creation of the so-called iron "plantations" in the Pines (approximately two dozen in all between 1760 and 1850) was triggered by growth in both the population and the self-sufficiency of the Colonies as they broke away from England. The Pinelands was a particularly appropriate place in which to found iron manufacturing because of its proximity to Philadelphia with her ready markets and capital, and its richness in the resources that were necessary for iron making, especially wood for fuel.

The period from 1810 to 1860 was a busy industrial time not only for the forest regions of the Pinelands, but also for the rural Northeast as a whole. The Industrial Revolution was just beginning, but meanwhile most industrial products were being produced in rural areas which used wood for fuel and water for power. In addition to the two dozen iron furnaces in the Pines, there appeared a smaller number of forges and slitting mills. Beginning about 1800, a series of 20 glass factories were founded. Most had died by 1870. Some lasted to 1930, and a few were transformed into the modern glass industry around Millville. There were also several paper mills, which used salt hay (*Spartina*) for raw material, and cotton factories, which produced cloth from rags.

The principal natural resource on which rural industry rested was wood, and because of this, people continually cut the woodlands, from the smallest pitch pine to the largest cedar. The enormous demand for wood,

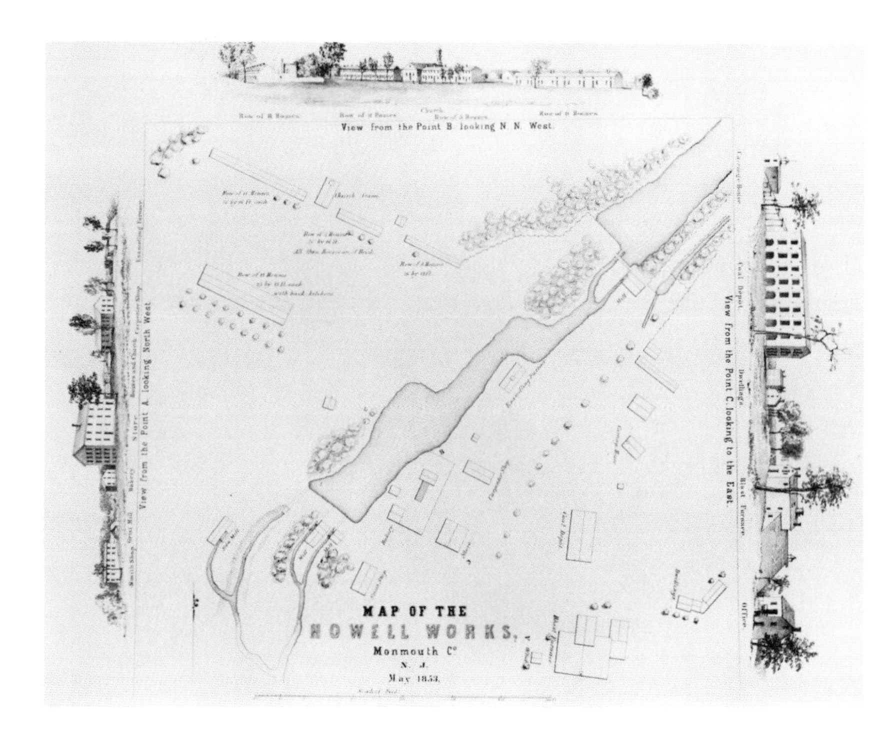

Map of the Howell Iron Works, Monmouth County, 1853. Courtesy of the Monmouth County Historical Association.

coupled with uncontrollable forest fires, drastically changed the landscape of the region from its earlier form.

A medium-sized iron furnace, such as the one at Batsto that produced some 900 tons of iron annually, required about 6,000 cords of wood per year for the 12,000 bushels of charcoal used to fire its furnaces. Forests in the Pinelands generally produce only one-third to one-half a cord of wood per acre per year. A conservative estimate is that the five furnaces in the Batsto area (the Wharton Tract) would have required 40,000 acres of woodland annually for their production. But in fact more wood was needed because in that same area were three forges, four glasshouses, a paper mill, and a cotton mill. Furthermore, oak had to be cut for domestic fuel and cedar for

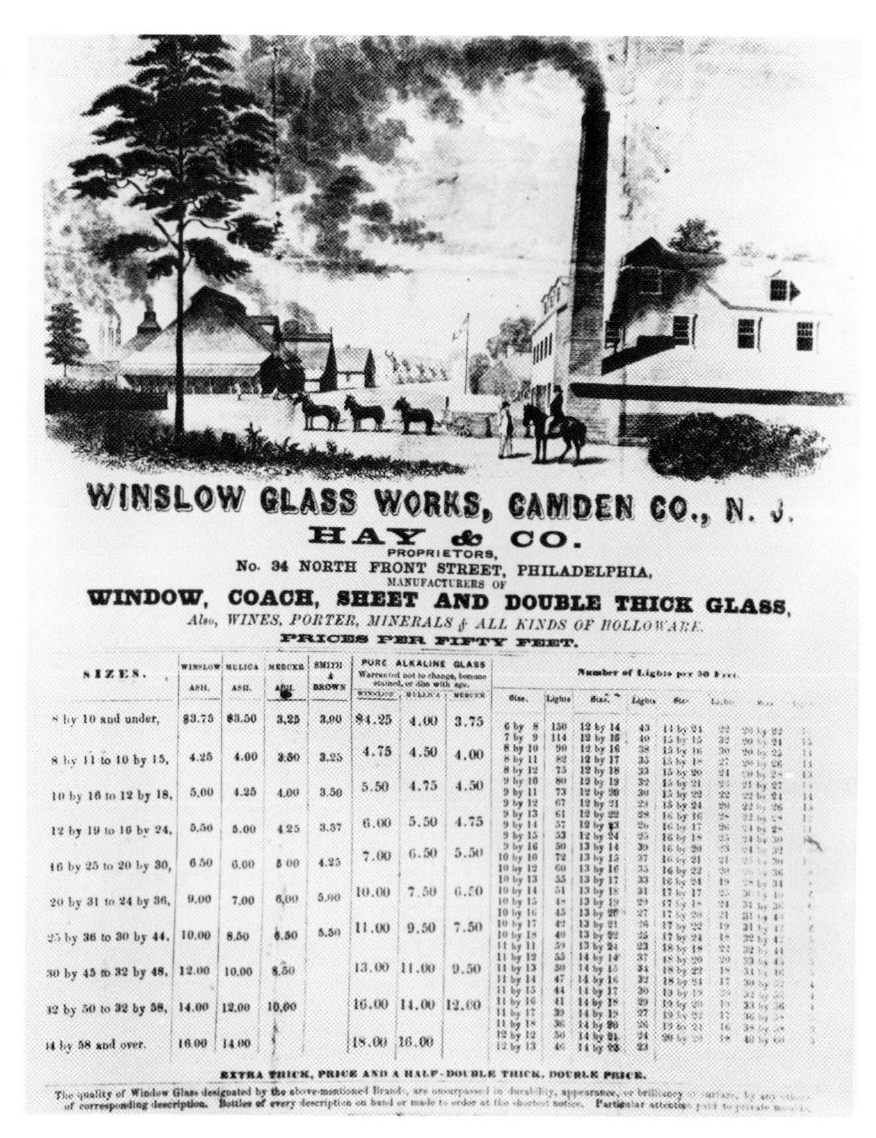

Winslow Glass Works, Camden County. Courtesy of the Donald A. Sinclair New Jersey Collection. Special Collections and Archives, Rutgers University Libraries.

George C. Rowand charcoal burner's cabin, MacDonald's Branch, Lebanon State Forest, 1927. Courtesy of George H. Pierson, New Jersey Bureau of Forest Management, Division of Parks and Forestry.

boatbuilding and shingle production. Coppice, or young oak growth, was cut for hoop-poles, for barrels, for pole wood, and for basketmaking. Most economic activities in the Pines depended on wood and hardly a man was born in the forest regions who, at one time or another, did not cut wood or fell trees.[7]

The Atlantic white cedar of the wetlands was and still is by far the most valuable tree in the Pines, as much because of its scarcity as because of its irreplaceable qualities of light weight and resistance to weather and disease. It is in high demand for roofing, fence posts, and boats. Oak is most valuable as a mature tree for construction and in small pieces for heating fuel. Pitch pine was, during the first historic period, the most useful wood because it was essential to industry. Small trees were cut mostly for charcoal, but some large pines were used for construction, poles, and later for railroad ties and newspaper pulp.

The insatiable demand for wood from the Pines throughout the nineteenth century required that the whole region be cut several times over. Each year the woodcutters and colliers combed their areas for new sources of wood and cut everything one inch in diameter and up. They then moved to the next woodlot, leaving slash and sand behind them, and sometimes even purposely burned the site so they could purchase it cheaply and harvest it 15 years later. Further, wildfires claimed tens of thousands of acres of forest every decade. Reports from the *New Jersey State Geologist* in the late nineteenth century are filled with descriptions of fires that burned 10, 20, or 30 thousand acres at a time. A hundred years ago much of the Pinelands looked barren indeed—a forest devastated by axmen and wildfires.[8]

Did nineteenth-century people understand that they were destroying their forests? Perhaps we need to understand what destruction and creation meant to them and to explore what was economically important to those people rather than aesthetically important to us. At what point in the life of a forest—in terms of ecological succession—could the rural industrialists expect to get the species of trees they needed to support their economy? Put most simply, how could people in the nineteenth century get the most cedar and pine in the shortest time possible?

Had the people of the Pines been ardent conservationists a hundred years ago and protected their forests from fire and cutting, they would have produced precisely the kind of woodland least useful to their economy, since climax conditions result in oak and mixed swamp-hardwood stands. People needed pine and cedar, which generally occur early in succession and, in fact, cannot regenerate extensive stands unless they are disturbed. By cutting forests and starting fires intentionally, the people of the Pines were, in ecological terms, keeping the forest in an early successional stage and, in economic terms, assuring themselves of the kinds of wood, especially pine, necessary to support their industries.

In 1860, after three generations of heavy cutting and burning, the Pines were certainly barren. Photographs from the turn of the century show desolate backgrounds, barren land, and slash piles. But the regenerative powers of even this poor soil were suufficient to supply nineteenth-century industrial needs. Out of the ashes and stumps and roots grew the pines that made the wood and charcoal that ran the furnaces, forges, and glasshouses. Every 20 years the wood choppers and colliers gathered their harvest on a given site, and the cycle of the phoenix continued until the industries died.[9]

Industrialization of the forest regions also meant drastic changes in settlement patterns, and for a hundred years the forest provided habitation for at least twice the number of people now living there. Following is a contemporary description of a typical iron town, this one in Weymouth near Mays Landing:

> Weymouth, blast furnace, forge and village, in Hamilton Township, Gloucester County, upon the Great Egg Harbour River, about five miles

Aerial view of Batsto. Courtesy of the Office of New Jersey Heritage, Division of Parks and Forestry.

above the head of navigation. The furnace makes about 900 tons of casting annually: the forge having four fires and two hammers, makes about 200 tons bar iron, immediately from the ore. There are also a grist and a saw mill, and buildings for the workmen, of whom 100 are constantly employed about the works, and the persons depending upon them for subsistence, average 600 annually. There are 85,000 acres of land pertaining to this establishment. . . . The works have a superabundant supply of water, during all seasons of the year.[10]

All the connotations of the word "plantation" were appropriate for the iron towns which the Philadelphia Quakers established. The most important plantation owners were the Richardses of Philadelphia, who created, in Arthur Pierce's words, an "Empire in the Pines." The family bought huge tracts of forest land, the borders of which were contiguous, so that a pattern of large-sized land holdings was established. The result was to create outside ownership of much of the forest region, some of which still exists today in the hands of speculators and developers. Other large forest tracts are held by local residents who are cranberry farmers.

The central place in these tracts was the iron village, and in the center of the iron village was the owner's mansion.

Batsto was the seat of the Richards's paternalistic empire, and their rambling, wooden mansion was the center of authority and social life in the village. In addition to the family, the mansion housed servants and unmarried workers, who lived in basement quarters. Near the mansion were a gristmill, sawmill, spring house, summer house, garden and orchard, smokehouses, and barns. Across the millpond were workers' cottages, gardens, and outbuildings. The workers, who came from Europe, other forges, and employees' families, enjoyed secure, well paid jobs and active social lives. Long tenure of skilled workers was common.[11]

The residents of the rural industrial towns were individuals from families who had settled there in the eighteenth century as well as immigrants from northern Europe and a few blacks whom the Quakers had helped to come north on the Underground Railway. While there were a few Catholics in the Pines, chiefly Irish immigrants, most of the residents had become Methodists by the first quarter of the nineteenth century. The Methodist circuit riders of the early 1800s had achieved a clear success in the Pinelands against the rival Episcopalians and Presbyterians. To this day, Methodism exerts a strong influence in Pinelands towns, and the church continues to play the principal role in the social life of women there.

By mid-century the tide of the Industrial Revolution caught up with the rural industries, not only in the Pinelands, but throughout the Northeast. New technologies in industry and transportation emptied all the rural industrial areas of the region as well as those of New York and New England. Within a generation, the population of the Pinelands was cut in half as workers and their families moved to the cities. The pine woods, sand, and bog iron were no longer useful to an industrializing society.

When revolutionary technological, economic, and social changes occur, however, some of the old ways of life are always left to mix with the new. Some activities from the first historic period continued into the second, which spanned the years from 1860 to 1950: fishing, gunning, trapping, salt haying, and to a lesser extent, boatbuilding. Some activities increased, such as agriculture, and some new activities appeared: plant gathering and collecting, cranberrying and commercial blueberrying, working on military installations, small manufacturing (mostly of clothing), and catering to tourism.

The impetus for economic growth came chiefly from railroads, which opened access to urban markets for goods and services from the Pinelands region. The most lasting impact was to cut the Pines into two sections, north and south of the Mullica River. Because the two major east-west lines ran to the tourist resorts of Atlantic City and Cape May, a rail net developed

Apple harvest, Cumberland County. Courtesy of the Donald A. Sinclair New Jersey Collection. Special Collections and Archives, Rutgers University Libraries.

throughout the southern section, while the northern section had no east-west link, only one that ran north-south through the central, least populated, areas.

The rail net south of the Mullica spawned truck farms and orchards, health spas and boondoggle real-estate developments, small clothing industries, and ethnic settlements. Although large tracts of forest remained for charcoal makers and loggers to work, much of it was turned into fields—the higher ground for peaches, apples, tomatoes, cucumbers, and sweet and white potatoes, and the wetter ground for blueberries, which were cultivated after World War I by Elizabeth White at Whitesbog near Browns Mills. In addition, cranberry fever hit the region in the 1860s, and landowners and lessees turned old millponds and swamps into cranberry bogs. An enormous amount of cedar was lost as the trees were drowned and cut to make way for new bogs. The map of the Tuckahoe region illustrates these patterns from the second historic period with its cranberry bogs, blueberry fields, and rail lines.

As it did throughout America, the railroad also created new speculative and settlement opportunities in the Pinelands. Italians, Germans, and Eastern European Jews settled in ethnic communities in the southern region. Some came as railroad workers, while others bought or were given land by German Protestant and Jewish philanthropists. The Jews of Woodbine, the Germans of Egg Harbor City, and the Italians of Hammonton built self-contained agricultural and industrial communities and thrived; their economic preeminence held until after World War II. Speculators, including railway companies, bought cheap forest land and advertised

Jewish farm family, Woodbine, circa 1900. From William Stainsby, The Jewish Colonies of South Jersey. *Camden: Chew and Sons, 1901.*

small lots to anyone foolish enough to buy them. Speculative development littered the maps of the southern Pinelands, almost all of them failures like Montefiore, Gigantic City, and Little Italy. In the late nineteenth century, New York and Philadelphia newspapers commonly gave away small pieces of ground in the Pinelands with new subscriptions.

North of the Mullica River, speculators and cranberry operators also flourished. As happened in the southern Pines, few developments succeeded. Paisley and Fruitland, for example, are now mere curiosities. Because the watersheds were somewhat larger there than in the south, and competing land uses fewer, extensive cranberry bogs still exist. The old rural industrial towns and outlying villages, however, are now lost in the woods. No railroads came to reinvigorate the economy, so the northern Pinelands slowly returned to woods, swamps, and bogs after the Civil War. Woodcutters, colliers, gatherers, hunters, and berry cultivators exploited the section of the Pines no one else wanted.[12] These same patterns set the stage for most of what a traveler can see today on Pinelands landscapes.

And what happened to the people of the Pines during this period? As the saying goes: "You can always make a buck from the bay or a dollar in the woods." Those families who elected to remain lived with the rhythm of the seasons and followed the complex patterns of resource exploitation established by their predecessors. As the processes of industrialization and urbanization sped forward, the gap between rural and urban cultures widened to the point where metropolites looked on rural people as odd and backward. It was during this period that the stereotype of the "Piney" was born: the stories of strange and isolated people deep in the Pine woods, all of whom would shoot strangers on sight.

The root of the stereotype lies in the study of the Kallikak family that was conducted at the Vineland Training School in the first decade of this century by H. H. Goddard and his research assistant, Elizabeth Kite. Stephen Jay Gould studied Goddard's work and found that:

> Goddard discovered a stock of paupers and ne'er-do-wells in the Pine Barrens of New Jersey and traced their ancestry back to the illicit union of an upstanding man with a supposedly feeble-minded tavern wench. The same man later married a worthy Quakeress and started another line composed wholly of upstanding citizens. Since the progenitor had fathered both a good and bad line, Goddard combined the Greek words for beauty (*kallos*) and bad (*kakos*), and awarded him the pseudonym Martin Kallikak.[13]

In fact, the family on which the Kallikak study was based was from Hunterdon County, not the Pines, as noted in a recent book by J. David Smith.[14]

In an absurd attempt to support the budding eugenics movement, Goddard and Kite traced the *kakos* family history through nine generations of alcoholism, blindness, criminality, deafness, epilepsy, insanity, syphilis, sexual immorality, tuberculosis, and feeblemindedness. The Goddard study was a hoax. Goddard's photographs, which he used as supporting evidence, "were phonied by inserting heavy dark lines to give eyes and mouths their diabolical appearance."[15] Even the tongues hanging out of the depraved mouths had been forged. Still, the stigma of the *kakos* line attached itself to all residents of the Pine Barrens.[16]

In fact, the rural people of the Pines lived not much differently from people in rural Pennsylvania, New York, or New England. They led independent lives in small communities centered on the Methodist church, and, in the early twentieth century, on gun clubs and volunteer fire companies. In the north, the pattern of large-scale land holdings continued. The Richards family was replaced with cranberry owners and another Quaker from Philadelphia, Joseph Wharton, who bought more than a hundred thousand acres of land in hopes that he could use the Mullica River basin as a source of potable water for Philadelphia.[17] When the state of New Jersey made out-of-state water shipments illegal, Wharton simply held the land in single ownership, and his heirs sold it in 1950 to the state, which established Wharton State Forest.

In other forest sections, speculators bought hundreds of thousands of acres in both large and small parcels, regardless of whether their titles were clear. Land speculation and confusing small landholdings are now part of the contemporary heritage of the Pines.

Still other parts of the Pinelands became cultural refuges for groups fleeing Europe. A great brick synagogue bears witness to the Jewish settlement in Woodbine during the 1890s, although few Jews are left, most of them having gone to nearby urban areas. Egg Harbor City was established by German immigrants in the 1850s, and the pattern of the town and its architecture reflect typical German burgher attitudes. Hammonton became a predominantly Italian town settled by emigrants from several communities in the Neapolitan region and Sicily during the 1860s and 1870s. Cassville in the northern Pines was settled after the Russian Revolution by Russian Orthodox groups and still serves as a major regional religious center. So long as land was cheap and immigrants plentiful, the Pinelands served them as a refuge, and the refugees added a great deal of cultural diversity to a region which, up until the mid-nineteenth century, had been culturally homogeneous.

Lastly, the railroad brought significant growth to coastal sections of the Pines. Atlantic City burgeoned in the 1870s as the playground for blue-collar workers from Philadelphia. Along the coast of northern Ocean County, recreational activities and summer home development grew slowly, so that many coastal residents were able to earn cash in the building trades as well as in older seasonal activities. The slow pace of change, however, would be threatened by new technological and social developments after World War II.

In the period following the war, from 1950 to 1985, two major shifts occurred: the small industrial complexes of the southern section began to die as industry continued to concentrate in metropolitan areas, and suburban, second-home, and retirement developments began to grow with the advent of the population boom and high-speed roads and automobiles.

The northern fringes of the Pinelands have borne the brunt of suburbanization since 1950, not so much because land was cheaper, but because the area was closer to Philadelphia and New York. Ocean County was the fastest-growing county in the United States between 1960 and 1980. Although the north-central section, the so-called "core" of the region, still displays some landscapes left from rural industrial and railroad periods, suburbs and retirement communities have nibbled off the borders all along the coastal and Delaware Valley sections. At times suburbs have leap-frogged over existing developments into the heartland, as has the retirement community of Leisure Village.

The southern section of the Pines, being farther from any metropolis, witnessed little suburbanization until recently and has seen its small industrial base die. Woodbine lost all its industries, and Egg Harbor is but a shadow of its old self; only Hammonton retains a solid but diminished industrial base. Along the coastal strip are miles of vacation homes, but few such developments have appeared inland, although significant tracts are now being set aside for camping. The southern section still bears the imprint of the second wave of settlement from the railroad, but, because of its proximity to Atlantic City casinos, it is finally under the pressure of suburbanization.

It is, nonetheless, surprising how similar a twentieth-century land use map is to one of the nineteenth-century. Today, residents of the Pinelands still use almost all the resources of the region that were used in the eighteenth and nineteenth centuries, from the shellfish of the bays to the wood of the forests to the water underground, and ties to land and resources are still important.

What is so valuable about the old ways, seasonal activities, and family and community ties? Though intrinsic interest in the old way of doing things is part of the value, the mutually advantageous balances between people and environment that are maintained in seasonal lifestyles are the greater value. The balances have evolved over time and have mediated changes in resources and demand. They have supported subsistence lifestyles independent of highly technological processes. But they are also based on interdependencies that suburbanization and industrialization tend to break apart. Once this happens, a whole series of adaptations also fall apart.[18]

What threatens the people of the Pines now is not change itself, but change that does not allow time for people's institutions and sense of place to settle into new patterns, and that, furthermore, may destroy crucial resources, whether clams or cedar. The attitudes of Pinelands residents do not differ significantly from those of residents of other areas. Some search for money, others for excitement and material goods, while most, like us, would like a whole mix of values even if the mix is unobtainable. The ways in which the people of the Pines work out their value systems and deal with their resources will decide the future of the region's landscapes and human ecology.

NOTES

1. John B. Jackson, *Discovering the Vernacular Landscape* (New Haven: Yale University Press, 1984), 12, 148.

2. Jonathan Berger and John W. Sinton, *Water, Earth, and Fire: Land Use and Environmental Planning in the New Jersey Pine Barrens* (Baltimore: Johns Hopkins University Press, 1985), 6.

3. Thomas Gordon, *A Gazetteer of the State of New Jersey* (Trenton: D. Fenton, 1834), 254.

4. Glenn S. Gordinier, "Maritime Enterprise in New Jersey: Great Egg Harbor During the Nineteenth Century," *New Jersey History* 97 (1979): 105–117.

5. Berger and Sinton, *Water, Earth, and Fire*, 62–63.

6. For more information on the South Jersey house, see Elizabeth Marsh, "The South Jersey House," in *History, Culture, and Archeology of the Pine Barrens*, ed. John W. Sinton (Pomona, NJ: Stockton State College, 1982), 185–187.

7. Berger and Sinton, *Water, Earth, and Fire*, 116.

8. Ibid., 117.

9. Ibid., 118.

10. Gordon, *Gazetteer*, 263.

11. Berger and Sinton, *Water, Earth, and Fire*, 64.

12. Ibid., 8–9.

13. Stephen Jay Gould, *The Mismeasure of Man* (New York: W. W. Norton, 1981), 168.

14. J. David Smith, *Minds Made Feeble: The Myth and Legacy of the Kallikaks* (Rockville, MD: Aspen, 1985).

15. Gould, *The Mismeasure of Man*, 168.

16. Berger and Sinton, *Water, Earth, and Fire*, 11.

17. Ibid., 66–67.

18. Ibid., 41.

LIVING WITH THE LANDSCAPE:

Folklife in the Environmental Subregions of the Pinelands

THROUGHOUT THE HISTORICAL PERIODS OF LAND USE IN the Pinelands, the traditional knowledge of place has enabled natives to shape the environment and harvest its resources, as well as to adapt to change. An important component of that knowledge has been the character of each of five major environmental subregions in the Pines: the pine and oak forests, the cedar and hardwood swamps, the meadows (salt- and fresh-water marshes), the rivers and bays, and the farmlands. Each of the first four has a special relationship to the underlying Cohansey Aquifer, and supports a characteristic array of flora and fauna. The farmlands are distinguished chiefly by their soils and type of use.

Opposite: *Detail. Cranberries being scooped, Hog Wallow, circa 1940. Photograph by William Augustine. Courtesy of the Donald A. Sinclair New Jersey Collection. Special Collections and Archives, Rutgers University Libraries.*

In the woodlands, where the water table is two feet or more beneath the sandy soil, pine and oak trees compete for light, water, and nutrients. In pine-oak forests, pines are more abundant than oak trees, and smaller scrub oaks and shrubs of black huckleberry and lowbush blueberry crowd beneath them. In oak-pine forests, on the other hand, leafy canopies of 50-foot oaks throw deep shadows over pine trees and shrubs. A wide variety of birds, as well as rabbit, fox, raccoon, squirrel, and deer, dwell in these forests. In the Pine Plains near Warren Grove, pygmy pines of less than six feet huddle with scrub oaks and mountain laurel. These dwarves, which hold their seed tightly in closed (serotinous) cones that only open in forest-fire temperatures, have survived frequent fire through their ability to sprout from the stump after their crowns are burnt.

Rita Moonsammy / David S. Cohen / Mary T. Hufford

Interspersed throughout the forests and ribboned along rivers and streams are the swamps, the second type of environment. In a cedar swamp, the land dips near the aquifer, often along a stream bed, and acid soil and water nourish the tall, straight Atlantic white cedar. The water and soil are carpeted with spongy mats of sphagnum moss, and decorated with insectivorous plants such as sundews. In swamps where cedar has been cut off, hardwoods such as red maple and black gum now flourish.

In the meadowlands that skirt the woods and edge the rivers and bays, the water table reaches the surface. Plants such as broomsedge and bullsedge form tussocks in the flat, treeless freshwater marshes locally known as savannas. On the salt water meadows lining the coast, *Spartina* grasses are the dominant lifeform, with *Phragmites* (common reed), sedges, and rushes also growing along the brackish upper margins. Muskrat, mink, otter, beaver, and weasels inhabit the meadows, and waterfowl and shorebirds visit on their journeys up and down the Atlantic flyway.

The rivers that drain the Pinelands, and the bays of the Atlantic Ocean, into which they empty, are the fourth type of environment. They serve as transit routes for humans and as homes for many kinds of fish and shellfish. In the lower portions of rivers such as the Maurice and the Toms, marine fish such as the American shad and the striped bass spawn, while clams, crabs, and oysters populate the bays.

These four are the environments that nature created in the Pinelands. Humans created the fifth, the farmlands that have, over the years, been carved out of woodlands, swamps, and meadows. Their soils range from gravelly sand to sandy loam. In sandy, acid soils such as those classified as the Lakewood series, berries grow well. In those with more clay, such as the Sassafras series, vegetable farms prosper. Over time, people have learned to match soils and crops.

The folklife expressions that have been shaped by these environments, and that in turn have shaped them, are many. They range from the tools and processes that have been used to work the land, to the stories and pictures that reflect it. In all of these, as well as in the words of its natives, we may read the story of the Pinelands.

PINE AND OAK FORESTS

Lumbering

"Sawmills" wrote archeologist R. Alan Mounier, "are among the earliest and most durable sites of historic settlement in the Pine Barrens of New

Jersey."[1] Local wisdom underscores this scholarly observation in the saying "All the crossways lead to Candlewood." Located in Barnegat Township, Candlewood was one of the earliest sawmills. The remaining network of crossways (wooden roads) that surrounds it attests to the importance of sawmills in the early settlement of the woodlands. Many of these mills were built in the early eighteenth century to provide lumber for local ship- and house-building, as well as for export. Although they were scattered in isolated locations throughout the Pines, sawmill sites became centers for rural settlements and were followed over time by commercial activities such as gristmills, iron plantations, and cranberry bogs. Today the Pines is crisscrossed with roadways and dotted with place names and ruins that reflect that past: Braddock's Mill, Browns Mills, Pleasant Mills, Jones' Mill.

The two most important types of wood harvested in the uplands have been oak, used for ship- and home-building and basketmaking in the past, and for firewood in the present, and pine, used as fuel for Pinelands industries in the form of charcoal and cordwood. In the swamps that line stream beds in the woodlands grows the white cedar that has also been an important wood in the lumbering industry.

Because early sawmills were water powered, they had to be located near a water supply that could be dammed to make a "head of water," or millpond, from which a sluiceway would convey the water to the waterwheel. Because the streams of the Pinelands are generally shallow and slow moving, the waterwheels were probably "undershot," meaning that the water passed below, rather than above, the wheel. In about 1825, the circular saw replaced the vertical frame saw, and steam power replaced water in the latter part of the century. Early in the twentieth century, the internal combustion engine was introduced.

Today, the logging of numerous species of trees is usually done in small operations that include a wide range of technology. Jack Cervetto describes the engine that used to power his mill:

> That's a straight A Studebaker. '36. And that'd purr like a kitten when I was runnin' the mill. Run a belt from the driveshaft to the main pulley and that would run everything here. I had four or five saws there. And I got a well dug right alongside the motor, and then with a hose into the motor and the water would go in and come out the other side and keep the motor cool.[2]

George Brewer, Jr., of Dennisville eschews a highly automated setup and compares his mill with that of his father, George Sr., in an earlier time. "Our sawmill is very elementary. It's about as simple as you can get." It is

Interior of sawmill of Leroy Creamer of Dennisville. Photograph by Jens Lund. PFP216898–5–3.

diesel powered, and everything is cut on a head saw, because "with too much investment, you'd have to run a lot more material through to make it pay, and there's not that much available."[3]

Thorough knowledge of both the forest and the trees enables the woodsman to maintain himself. He can "read" in a stand of trees both its history and the kind of "bottom," or soil, that supports it. "If you find a bunch of pine in oak bottom, man has been there," explains Cervetto.[4] Early settlers once cleared oak stands to use the rich, loamy soil. Left untilled, however, the soil was quickly taken over by pines.

Fire, as well as humans, has also cleared land, and knowledge of its history can pay off:

> See, there's a place back there. Now a forest fire hasn't been in there and see, those oaks are pretty good size. But this was all burned out in 1936. Clean killed all that. I cut that dead wood off for firewood. And that growed since 1936. There's nothing there to cut anything out of.[5]

Bottoms can affect safety and business as well as vegetation. "Sugar sand bottom" has caused travelers problems for years.

> There's three hills below my house going to Tuckerton and two of 'em is sugar sand. So an automobile or even in horse and wagon days, they'd never go to Tuckerton on account of that. They'd cut right down in that sugar sand, of sand bottom. And you'll find pieces of it in different old roads through the woods. It's bad as clay bottom. It dries like concrete. Gets wet and it's like mud. Go right down to the axle.[6]

A woodsman must know the characteristics of different woods and their uses. Cervetto explains that red cedar is useful for smaller items such as hope chests and gun stocks but is too knotty for large lumber. Pine, on the other hand,

> . . . is a valuable wood. Pine is used where it don't get wet. . . . Pine was used to frame a house, for floor beams, for ceiling, two-by-fours, rafters, but it also has to be covered away from dampness. Because the drier it gets, the harder it gets. For the outside wood they used cedar, for clapboard, cedar shingles. That withstood the weather a lot better than pine. All the old homes are built that way. Pine frame and covered with cedar.[7]

Many people in the Pinelands have built their own homes, including Cervetto.

Whether armed with an ax which he sharpens daily, as Cervetto does, or different sizes of chain saws, a woodsman will have to use the wind as another tool. "The main thing is that you got to catch the breeze," says Cervetto. "You see the wind controls them if the wind's blowing. That green top up there, if a little wind blows that, you couldn't hold it no how. I always watch that wind and then get to cutting where the wind helps you drop it where you want to."[8]

While today, independent woodsmen like Brewer and Cervetto feel that they make a comfortable living by combining lumbering with other seasonal work, in the nineteenth century, at least some of the woodsmen who worked for the owners of iron foundries felt less content. Their folksongs reflect their dissatisfaction with their lot in life. In 1937, John Youmans of Lakehurst sang "This Colliers Mill's a Very Fine Place" for folklorist Herbert Halpert. The song had been composed earlier by Rasher Moore of Colliers Mill to mock an owner who paid his workers in scrip money which was only good at the company store. In winter, when no work was available, he cut off credit at the store and made the workers either pay rent or vacate the company houses in which they lived. Sang Youmans:

Basketmaker named Dilks near Pennypot, Atlantic County, 1940. Photograph by Nathaniel R. Ewan. Courtesy of the Burlington County Historical Society.

This Colliers Mill's a very fine place,
a sawmill and a store,
And a hundred-acre cranberry bog
lie just before the door.
As I went down to Colliers Mill,
the snow bein' very deep,
As I went down to Colliers Mill
to get something for to eat.
There I received a notice and plainly
it did say,
"Vacate your house, old Rasher,
or I'll have you right away."[9]

Basketmaking

White oak (*Quercus alba*), which grows in the uplands, was used in the nineteenth century and in the first half of the twentieth century for basketmaking. There are two European-derived styles of traditional, wooden-splint baskets in the eastern United States. In the Adirondack Mountains and throughout New England, the splints were made by pounding ash logs to separate the yearly rings. Throughout the southern Appalachian Mountains, oak logs were split, and the splints were shaved on a shaving horse.

According to folklorist Henry Glassie, the occurrence of the pounded-ash and the split-oak traditions helps define the boundary between the southern and northern Appalachian Mountain culture areas.[10] The South Jersey baskets were in the split-oak tradition and were made in a variety of shapes and sizes, depending on their use. They included charcoal baskets, berry baskets, farm baskets, egg baskets, and eel traps. In the twentieth century, factory-made peach baskets began to replace the handmade, traditional baskets.

Perhaps the most famous basketmaker in the Pinelands was "Indian Ann," who lived near Indian Mills. She was said to be the daughter of a Delaware Indian named Ash Tamar or Elisha Moses Ashatama. Ashatama was indeed a Delaware Indian surname. The 1802 agreement to sell lands at

Hires
R-J
ROOT BEER
WITH REAL ROOT JUICES
NOW
MILD
LaAZORA
CIGAR
NOW 5¢

Clarence Morgan using splint-shaving horse. Courtesy of George H. Pierson, New Jersey Bureau of Forest Management, Division of Parks and Forestry.

the Brotherton Indian Reservation was signed by Elias Ashatama and Ann Ashatama, among others.[11]

"Indian Ann" married a man named Roberts. She was listed in the 1880 census as an Indian, age 87, whose occupation was basketmaking.[12] Her will, dated 1894, mentions that she had three sons and three daughters and that she owned a house and land in Shamong Township, Burlington County.[13] According to a local newspaper article dated 1932:

> Indian Ann outlived her husband by many years and supported herself by making baskets which she sold in the neighborhood and in nearby towns. In summer she gathered blackberries and huckleberries and frequently walked as far as Vincentown, eleven miles distant, to sell them. She was a familiar figure in Vincentown and Medford, and many interesting anecdotes are told about her visits to these villages. She was usually accompanied by two small dogs, to which she was greatly attached.[14]

Opposite: *Clarence Morgan of Dividing Creek, basketmaker, 1944. Courtesy of George H. Pierson, New Jersey Bureau of Forest Management, Division of Parks and Forestry.*

Several local historical societies have baskets attributed to Indian Ann, and she is a popular figure as well in efforts to evoke the past. Indian Ann applehead dolls are made by residents and sold at festivities such as the Medford Apple Festival. Baskets that she is believed to have made also appear in decorative assemblages in area homes.

Charcoal Burning

The abundant pine of South Jersey woodlands was used to make charcoal, the fuel which, starting in the fourteenth century in Europe, was used in iron foundries. It also was used as the fuel in the New Jersey iron industry, and developed in the Pinelands along with that industry. After the decline of the iron industry here, charcoal was put to use as fuel for Delaware River steamboats and home furnaces in urban New Jersey.[15]

In the twentieth century, rural colliers had to compete with large commercial operations using brick kilns, but the small charcoalers continued to work in New Jersey until about 1970. In recent years, colliers Harry and Herbert Payne of Whiting made charcoal for a variety of users, including roofers who heated their tar pots with charcoal, restaurants that used it for

Split-oak baskets made by Clarence Morgan of Dividing Creek. New Jersey State Museum Collection, 175.1.2, 175.24, 82.50.19, 175.26 and 77.80.7. Photograph by Joseph Crilley.

"hearth" cooking, and local artists who used it for drawing. The brothers recalled to folklorist Mary Hufford that it "got to be quite an industry" in Whiting, where "you can't walk more than a half a mile in any direction without stumbling across evidence of an old kiln." During World War II, they sold much of their charcoal for use in the manufacture of gunpowder.[16]

In large-scale operations during the heyday of the industry, six to ten pits were often fired at the same time, spaced approximately 200 feet apart in an arrangement known as the "ring." A crew of men with specialized jobs were involved in the process: woodchoppers cut the cordwood; setters constructed the pits; scalpers placed the floats (pieces of turf) on top; blackers covered the floats with sand; firemen kept watch on the burning pits; drawers raked the floats off the pits; and teamsters transported the wood and charcoal.

These colliers often lived in temporary log cabins, known as "float cabins" because they were covered with floats and sand, like the charcoal pits. While the firing was underway, smaller lean-to structures, known as "watch

Indian Ann's house, Shamong Township. Photograph by William Augustine. Courtesy of the Donald A. Sinclair New Jersey Collection. Special Collections and Archives, Rutgers University Libraries.

A coaling site in the Pinelands, circa 1940. Courtesy of George H. Pierson, New Jersey Bureau of Forest Management, Division of Parks and Forestry.

Charcoal burning tools. Courtesy of George H. Pierson, New Jersey Bureau of Forest Management, Division of Parks and Forestry.

cabins," provided shelter and continuous visibility of the pit. Some of these watch cabins had handles so that they could be carried from one ring to another.[17]

Two types of pits were generally used in the Pinelands: the chimney pit and the arch pit. The pit was constructed in tiers around a "fagan," or guide pole, stuck vertically in the ground. In a chimney pit, the tiers were built around an interior triangular stack of cordwood that served as the chimney. In an arch pit, the wood was stacked in a V-formation with the fagan at the point, until the arch reached a height at which logs could be piled upright to form the first tier. When finished, the resulting pit had a side opening, or "arch," through which it could be fired and additional kindling inserted. This was also called a "female" pit, presumably in contrast to the chimney pit.

The second step was "turfing and blacking." It involved covering the pit four to five inches deep with "floats" of turf, which were usually cut from sandy soil containing sheep laurel, black huckleberry, and teaberry. Then

George Crummel, "the Indian Collier of Jenkins Neck," circa 1950. Photograph by William Augustine. Courtesy of the Donald A. Sinclair New Jersey Collection. Special Collections and Archives, Rutgers University Libraries.

Covering the pit with floats. Courtesy of George H. Pierson, New Jersey Bureau of Forest Management, Division of Parks and Forestry.

the pit was "blacked" by covering it with a four- to six-inch-deep layer of sand which formed an airtight seal.

To "fire the pit," kindling was inserted in the chimney or through the arch and set afire. Then the chimney or arch was filled with two-foot lengths of cordwood and covered with floats and sand. Flue holes were then punched through the floats and sand about one foot apart and one to two feet above the base of the pit. By covering and opening these flue holes, the collier could control the speed and evenness of the charring process. Too much air caused the wood to decompose into gases, vapors, and solids, leaving only ashes instead of charcoal.

It took eight to ten days to burn an eight-cord pit. Many pits burned for as long as two weeks, during which time they had to be watched night and day. The collier could monitor the burn from the appearance of the smoke. If white steam came out of the flue holes, the wood was being carbonized properly. But if blue smoke emerged, the wood was burning, and a "soft spot" would develop where the wood caught fire and burned a hole. In this

A burning charcoal pit, circa 1940. Courtesy of George H. Pierson, New Jersey Bureau of Forest Management, Division of Parks and Forestry.

Bagging the charcoal. Courtesy of George H. Pierson, New Jersey Bureau of Forest Management, Division of Parks and Forestry.

case, the soft spot had to be "dressed" immediately by cleaning it out, filling it with chunk wood, and covering it with new floats and sand.

After the pit had slowly burned for eight to fourteen days, it would settle, and sand would sift unevenly down through the pit. The fire had "come to foot" when the pit finished settling. It would then be "keeled" by closing the flue holes and turning the floats so that sand would sift down and the pit would cool. Finally, the pit was "drawn" when the turf and sand were raked off. The cooled charcoal was then raked into a ring, and the "braise," or charred bark, was flaked off and left in the pit. The large chunks, known as "ring coal," were broken into smaller sizes. The charcoal was then loaded into boxes or sacks.[18]

An entry of July 1812 from the *Martha Furnace Diary and Journal* reflects this scene:

> 22. Joseph Camp carting floats in coaling with three horses.
>
> 23. Camp continues coaling. Abbot and Sermon ariv. last evening.

Cranberry label. Photograph by Anthony Masso.

Birds-eye view of Egg Harbor City, 1865. Lent by the New Jersey Historical Society. Photograph of lithograph by Anthony Masso.

A Pinelands Mosaic. Photograph by Joseph Czarnecki. PFP 83–CJC045–A/3.

Blueberry Packing House. *Painting by Margaret Bakely of Vincentown. Photograph by Anthony Masso.*

Roadside stand, New Egypt. Photograph by Joseph Czarnecki. PFP 83–JC009–5.

Bill Wasiowich of Woodmansie. Photograph by Joseph Czarnecki. PFP 83–CJC016–07.

Bucci Farm Market, Highway 40. Photograph by Elaine Thatcher. PFP 83–CET007–17.

Grave blanket, Egg Harbor City Cemetery. Photograph by Susan Samuelson. American Folklife Center, PFP 83–CSS047–19.

Railbird hunting on the Maurice River. Photograph by Dennis McDonald. PFP 84–CDM030–18.

NEW JERSEY
PINELANDS
MILES
0 2 4 6 8
KILOMETERS
0 2 4 6 8 10 12
TRENTON
PHILADELPHIA
Burlington
Willingboro
Mount Holly
Cherry Hill
Medford
Berlin
Hammonton
Batsto
Chatsworth
Browns Mills
Lakehurst
Lakewood
Toms River
Manahawkin
Allentown
Pt. Pleasant
MERCER
MONMOUTH
BURLINGTON
CAMDEN
OCEAN
GLOUCESTER
DELAWARE RIVER
BARNEGAT
BAY
NEW JERSEY TURNPIKE
GARDEN STATE PARKWAY
ATLANTIC
OLD MAPPING
OLD MAPPING
Mullica River
Batsto River
Toms River
Metedeconk River
Manasquan
Crosswicks Creek
Rancocas Creek
Cedar Creek
Oyster Creek
Oswego River
Wading River
Westecunk Creek
Mill Creek
Nescochague Creek
Sleeper Branch
Cooper River
Timber Creek
Assiscunk
PLATE 10

Vegetation map, showing the Pinelands environments. New Jersey Pinelands Commission.

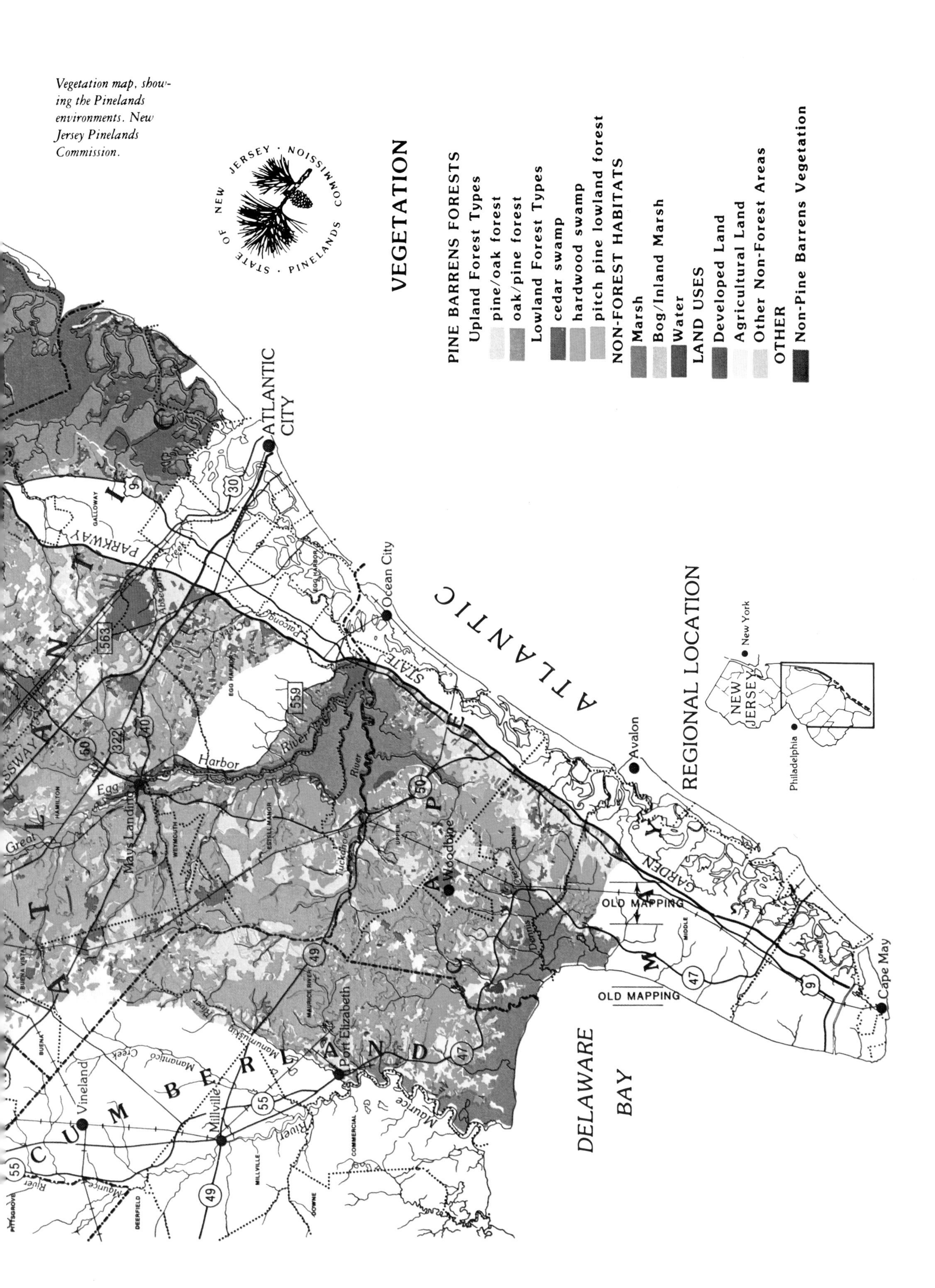

Cranberries floating in a flooded bog, Haines and Haines cranberry bog, Hog Wallow. Photograph by Susan Samuelson. PFP 83–CSSD05–12.

A row of beating machines in a flooded cranberry bog, Hog Wallow. Photograph by Joseph Czarnecki. PFP 83–JC004–5.

Beating machines in action, Hog Wallow. Photograph by Joseph Czarnecki. PFP 83–CJC004–13.

Cranberries being loaded by a conveyor belt onto carts for transportation to the packing house, Hog Wallow. Photograph by Joseph Czarnecki. PFP 83–JC005–12.

Pair of merganser decoys by Hurley Conklin of Manahawkin. New Jersey State Museum, 83.64.1,2. Photograph by Anthony Masso.

Glass items made by Ted Ramp of Egg Harbor City. Photograph by Anthony Masso.

Batsto. *Painting by Win Salmons of New Gretna. Lent by Anne Salmons. Photograph by Anthony Masso.*

Cranberry Reservoir. *Painting by Margaret Bakely of Vincentown. Photograph by Anthony Masso.*

Picking Cranberries at Peterson's Bog. *Painting by Bluma Bayuk Rappoport Purmell. Lent by Michèle Rappoport. Photograph by Anthony Masso.*

Signature quilt in the tumbling block pattern made by the Sew and Sews of the Tabernacle Methodist Church. Lent by the Tabernacle Historical Society. Photograph by Dennis McDonald.

24. ditto ditto ditto Job Mathis has gone to Tuckerton.

25. ditto ditto ditto All going well. Friend Pedler arrived here last evening with books and skins of various kinds.

27. Thomas Anderson drunk. Craig on turn all night.

28. All going on well. Steward carting logs for I. Cramer. Camp in the coaling.

29. Camp carting floats for Patrick Hamilton.

30. Garner Crane went home at noon on account of his wife being indisposed. Camp in the coaling.

31. Camp carting floats in the coaling.[19]

Individual colliers modified the process of charcoal burning. Herbert Payne, who "coaled" until 1974, experimented with both materials and methods. He developed his own personal technique by combining the chimney and arch pits. In order to meet a demand for extra-hot charcoal, he began using old railroad ties. He found that the ties from the defunct Tuckerton Railroad worked best, so that he eventually could claim that he'd turned the whole Tuckerton Railroad into charcoal!

Payne recalled that in the old days, the "firewalker" would walk on the pit as it smoldered, packing soft spots so that the wood would smolder evenly. On one occasion, Payne burned his leg while firewalking, but his skill in judging the burn eventually became so great that he was able to moonlight at another job in Toms River while his pit was smoldering in Whiting.[20]

Ironmaking

According to historian John Stilgoe, from the time of the Middle Ages, mines and furnaces were associated in the European folk mentality with Satan and Hell. The "husbandman," or farmer, on the other hand, was associated with the positive values of hearth and home. Such attitudes did not interfere with commerce, however. The farmers bought the stoves, plows, skillets, and hoes that rural blacksmiths made from the products of the furnace.[21]

Attitudes derived from such ancient European roots may be implicated in regional prejudices. Many of the people who settled in the Pinelands in the 1800s were employed in the iron industry with its blazing furnaces, and so may have been regarded suspiciously by those outside the Pines. This bias, then, may have contributed to the stereotype of Pinelanders that survived long after the demise of the iron furnaces in the mid-nineteenth century.

The South Jersey iron industry began about 1765. It drew not only upon the bog iron, or limonite, found in the swamps along the streams of the Pinelands, but also upon the available water for power, the extensive woodlands for charcoal, the sandy soil for castings and insulation, and the lime in oyster shells found along the rivers and bays for smelting. The main products of the furnaces and forges were munitions, and cast-iron stoves, pots, pans, axe heads, shovels, and water pipes.

The iron ore was dug up and transported in large ore boats to the furnace. First, the ore had to be prepared in the "stamping mill," where large, water-powered trip hammers crushed it into smaller particles. Then the ore was roasted (by placing the ore on a grate and igniting a fire under it) to remove excess water and to convert any carbonates or sulfides into ferric oxide.

The smelting process required a flux and a reducing agent. In South Jersey, the flux was usually lime (in the form of oyster shells), which combined with the impurities in the iron to form slag. The reducing agent was carbon (in the form of charcoal), which combined with the oxide in the iron oxide to form carbon monoxide.

Small iron fireplace kettle, Allaire Iron Works (Howell) Monmouth County, 1823–1846. New Jersey State Museum Collection, 77.80.6. Photograph by Joseph Crilley.

A fire was kindled in the hearth until the stack was thoroughly heated. The batch of ore, flux, and charcoal combined in specific proportions was transported in wheelbarrows over a trestle bridge to a platform around the top of the furnace and dumped into the top of the stack in layers. Blasts of cold air kept the molten mass in a state of agitation and provided oxygen to complete the combustion. As the mass melted, the slag floated to the top and the purified metal sank to the bottom. The slag was drawn off periodically, and the iron drawn out every nine to ten hours.

Forges usually were located near the furnaces. The forge refined and resmelted the pig iron into malleable wrought iron, or "bar iron." Pig iron had limited use as stoves, hollow ware, kettles, sash weights, and firebacks; wrought iron could be made into such items as tools, horseshoes, and wagon tires.

At the forge, the pig iron was reheated in a blast fire and pounded by

Howell Iron Works, Monmouth County, circa 1850. Courtesy of the Monmouth County Historical Association.

heavy, water-powered hammers into bars called "anconies." Some furnaces combined the operations of forge and furnace. Some ironworks also had "rolling mills" that made sheet iron, and "slitting mills" that made iron rods, nails, and tires for wagon wheels.

The production of iron in the late eighteenth century was a rural industry, but it required considerable capital and a large workforce. Most of the "iron masters," such as Charles Read, Isaac Potts, John Cox, and William Richards, were wealthy men, but they required sometimes two or more partners to raise the required capital.

Around the forge or furnace was a village usually consisting of the iron master's mansion, the workers' houses, a sawmill, a gristmill, a store, a school, a church, and sometimes a stamping mill. Martha Furnace, for example, founded in 1793 by Isaac Potts, had 40 to 50 structures and a population of 400 in its heyday.

Iron plantations required a workforce that included founders; fillers or bankmen, who loaded the furnace; guttermen and molders, who handled the molten metal; and blacksmiths and patternmakers. In addition, there were workers outside the furnace, including "ore raisers," who dug the ore out of the bogs, charcoal burners, lumbermen, and teamsters, who hauled

Iron fireback set, Atsion Furnace, Burlington County, 1790–1840. New Jersey State Museum Collection, 72.129a, b,c. Photograph by Joseph Crilley.

the raw materials to the furnace and the finished product to nearby river landings.[22]

The furnaces operated 24 hours a day continuously from spring to December or January. Some of the forge workers were slaves, but most were white, indentured servants who agreed to work for a specified number of years. Nevertheless, it was not uncommon for them to run away.

The entries from the *Martha Furnace Diary and Journal* reflect the round of activities associated with the iron industry:

April 1808

26. John Lynch helping P. Applegate about the furnace wheel. S. Stewart hauled 6 loads Ore from the Pond.

27. Commenced hauling coal. Owen Hedger driving John Hedgers Team. Stewart hauling sand to Furnace. . . .

29. . . . Peter Cox began to fill the furnace.

30. At 11 o'clock A.M. put the Furnace in Blast. Asa Lanning filled 1 turn. J. Hedger Banksman. John Cunning doing Gutterman's duty. . . .

November 1808

5. John Bodines Team hauled 1 Ld. Shells & ½ Ton Iron. Luker hauled Logs in forenoon. Stewart Cuttg. Logs.

6. J. Bodines Team hauled 2 Lds. Hay & 1 Ton stoves. Isaac Cramers Team hauled wood to Ventling.

7. Forenoon. Hauled 2 Tons Stoves in afternoon. . . .

April 1809

20. At 25 M past 2 o'clock P.M. put the Furnace in Blast. Delaney & Cox Fillers. Hedger putting up the Ore & Donaghou Banksman. Ventling & Townsend working in the Blacksmiths Shop.

21. Teams hauled Moulding sand in forenoon (Bennett & Brown). Ventling ½ day in Blksmh. Shop. . . .

22. Teams hauling from Gravelly Point. King & Cos. Team hauled Mdg. Sand in the afternoon. Albertus King arrived from Philadelphia.

23. The furnace working well. . . .

July 1812

15. Craig & Anderson filling up the furnace with coal.

16. Furnace went some time last night. . . .

17. All hands at their usual business.

18. Teams carting moulding sand. . . .

20. Furnace made a great puff on Sunday night last, but fortunately done no damage.[23]

In the mid-nineteenth century, the South Jersey iron industry declined. Experts disagree on the reasons for this decline, but one factor was that the ore began to be depleted and furnaces such as Batsto had to import ore from other places. Also, the discovery of anthracite coal near the magnetite, or rock iron, ore in Pennsylvania—coupled with the development of the hot-blast method—made the charcoal, limonite, cold-blast complex of South Jersey obsolete. Because of its high phosphorous content, bog iron could not be used to make steel.

The history of Batsto Furnace illustrates the rise and fall of the iron industry. Located near the site of a sawmill on Batsto Creek, a tributary of the Mullica River, Batsto was built in 1766 by Charles Read and his business associates, who also built forges and furnaces at Etna, Taunton, and Atsion. Under Read's ownership the principal product of Batsto was pig iron, some of which was transported to the forge at Atsion to be refined into bar iron.

At the time of the Revolutionary War, Batsto was owned by John Cox, a Philadelphia merchant, and his associates. Cox expanded the furnace's products to include pots, kettles, Dutch ovens, skillets, stoves, mortars and pestles, sash weights, forge hammers, and evaporating pans for salt works. Batsto also made cannon balls for the Continental Army.

In 1778 Batsto was purchased by Joseph Ball, who had been the manager of the works under Cox and his partners. Ball constructed a forge and slitting mill about half a mile from the furnace. In 1784 Ball and his partners sold Batsto to William Richards and his associates. Richards rebuilt the furnace.

Batsto flourished under the ownership of William Richards's son, Jesse Richards, and his partners. The principal reason for the boom was the demand for munitions during the War of 1812. During the 1840s Batsto had two sawmills and a brick factory; its products included pipe, ornamental iron, and firebacks. In 1841 Jesse Richards built a "cupola," or resmelting furnace, that refined pig iron into products of higher quality.

When nearby iron ore began to be depleted, Richards imported ore from New York and Pennsylvania. As the iron industry in New Jersey began to decline, Richards decided to diversify. In 1846 he built a factory to make window glass. A second glassworks was also built, but the two were not successful.[24]

Over the latter half of the nineteenth century, Philadelphian Joseph Wharton purchased the Batsto mansion and more than 100,000 acres of Richards's land. He planned to divert the region's abundant water supply to Philadelphia. In 1884 the New Jersey state legislature thwarted his plan, but over the next 75 years, the area continued to be the target of development proposals. During the 1950s the state purchased the Wharton tract, and began restoration of the village soon after.

Today, as a state historic site, Batsto encompasses reconstructed operating mills, workers' houses, and a nature center; various programs interpret the relationship between culture and nature. Even the sight of unrestored ruins can powerfully evoke the past. Such an encounter seems to have taken place as early as 1823, when the traveler J. F. Watson wrote:

> Was much interested to see the formidable ruins at Atsion iron works. They looked as picturesque as the ruins of abbeys, etc., in pictures. There were dams, forges, furnaces, storehouses, a dozen houses and lots for the workmen, and the whole comprising a town; a place once overwhelming the ear with the din of unceasing, ponderous hammers, or alarming the sight with fire and smoke, and smutty and sweating Vulcans. Now all is hushed, no wheels turn, no fires blaze, the houses are unroofed, and the frames, etc., have fallen down and not a foot of the busy workmen is seen.[25]

Reconstructed villages such as Batsto and Allaire, the site of the old Howell Iron Works, make the past accessible to visitors and residents alike, just as do the re-enactments of historical events that take place in costume and pageantry at local festivals.[26]

The element of personal history that connects place and past may be the impetus for formal expressions of sense of place. The families of artist Win Salmons and his widow, Anne, lived in the New Gretna area for generations. Salmons's paintings depict the multiple connections between their lives and the Pinelands. Mrs. Salmons's grandparents were caretakers of the Batsto mansion for years. Win, she recalls, loved to go up and sketch scenes, and through some of his paintings she unravels the thread of the personal past that is woven into public history. Of Salmons's painting of the Batsto mansion, she explains:

> This is a picture that is of the Wharton tract. It's the Batsto mansion. My grandfather and grandmother lived here till just before the state bought

The Pines. *Painting by Win Salmons of New Gretna. Photograph by Anthony Masso.*

it in '54. In fact, they stayed right here in this room which is the oldest part of the mansion, see? This part was built first by Read.[27]

Another of Salmons's paintings, depicting a clearing in the Pines, prompts her to recall special encounters with nature:

> We used to talk when we'd like pick up things to make wreaths. You know—for Christmas for cemeteries and things. The Pines sing, you know. They just kind of sing. It's like we—most everybody knows—around here. You can tell there's going to be a storm. They sigh. It's like a pretty music in the trees.[28]

Iron forges have long provided the context for verbal genres in the Pines, as well. In the late 1930s and early 1940s, folklorist Herbert Halpert collected a cycle of stories about a wizard or magician named Jerry Munyhun. Many of these stories were centered on Old Hanover Furnace and its vicinity. No one knew where Munyhun originated, or even whether he really lived. Some people thought Munyhun was Irish, others thought he was black, and still others called him "the Old Hanover Hessian." Harry Payne recalls his father talking about Munyhun:

> Was an old fellow from down at Waretown, fellow by the name of Acton Bunnel. He was a fisherman. . . . He was tellin' about Jerry Munyhun. Said that Jerry had sold his soul to the devil so that he would have unusual powers over other human beings. He was a man about my father's age, and I guess Pop met him through buying salt hay for his horses. Once in awhile he'd come by and he'd sit down and talk and have dinner together. He'd say "Charlie, remember old Jerry Munyhun? Remember?" Pop'd say, "Yeah, I heard about him," and tell the story all over again.[29]

Munyhun, it was said, could stop a person in his tracks and chop large amounts of wood by making the axe do the work. He could turn oyster shells into money or hogs into corn cobs. He could not be shot or held in jail.

On June 26, 1941, George White of Lakehurst told the following Munyhun tale to Halpert:

> Jerry Munahun—not Monahan. My father-in-law, Tom Luker, used to tell us these yarns about him. They told it for the fact. He worked at Old Hanover Furnace—and they owed him quite some money. So, they wouldn't pay him; so they couldn't do anything with their plant—couldn't get no steam with their boilers. And he went away and was gone awhile. And he come back again. And he asked them what the trouble was—and they said they didn't know. Couldn't get no steam; couldn't make the fire burn. And he told them he knowed what the trouble was. He says, "If you pay me, they'll all fly out." Which they paid him, and they did—went out. They claimed they flew out of there one after the other. That's the way he made it look to them. They told that for the truth. These real old people, I've had more than one tell me these things he done. They claimed he was a great man.[30]

Papermaking

With the decline of the iron industry, other uses were found for the resources left by the numerous ironworks. These resources included mills and lakes which could be used for other manufacturing processes.

In 1832, William McCarty bought the forge and slitting mill that had been built in 1795 by Isaac Potts on the Wading River near the site of Harris-

South Jersey wagon in front of the Harrisville paper factory. William Augustine Collection, Donald A. Sinclair New Jersey Collection. Special Collections and Archives, Rutgers University Libraries.

ville. The Wading River Forge and Slitting Mill had used pig iron from nearby Martha Furnace, which Potts had also owned.

McCarty renamed the place McCartyville and decided to make paper rather than iron. He enlarged the slitting mill dam and constructed a system of canals and millraces. By 1835, McCarty had in operation a double paper mill, with one mill manufacturing nearly a ton of paper a day.

In 1851 William D. and Richard C. Harris bought the property and renamed it Harrisville. They expanded the paper mill, enlarged the canals, and built a gas plant to supply the factory and village. The main product at Harrisville was a thick, heavy, brown paper known as "butcher's paper." Despite attempts to get the Raritan and Delaware Railroad (later the New Jersey Central) to build a line to Harrisville, the railroad was built eight miles to the west, leading to the demise of the papermaking operation at Harrisville.

The raw materials for papermaking included old rags, rope, scrap paper, bagging, and salt hay. The salt hay was harvested at nearby marshes and transported in wagons to the factory. The other raw materials were shipped from Philadelphia or New York.[31]

Glassmaking

Glassmaking began in South Jersey in 1739, when Caspar Wistar built a glasshouse on Alloways Creek in Salem County. He imported experienced German glassblowers. While most of the early South Jersey glass factories were outside the boundaries of the Pinelands, in the nineteenth century there were more than 20 glasshouses in the Pines. Most of these factories made bottles and window glass; a few made cut glass and tableware.

The Pinelands offered many resources to Wistar and the glassmakers who followed him. Glass was made from a "batch" consisting of silica, lime, and soda which had been mixed with broken glass. Silica was available in the sandy soil of the Pinelands. The lime came from limestone brought from Staten Island and Valley Forge. The soda was also imported from outside the region. The furnaces were fueled by cordwood, which was readily available in the Pines. Salt hay was used for packing.

In the nineteenth century, hollow ware factories were those which pro-

Sand pits owned by Samuel Hilliard, on the Maurice River near Millville, circa 1876. From New Historical Atlas of Cumberland County, New Jersey. *Philadelphia: D. J. Stewart, 1876.*

Interior of the Star Glass Works, Medford, circa 1910. Courtesy of Everett Mickle and the Medford Historical Society.

duced bottles. They were divided into "shops." A shop was a group of men working around a single "glory hole" (opening in the furnace). Each shop consisted of a gatherer, a blower, or "gaffer," and sometimes a finisher. There were also helpers called "boys." The gatherer placed the head of the blowpipe into a pot of molten glass, called the "metal." He worked the metal into an elongated ball, and he "marvered," or flattened, it on a flat plate to remove impurities and to consolidate the mass. The blower then placed this mixture, called a "parison," into a wooden mold and blew

through the blowpipe to form the desired shape. Next, the finisher took it to the gaffer's bench, where he placed a pontil rod on the bottom of the bottle and broke off the blowpipe. The gatherer then brought a small amount of metal to the finisher, who attached it to the neck and then shaped the top and the lip. Once the bottle was finished, it was placed in a lehr to cool gradually. In the first decade of the twentieth century, an automatic bottle-blowing machine was introduced, making the earlier method obsolete.

There were two methods for making windowlight. The earliest was known as the crown method. The blower took a pear-shaped globule of melted glass from the oven and twirled it into a disk about 36 inches in diameter. After it was cooled in an annealing oven, small panes were cut from the disks. Most of the windowpanes made in New Jersey, however, were produced by the cylinder method. The gatherer gathered 80 to 100 pounds of melted glass in two or three stages and turned it into a hemispherical shape. Then the blower blew the glass into a balloon shape several feet long and took it to a pit. Swinging his blowpipe like a pendulum, the blower stretched the glass into a six-foot length. After the glass was cooled, the ends were cracked off, leaving a glass cylinder which was grooved, split, and then flattened into sheets which were cut into windowpanes.

Around the turn of the twentieth century, machines were developed that could draw out flat sheets of glass, thus making the cylinder method obsolete. New Jersey factories did not adopt this new technology and instead ceased to produce windowlight.[32]

Lampwork, also known as flameworking, torch working, or scientific glassblowing, has been used in factories since the late nineteenth century to make laboratory apparatus. A glass rod or tubing is manipulated in the intense heat of a small burner to the desired shape. Glassworkers have used the same technique to make their own folk objects, such as toys and miniatures.

According to former lamproom blower Dorothy Lilly, "If you grew up in Millville, you were raised on glass."[33] The statement could be applied as well to towns of the Pinelands whose economies and societies were shaped and sustained by glasshouses. A glass factory required a wide range of auxiliary industries, including sand mining, woodcutting, and potmaking. Within the glasshouse itself were many shops filling different needs: mold shops, art shops, cutting and grinding shops, and packing shops. Therefore, a sizable population could be sustained by a few factories.

The annual cycle in these towns revolved around the fiery furnaces of the glasshouse. In the summer, they would shut down. Lilly, whose grandparents and parents worked in the industry, recalls that her family would go to Wildwood for the summer, where her father worked as a trolley conductor.[34]

Social hierarchies were established according to occupational hierarchies. Glassblowers were well paid and always in demand. They often became a social elite within the town as a result.

Many families counted many members and several generations among glasshouse employees. Lilly's grandparents, parents, and husband all worked in glass. Women often did decorative work and packing in the factories. Family members passed on skills in formal and informal ways. Malcolm Jones, whose father, like himself, was a moldmaker, recalls that it was difficult to get an apprenticeship in the mold shop unless one had a relative there.[35] Lilly began learning in a less formal way. She would take lunch to her father, who would let her try to blow a gather on his blowpipe. She quickly learned that if she blew too fast, the glass would burst and scatter "blovers" around the shop. During World War II, Lilly developed her skills further doing lamproom blowing at the Frederick Dimmick plant.[36]

Within the many shops of the glasshouse, communities of workers developed their own cycles, hierarchies, and customs. The more skilled craftsmen had to go through an apprenticeship of up to five years. A young boy might begin his apprenticeship as young as age 12. During the years until he was "out of his time," as completion of apprenticeship was called, he would work at many tasks within the shop, and learn by observation as well as by practice. In the moldmaking shop, an apprentice would be as-

Left: *Engraved vase and decanter. Lent by Dorothy Lilly. Photograph by Anthony Masso.*

Right: *Engraved vase made from a chemical beaker by Walter Earling of Millville. Lent by Walter Earling. Photograph by Michael Bergman.*

signed either to the machine or to the bench in his third year. The latter spot was much coveted, for it meant that he would become a letter cutter, the highest paid and most skilled man in the shop. A letter cutter would do the fine handwork, such as carving letters or designs in the iron mold after the machine work had been completed.

Apprentices quickly learned the physics of glassmaking with some traditional pranks. Ted Ramp recalls, "You'd blow a long piece of glass, like a tube, and pinch it shut, and that air'd get so hot in there, it would blow all to pieces."[37] In his reminiscences of the glasshouse in *Tempo: The Glass Folks of South Jersey,* Roy C. Horner recalls similar items called "snappers," "Dutch tears," or "Prince Rupert's drops" that were formed by dropping molten glass into a bucket of water. The glass would harden into a tear-shaped drop with a tail. When a wily journeyman would place the snapper in the hand of the apprentice and quickly snap off the tail, it would explode into fine dust.[38]

More pleasant traditions often drew on the materials in the shop and the special skills of the workers. In the glasscutting shops of the Whitall–Tatum factory, it was traditional to present each girl in the art shop with an

Incised glass certificate presented to Romine B. Butler, potmaker, made by Joseph Marshall, Cohansey Glass Manufacturing Company, 1883. Lent by Wheaton Historical Association. Photograph by Anthony Masso.

engraved decanter or vase on her sixteenth birthday. In Lilly's family, there are two of these. One is a decanter engraved with an "E," which was presented to her grandmother, Ella Estlow, around 1880. The other is a vase, created from a chemical beaker by cutting the top off and polishing the edge, which is engraved with "Hazel B. Estlow," and which was presented to Lilly's mother.[39]

Apparently, in some shops a certificate was presented on the completion of apprenticeship. At the Museum of American Glass at Historic Wheaton Village in Millville, an elaborately engraved pane of glass commemorates that occasion for Romine B. Butler, who "served his full apprenticeship for the art of potmaking at Cohansey Glass Manufacturing Company of Bridgeton." The certificate was engraved by Joseph Marshall in June 1883.

Glasscutter Walter Earling recalls other items that he and other cutters and grinders made both to hone skills and to create "whimsies," as such folk art is often called. Earling once ground lapidary cuts onto a glass doorknob, then mounted it on a base. Dentists' dapping dishes became salt cellars by such a process, and he once engraved his name and a sampler of designs on a chemical beaker to create a vase.[40]

South Jersey jar, made by the Moore Brothers Glass Company, Clayton, circa 1870. New Jersey State Museum Collection, 324. Photograph by Joseph Crilley.

Ted Ramp of Egg Harbor City blows through a blowpipe to inflate a parison, or bubble, of molten glass, 1970. Photograph by William Augustine. Donald A. Sinclair New Jersey Collection. Special Collections and Archives, Rutgers University Libraries.

Ted Ramp places the parison into a ribbed mold, or crimp, which will produce a ribbed effect in the finished vessel, 1970. Photograph by William Augustine. Donald A. Sinclair New Jersey Collection. Special Collections and Archives, Rutgers University Libraries.

Much of this was done before and after work or during breaks, and has come to be called "end of day" work or "tempo" work. It often went home for use as sugar bowls, pitchers, rolling pins, and vases. Other items were used decoratively, such as chains, batons, canes, hats, lilies, knitting needles, button hooks, Christmas ornaments, and paperweights.[41]

The paperweight became an important emblem of skill among glassworkers. While some were made using a metal die, they were not mass produced here. In southern New Jersey, characteristic shapes became localized in such designs as the "Millville Rose." The men who perfected these designs, such as Ralph Barber and Emil Larsen, were highly regarded, and the items eagerly sought after. Yet, glassblowers were never allowed to make paperweights on company time because they were not a profitable product for the factory. Most glassblowers at one time or another tested their skills by making paperweights. Often they incorporated into the designs the traditional community values of home, church, and country.

Many glassworkers took the skills they had acquired in the large factories

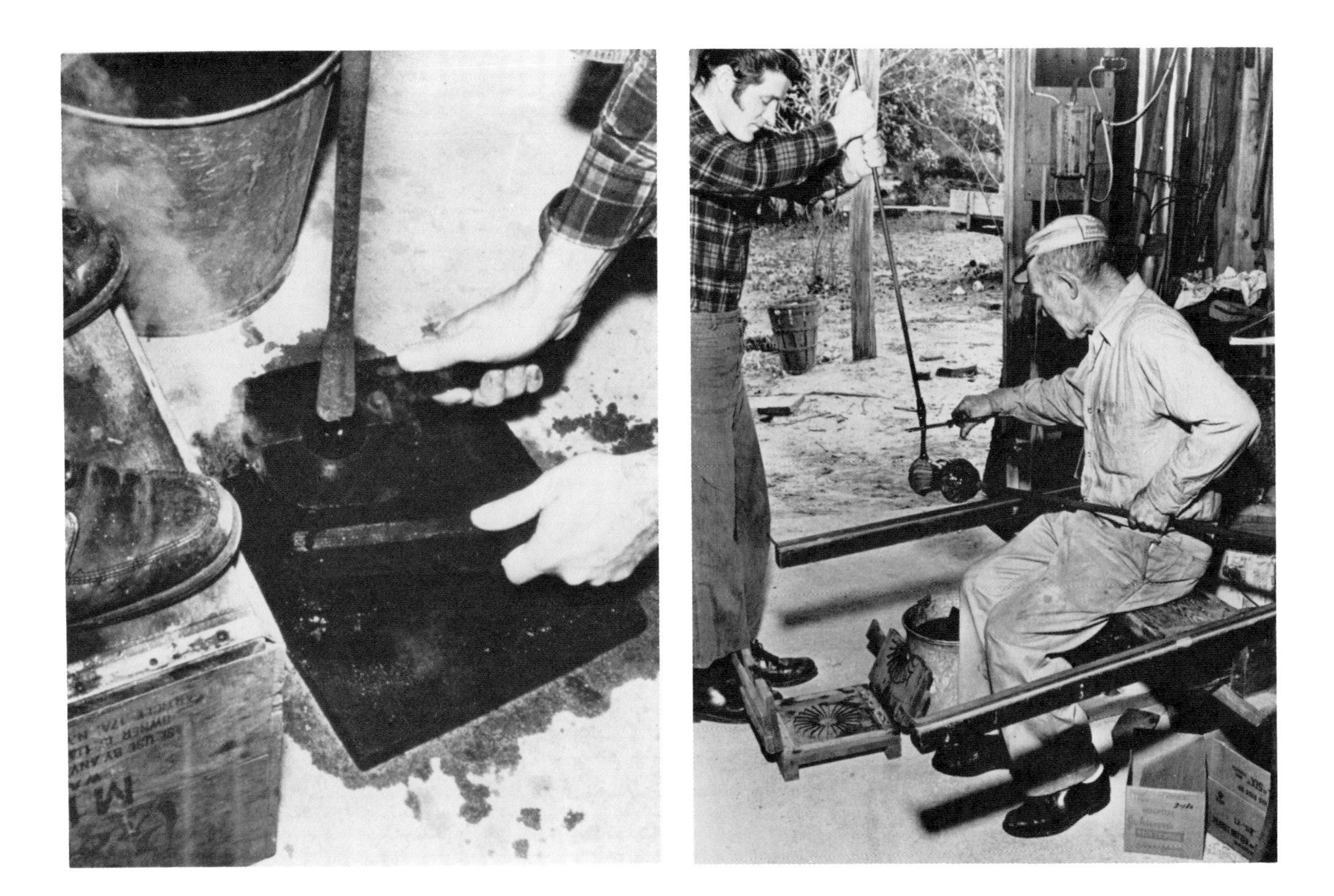

Ted Ramp's son, William, closes the wooden mold, which has been soaked previously in water, around the ribbed parison, while Ted rotates the blowpipe and blows into it to inflate the bulbous base of the vessel, 1970. Photograph by William Augustine. Donald A. Sinclair New Jersey Collection. Special Collections and Archives, Rutgers University Libraries.

William Ramp holds small amount of molten glass to the vessel now affixed to a pontil rod, while Ted, using shears, guides its application to form a handle, 1970. Photograph by William Augustine. Donald A. Sinclair New Jersey Collection. Special Collections and Archives, Rutgers University Libraries.

and began their own small shops. Among these were the Clevengers, whose family involvement in glass reached back to Batsto, and Ted Ramp, whose occupational and familial roots are German.

William Clevenger worked in the windowlight furnace at Batsto between 1844 and 1866. When the Batsto works declined, he moved to Clayton to work in the Moore Brothers' glass factory; his sons, Thomas (Tommy), Lorenzo (Reno), and William Elbert (Allie), went to work in the factory also. When the Moore factory closed down just before World War I, the Clevengers, like other glassworkers, had to resort to odd jobs.

According to a South Jersey antiques expert named Ernest C. Stanmire, he accompanied the Clevenger brothers to the 1926 Sesquicentennial Exposition in Philadelphia, where they saw the original log cabin Booz bottles made by the Whitney Glass Works being sold as collectibles. With money borrowed from Stanmire, the Clevengers built a one-pot furnace and began producing handblown glass. Many collectors consider their glass to be "reproductions" of colonial glass to supply the fashionable Colonial Revival market of the 1920s. But they can also be considered as reviving the folk tradition of "free-blown" glass that was on the verge of dying out.

Allie Clevenger continued to blow glass until his death in 1960. The Clevenger Works continues today under the proprietorship of James Travis, Jr.

Ted Ramp of Egg Harbor City recalls that "old man Larsen was a dandy" when it came to making paperweights. Larsen was one of the many glassblowers with whom Ted worked over his long career.

Ramp began working as a gathering boy at the Liberty Cut Glass Works in Egg Harbor. He eventually worked as a gaffer in shops throughout South Jersey and in West Virginia, as well as in the shop he ran behind his own home for approximately 15 years.

Ramp's years at the Hofbauer glass factory in Vineland had a great impact on his work. Hofbauer came from Austria, opened his factory in 1932, and produced art glass that was sold through department stores like Gimbels. While he worked there, Ted acquired many of the stylistic traits which continued to characterize his work, including crimped edges, optical effects, and swirled and fluted motifs.

But the moldmaking that Ramp learned at Hofbauer's may have been as important as anything else he learned there. As he tells it:

> Hofbauer made the wooden molds himself, and I'm watching him, and he said, "You think you could do that?" And I said, "Yeah, maybe a little

Glass pitcher made by Ted Ramp of Egg Harbor City. Lent by Helen Zimmer. Photograph by Anthony Masso.

> better than you." He'd cut a piece of wood, round like that, take an axe and split it. And you know how rough everything is and that's why I told him. He throwed a hatchet at me! He was a crazy Dutchman! . . . It took me all day to make [my first mold]. You had to chip it out by hand, you know, on a gouge. And then he give me hell why didn't I tell him I could make it before! I made prit' near all his molds after that.[42]

He cut cherry and apple trees and carved the molds out in his basement for five dollars each.

These handmade molds became a hallmark of Ramp's own work, for few people were using wooden molds by that time. He also made molds for other factories later, including Clevenger Brothers and Wheaton.

While Ramp ran his own shop behind his house, he was working full time elsewhere. He would start the oven up on Thursday evening and stay up all weekend blowing glass and tending the fire so that the "batch" wouldn't be ruined. Customers would arrive by five in the morning, his wife recalls; she had to hide pieces she wanted to keep! His younger son, and later his granddaughter, sometimes helped him in the shop.

Most of Ramp's work is in amber, amethyst, cobalt, green, or bluestone. He would create optical effects by first blowing the gather into wire, then finishing it in a wooden mold. Stripes were applied on hot glass in the following manner: "When you got the hot glass—that was white. You stick that on the pipe and hold it there till it starts to melt, then you twist it—twist it around. And then you take a pair of pincers and pull the color up."[43] Though he made many pitchers and vases, it is the small, fluted basket that is distinctively "Ramp glass."

Hunting, Trapping, and Gathering

With the decline of great industries in the Pines in the late nineteenth century, much of the landscape which had bustled with activity for a century reverted to wilderness. This change coincided with changes in the general perception of wilderness.

Geographer Yi Fu Tuan points out that in the late nineteenth century, under the influence of the conservation movement, the wilderness began gaining value as an "edenic" retreat from the chaos of the city. Twentieth-century values and views of the wilderness as threatened by development have reinforced this notion.[44] Recreational uses and suburban settlements have resulted from it.

HARMONY
GUN CLUB

Opposite: *Harmony Gun Club. Photograph by Mary Hufford. PFP218973-1-25.*

Many of the members of the innumerable Pinelands hunting clubs view that environment as a kind of ritual homeland where men gather to share each other's company and to forget the rigors of modern life in the city.

In 1980, there were more than 300 such clubs in the Pines, most of them maintained by men living outside the area. Many of the groups first formed their clubs and built their lodges in the 1920s. Membership usually reflects family and occupational networks. Members may use a lodge for fishing or family events, but its main function is to house the group during the week-long annual deer hunt in December.

While the week in the woods is a ritual itself, it also serves as a frame for many other carefully constructed and observed customs. Typically, club members arrive several days before the season opens to prepare the lodge, the woods, and themselves. Some members scout the woods to identify recent deer movement and to plan the hunt. On the Sunday of deer week, each club has a party and members of different clubs visit each other in a formal and congenial recognition of their shared space in the woods. This is honored as well in the structure of the hunt. The "woods captain" of each club has a "woods schedule" which reflects the silent agreement among the clubs to manage shared space equitably: "You don't push onto us and we won't push onto you."[45] During a hunt, some club members will "drive," that is, move in a tight line through the woods to flush the deer, and others will "stand" a good distance upwind of the drivers to shoot deer that run by.

Traditionally, a hunter who misses a shot will have his shirttails cut off. Mementos of such infamy are tacked on walls, framed in pictures, and chartered into "shirttail clubs." At the annual banquet held after the deer season, most clubs re-enact or satirize particular events of the year's hunt.

While most club members live outside the Pinelands, their long years of association with the region and the natives, as well as their caretaker attitudes towards the woods, usually earn them a special "resident nonresident" status. Much of the hunting is done on privately held lands by agreement with the owners. Natives often regard hunting as a form of deer population control that benefits them personally, and club members are generally careful to safeguard the owner's property as they hunt. Reciprocally, natives watch over the hunting lodges, which are targets for vandalism by "outsiders."[46]

Even for the man who hasn't gotten a deer in nine years, the challenge of the hunt and the camaraderie of the club make the hunting lodge a regenerative woodland shelter. According to one: "You're down here for a week; it is another world. No cars, no nothing. And you don't realize there's cars until you get up to the White Horse Pike, and then it's the old hubbub again."[47]

Like many other Pinelands woodsmen, Jack Cervetto has guided hunters for many years. Hunter guiding serves as an important link in two chains of activity. With their knowledge of place, native guides map the otherwise indecipherable environment for visitors, helping them use the area as a recreational resource. On the other hand, guiding itself is an important occupational resource for men who "work the cycle."

> I used to take deer parties, big clubs, out here years ago for deer season. At one time, I had thirty, forty men. Put thirty of 'em on a stand and ten for drivers. Sometimes you have twelve drivers. It all depends on the bottom, how you can see. If it's thick bottom, you have to get closer together. If it's open bottom, you'd take less. And my drives were already laid out on paper, you know, from one drive to the other, and eventually the boys got to know the drives, too.[48]

Knowing the life cycles, habits, and habitats of the flora and fauna, a woodsman can gather in the spring and summer, guide in the fall, and hunt and trap in the winter. These activities are performances of the skill of woodsmanship, an intricate knowledge of the environment and the ability to maneuver in it.

The skills of woodsmanship are developed from an early age, and they serve as markers on the road to manhood. Trapper Tom Brown, who harvests from both woodlands and wetlands, recalls first learning to trap rabbits with handmade snoods, snares, and deadfalls when he was 11. He received his first knife the next year, on his twelfth birthday.

Family members are often important teachers, and the development of camaraderie is as important as the acquisition of skills. Recalls Brown:

> The ones that are killed, that's it, but the ones that are missed, that's the ones you really had the fun over 'em. I was with my grandson when—well, both of 'em—when they killed their first rabbit. But Bucky, when he got his first trout I was with him. When he got his first rabbit, when he got his first deer. So I told him week after he got his grouse—I was tendin' traps out at Coleman—so I said "Buck, there's another grouse out there." I said, "You go with me on Saturday morning. You'll be out of school. You'll get a crack at it." So we walked down the road, and lo and behold! The grouse kicks up. It went so fast he didn't even get a shot at it. I said, "Well, you got another chance, Buck." So we go around, and we get him in another open spot, and he missed him again! I said, "Buck,

> did you hear what that grouse said when he took off?" He said, "No, but I'm gonna hear it." I said, "Well, I heard what it said. It said 'Bucky, you mighta killed my mate before but you couldn't hit my rear end with a barn door!'"[49]

Pelts and game meat have always been an important commodity in the Pinelands, especially for those who work the cycle. Although beaver, otter, mink, and skunk were trapped in the past, now they either require a special permit or are considered unmarketable. Fox, raccoon, and muskrat are still important quarry.

The woodsman's keen knowledge of the environment is his first tool for successful hunting and trapping. He knows both the habits and the routes of his prey, and that knowledge shapes his technique. One of the first tenets a trapper learns, in Tom Brown's words, is to "set down a great attraction and remove all suspicion."[50] Brown has a variety of methods for doing so, including dyeing traps in a walnut bath and removing human scent with red-cedar shavings. This removes suspicion. Among the tricks to create attraction is the use of liquid lures. Such potions as skunk essence, claims Brown, are responsible for the saying "Old trappers never die, they just smell that way." Finally, the trapper tries to create "eye appeal" by putting feathers, eggshells, or a piece of dry fur by the set to catch the attention of the prey.[51]

Brown formerly sold his pelts directly to Sears, Roebuck, and Company; today, trappers sell directly to fur buyers or to one of the large fur auctions.

The woodsman's ability to maneuver in the environment opens another occupational avenue to him: gathering. A vast array of plants grow in both dry uplands and wet lowlands of the Pinelands. Everywhere there is something worth collecting: pine cones in the Plains, Indian grass in the cranberry bogs, acorn sprays in the pitch pine lowlands, cattails and statice in the salt marshes, grapevines in the woods, bayberry at the seashore, sweetgum balls in domestic yards, sphagnum moss in the swamps, "cornettes" in corn fields, and a wide variety of huckleberries and blueberries in many places.[52]

Foragers know the worth of plants that are "just weeds" to other people. Farmers are often grateful to Leo and Hazel Landy when they ask permission to pick the podded mustard plant that clutters vegetable fields, or the okra that has gotten too woody to sell. To the Landys, the woodier the better.

Gathering has been an important part of the Pinelands economy for generations. Berger and Sinton point out that Americans have decorated their homes with dried and fresh flowers since Colonial times, and that the proximity of the Pines to major cities has provided a constant market. The fact that residents were supplying those markets abundantly as early as 1916 is underscored by botanist John Harshberger's worry that those who made bouquets and wreaths from Pinelands flora and sold them in Philadelphia were going to exhaust the supply of mistletoe, water lily, sweet-bay, holly, pink azalea, mountain laurel, and arbutus.[53]

In addition to the ready supply of plants, the independent lifestyle favored by residents has made foraging a widespread and well-favored industry. The variety of plants and plant life cycles offer a variety of occupational options to the woodsman. He can choose to concentrate on gathering certain plants from May through December, piecing the work year together with trapping or lumbering during the winter. He can gather occasionally to plug up smaller gaps in a work year that concentrates on hunting, trapping, and fishing. He can "moonlight" at gathering to supplement the income of a full-time job in business or industry. Or, like the Landys, he can rely on gathering, processing, and marketing to wholesalers for his entire subsistence.

Though well-equipped with his knowledge of plant types and cycles, the gatherer must still be willling to "get mud up to your eyebrows," according to Leo Landy.[54] His point might be broadened to mean that many plants are found in hard-to-reach places, and, further, that a gatherer must be prepared for a host of other inconveniences.

He must, first of all, be willing to let nature set his alarm clock. Many plants are marketable only at a certain stage of growth, and a good woodsman will recognize that often brief period and time his work schedule to it. If dock (*Rumex crispus*), for example, is picked when it's too green, it will turn a dull brown instead of the bright russet color that florists prefer. On the other hand, if it is "too far gone," it will shed its seeds and lose its shape.[55]

Various plants get specialized handling in order to bring out their best colors and textures. For instance, the Landys gather the dark brown pod of sensitive fern in low spots in the woods after first frost. They cut them with a sickle, leaving a 14-inch stem, tie them loosely in bunches, and place them, pod end up, in baskets to dry. Because the pods will lose color if they are dried in high temperatures such as those in a hothouse, they set them in a storehouse for three to four days. Only after they are fully dry are they boxed for sale.

Bill Wasiowich gathering pine cones. Photograph by Joseph Czarnecki. PFP219558–9–12A.

For many people, the pine cone is a natural symbol of the region. Its importance to gatherers makes this especially appropriate. Though sales usually fluctuate from year to year, the Landys may process and sell as many as 70,000 pine cones in a year.[56]

Although both Virginia pine and pitch pine grow in the region, most gatherers prefer to pick the cones of dwarf pitch pines because the trees are shorter. They are also thickly covered with cones that produce a lot of seed, an important trait to processors who sell the seed as well as the cones.

A woodsman will recognize the age of a cone by its color: this year's crop is green; last year's is grayish-brown; and the previous year's is grayish-black. Although second-year cones are saleable to dealers who paint cones, the green cones are preferable. They are still tightly closed and will open into a shiny brown rosette when heated in a "pine cone popper." Third-year cones are often wormy and therefore disintegrate in the heat.

Portrait of James Still, the "Doctor of the Pines," Medford, New Jersey, circa 1877. From Early Recollections and Life of Dr. James Still, *Philadelphia: J. B. Lippincott, 1877.*

A gatherer times his harvest for both ease and conservation. "You wait till frost," says Jack Cervetto. "The cone is mature then and it's kinda dryin' up and it snaps off the vine easier. You go to pull that off [out of season], that's so sticky with pitch now you have to get half the branch with it. Do an awful lot of damage to the tree, lettin' that sap right out of it. Then for about five years you won't get any more pine cones out of that tree. You got the best nourishment out of it."[57] Moreover, woodsman Bill Wasiowich observes that trees produce cones more abundantly when they are picked regularly.[58]

Although today most gatherers sell their cones to processors such as Landy, the "pine cone popper," a crude brick or cement building with a tin roof, is common to the Pinelands because many woodsmen at one time or another gathered and processed cones. The "popper" acts as a giant oven in which to heat cones with slow, wood-fired heat; this dries and opens them.

In addition to their commercial value, forage plants have personal value for Pinelands residents who use them as a medium for aesthetic expression in traditional forms such as wreaths, bouquets, and grave blankets. The grave blanket is a highly localized form, according to folklorist Susan Samuelson. It is especially popular in the southern agricultural area of the Reserve. While it has become an important commercial item for produce merchants, it continues to be what it began as: a traditional Christmas decoration created by families for their burial plots.

Generally, a grave blanket is constructed on a base of wood and sphagnum moss. Evergreen boughs such as spruce or scotch broom are arrayed on the base to form a flat spray. The size varies, but two feet by three feet is standard. Decorations of either natural materials, such as pine cones, laurel, or red ruskus, or artificial materials, such as ribbon and plastic ornaments, are usually added.[59] The blanket is placed on the grave around the first Sunday in December.

Although there are analogues in other places, the grave blanket in southern New Jersey most clearly expresses the close association of human society and natural environment. Hazel Landy recalls that her father would always add a few pieces of princess pine to the blankets he made. Now, she says, she always does the same on the blankets she makes for the family plot.[60]

Probably more pervasive, but less well documented, has been the use of Pinelands flora in foodways and in folk medicine. Poke, for instance, has traditionally been gathered in spring and cooked like tender greens. Wild huckleberries, blueberries, and cranberries have been gathered for pies and jellies.

His use of gathered ingredients was part of the reason for the fame of Dr. James Still, the son of two slaves from Maryland who bought their freedom and settled in South Jersey. He was born in 1812 at Indian Mills in Burlington County. His brother, William, was active in the abolitionist movement and was the author of a book about the Underground Railway.

Still was not formally trained in medicine, but learned about medicine by observing druggists on visits to Philadelphia and by reading books on botany. He adapted this knowledge to the woodland environment around him when he moved in 1827 to Medford and began practicing folk medicine. Other physicians in the area disapproved of his practice, but many people in the Pines near Medford attested to the effectiveness of his cures. In 1877, he published an autobiography entitled *Early Recollections and Life of Dr. James Still.* The following recipe used some of the plants he gathered in the western fringes of the Reserve at Medford:

Take

Spikenard root	8 oz.
Comfrey root	8 oz.
Horehound tops	8 oz.
Elecampane root	8 oz.
Bloodroot	8 oz.
Skunk-cabbage root	8 oz.
Pleurisy root	8 oz.

All bruised; then boil in two gallons of soft water down to one gallon; express and strain the liquid, and see that you have one gallon. Then add ten pounds of white sugar, and boil to form a syrup. When done, strain again into something to cool, and when nearly cool take two drachms oil anise and four ounces alcohol, mix and pour into the balsam; also one pint tincture of lobelia. Let the whole stand twenty-four hours to settle, then bottle up in half-pint bottles. Dose:—One teaspoon three, or five times a day. This balsam far excels anything I have ever known used for pulmonary affections and coughs of long standing. It is admirably calculated to relieve that constricted state of the lungs which is so often met with in consumption. It assists expectoration and invigorates the whole system, and is seldom or never given without benefit. This is an excellent remedy for asthma or any bronchial affection attended with difficulty of respiration.[61]

Lake Communities and Ethnic Resorts

Since the nineteenth century, there have been various attempts to develop resorts, retirement communities and suburbs in the Pinelands. Some of these schemes, like the plan of a Long Island real estate promoter to develop the "Magic City" of Paisley between 1888 and 1891, came to naught.[62] Other plans, such as the development by Gloucester Farm and Town Association of Egg Harbor City and its vicinity as an agricultural center, were very successful. Many of these land developments, which centered on the adaptation of old millponds and former cranberry bogs to new uses as recreation centers, grew with the change in attitudes toward wilderness.

The history of Lakewood traces such an evolution. Its first use was as the site for the Three Partners' sawmill, erected circa 1786, on a branch of the Metedeconk River. The mill was replaced first by the Washington Furnace, which was in operation between 1814 and 1832, and then by the Bergen Iron Works, which operated there between 1833 and 1854. Joseph W. Brick was the proprietor of the latter until his death in 1847. In 1865 the place was named Bricksburgh, and in 1866 a company named Bricksburgh Land and Improvement Company was formed to develop the land into fruit farms. In 1879 the company sold its interests to a land association that renamed the place "Lakewood" and attempted to develop a winter resort there. Apparently the effort was successful, for in 1889 Gustave Kobbé described Lakewood in the following words:

> *Lakewood* is a little winter paradise created by good taste and sound judgement backed by the necessary capital, on the site of one of the iron furnaces which formerly reared their stacks among the pines. . . . There are few places which one recalls with as much affection as Lakewood. It has the tranquility of a refined home while affording a varied range of amusements. Though a health resort, it is not over-run with invalids, so that a person who goes there for relaxation does not have his spirits dampened by silent but no less piteous appeals to sympathy. In fact, Lakewood is a place of rest rather than a health resort. People go there to recuperate after a rapid social season or to tone up the nerves after they have been subjected to an unusual strain.[63]

The site was laid out around Lake Carasaljo, which was named after the three daughters of Joseph W. Brick—Caroline, Sarah, and Josephine. Scenic roads intended for driving, riding, cycling, and walking wound through the woods around the lake, and the Laurel House Hotel graced the town. Kobbé described the hotel in glowing terms:

> The Laurel House is old-fashioned in that it is home-like; and modern in that it lacks none of the latest improvements. Among its sources of comfort are the ample hearth in the sitting hall . . . ; its parlors and reading room; its spacious and well-equipped nursery; a smoking room in which whist, a game as aristocratic as the gout, is assiduously cultivated; its *cuisine,* admirable and abundant; its piazzas of 380 feet, which are glassed over and kept at an agreeably warm temperature so as to form a pleasant promenade in wet weather; and the open fireplace in all the bedrooms, wood fires being supplied free of charge. There is also a large hall for music and dancing.[64]

In the twentieth century, Lakewood became a Jewish winter resort, perhaps because of its proximity to the Jewish poultry farmers in Farmingdale. Today a number of ethnic groups have settled there, including Orthodox and Hasidic Jews, and Estonians.

Medford Lakes grew from similar roots, but in a different direction. On the tributaries of the Upper Rancocas Creek, it, too, began as the site of a sawmill. In 1766, Charles Read established Etna Furnace in the vicinity. By 1770, there was a small stamping mill, a gristmill, and a furnace on the site. Read also built the furnaces at Taunton, Atsion, and Batsto. But he experienced both health and financial difficulties in the early 1770s, and he moved first to Antigua and then to St. Croix. His son, Charles Read, Jr., inherited Etna Furnace in 1773, but with his death in 1784, the land passed out of the hands of the Read family. Under the new owners, who renamed it Etna Mills, the two sawmills and the gristmill continued. Some cranberry cultivation was tried, too, but it was unsuccessful. Finally, in 1927, a Texas land speculator named Captain Clyde W. Barbour bought the property with the idea of developing a summer colony around the 21 lakes or potential lake sites it encompasses. Barbour's representative was a real-estate man named Leon Edgar Todd, who eventually bought out Barbour's interests. Todd moved to Medford Lakes, as it was renamed, and supervised the laying out of roads, trails, lakes, beaches, parks, and building lots.[65] The houses were built in a log-cabin style, a practice that persists today. In 1928 a community house named "The Pavilion" was constructed, and in 1930 the Medford Lakes Lodge, said to be the largest log-cabin hotel in the United States, was built. Soon the town was transformed from a summer resort to a year-round community of permanent residents.

Today residents belong to the Colony Club, which oversees care of the lakes and recreation facilities and activities in the borough.

Many of the permanent residents of Medford Lakes first used the area as a summer retreat. For them, the lake community offers the chance to maintain both a sense of personal isolation from city hubbub and the security of a readily available social network.[66] The lake becomes the focal point of community interaction as neighbors participate in the care, management, and surveillance of the lake. They also share special social events, perhaps best represented by the annual "canoe carnival" held every August.

The canoe carnival has been held for more than 50 years. Individuals, families, neighbors, and members of organizations who live in Medford Lakes compete with each other in the construction of large and brilliantly lit floats that are supported by canoes. Each entry may use no more than two canoes and two paddlers. The challenge is in moving the canoe, with its elaborate assemblage, around Lake Aetna without capsizing.

Power boats are not allowed on the lakes in Medford Lakes for ecological reasons. However, along with the "Indian trail" names that mark the streets, and the rustic architectural style of the homes, the use of canoes is a conscious attempt to link the modern community with a mythical past.

In Cassville, in the northwestern reaches of the Pinelands, a golden onion dome gleams incongruously among the green pine tops. It crowns St. Vladimir's Russian Orthodox Church, one of the two churches that serve and symbolize the Russian community of Rova Farms. The Pines offered to this group both a healthful resort area and a rich supply of environmental features which could support ethnic folklife.

In 1926, Russians who had migrated in the late nineteenth and early twentieth centuries to big cities in New York, New Jersey, and Pennsylvania, founded the Russian Consolidated Mutual Aid Society to provide support and assistance to each other in times of need. One perceived need was to get people out of the crowded cities and into the healthful countryside, so in 1934, the association, by then known as ROOVA, bought 1600 acres of land, including a mill lake, just north of Lakewood. It was then resold in smaller plots to ROOVA members for vacation and retirement homes. Eventually a community center, restaurant, gift shops, cemetery, home for the aged, and churches were built. Many families settled there permanently.

At Rova, the residents have used the natural resources of the Pinelands to re-create old world ambience. Birch trees, a prominent feature of both Eastern European and Pinelands forests, have been annexed literally and figuratively into the landscape of this Russian community. A row of birches flanks

Following overleaf:
Blessing of the water, St. Vladimir's Day Russian festival at Rova Lake in Cassville. Photograph by Dennis McDonald. New Jersey State Council on the Arts.

the walkway to the center, and small trees and branches are often used for decoration on holy days such as Pentecost and at community festivities.

Cedar miner holding a "progue," 1937. Courtesy of George H. Pierson, New Jersey Bureau of Forest Management, Division of Parks and Forestry.

During the celebration of St. Vladimir's Day in July, ceremonies commemorate the Christianization of Russia by St. Vladimir. The blessing of the waters is a traditional element of the day, and at Rova, the ritual takes place beside the lake.

Another important resource that the Pinelands offers the Russian community is the panoply of mushrooms that grows there. Mushrooms are an important ingredient in Russian cooking. After being dried or pickled, they appear in many dishes, especially during the meatless Orthodox Lenten season. Mushroom hunting is a skill that Russian girls learn from their mothers. Pinelands forests offer a wide variety of microenvironments to nurture the different "communities" of mushrooms that women such as Dusia Tserbotarew and Valia Petrenko have learned to recognize in particular places such as under nut trees and on old stumps.[67]

Two other Russian communities have made their homes in the Pinelands. After the Russian Revolution, a number of Byelorussians who had been officers in the Tsarist army migrated to New York City. In the 1940s some of them moved to Lakewood Township to engage in chicken farming. A third group of Russians came to America as displaced persons after World War II. This wave included a group of Kalmyks, Mongolian Buddhists who had migrated to southwestern Russia in the sixteenth century. They were temporarily housed in barracks in Vineland, and some of them later settled in Freewood Acres north of Lakewood.

SWAMPS

Cedar Shingle Mining

Over the years, Pinelands swamps have offered exotic products to woodsmen resourceful enough to find uses for them. Surely one of the most unusual was Atlantic white cedar that had been buried in the rich muck for centuries. In the eighteenth century, it was discovered that thousands of these logs lay deep in the swamps of Dennisville. Until the early twentieth century, they were excavated, or "mined," from the swamps and used, because of their preserved state, to make shingles. Thus the men who mined cedar were known as "shingle miners."

Raising, or mining, cedar timber, circa 1857. From Geological Survey of the County of Cape May. *Trenton: The True American, 1857.*

The shingle miner first probed the muck with an iron rod, known as a "progue," to locate the cedar log. Then he dug into the tangle of roots with a "cutting spade" and sawed off a section about one foot long, called a "cut-off," which would pop to the surface like a cork. If he determined that the log was of good quality, the miner then cut away the roots and dug out the muck, exposing a hole that quickly filled with water. He sawed off the ends and, using levels, raised the log. When free, the log floated to the surface and turned over.

The shingle miner used a crosscut saw to cut the log into shingle lengths. Then he used a mallet and froe to split it into blocks called "bolts." Each bolt was then split into shingles. The rough shingles were dried in the sun and then shaved with a drawing knife on a shaving horse.[68]

Charles Pitman Robart was the last shingle miner in Cape May County. He was born at Dennisville in 1828, and he died there on September 29, 1907, in his seventy-ninth year. Edwin Robart, his son, wrote the following reminiscences about his father and shingle mining in a letter quoted by Charles Tomlin in 1913:

> My experience with my father in removing these logs from the soil and converting them into shingles was, from the time I arrived at the age of

Sawing cedar logs and making shingles. From Geological Survey of the County of Cape May. *Trenton: The True American, 1857.*

ten until I was sixteen years of age, always in the summer when I was out of school.

. . . No doubt the greatest place where these logs were found was in what is known as Robbin's Swamp. This swamp was cut off about 1864, enabling miners to investigate the bottom. The result was that hundreds of thousands of these shingles were taken out as there were several shingle miners at the time. Roads had been made of poles and bark to get the live timber out, and these same roads were used to cart out these shingles. Shingles secured in most places had to be carried out on the backs of men and boys to the creek and then taken by boat to the landing.

. . . Father secured thousands of these shingles in what is known as Hawk Swamp. . . . I being a boy could carry but twelve of these shingles if they were shaven as soon as they were dug, which was quite frequently the case. If the shingles could be dug and allowed to dry thoroughly, a man could make much more headway in shaving them. Father was considered a fast worker and could shave six hundred a day. With things favorable, he could mine and get ready for market one thousand shingles per week which usually sold for $16. In later years when sawed shingles came on the market, the price was as low as $12, which was very poor pay for mined shingles.

> . . . For a number of years after the Civil War, in which he served, father was engaged in the milling of grain, but after the mill blew out he returned to shingle mining, following it until he was 76 years of age.[69]

Pinelands residents continue to be fascinated by the phenomenon of mined cedar and its products. Charles Pomlear recalls logs that were dug up when he worked at the Durell plantation: "I could show you cedar trees in there that you and I, if they were standing, couldn't reach around them, three or four of us."[70] They conjecture about how the smell of the cedar chip conveyed information, suggesting that a windfall would contain sap. Cedar shingles, and homes that still wear them, are valued pieces of local history. One man, it is said, has refashioned them into cabinets for his home.

Cedar Farming

"That's our wood, Tom," was Ed Hazelton's response to folklorist Tom Carroll's query about why he uses Jersey white cedar for his miniature boats.[71]

Indeed, the tree and its wood have been distinctively a part of both the economy and the identity of the Pinelands throughout its history. Cedar's lightness and durability have made it a favorite for export and for regional uses. In 1749, traveler Peter Kalm described it:

> A tree which grows in the swamps here, and in other parts of America, goes by the name of the white juniper tree. Its trunk indeed looks like one of our tall, straight juniper trees in Sweden: but the leaves are different, and the wood is white. The English call it white cedar, because the boards which are made of the wood, are like those made of cedar. But neither of these names is correct, for the tree is of the cypress variety. It always grows in wet ground or swamps; it is, therefore, difficult to get to it, because the ground between the hillocks is full of water. The trees stand both on the hillocks and in the water: they grow very close together, and have straight, thick and tall trunks; but their numbers have been greatly reduced. In places where they are left to grow up, they grow as tall and as thick as the tallest fir trees. They preserve their green leaves both in winter and summer; the tall ones have no branches on the lower part of the trunk.[72]

More recently, boatbuilder Joe Reid called it ". . . the best wood that grows for boats," and described its virtues for boatbuilding. It planes well, doesn't split off, and has a long grain that can be steamed and shaped to the forms of Jersey boats.[73]

Cedar grows crowded together in small scattered stands throughout the Pinelands, taking from 80 to 100 years to mature. Although the crowding slows down its growth, it enhances its survival. Cedar trees are easily toppled by the wind, and Pinelands cedar farmers have classified the types of wood produced by "windthrow." "Boxy" cedar is the dense wood produced when a thrown tree eventually rights itself. "Shook" timber has inner cracks caused by twisting of the tree in the wind.[74] "Brazile" or "brasilie" is the exceptionally hard wood formed by cedar that has been in too much water. Oldtimers would detect it by putting a suspect board in the sun; it would quickly curve up if it were brasilie.[75] "Brassy" cedar is full of knots from the low branches that grow after the top of a cedar tree is burned.[76]

Many cedar farming methods aim to produce "clear" cedar that is free from knots. Adequate crowding inhibits the formation of branches, which cause knots. But all seedlings and saplings will not survive, so cedar farmers wisely thin the crop of marketable poles.

The uses of cedar have been many and varied over time. Kalm catalogued some of them:

> The white cedar is one of the trees which resists decay most; and when it is put above ground, it will last longer than underground. Therefore it is employed for many purposes; it makes good fence rails, and also posts which are to be put into the ground; but in this point, the red cedar is still preferable to the white. It likewise makes good canoes. The young trees are used for hoops, round barrels, turns, etc., because they are thin and pliable; the thick, tall trees afford timber and wood for cooper's work. Houses which are built of it surpass in duration those which are built of American oak.[77]

Today, Jersey white cedar fortifies both artifacts and traditions throughout the Pinelands. Boatbuilders make Jersey garveys and Barnegat Bay sneakboxes, and carvers shape decoys and boat models from it. Delaware Bay oyster schooners are kept alive with cedar planks, and coastal docks and bulkheads are kept afloat with cedar posts. Cedar clam stakes mark underwater beds in the bays, and cedar stales (long handles) move clam tongs onto those beds. Pole limas climb cedar poles in South Jersey gardens.

"You could use the regular two-by-twos from the lumber yard, but the cedar has a rough bark that makes it very easy for the beans to climb on," said George Heinrichs. "Yeah, like Jack and the beanstalk, if you had one twenty feet in the air, he'd go clear to the top of it." [78] Cedar trees are often used as Christmas trees, and many residents understand why Helen Zimmer ". . . couldn't live in a place where there's no Jersey white cedar," so thoroughly has the culture incorporated the wood into its existence. [79]

Although other methods of entering the swamps are used by most cedar farmers today, the building of "corduroy roads" to carry horses and wagons into the swamps was long an important activity. Pinelands swamps are replete with these paths, also known as "pole roads," "causeways," and "crossways." So important are they that they have names. One of the oldest, Frankie's crossway, dates back to the eighteenth century. One of the newest, Collins's crossway, was built by Clifford Frazee and his son, Steve, recently.

The Frazee family has lived in Forked River for generations, and the water in the area has always been vitally important to the family. Cliff's father oystered in the Barnegat Bay until the water became too saline in the 1930s. The oysters stopped growing in the bay, and Cliff turned to the

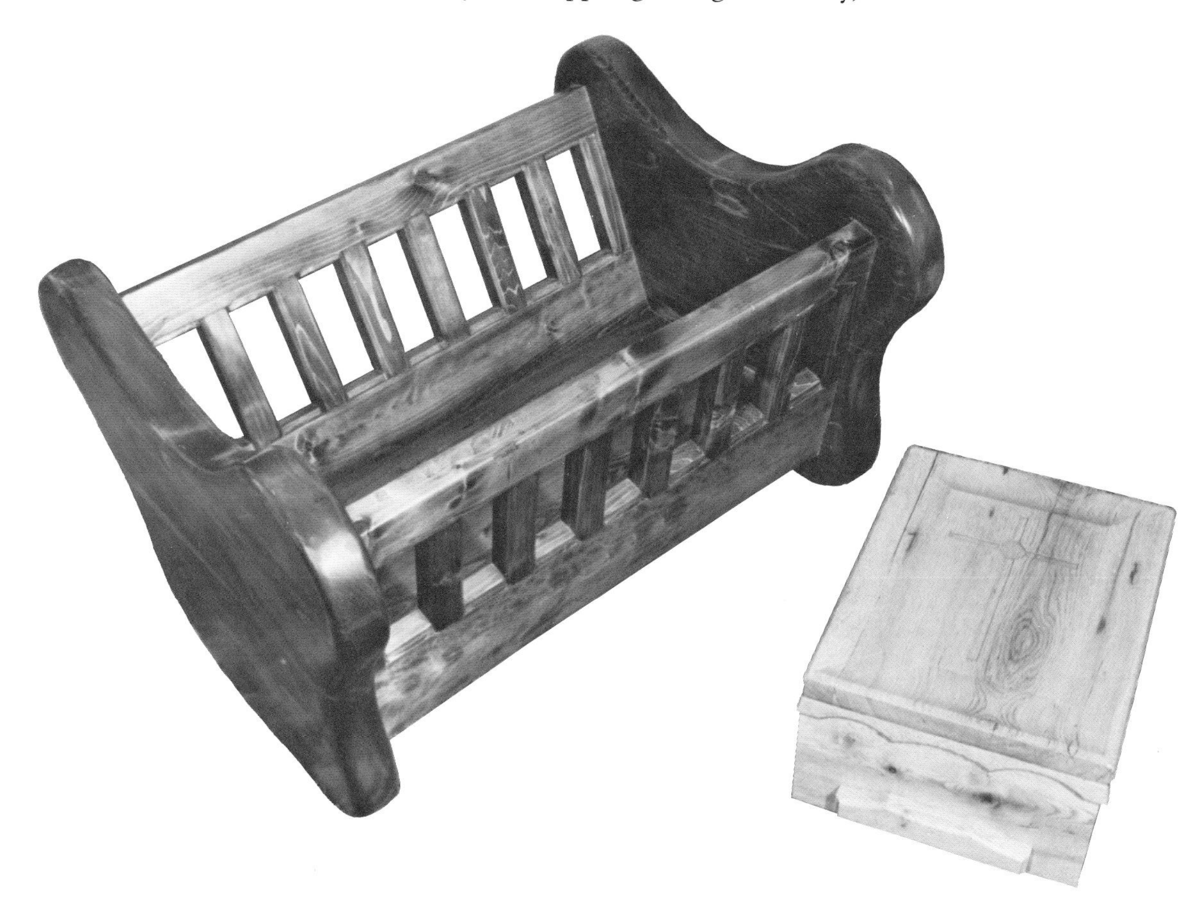

Cedar-lined Bible chest made by Cliff Frazee, Jr., and cedar cradle made by Steve Frazee. Lent by Cliff Frazee, Sr., and Steve Frazee. Photograph by Anthony Masso.

woods and swamps for his living. He worked a cycle that included scooping cranberries, clamming, cutting wood, and working for the state.

In the 1950s he started buying cedar swamps. Eventually, he bought a 600-acre tree farm and a mill, which Steve now runs. Since that time, wood, especially cedar, has been an integral part of both work and family.

Cliff's home grew with his family, acquiring cedar closets and bathroom, sassafras trim, and pine-paneled walls. In the living room, an oak Bible chest lined with cedar protects the family Bible, a two-century-old heirloom inscribed with the genealogy of the family. When the family inherited the book, Cliff, Jr., the eldest son, made the chest.

Steve Frazee lives nearby with his wife and four sons in a home that he built, principally from cedar. His esteem for wood is reflected in the many ways he finds to display its grain and color. Much of the furniture was also built by Steve, and some items seem particularly invested with a sense of family continuity. Before they married, Steve made a cedar hope chest for his wife. All of Steve's sons spent their first months in a cedar cradle he made, as well. "When I build things," says Steve, "I like to build them right. It takes a little longer, but I wanna build so it'll last forever. See, that's the way I think."[80] Individual trees prompt imaginative expressions of sense of place: "One tree I saw down right by Frankie's crossway. The tree was about this big. But when the tree was this big, they cut the surveying line across it. You could see, as I sawed it, I could see where they cut each branch off, a long time ago. And I took the one tree, and I was thinking about makin' a wall out of it, then this tree would be a history of when that surveying line was cut."[81]

Two days a week, most weeks of the year, the Frazees go into the swamps to cut cedar. During the dry season, they work each day on the crossway that carries them into the swamp and allows them to carry cedar out. To build a crossway, or corduroy road, they first mark the borders of the road, using a chain saw. They then level the roadbed by removing the stumps, first cutting horizontally along the bottom of the stump with a chain saw and then cutting vertically to the level of the horizontal cut. They discard the roots and compress the moss and dirt into the spongy ground. Next, they lay cedar, maple, and gum logs of less than four inches diameter horizontally along the road, with the thinnest logs in the middle and the thicker logs at the outer edges where the wheels of the truck will run. In order to level the road, they place a load of slabs from the mill between the logs already laid. Then they place more slabs perpendicular to the first level. Finally, they place on top trimmed limbs from cedar trees with the crotch ends facing outward and the brush ends facing inward.[82]

Cliff Frazee and his son, Steve, loading cedar logs on his truck. Photograph by Joseph Czarnecki. PFP215661–2–8A.

Nature has much to say about how and when they work. The swamp is full of holes and soft spots that can practically swallow heavy equipment. The sawyer must maintain his balance as he cuts. Swamps that have not been burned for a long time become nearly impenetrable with undergrowth. Woodsman George Brewer facetiously named one particularly dense and dangerous area "Laurel City." Trees are felled toward the crossway, so cutting can only be done when the wind is blowing in the direction in which they want the trees to fall.

Cutting a tree takes about 30 seconds. Steve cuts and notches the tree and Cliff pushes it to fall into the clearing. Both men then trim the branches from the entire trunk with axes. They cut the logs into 16-foot lengths and drag them to the roadside where they will be picked up the next day and hauled out of the swamp.[83] The remaining days of the week, the Frazees mill the wood at the sawmill near Double Trouble.

Cliff Frazee of Forked River standing on a crossway. Photograph by Joseph Czarnecki. PFP216395–6–34.

In 1749, Peter Kalm fretted about the depletion of cedar.

> In many parts of New York province where the white cedar does not grow, the people, however, have their houses roofed with cedar shingles, which they get from other parts of New Jersey, to the town of New York, whence they are distributed throughout the province. A quantity of white cedar wood is likewise exported every year to the West Indies, for shingles, pipe staves, etc. Thus, the inhabitants here, are not only lessening the number of these trees, but are even extirpating them entirely. People are here (and in many other places) in regard to wood, bent only upon their own present advantage, utterly regardless of posterity. By these means many swamps are already quite destitute of cedars, having only young shoots left.[84]

Today Cliff Frazee worries, too, but, knowing that clear-cutting a swamp provides the best chance for regrowth of young seedlings, he worries more about the water—its depth, flow, and salinity. As long as there is sufficient moisture, cedar will regenerate and seedlings will thrive. With too much water, cedar saplings will drown. Although natural damming often maintains an adequate water level, Cliff believes that a good cedar management plan would allow damming to maintain depth. If the water does not become brackish, cedar will continue to grow in the Pinelands swamps, and future cedar farmers may one day mark Cliff Frazee's handiwork on an old corduroy road as they harvest a 70-year-old stand of cedar.

Moss Gathering

"There is no rug prettier, with its golden fringe around it," Jack Cervetto told ethnobotanist Eugene Hunn.[85] The rug Cervetto refers to is sphagnum moss. It carpets Pinelands swamps, where tall cedars crowd, their dark green tops a canopy over soggy depressions furnished with cedar hassocks and unusual plants. "Hassock" is the name that natives have given to the gnarled and knobby root systems of cedars that poke out of the water at the base of the tree.

Because swamps punctuate the woodlands, moss was for many years a staple in the gathering calendar of most woodsmen. Sterile and highly absorbent, it was used as a base for floral arrangements and in surgical dressings. In the 1950s, however, a shift in Pinelands' manpower to heavy industry and the invention of styrofoam simultaneously reduced supply

Jack Cervetto of Warren Grove. Photograph by Joseph Czarnecki. PFP222306–3.

and demand. Today, a few gatherers still harvest and process the moss, including Bill Wasiowich of Woodmansie.

Wasiowich takes wheelbarrow, bucket, and pitchfork into swamps over old crossways abandoned by cedar cutters. He "pulls" the moss, hauls it out of the swamp, and takes it back to his homestead, where he spreads it on the sandy ground. With rakes that he makes himself, Bill turns the moss until it is dry. Then he bales it in a moss press, usually a homemade contraption.

Methods of processing are like moss presses: highly varied but achieving the same end. Because wet moss is so heavy, Jack Cervetto used to drape it over the cedar hassocks in the swamp. After two or three days, the moss, which could weigh as much as 200 pounds wet, would dry to a mere 15 pounds. He then would haul it out and press it into two-foot bales.

Woodsmen in the Pinelands have long visited the swamps to harvest moss and cedar, but for men such as Cervetto, the swamp offers greater rewards: "I like other woods, yeah, the Plains. But if you get right in the

middle of a cedar swamp with a good growth of cedar, you really think that you're the only person in the world. You're closed in there in a way that gives you the feeling there's nobody else around. And the smell! Yeah, I love that sphagnum smell. Boy, that's somethin'. I found out that's why I like workin' moss and cedar."[86]

MEADOWS

Salt Hay Farming

In 1879 the Reverend Allen H. Brown, a local historian, described the New Jersey meadows:

Following overleaf: *Charlie Weber harvesting salt hay with a team of horses, a South Jersey wagon, and a scow at Lower Bank, circa 1940. Photograph by William Augustine. Courtesy of the Donald A. Sinclair New Jersey Collection. Special Collections and Archives, Rutgers University Libraries.*

> The salt marshes or salt prairies of the coast may be reckoned among the natural privileges, as they produce annually, without cultivation, large crops of natural grasses. The arable land comes down to the sea in the northern portion of Monmouth County, and again at Cape May; but in the long interval the sea breaks upon a succession of low sandy beaches. Between these long narrow islands, and the mainland, which is commonly called "The Shore," are salt meadows extending for miles, yet broken and interrupted by bays and thoroughfares. More than 155,000 acres of salt marshes are distributed along the coast from Sandy Hook to the point of Cape May, including also the marshes on the Delaware Bay side of that county. As of old, so now, they furnish good natural pastures for cattle and sheep all the year round, and are highly esteemed by the farmers whose lands border on them, as they constitute also an unfailing source of hay for winter use and a surplus for exportation.[87]

Local people call these wetlands "meadows." Like the bogs and swamps, the saltwater meadows bear crops that resourceful natives have found many uses for. Chief among these is "salt hay," a type of cordgrass (*Spartina*) that, like its upland namesake, sweet hay, was considered useful in the keeping of livestock.

Opposite: *Bill Wasiowich gathering sphagnum moss. Photograph by Joseph Czarnecki. PFP219558-9-35A.*

Salt hay actually comprises three plant species which grow on slightly different elevations and are harvested together. Salt hay farmers recognize their subtle differences as keenly as they do the one- or two-inch variations in meadow height. "Black grass" (*Juncus gerardi*), a rush, grows earlier than

the others and on higher meadows. It must be cut by July or it will become oily and black. *Distichlis spicata,* which salt hay farmer George Campbell calls "rosemary," has a slightly hollow stem and grows straight up on slightly lower elevations. "Yellow salt" (*Spartina patens*) is the very fine grass that grows at still lower elevations. Campbell calls it "the real salt hay" because of its excellent qualities.[88]

Over the years, salt hay has been put to many uses, and residents enjoy competitively cataloging those uses. In the eighteenth and nineteenth centuries, many farmers grazed their cattle on the salt hay in the meadows and also baled it for winter feed and bedding. At Harrisville, in the 1840s, it was used to produce as much as a ton of wrapping paper and "butcher" paper a day.[89] Farmers favor it for mulching sweet potatoes and strawberries because it will not sprout in upland soil. Salt hay farmer George Campbell's father even sold it to banana boats in Philadelphia.

In the nineteenth and early twentieth centuries, the sturdy bulk of salt hay made it a favorite for packing material in nearby glasshouses, pottery plants, and brickyards. It once held down the shifting sands of dunes that had been disturbed by real-estate development.[90] Because it doesn't rot, it has also been used in sewers. In Port Norris, factories used to make it into rope used chiefly in the casting of iron pipe. Jack Carney, whose father ran such factories, still supplies a small market from a backyard building with two machines.

Much of the salt hay farmer's work is an attempt to maintain a balance or rhythm with natural forces. Salt water nourishes salt hay, but if tides wash the meadows too often, according to George Campbell, sedges grow. The farmer must maintain his dikes and adjust his floodgates to control tidal wash on his meadows. The growth of grasses provides a sturdy surface to the undulating meadow. If left too long undisturbed, it becomes "rotten marsh" that can mire machines. Harvest and burning must be timed to deter this. Even with its protective mat, the meadow can be treacherous terrain. Knowledge and equipment enable the farmer to keep his balance on it.

In the early nineteenth century, salt hay was mowed with scythes and piled on parallel poles that two men would carry to a stack or boat. Later, both oxen and horses were used to pull mowing machines and sometimes to power hay-balers.[91] Campbell recalls:

> We mowed it with a horse, and then we took another horse and we raked it with the old dump rake. We'd rake it into windrows, and then you'd

> turn around and reverse and put in what we called "cocks." Then we come in with the horses again, with the old wooden "shovel" ones that, you've seen pictures of them, years ago. And we'd pitch it all on the wagons and bring it ashore and then we'd put it in big stacks, loose. Well, there we got pretty mechanized. We had a boom with a cable on it with an old Hettinger engine that we had forks on. We would lift it off of the wagons in big forkfuls, okay? And stack it, and then all winter we had old stationary balers. We'd pull along from bent to bent and bale it in the big old three-wire stationary bales. And back then we'd still used to sell quite a bit loose, also . . . and we started on what I call my Cabin Marsh. My father used to go down there and bring it up on scows and boat it up creeks and, you know, to get it ashore.[92]

Farmer Ed Gibson of Port Norris remembered that it was necessary to keep two teams of horses, for one would sometimes have to pull the other out.[93] To avoid this, draft animals were specially outfitted. Oxen wore special collars; horses wore "mudboots." These shoes worked like snowshoes, giving the horse a broader base on the soupy meadow surface. Such shoes were made of wood or leather, and had "uppers" which came up on the front and sides of the hoof. They were secured with heavy straps and buckles. Other types modified the iron shoe, adding side loops or a wider base to it.[94]

Removing the hay required pole roads or scows. Pole roads were constructed by laying large, heavy poles, called "deadmen," lengthwise as a solid base, and then placing smaller poles or planks crosswise atop them. The deadmen were often made of cedar because it could withstand moisture. A hay scow was a bargelike boat that was pushed with a 15-foot cedar pole.[95]

Salt-meadow horseshoe. Lent by the Ocean County Historical Society. Photograph by Anthony Masso.

Harvesting salt hay, 1943. New Jersey Department of Agriculture Collection, Archives and Records Management, New Jersey Department of State.

Today, upland haying equipment is adapted for use on the meadows with balloon tires, traction belts, sleds, and flotation rings. A tractor-pulled windrower, baler, and wagon cut, bale, and move the hay. Where once a salt hay operation required 10 or 12 men, it now takes only 3 or 4.

The main task in salt hay farming has always been water control. Salt water is necessary for salt hay. It provides nutrients as well as irrigation. Campbell floods his meadows for six to eight weeks in the spring to control weeds and to fertilize for summer growth, but usually keeps tides out with dikes the rest of the year. If the meadows become too dry, however, upland weeds crowd out the salt hay. A system of drainage ditches and sluice gates facilitates the control of water.

Harvesting is usually done from June to January. Some spots, however, can only be gotten into during a hard (but snowless) freeze or a dry summer. If a field is uncut for longer than two years, it will be burnt over to destroy dead vegetation and encourage new growth. Drainage ditches, which must be kept clear, are also burnt.

The principal difference in equipment and methods for the salt hay farmer today is not the absence of the draft horse, but the presence of a giant crane and backhoe. While changes in equipment are usually caused by developments in technology, this change was caused by nature. For all the

George Campbell, salt hay farmer. Photograph by Joseph Czarnecki. PFP222306-6-35.

Delaware Bay salt hay farmers, the storm of 1950 marked the end of an era. The storm destroyed the natural banks of the meadows and upset the natural rhythm that had governed salt hay farming for generations.

Before 1950, the farmer's main task was cleaning the ditches and adjusting the gates. Since then, however, his major task has been building and maintaining the artificial banks against a rising sea level. ". . . So we're having to hold more and more water out of the meadows with the banks and it's getting, the economics is getting to the point that we're not gonna be able to do it much longer. It's going fast. I just keep running from one hole to another now on my cranes," Campbell says.[96] To compound the problem, tides have been higher in recent years.

Indigenous combatants in Campbell's fight to maintain a balance are wetlands animals and plants. Since 1950, *Phragmites communis* has claimed more territory in the meadows. This tall, plumed reed, while desirable to gatherers for use in dried arrangements, can outcompete *Spartina* under certain conditions. Campbell tried killing it with herbicide for awhile, but concluded that, like the mosquito, *Phragmites* will adapt to any adversity. "When something did come back," he told folklorist Jens Lund, "the *Phragmites* come back before anything else."[97]

Muskrats and otters can destroy banks, the "marsh bunny" with its burrowing and the otter with its play. Ed Gibson says otters "play like kids," sliding down the muddy banks.[98] Trappers are usually a welcome form of pest control.

Other creatures serve as alarm systems. While perch are normally found in the drainage ditches on Gibson's meadows, a carp is a signal that there is a break in the dike. Carp, which normally enter the meadow only during spring flood, and migrate out after spawning, may enter the ditches from the bay in search of roots if there is a break in the dike.

Though he may be fighting a losing battle with the bay, Campbell views salt hay farming as "a way of life," one followed by his father and grandfather before him, that he will not easily let go.[99]

Pest Control

For George Campbell, "half the enjoyment of working outside is going down in the meadow and seein' the osprey and the eagle taking the fish."[100] With the fiddler crab and the mosquito, the deer and the snow goose, they are familiar co-habitants of his environment. Not all of them elicit such favorable responses from Campbell, but their presence is part of the place and their absence a sign that something is amiss. When the meadows were sprayed with DDT to control mosquitoes, the crabs, dragonflies, fish, and mussels disappeared, too. Their return is welcomed as a sign that balance is back, even if it does mean more mosquitoes.

Pinelands residents know which creatures control others. They use their knowledge of these relationships in indigenous systems of pest control. Box turtles are kept in basements as silent snail catchers. Skunks are allowed to rid gardens of bugs and mice. Perhaps the most widely celebrated set in the Pinelands is the mosquito and the purple martin. The Jersey swamp mosquito ". . . is still bitin' you when everything else is dead and gone," according to local experience.[101] It long ago became part of regional identity. In 1685, a newcomer wrote: "In the Marshes are small flies called Musketoes, which are troublesome to such people as are not used to them."[102] Peter Kalm's description in 1748 was more graphic:

> The gnats, which are very troublesome at night here, are called mosquitoes. . . . In daytime or at night they come into the houses, and when the people have gone to bed they begin their disagreeable humming, approach nearer and nearer to the bed, and at last suck up so much blood that they can hardly fly away. When the weather has been cool for some days the mosquitoes disappear; but when it changes again, and especially after a rain, they gather frequently in such quantities about the homes that their numbers are legion. The chimneys of the English, which have no dampers for shutting them up, afford the gnats a free entrance into the houses. On sultry evenings they accompany the cattle in great swarms from the woods to the houses or to town, and when they are driven before the houses, the gnats fly in wherever they can. In the greatest heat of summer, they are so numerous in some places that the air seems to be full of them, especially near swamps and stagnant waters, such as the river Morris [Maurice] in New Jersey. The inhabitants therefore make a fire before their homes to expel these disagreeable guests by smoke.[103]

Later, perhaps as proof of the power of humor to help people accept the inevitable, jingles, tales, and jokes featured the Jersey mosquito.

Kalm mentions one of the many methods of mosquito control that have been tried over the years—smoking. Another, suggested by a local newspaper, involved hanging over one's bed a rag that had been soaked in a mixture of camphor and whiskey.[104] Early in the twentieth century, the growth of tourism in shore areas prompted organized efforts at mosquito control.

Purple martin bird-houses. Photograph by Jens Lund. PFP217721–14–3.

These included glossing breeding grounds with kerosene, and eventually ditching and draining thousands of acres of salt meadow. These ditches, unfortunately, also contributed to the loss of many salt hay meadows. The spraying of DDT in the 1950s killed more mosquitoes but sterilized the meadows of other wildlife, as well.

While Pinelands residents are acquainted with a variety of natural enemies of the mosquito (including the scents of sphagnum moss and cedar), the traditional favorite is the purple martin, an insect-eating bird that summers in New Jersey. Long ago, the Lenape Indians, it is believed, removed most of the branches from tall saplings and hung gourds in them as homes for the martins. For over a century now, the square, gable-roofed martin house set atop a tall pole has been a familiar part of the landscape, especially in the open spaces near meadows, bogs, and blueberry farms. The martins' helpfulness with the mosquito problem, as well as their interesting habits, have earned them special status. This is reflected in the way people speak about the martin. In stories, it gains personhood and human characteristics: "They're your friend," says Leslie Christofferson. "In other words you can go out there, and when you'd shoot a starling, well, they'd be tickled to death. Oh, they'd whistle and talk to you. That's right! They knew you were helping 'em, believe it or not." [105] Starlings are a serious enemy of the martin, usurping nest compartments and sometimes even stealing eggs.

Christofferson, who lived most of his life in Whiting, New Jersey, has built many martin houses in his years as a carpenter and blueberry farmer. His observations of this species' social nature over the years prompted him to build a "purple martin palace." This masterpiece is more than six feet high and two feet in diameter, and weighs about 500 pounds. Though the structure adheres to the specifications requisite for a martin house, it elaborates on them in grand fashion. Each of its 112 apartments is reached through a hole that is about one and seven-eighths inches wide and has an inner dimension of approximately seven inches by six inches to give the bird space to build a nest and to turn around. "Just what makes common sense to you, you know, when you look at the bird." [106]

But more subtle aspects of the martins' social nature are catered to, with individual verandas on the eight-storied skyscraper:

> You'd be surprised what personality they've got! And you'd have the same ones year after year on the same date. I could look out April the fifteenth, and I wouldn't see a bird. Maybe I'd start looking the fourteenth—I mean there within two or three days. After a bit, you'd look out there and you'd see waaay up there, maybe one or two birds. Way up in the air. Then the next day, half a dozen. Next day, twenty-five would come in. See, they'd go back and give them the signal—all's clear. They'd bring their families with them. And you'd go out there in the evening. The males sit out on the porches on a hot night. They were never in the houses, you know. You'd got out there, say, when the moon was bright, . . . and they'd lean over and talk to you. Oh, beautiful songs they have! [107]

The purple martin palace was crafted through Christofferson's delight in both the creatures and the process of creation. It began simply enough: "I said, 'I guess I'll build a birdhouse,'" but grew, ". . . as a little challenge. I could build just about anything. I was going to build a round one, you know, but it turned out fourteen-sided. Fourteen sides. I divided it up and it came out just about right." And it ended with satisfaction: "Well,

Opposite: *Purple martin palace. Lent by Leslie Christofferson. Photograph by Anthony Masso.*

when I stood there, I said, 'Well, that looks like that should be it!' I built it right. There wasn't a joint you could get a razor blade in. It was built to perfection." [108]

Wetland Trapping

The meadows that fringe the rivers and bays are a transitional zone where a tug of war between earth and water goes on. In many places, meadows have been pulled violently back into the water, as along the Campbell's bayside property. In others, the change was more gradual. Many of the meadows along the Maurice River were once fertile farmlands whose banks were reinforced by huge floating mud machines. Eventually, the river reclaimed them.

The amount of rainfall also affects these wetlands. When rain is scarce, the saltwater creeps landward, causing critical changes in both fresh and salt water bodies. For example, as the salinity increases on Delaware Bay oyster beds, the oysters are less able to resist disease. As freshwater meadows along the Maurice River become brackish, the "wild rice" that attracts the sora railbird disappears, and so does the railbird.

Yet residents of the wetlands have successfully adapted to change, both normal and cataclysmic. They have structured their lives around the daily fluctuations of the tides as well as the changes of the seasons. They've even met and mastered the considerable challenges of harvesting the resources of the muddy marshes, the muskrat and the snapper. People in Cumberland County have coined a name that celebrates this skirmish between man and nature and seems to declare man the victor. While those who trap, fish, and probe for turtles in other places are usually called "proggers," around Mauricetown they are sometimes called "mudwallopers." Chief among a mudwalloper's skills is maneuvering through the muddy areas where muskrats and snapping turtles can be found.

Albert Reeves of Mauricetown and his eight brothers learned these skills from their father, "Gummy" Reeves (so named because of his ever-present gumboots). According to Reeves, each man must know well the meadow he's trapping, watch for soft spots, and proceed with caution: "You wouldn't go along haphazard. You wouldn't just go along head up and not looking. You'd just sort of skid along, sort of feel your way along, feelin' for how much the mud's givin'. You can tell the soft spots. There's some water standin' there. The level spots would usually be the softer spots. When you get in a soft spot, you're out of luck." [109]

Tom Brown of Millville, trapper. Photograph by Dennis McDonald. PFP219500–06–20.

The consistency of soft spots varies greatly, calling for a variety of manuevers: "I've seen my brother actually get down and swim in it! Thick, like a real thick puddin', it's so soft. He leaned over and actually swam to the boat, it was so soft."[110]

"Blue mud," recognized by its bluish cast, is more troublesome. It is more solid and sticky, with a consistency Reeves compares to putty. "It's terrible! Once you get in, you can't get out. That's how I hurt my hip, in blue mud, trying to pull my leg out. I've had trouble with it ever since."[111]

One animal that makes such hazardous pursuit worthwhile is the muskrat, or "marsh rabbit," a small furbearer that is as abundant and valuable now in South Jersey as it was in 1749 when Peter Kalm made the following observation.

> The muskrats, so called by the English in this country on account of their scent, are pretty common in North America; they always live near the water, especially on the banks of lakes, rivers and brooks. On traveling to places where they are, you see the holes which they have dug in the ground just at the water's edge, or a little above its surface. In these holes they have their nests, and there they stay whenever they are not in the water in pursuit of food. . . . Their food is chiefly the mussels which lie at the bottom of lakes and rivers. You see a number of such shells near the entrance holes. I am told they likewise eat several kinds of roots and plants. . . . They make their nest in the dikes that are erected along the banks of the rivers to keep the water from the adjoining meadows; but they often do a great deal of damage by spoiling the dikes with digging and opening passages for the water to come into the meadows; whereas beavers stop up all the holes in a dike or bank. They make their nests of twigs and such things externally, and carry soft stuff into them for their young ones to lie upon. . . . As they damage the banks so considerably, the people are endeavoring to destroy them when they can find their nests. The skin is sold and this is an inducement to catch the animal. The skin of a muskrat formerly cost but threepence, but at present they bring from sixpence to ninepence in the market. The skins are chiefly used by hatters, who make hats of the hair, which are said to be nearly as good as beaver hats. The muskrats are commonly caught in traps, with apples as bait.[112]

Muskrat is an important "cash crop" to many part-time trappers as well as to men who work the cycle full time. Boys who go to high school and men who work in offices and factories alike draw income from the traps they set and empty in the early hours each day between December 1 and March 15.

205
'83

Tom Brown taking a muskrat from a trap. Photograph by Dennis McDonald. PFP219500–03–24.

On state-owned as well as privately held brackish tidal meadows that muskrats prefer, trappers set and tend traps on "low water," marking them with a pole. An experienced trapper may also time his activity to the weather and the mating cycle, therefore starting later and stopping earlier than the legal dates. Animal pelts "prime up" in the colder weather of January and thus are worth more when sold. Conversely, damage to fur caused by the fighting that occurs between males during the mating season in March lowers a pelt's value.[113]

To find muskrats, trappers may look for the "three-cornered grass" that they favor, and observe the maxim that "every animal has its highways"; the muskrat's are called "runs." Traps are set in the run about 50 feet from a "lodge," the mound of sticks, grass, and mud that houses the animal. A careful trapper will not step in the run. "Step your foot in it or mux it up in some way, and as a rule, he'll go somewhere else. That's the art of trapping."[114]

Both law and logic require trappers to visit their traps daily. An animal left unattended for a longer time may be damaged or stolen.

Little of the crop is wasted. Animals are first "skinned out," with the meat being frozen, eaten, or sold fresh by the trapper. Muskrat meat is often likened to a cross between wild duck and rabbit, and is a popular dish in the region. In Hancock's Bridge, just outside the Reserve, the annual Muskrat Dinner is always sold out. Trappers who also farm often use waste parts such as blood and bone, as fertilizer.

But by far the most valuable part of the animal is the fur. Value is expressed in many ways. The year 1980 is remembered as a good season because brown pelts brought $9.25 each. But an even more meaningful measure is often recorded in a statement such as, "My dad paid my college tuition that year with what he made trapping 'rats."

Fleshing out furs is therefore an important skill, for unremoved bits of fat can turn a fur rancid. Knife or push-pin holes can mar it. Tom Brown has developed equipment and methods to do this job to his high standards. A skinning bench with a "fleshing beam" mounted on heavy sassafras stumps provides a sturdy work base. After the pelt has been thoroughly scraped, it is slipped inside-out over a "stretcher" and hung to dry. Stretchers used to be made of wood, but are now most often made of wire.

Another denizen of the Pinelands—one more grotesque and cantankerous than the muskrat—is the snapping turtle, which lurks on the bottoms of brackish and freshwater ponds and streams from April through October and hibernates in the mud the other half of the year. Though the season is unregulated, it is difficult to capture the creature in months other than May to July in the traps that are most commonly used now.

Two other methods, which are only occasionally used today, were popular in the past. Telltale air holes on muddy flats indicate the presence of a snapper. The trapper probes with a large snapper hook until he hits a shell, then hoists the turtle up. The other method is called the "choker set," in which wire is wrapped in a groove around the middle of a four-inch piece of quarter-inch wooden dowel. The trapper tucks the stick into a piece of salt eel, fastens the choker to a line, drops it into a pond, and ties the line to a bush or a stake on shore. After the snapper swallows the bait, the choker turns crosswise so that it cannot be disgorged, allowing the reptile to be caught alive.[115]

Today, snappers are most often trapped in fykes. A fyke is a cylinder approximately five feet long and two feet in diameter. When made of net, the cylinder is supported by three evenly spaced wooden hoops. One end of the cylinder is entered through a net funnel that leads to a bait box. The other end is closed. Until fairly recently, the body of the cylinder was made of hand-knit net. Today it is usually made of wire fencing.

Whether in a pond or on a tide-washed meadow, the fyke is set so that one end of it will always be above water, and the turtle will not drown. This

Exterior of Tom Brown's hunting and trapping lodge. Photograph by Joseph Czarnecki. PFP221130–4–4.

may require anchoring the trap with sturdy poles or floating it with plastic buoys. On tidal meadows, the fyke is set on high tide and returned to on the next high tide.

Though snappers can be easily enough coaxed into a fyke with rank meat or fish, they can be safely gotten out only with caution and subterfuge. Tom Brown "bridles" the turtle by getting it to bite on a stick, then pulling the head back and tying it.[116] Other trappers spin the turtle to disorient it.[117]

Once out of the fyke, the turtle can be held by the tail, but only with the top of his shell facing away from the trapper. A snapper can only bend its head backward. The unwary trapper who gets bitten by a snapping turtle will testify to the power of his jaws, as do the sayings about the turtle. Some people chop off the head and cut it in two to prevent the snapper from

Interior of Tom Brown's hunting and trapping lodge. Photograph by Joseph Czarnecki. PFP217721–14–3.

biting for the two days after its death during which it is reputed to be capable of biting.

Most turtles are sold live to restaurants that may serve snapper soup with sherry on the side. Those that are used locally more than likely are served at parties at which a big pot of snapper soup with hard-cooked eggs in a vegetable-laden broth is the focus of enjoyment.

Railbird Shooting

Kenneth Camp's narrative on the origin of the name "railbird" recalls a time when gunning the birds was a far easier task than the one guides now experience:

Herbert Misner of Medford turtle trapping. Photograph by Dennis McDonald. PFP235200-7-8.

> This was a dairy farm in here. They used to have the split rail fence. And when the meadows went out the fences went underwater, see, and started floating all over. And these rails would get on them and line up. They'd shoot them right down the rail. Yeah. That was more or less years ago, you'd market hunt them. Wouldn't be nothing to come out and shoot a hundred shells.[118]

Wildfowl are seldom so cooperative, and his knowledge of their habits and habitats is a guide's stock in trade.

Guiding gunners of the railbird was an important part of the yearly round for families such as the Reeves, who combined it with trapping and fishing. Ken Camp, his father, and his son work things slightly differently; they farm much of the year and guide gunners in the fall. Ken is also employed by the state as a forest fire warden.

Ken Camp of Port Elizabeth, railbirding guide. Photograph by Dennis McDonald. PFP235042-7-6.

In September and October, the "wild rice" (*Zizania aquatica*), or "wild oats," as locals sometimes call it, attracts flocks of the tiny sora rail to the freshwater marshes along the Maurice River in their migration south. These

NRA

meadows have names that recall the farmers who farmed them when they were diked: Hampton's Meadow, Coxe's Meadow, the Boudelier Meadow.

The bird, in turn, attracts gunners who rely on local guides to push them through the meadows atop the double-ended skiffs specially built for the hunt. The railbird skiff has two platforms, the rear one for the "pusher," who propels the boat with a long pole. Its relatively flat bottom, shallow draft, and pointed ends allow it to cut through the grassy meadows where the birds hide. The gunner stands near the front platform, gun poised.

"You gotta have a double end boat," according to Camp:

> It has to be pointed so that you can push it through the reeds. . . . You gotta have them wide enough to make them stable, and long enough. The balance. You can't have a round bottom boat. You can't have a flat bottom boat. You gotta have more or less an in between, and you got to have a bow to them. The railbird skiff, mostly—the old ones were all made out of cedar. Cedar planking on the bottom. We modernized them by taking that off and putting plywood and fiberglass in them in the last

Railbirding skiff and push poles. Photograph by Dennis McDonald. PFP235042-2-10.

> twenty years. Any of them that's built today is mostly built out of plywood and fiberglass. But there's not too many of them being built.[119]

The guide and the gunner go out about an hour before high tide in search of the seven-inch bird. It is easier to flush the rail during high water because it has no stable surface underfoot. Therefore, the hunt ends as the tide recedes.

As he poles through the meadows, the guide shouts the location of a bird for the gunner. If the shot is successful, the guide then poles the boat to the spot and retrieves the bird.

For a pusher, four critical aspects of a hunt are summarized in Reeves' maxim about what a guide hopes for: "Big tide, small man, good shot, big tip."[120] A big tide is one which comes in high and stands for a long time. The added depth it brings not only makes it easier to push the boat, but also causes the birds to flush more easily, producing "good shot," or many birds for the gunner.

It is easy to understand why a "small man" makes an easier day for a pusher. Yet it goes beyond lightness of load. As Camp explained to folklorist Gerald Parsons: "If you get a heavy boat and it's balanced out with a gunner, you can push it just as easily as you can another boat. The balancing out is the main thing, getting them going."[121]

Many of the people who have come to the Maurice River to shoot railbirds have been wealthy and prominent. Names such as Kean, Strawbridge, and Pugh are common in Reeves's repertoire of railbirding tales. A camaraderie between Reeves and the visiting gunners often developed and resulted in the "big tip" that could happily conclude a hunt for the pusher.

The good pusher needs numerous skills and reliable equipment to garner that big tip. In addition to a stable skiff, he wants a pole with a smooth, lengthwise grain that won't snap under stress.[122] He uses the pole with a side-to-side motion, creating a windmill effect.

The push poles used to be made of cedar, "but we get lazy, you know, and we go buy bannister rails out of the lumberyard." The pole has three blades on the end that prevent it from sinking into the mud. "The trick to pushing one of these boats," says Camp, "is really experience. It takes you a few trips before you can get on to it. You use your leg to steer with, and [learn] where to place the pole in the back of the boat."[123]

Because "you can do more in a half an hour at the right time than you can in three hours at the wrong time," a pusher must be able to read tides and weather fluently.[124] He'll avoid "pogee" (apogee) tides that are "lazy": they don't come in very far and they don't go out very far.

The two most important qualities of a good pusher are visual acuity and

Opposite: *Railbirding on the Maurice River. Photograph by Dennis McDonald. PFP235042–8–36.*

strength, which enable him to spot birds and to push quickly and smoothly. Reeves, who hired many a local man to push gunners in his boats, explains: "Sometimes you have a real good pusher, but he can't mark so good. Then some guy could mark better, but couldn't cover as much territory, couldn't get around as fast. But you get the two together is what you got to do to get what you call a top pusher."[125]

The ecology of the region has changed as salt grass replaces the wild rice, which needs fresh water. "The salt grass is moving up more every year," Camp says. "Everything below the [Mauricetown] bridge is all salt grass now. Years back—say twenty years ago, twenty-five years ago—they used to hunt below the bridge. But that's all gone."[126]

RIVERS AND BAYS

Shipping

In 1862, after the death of his father the previous year, Alonzo T. Bacon of Mauricetown shipped out on the schooner *Robbie W. Dillon* which was owned by Captain George S. Marts, a neighbor and friend of his father:

> The occupations open to a boy, in which no capital was necessary, in any of the towns or villages in southern New Jersey were oystering, seafaring, or farming. I chose the seafaring life as being one by which, with determination to put forth my very best efforts, success would be achieved, and more remunerative occupation obtained.[127]

Advancement was quick for the right kind of person. Before he was 21 years old, Alonzo Bacon was a mate on the schooner *Thomas G. Smith,* and by the age of 23 he was the master of this same vessel. During his career as a coastwise schooner captain, he sailed to Boston, Charleston, Savannah, Baltimore, New York, Philadelphia, and also to Maine, Newfoundland, Nova Scotia, Texas, Louisiana, Florida, and the West Indies. In 1897, he sold his vessel to Captain William Crawford and bought the ship chandlery in Bivalve.

Until the building of the railroads in the late nineteenth century, and even a bit after, the products of the Pinelands were shipped by water from various landings along the rivers and bays.

Somers Point was the official port of entry for the entire Great Egg Harbor District. Each landing had a wharfmaster who maintained the wharf and collected a fee from boats using the dock. Jeffries' (also known as Jeffer's) Landing, on the Great Egg Harbor River, was named after John Jeffrey, who became wharfmaster in 1819. The goods shipped from his wharf included lumber, cordwood, charcoal, iron from Weymouth Furnace, farm produce, and salt hay from the meadows around the bay.

Various timber products were also shipped from English Creek Landing. In the last half of the century, cordwood was shipped to Haverstraw, New York, for use in brick manufacturing, and cordwood was carried to Philadelphia in small sloops called "bay craft."

By the first decades of the twentieth century, the railroad had completely eclipsed the sailing vessel, and in 1915 the customs district was officially closed.[128]

Boatbuilding

The importance of boats, their building and use, to southern New Jersey commerce and society were well captured by a visitor to Tuckerton who wrote in *Watson's Annals* in 1823:

> Little Egg Harbor was once (in my grandfather's time, when he went there to trade) a place of great commerce and prosperity. The little river there used to be filled with masted vessels. It was a place rich in money. Farming was but little attended to. Hundreds of men were engaged in the swamps cutting cedar and pine boards. The forks of Egg Harbor was the place of chief prosperity. Many shipyards were there. Vessels were built and loaded out to the West Indies. New York and Eastern cities received their chief supplies of shingles, boards and iron from this place.[129]

A wide range of boats have come from South Jersey boatyards in the past two-and-a-half centuries. Great schooners, coasting vessels, sloops, and shallops were crafted for work in the bays and ocean, while garveys, bateaux, sharpies, sneakboxes, and skiffs were, and in some cases still are, constructed for tasks such as sturgeon fishing and railbirding in the bays, rivers, and meadows.

These boatyards drew on the resources of the Pinelands. Many of them lay along the network of rivers and creeks which empty into Great Egg Harbor and drain the southeastern section of the Pine Barrens. They included two shipyards at Bargaintown on Patcong Creek and the Israel Smith yard on English Creek. Below Mays Landing, at the head of navigation on the Great Egg Harbor River, there were shipyards at Kats Landing, Pennington Point, Huggs Hold, Junk House Wharf, and Coal Landing, among others.[130]

Shipbuilding was also an important activity on the Dennis Creek in northern Cape May County. Dennisville, located five miles up the creek, produced at least 50 vessels between 1849 and 1901. Twenty-three of them were three-masted schooners up to 150 feet long. They had to be launched sideways and kedged down the creek by hauling in lines secured to anchors. It took several tides for them to reach the bay.[131]

On the Maurice River in Cumberland County, schooners were built both for coastwise trade and for oystering on Delaware Bay. The Lee brothers, two ship carpenters from Egg Harbor, founded a shipyard at Leesburg in 1795. A three-masted schooner was built at the Banner and Champion Yard in Dorchester in the 1880s.[132]

A large workforce was necessary for the many steps involved in constructing a wooden schooner. Skilled craftsmen, such as ship carpenters, caulkers, blacksmiths, and sailmakers, were drawn from the local population. John DuBois, a retired boatbuilder and oysterman, recalls that, "When I was young, the only thing to do in our town was either to work in the shipyard or go on to the boats."

Though workers were often motivated in their work by payment of a share in the ship's worth, DuBois cites the equally important reward of pride of workmanship: "You start with a pile of flitch lumber with bark on it, but when you get done you started from something with bark on [it] and you wind up with something that's finished, and you're proud of. There's a lot of satisfaction in building a nice boat."[133]

In the building of a schooner, the dimensions of the boat were first drawn on the mold loft floor. A carved half-model of the hull, which could be separated into cross-sections, was used. The framing timbers were cut from these patterns in symmetrical pairs. The keel, the backbone of the boat, was assembled, and the sternpost hoisted into position. Then the frames were also hoisted into position and fastened with wooden pegs known as "treenails" (pronounced "trunnels"), which would swell once the boat was in the water. The planks were steamed, fitted, and nailed to the frame, strake by strake, and the seams between the planks were caulked (made watertight) with oakum (hemp) and pitch. Then the masts, booms, gaffs, mast hoops, halyards, sheets, and blocks were installed. Typically a sailing schooner had two masts. The sails consisted of a jib, a staysail, a foresail, a mainsail, and topsails.

The age of steam power and of iron ships constructed in urban shipyards brought an end to the wooden shipbuilding industry in remote places such

Ship caulker at the Delaware Bay Shipbuilding Company, Leesburg, circa 1928. Photograph by Graham Schofield. Courtesy of John DuBois and John T. Schofield.

as the Pinelands. In the 1920s, the last wooden schooner was built in the Dorchester Shipyard on the Maurice River.

Boatbuilding in smaller operations continues today along the coast, however. Fiberglass pleasure craft are manufactured in Egg Harbor City, for instance, and much of the work on wood and engines is done by members of old coastal families.[134]

More common still is the backyard boat shop where garveys and sneakboxes are constructed by boatbuilders and fishermen who make boats for their own use.

Folklorist Tom Carroll's observation about the garvey, that it "results from the conjunction of local resources and local conditions," applies as well to other indigenous boat types—the schooner, sneakbox, and railbird skiff.[135] Each of these has been shaped by the local environment in which it is used, the particular tasks for which it is created, and the locally available materials and craftsmen.

Joe Reid, who knows boats from the standpoints of both use and construction, speaks for many residents when he says that Jersey cedar is best for Jersey boats. It has a long grain and can be steamed. This is in contrast to the "southern" cedar that he once tried to use. He explains that, "Every night we'd steam a board up in the bow, and clamp it, and leave it overnight. Next morning it'd be broke right in two. We didn't know that until we started using it." Fiberglass, another frequently used material, doesn't handle as well in water as cedar does: "Cedar takes in just the right amount of water. When it's first put there, it tends to sit right on top of the water. In a couple of weeks it soaks up the right amount of water and settles down. Then it handles really well. You can't beat cedar for a boat." [136]

The garvey, sneakbox, and railbird skiff are all tailored to the shallow waters they ply. Barnegat Bay, where garveys are used for clamming, ranges in depth from two to twelve feet, and the flat-bottomed garvey can pass in less than three feet of water. Water depths in grassy meadows where railbird skiffs are pushed and sneakboxes secreted are sometimes as little as two feet; consequently, those boats also have very shallow drafts.

The waterman's need for steady footing on the water is answered in the design of several of these boats. The garvey and the much larger schooner have each been described by users as "a sturdy working platform." The Jersey schooner is wide for its length, and the garvey has become wider over the years. Such proportions provide greater stability as well as ample deck space for the shellfish that they carry. The rail skiff provides a raised platform at each end. Its proportions and weight must be finely balanced in order for the pusher to maintain his balance. Albert Reeves, who pushed hunters on dozens of rail skiffs in his years, recalled one boat that missed the mark: "It had too heavy a frame. And not the right kind of flare. It was top heavy. I don't know how the hell I ever stayed on it!" [137]

Though they are all shaped by traditional templates, no two boats are alike, builders say. For one thing, the boat is often the collaborative product of the customer and the builder. John DuBois, who worked on the schooners, recalls that even oystermen sometimes played "keep up with the Joneses" by ordering the same boat a bit longer, larger, or grander. Others would have the boat tailored to their own way of sailing and dredging. [138]

James Reid, who works with his father, emphasizes that a craftsman continues to learn as he works: "I can't think of a single boat that we've done in the last five years that we haven't thought of something that coulda been done a little different. Only thing, my father would agree and say that you never really stop learning how to do it until you die." [139]

Yet good design, and the craftsmanship to execute it, is elusive, according to Albert Reeves: "It's hard to explain. I took one of my boats I loved so well—the design of it and everything—I took it to an expert, and took it there so he could make a copy of it, see. But it wasn't like it! He had the damn boat right there, but it still wasn't like it! He didn't know his stuff!" [140]

Les Hunter, who lived in Haleyville and built Reeves's favorite railbird boat (which he eventually sold to Frank Astemborski), did know his stuff.

Opposite: *Model schooner* Charles Stowman *made by the crew at the Stowman Shipyard, 1929. Lent by Albertson Huber. Photograph by Dennis McDonald.*

For some boatmen, change is a problem. ("Well, sometimes you can't make things better than they are," said Albert Reeves.) Yet a skilled craftsman alters his product within acceptable limits to suit a changed market or technology. Even though a tonger needs straighter sides on his garvey, Reid will flare the sides of pleasure garveys to meet the aesthetics of his customer. His work garveys have gotten wider to increase the load capacity, and the bow has been raised so that the boat rides higher in the water.[141] In such matters, Reid's judgment is so well regarded that people bring their boats to his yard so they can work on them under his practiced eye.[142]

Boats have a strong significance in the bay areas because of the personal and economic dependence on them; they are closely tied to many aspects of human life cycles. A waterman always remembers his first boat, whether it was an old garvey pulled out of the mud or the first oysterboat in which he "went up the Bay."

A boat is often the site of learning to work as well as becoming an adult in the community. Most men along the Delaware Bay have worked both in the boatyards and on the boats. In the words of Belford Blackman, a former oysterman, a youth would begin learning "how to work and how to work the boats" in the middle deck, where he would cull and clean up.[143] On the other hand, achieving the role of captain represented maturity and manhood in the work community. In the past, recalls Fenton Anderson, an oysterboat captain was called "the old man," and the crew were "the boys," no matter what their ages.[144]

Captain Louis Peterson's recollection of his grandfather underscores the importance of boats in men's lives: "If it could be done with a boat, he did it. If it couldn't be done with a boat, it wasn't worth doing."[145] It is a sad occasion, therefore, when a man ends his work on the water and sells his boat: "I sorta hated to let it go. It was part of my life and everything. It was my last boat that I ever owned."[146]

Men are identified with boats and vice versa. In young Captain Todd Reeves's words, "If the boat don't do nothin', people will think of you as you don't do nothin'."[147] This attitude is reflected in the fact that most of the Delaware Bay oysterboats bear family names. Stories often refer to a man and his boat interchangeably:

> Tarburton was the captain of this boat, and this boat meant more to him than any boat he'd ever had 'cause she was the nicest and the biggest boat. And the first day of the season, we had an awful blow up here. I think he was the only boat that worked. Tarburton was the only boat that worked with that boat, and he planted two loads of oysters with her over in the state of Delaware.[148]

CHAS H STOWMAN

Miniature garvey The Gladys *made by Joe Reid. Lent by Joe Reid. Photograph by Anthony Masso.*

Because of this importance of boats, the schooner has been appropriated as a symbol. A sailing schooner rides on the door of the Port Norris fire truck. In many homes, half models and pictures of family boats hang beside those of family members. A fully equipped scale model of a Jersey dredgeboat (oysterboat) was built by the workers in the Stowman Shipyard in 1929 to represent them in a Vineland parade.

The miniatures made by men such as Joe Reid, therefore, are meticulous statements of both sense of place and sense of person. Reid has spent his life in Waretown using and building boats. Now in his seventies, he seems to know everything there is to know about the garvey. Joe knows the subtle variations that occur in boats even within a limited region. When he was a boy, he recalls, the garveys in Tuckerton were lower and narrower than those in Waretown, and in Barnegat they used sharpies, a special rowboat that one rarely sees now.[149]

When he builds a boat, Reid says, "what I like is to let the wood more or less shape itself. Not try to force it. Have a graceful curve. No flat places in it."[150] His respect for the wood and his knowledge of the place have been crafted into the model garveys that Reid has made in recent years. He builds them of the same materials and according to the same methods as the full size boats, planing the cedar down to proportionate thickness and whittling by hand some tiny pieces such as the water scoop that he included in one model of a working garvey.

Though Joe attributes his first model to no grander a motive than "foolin' around," his miniatures really are grand statements writ small. They encapsulate a lifetime of experience, knowledge, and concern.

Fishing

The rivers and bays have enriched folklife in numerous other ways over the years. Along the Maurice River, proggers who trap muddy meadows for snapping turtles in spring and summer have also worked the river and bay for livelihood by fishing, shellfishing, and bird hunting.

Many fishermen believe that different species of fish come and go in great natural cycles. That seems to be the case on the Maurice River. Earlier in the century, striped bass (locally called stripers and rockfish) were so abundant that fortunes were made fishing them. By the 1970s their numbers had dwindled considerably.[151] Today, commercial fishermen are restricted from catching them in nets.

Another fish formerly caught in the Maurice was the sturgeon. Its value was greatly enhanced by its roe. Albert Reeves likes to recall being "up to my elbows" in caviar. Local men would process the roe of sturgeon before selling it to caviar companies in New York. The sturgeon would first be bled through the tail to avoid polluting the roe. The roe was then rubbed through a wire mesh to remove the fat and put in a clean wooden bucket, where it was salted in proportions of one pound of salt to ten pounds of roe. After several days, the roe was drained overnight on netting. Much of the roe that Reeves processed was sold to Ferdinand Hansen, a caviar dealer from the Fulton Fish Market in New York City.

The delicacy inspired little enthusiasm among the Reeveses, however, who sometimes used it in eelpots. Pickled sturgeon, though, is recalled with much relish.[152]

Menhaden ("bunker"), a valuable and prolific fish that has numerous industrial uses, was abundant along the Jersey coast until the 1950s. The huge schools of menhaden were encircled in purse seines (bag-bottomed nets) that were then drawn up by two dozen men in two striker boats. Their movements were coordinated with traditional songs called menhaden chanties.[153] Today, menhaden are caught in gill nets in smaller numbers and sold as bait.

Pound nets, which were actually huge fish traps that were often staked on the shore, were most often used in the past. Today, this and other methods, such as the fencing of carp as they emerge from meadowland ditches, are illegal. Still other methods, such as hauling seine and drifting gill nets, have changed only in their materials.

Before the invention of synthetic textiles, most fish nets were hand knit of cotton or linen with a large, flat wooden "needle" during idle winter months. Fisherman Nerallen Hoffman of Dividing Creek recalls the chant "Row upon row, watch it grow" that accompanied knitting sessions (which might include whole families).[154] The care of such nets was equally time consuming. After each use, a linen net had to be dragged through a lime box and put on a pole to dry. The coast in those days was lined with cedar drying poles spiked with nails.[155]

Today the weakfish, which was scarce in the past, is the "money" fish. Along with flounder, perch, bluefish, and shad, they occupy fishermen who haul seines or drift or stake nylon gill nets. Making a net today involves "tying in" purchased "lease" (machine-made netting) to nylon top lines. First, the fisherman must choose lease with appropriate-size holes for the type of fish he wants to catch. Three-and-a-half-inch lease catches most desirable fish, and few undesirable ones. However, a fisherman who wants to have some "fun" catching the giant drum fish may choose a 10-inch lease. The fish swims part way through the hole and catches its gills in the net.

The fisherman next decides how much slack he wants in the net; generally, he wants more slack in cold weather. Then he will decide how much he wants the net to sink or float, and will add corks to the top and leads to the bottom accordingly.[156]

Knowledge of fish, tides, and weather will then determine when and where the fisherman uses the net. Many fishermen believe that different fish come on different winds. For commercial fishermen, Reeves recalls, a northwest wind generally meant a poor catch. A southerly or easterly wind, however, would bring, among other varieties, shad.

Fishermen still may "make a haul" with a hauling seine, but this method is less used than in the past, perhaps partly because, as Reeves recalls, it means "simply pulling your guts out!"

Either working alone or with one or two others, Reeves would put the seine out in a half circle from the shore and immediately pull it in. A strong ebb tide would make the haul more difficult by pulling the net away from him.

A good fisherman, Reeves claims, must stay ahead of the net. To minimize the tug of war with the tide, fishermen may haul seine "on top of the tide" at "high water slack," when the tide is not moving. Whenever they haul, however, if the fish are not moving they'll get only a "water haul."[157]

The final vagary with which the fisherman must deal is the market. "It's like I said, there's no happy medium for the buyers and the fishermen. He knows you got a product's gonna go bad. You can't eat it and you can't keep it, so you gotta take what you can get for it. I always said, they had a little shelf about six inches wide and they take their hand in their pocket, when you got a lot o' fish, and they throw the change up there and what hits there's what you get."[158]

Another creature that fishermen may pursue is the eel, and the pursuit is a full-time task from June through November. Eeler Lynwood Veach thinks nothing of following fish from 6 A.M. to 8 P.M. in the months when they are moving.

During that time, the eels feed and travel 24 hours a day, moving so quickly that an eeler needs a fast boat, a nimble body, and a good memory. Eels return annually to the same feeding spots, but the eeler will keep his eel pots well spread out and retrieve them regularly in his efforts to keep up with the eels.

Eels were caught more in the past, when spearing, dredging, and bobbing were legal. Searching the mud bottoms of clear streams for small entry holes, or probing in murkier waters with a rod, eelers could spear more than one eel at a time.

Handmade, split-oak fykes were also much used in the past. Today, most eels are caught in "pots" constructed on the same principle but of different materials, usually hardware cloth or wire mesh. Because the eel is "tem-

Left: *Eel spear, circa 1900. New Jersey State Museum Collection, 174.5. Photograph by Anthony Masso.*

Right: *Eel trap basket. New Jersey State Museum Collection, 174.6. Photograph by Anthony Masso.*

Opposite: *Making eel baskets, Greenwich. Photograph by William Augustine. Donald A. Sinclair New Jersey Collection. Special Collections and Archives, Rutgers University Libraries.*

peramental," according to Veach, he uses a wide range of variations in his pots: "Well, there's square ones; there's round ones. There's what I call the sea bass pot that are flat on one side, and there's quite a different variety. There's double funnel, with a long wire to go into where the net mesh is. Listen, if you could make a pot that would work everywhere and work all the time you could be a millionaire in a couple o' years." [159]

Like the lobster, the eel enters the pot tail first. To entice him to do so, eelers bait their traps with cut up horseshoe crabs containing eggs, or waste parts that local clam factories discard. Small eels, caught early in the season, are salted and sold as bait to crabbers. Larger ones are kept in a "live car," a small, boatlike box with an opening on top and screened ends to promote water circulation, until they are shipped by "live truck" to buyers in large cities. Locally, eels are usually smoked by individuals in makeshift smokehouses. Some oldtimers used to smoke fish in an old refrigerator mounted on cinder blocks, with a fire below, usually made of hickory, occasionally of sassafrass. [160]

Oystering and Clamming

Right by the Delaware Bay sits the small town of Bivalve, its name a reminder of the importance of shellfish to the area. As recently as the 1930s, residents claimed that "the whole world meets in Bivalve." Since the earliest days of settlement, coastal residents have built their lives around the harvest of oysters and clams. Oystering, particularly, has generated a distinct and elaborate folklife.

In Colonial days, oysters were gathered by hand with rakes or tongs. Some people continued tonging for oysters into the twentieth century, especially in the rivers. However, in the early nineteenth century the introduction of the oyster dredge launched the commercial oyster industry in Delaware Bay. The dredge originally consisted of a wooden crossbar with iron teeth to which was attached a rope mesh bag. After the Civil War, the dredge frame and mesh bag were made of iron. It was "drudged" along the bottom of the bay and then hauled in. The oysters were dumped on the deck and "culled," that is, separated into different grades by the crew. Crews were usually made up of men from the immediate region or immigrants who came from Philadelphia or who followed the dredging season up the coast.

The completion of the railroad to the Maurice River in 1876 made possible the shipping of oysters in refrigerated cars to markets in Philadelphia and New York. Prior to the 1920s, the oysters were shipped unshucked.

They were "floated" in wooden frames to keep them fresh and plump, and then moved with "oyster forks" by float crews into handmade wooden oyster baskets when they were sold. Carloads of the bivalves were shipped out daily by rail.

An oysterman's yearly cycle has always been built around the Delaware Bay practice of "planting" oysters. Small "seed" oysters that are raised on state-owned beds in less salty waters are dredged freely by licensed oyster boats for four weeks in May and transplanted to leased beds farther down the Bay. The seed oysters are left to grow until they reach a marketable size, which may take three months or a couple of years. Many of these leases have been held by one family for generations and may be named for an early owner. Fenton Anderson, for instance, who bought his leasing rights from Tulip Westcott, calls his lot the "Tulip Grounds." [161]

Dredging season traditionally began in September, quickened as the holiday season approached, and tapered off in January and February. In recent years, some oystermen dredge year round. Many oystermen, as befits the local title "oyster planter," farmed during the off season, while others worked in boatyards. Still others used their boats to fish, take out fishing parties, and haul freight, practices still followed today.

Under sail, boats would go out on Monday morning and return on Friday evening. Crew members, recalls John DuBois, would never bring a black suitcase aboard or use the word "pig," for fear of jeopardizing their journey. [162]

The crew was responsible not only for "culling" and shoveling discards over the side, but also for working the boat and dumping the dredges. The captain was a man well practiced in the art of tacking up and down as he dragged the heavy iron dredges over the oyster beds and raised them to the deck. His skill in maneuvering was a part of his reputation, as was his boat. Many oystermen reveled in the image of being tough, and this is reflected in their stories, such as the one about Elmer Tarburton.

Most of the towns along the Maurice River harbored oyster boats. A fleet of small sailing vessels called the "Mosquito Fleet" sailed out of Dividing Creek. These smaller boats could tong or dredge smaller beds and shallower waters.

In 1945, a prohibition against using power boats on oyster beds was lifted. Masts were removed, decks were rearranged, and a way of life was altered.

Today, though most of the oyster boats are survivors of the era of sail, they work in different ways. The crew usually numbers no more than four or five, for automatic dumpers and culling machines now separate most of the oysters and remove debris. The captain can run a power boat in slow circles to dredge the oysters. Boats go out early in the morning and return by

Tonging for oysters, Cedar Creek, circa 1910. Donald A. Sinclair New Jersey Collection. Special Collections and Archives, Rutgers University Libraries.

4 P.M. Because the oyster crop is smaller, dredging is done less often, and is more keyed to demand.

Since the 1920s, oysters have been "shucked" (opened) in shucking houses by oyster shuckers, who come mostly from Maryland and Virginia. After they are removed from their shells by hand, the oysters are "blown" to clean them in a vat, graded, and packed in metal containers that are then refrigerated.

Whereas oysters have been especially important to baymen on Delaware Bay, clams have been an important resource to their counterparts on Barnegat Bay.

Joe Reid of Waretown grew up on Barnegat Bay. He used to clam in the summer and build garveys in the winter, but for the past 10 years he has been building boats full time. In his day he clammed at numerous places in the bay. "I worked from Forked River to the Mullica River, which is at New Gretna, according to where the clams are and according to the weather. We don't stick to one spot."[163] Reid didn't use navigational charts; he knew the bay both from his own experience and from the traditional knowledge of other baymen, who have given places in the bay names such as Sammy's

Charles DeStefano of Pleasantville clamming on Barnegat Bay. Photograph by Joseph Czarnecki. PFP216651–11–14.

Slough (a place of deep mud), Buck's Channel, and Diamond Flat. "I don't know how they originally got their names. Since I was a kid, we just knew the names." [164]

Clamming is quite different from oystering. While oystermen transplant seed oysters onto their own leased beds, clammers go wherever they want to clam. Some clammers, however, lay their surplus in staked lots when the market for clams is poor. These lots are leased from the state and marked with corner stakes bearing the name of the clammer.

The Lenape Indians used to "tread" for clams and oysters; that is, they waded out into the shallow water to gather them. Today, clammers use two methods: raking and tonging. The most common rake is the Shinnecock, named after Shinnecock Bay on Long Island, New York. The rake is dragged along the bottom, often from the stern of the boat. Tongs, on the other hand, work like large scissors. They dig into the bottom to grab the clams, for clams burrow just below the surface, unlike oysters, which lie on the surface. The oldest tongs had wooden heads. Today the heads are made of iron and only the handles are wooden. Reid acknowledges that the iron heads are superior.

The garvey is essential for clamming with tongs, but not necessarily for raking. Reid explains why:

> The main object of a garvey for clamming is the fact that it lays still in the water. If you're tonging, for instance, you can't have a boat moving back and forth. You drop your tongs down and you feel clams, and the boat has to stay there so that you dig the clams out. If the boat moves, you can't get them out. Now with a rake—some of them use rakes off the stern—it doesn't matter. They drag these big rakes. Sort of a jerking motion. And then it doesn't matter what kind of boat you've got.[165]

Legendary tales are told about clammers, celebrating their skills, their knowledge of the bay, and their shrewdness. Mr. Reid tells the following story about his brother.

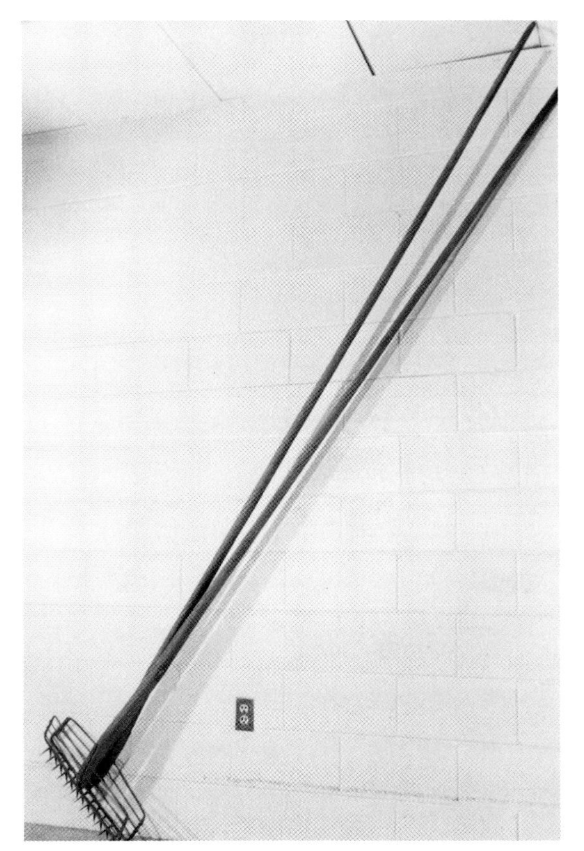

Clam tongs used by the Ridgway family in the early twentieth century. New Jersey State Museum, 81.99.4. Photograph by Anthony Masso.

Opposite: *Joe Reid of Waretown, garvey builder. Photograph by Joseph Czarnecki. PFP216395-1-22.*

> My brother was always very particular about having his own spots, where he clammed. He didn't encroach on anybody. And he didn't like anybody to do it to him. Though this young fellow came out of the army and decided to go clamming. So they told him my brother was one of the best clammers. So naturally he followed him. And he followed him all day, and my brother would clam ten minutes here, ten minutes there, and he would move. So when they came in that night, he said to him: "How come they tell me you're the best clammer in the bay, and I followed you all day, and I didn't find anything." So my brother said to him, "Well, I know where they are, and I know where they ain't." And he said, "Today I've been clamming where they ain't." [166]

Fishermen, whether tonging shellfish or netting finfish, often cite their love of independence and the outdoors when they explain their attachment to their lifestyle. For some, it is an almost organic attachment, expressed in the desire to "die with salt in my ears."

Duck Hunting

The Barnegat Bay decoy and the duck hunting boat known as the "sneakbox" are both made of white cedar and are both adapted to the specific conditions of Barnegat Bay, a wide expanse of shallow water. "Many of the earliest late-nineteenth-century decoys from the Barnegat Bay area have long been admired for their similarity to small craft themselves," wrote folklorist Bernard Herman and archeologist David Orr.

> The decoys of the Jersey shore, centering on Barnegat Bay, form a marvelous assemblage of artifacts. . . . They closely parallel the corresponding evolution in hunting boats developed in the same area. The premium was on weight and size since these commodities were in sparse supply on a twelve foot craft. . . . There was an additional desire for naturalistic realism for the hunting was on open expanses of water. . . . Outstanding examples of original design, the Jersey hollow duck or goose is not only formally symmetrical and cleanly carved and ornamented, but it is also an excellent solution to the combinative determinants of Bay ecology and small craft architecture. [167]

The sneakbox is a shallow-draft hunting boat, usually 12 feet long and 4 feet abeam, with a spoon-shaped bow and a melon-shaped hull. It is thought that the first sneakbox was made about 1836 by Hazelton Seaman

Barnegat Bay sneakbox, circa 1910. Made by J. Howard Perrine. New Jersey State Museum Collection, 70.156. Photograph by Joseph Crilley.

of West Creek. Originally they were either rowed or sailed with a sprit-rigged sail to the gunning grounds. Today they are powered by outboard motors or towed to the grounds by larger power boats. Some are still built as sailboats, used more for pleasure than for gunning. Those built for sailing have a centerboard trunk and centerboard. Those built for gunning have a rack on the stern to hold the rig of decoys.

Sam Hunt of Waretown makes his sneakboxes with white-oak timbers (frames), which he first softens by steaming them in a homemade steambox and then clamps into place on a model of the hull called a "jig." But, he explains, the original sneakboxes didn't have steamed timbers. Instead, they had frames which were sawed from large cedar logs. Hunt uses white cedar for the planks, because, he says, "I never seen a piece of it really rot in a boat. Never. It's light, and it will swell. You can put it together and leave the seams that far apart and it still comes tight. It won't leak." [168]

According to Hunt, the boat is called a sneakbox because it enables the gunner to "sneak up" on ducks. Actually, they were used quite differently, as decoy carver Ed Hazelton of Manahawkin told fieldworkers Tom Carroll and Nora Rubinstein. The hunter cut a V-notch in the meadow along the bay, and pulled his boat into it. The sneakbox was covered with brush, marsh elders, or hay, and the gunner sat facing the stern, waiting for the ducks to be lured down by the decoys. Today, Hazelton explained, it is more common for the gunners to use "gunning boxes" built with seats and shell shelves. They are more comfortable and can seat as many as six men.

Two different gunning traditions came out of the nineteenth century. One was that of the sportsmen gunners, usually wealthy individuals from outside the region, who hired local guides. The other was that of the market gunners, local people who made a living, at least in part, by supplying game for restaurants in nearby cities. The Federal Migratory Bird Act of 1918, which restricted wildfowl gunning to strictly regulated seasons, brought market gunning to an end.

In the nineteenth century, according to local historian Patricia H. Burke, the sportsmen traveled to Barnegat Bay by boat or stagecoach. After 1870, railroad connections to New York and Philadelphia made the trip much easier. The accommodations included boarding houses, hotels, and inns, such as the Chadwick House at Chadwick Beach, south of Mantoloking, and the Harvey Cedars on Long Beach Island. For the very wealthy, private gunning clubs were organized, including the Marsh Elder Gunning Club on Marsh Elder Island, the Manahawkin Gunning Club, and the Peahala Club.

Sometimes businesses maintained facilities for the use of their clients. George Heinrichs of New Gretna worked in a menhaden fish processing plant that maintained such a hunting lodge, and he sometimes would act as a guide. "We'd take them out for anywhere from a day to a week. They'd sleep at the lodge. We had a cook for the meals. . . . The lodge is torn down now. I went there in 1959 as a hunting guide, and I left in 1964–65. The lodge was torn down then."[169]

The local men who served as guides for visiting hunters would sometimes maintain small lodges for their own use. Milton R. Cranmer of Manahawkin, who was a guide from 1943 to 1957, leased the Flat Creek meadows near Marsh Elder Island for hunting. His father built a small clubhouse there and they nicknamed it "Cream Puff Castle."[170] Local hunters both built their own boats and carved their own decoys, because they couldn't afford the decoys made by the more prestigious carvers.

All Barnegat Bay decoys had certain characteristics in common. They all were made of white cedar, they all were hollow, and they all had painted, rather than carved, tail feathers. Within these parameters, four different styles developed along Barnegat Bay. The decoys from the "Head of the Bay" had a high tail with a small knob on it, a rounded bottom, a tight eye groove, and a rectangular lead pad on the bottom for ballast. The decoys from the town of Barnegat were similar, except that the tail was not as high, the bottom was not rounded, and the eye groove was less pronounced. The carvers from Parkertown made their decoys with a low tail and flat head. The ballast was made by pouring molten lead into a rectangular opening carved into the bottom, and an "ice-catcher" was carved behind the neck. The decoys from Tuckerton had a low tail, a round bottom, and ballast similar to that of the Parkertown decoys.[171]

Ed Hazelton said that the old carvers made all species of ducks, but mainly Canada geese, brant, black ducks, broadbills (scaups), canvasbacks, and redheads. The species depended on what the gunner wanted to gun for. The clubs gunned for broadbills, black ducks, and geese, because they were plentiful.

"The decoy as a historic artifact is basically a social tool," folklorist Bernard Herman has said. "It was used not [only] as an implement to lure birds, but as a status statement. One went into a resort hotel, got the finest guides, the finest sneakboxes, and obtained the best birds made by the best carvers in the area."[172]

Some of these prestigious carvers were Eugene Birdsall (c. 1857–1917) of Point Pleasant, Percy Grant (1864–c. 1960) of Brick Township, Jesse Birdsall (1843–1927) and Henry Grant (1860–1924) of Barnegat, Nathan Horner (1881–1942) of West Creek, Jay Cooke Parker (1882–1967) and Lloyd Parker (1858–1942) of Parkertown, and Harry V. Shourds (1861–1920) of Tuckerton.[173]

Harry Shourds's grandson, Harry V. Shourds, II, of Seaville, is carrying on the family tradition. He argues, however, that his decoys are different from his grandfather's and his father's decoys. He distinguishes between working decoys and decorative decoys, but he acknowledges that most of his decoys are made for collectors, rather than gunners. "Most people today buy them for decoration, but they're really a working decoy. They're what I think a duck looks like. I don't copy off a real duck, and none of the old-timers did. They hunted ducks and saw ducks in the wild, and took their memories with them. They worked from those memories."[174]

Harry Shourds judges a good decoy by how it behaves in the water:

> For a decoy to be a working decoy, it has to look like a duck, float like a duck, and act like a duck in the water, the way it bounces and swims around. I think a decoy has to have enough roundness so that it will rock a little. It's got to swim back and forth. It has got to be weighted right so that it sits the right depth in the water, so when a duck comes over, he sees that more than he would see the shape of the duck. He sees it at a

> distance, say one hundred yards, and looks down and says, "That's a nice flock of ducks down there. Think I'll go down and see what they're doing." [175]

Carver Hurley Conklin was born in 1913 in Cedar Run, but for the past 40 years he has lived in Manahawkin. He made his first decoy in 1928, mainly for his own use. He reserves the term "decoys" only for those used by gunners. But, he says, "Today carvers, including myself, do not make decoys for gunners; they're too expensive."

For John A. Hillman, a gunner and decoy collector from Sea Girt, the traditions surrounding gunning are very much a part of place: "Each place, we gun different, just like neighborhoods. People are different in each neighborhood, and around Barnegat Bay this is what they've done over a hundred years." [176]

FARMLANDS

The following tall tale told by Tom Brown contradicts the image of the Pinelands as a "barrens" with poor soil for agriculture.

> You know, that ground down there to Linwood is very rich. My friend Elwood Ford used to tell this story. "When we'd plant corn," he said, "Grandmother would make the hole for the hill of corn. Us children would drop it, then Dad would cover it up. Then he'd jump out of the way, the corn came up so quick. One day I was real brave. I grabbed ahold of it. I rode up in the air. I'd of starved to death, if Grandpa hadn't shot biscuits up there with a shotgun." [177]

Actually, there are a variety of Pinelands soil types, ranging from gravelly sand with almost no humus, too porous to retain water and fertilizers, and thus too poor for conventional agriculture, to fine sandy soil that holds fertilizers and is well suited for crops such as sweet potatoes, to sandy or gravelly loam with some heavy clay, that is good for fruit trees, but not wheat, rye, or other grain crops. Agriculturally, the region is divided into broad subregions: north of the Mullica River, where cranberries and blueberries have predominated; and south of the Mullica, where grapes, strawberries, sweet potatoes, and fruit trees have been grown.

In the eighteenth century, agriculture was a part-time activity in the Pinelands, second to lumbering, shipbuilding, and iron manufacturing. Farming was restricted primarily to the southern and coastal regions. The land was cleared by burning, a practice that was criticized by Count Julian Niemcewicz, who traveled through the region in 1799:

> Again sand and pine forests, the more sad because it was all burnt over. This tremendous damage is caused by indigent inhabitants who, having no meadows in which to feed their cattle, burn the woods. The fire running along the ground, turns the lower bushes to ashes; with this the earth is enriched and puts forth grass and other plants—in a word excellent pasturage for cattle. This advantage does not compensate for the harm done by the fires which rise from the lower growth to the tops and burn the taller and more useful trees.[178]

Cattle were allowed to graze on woodland pastures and on salt hay in meadows and on barrier islands. Early crops there included corn (maize) and rye, which were grown primarily for self-consumption.

Cranberry Farming

J. J. White saw the potential for agriculture in what had been previously thought of as "barrens." In 1916, he wrote:

> From the position it occupies, between the two great cities of the nation, it may be a marvel to some that this region should have remained so long uncultivated, but it is explained in few words. The soil is light and sandy, not suited to growing grass or the cereals, but yielding good crops when planted in small fruits. These, with the exception of cranberries, require easy and rapid facilities for marketing; such as are only obtained in the interior by the use of railroads, and those, until recently have been withheld. Hence, the swamps were left to make cedar, and the uplands to produce pine timber. But now, railroad facilities are being afforded, and large portions of "The Pines" are destined to become a fruitful garden under the skillful management of the fruit grower.[179]

The fruitful garden which White predicted, and which he himself helped to create, dazzles the eye with its color and the mind with its ingenuity. Out of acidic streams and sandy soils, cranberry farmers have built intricate waterways and fertile bogs. Cranberry agriculture is perhaps the best example of the resourcefulness and judicious adaptation that characterize Pinelands folklife.

The history of the development of the industry is a story of environmental recycling. The cranberry (*Oxycoccus macrocarpus*) is a vining evergreen that grows naturally in sandy swampy areas of eastern Massachusetts, the upper Midwest, and southern New Jersey. Its fruit ripens in the early fall and can be stored until spring. Natives gathered wild cranberries regularly before the 1800s, but the decline of other Pinelands industries—most notably the iron foundries—encouraged residents to adapt these natural resources to different uses in the cultivation of cranberries.

The first cultivated cranberry bog in the Pines was planted in the 1830s by Benjamin Thomas at Burrs Mill in Southampton Township.

In 1857, when James A. Fenwick purchased the 108 acres that would become the nucleus of the Whitesbog agricultural plantation, he was recycling the water supply and work force from nearby Hanover Furnace. As Mary Ann Thompson points out, the bogs themselves are "remodeled cedar swamps," the components of which—water, turf, and cedar—became reservoirs, roads, and flood gates.[180]

Eventually, Whitesbog would encompass 3,000 acres, most of them purchased between 1884 and 1909 by Joseph H. White, Fenwick's son-in-law. At the height of its development, Whitesbog consisted of cranberry bogs, blueberry fields, a water supply system, the village of Whitesbog that grew up around the sorting house, and the migrant workers' villages of Florence and Rome, named in honor of the Italian workers at the turn of the century.

Four parallel streams that flow from east to west into the Rancocas Creek, a tributary of the Delaware River, provided the water supply for Whitesbog. Gaunt's Brook and Antrim's Branch are located to the north and Cranberry Run and Pole Bridge are to the south. To the east and west are swamps. The main water supply came from the Upper Reservoir. The second water supply system was created by damming Pole Bridge Branch to form Canal Pond.

Between 1857 and 1912, 40 bogs covering 600 acres were designed and built at Whitesbog. Lower Meadow Bog, Old Bog, and Little Meadow Bog were the original bogs built by Fenwick. White built five other groupings of bogs, named the Cranberry Run Bogs, the Ditch Meadow Bogs, the

Cranberry pickers' village at Florence, circa 1910. Courtesy of William C. Bolger and the New Jersey Conservation Foundation.

Pole Bridge Bogs, the Upper Reservoir Bogs, Antrim's Branch Bogs, and a grouping consisting of Big Swamp Bog, Billy Bog, and John Bog. Today it is more common for the bogs simply to be numbered.[181]

At Whitesbog, in the early twentieth century, J. J. White helped develop the process of adaptation to an art. In his manual on cranberry culture, he sketched out the recycling of "muck" into bogs.[182] And over the years, even the wild vines have been given second life, as growers continue to hybridize them, often sowing their own identity into the fruit with names such as Howard Bell, Richard, Garwood, Bozarthtown Pointer, Braddock Bell, Applegate, and Buchalow. The "bell" in some of the names refers to the bell-like shape of the cranberry; the names refer to the growers.

The engineering of land and water that are basic to cranberry farming begin with the location of the bogs on a good water supply. Most planters seek a ten-to-one ratio of headlands, or upland water supply areas, to bogs. Each of the cranberry plantations in the Pines today is located on a branch of a major waterway. A system of reservoirs, dams, canals, and dikes connects the bogs with the main water supply and acts as a system of valves and pipes which can be regulated as necessary with the gates that farmers build.[183]

The bog itself is a structure of symmetry and precision, and the present bog landscapes of the Pines represent many lifetimes of work. Today, cranberry farmers assess the fine points of various bogs and bog-building techniques, and even refer to some bogs, such as Buffin Meadows at the Darlington plantation, as "state of the art."[184]

As described by cranberry farmers Stephen Lee and Henry Mick, the building of a bog begins with planning for efficient water management. At a spot near a stream, the topography of the land is first surveyed, and the reservoir placed at the highest point.[185] The area in which the bog was to be built was formerly either scalped of its vegetation or flooded to kill the vegetation. Today the trees are usually removed with heavy machinery, and useless vegetation is burned off. Cedar and pine are kept for gates and equipment, cedar for things to be partly submerged and pine for things fully submerged. Irrigation channels, consisting of a main center ditch with perpendicular side ditches, were used in older bogs; today, underground pipes are used for irrigation. A dike surrounds the bog. It is made of a core of sand from the ditches that are dug around the bog, and it is covered with turf cut from the bottom of the bog to prevent erosion. These dikes also serve as narrow roads between the bogs. Wooden gates at key places in the dike walls control the flow of water.

The bottoms of older bogs may be uneven and rough compared with those of recently built bogs. Today, bogs are graded so that they are within a half inch of level for conservation of water and ease of harvest.

It may take three years to build a bog, and several more to develop the plants. Even then, a bog, like a house, may "settle" in places within the first year or two and require regrading.

Most new bogs are planted with cuttings from other bogs. A cover of sand is spread over the muck at the bottom of the bog. The vines are inserted in small holes in a grid pattern made with a long-handled tool sometimes called a "dibble."

Water is essential to the cranberry, not only as sustenance but also as protection. Thus, water levels in the bogs follow a pattern of response geared to both regular growth cycles and irregular temperature fluctuations. The "winter flood" protects vines from the cold of winter months, and the "spring draw" uncovers them for the growing season. At full fruition, the "fall flood" buoys the fruit for easy harvest. In between, water may play havoc if it turns to heavy ice and chokes off oxygen or if it appears as a spring rainstorm that inhibits pollination. Moreover, unseasonable drops in temperature can

Cranberries being picked by hand, Ocean County, 1877. Lithograph by Granville Perkins. From Harper's Weekly, *1877.*

damage uncovered vines. During frost season, according to Stephen Lee, Jr., "you don't plan to do anything in the evening. You have to watch the thermometer." [186]

The management of these floods and the surveillance of the dikes is entrusted to a "water foreman" whose credentials usually include mental maps of complex water systems compiled through a lifetime of living in the Pines.

Cranberry farmers have a history of independently crafting technology for their small, rather esoteric branch of agriculture. A. S. Doughty's description of hand picking in 1878 picturesquely illustrates the circumstances that would encourage the development of technology in the twentieth century.

> A few days since it was the privilege of the writer to visit the cranberry bog of D. R. Gowdy, Esq., at Stafford Forge, distant one and one-half miles from West Creek, in Ocean County. . . . They were in the midst of gathering the crop, and the scene presented some novel features. Six or

> seven hundred pickers, comprising both sexes, were gathered from the surrounding neighborhood, some coming a long distance, and with their wagons scattered around suggested the idea of a gipsy encampment. The pickers formed a line, and each provided with handled basket of half bushel capacity, would gather the berries with remarkable dexterity, and average, for five or six hours, from two to four bushels in as many hours, and on a wager could pick twelve bushels in ten hours. The price paid per bushel for picking is forty cents.[187]

In the early twentieth century, the cranberry scoop was introduced in the Pinelands. The Makepeace scoop, which came from Cape Cod, and the Applegate scoop, which was designed by David Applegate of Chatsworth, provided the types on which other local scoop makers worked many variations. Basically, the scoop is a wooden box with one open side. Using the scoop requires bending low to the ground and "combing" the vines with light short strokes that do not tear the vines as the berries are pulled off. The operation was carried out as Doughty described, with scoopers strung across the bog, each worker having a one-peck box. When the box was filled, he would dump it into a one-bushel packing box and receive a chit from a field supervisor.

Cranberries being scooped, Hog Wallow, circa 1940. Photograph by William Augustine. Courtesy of the Donald A. Sinclair New Jersey Collection. Special Collections and Archives, Rutgers University Libraries.

Cranberry scoops were used until the 1960s, and they have now joined the ranks of other utilitarian objects, such as the decoy and the sneakbox, which have become aesthetic symbols of place. Joe Reid, whose wife, Gladys, is the grand-niece of David Applegate, makes replicas of the scoop to be used as flower holders and magazine racks.

Mark Darlington, great-grandson of J. J. White, says his father, Tom, can remember "back in the very old days" when everyone was on hands and knees raking with their fingers. Cranberry scoops represent a variation on that theme, as do many of the automated dry-pickers that have been used in this century, including one his father designed: "The way we picked these, now, before the wet-harvester became available, he had a small machine about the size of a desk that you walked behind that had rows of cones that would work in a dry bog, and it would sort of rake through the vines and flop up in a rotating bucket arrangement and then dump 'em in a bag in the back."[188]

Today only a few growers dry-harvest because it is less efficient, more time consuming, and more destructive to the vines than the wet-harvest. However, dry-harvested berries are preferable for fresh sales, so there is still a market for them.

In the early 1960s the "wet method" of harvesting cranberries was developed. The fields are flooded to a depth of at least two inches above the top of the vines, and special machines—either "walk-behinds" or "ride-ons"—agitate the vines, knocking the berries off. The berries float to the surface and are "hogged," or gathered, toward a conveyor belt that moves them onto a truck.

The wet-harvester is the subject of endless improvisation and adaptation which Darlington describes as "sort of a communal project." Because there are simply not enough farms to make the development of special cranberry machinery worthwhile to companies such as John Deere, the cranberry farmers pool their resources. Darlington describes how it may unfold:

> It's a very tight knit industry for the most part in terms of all the other locals. We can borrow equipment back and forth. . . . Dave Thompson, machining and fabricating, is fantastic at coming up with stuff that my dad can conceive of. . . . My dad will work out something and Dave will make it, and what'll happen is, somebody'll come up and look at it or will hear about it and we'll loan it to 'em and they'll try it out for awhile and they'll come back and say, "Well, we need this change or that change on it." And some grower can go to Thompson and . . . he can just make it up for them.[189]

Cranberry sorting machine in operation. Courtesy of Everett Mickle and the Medford Historical Society.

Other exercises of ingenuity may involve the adaptation of commercial machines. Abbott Lee developed a cranberry pruner by adapting a John Deere Model 640 rake.

The packing house is the focal point of a cranberry plantation and the final point for technological innovation, set within a traditional frame. In the nineteenth century, harvested cranberries were transported from the bogs to the packing houses in covered wagons. The packing house was often located in a town, such as Medford or Vincentown, that was a good distance from the bogs. Later, as the workforce grew, so did villages, such as Whitesbog and Double Trouble, near the bogs and around the packing house.[190]

At the packing house, the berries were separated from the chaff. Then they were sorted by machines which operated on a "bounce" principle: good berries bounce; bad ones don't. Sorting machines bounce the berries as many as seven times. Legend has it that the bounce principle was discovered by Peg Leg John Webb when he dropped a box of berries at the top of a staircase.[191]

Cranberry pickers. Courtesy of Mary Ann Thompson, Virginia Durell Way, and Cranberry History Collection.

A final sorting under sunlight was done before the berries were packed for shipping in barrels emblazoned with the special label of the grower. Often the label carried family nicknames. The American Cranberry Exchange classified berries and often assigned names of indigenous plants and places. These would appear on the label, also.

The traditional structure of packing houses evolved to accommodate these tasks. Their high-peaked, multiwindowed buildings allowed maximum light and ventilation.

Today, the berries arrive in trucks and depart in large, square boxes. In between, they must still be cleaned and sorted. Local ingenuity continues to rework machines to bounce the berries.

The development of cranberry agriculture dramatically affected the Pinelands population. In the nineteenth century, cranberries were harvested by local families. By the turn of the century, however, growing manpower needs brought about the recruitment of Italian workers from Philadelphia. Entire families were hired in September by a "padrone," or boss, who orga-

Abbott Lee, cranberry and blueberry grower of Speedwell, 1985. Photograph by David S. Cohen.

nized, transported, and oversaw the workers. Some Jews from the nearby Jewish agricultural colonies were also hired.

Rev. E. H. Durell with a cranberry bouquet. Courtesy of Mary Ann Thompson, Virginia Way Durell and Cranberry History Collection.

With the influx, villages grew up near the bogs. In recent times, however, the practice of housing migrant workers has given way to the hiring of day labor. Migrant blacks, Puerto Ricans, Haitians, and Cambodians have all worked the bogs, along with Anglo natives of the region. Many of the Puerto Ricans who used to migrate to the bogs have now settled into permanent communities at Woodbine, Hammonton, and Vineland. With the natives, they form the core of experienced bog workers.

George Marquez and Orlando Torres represent two different patterns in the lives of Puerto Rican agricultural workers. Mr. Marquez has worked for the past 20 years or more at the Birches Road Cranberry Farm in Tabernacle Township. He found the job through a friend who worked at the agricultural employment office. When he started, he told folklorist Bonnie Blair, he drove about 30 workers each day in a bus from Philadelphia. They hand-harvested cranberries using scoops. Only four or five of them were Puerto Rican. Today Marquez lives in Philadelphia. He works in the cranberry packing house only during September and October, using vacation time from his regular job in Philadelphia to work in New Jersey.[192]

Orlando Torres was born and raised in Chatsworth and now lives with his wife Hazel in Vincentown. His father was a crew foreman at the Haines's cranberry bogs, and he began working in his father's crew. Eventually, he himself became a crew foreman because he could speak both Spanish and English.

Most of Torres's crew are seasonal workers. They work for six months in New Jersey, picking blueberries in the summer and cranberries in the fall, and then they return to Puerto Rico. In Chatsworth, however, there is a permanent Hispanic population, which includes Torres's parents. Though his parents think of themselves as "Puerto Rican," Torres thinks of himself as a "Piney," a sentiment he proclaims on his "Piney Power" cap. He explained his feelings about the Pines to folklorist Bonnie Blair by comparing himself to the bird that leaves its nest before its time and will always come back, for, he says, quoting his father-in-law, "You can take the Piney out of the Pines, but you can't take the Pines out of the Piney."[193]

The cranberry families have created communities. At the center of the community of the growers are the families, many of which involve several generations working together. Often different aspects of the business are divided among members. His father, according to Mark Darlington, is the creative mastermind of machinery at Whitesbog. His brother, Joe, is the

manager, and Mark is the mechanic.[194] In the Lee family, Abbott oversees tasks such as development of techniques and machinery, and Stephen III handles marketing and management.[195] Wives and children are also integral to the farm. In smaller operations, they help with many of the chores of cranberrying.

The community of growers emerges from this nucleus of local families. The methods and machines that they trade are symbolic of the larger concerns that they share as growers. While each is guardian of the specific waterways that feed his bogs, all watch over the Pinelands water system at large. Their common problem of handling processed berries led many of them to form the Ocean Spray co-op 50 years ago. The practices of planting new bogs with shared cuttings and of burying a plantation owner with a bouquet of cranberries speak metaphorically of their shared identity.

Blueberry Farming

Many of the cranberry growers also raise blueberries. "Blueberries was a way of supplementing your income as it first came on," grower Brad Thompson of Vincentown told ethnobotanist Eugene Hunn, "so people would get enough money together to put in three rows of blueberries, and then they would get three more, and then they would get an old cranberry bog annd plant all that. And my grandmother's generation started that way."[196] Stephen Lee and his two sons Abbott and Stephen grow both cranberries and blueberries at Speedwell. "Our blueberry fields and cranberry bogs here are practically the same type of ground," says Mr. Lee. "In fact, we can turn any of the blueberry fields into cranberry bogs the way we're located. It's the same ground. If you've got good blueberry ground, you've also got good cranberry ground."[197]

Two varieties of blueberries grow naturally in the Pinelands: the highbush (*Vaccinium corymbosum*) and the lowbush (*Vaccinium pennsylvanicum*). Local residents refer to them as "huckleberries." In 1916, Dr. Frederick V. Coville of the United States Department of Agriculture, and Elizabeth White, the daughter of J. J. White, the founder of Whitesbog, attempted to cultivate blueberries in New Jersey. They enlisted the help of local gatherers to locate the best wild blueberry shrubs, and named many of the cultivated blueberries after the woodsmen who found them. The "keystone" of the cultivated blueberry, Miss White said in a 1953 interview, was the Rubel, named for Rube Leek of Chatsworth.[198]

Since that time many more varieties of blueberries have been developed through crossbreeding the original varieties. Some of the newer varieties are the June, introduced in 1930; the Weymouth, introduced in 1936; the Berkeley and the Coville, both introduced in 1949; and the Bluecrop, introduced in 1952.[199] Brad Thompson has grown all of these, but he took out the Covilles, because they weren't producing much, and he put in Weymouths, because, he says, they're sweeter than Bluecrop and Berkeley, they're a "good cropper," they get "good money," and you can get a tractor between the bushes.[200]

The propagation of blueberries is usually done from cuttings. The 12- to 30-inch-long shoots from which the cuttings are made are called "whips." The cuttings are collected in the spring, before the beginning of bud growth (usually between March 15 and April 10), and are planted in sphagnum-lined boxes containing peat and sand. In the early fall, the rooted cuttings are planted in the fields.

When the plant is one year old, it will flower and produce fruit. The flower buds are initiated in the summer and develop through the fall and winter. Coville advocated planting different varieties of blueberry bushes in alternating rows, because when the flowers are pollinated with pollen from the same variety, the berries are smaller and mature later.[201]

The pollination is done by honeybees hired from beekeepers. Bob Reeves, a blueberry grower from New Lisbon, figures that it takes one hive per acre.

Blueberry scoop and picker's box. Lent by Mary Ann Thompson and Cranberry History Collection. Photograph by Anthony Masso.

The beekeeper who supplies him takes his bees to Florida in the winter to pollinate orange and grapefruit trees, follows the blueberry blossoms from Carolina into New Jersey, and then the cranberry blossoms to Cape Cod. "You don't try to bring [the bees] in until there is bloom showing," says Reeves, "because if you bring them in too quick, they'll go out on to wild flowers." [202] The bees are left in the fields for three to four weeks. Stephen Lee notes that he hires bees first to pollinate his blueberrry bushes and a second time to pollinate his cranberry vines. "In other words," he says, "the same colony of bees is hired twice." [203]

Blueberries being picked by hand. Photograph by Dennis McDonald. PFP235203-1-35.

The blueberries are usually harvested between June 20 and August 15. In the past, lowbush blueberries were harvested using a hand rake, sometimes called a "huckleberry knocker," similar to a small cranberry scoop. The highbush variety has usually been harvested by hand picking. Since the 1960s, growers have also used a mechanical harvester that straddles the row of blueberry bushes while hydraulically operated rods vibrate the bush to detach the fruit. Today, the hand pickers are either contract migrant workers or local high-school students.

There are two methods of hand picking. In the first, the berries are put into a small pail fastened to the picker's waist. The pail is then dumped into larger trays. In the second method the berries are picked directly into market boxes, which are transported on pallets to the packing shed to be processed for the fresh market. In the shed, the green berries, dirt, leaves, and sticks are sorted from the marketable berries by rolling them down a screen. Electric fans are now used to blow the leaves off the screens.

Impressions of the blueberry fields gathered while picking are captured in Margaret Bakely's painting of a blueberry packing house and field. Bakely and her husband Vernon live on the Thompson cranberry plantation in Vincentown. Years ago, Margaret tried to pick blueberries, but found that there is a knack to it that she didn't have:

> You've got to have big hands, or the berries will fall right through your hands. That's what happened to me. Some people could pick twenty trays a day, but I could only do a couple. You pick not with your fingers, but with your thumbs. You hold your hands like a cup, and go up to the bush and work your thumbs on them, and they drop into the cup. So the bigger hands you've got, the bigger the cup! [204]

Even more taxing than the task is the heat in a blueberry field, Bakely says: "You have to be able to stand the heat, and I couldn't."

Such strong impressions seem to have motivated by contrast the sense of

Opposite: *Blueberries being unloaded at packing house. Photograph by Dennis McDonald. PFP235203-2-33.*

relief that Bakely reads in her painting. "I think of a warm autumn day—Indian summer—with the yellow sky reflecting the colors. And the crows. There are always lots of crows around. It's a feeling of rest after a busy summer." [205]

"When we're finished picking blueberries," said Stephen Lee, "we take about a week or ten days as a break, and then we bring in a crew of people that will prune blueberries for us." [206] The purpose of pruning is to balance the size of the fruit with the production of new fruit the following year. Too much pruning increases the size of the blueberries, but decreases total production; light pruning will result in more, but smaller, berries. Those who wield the shears have to know how to strike a balance. Vernon Bakely "used to be a great pruner," his wife recalls. "He picked it up from his mother. They'd go around and prune off patches together." [207] The pruning is done while the bush is dormant, and it is interrupted by the cranberry harvest in October. In November, pruning resumes and continues through the first week of April.

Many of the New Jersey blueberry growers are associated with the TRU BLU Cooperative, which was established in 1927. It accounts for about 20 percent of the total blueberry crop in the state. It represents both the large growers, such as the Haines, White, Reeves, Lee, and DeMarco families, and the small growers. Because blueberry growing is labor intensive, the small growers can continue to compete with the large growers. Brad Thompson feels that he has made a good living growing blueberries. "Of the natives," he told Eugene Hunn, "I know very few instances of abject poverty. There are still enough resources to keep a fairly good level of living, from what I've seen." [208]

Vegetable and Fruit Farming

In the last half of the nineteenth century, improved transportation led to the growth of agricultural towns at Egg Harbor City, Woodbine, Hammonton, Vineland, and Bridgeton, where crops were loaded on railroad cars or boats for shipment to markets and food processing facilities in Camden, Philadelphia, and New York. New crops were grown, including strawberries, blackberries, raspberries, grapes, sweet potatoes, apples, and peaches.

Strawberries were the first money crop in the region south of the Mullica River, where berry growing began on a commercial scale in 1861. By the turn of the century, blackberries, or dewberries, became more important than strawberries. Raspberries were a third important crop. The three were compatible, because they were grown in the same kind of soil, and because blackberries mature earlier than raspberries but later than strawberries.

Sweet potato harvest. Photograph by Joseph Czarnecki. PFP216422–1–6.

Hammonton became the strawberry center of New Jersey. Out-of-town buyers and commission agents went there to buy the berries directly from the farm wagons. In 1909 a farmers' canning company was organized in Vineland for canning strawberries, blackberries, tomatoes, and sweet potatoes.[209]

Grapes were another nineteenth-century crop. They were introduced by German farmers near Egg Harbor City and by Italians near Hammonton. The Italians grew them mainly for home consumption in a sour wine made by drawing off the liquor from pressed grapes. A grape juice factory was established in Vineland, but in 1911 only about 25 percent of its grapes were supplied by local farmers.[210] The Germans also initially raised grapes for home consumption, but by the 1870s wine production became the most important industry in Egg Harbor City. At that time there were more than 700 acres of vineyards in the Egg Harbor vicinity. Large stone vaults were built to store the wine.[211]

Unlike white potatoes, sweet potatoes (sometimes called "Vineland Sweets") were well suited to the sandy Pinelands soil. In the nineteenth century, potato fields were plowed with a one-horse plow and planted by hand, using a trowel or dibble. Harvesting, using potato forks, began about September first. The potatoes were gathered by hand and stored either in barns or cellars. There they were sweated by wood fires to raise the temperature to about 80° F. After several days the temperature was lowered, and the seed potatoes could be stored all winter in a dry place.

In the first decades of the twentieth century, new methods replaced old. The new farm machinery was at first horsepowered, and included two-horse plows, disc and acme harrows, and sulky cultivators. Today, tractors have replaced horses, but the equipment is similar.[212]

Today, Leonard Maglioccio of Middletown in Upper Township, Cape May County, explained to folklorist Jens Lund, a tractor pulls a harvesting machine called a "potato digger"; it digs into the ground and forces the yams to the surface. A farm worker sits on the potato digger and clears the vines away from the harvesting machines, while two other laborers on foot pick up the yams. Most of the yams are sold to a packing plant in East Vineland.[213]

In the 1930s and 1940s the peach crop was so important that Hammonton called itself "Peach City." Fruit farmers in South Jersey usually grew both peaches and apples, which are compatible crops. Peaches are harvested in the summer; apples in the fall. Compatibility is important, for it enables

Ed Lamonaca sorting peaches in packing house, Hammonton. Photograph by Dennis McDonald. PFP233374-4-10.

the farmer to use the same machinery, keep the labor supply constantly active, increase his income, utilize land not good for peaches, and reduce the risk of relying on a single crop.

The peach harvest begins July 1 and lasts until September 10. Most peach farmers harvest their crop two or three times during the season. The picking is done by hand, using ladders and bags, and the peaches are put either in baskets or bins. Forklifts move the bins to the packing house.[214]

Peach farmer Ed Lamonaca's orchards are located in Hammonton near the farmlands his grandfather purchased after he immigrated from Italy. Lamonaca grew up there in a 15-room farmhouse shared with a large extended family. The families also shared the work of the farm.

They first raised peaches, then changed to apples, and then switched back to peaches. Lamonaca explained their reasons. A peach tree will survive 15 years, while an apple tree can survive 30 years or longer. However, apples are grown all over the country, whereas peaches require particular conditions that are not found everywhere. Furthermore, it is hard for South Jersey apple growers to compete with those of New York and Pennsylvania, and apples are picked through Christmas, while peaches bring in more money and sooner.

"The best thing for peaches," according to Lamonaca, "is a lot of fertilizer, high ground with a lot of good drainage, and constant water," the opposite conditions needed for blueberries and cranberries. Pruning is also good for peaches. "You don't want to put the energy back into making leaves. . . . You want your peaches on the outside of the tree, where the sun is." What's more, "you want to open the inside of the tree . . . and let the sun in." Both the peach and the tree are very delicate. The sun in the crotch of the tree and on the leaves makes the tree grow, he explained. "The tree likes hot weather. It likes Miami."[215]

Lamonaca thinks of the seasonal variations in the business as Mother Nature's way "of weeding out the amount of boxes as opposed to the amount of people." To avoid relying too heavily on a single crop, he plants cucumbers between his peach trees.[216]

Lamonaca believes that it would be impossible for someone to start an orchard from scratch today because peach orchards involve a great initial investment and it takes at least three years for the trees to produce. It's a long term investment: "The only reason I'm farming is because my family had it," he says. "What you're doing is you're trying to create something for your children, so that you can pass it down to them . . . even though they might not be interested in farming."[217]

PUMA

The South Jersey Wagon

Crops were brought to regional markets in South Jersey wagons, a wagon type similar in appearance to the more famous Conestoga wagon of Pennsylvania, which became the covered wagon of the American West. Both can be traced back to the English farm wagon, which, according to folklorist J. Geraint Jenkins, is derived from the Dutch wagon.[218] Both have a two-part undercarriage, consisting of a forecarriage and rearcarriage connected by a shaft called the "coupling pole." In both, the rear wheels are larger than the front wheels.

South Jersey wagons were used in the Pinelands to haul iron products from the furnaces, salt hay to the glass and paper factories, and cranberries to the sorting houses. The body of the wagon could be changed to fit the job. Often they were covered with canopies, like the Conestoga wagons. In 1830 it was reported that Burlington County had 977 "covered wagons."[219] In 1857 Colonel James A. Fenwick, the original owner of the cranberry bogs that became Whitesbog, described how these wagons were lined up during the cranberry harvest. "There have been seen at one time as many as sixty covered wagons with horses hitched to trees around the edges of these meadows. These wagons brought farmers' families who were busily engaged in picking cranberries."[220]

Most South Jersey wagons were made by local wagonmakers, wheelwrights, and blacksmiths. Wagons were usually named after the person who made them or the town in which they were made. In 1845, local historian Isaac Mickle wrote about the famous Pine Robber William Giberson: "It was said that at a running jump he could clear the top of an ordinary Egg Harbor wagon."[221]

Wagons were still being made in the first decades of the twentieth century. One of the many South Jersey wagonmakers was H. B. Ivins of Medford. In 1913 he advertised in a Burlington County farm directory as follows:

H. B. IVINS

Practical Wagon Builder and Horse Shoer

JOBBING A SPECIALTY

Painting and Lettering by Experienced Hands,
Trimming and Wheelwright Work.
Agent for Handford's Balsam
of Myrrh and Snow Flake Axle Grease.

Bell Phone 64–2 Branch Street, Medford, N.J.[222]

Ivins's wagonmaking shop, Medford, circa 1910. Courtesy of Everett Mickle and the Medford Historical Society.

The farm market wagon owned by Byron T. Roberts of Moorestown was similar to wagons made by C. T. Woolston of Riverton and William Frech of Maple Shade. Frech made all kinds of wagons, a potato sorter, and commercial car bodies. His farm market wagon was called a "Frech Underslung." He advertised that "the word 'Frech' on a wagon is like 'Sterling' on silver, it is a mark of quality, it stands for the best of everything."[223]

Ethnic Agricultural Communities

A long succession of ethnic groups has been associated with farming south of the Mullica River, including Germans, Russian Jews, and Italians in the nineteenth century, and blacks and Puerto Ricans in the twentieth.

The Germans settled near Egg Harbor City. The city was planned by the Gloucester Farm and Town Association, a Philadelphia-based organization founded in 1854 to resettle German-Americans from the cities in rural areas. The association purchased 30,000 acres in the Gloucester Furnace Tract, 5,000 in the Batsto Tract, and 1,000 additional acres.

Located on the Camden and Atlantic Railroad line in Atlantic City, Egg Harbor City was laid out in 1854. Each share of stock entitled the holder to a 20-acre farm and a building lot in town. The town's German heritage was evident in street names such as Cologne and Heidelberg and calendrical festivals, including a Weinfest, a Winzerfest, a Weinlesefest, and an Oktoberfest. In 1856 a monthly newspaper, the *Independent Homestead,* began publication in both German and English.[224]

Between 1880 and 1890 a large number of Italians settled in the Hammonton area. Located on the Camden and Atlantic Railroad, Hammonton was developed by Charles K. Landis, a lawyer, and Richard J. Byrnes, a banker. Landis went on to found the city of Vineland, but Byrnes remained in Hammonton. Believing that Italians were the best farmers, they sought to interest Italian immigrants in settling in both Vineland and Hammonton by means of agents and advertisements. Some of the immigrants came from cities such as New York and Philadelphia; others came directly from Italy.

Sometimes a large number of immigrants would come from a single town. After the Campanello brothers from Gesso, Sicily, settled in Hammonton, for instance, they were followed by many other immigrants from the same town. Neapolitans and Calabrians also settled in Hammonton. The Tell family, for instance, came from Naples in about 1870 and became prosperous berry farmers. At first the Italians from these various regions of Italy remained apart from each other, but eventually they developed an Italian-American community that shared such traditions as the Festival of Our Lady of Mt. Carmel, which is still held on July 16th in Hammonton.

Many of the farm laborers around the turn of the century were Italians brought in from Philadelphia and New York by padrones, who were paid a certain amount per laborer. The farmers met them at the train station and took them to the farms in wagons. The padrone usually remained in the city and left the supervision of the work to a crew member called a "row boss."

Many other Italians bought farms. Most of them grew the same crops as those grown by other groups, except for Italian beans and peppers. The pepper industry began, in fact, when Italian farmers started shipping a few of the peppers they had raised in their gardens to friends in Philadelphia. By the 1890s it became an important crop, especially in the vicinity of Newfield in Gloucester County.

These farmers marked the landscape as Italian in various ways. To bake bread, Italians in the Hammonton vicinity built outdoor beehive bake ovens fired by brush, roots, and even raspberry canes cut from the berry patches.[225] Around Vineland, farmers sometimes marked the boundaries of their fields with grapevines.

In the late nineteenth century, several Jewish agricultural colonies were established in South Jersey. The experiment was motivated partly by the Russian Jewish tradition of agrarian idealism, out of which Zionism developed, and partly by the desire of Jewish philanthropists to resettle Jewish immigrants from the urban ghettos in rural areas.

The first colony was Alliance, founded in 1882 in Salem County, three miles west of Vineland. It was organized by the Jewish Emigrant Aid Society, based in New York. Its members established a cigar factory and a shirt factory as well as farms. Alliance was followed by Rosenhayn, Norma, Carmel, Mizpah, and Woodbine. Only Woodbine, in northern Cape May County, is located within the boundaries of the Pinelands National Reserve.

Established in 1891 by the Baron de Hirsch Foundation, the philanthropic organization that supplied the funds for the Jewish Emigrant Aid Society, Woodbine became the largest and most widely known of the colonies. Although it was intended to be an agricultural colony, it developed instead as an industrial center with factories making clothing, hats, hardware, and machinery. In 1894 an agricultural school was also established there.

By 1901 there were approximately 3,300 Jewish farmers in these agricultural colonies.[226] They raised the same kinds of crops as their German and Italian neighbors, including strawberries, blackberries, and, later, grapes, sweet potatoes, and fruit.

These new immigrants faced not only the difficult circumstances of agricultural life familiar to all farmers, but also some unique to the new colonies. Many of these settlers had never farmed before, and so were just learning agriculture at the same time that they were trying to support themselves with it. Moreover, the acreage on which they were settled was in many cases uncleared and uncultivated.

Enterprises such as the clothing factories and institutions such as the Woodbine Agricultural School were established to help the colonists sustain themselves. While developing their knowledge of farming, they worked for others to bolster their often meager farm income.

"I started to earn my bread and butter when I was two years old," claims Bluma Purmell, who is the daughter of Moses Bayuk, one of the original founders of Alliance. Life on the farm was very hard, and Bluma recalls that Jewish families would often do day labor on other farms. This work included scooping cranberries in nearby bogs, a scene she has depicted in one of her many paintings.

Bluma's paintings provide a picture of life in the Jewish agricultural colonies, filtered through memories in old age. Vivid details of both the daily life and its larger meanings are distilled in her pictures. The cranberry bog, she recalls, belonged to the Petersons. While helping pick cranberries there with her father and sister, Bluma received her first lesson from her father on how to properly take care of things: "I ran in front of my father. I had my little tin cup, and I started picking berries. Then my father said, 'You go follow behind me and get the ones I miss.' From that I learned to always finish things up carefully."[227]

This philosophy of her father motivated Bluma in the many activities of her full life. At the age of 83, after a long career as a nurse and a nursing home administrator, Bluma took up painting. Her subjects include the family farm and farm chores. In 1981, with the help of Felice Lewis Rovner, she published her memoirs under the title *A Farmer's Daughter: Bluma.*

Other forms of the visual arts also depict the circumstances of the early years in the Jewish colonies. The Crystal family, who farmed in Norma, named their homestead "Tall Weed Farm," good-naturedly recalling the poor condition it was in when they acquired it. Martha Crystal has commemorated that name and scenes of their daily life there in an embroidered picture that hangs in the kitchen of their farmhouse.

Like other immigrants, those of the Jewish agricultural colonies became

Recollection of My Homestead. *Painting by Bluma Bayuk Rappoport Purmell, formerly of Alliance. Lent by Michèle Rappoport. Photograph by Anthony Masso.*

Poultry farmers, Cumberland County. Photograph by Harvey W. Porch. Courtesy of the Donald A. Sinclair New Jersey Collection. Special Collections and Archives, Rutgers University Libraries.

experts at adapting. When they were joined after World War II by approximately 1,000 Jewish survivors of the Holocaust, many of them turned to the burgeoning field of poultry farming. The new immigrants settled in Farmingdale, Toms River, Lakewood, and Vineland. In 1951 the Jewish Poultry Farmers' Association was founded, and Vineland became known as the "Egg Basket of the East." [228]

By the 1970s, however, two major changes brought the era of the colonies to an end. First, the huge technological poultry operations of the South forced the smaller New Jersey operations out of business, and second, many of the younger generation had become well educated, as their parents had hoped, and entered professions. Another migration took place as they moved from the farms to the suburbs and cities.

Today, when reunions are held for those who lived in the Jewish agricultural colonies, they come from all over the country.

Beginning in the 1930s, blacks from Florida, South Carolina, and Georgia replaced Italians as seasonal agricultural workers. About half of them brought their families with them as they followed the harvest northward through the South and into the Middle Atlantic States. They generally traveled with a crew leader in the crew leader's bus; however, some, known as "free-wheelers," came in their own cars. Their stay in New Jersey typically lasted four months. Accommodations were provided for them on the farms on which they worked. Most returned south for the winter.

By 1967, southern blacks were no longer the major component of the seasonal farm work force, having been replaced by Puerto Rican workers. In 1947 the Puerto Rican Department of Labor initiated a migration plan with the federal, state, and local authorities that established a uniform contract for agricultural workers from Puerto Rico. Most of these workers were men between the ages of 20 and 34. They were recruited by the Puerto Rican Department of Labor and flown on chartered planes to either New York City or Millville, from which they were transported to the Glassboro Service Association Camp, which had been established at an old WPA camp in 1947. Within 24 hours, they were usually assigned to a grower, generally the same grower each year.

Unlike the black migrant workers, who brought their families with them, the Puerto Rican workers usually left their families at home, returning to them after a stay of about five months in New Jersey.[229] Some of the Puerto Ricans have stayed in this country and established communities in Hammonton, Woodbine, and Vineland.

Domestic Life of the Farm

"Just because you plant it doesn't mean it's going to grow. . . . It means you're going to be out there working all the time!"[230] Hard work is a common theme of the oral histories of farming people such as Rae Gerber, whose family tree springs from far back in the history of Burlington County agriculture. Many of the traditional aspects of farm life derive from the demands and risks involved in farming. The structure of the yearly round, methods of thrifty resource use, and participation in communal work events are all ways of surmounting the difficulties of an agricultural lifestyle.

Just as a trapper organizes his year around the seasons of the animals he harvests, a vegetable farmer follows a calendar charted by the growing seasons of his crops. In South Jersey, preparation of fields for planting begins as soon as the soil thaws, usually in March. Planting follows, closely keyed to weather. Modern agricultural science records frost dates to guide planting schedules. Most farmers, however, also pay close attention to lunar cycles. Long experience shows that frost often occurs around a full moon, therefore they will avoid planting especially tender crops at that time. A farmer is guided in his planting routine by the accumulated wisdom of his own and his parents' experience in farming. Sometimes, other domains of life are incorporated into these practices; for instance, Italian farmers around Vineland used to say that peas planted on St. Guiseppe's Day (March 19) would flourish.

With the spring planting season, the busiest part of the year begins. During the summer and fall harvests, other activities are of secondary importance. Once winter cold sets in, the farmer's life slows down, but it is still busy with maintenance and repair of buildings and equipment.

Like Rae Gerber, who knows that "just because you plant it, doesn't mean you're going to get a crop," those who produce most of what they live on take care to see that resources are fully used.[231] This philosophy has traditionally been couched in the saying that is used to describe the preparation of meat from a butchered hog: "We use everything but the squeal." It is equally apparent in other foodways practices in the region.

Home canning and freezing have been important among many families, both those currently involved in farming and others, such as George and Helen Zimmer, who no longer farm but who have relatives and friends who still do. Folklorist Susan Samuelson observed that a cycle of reciprocity sees to it that nothing is wasted. Helen Zimmer receives fresh produce from members of her wide family network in the region. She spends a good deal of time processing the food in summer and fall, with the result being "well stocked freezers and cellar shelves lined with jellies, canned fruits and vegetables, stew mixes, purees, homemade wines, pickles, sauerkraut, and the like." These preparations are then shared with the original benefactors in an efficient meshing of natural and cultural resource management.[232]

Italian families often have "can houses," summer kitchens used especially for cooperative sessions of canning when all the women of the family work together to process large amounts of produce and to share it.

Communal work sessions provide two important resources: many hands to lighten the task and companionship to brighten it.

The hog killing was an annual late fall event in the past. At a hog killing, families would help each other butcher and prepare the meat from a hog. Each age and sex group had a job. The men did the work of butchering, with the younger men responsible for lifting and moving the heavy carcass. The women took care of the preparation of the meat and of a meal for the participants, with the oldest women doing the job of preparing the sausage casings since they could do it seated. The children ran errands.

The slaughtered hog was dipped into a tub of scalding water before the

Opposite: *Helen Zimmer of Egg Harbor City preparing chow-chow. Photograph by Elaine Thatcher.* *PFP215661-3-25.*

hair and bristles were scraped off the skin. Then the various parts were prepared. The liver of the hog was often cooked that day to feed the group. Fatty scraps were rendered for lard, and hams were cut to be cured. Ground sausage mixture was prepared from scrap parts and stuffed into casings made from intestines that had been scraped, rinsed with running water several times, and turned inside out. Often, the sausage was divided among the participating families. Other parts were salted and stored on an upper floor of the house.[233]

Following overleaf: *The Sew and Sews of the Tabernacle Methodist Church. Photograph by Dennis McDonald.*

Quilting has traditionally been another important activity whose rhythm is set by natural cycles. In South Jersey, where asparagus is a first crop for many farmers, the quilting season spans the winter and halts in April or May when it is time to pull asparagus. The Sew and Sews quilting group of Tabernacle Methodist Church has met weekly for 50 years in Rae Gerber's home in Tabernacle. They quilt the pieced tops that members make or others bring them, often completing as many as 25 quilts a season. The income from their work helps support the church.

However, such "bees" are a remnant of an era when all bedcovers were handmade and many hands were needed to get the task done. Other traditions surrounding quilting are recalled by Mrs. Gerber, who counts several seamstresses among her great-aunts.

To hone her sewing skills, a young girl made 12 quilts before her marriage. Upon her engagement she began the thirteenth, her bridal quilt. Mrs. Gerber still has the bridal quilt made in 1856 by her grandmother, Rachel Evans, for her marriage to Jesse S. Braddock. Like most of her other historic quilts, it is large—approximately three yards square—thin, and light. A thin, warm layer of wool forms the interfacing. Rachel Evans's wedding quilt is in a pattern similar to one called "Ocean Waves," yet is distinct in that the pieces are square rather than triangular. The quilting is in a feather and rose stitch.[234]

According to Mrs. Gerber, other patterns popular in the region in the past were the Oakleaf and Reel and the Album. Her collection includes one of each of these made by her forebears. The Album quilt is signed by many friends and relatives of Rachel Evans from the Medford area. In the 1970s, the Tabernacle group made a signature quilt from an historic Tumbling Block pattern. It bears the names of all the donors to the Tabernacle Historical Society fund drive, making it a colorful statement of people and place. Most of the names are of Tabernacle families.

Rachel Evans's quilts were probably stitched on the same frame that Rae Gerber and her friends have worked around for so long. It was made in the 1840s by one of the members of the Braddock family, with sturdy walnut posts and "Jersey iron rings that will never rust."[235]

In addition to making good use of womanpower, quilting makes economical use of materials. In the past, scraps from other sewing projects were stitched into quilt tops. Mrs. Gerber still has boxes of such materials, from which she would choose colors and patterns most pleasing to her taste. Other quilts made use of the colorful cloth used for feedbags before World War II. In more recent times, she has purchased fabric specifically for her quilts.

For Rae Gerber, the best patterns are the more difficult ones, such as the Rolling Star. These occupy the hands and entertain the eye. "My hands have to be busy," she says. But undirected or imprecise busy-ness is not sufficient. As if speaking for all women who have made beautiful and precise functional pieces, Rae Gerber says, "If you're going to put something together, why not make it beautiful while you're doing it, rather than slap it together haphazardly."[236]

Though a quilt warms, decorates, and occupies, its most eloquent function is expressiveness. The quilts Rae Gerber has made for each of her grandchildren on the occasion of his or her wedding, each of different design, speak not only about the individuality of the offspring and the creativity of the grandmother, but also about the continuity and importance of the family.

CONCLUSION

Environmentally, the New Jersey Pinelands is a special place. It is a wilderness in the midst of the megalopolis, and the repository of a large underground supply of fresh water. It marks the northernmost range limits for certain animal and plant species, and the southernmost limit for others, and is the habitat of some endangered species, such as the Pine Barrens tree frog. It is the first National Reserve in the United States. This specialness is recognized in the words of the Pinelands Commission Management Plan: "No other area has the same pattern of natural habitats, distribution of plant and animal populations, and unusual variety of plant and animal species."[237]

Yet in other ways it is similar to other places. Many of the products and industrial activities of the Pinelands can be found elsewhere. Lumber and shipbuilding supplies have been produced from the pine woods of North Carolina, and iron has been forged in the forests of Pennsylvania. Cranberries have been harvested from the bogs of Cape Cod, and clams have been gathered commercially from the bays of Long Island. Some of the methods used in those places have been adopted here, as well—the wet-harvesting of cranberries, for instance, which was developed in Massachusetts.

Even the stereotype of the "Piney" is not unlike that of isolated people elsewhere in the United States, such as the Southern Appalachian mountain people, and it has the same roots as that stereotype. Conversely, the reversal of that image, embodied in such slogans as "Piney power," is similar to the way that Americans of an earlier day took the comic image of the Yankee and turned it into a positive figure of national pride.

What makes the Pinelands distinct is the particular interaction of the people with their special environment. They have gained intimate knowledge of it over the years and have used their knowledge both to shape the landscape itself and to create their folklife.

The landscape has not always appeared as it is today. A rural industrial area became a wilderness, barrens became rich farmlands, and a hinterland became a popular recreational center. Each of these changes was wrought by residents using their knowledge of place to develop resources. While some of the uses have been destructive, such as the depletion of the woodlands to fuel nineteenth-century industries, others have been constructive. Sand has been processed into glass; cranberry bogs have been built from swamps. Folk technologies have made some of these transformations of the landscape possible.

In addition to creating such technologies, the people of the Pinelands have used their intimate knowledge of place to create many other folklife forms. They have shaped artifacts tailored to their environments: schooners designed for the special conditions of Delaware Bay and sneakboxes superbly suited to Barnegat Bay. Some artifacts are themselves bits of the environment reshaped, such as evergreen grave blankets and decoys. The people shape and transform the environment in other ways when they create art forms. Their special vision and deep attachment infuse their songs and stories, their paintings and carvings. These forms reveal the Pinelands that residents experience.

While there are many ways to understand the Pinelands, its folklife is one of the best.

NOTES

1. R. Alan Mounier, "A Study of Waterpowered Sawmills in the Pine Barrens of New Jersey," in *Historic Preservation Planning in New Jersey,* ed. Olga Chesler (Trenton: Office of New Jersey Heritage, 1984), 93.

2. Interview, Eugene Hunn with Jack Cervetto, June 22, 1984, PFP84–AEH014, Pinelands Folklife Project Archive, American Folklife Center, Library of Congress. Hereafter, references to archive materials will be listed with names, data retrieval number, and date only.

3. Interview, Jens Lund with George Brewer, October 15, 1983, PFP83–RJL008.

4. Interview, Eugene Hunn with Jack Cervetto, March 18, 1984, PFP84–AEH006.

5. Hunn with Cervetto, PFP84–AEH014.

6. Interview, Eugene Hunn with Jack Cervetto, March 18, 1984, PFP84–AEH002.

7. Hunn with Cervetto, PFP84–AEH006.

8. Hunn with Cervetto, PFP84–AEH002.

9. Herbert Halpert, "The Piney Folk Singers," *Direction* 2 (1939):15.

10. Henry Glassie, "William Houck: Maker of Pounded Ash Adirondack Pack-Baskets," *Keystone Folklore Quarterly* 12 (1967):40–42.

11. Agreement of Brotherton Indians to Sale of Land at the Brotherton Reservation, January 15, 1802, New Jersey Division of Archives and Records Management, Trenton, N.J.

12. U.S. Census, 1880, Burlington County, Shamong Township (manuscript).

13. Will of Ann Roberts, August 7, 1894, Will 21042C, N.J. Division of Archives and Records Management, Trenton, N.J.

14. *Mt. Holly Herald,* 15 April 1932, 3.

15. Harry B. Weiss and Robert J. Sim, *Charcoal Burning in New Jersey from Early Times to the Present* (Trenton: New Jersey Agricultural Society, 1955), 9–12.

16. Interview, Mary Hufford with Harry and Gladys Payne, November 14, 1983, PFP83–RMH032.

17. Weiss and Sim, *Charcoal Burning,* 20–33.

18. Ibid.

19. Henry H. Bisbee and Rebecca B. Colesar, eds., *Martha: 1801–1815: The Complete Furnace Diary and Journal* (Burlington, NJ: H. H. Bisbee, 1976), 63.

20. Mary Hufford, "Folk-Artists-in-the-Schools Final Report," (New Jersey State Council on the Arts, 1980).

21. John R. Stilgoe, *Common Landscape of America, 1580 to 1845* (New Haven and London: Yale University Press, 1982), 268–274.

22. Charles S. Boyer, *Early Forges and Furnaces in New Jersey* (Philadelphia: University of Pennsylvania Press, 1931), 2–9; Arthur D. Pierce, *Iron in the Pines* (New Brunswick: Rutgers University Press, 1957), 10–19.

23. Bisbee and Colesar, *Martha,* 12–13, 17–18, 25, 62–63.

24. Boyer, *Early Forges and Furnaces,* 174–190; Pierce, *Iron in the Pines,* 117–155.

25. Quoted in Pierce, *Iron in the Pines,* 35–37.

26. Mary Hufford, "Navigators in a Sea of Sand" (Pinelands Folklife Project, 1985), 210.

27. Interview, Rita Moonsammy with Anne Salmons, March 15, 1985.

28. Ibid.

29. Hufford with Payne, PFP83–RMH032.

30. Herbert Halpert, "Folk Tales and Legends from the New Jersey Pines: A Collection and a Study" (Ph.D. dissertation, Indiana University, 1947), I:240–241.

31. Michael Fowler and William A. Herbert, *Papertown of the Pine Barrens: Harrisville, New Jersey* (Eatontown, NJ: Environmental Education Publishing Service, 1976).

32. Adeline Pepper, *The Glass Gaffers of New Jersey* (New York: Charles Scribner's Sons, 1971), 60–65, 313–315.

33. Interview, Rita Moonsammy with Dorothy Lilly, April 13, 1982.

34. Ibid.

35. Interview, Rita Moonsammy with Malcolm Jones, April 26, 1982.

36. Moonsammy with Lilly.

37. Interview, Rita Moonsammy with Theodore Ramp, March 18, 1985.

38. Roy C. Horner, *Tempo: The Glass Folks of South Jersey* (Woodbury, NJ: Gloucester County Historical Society, 1985), 5–6.

39. Moonsammy with Lilly.

40. Interview, Rita Moonsammy with Walter Earling, April 21, 1982.

41. Donald Pettifer, "Glass Folk Art in New Jersey" in *The Challenge of Folk Material for New Jersey Museums,* (forthcoming).

42. Moonsammy with Ramp.

43. Ibid.

44. Yi Fu Tuan, *Topophilia: A Study of Environmental Perception, Attitudes, and Values* (Englewood Cliffs, NJ: Prentice-Hall, 1974), 105.

45. Interview, Mary Hufford with members of Harmony Gun Club, December 10, 1983, PFP83–AMH024.

46. Interview, Rita Moonsammy with Jack Cervetto, May 29, 1985.

47. Interview, Mary Hufford with members of the Spartan Gun Club, November 17, 1983, PFP83–RMH044.

48. Hunn with Cervetto, PFP84–AEH006.

49. Interview, Eugene Hunn with Tom Brown, December 1983, PFP83–AEH008.

50. Interview, Rita Moonsammy with Tom Brown, February 17, 1982.

51. Interview, Eugene Hunn with Tom Brown, October 10, 1983, PFP83–AEH002.

52. Jonathan Berger and John W. Sinton, *Water, Earth, and Fire: Land Use and Environmental Planning in the New Jersey Pine Barrens* (Baltimore: Johns Hopkins University Press, 1985), 123.

53. John Harshberger, *The Vegetation of the New Jersey Pine-Barrens: An Ecologic Investigation* (New York: Dover Publishers, 1970), 30.

54. Fieldnotes, Eugene Hunn, PFP84–FEH0618.

55. Ibid.

56. Interview, Rita Moonsammy with Hazel Landy, December 19, 1985.

57. Hunn with Cervetto, PFP84–AEH014.

58. Fieldnotes, Eugene Hunn, PFP83–FEH1011.

59. Susan Samuelson, "Christmas in the Pines" (Unpublished ms.).

60. Moonsammy with Landy.

61. James Still, *Early Recollections and Life of Dr. James Still, 1812–1885*, facsimile edition (1877; reprint, Medford, NJ: Medford Historical Society, 1971), 119–120.

62. Alvin T. M. Lee, *Land Utilization in New Jersey: A Land Development Scheme in the New Jersey Pine Area*, New Jersey Agricultural Experiment Station Bulletin no. 665 (New Brunswick: New Jersey Agricultural Experiment Station, 1939).

63. Gustav Kobbé, *The New Jersey Coast and Pines* (Short Hills, NJ: Gustav Kobbé, 1889), 86–88.

64. Ibid., 88.

65. Pierce, *Iron in the Pines*, 156–173.

66. Berger and Sinton, *Water, Earth, and Fire*, 80–83.

67. Interview, Rita Moonsammy and Mary Hufford with Valia Petrenko, October 21, 1983, PFP83–RRM001.

68. Robert C. Alexander, "The Shingle Miners," *Cape May County Magazine of History and Genealogy* 4 (1957): 99–104.

69. Ibid.

70. Interview, Jens Lund with Charles Pomlear, November 10, 1983, PFP83–FJL028.

71. Tom Carroll, "Final Report, Phase I" (Pinelands Folklife Project, 1983).

72. [Peter Kalm,] *The American of 1750: Peter Kalm's Travels in North America . . .*, ed. Adolph B. Benson (New York: Wilson-Erickson, 1937) I:298–299.

73. Interview, Tom Carroll with Joe Reid, October 8, 1983, PFP83–ATC002.

74. Mary Hufford, "Culture and the Cultivation of Nature in the Pinelands National Reserve," in *Folklife Annual* I (1985): 10–39.

75. Interview, Tom Carroll with Joe Reid, October 8, 1983, PFP83–RTC001.

76. Fieldnotes, Eugene Hunn, PFP84–FEH0318.

77. Kalm, *Travels*, 299.

78. Hufford, "Navigators," 145.

79. Interview, Rita Moonsammy with Helen Zimmer, March 18, 1985.

80. Interview, Rita Moonsammy with the Frazee family, March 21, 1985.

81. Ibid.

82. Fieldnotes, Mal O'Connor, PFP83–FM01014.

83. Ibid.

84. Kalm, *Travels*, 300.

85. Hunn, PFP84–FEH0318.

86. Ibid.

87. Rev. Allen H. Brown, "The Character & Employments of the Early Settlers of the Sea-Coast of New Jersey," 1879, reprinted in New Jersey Historical Society *Publications*, 2nd ser., 6 (1879–1881): 41.

88. Interview, Jens Lund with George Campbell, November 7, 1983, PFP83–FJL023.

89. Harold F. Wilson, *The Jersey Shore: A Social and Economic History of the Counties of Atlantic, Cape May, Monmouth, and Ocean* (New York: Lewis Historical Publishing Company, 1953) II: 747.

90. Ibid.

91. Ibid., 748–51.

92. Interview, Jens Lund with George Campbell, November 7, 1983, PFP83–FJL024.

93. Interview, Rita Moonsammy with Ed Gibson, May 10, 1982.

94. Wilson, *The Jersey Shore*, 751.

95. Ibid., 748.

96. Lund with Campbell, PFP83–FJL023.

97. Ibid.

98. Moonsammy with Gibson.

99. Lund with Campbell, PFP83–FJL023.

100. Lund with Campbell, PFP83–FJL024.

101. Ibid.

102. Wilson, *The Jersey Shore*, 889.

103. Kalm, *Travels*, 76–77.

104. *National Standard and Salem County Advertiser*, 14 September 1853, p. 2, as quoted in Wilson, *The Jersey Shore*, 889.

105. Interview, Rita Moonsammy with Leslie Christofferson, March 6, 1985.

106. Ibid.

107. Ibid.

108. Ibid.

109. Interview, Rita Monsammy with Albert Reeves, June 3, 1982.

110. Ibid.

111. Ibid.
112. Kalm, *Travels,* 239–240.
113. Moonsammy with Brown.
114. Ibid.
115. Interview, Rita Moonsammy with Tom Brown, December 17, 1985.
116. Interview, Eugene Hunn with Tom Brown, June 21, 1984, PFP84–AEH009.
117. Berger and Sinton, *Water, Earth, and Fire,* 47.
118. Interview, Gerald Parsons with Kenneth Camp, October 2, 1984.
119. Interview, David Cohen and Rita Moonsammy with Kenneth Camp, September 19, 1985.
120. Interview, Rita Moonsammy with Albert Reeves, February 12, 1982.
121. Parsons with Camp.
122. Interview, Rita Moonsammy with Kenneth Camp, September 23, 1982.
123. Cohen with Camp.
124. Interview, Gerald Parsons with Albert Reeves, September 24, 1984, PFP84–RGP006.
125. Ibid.
126. Cohen with Camp.
127. Alonzo T. Bacon, *Recollections Pertaining to the Seafaring Life* (Bridgeton, NJ: The Evening News, 1970).
128. Glenn S. Gordinier, "Maritime Enterprise in New Jersey: Great Egg Harbor During the Nineteenth Century," *New Jersey History* 97 (1979): 105–117.
129. Brown, "Character & Employments," 17.
130. Gordinier, "Maritime Enterprise," 105–117.
131. Alan Frazer and Wayne Yarnall, "New Jersey Under Sail," *New Jersey History* 99 (1981):196.
132. Donald H. Rolfs, *Under Sail: The Dredgeboats of Delaware Bay* (Millville: Wheaton Historical Associations, 1971), 11.
133. Interview, David Cohen and Rita Moonsammy with John DuBois, October 19, 1983.
134. Berger and Sinton, *Water, Earth, and Fire,* 63.
135. Carroll, "Final Report," 4.
136. Interview, Tom Carroll with Joe and James Reid, October 8, 1983, PFP83–RTC001.
137. Moonsammy with Reeves, June 3, 1982.
138. Interview, Rita Moonsammy with John DuBois, February 19, 1982.
139. Carroll with James Reid, PFP83–RTC001.
140. Moonsammy with Reeves, June 3, 1982.
141. Fieldnotes, Tom Carroll, PFP83–FTC1007.
142. Carroll with Reid, PFP83–RTC001.
143. Interview, Rita Moonsammy with Belford Blackman, September 17, 1984.
144. Interview, Rita Moonsammy with Fenton Anderson, September 17, 1984.
145. Interview, Jens Lund with Louis Peterson, October 22, 1983, PFP83–RJL017.
146. Moonsammy with Reeves, June 3, 1982.
147. Interview, David Cohen with Todd Reeves, December 15, 1983.
148. Interview, Rita Moonsammy with Norman Jeffries, September 17, 1984.
149. Interview, Rita Moonsammy with Joe Reid, March 13, 1985.
150. Ibid.
151. Berger and Sinton, *Water, Earth, and Fire,* 58.
152. Moonsammy with Reeves, February 12, 1982.
153. Mary Hufford, "Maritime Resources and the Face of South Jersey," in *Festival of American Folklife Program Book,* ed. Tom Vennum (Washington, D.C.: Smithsonian Institution, 1983), 12–15.
154. Interview, Rita Moonsammy with Nerallen Hoffman, February 11, 1983.
155. Interview, Jens Lund with Louis Peterson, October 22, 1983, PFP83–RJL016.
156. Ibid.
157. Interview, Rita Moonsammy with Albert Reeves, December 19, 1985.
158. Interview, Jens Lund with Veach family, November 11, 1983, PFP83–FJL032.
159. Interview, Jens Lund with Veach family, November 11, 1983, PFP83–RJL029.
160. Lund with Peterson, PFP83–RJL016.
161. Interview, Rita Moonsammy with Fenton Anderson, February 8, 1982.
162. Moonsammy with DuBois.
163. Interview, Mary Hufford and David Cohen with Joe Reid, November 4, 1985.
164. Ibid.
165. Ibid.
166. Ibid.
167. Bernard Herman and David Orr, "Decoys and Their Use: A Cultural Interpretation," Academy of Natural Sciences of Philadelphia *Frontiers* 1 (1979):6–7.
168. Interview, Christopher Hoare with Sam Hunt, April 28, 1978, quoted in David Steven Cohen, *The Folklore and Folklife of New Jersey* (New Brunswick: Rutgers University Press, 1983), 122.
169. Interview, Elaine Thatcher with George Heinrichs, October 1, 1983, PFP83–RET009.
170. Patricia H. Burke, *Barnegat Bay Decoys and Gunning Clubs* (Toms River, NJ: Ocean County Historical Society, 1985), 25–37.

171. Ibid., 11.

172. Quoted in the documentary film, *In the Barnegat Bay Tradition*, produced by the New Jersey Historical Commission and New Jersey Network (1982).

173. Burke, *Barnegat Bay Decoys and Gunning Clubs,* 12–23.

174. Quoted in *In the Barnegat Bay Tradition.*

175. Quoted in ibid.

176. Quoted in ibid.

177. Interview, Rita Moonsammy with Tom Brown, June 4, 1985.

178. Julian Ursyn Niemcewicz, *Under Their Vine and Fig Tree: Travels Through America in 1797–1799, 1805* . . . New Jersey Historical Society Collections, vol. 14 (Elizabeth, NJ: Grassmann, 1965), 217–218.

179. J. J. White, *Cranberry Culture* (New York: Orange Judd, 1916), 24.

180. Mary Ann Thompson, "The Landscapes of Cranberry Culture," in *History, Culture, and Archeology of the Pine Barrens: Essays from the Third Pine Barrens Conference,* ed. John W. Sinton (Pomona, NJ: Stockton State College, 1982), 193–211, as cited in Hufford, "Navigators," 112.

181. William Bolger, Herbert J. Githens, and Edward S. Rutsch, "Historic Architectural Survey and Preservation Planning Project for the Village of Whitesbog, Burlington and Ocean Counties, New Jersey" (Morristown, NJ: New Jersey Conservation Foundation, 1982).

182. White, *Cranberry Culture,* 29–35.

183. Berger and Sinton, *Water, Earth, and Fire,* 76.

184. Interview, Mary Hufford with Mark Darlington, November 1983, PFP83–AMH015.

185. Interview, Christine Cartwright with Haines, Henry, and Francis Mick, September 29, 1983, PFP83–RCC011; Interview, Mal O'Connor with Stephen, Jr., Stephen III, and Abbott Lee, November 12, 1983, PFP83–RMO015, as cited in Hufford, "Navigators," 114–115.

186. O'Connor with Lee, PFP83–RMO015.

187. A. S. Doughty, quoted in T. F. Rose and H. C. Woolman, *Historical and Biographical Atlas of the New Jersey Coast* (Philadelphia: Woolman and Rose, 1878), 308.

188. Hufford with Darlington, PFP83–AMH015.

189. Ibid.

190. Thompson, "The Landscapes of Cranberry Culture."

191. Hufford, "Navigators," 111–112.

192. Interview, Bonnie Blair with George Marquez, October 14, 1983, PFP83–RBB001.

193. Interview, Bonnie Blair with Orlando and Hazel Torres, October 20, 1983, PFP83–RBB003.

194. Hufford with Darlington, PFP83–AMH015.

195. Mal O'Connor, "Time, Work, and Technology: A Multi-Generational Approach to Family Business in the Pinelands" (Paper presented at American Folklore Society 1984 Annual Meeting, San Diego, California, October 14, 1984).

196. Interview, Eugene Hunn with Brad Thompson and Bob Reeves, June 22, 1984, PFP84–AEH011.

197. Interview, David Cohen with Stephen Lee, Jr., October 16, 1985.

198. Quoted in Bolger, Githens, and Rutsch, "Historic Architectural Survey," 47.

199. Paul Eck and Norman F. Childers, *Blueberry Culture* (New Brunswick, NJ: Rutgers University Press, 1966), 65.

200. Hunn with Thompson, PFP84–AEH011.

201. Eck and Childers, *Blueberry Culture,* 183.

202. Interview, Eugene Hunn with Brad Thompson and Bob Reeves, June 22, 1984, PFP84–AEH012.

203. Cohen with Lee.

204. Interview, Rita Moonsammy with Margaret Bakely, December 23, 1985.

205. Ibid.

206. Cohen with Lee.

207. Moonsammy with Bakely.

208. Hunn with Thompson and Reeves, PFP84–AEH012.

209. United States Immigration Commission, *Immigrants in Industries,* Part 24, *Recent Immigrants in Agriculture.* U.S. Senate Document No. 633, 61st Congress, 2d Session (Washington, D.C.: U.S. Government Printing Office, 1911) I: 72–73, 97.

210. Ibid.

211. Dieter Cunz, "Egg Harbor City: New Germany in New Jersey," New Jersey Historical Society *Proceedings* 73 (1955): 89–123.

212. U.S. Immigration Commission, *Recent Immigrants in Agriculture,* I: 63–65.

213. Interview, Jens Lund with Leonard Maglioccio, PFP83–RJL006.

214. Janet L. Donohue-McAdorey, "An Economic Analysis of Peach Production by Southern New Jersey Growers" (M.S. thesis, Rutgers University, 1984), 53–84.

215. Interview, Mal O'Connor with Ed Lamonaca, June 20, 1985, PFP85–AMO001.

216. Ibid.

217. Ibid.

218. J. Geraint Jenkins, *The English Farm Wagon: Origins and Structure* (Reading, England: Published by the Oakwood Press for the University of Reading, 1961).

219. Thomas F. Gordon, *A Gazetteer of the State of New Jersey* (Trenton: D. Fenton, 1834), 111.

220. Quoted in Bolger, Githens, and Rutsch, "Historic Architectural Survey," 17.

221. Isaac Mickle, *Reminiscences of Old Gloucester* (Philadelphia: Townsent Ward, 1845), 85.

222. *The Farm Journal and Farm Directory of Burlington County, N.J. 1913* (Philadelphia: Wilmer A. Kinson Co., 1913), 176.

223. Ibid.

224. Cunz, "Egg Harbor City."

225. U.S. Immigration Commission, *Immigrants in Agriculture,* I:51–60, 70–72, 90–92, 125, 129.

226. Ibid., II:89–144; Joseph Brandes, *Immigrants to Freedom; Jewish Communities in Rural New Jersey Since 1882* (Philadelphia: University of Pennsylvania Press, 1971); Edward S. Shapiro, "The Jews of New Jersey," in *The New Jersey Ethnic Experience,* ed. Barbara Cunningham (Union City: William S. Wise and Co., 1977), 294–311.

227. Interview, Rita Moonsammy with Bluma Purmell, July 1, 1985.

228. Shapiro, "The Jews of New Jersey," 303–304.

229. Pearl J. Lieff, "Types of Family Structure of Migrant Agricultural Workers" (Ph.D. dissertation, Rutgers University, 1971); Isham B. Jones, "The Puerto Rican Farm Worker in New Jersey," in *The Puerto Rican in New Jersey* (Trenton, NJ: New Jersey Department of Education, Division Against Discrimination, 1955), 14–20; Governor's Migrant Labor Task Force, *Seasonal Farm Workers in the State of New Jersey* (Seattle, WA: Consulting Services Corporation, 1968).

230. Interview, Christine Cartwright with Rae Gerber, November 1983, PFP83–RCC030.

231. Ibid.

232. Susan Samuelson, "The Landscape Experienced, Summary Statement for Phase II" (Pinelands Folklife Project, 1983), 5.

233. Interview, Christine Cartwright with Ephraim and Alice Tomlinson, October 18, 1983, PFP83–RCC016.

234. Interview, Rita Moonsammy with Rae Gerber, March 6, 1985.

235. Ibid.

236. Interview, Christine Cartwright with Rae Gerber, November 1983, PFP83–RCC029.

237. Pinelands Commission, *Comprehensive Management Plan for the Pinelands National Reserve . . .* (New Lisbon, NJ: Pinelands Commission, 1980), xvii.

INDEX

Living
Vegetarian

DISCARD

by Suzanne Havala Hobbs, DrPH, MS, RD

Foreword by Michael F. Jacobson, PhD
Executive Director, Center for Science in the Public Interest

Wiley Publishing, Inc.

Living Vegetarian For Dummies®

Published by
Wiley Publishing, Inc.
111 River St.
Hoboken, NJ 07030-5774
www.wiley.com

Published by Wiley Publishing, Inc., Indianapolis, Indiana

Published simultaneously in Canada

Library of Congress Control Number: 2009939781

ISBN: 978-0-470-52302-5

Manufactured in the United States of America

10 9 8 7 6 5 4 3 2 1

About the Author

Suzanne Havala Hobbs, DrPH, MS, RD, is a registered, licensed dietitian and nationally recognized expert on food, nutrition, and dietary guidance policy. She holds a doctorate in health policy and administration from the University of North Carolina at Chapel Hill, where she is a clinical associate professor in the Gillings School of Global Public Health, the nation's top public school of public health. There she directs the doctoral program in health leadership and serves on the faculty of the Department of Health Policy and Management and the Department of Nutrition.

Sue was the primary author of the American Dietetic Association's 1988 and 1993 position papers on vegetarian diets and the founding chair of the association's Vegetarian Nutrition Dietetic Practice Group. She serves on the editorial board of *Vegetarian Times* magazine and advisory boards of the nonprofit Vegetarian Resource Group and the Physicians Committee for Responsible Medicine.

A vegan-leaning, lacto ovo vegetarian for 35 years, Sue explores topics related to food, nutrition, and policy issues in her popular newspaper column, *On the Table.* The column reaches more than 400,000 readers weekly in *The News & Observer* of Raleigh, North Carolina, and in *The Charlotte Observer.* An archive of *On the Table* columns, as well as Sue's blog, may be found at `www.onthetable.net`.

She has written 11 books, including *Get the Trans Fat Out* (Three Rivers Press), *Vegetarian Cooking For Dummies* (Wiley), *The Natural Kitchen* (Berkley), *Good Foods, Bad Foods: What's Left to Eat?* (Wiley), and *Shopping for Health: A Nutritionist's Aisle-by-Aisle Guide to Smart, Low-Fat Choices at the Supermarket* (Harper Perennial). She is a contributing writer for the "Bottom Line/Personal" newsletter and has been a regular writer for *Vegetarian Times* and *SELF* magazines and other national publications.

Sue is a member of the American Public Health Association, American Dietetic Association, Association of Health Care Journalists, Association of Food Journalists, and the American Society of Journalists and Authors. She served on the board of directors of the Association of Health Care Journalists and the Center for Excellence in Health Care Journalism. She also serves on the board of trustees of the North Carolina Writers' Network.

She lives in Chapel Hill, North Carolina. Her family includes her husband, Michael R. Hobbs; their children, Barbara and Henry; and dogs Kailani and Sperry and cat Kodak.

Dedication

This book is dedicated to people everywhere who strive to eat well to support their health and to protect the well-being of our environment and the other living things with whom we share our beautiful planet.

Author's Acknowledgments

Heartfelt thanks to the kind, competent, and hardworking team at Wiley Publishing who made this book possible: to Acquisitions Editor Michael Lewis, Project Editor Sarah Faulkner, Copy Editor Todd Lothery, and intern Beth Staton, who expertly guided this book from concept to completion, and to their very talented colleagues, who worked their magic in the design and production departments. I'm grateful to Patricia Santelli for her assistance with the nutritional analyses of the recipes, and to expert recipe tester Emily Nolan for her good work. I'm especially indebted to my longtime friend and colleague, Ginny Messina, for her help with the technical review. Many thanks as well go to my agent, Mary Ann Naples, and her colleagues at The Creative Culture, as well as to my former agent and good friend, Patti Breitman, with whom I worked on the predecessor to this book, *Being Vegetarian For Dummies* (Wiley). It is such a privilege and joy to be part of a team of so many outstanding professionals.

Many of my colleagues in the U.S., in Canada, and around the world have dedicated their lives and careers to advancing knowledge in nutrition science, the links between diet and health, and the practice of diet and health policymaking. My work builds on theirs, and I salute the collective efforts of this community of scholars and practitioners.

I am grateful for my family and friends and their continued support and good humor. Special thanks to my sisters-in-law, Laura Bridges and Karen Bush, for lending me a beautiful beachfront getaway on the North Carolina coast, where I wrote several chapters in record time under the spell of ocean breezes, sea oats, and swooping pelicans. My husband, Mike, helped me day-to-day with his encouraging words, brilliant ideas, and the occasional caipirinha on the back deck.

I am indebted, too, to readers of my newspaper column, *On the Table*. Their feedback and encouragement help me stay in touch with issues of primary concern to people trying to do their best to make wise food choices.

Publisher's Acknowledgments

We're proud of this book; please send us your comments at http://dummies.custhelp.com. For other comments, please contact our Customer Care Department within the U.S. at 877-762-2974, outside the U.S. at 317-572-3993, or fax 317-572-4002.

Some of the people who helped bring this book to market include the following:

Acquisitions, Editorial, and Media Development

Project Editor: Sarah Faulkner

Acquisitions Editor: Michael Lewis

Copy Editor: Todd Lothery

Assistant Editor: Erin Calligan Mooney

Editorial Program Coordinator: Joe Niesen

Technical Editor: Virginia Messina, MPH, RD

Recipe Tester: Emily Nolan

Nutrition Analyst: Patricia Santelli

Editorial Manager: Christine Meloy Beck

Editorial Assistants: Jennette ElNaggar, David Lutton

Art Coordinator: Alicia B. South

Cover Photos: © SoFood / Alamy

Cartoons: Rich Tennant (www.the5thwave.com)

Composition Services

Project Coordinator: Lynsey Stanford

Layout and Graphics: Carl Byers, Ashley Chamberlain

Illustrator: Elizabeth Kurtzman

Proofreaders: Evelyn C. Gibson, Dwight Ramsey

Indexer: Dakota Indexing

Special Help Elizabeth Staton

Publishing and Editorial for Consumer Dummies

Diane Graves Steele, Vice President and Publisher, Consumer Dummies

Kristin Ferguson-Wagstaffe, Product Development Director, Consumer Dummies

Ensley Eikenburg, Associate Publisher, Travel

Kelly Regan, Editorial Director, Travel

Publishing for Technology Dummies

Andy Cummings, Vice President and Publisher, Dummies Technology/General User

Composition Services

Debbie Stailey, Director of Composition Services

Contents at a Glance

Recipes at a Glance

Appetizers

Beverages

Breads

Breakfast Foods

Desserts

Entreés

Salads

Side Dishes

Soups

Table of Contents

Foreword

Simply put, this book may be the most important book you read this year . . . or this decade.

When I was in college, I certainly could have used a book like *Living Vegetarian For Dummies.* Somehow, I had heard that a vegetarian diet was healthful, or perhaps just cool, so I tried it. The first evening I ate a pound of broccoli. The next night I ate a pound of cauliflower. And so on. Actually, my first bout with vegetarianism may not have lasted more than those first two days.

I thought that "vegetarian" meant eating only vegetables. I only wish that Suzanne Havala Hobbs had been around to hold me by the hand (as this book will do for you) and show me that vegetarian diets are typically more varied, more healthful, and more delicious than the typical steak-and-potatoes (or burger-and-fries in my case) American diet.

Truth be told, I never became a full-fledged vegetarian, let alone a vegan, and most casual claimants to vegetarianism probably also cheat a bit. Flexitarian is what we say we are, which often means eating mostly vegetarian, but dining on fish, chicken, or even beef or pork occasionally. Such folks gain most of the health benefits of well-constructed vegetarian diets, but can't claim to be free of any responsibility for the maltreatment of farm animals (especially layer hens, dairy cows, and veal calves). At the other end of vegetarianism are vegans, who, notwithstanding all the temptations of daily life in North America, eschew even the lacto ovo foods that regular vegetarians eat plenty of. Fortunately, Sue Hobbs provides sensible guidance for people at every point on the vegetarian spectrum.

Some people (including me) have moved toward vegetarian diets mostly for health reasons, and those reasons are ample. Vegetarians (and I don't mean people whose notion of vegetarianism is chowing down on soft drinks, cookies, quiches, and chocolate bars) have lower risks of obesity, diabetes, heart disease, high blood pressure, and cancer. (Because most studies include few vegans, it's unclear whether vegans fare even better than run-of-the-mill lacto ovo vegetarian.)

Probably more people are attracted to vegetarianism for moral reasons. Eating animal products inevitably means that one is contributing to the miserable circumstances in which most farm animals spend their lives. Raising cattle, pigs, and chickens on grass and grubs certainly reduces the misery, but most vegans end up vegans because they don't want to feel culpable for any part of the raising and killing of animals.

If better health for yourself and avoidance of cruelty to animals aren't reason enough to eat a more plant-based diet, consider the benefits to the environment. Raising animals means using more energy-intensive fertilizer, much of which ends up polluting waterways. It means using huge quantities of water to irrigate fields of feed grains. Questionably safe pesticides endanger farm workers and wildlife. And the animals themselves emit greenhouse gases in the form of manure (which may also pollute rivers and streams) and cows' belching of methane gas. Eating fewer animal products and more fruits, vegetables, beans, whole grains, and nuts will help protect our increasingly crowded and polluted planet.

But back to basics: your taste buds will thank you, day after day, for moving in a vegetarian direction. Enjoy!

Michael F. Jacobson, PhD
Executive Director
Center for Science in the Public Interest
Washington, DC

Introduction

Vegetarianism has come a long, long way.

As a child, I wore a button that said, "Real People Wear Fake Furs." I'd picked it up at the Ann Arbor Street Art Fair when my older sister was in college at the University of Michigan. It was the late '60s, and it wasn't much longer before my mother announced to our family that from then on, she would be a vegetarian. She never said why, but for the next several years, the former Wisconsinite ate cheese omelets or cheddar-cheese-and-pickle sandwiches on whole-wheat toast for dinner while the rest of us ate the meat she prepared for us. That is, of course, until we kids followed her lead and, one by one and without fanfare, became vegetarians ourselves.

My dad worried we'd miss vital nutrients. He chided my mother for planting the idea. Mom, a registered nurse, was considered a bit odd by her hospital colleagues. By now, it was the early '70s, and vegetarians lived on communes or wore Birkenstocks and long hair on college campuses. They weren't kids and working, middle-aged moms.

A competitive swimmer in high school, I hoped that a vegetarian diet would boost my endurance and athletic performance, as Olympic gold medalist Murray Rose claimed it had for him. It didn't help enough, but it did pique my interest in nutrition and set me on the path to a career in dietetics. It would be many years, however, before the scientific community came around to the idea that a diet of grains, fruits, vegetables, legumes, seeds, and nuts can be adequate — never mind superior — to a diet centered on animal products.

In college, I learned about vegetarianism in a lesson on fad diets. At that time, in the early 1980s, a blood cholesterol level of 300 mg/dl was considered normal, and patients in the coronary care unit in the hospital got bacon and eggs and white toast for breakfast.

My grandmother worried that I wouldn't get enough iron if I didn't eat red meat. She thought that my slender body wasn't "healthy" enough in size as compared to her old-world, European standards. For baby boomers like me, this was the environment for vegetarians in North America 30 years ago.

Everything is different now.

In the last 20 years, the American Dietetic Association — long the conservative holdout on such matters — went from cautious at first, to later tentative at best, to now clearly stating in its position papers that vegetarian diets confer health advantages. U.S. government dietary recommendations now explicitly acknowledge the vegetarian alternative and advise all Americans to make fruits, vegetables, grains, and legumes the foundation of a healthy diet. It's as close as the government can come to a stamp of approval for a plant-based diet as it balances science with the economic interests of the powerful meat and dairy industries.

As a practicing nutritionist and vegetarian, I've observed these changes taking place over decades. The scientific rationale for eating a plant-based diet is well-documented. The advantages for everyone and everything on our planet are compelling. The next task is helping people everywhere make the transition to an eating style that, at this time, is still outside the cultural norm in many countries. Accomplishing this requires education and the political will to initiate and enforce policies to create an environment that makes it easier for you and me to sustain lifestyles that support health.

Living vegetarian is an excellent way to meet today's dietary recommendations for good health. This book is for everyone who wants to understand the future of preventive nutrition and get a head start on making the switch.

About This Book

This book is for vegetarians and prospective vegetarians, too — for anyone curious about what a vegan is, for those who still have questions about where vegetarians get their protein, for parents who are wringing their hands because Junior has "gone vegetarian," and for Junior to give to Mom and Dad so that they won't worry.

This book is for vegetarians and nonvegetarians alike. Whether you want to control or prevent diseases such as diabetes and coronary artery disease, manage your weight, save money, or help keep the planet healthy and the animals happy, this book has what you need. That's because the secret to living well is eating well, and to eat well, you need to make plant foods the foundation of your diet.

It's the simple truth.

Don't feel you need to read the chapters in this book in order or read the book from cover to cover. It's designed to make sense and be helpful whether you surf it or read it in its entirety. Throughout the text, you'll find cross-references to guide you to other parts of the book where you can find related information.

Conventions Used in This Book

To make this book easier to read, I adhere to the following conventions or rules throughout:

- When I use the term *vegetarian,* I'm using it generically. In other words, it includes all the various subtypes — vegans, lacto, lacto ovo, and other variations of a vegetarian diet. When I want you to know something unique to a particular form of vegetarianism, I refer to the specific diet subtype. (Chapter 1 gives you definitions and explanations of each of these types of vegetarianism.)
- I use *italics* to introduce new terms, and I give you definitions of the new terms shortly thereafter.
- **Bold** text makes it easy to spot keywords in bulleted lists.
- Web addresses are printed in monofont, like `this`.
- When this book was printed, some Web addresses may have needed to break across two lines of text. If that happened, rest assured that I haven't put in any extra characters (such as hyphens) to indicate the break. So, when using one of these Web addresses, just type in exactly what you see in this book, pretending the line break doesn't exist.
- I don't specify most recipe ingredients as being organic, conventional, low-sodium, or other possible variations. When you shop for ingredients, feel free to make these choices as you see fit.
- Some recipes note the substitutions to make the dish suitable for vegans. In cases where I don't provide that information, feel free to experiment and make the substitutions yourself. I provide lots of information about recipe substitutions in Chapter 8.
- All margarine in the recipes is trans fat-free.
- All temperatures in the recipes are in Fahrenheit.

What You're Not to Read

It's great if you read the entire book. You won't miss any helpful hints and information that way. On the other hand, some information I include isn't as critical for you to know as the rest. If you need to pare down your reading, here's what you can save for later:

- **Material flagged with the Technical Stuff icon:** These paragraphs contain information that, while interesting, isn't vital to your understanding of the topic.
- **Sidebars:** This information is scattered throughout the book in shaded boxes. It's similar to the Technical Stuff: great if you have the time, but not critical for you to read.
- **Recipes:** I include a collection of good starter recipes for anyone who wants to give them a try. No need to read these unless you're ready to get started in the kitchen.

Foolish Assumptions

If you're holding this book, you or someone who loves you bought or borrowed this book to gain a better understanding of how to live a vegetarian lifestyle. I'm assuming that this book is appropriate for a variety of purposes, including:

- Dipping your toe into the topic. If you just want a little more information to help you decide whether living vegetarian may be something you'd like to consider doing, this book is appropriate for you.
- Digging in deeper. You may already have a general sense of what's involved in living vegetarian, but you want more in-depth advice and understanding of how to go about it. This book is for you.
- Sharing the knowledge. If you know someone with an interest in going vegetarian — or someone who may simply be curious and interested in finding out more — this book is a reliable resource.
- Refreshing your own knowledge. Longtime vegetarians may benefit from the up-to-date information in this book.
- Having a reference on hand. Health professionals often encounter vegetarians in their work and have to give them medical or dietary advice. If you're a health professional and you have no personal experience with a vegetarian lifestyle, this book may be helpful as an accurate and quick reference.

You can make some assumptions about me, too:

- I know what I'm talking about. I'm a licensed, registered dietitian with a master's degree in human nutrition and a doctorate in public health. I'm a leading expert on vegetarian nutrition and have lived a vegetarian lifestyle myself for 35 years.

- My advice is practical. It's informed by my own experience of living vegetarian for more than three decades, as well as many years of experience counseling individuals on special diets, including both vegetarians and nonvegetarians.
- I'm not giving individualized advice. As much as I wish it were possible, books aren't an appropriate means of dispensing medical or dietary advice tailored to individual needs. I can give you general information that provides you with a good foundation of knowledge about the topic. However, if you have specific issues you need help with — particularly medical conditions that require you to follow a special diet — you should get additional, individualized guidance from a registered dietitian. I include information in Chapter 1 about how to locate a dietitian with expertise in vegetarian diets.

How This Book Is Organized

Living Vegetarian For Dummies is divided into six parts. The book is organized to take you through a logical progression of information, moving from basic to more in-depth, depending on your level of interest and experience.

Parts I and II provide fundamental information that you should know if you're contemplating going vegetarian full time or part time. Part III includes recipes to get you started. Parts IV, V, and VI are important for anyone ready to dig a little deeper who wants more advanced-level skills.

Each part focuses on a different aspect of vegetarianism, from the basic who, what, and why to the nutritional underpinnings of a diet without meat, strategies to help you make the transition, and tips on how to maintain the lifestyle over time. Together, the six parts of this book lay the foundation for understanding the vegetarian lifestyle and building the skills necessary to successfully adapt.

Part I: Being Vegetarian: What It's All About

This part peels away the first layer of mystery around issues of vegetarianism. It gets to the bottom of the various definitions of vegetarian diets, revealing once and for all what the word *vegan* means and how to pronounce it. It looks at what vegetarians *do* eat, including vegetarian traditions around the world, rather than stopping at what they *don't* eat. This part also discusses the reasons people adopt a vegetarian diet and the nutritional aspects of vegetarian diets. It also guides you with good-sense advice and strategies for making the transition.

Part II: Planning and Preparing Your Vegetarian Kitchen

This part explains how to set up a vegetarian-friendly kitchen so that you can make more meals at home. It covers what you need to know about common and versatile ingredients, where to shop for them, and strategies for getting the best values. This part also focuses on practical equipment and basic cooking techniques you should know to help you get started.

Part III: Meals Made Easy: Recipes for Everyone

I provide a good set of starter recipes in this part, covering the major food categories and including recipes that are versatile and practical. Ingredient lists are short, and basic cooking skills are all that's necessary to follow the simple instructions. You can modify most of the recipes to add or subtract animal ingredients, depending on the extent to which you want to include or exclude them.

Part IV: Living — and Loving — the Vegetarian Lifestyle

This part provides advanced advice for anyone who's ready for intermediate- to advanced-level skills in living vegetarian. It includes strategies for families that have only one vegetarian in the household and tips for getting along in social situations outside your home. This part also includes information about how to maintain a vegetarian lifestyle when eating out at restaurants and traveling away from home.

Part V: Living Vegetarian for a Lifetime

Part V takes a life-course view of living vegetarian, with advice that's specialized for whatever stage you're in. I include information about living vegetarian during pregnancy, infancy, childhood, and the teen years, as well as into adulthood and older adulthood.

Part VI: The Part of Tens

All *For Dummies* books end with The Part of Tens, a collection of handy tips, lists, and fun facts that are easy to read at a glance. The chapters in this part provide you with a quick list of reasons why it makes sense to go vegetarian, as well as practical advice about how to make it happen, including simple ingredient substitutions and easy lunchbox ideas.

Icons Used in This Book

Another fun feature of *For Dummies* books is the clever icons that flag helpful nuggets of information. Each icon denotes a particular type of information. Here's what each icon means:

Tips are insights or other helpful clues that may make it more convenient or hassle-free for you to follow a vegetarian diet.

When you see this icon, the information that follows is a rule-of-thumb or another truism you should keep in mind.

If you see this icon, the information is meant to help you avoid a common pitfall or to keep you from getting into trouble.

This is information that, while interesting, isn't vital to your understanding of the topic. In other words, some of you may skip it, but it's there if you care to find out more.

Where to Go from Here

The science of nutrition is complicated, but being well-nourished is a relatively simple matter. It's even easier to do if you eat a wholesome, plant-based diet. That's where this book comes in.

If you want a clearer understanding of what vegetarianism is, start with the foundational information in Chapter 1. If you have a child or teenager who's interested in becoming vegetarian, check out Chapter 21. If you're ready to whip up some tasty vegetarian meals, head straight to Part III — you can start with the breakfast recipes in Chapter 9 or skip straight to the dessert recipes in Chapter 14 (I won't tell!).

Whether you go vegetarian all the way or part of the way, moving to a more plant-based diet is one of the smartest moves you can make. I hope this book helps. Best wishes to you as you take the first step!

Part I

Being Vegetarian: What It's All About

"This isn't some sort of fad diet, is it?"

In this part . . .

To change the way you eat, you not only have to gain knowledge and develop and practice new skills, but you also have to change your mind-set. That includes replacing old traditions with new ones. That's the fun of it, and that's the challenge, too.

In this part of the book, I cover basic information you need to help you get started. I define the various types of vegetarian diets and explain the reasons many people make the switch. I give you the background you need to understand nutrition issues pertaining to meatless diets, including how to ensure you get what you need from whole foods. I also discuss the pros and cons of taking vitamin and mineral supplements.

I also share some good-sense advice about living the vegetarian lifestyle. I explain how to plan for meatless meals, and I coach you on practical ways to master the behavioral changes that are a part of the transition to a new eating style.

It's exciting! Let's get started. . . .

Chapter 1

Vegetarianism 101: Starting with the Basics

In This Chapter

- Defining different types of vegetarianism
- Explaining why meat-free makes sense
- Fixing meatless meals
- Adopting a new mind-set about food

Mention a vegetarian diet, and many people visualize a big hole in the center of your dinner plate. They think that to be a vegetarian, you have to like lettuce and carrot sticks — *a lot.* Just contemplating it leaves them gnawing on their knuckles.

Nothing could be further from the truth, however.

Vegetarian diets are diverse, with an abundance of fresh, colorful, and flavorful foods. For anyone who loves good food, vegetarian meals are a feast. That may be difficult for nonvegetarians to understand. Vegetarian diets are common in some parts of the world, but they're outside the culture and personal experience of many people.

That's why I start with the basics in this chapter. I tell you about the many forms a vegetarian diet can take and the reasons people choose to go meat-free. I give you a quick overview of what's involved in planning and fixing vegetarian meals, and I introduce some important considerations for making the transition to meat-free a little easier.

Vegetarian Label Lingo: Who's Who and What They Will and Won't Eat

Most of us are pretty good at describing a person in just a few words:

> "He's a liberal Democrat."
>
> "She's a Gen-Xer."

It's like the saying goes: "A picture (or label) paints (or says) a thousand words."

People use labels to describe vegetarians, with different terms corresponding to different sets of eating habits. A lacto ovo vegetarian eats differently than a vegan eats. In some cases, the term used to describe a type of vegetarian refers to a whole range of lifestyle preferences, rather than to just the diet alone. In general, though, the specific term used to describe a vegetarian has to do with the extent to which that person avoids foods of animal origin. Read on for a primer on vegetarian label lingo, an explanation of what I call the vegetarian continuum, and an introduction to vegetarian foods.

From vegan to flexitarian: Sorting out the types of vegetarianism

In 1992, *Vegetarian Times* magazine sponsored a survey of vegetarianism in the United States. The results showed that almost 7 percent of Americans considered themselves vegetarians.

However, a closer look at the eating habits of those "vegetarians" found that most of them were eating chicken and fish occasionally, and many were eating red meat at least a few times each month. Most vegetarian organizations don't consider occasional flesh-eaters to be vegetarians.

As a result, the nonprofit Vegetarian Resource Group (VRG) in 1994 began sponsoring national polls on the prevalence of vegetarianism, wording the interview questions in such a way as to determine the number of people who *never* eat meat, fish, poultry, or byproducts of these foods. (The organization continues to conduct periodic polls, and you can find the results online at `www.vrg.org`.) Over the years, the number of people who fit the VRG definition of vegetarian has remained relatively stable at between 2 and 3 percent of the adult population in the U.S.

The fact is, people interpret the term *vegetarian* in many different ways.

Many people use the term loosely to mean that they're consciously reducing their intake of meat. The word *vegetarian* has positive connotations, especially among those who know that vegetarian diets confer health benefits. But what about the true vegetarians? Who are they and what do they eat (or not eat)?

The definition of a vegetarian most widely accepted by vegetarian organizations is this: A vegetarian is a person who eats no meat, fish, or poultry.

Not "I eat turkey for Thanksgiving" or "I eat fish once in a while." A vegetarian consistently avoids all flesh foods, as well as byproducts of meat, fish, and poultry. A vegetarian avoids refried beans made with lard, soups made with meat stock, and foods made with gelatin, such as some kinds of candy and most marshmallows.

The big three: Lacto ovo vegetarian, lacto vegetarian, and vegan

Vegetarian diets vary in the extent to which they exclude animal products. Historically, the three major types of vegetarianism have been:

- **Lacto ovo vegetarian:** A *lacto ovo vegetarian* diet excludes meat, fish, and poultry but includes dairy products and eggs. Most vegetarians in the U.S., Canada, and Western Europe fall into this category. Lacto ovo vegetarians eat such foods as cheese, ice cream, yogurt, milk, and eggs, as well as foods made with these ingredients.
- **Lacto vegetarian:** A *lacto vegetarian* diet excludes meat, fish, and poultry, as well as eggs and any foods containing eggs. So a lacto vegetarian, for instance, wouldn't eat the pancakes at most restaurants because they contain eggs. Some veggie burger patties are made with egg whites, and many brands of ice cream contain egg. A lacto vegetarian wouldn't eat these foods, either. A lacto vegetarian would, however, eat dairy products such as milk, yogurt, and cheese.
- **Vegan:** Technically, the term *vegan* (pronounced *vee*-gun) refers to more than just the diet alone. A vegan is a vegetarian who avoids eating or using all animal products, including meat, fish, poultry, eggs, dairy products, and any foods containing byproducts of these ingredients. Vegans also use no wool, silk, leather, and any nonfood items made with animal byproducts. Some vegans avoid honey, and some don't use refined white sugar, or wine that has been processed using bone char or other animal ingredients. Needless to say, vegans also don't eat their dinner on bone china. (For more details on veganism, see the nearby sidebar.)

In academic nutrition circles, *strict vegetarian* is the correct term to use to describe people who avoid all animal products but who don't necessarily carry animal product avoidance into other areas of their lives. In practice, however, the term vegan is usually used by both strict vegetarians and vegans, even among those in the know. In other words, technically, the term strict vegetarian refers to diet only. The term vegan encompasses both food and other products, including clothing, toiletries, and other supplies.

More than a diet: Veganism

Maintaining a vegan lifestyle in our culture can be difficult. Most vegans are strongly motivated by ethics, however, and rise to the challenge. A large part of maintaining a vegan lifestyle has to do with being aware of where animal products are used and knowing about alternatives. Vegetarian and animal rights organizations offer information and materials to help.

Sometimes vegans unwittingly use a product or eat a food that contains an animal byproduct. Knowing whether a product is free of all animal ingredients can be difficult at times. However, the intention is to strive for the vegan ideal.

So a vegan, for instance, wouldn't use hand lotion that contains lanolin, a byproduct of wool. A vegan wouldn't use margarine that contains casein, a milk protein. And a vegan wouldn't carry luggage trimmed in leather. Vegans (as well as many other vegetarians) also avoid products that have been tested on animals, such as many cosmetics and personal care products.

The list goes on: Semi-vegetarian, flexitarian, and others

Lacto ovo vegetarian, lacto vegetarian, and vegan are the three primary types of vegetarian diets, but there are more labels for vegetarians, including the following:

- A *semi-vegetarian* is someone who's cutting back on his intake of meat in general.
- A *pesco pollo vegetarian* avoids red meat but eats chicken and fish.
- A *pollo vegetarian* avoids red meat and fish but eats chicken.

These terms stretch the true definition of a vegetarian, and in practice, only the term *semi-vegetarian* is actually used with much frequency.

In recent years, another term has been introduced as well. A *flexitarian* is basically the same thing as a semi-vegetarian. It refers to someone who's generally cutting back on meat but who may eat meat from time to time, when it's more convenient or on a special occasion.

Don't leave out: Raw foods, fruitarian, and macrobiotic diets

The list actually goes even further. One adaptation of a vegetarian diet is a *raw foods diet,* which consists primarily of uncooked foods — fruits, vegetables, sprouted grains and beans, and vegetable oils. Though raw foodists never cook foods in an oven or on a stovetop, some of them eat ingredients that have been dehydrated in the sun.

In practice, most raw foodists in North America actually eat a *raw vegan* diet. The proportion of the diet that comes from raw foods is typically anywhere from 50 to 80 percent. Most raw foodists aim for a diet that's 100 percent raw, but what they can realistically adhere to still includes some amount of cooked food.

Another adaptation, the *fruitarian diet,* consists only of fruits, vegetables that are botanically classified as fruits (such as tomatoes, eggplant, zucchini, and avocados), and seeds and nuts. Planning a nutritionally adequate fruitarian diet is difficult, and I don't recommend the diet for children.

Macrobiotic diets are often lumped into the general category of vegetarian diets, even though they may include seafood. This diet excludes all other animal products, however, as well as refined sugars, tropical fruits, and "nightshade vegetables" (for example, potatoes, eggplant, and peppers). The diet is related to principles of Buddhism and is based on the Chinese principles of yin and yang. Therefore, macrobiotic diets include foods common to Asian culture, such as sea vegetables (including kelp, nori, and arame), root vegetables (such as daikon), and miso. Many people follow a macrobiotic diet as part of a life philosophy. Others follow the diet because they believe it to be effective in curing cancer and other illnesses, an idea that has little scientific support.

The vegetarian continuum: Going vegetarian a little or a lot

Pop quiz: What would you call a person who avoids all flesh foods and only occasionally eats eggs and dairy products, usually as a minor ingredient in a baked good or dish, such as a muffin, cookie, or veggie burger?

Technically, the person is a lacto ovo vegetarian, right? But this diet seems as though it's leaning toward the vegan end of the spectrum.

As a nutritionist, I see this kind of variation — even within the same category of vegetarian diet — all the time. One lacto ovo vegetarian may eat heaping helpings of cheese and eggs and have a high intake of saturated fat and cholesterol as a result. In fact, this type of vegetarian may have a nutrient intake similar to the typical American's — not so good. Another lacto ovo vegetarian may use eggs and dairy products, but only in a very limited fashion — as condiments or minor ingredients in foods. This person's nutrient intake more closely resembles that of a vegan.

What am I getting at? That labels are only a starting point, and they have limitations. Even if you know generally what type of vegetarian a person is, you may see a lot of variation in the degree to which the person uses or avoids animal products.

Many new vegetarians find that their diets evolve over time. At the start, for example, many rely heavily on cheese and eggs to replace meat. Over time, they learn to cook with grains, beans, and vegetables, and they experiment with cuisines of other cultures. They decrease their reliance on foods of animal origin, and gradually, they consume fewer eggs and dairy products. Some eventually move to a mostly vegan (or strict vegetarian) diet.

You might say that vegetarian diets are on a continuum, stretching from the typical American, meat-centered diet on one end to veganism on the other (see Figure 1-1). Most vegetarians fall somewhere in between. Some may be content staying wherever they begin on the continuum, while others may progress along the spectrum as they hone their skills and develop new traditions, moving from semi-vegetarian, or lacto ovo vegetarian, closer to the vegan end of the spectrum.

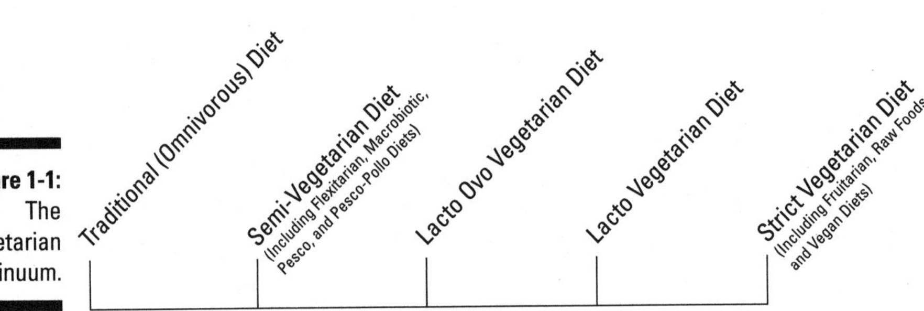

Figure 1-1: The vegetarian continuum.

Common foods that happen to be vegetarian: Beyond mac and cheese

Your eating style is a mind-set. For proof, ask someone what she's having for dinner tonight. Chances are good she'll say, "We're grilling steaks tonight," or, "I'm having fish." Ever notice how no one mentions the rice, potato, salad, vegetables, bread — or anything other than the meat?

Many vegetarians eat these common foods — side dishes to nonvegetarians — in larger quantities and call them a meal. Others combine them in new and delicious ways to create main courses that replace a burger or filet. Your skills at assembling appealing vegetarian meals will improve over time.

Until they do, going vegetarian doesn't have to mean a whole new menu. Many vegetarian foods are actually very familiar to nonvegetarians as well. Some examples include:

- Falafel
- Pasta primavera
- Salad
- Tofu
- Vegetable lasagna
- Vegetarian chili
- Vegetarian pizza
- Veggie burger

When meat-free isn't vegetarian: Bypassing meat byproducts

Living vegetarian means avoiding meat, fish, and poultry, and it includes eliminating ingredients made from those foods, too. Vegetarians don't eat soups that contain beef broth or chicken stock. They avoid meat flavoring in pasta sauces, Worcestershire sauce (which contains anchovies), and many stir-fry sauces, which contain oyster sauce. They don't eat marshmallows and some candies, which contain gelatin made from the cartilage and skins of animals. In Chapter 5, I cover all this in more detail, listing foods that may contain hidden animal products.

Going Vegetarian Is Good for Everyone

Some people go vegetarian for the simple reason that they don't like meat. They chew and chew, and they still have a glob of aesthetically unpleasant flesh in their mouth. Some people just like vegetables better.

Others go vegetarian because they recognize the link between diet and health, the health of ecosystems on our planet, the welfare of animals, or the ability of nations to feed hungry people. Whichever issue first grabs your attention, the other advantages may reinforce your resolve.

Eating for health

Many people view their health (or lack thereof) as something that just sort of happens to them. Their bad habits catch up with them, or they have bad genes. Their doctor just gave them a clean bill of health, and then they had a heart attack out of the blue. (Well, we all have to die of *something*.) Who could have foreseen it? They lived reasonably — everything in moderation, right? What more could they have done?

A lot, most likely. You may be surprised to discover how much power you wield with your knife and fork. The fact is that vegetarians generally enjoy better health and longer lives than nonvegetarians.

In comparison with nonvegetarians, vegetarians are at lower risk for many chronic, degenerative diseases and conditions. That's because a diet composed primarily of plant matter has protective qualities. I cover the diet and health connection in more detail in Chapter 2.

Protecting our planet

A disproportionate amount of the earth's natural resources is used to produce meat and other animal products. For example:

- It takes about 25 gallons of water to grow 1 pound of wheat, but it takes about 390 gallons of water to produce 1 pound of beef.
- A steer has to eat 7 pounds of grain or soybeans to produce 1 pound of beef.

Animal agriculture — the production of meat, eggs, and dairy products — places heavy demands on our land, water, and fuel supplies, and in some cases, it contributes substantially to problems with pollution. You should understand how your food choices affect the well-being of our planet. I discuss the issues further in Chapter 2.

Compassionate food choices

Many people consider a vegetarian diet the right thing to do. Their sense of ethics drives them to make very conscious decisions based on the effects of their food choices on others. You may feel the same way.

In Chapter 2, I describe more fully the rationale for considering the feelings and welfare of animals used for their flesh, eggs, or milk. I also discuss the implications of food choices for world hunger. A strong argument can be made for living vegetarian as the humane choice — not just in terms of the effect on animals, but also because of what it means for people, too.

Meatless Meals Made Easy

Making vegetarian meals doesn't have to be time-consuming or difficult. Despite all the gourmet cooking magazines and high-end kitchen supply stores around, you and most other people probably don't anticipate spending much of your free time fixing meals.

Not to worry. You can make the best vegetarian meals quickly, using basic ingredients with simple techniques and recipes.

Mastering meal planning and prep

After nearly 30 years of counseling individuals on many types of diets, I've found one thing to be universally true: Nobody follows a structured meal plan for very long. Though it may be helpful for some to see a sample meal plan, following rigid diet plans doesn't work well for most people. That's because you, like me, probably juggle a busy schedule that requires a fair amount of flexibility in meal planning.

You find a good deal of advice in this book that pertains to planning and preparing meatless meals in the most efficient way possible. In general, though, your meals should follow the guidelines I present in Chapter 7.

The best way to prepare meals with a minimum of fuss is to remember the key word — simple.

You need very little equipment and only basic cooking skills — boiling, baking, chopping, and peeling — to prepare most vegetarian foods. The best recipes include familiar, easy-to-find ingredients and have short ingredient lists.

The recipes I include in this book in Part III are a great place to start. You may also find, as you dig a little further into living vegetarian, that you don't have to rely on recipes at all to fix great meals. My hope for you is that you gain confidence in your ability to put ingredients together in simple and pleasing ways so that you can quickly and easily assemble delicious, nutritious vegetarian meals.

Shopping strategies

Grocery shopping doesn't take exceptional skills, but smart shopping habits can help ensure that you have the ingredients you need on hand when you need them. Because you probably don't have lots of free time to spend roaming the supermarket, you want to shop efficiently, too.

A few tips to keep in mind:

- **Keep a list.** Post on your refrigerator a running list of ingredients you need to pick up the next time you're at the store. You'll be less likely to forget a key item, and you'll be less likely to spend impulsively, too.
- **Shop for locally grown, seasonal foods.** Stop at your local farmer's market or roadside vegetable stand. Fruits and vegetables grown near your home taste better and retain more nutrients than foods that spend days on a truck being shipped across the country after being picked.
- **Mix it up.** Visit different grocery stores from time to time to take advantage of new food items and varied selections across stores. Ethnic markets and specialty shops can give you good ideas and offer some interesting new products to try.

In Chapters 6 and 7, I describe various commonly used vegetarian ingredients, and I give advice about shopping and stocking your vegetarian kitchen.

Mixing in some kitchen wisdom

If the idea of fixing meatless meals is new to you, harboring some concerns about your ability to plan and prepare good-tasting meals is understandable. Until you've had some experience, you may mistakenly believe that living vegetarian means buying lots of specialty products or spending hours slaving over the stove.

Not true. Going vegetarian — done well — will simplify your life in many ways. You'll have fewer greasy pans to wash, and your stove and oven should stay cleaner longer. Foods that contain fewer animal ingredients are likely to be less of a food safety concern than those that contain meat, eggs, and dairy products.

And living vegetarian costs less.

Of course, you can spend as much time and money as you want to on your meals. Living vegetarian, though, is all about basic foods prepared simply. The staples — fruits, vegetables, grains, and beans — are generally inexpensive and easy to use.

Cooking creatively

Vegetarian meals invite creativity.

After you're free of the idea that a meal has to be built around a slab of meat, the variety really begins. After all, you can put together plant ingredients to make a meal in an almost endless number of ways. Plant ingredients come in varied colors, textures, and flavors. The sampling of recipes I include in this book (see Part III for these recipes) are an introduction to what's possible. Don't hesitate to experiment with these — add a favorite herb or substitute Swiss chard for kale.

You'll soon be spoiled by the variety and quality of vegetarian meals. After you've practiced living vegetarian for a while, you'll find that you don't have space on your plate for meat anymore, even if you still eat it! The vegetarian foods are so much more interesting and appealing.

Embracing a Meat-Free Lifestyle

At this point, you may be distracted by such thoughts as, "I need a degree in nutrition to get this right," or, "I wonder whether I'll be vegetarian enough." Your mind may be leaping ahead to such concerns as, "Will my family go along with this?" In this section, I help you put issues like these into perspective.

Taking charge of your plate

You have no reason to be afraid to stop eating meat. I haven't touched a hamburger in more than 35 years, and I'm alive and well.

In the opening of his book *Baby and Child Care,* Dr. Benjamin Spock wrote these famous words: "Trust yourself. You know more than you think you do." You may not have studied the Krebs cycle or be able to calculate your caloric needs, but if you're reading this book, you probably have enough gumption to get your diet mostly right.

Mostly right is usually good enough.

The science of nutrition is complicated, but being well nourished is a fairly simple matter. It may sound strange coming from a nutritionist, but you really have no need to worry about nutrition on a vegetarian diet. Vegetarian diets aren't lacking in necessary nutrients.

Poor nutrition is a function of lifestyle — vegetarian or not. The greatest threats? Junk foods and a lack of time to fix the foods you need most to support health — particularly fruits, vegetables, whole grains, and legumes.

Your public persona: Affirming your choice

Vegetarianism has become mainstream in many places in recent years. If you live in an urban setting or near a college town, chances are good that a number of your friends or coworkers are vegetarians of one sort or another.

Other people aren't so lucky. If you find yourself feeling isolated because you're the only vegetarian you know, keep a few points in mind:

- You're 1 in 40, more or less — only about 2 to 3 percent of the population is truly vegetarian. If you feel a little different, it's because you are. You'd fit right in if you lived in India or any other culture in which vegetarianism has been a tradition for thousands of years.
- You alone decide what to eat. You may be different from most people, but you should feel confident in your decision to live a vegetarian lifestyle. It's better for your health and the environment, and it's a nonviolent choice. A vegetarian diet is the thinking person's diet.
- You owe no one an explanation. Stand tall and take comfort in knowing that you're on the cutting edge of the nutritional curve.

Cohabitating harmoniously

The most important things to do if you want your partner, children, parents — or anybody else you live with — to be happy in a vegetarian household are:

- **Take a low-key approach.** Arguing, chastising, and cajoling are seldom effective in gaining buy-in from other people. In fact, being pushy is likely to give you the opposite result. Explain your rationale to adults and older children, and then let them decide for themselves what they will or won't eat.
- **Model the preferred behavior.** The choices you make and the way in which you live send the most compelling — and noticeable — message.

Vegetarian diets are a great idea for kids. Establishing health-supporting eating habits early in life can help your children maintain good eating habits into adulthood. I include detailed information about managing the nutritional aspects of vegetarian diets for kids of all ages in Part V.

Setting realistic expectations

Making any lifestyle change requires you to master new skills and change longstanding habits, and that takes time. Don't be too hard on yourself if you experience a few occasional setbacks.

Experiment with new recipes, try vegetarian foods in restaurants, and invite friends to your house for meals. Read all you can about meal planning and vegetarian nutrition. With practice and time, you'll gain confidence and get comfortable with your new lifestyle. Think of this as a long-term goal — it took me several years before I felt like an expert at it.

Educating yourself with reliable information

This book gives you a solid introduction to all things vegetarian, but don't stop here. It's helpful to hear (or read, or view) the same subject matter presented differently by different sources. It takes most people several rounds before they absorb and understand a new subject well.

I mention good resources throughout this book, but I also list several of the best all-around resources for vegetarians of any skill level in the upcoming sections.

The Vegetarian Resource Group

The Vegetarian Resource Group (VRG) is a U.S. nonprofit organization that educates the public about the interrelated issues of health, nutrition, ecology, ethics, and world hunger. The group publishes the bimonthly *Vegetarian Journal* and provides numerous other printed materials for consumers and health professionals. The organization provides the materials free of charge or at a modest cost in bulk, and many are available in Spanish. VRG's health and nutrition materials are peer-reviewed by registered dietitians and physicians. VRG also advocates for progressive changes in U.S. food and nutrition policy. (Full disclosure: I've served as a nutrition adviser to VRG for about 25 years.)

You can reach VRG at P.O. Box 1463, Baltimore, MD 21203; phone 410-366-8343, fax 410-366-8803; e-mail `vrg@vrg.org`, Web site `www.vrg.org`. From VRG's Web site you can download and order materials, including handouts, reprints of articles, recipes, and more.

The North American Vegetarian Society

The North American Vegetarian Society (NAVS) is best known for its annual vegetarian conference, Summerfest, which is often held in Johnstown, Pennsylvania in July. This casual, family-oriented conference draws an international crowd of 400 to 600 people with diverse interests. Nonvegetarians are welcome. Summerfest is an excellent opportunity to sample fabulous vegetarian foods, meet other vegetarians, attend lectures, and pick up materials from a variety of vegetarian organizations. The group also publishes *The Vegetarian Voice,* a newsletter for members.

You can contact NAVS at P.O. Box 72, Dolgeville, NY 13329; phone 518-568-7970; e-mail `navs@telenet.net`, Web site `www.navs-online.org`.

Vegetarian Nutrition, a Dietetic Practice Group of the American Dietetic Association

The Vegetarian Nutrition Dietetic Practice Group (VNDPG) is a subgroup within the American Dietetic Association (ADA) for dietitians with a special interest in vegetarian diets. The group publishes a quarterly newsletter, and other consumer nutrition educational materials are available online at `www.vegetariannutrition.net`. (Full disclosure: I was VNDPG's founding chair in 1992 and have been a member ever since.)

The ADA also offers a referral service for people who need individual nutrition counseling. To find the name and contact information of a registered dietitian in your area with expertise in vegetarian diets, go online to `www.eatright.org/cps/rde/xchg/ada/hs.xsl/home_4875_ENU_HTML.htm`.

Other great resources

If you're looking for even more information on the vegetarian lifestyle, consider the following:

- **Vegetarian Nutrition** at `www.vegetarian-nutrition.info`. This Web site is devoted to information about vegetarian diets and healthful lifestyles.
- **A Dietitian's Guide to Vegetarian Diets** at `www.vegnutrition.com`. This site includes basic information for newcomers to the vegetarian lifestyle.
- **VeganHealth.org.** This is a project of the nonprofit Vegan Outreach. The Web site provides information about diet and health aspects of vegetarian lifestyles.
- **Physicians Committee for Responsible Medicine** at `www.pcrm.org`. PCRM is a nonprofit organization of physicians and others who advocate for compassionate and effective medical practices, research, and health promotion, including vegetarian diets.

Chapter 2

Vegetarians Are Sprouting Up All Over: Why Meatless Makes Sense

In This Chapter

- Choosing a vegetarian diet for your good health
- Rejecting meat to help save the planet
- Becoming a vegetarian for ethical reasons

Sometimes, what you don't know won't hurt you. Other times, what you *do* know can help you and others . . . a lot.

Vegetarian diets are like that. You can find many compelling reasons why living vegetarian is a good thing to do. One of those reasons may pique your interest, and after you discover more, other reasons may reinforce your resolve. When you think about it from all angles, a vegetarian diet makes a lot of sense.

What you eat is a highly personal decision. Many factors, including a variety of psychological and social issues, affect your capacity to make changes in your lifestyle. That's why many people who adopt a vegetarian lifestyle see their diet evolve over time. Consider the rationale for adopting a vegetarian diet, and do what you can to move in that direction at a pace that's right for you.

In this chapter, I share some details about the prevalence of vegetarianism. I describe how eating a vegetarian diet promotes your overall health and the health of the planet. I also explain why ethics and compassion for animals compel many people to make the switch. In Chapter 5, I share some additional advice on how to help you along in the transition to being meat-free.

You're in Good Company

Contemplating a move to a more plant-based diet? You're not alone. In fact, you may be among an elite group of deep thinkers with the determination it takes to live a lifestyle outside cultural norms. That takes courage and tenacity, but it's a decision more people are making.

The percentage of people in the U.S. who are consistently vegetarian has remained relatively constant for decades. That figure stands at about 2 to 3 percent of the adult population, according to polls conducted for the nonprofit Vegetarian Resource Group (VRG).

Within those figures, though, the number of people eating a vegan diet continues to rise, especially among younger vegetarians, according to survey data collected for more than ten years for the VRG. The VRG estimates that one third to one half of all vegetarians in the U.S. eat a vegan diet free of all animal products, including meat, fish, poultry, eggs, dairy products, and byproducts of these foods. For information about these surveys, check out `www.vrg.org/nutshell/faq.htm#poll`.

Among nonvegetarians, the number of people actively cutting back on their meat intake is also increasing, fueling the explosion of meatless food products in supermarkets and the regular sight of vegetarian options on mainstream restaurant menus. A 2006 poll conducted for the VRG found that nearly 7 percent of people in the U.S. say they never eat red meat.

Famous vegetarians yesterday and today

If you're thinking about going vegetarian, you may be interested to know that you're joining an illustrious group of like-minded individuals.

Considering the small number of people who are consistently vegetarian, it's remarkable how many well-known personalities throughout history have chosen the vegetarian lifestyle. The list includes philosophers, artists, intellectuals, entertainers, political leaders, sports figures, and many more. (I cover vegetarian diets for athletes in Chapter 22.)

The following list includes some of the better-known people throughout the centuries who have advocated a vegetarian lifestyle:

- Hank Aaron: Professional baseball player
- Louisa May Alcott: Author
- Susan B. Anthony: Suffragist
- David Bowie: Rock musician
- Berke Breathed: Cartoonist
- Leonardo da Vinci: Artist, scientist, inventor
- Isadora Duncan: Dancer
- Thomas Edison: Inventor
- Albert Einstein: Physicist

- Benjamin Franklin: Founding Father, scientist, inventor, philosopher, printer, and economist
- Mahatma Gandhi: Spiritual leader
- Jerry Garcia: Musician
- Philip Glass: Composer
- Dustin Hoffman: Actor
- Desmond Howard: Professional football player
- Steve Jobs: Founder of Apple Computer
- Carl Lewis: Olympic runner
- Martina Navratilova: Tennis champion
- Sir Isaac Newton: Physicist
- Bill Pearl: Bodybuilder and four-time Mr. Universe
- Plato: Philosopher
- Pythagoras: Philosopher
- Fred Rogers: Television's "Mister Rogers"
- Albert Schweitzer: Physician and Nobel Peace Prize winner
- Alicia Silverstone: Actress
- Socrates: Philosopher
- Benjamin Spock: Pediatrician and author
- Biz Stone: Twitter co-founder
- Carrie Underwood: Singer

Supporting Your Health with a Plant-Based Diet

Scientific consensus today — supported by a large body of research — indicates that vegetarian diets generally support health and confer health benefits. That doesn't mean vegetarian diets are foolproof or the Fountain of Youth; if your vegetarian diet largely consists of soda and French fries, you likely won't see any health benefits. But if you follow my basic guidelines for a healthful, meat-free diet, you'll decrease your risk for disease and promote your overall health.

People often worried in the past about whether vegetarians could get the nutrients they need from a diet without meat, but the tables are turned today. The riskier diet is the one that *isn't* vegetarian.

Protecting yourself from disease

In recent years, the U.S. government, in its "Dietary Guidelines for Americans," as well as Health Canada and leading health organizations, have issued statements acknowledging that well-planned vegetarian diets meet nutritional needs and support health. In fact, vegetarian diets are associated with lower risk for many chronic diseases and conditions, including coronary artery disease, diabetes, high blood pressure, some forms of cancer, obesity, and others.

It's not hard to understand why.

Vegetarian diets tend to be lower in *saturated fat* than nonvegetarian diets. Although you can find saturated fat in some plants, saturated fats come primarily from animal products, particularly high-fat dairy foods and meats. In fact, two-thirds of the fat in dairy products is saturated fat. Even so-called low-fat dairy products contain a substantial amount of saturated fat.

Saturated fats are usually firm at room temperature, like a stick of butter. That fact can help you identify foods that may be high in the artery-clogging fats. Foods that are high in saturated fat include red meats, the skin on poultry, butter, sour cream, ice cream, cheese, yogurt made with whole milk, 2 percent milk, and 3.3 percent whole milk.

Saturated fats stimulate the body to produce more cholesterol. *Cholesterol* is a waxy substance found in plaque in diseased arteries. Everyone needs some cholesterol, but the human body manufactures a sufficient amount. You don't need more from outside sources, and for people with a predisposition for coronary artery disease, high blood cholesterol levels may contribute to hardening of the arteries.

Cholesterol is produced in the liver, so it's found only in animal products. Foods of plant origin contain no cholesterol. (Have you ever seen a lima bean with a liver?)

WARNING!

Lacto ovo vegetarian diets have the potential to be high in saturated fat and cholesterol if you don't limit the eggs and high-fat dairy products you eat. If you switch to a lacto ovo vegetarian diet, try not to rely too heavily on these foods to replace the meat you once ate. (Chapter 1 has definitions of lacto ovo and other kinds of vegetarianism.)

Getting more of what you need — and less of what you don't

Vegetarian diets are slimming. In general, vegetarians eat more filling, low-calorie foods, such as fruits and vegetables, than nonvegetarians do. Dietary fiber makes vegetarian foods bulky. *Dietary fiber* is the part of a plant that's only partially digested by your body, or not digested at all. When you eat fiber-rich foods, you tend to get full before you take in too many calories. In that way, foods that are rich in fiber help to control your weight. Read on for more details about the benefits of dietary fiber as well as protein and phytochemicals! All these nutrients play important roles in supporting your health.

Fiber

Dietary fiber does more than help control your weight. It brings several other health benefits, too. For example, fiber can bind with environmental contaminants and help them pass out of the body. Fiber also decreases the amount of time it takes for waste material to pass out of the body, so potentially harmful substances have less time to be in contact with the lining of your intestines.

If you get plenty of fiber in your diet, you're less likely to have constipation, hemorrhoids, and varicose veins. Getting plenty of fiber (and water) in your diet also keeps your stools large and soft and easy to pass. You don't have to strain and exert a lot of pressure to have a bowel movement.

TECHNICAL STUFF

Eating a high-fiber vegetarian diet may prevent you from developing *diverticulosis,* a condition marked by herniations or small outpocketings in the large intestine. These pouches can fill with debris and become inflamed, a serious and painful disease called *diverticulitis.*

Diets high in fiber are associated with lower rates of colon cancer and coronary artery disease than diets low in fiber. If you have diabetes, you can better control your blood sugar by eating a diet that's high in fiber, too.

On average, Americans get about 16 grams of fiber in their diets each day — that's about half the recommended amount. Vegetarians typically get much more — up to double the amount. Vegans get more fiber than most lacto ovo vegetarians. One cup of oatmeal contains 8 grams of fiber, a medium pear with skin has 4 grams of fiber, a cup of vegetarian chili has 14 grams of fiber, a slice of whole wheat bread contains 2 grams of fiber, and a cup of chopped, steamed broccoli has 6 grams of fiber.

Current dietary recommendations call for fiber intakes of at least 14 grams for every 1,000 calories you eat each day. That means most people need at least 25 to 35 grams of fiber daily. It's also important to drink plenty of fluids, particularly water, when your fiber intake is high.

Protein

Most vegetarians get enough protein, but they don't overdo it, and this moderation has its benefits. When you moderate your protein intake, you help to conserve your body's stores of calcium (I tell you more about the relationship between protein and calcium in Chapter 3). Diets that are too high in protein, especially protein from animal sources, cause the kidneys to let more calcium pass into the urine. Meat protein is high in sulfur-containing amino acids that, when metabolized, result in higher acid levels in the body. To neutralize or buffer some of the excessive acid, calcium is leached from the bones and excreted in the urine.

Speaking of your kidneys, when you moderate your protein intake by eating a vegetarian diet, you also cause less wear and tear on your kidneys. Vegetarians have fewer kidney stones and less kidney disease than nonvegetarians. A good rule of thumb is to aim for a little less than a half gram of protein for every pound you weigh. A 120-pound person, for example, should get about 44 grams of protein per day. High intakes of animal protein are linked with higher blood cholesterol levels and more coronary artery disease, as well as a greater incidence of some types of cancer.

Phytochemicals

Vegetarian diets are rich in the plant components called *phytochemicals* that promote and protect human health. Conversely, meat is low in beneficial phytochemicals such as antioxidants and high in the oxidants that cause your body to produce detrimental *free radicals.*

The more animal products you include in your diet, the less room you have for plant matter. Whether or not you make the transition to a fully vegetarian diet, you can benefit greatly from dramatically increasing the ratio of plant to animal products in your diet (Figure 2-1 shows a typical American diet and the vegetarian goal).

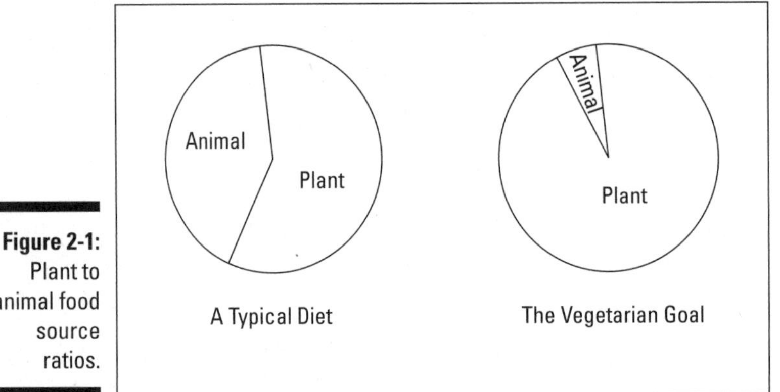

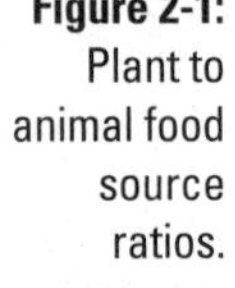

Figure 2-1: Plant to animal food source ratios.

In case you're wondering, *free radicals* aren't renegades from the '60s; they're molecules produced as a byproduct of your body's normal metabolism. They're also produced when you're exposed to environmental contaminants such as air pollution and ozone, sunlight and X-rays, and certain dietary components such as fat and the form of iron found in meat. Free radicals speed up the aging process by damaging your cells. They can impair your immune system and cause numerous diseases and illnesses.

Worrying less about food safety

Vegetarians are less likely than nonvegetarians to be affected by food-borne illnesses carried by animal products. Listeria and salmonella are types of bacteria found in meats and other animal products. A good dose of either can cause severe illness at best and death at worst. Many cases thought to be the "24-hour flu" are actually caused by a food-borne pathogen from an animal product. Creutzfeldt-Jakob disease, also known as CJD or mad cow disease, is a progressive and fatal disease that causes brain tissue in humans and animals to become spongy and porous; the animal or person literally loses its mind. Mad cow disease has been found in U.S. and Canadian cattle. Safety measures to protect other cattle from becoming infected are routinely disregarded and aren't effectively monitored or enforced by governments.

Saving the Planet One Forkful at a Time

Choosing to live vegetarian doesn't just benefit your body. When you choose a vegetarian lifestyle, you contribute to the health of the whole planet.

All things considered, vegetarian diets make more efficient use of land, water, and fossil fuel resources than diets that give prominence to meat, eggs, and dairy products.

Soil sense

You and I have to take care of the land, or our land won't be able to take care of us. *Animal agriculture* — raising animals for human use — places a heavy burden on the land.

Unfortunately, the amount of land resources needed to raise animals for their meat is far greater than the amount needed to grow enough plant matter to feed the same number of people directly.

Livestock grazing causes *desertification* by eroding the topsoil and drying out the land, preventing it from supporting plant life. Topsoil is being depleted faster than it can be created. Healthy, abundant supplies of topsoil help ensure we have enough arable land to grow the food we need to survive.

The appetite for meat and other animal products also consumes trees and forests, although these trees may not be the ones in your own backyard or your neighbor's. Much of the deforestation happens in Central America and South America, where the number of acres of tropical rain forest cleared to make way for grazing cattle is so big that it's hard to comprehend its enormity.

Meat costs even more than you think!

If the true cost to society of producing animal products for human consumption were passed on to consumers, fewer people could afford to put these foods on the dinner table as often as they do today. Fortunately, though, for people with a hankering for ham and eggs, government agricultural policies subsidize and protect the production of many of these foods to keep businesses and consumers satisfied with low prices. Of course, the costs don't go away — they're only deferred or shifted. As I point out in this chapter, the production and distribution of animal products makes intensive use of our nonrenewable natural resources and creates byproducts that contaminate our water, air, and soil and harm our health. The cost of meat is sort of like the cost of tobacco to society, which everyone shares in the dollars we spend on healthcare. As someone once said, "You can pay me now, or you can pay me later."

Think of tropical rain forests as the planet's lungs. One of the many ways that trees help keep the world healthy is by exchanging the carbon dioxide emitted by humans for the fresh oxygen we need to survive. If you didn't have trees, you'd be huffing and puffing as though you had reached the summit of Mount Everest; "breathing through a straw" is how mountain climbers describe the feeling of thinning oxygen. Losing a large percentage of the earth's rain forests has repercussions for every corner of the planet.

In deforestation, many species of plants and animals are wiped out. These plants and animals hold the keys to scientific discoveries that may benefit humankind. Plants and animals of the rain forests are on the front line, but eventually, the assault on them will reach all of us.

Wasting water

When you look at a map of the earth, you see a lot of blue. It's hard to imagine that the earth could be short of water anytime soon. The planet does have a lot of saltwater in the oceans, but one of the greatest threats of animal agriculture is to supplies of *fresh* water.

You can't drink saltwater. The aquifers that lie deep below the earth's surface hold the fresh water needed to irrigate farmland and for drinking. Those giant pools of fresh water are dwindling rapidly because people are sucking up great quantities of the stuff to irrigate huge tracts of land to graze animals that provide a relatively small amount of food. What an inefficient use of a resource as precious as our fresh water!

Furthermore, animal agriculture is polluting both fresh water and saltwater. Pesticides, herbicides, and fertilizers used to grow feed for animals are contaminating water supplies, and nitrogenous fecal waste produced by animals washes into streams, rivers, lakes, and bays. The planet has less clean water than it used to, and the creatures that live in that water are becoming contaminated or are being killed off by the pollution.

Filching fossil fuels

Animal agriculture is ransacking our planet of fossil fuels. The production of meat, eggs, and dairy products requires intensive use of fossil fuels — including oil, coal, and natural gas — for everything from transporting animal feed and the animals themselves to running farm machinery and operating the factory-homes where animals are raised. Fossil fuels are also used to make synthetic fertilizers used on factory farms.

Unlike wind, solar, and geothermal sources of energy, fossil fuels aren't readily renewable. After we use our supplies, they're gone. Byproducts such as carbon dioxide and nitrous oxide created by burning fossil fuels for energy also contribute to global warming. Other pollutants, such as methyl mercury from coal-burning power plants, find their way into our water supply, air, and soil and harm the health of humans and other living things.

Considering the Ethics

Of course, choosing a vegetarian diet isn't just about improving your own health and well-being. When you adopt a vegetarian diet, you also help to diminish the pain and suffering of those without a voice.

Philosophically speaking

Some of the world's greatest thinkers and philosophers chose or advocated a vegetarian lifestyle, including Pythagoras, Socrates, Leonardo da Vinci, and Benjamin Franklin.

For some people, a vegetarian lifestyle stems from religious or spiritual teachings. Some Christians, for example, interpret the passage Genesis 1:29 from the Old Testament to mean that humans should eat a vegetarian diet: "And God said, behold, I have given you every herb-bearing seed, which is upon the face of all the earth, and every tree, in which is the fruit of a tree yielding seed; to you it shall be for meat."

Members of the Seventh-day Adventist Church are encouraged to follow a vegetarian diet, and about half of them do. The Trappist monks, who are Catholic, are also vegetarians, and numerous Eastern religions or philosophies, including Buddhism, Jainism, and Hinduism, also advocate vegetarianism.

Vegetarianism is also an ethical choice. Hunger is a good example. For most people in the U.S., being hungry is a temporary state of being — the result of skipping a meal because you were in a rush to get out the door in the morning, for example. For less fortunate people in the world, however, hunger is a state of slow starvation with no next meal or snack in sight.

World hunger is complicated in terms of ethics, politics, and economics. Many people choose a vegetarian lifestyle as a way of contributing to the fight against world hunger. In the same sense of knowing that one's lone vote really does count, many people choose to cast their vote in favor of a diet and lifestyle that can sustain the most people.

In the simplest sense, eating foods directly from the soil — fruits, vegetables, grains, legumes, nuts, and seeds — can nourish many more people than can be fed if the plant matter is fed first to animals and then people eat the animals.

But the appetite of affluent nations for meat and other animal products also creates a market in poor countries for the resources needed to produce those foods. The result is that in developing nations, those with power often grow cash crops or livestock feed for export, instead of growing less profitable crops that might feed the local people.

Understanding animal rights and animal welfare

Inhumane treatment of animals compels many people to adopt a vegetarian lifestyle. They consider a meatless diet to be the compassionate choice.

If you share your life with dogs or cats, you know that animals have feelings. You probably also know that cattle, chickens, and pigs don't live in idyllic pastures and barnyards anymore. They live and die in factory farms and slaughterhouses, the likes of which would rival any horror movie.

Male calves born to dairy cows, for instance, are taken from their mothers and raised in confinement for veal. Chickens raised for their eggs live in factory-farm conditions in which they are routinely subjected to *debeaking* — the practice of using a machine to cut off the end of a chicken's beak to reduce the pecking damage caused when crowded chickens lash out at one another.

The part we see is the neat and tidy packages of legs and shoulders on Styrofoam plates wrapped in plastic. As morbid as it sounds, the meat counter is the grocery store morgue.

Suffice it to say, some people do allow themselves to think about how animals are treated. They put themselves in the hooves and claws of other creatures, and they take a stand for nonviolence.

In addition to avoiding meat, fish, and poultry, many vegetarians avoid the use of all other animal products. In part, that's because many animal byproducts, such as leather and wool, subsidize the meat industry. The production of other animal products, such as eggs, milk, and other dairy products, is also seen as exploiting those animals, subjecting them to inhumane conditions and treatment, and supporting the meat industry.

For more information about animal rights, including sources of cruelty-free products, check out the Web site of People for the Ethical Treatment of Animals (PETA): `www.peta.org`.

Chapter 3

Nutrition Know-How for Living Vegetarian

In This Chapter

- Putting protein needs into perspective
- Optimizing plant sources of calcium
- Understanding the basics about iron
- Getting enough vitamin B12
- Identifying sources of omega-3s
- Managing other vitamins and minerals

Do you have the old Basic Four Food Groups branded on your frontal lobe? When someone asks you to name foods that provide protein, calcium, and iron, does your mind reflexively retrieve images of rib-eye steaks, cheese wheels, and tall glasses of frothy white cow's milk?

I understand.

You may be savvy enough to know that you can find other food sources of key nutrients, but it's hard to change your mind-set in a culture in which animal products have held center stage for generations.

That's why I devote this chapter to some basic nutrition issues. The information that follows helps to clear up questions and concerns you may have about the nutritional adequacy of a diet that limits or excludes foods of animal origin.

Consuming Enough Protein on a Vegetarian Diet

In all my years of practice as a nutritionist, the number one question people ask about vegetarian diets hasn't changed. They want to know, "Where do you get your protein?"

And you know what? Protein is the nutrient about which vegetarians have the least reason to worry.

Even so, it's not hard to understand why people do worry about protein. In the first grade, your teacher probably made you cut out pictures of protein-containing foods from magazines and paste them onto poster board. You and your classmates collected pictures of hamburgers, hot dogs, pot roasts, and pizza. Likely absent from that poster were pictures of baked beans and brown rice.

In North America and much of the developed world, a cultural bias exists that places the highest value on protein from animal sources. The food industry — in close collaboration with the government — perpetuates policies and practices that maintain the status quo for complex political and economic reasons.

However, getting most or all of your protein from plant sources instead of from animals is easy and desirable.

In this section, I help you loosen your grip on some misconceptions about the role of protein in human health and tell you all you need to know about protein in meatless diets.

Examining protein facts

To understand how vegetarian diets deliver the protein you need, you first need a solid grasp on what protein is and how your body uses it. That will help illuminate the misunderstandings that persist concerning vegetarian diets and protein.

The word *protein* comes from a Greek word meaning "of first importance." Protein was the first material identified as being a vital part of all living organisms. Proteins make up the basic structure of all living cells, and they're a component of hormones, enzymes, and antibodies.

Most of the protein in your body is in your muscles, but you also have protein in your bones, teeth, blood, and other body fluids. The collagen in connective tissue is a protein, and so is the keratin in hair. The casein in milk, albumin in eggs, blood albumin, and hemoglobin are all examples of proteins.

Protein is a vital part of all living tissues. Proteins are nitrogen-containing compounds that break down into *amino acids* during digestion. Amino acids are the building blocks of proteins and have other functions in the body as well. *Essential amino acids* are amino acids that the body can't manufacture. There are nine of them, and you have to get them from your food.

Most plants contain protein, and some are rich sources. Vegetables, grains, legumes, seeds, and nuts all contain protein. Examples of vegetarian foods that are good sources of protein include

- Barbecued tempeh
- Bean burritos
- Bean soup
- Cereal and soymilk
- Falafel (garbanzo bean balls)
- Lentil soup
- Oatmeal
- Pasta primavera
- Red beans and rice
- Tofu lasagna
- Vegetable stir-fry
- Vegetarian chili
- Vegetarian pizza
- Veggie burgers

Plants contain all the essential amino acids in varying amounts. Some plants are high in some essential amino acids and low in others. Getting enough of what you need is easy to do, even if you eat nothing but plant-based foods and your diet contains no meat, eggs, or dairy products at all.

In fact, as long as you get enough calories to meet your energy needs, it's nearly impossible to be deficient in protein. If you ate nothing but potatoes but ate enough potatoes to meet your energy needs, you'd still get all the essential amino acids you need. Now, I don't recommend a potato-only diet.

You need other nutrients from food besides protein, and no one food has them all. That's why it's important to include a reasonable variety of foods in your diet.

However, the practical fact about protein and essential amino acids is that if you eat enough vegetables, grains, legumes, nuts, and seeds (or a reasonable variety of most of those foods), you'll meet or exceed your need for all the essential amino acids.

Debunking old rules about complementary proteins

Despite what you may have heard or read, you don't have to consciously combine foods or *complement proteins* to make sure you get enough protein on a vegetarian diet. The old practice of complementing proteins was based on solid science, but the conclusions that nutritionists drew — and the recommendations they gave — were faulty.

Here's how the idea of complementary proteins works:

Because plant foods are limited in one or more of the essential amino acids, vegetarians have to combine a food that's limited in an amino acid with a food that has an abundance of that same amino acid. The concept is to complement one plant food's amino acid profile with another's, fitting the foods together like puzzle pieces. That way, you have a *complete protein,* with adequate amounts of all the essential amino acids available to the body at the same time.

Nutritionists created complex protein complementary charts that detailed the way foods should be combined, and vegetarians had to be careful to eat their beans with rice or corn and to add milk or cheese to their grains (macaroni and cheese was a big favorite). It was complicated, and the practice made it seem as though vegetarians lived on the brink of nutritional peril if they didn't get the balance right.

Welcome to the new millennium.

Combining the amino acids in foods is no longer recognized as something vegetarians have to do consciously. Your body can do that for you, just like numerous other nutrient interactions that occur without your active intervention. Your job is to do two things:

- Make sure you get enough calories to meet your energy needs.
- Eat a reasonable variety of foods over the course of the day.

That's really all there is to it.

Even though it's not necessary to get all essential amino acids in one food, it's interesting to note that soybeans do contain all the essential amino acids. Soy can actually serve as the sole source of protein in a person's diet if that were necessary for some reason.

Getting the protein you need: It's easy to do

Really, you don't have to worry about your protein intake.

Okay, if you live on snack cakes and soda, maybe you can end up protein-deficient. But you'd be lacking many vitamins and minerals and other phytochemicals, too. Nevertheless, I know that some of you want to know precisely how much protein you need, so here's a simple way to figure it out:

The rule of thumb for determining recommended protein intake is to aim for 0.8 gram of protein for every kilogram of your body weight. One kilogram is equal to 2.2 pounds. That's body weight (in kilograms) times 0.8 = grams of protein needed. This formula has a generous margin of error worked into it. In reality, your body actually requires less than this amount. (An exception may be protein requirements for vegans. Although there is no consensus on this point, many vegan nutritionists think vegans should use a factor of 0.9 or even 1.0 to compensate for the lower digestibility of plant proteins.)

So, for example, if you weigh 120 pounds, that equals about 54.5 kilograms (120 pounds divided by 2.2 kilograms). Multiply 54.5 kilograms by 0.8 grams of protein, and you get 43.6 grams of protein, or approximately 44 grams of protein. That's how much protein your body would need each day, and it isn't hard to get.

To see how easy it is to get the protein you need, take a look at the list of foods and their protein contents in Table 3-1. Think about what you eat each day and the portion sizes. Calculate your own protein requirement using this formula and compare it to the amount of protein you eat in a typical day:

> Your weight in pounds _____ ÷ 2.2 kilograms = _____ your weight in kilograms × 0.8 grams protein = _____ your daily protein requirement

Table 3-1 Protein Values of Common Foods

Food	*Grams of Protein*
Animal Products	
1 ounce any type of meat	7
1 ounce cheese	7
1 egg	7
1 cup milk	8
Quarter-pound hamburger (no bun)	28
4-ounce chicken breast	28
10-ounce rib-eye steak	70
3-egg omelet with 2 ounces of cheese	35
Animal Product Alternatives	
1 typical veggie burger	5–25
1 typical veggie hot dog (1 link)	8
4 ounces tempeh	20
4 ounces tofu (depending on type)	5–9
8 ounces soymilk (plain)	10
Legumes (Dried Beans and Peas)	
½ cup most legumes (depending on type)	5–9
½ cup bean burrito filling	6
1 cup black bean soup	16
1 cup vegetarian chili	24
½ cup vegetarian baked beans	6
½ cup garbanzo beans on a salad	6
Nuts and Seeds	
1 ounce nuts (depending on type)	4–7
1 ounce seeds (depending on type)	4–11
2 tablespoons tahini (sesame seed butter)	6
2 tablespoons cashew, almond, or peanut butter	8
Grains and Grain Products	
1 slice whole-wheat bread (dense style)	2–3
1 bran muffin	3
½ cup whole-grain flake cereal	2
½ cup cooked oatmeal	3
1 whole bagel	6

Food	*Grams of Protein*
½ cup cooked pasta	7
½ cup cooked rice	4
1 flour tortilla	2
Vegetables and Fruits	
1 cup most vegetables (for example, green beans, tomatoes, cabbage, broccoli)	4
Most fruits	trace amounts

The sample one-day menu that follows provides about 1,600 calories and 53 grams of protein. You can see that even if you don't include any animal products at all, it's easy to get all the protein you need.

Breakfast:

1 cup cooked oatmeal, with cinnamon, raisins, and 1 cup plain soymilk

1 slice whole-wheat toast, with trans fat-free margarine and jelly

6 ounces fresh orange juice

Lunch:

Mixed green salad with vinaigrette dressing

1 cup lentil soup

1 chunk corn bread

½ cup fresh fruit salad

Water

Dinner:

4 ounces bean curd mixed with Chinese vegetables and brown sauce

1 cup steamed rice

½ cup cooked greens with sesame seeds

Orange wedges

Herbal tea

Snack:

Bagel with nondairy cream cheese

Herbal iced tea mixed with fruit juice

Avoiding protein pitfalls

Given the attention protein gets, you'd think that there's a looming threat of deficiency. In truth, many people are at risk from getting too much protein.

An excessively high intake of *sulfur-containing amino acids* may cause you to lose calcium from your bones. Foods high in sulfur-containing amino acids include eggs, meat, fish, poultry, dairy products, nuts, and grains.

Eating red meat and processed meat is associated with a greater risk of colorectal cancer, and meat consumption has been found in some studies to increase the risk of breast cancer, too. Too much protein in your diet may also put undue stress on your kidneys.

Vegetarian diets typically contain enough protein, without providing too much. Avoiding excessive amounts of protein — and animal products rich in protein — has definite health advantages.

Moooove Over Milk

Dogs do it. Deer do it. Cows do it. Even chipmunks and raccoons do it. They all produce milk for their babies. So do humans.

But dogs don't drink chipmunk milk. And deer don't drink raccoon milk. That's because milk is species-specific. The milk of each species is tailor-made for its own kind.

So how did people start drinking milk from cows? Even adult cows don't drink cow's milk.

It's an odd state of affairs when you think about it.

Milk is a concentrated source of calcium, and calcium is an essential nutrient for human health. That part is true.

But humans, like all mammals, have no need for milk after infancy. Your body is designed after infancy to get your calcium from other sources in your diet, like plants. Young children gradually stop producing *lactase,* an enzyme present in infancy that permits babies to digest the milk sugar *lactose.*

By adulthood, most people — with the exception of Caucasians — no longer produce lactase and have a limited ability to digest milk. This natural condition is called *lactose intolerance.* Most of the world's adults are lactose intolerant to some extent, including the majority of blacks, Asians, Hispanics, Native Americans, and people of Mediterranean descent.

When people who are lactose intolerant drink milk or eat other dairy products, they often experience symptoms such as gas, bloating, nausea, and diarrhea. An exception is people of North European ancestry who are thought to have inherited a genetic mutation that allows them to digest milk even as adults.

In our culture — and throughout much of the developed world — powerful dairy industry interests, in collaboration with local, state or provincial, and federal governments, strongly advocate milk consumption for complex political and economic reasons.

They do it despite the fact that most of the world's adults don't digest milk very well. They also push milk despite other health consequences — milk is high in sodium and artery-clogging saturated fat, and it's devoid of beneficial phytochemicals and the dietary fiber largely absent in most people's diets.

In this section, I explain what you need to know about getting calcium the way nature intended — on a vegetarian diet.

Determining who needs milk: The bones of current dietary recommendations

Cow's milk is rich in several nutrients, including protein, calcium, potassium, vitamin B12, riboflavin, and others. With the exception of B12, these are all also found in ample amounts in plant foods.

But the push to drink cow's milk is most closely tied to recommendations for calcium intake. Government recommendations for calcium intake are relatively high to compensate for calcium you lose because of other dietary factors. In most affluent countries, for example, high protein intakes from animal products and high sodium intakes from junk foods and processed foods cause calcium to leach out of the body. That, in turn, causes scientists to ratchet up recommended calcium intakes, which reinforces the idea that you need calcium supplements or super-concentrated food sources of calcium, such as dairy products.

It's a Catch-22.

When you eat a healthful vegetarian diet (as opposed to a junk-food vegetarian diet!), you may protect yourself against the dietary factors that cause less thoughtful eaters to risk their calcium stores.

Understanding the calcium connection

Calcium is vital to a healthy body. Too little calcium in your diet can make you a prime candidate for health problems like *osteoporosis,* the condition that results when the bones begin to waste away and become porous and brittle. Bones in this condition are susceptible to fractures. In the most severe cases, the bones can break with the slightest stress, such as a sneeze or a cough. Osteoporosis is a major health problem that can have life-threatening consequences.

Vegetarians around the world appear to have rates of osteoporosis that are the same as, or lower than, those of nonvegetarians. Still, osteoporosis is a very complex disease influenced by many factors, including lifestyle, diet, and genetics. (Your specific risk for osteoporosis needs to be assessed on an individual basis.)

Factors associated with vegetarian diets — low intake of animal protein, moderation of protein intake in general, and higher intake of soy protein, potassium, and vitamin K, for example — may protect against osteoporosis among vegetarians. On the other hand, too-low intakes of protein, vitamin D, and calcium, and lower estrogen levels in vegetarian women, may be risk factors for osteoporosis for some vegetarians.

Making sure you get enough calcium

All vegetarians should err on the side of caution and aim for meeting the levels of calcium intake recommended by the Institute of Medicine, a unit of the National Academy of Sciences. Most adults can meet these recommendations by following the meal-planning guidelines I include in Chapter 7.

Those recommendations call for at least 8 servings of calcium-rich foods each day. That's not as much food as it may seem. Mountains of broccoli aren't mandatory, but heaping helpings of green leafy vegetables, beans, and other good sources of calcium should be your habit.

Good plant sources of calcium include broccoli, Chinese cabbage, legumes, almonds, sesame seeds, dried figs, and dark green, leafy vegetables such as kale, Swiss chard, collards, and mustard and turnip greens. Other sources are tofu processed with calcium and fortified soymilk or orange juice.

You have lots of delicious ways to work high-calcium foods into your meals. Here are some examples:

- Bean chili over rice with steamed broccoli and a chunk of cornbread
- Black bean dip with broccoli and cauliflower florets and carrot sticks
- Falafel (garbanzo bean balls) in pita pockets and a glass of orange juice

- Fig cookies with a glass of fortified soymilk
- Green beans with slivered almonds
- Steamed kale with garlic and sesame seeds
- Stir-fried Chinese vegetables with bean curd (tofu) over rice

Hanging on to the calcium you have

When it comes to your body's calcium stores, what's important is not only getting enough, but also hanging onto what you have.

Too much protein — especially from foods high in sulfur-containing amino acids — can make you lose calcium. Other aspects of your diet and lifestyle also make a difference, including

- **Sodium:** High intakes of sodium make your body lose calcium. Table salt and processed foods contain lots of sodium, as do canned foods such as soups and condiments such as ketchup, mustard, and soy sauce.

  You need some sodium, but you can get all you need from the sodium that naturally exists in the food supply. You don't have to add sodium to foods. Read the labels on packaged foods and try to limit your sodium intake to no more than about 2,000 milligrams each day (1,500 milligrams is even better) — good advice for vegetarians and nonvegetarians alike.

- **Phosphorus and caffeine:** Too much phosphorus (found in red meats and soft drinks) causes the body to lose calcium, as does the caffeine found in soft drinks, coffee, and tea. These factors have a lesser effect on calcium balance than do protein and sodium, though.
- **Phytates and oxalates:** Substances such as *phytates* and *oxalates,* found in plant foods, inhibit your body's ability to absorb the calcium in the foods that contain them. Whole grains are high in phytates, and raw spinach is high in oxalates, making most of the calcium in these foods unavailable to your body. (Boiling spinach causes some of the oxalates to leach out into the cooking water, which can make the calcium more available to your body.)

  Overall, though, plant foods contain plenty of calcium that your body can absorb. In fact, some research shows that the body better absorbs the calcium in such plant foods as kale, Chinese cabbage, and broccoli than the calcium in cow's milk.

- **Physical activity:** The more weight-bearing exercise you include in your daily routine, the more calcium you'll keep. People who walk regularly or engage in strength training by using a weight set have denser bones than people who are couch potatoes.
- **Sunshine:** Your body manufactures vitamin D when you're exposed to sunlight, and vitamin D helps your body absorb calcium.

I should mention one more factor that people often forget. Your body adjusts the absorption of many nutrients, including calcium, according to its needs. In other words, when you need more calcium, your body becomes more efficient at absorbing the calcium that's present in your food. When you need less calcium, your body absorbs less, even if you flood yourself with calcium from dairy products or supplements.

Iron Issues

"Can you list three good food sources of iron?"

When I ask most people that question, they say, "Red meat," and draw a blank after that. Like protein and calcium, most people associate iron with an animal product. But also like protein and calcium, iron is widely available in foods of plant origin.

In this section, I explain what you need to know about getting the iron you need from a vegetarian diet.

Ironing out the basics

Iron is a mineral that forms part of the *hemoglobin* of red blood cells and helps carry oxygen to the body's cells. Iron is also part of the *myoglobin* that provides oxygen to the muscles in your body. When your iron stores are low or depleted, you can't get enough oxygen to your body's cells. A form of *anemia* results, and you may feel very tired.

Before you begin yawning, you might like to know that vegetarians aren't more prone than nonvegetarians to iron deficiency. *Iron-deficiency anemia* happens to be one of the most common nutritional deficiencies around the world, but most of it occurs in developing countries, not in affluent countries, and the cause is usually parasites, not diet. In affluent countries, iron deficiency is most likely to affect children and young people, as well as pregnant and premenopausal women.

Vegetarians need more iron in their diets than nonvegetarians, because the iron from plants isn't absorbed as efficiently as the iron from meat. That's not a problem for vegetarians, though. Most get plenty of iron in their diets — in fact, they get more than nonvegetarians.

Introducing heme and nonheme iron

Iron comes in foods in two forms: *heme* and *nonheme* iron. The iron in meat, poultry, and fish is the *heme* form. The human body easily absorbs heme iron. In fact, when meat is present in the diet, it increases the amount of iron that you absorb from plant foods as well. This characteristic doesn't necessarily make heme iron a better source of iron, and it may even have drawbacks.

Iron is a potent oxidant that changes cholesterol into a form that's more readily absorbed by the arteries, leading to hardening of the arteries, or *coronary artery disease.* For this reason, men — most of whom are at a higher risk of developing coronary artery disease than most women — are now advised *not* to take daily multivitamin and mineral supplements containing iron.

The form of iron found in plant foods is called *nonheme* iron. Nonheme iron is absorbed less efficiently than heme iron. However, certain other food components — which I discuss in the "Enhancers of iron absorption" section later in this chapter — can radically improve the body's absorption of nonheme iron. These enhancers of iron absorption in plant foods help vegetarians absorb the iron they need.

The recommended dietary allowances for iron vary by age and sex. Generally, the recommendations are higher for premenopausal women than for men and postmenopausal women, because women who menstruate lose more iron than women who don't. The recommendations for adult men and postmenopausal women are only 8 milligrams per day, compared to 18 milligrams per day for premenopausal adult women.

It's not practical, though, for most people to count the milligrams of iron in their diets. Rather than live life with a calculator in the palm of your hand, pay attention to including iron-rich foods in your diet regularly and do what you can to make sure the iron you do eat is well-absorbed. Keep reading as I explain how to do that.

Finding iron in plant foods

Iron is available everywhere you look in the plant world. Rich sources include whole or enriched breads and cereals, legumes, nuts and seeds, some dried fruits, and dark green, leafy vegetables.

Many of the plant foods that are rich in calcium are also high in iron. The meal ideas that I list in the earlier section "Making sure you get enough calcium" also work if you're trying to ensure that you get enough iron in your diet.

If you're looking for additional iron-rich foods, consider the following:

- Broccoli
- Brussels sprouts
- Cabbage
- Cantaloupe
- Cauliflower
- Green peppers
- Onions
- Oranges
- Tomatoes
- Strawberries

Balancing inhibitors and enhancers of iron absorption

You may not realize it, but a game of tug of war over iron is going on in your body. That's because certain substances in the foods you eat enhance iron absorption in your body, and other substances inhibit it. Most of the time, the two even out and everything turns out just fine. It pays to understand how the game works, though, so that you can ensure the amount of iron your body is absorbing is just right.

Inhibitors of iron absorption

One of the substances that inhibits your body's ability to absorb iron is the tannic acid in tea. In poor countries where diets are low in vitamin C-rich fruits and vegetables (which enhance iron absorption) and low in iron, a tradition of tea drinking can tip the scales and cause iron deficiency. That doesn't commonly happen in Western countries, where people generally eat a wider variety of foods and have access to plenty of fruits and vegetables.

Other substances that decrease the availability of dietary iron include certain spices, coffee, cocoa, the *phytates* in whole grains, and the calcium in dairy products. The reality is, however, that if you eat a reasonable mix of foods, inhibitors and enhancers of iron absorption offset each other and no harm is done.

If you're an avid tea drinker, be careful. More than coffee, tea impairs your body's ability to absorb dietary iron. In the American South, it's common for people to drink half a gallon or more of iced tea each day, and quantities of tannins that large may be problematic for some people. Many kinds of herbal tea also contain tannins, though others, including chamomile and citrus fruit teas, do not. If you drink tea, consider drinking it between meals, rather than with meals, to lessen the inhibitory effect of the tannins.

Enhancers of iron absorption

One of the most potent enhancers of iron absorption from a plant source is vitamin C. Healthful vegetarian diets are full of vitamin C. As a vegetarian, if you eat a rich food source of vitamin C with a meal, you can enhance the absorption of the iron in that meal by as much as 20 times.

You probably eat vitamin C-rich foods with many of your favorite meals and don't even realize it. In each of the following vegetarian meals, a vitamin C-rich food helps the body absorb the iron in that meal:

- Bowl of black bean soup with sourdough roll; a mixed green salad with tomatoes, green peppers, and onions; and a slice of cantaloupe
- Bowl of cooked oatmeal with fresh strawberry slices and soymilk
- Pasta tossed with steamed broccoli, cauliflower, carrots, green peppers, and onions, with marinara (tomato) sauce or garlic and olive oil
- Peanut butter and jelly sandwich on whole-wheat bread, with orange quarters
- Tempeh sloppy Joe, with a side of coleslaw
- Vegetarian chili over brown rice with steamed broccoli or Brussels sprouts
- Veggie burger with tomato and lettuce and a side of home fries

Other plant components also improve iron absorption, but vitamin C is the most powerful. Using cast-iron skillets, pots, and pans can also increase the amount of iron you absorb, especially when you use them to cook acidic foods, such as tomatoes or tomato sauce.

Building B12 Stores

Your body needs vitamin B12 for the proper functioning of enzymes that play a role in the metabolism of amino acids and fatty acids. A vitamin B12 deficiency can cause a form of anemia and a breakdown of the nerves in your hands, legs, feet, spinal cord, and brain.

That's serious stuff for a vitamin that we need in only miniscule amounts. The recommended intake of vitamin B12 is a teeny-weeny 2.4 micrograms per day. Just a fraction of a pinch would be enough to last your entire life.

So what's the big deal about B12?

Getting educated about vitamin B12

For vegetarians, the issue is this: Vitamin B12 is found only in animal products. As far as anyone knows, plants contain none. So vegans could theoretically risk a vitamin B12 deficiency if their diets didn't contain a source of it.

Vegetarians who eat eggs or dairy products don't have as much to worry about. They can get plenty of B12 from those foods if they eat them regularly.

Vitamin B12 is an issue that vegetarians — particularly vegans — should take seriously, because deficiency can lead to severe and irreversible nerve damage. Despite the need for such a miniscule amount of the vitamin, the stakes are high if you don't get what you need.

On the other hand, it would probably take you a long time to develop a vitamin B12 deficiency. Your body hoards and recycles much of the B12 it gets, which means that your body conserves B12 and excretes very little of it. Most people have at least a 3-year supply of vitamin B12 stored in their liver and other tissues, and some can go without B12 in their diet for as long as 20 years before depleting their stores and developing deficiency symptoms.

In fact, the most common cause of vitamin B12 deficiency isn't a lack of the vitamin in the diet — it's the diminished ability of the body to absorb the vitamin. This problem happens to everyone and isn't strictly a vegetarian issue. Older adults are thought to be at greatest risk, because the body's ability to produce *intrinsic factor,* the substance needed to absorb vitamin B12, diminishes with age. For that reason, recommended intakes of B12 are higher for people over age 50 than for younger adults.

Despite the theoretical risk, few cases of vitamin B12 deficiency in vegans have been reported. That may be due to several reasons, including the fact that folic acid can mask the anemia caused by vitamin B12 deficiency. Folic acid is found in plant foods, especially green, leafy vegetables. Vegetarians, especially vegans, tend to get a lot of folic acid in their diets, so it's possible that a vitamin B12 deficiency could go undetected.

Finding reliable sources of B12

Vitamin B12 is being manufactured all around you — and inside you, too. The vitamin is produced by *microorganisms* — bacteria in the soil, ponds, streams, rivers, and in the guts of animals, including humans. So if you were living in the wild and drinking from a mountain stream or eating vegetables pulled directly out of the ground, you'd have natural sources of B12 in the water and soil clinging to your food. Today, you buy your vegetables washed clean and you drink sanitized water. That's best for public health, but it does eliminate a source of B12.

If you eat any animal products — eggs, cheese, yogurt, cow's milk, or other dairy products — you're getting vitamin B12. Nonvegetarians get even more vitamin B12 from meat, poultry, and seafood.

Your intestines also produce vitamin B12, but it's produced past the site of absorption. In other words, you make vitamin B12, but it's not a reliable source.

Take heed: If you eat few or no animal products, make sure you have a reliable source of B12 in your diet. It's easy to find reliable sources of vitamin B12, such as B12-fortified soymilk and rice milk, fortified breakfast cereals, or even a B12 supplement. Include them in your diet regularly, just to be on the safe side.

If you take a vitamin B12 supplement, choose one that can be chewed or left to dissolve beneath the tongue. Vitamin B12 isn't absorbed very well from supplements that are swallowed whole.

Distinguishing between the different B12s

Just when you thought you understood — here's a twist on the B12 story.

Vitamin B12 comes in many forms. The form that you need is called *cyanocobalamin.* Other forms of vitamin B12 are known as *analogs.* The analogs are inactive in the human body.

Why would I burden you with this information? Because a considerable amount of confusion exists over which foods are reliable sources of vitamin B12, especially for vegans. In the past, sea vegetables, tempeh, miso, and nutritional yeast were touted as being good sources of vitamin B12. When those foods were tested for vitamin B12 content for purposes of labeling the food packages, the labs used a microbial assay that measured for all forms of vitamin B12 — not just specifically for cyanocobalamin. So people bought the foods thinking they were getting the amount of B12 that the label listed, when, in fact, much of what was listed may have been analogs.

Now, scientists say that up to 94 percent of the vitamin B12 in these foods is actually in analog form, rather than cyanocobalamin. Some are concerned that analogs may compete for absorption with cyanocobalamin and promote a vitamin B12 deficiency. Of course, other people think that this idea is hooey.

When you read food packages, look for the word *cyanocobalamin* to be sure that the food contains the right form of vitamin B12. You shouldn't have any trouble finding products that have been fortified with cyanocobalamin. Many of these food products, such as breakfast cereals and meat substitutes, are mainstream brands that you can find in any supermarket. Make it a habit to check the food labels regularly, though, because food manufacturers have been known to change the formulation of products from time to time. Other fortified products that are easy to find are soymilk and other milk substitutes. If you can't find them in your neighborhood supermarket, you'll find lots of choices at a natural foods store.

A product that many vegetarians — especially vegans — enjoy using is nutritional yeast. It has a savory flavor similar to that of Parmesan cheese, and many people use it the same way. You can sprinkle it over salads, baked potatoes, pasta, casseroles, cooked vegetables, and popcorn.

If you use nutritional yeast as a source of vitamin B12, be sure to buy the right kind. The regular yeast used for baking that you commonly find in the supermarket isn't what you want. Many brands of nutritional yeast in natural foods stores contain vitamin B12 analog. You don't want those, either. Look for Red Star Vegetarian Support Formula (T-6635+) at your natural foods store. If the store doesn't have it, ask the manager to order some for you.

Many vegetarians think that tempeh, miso (a fermented soy product used as a soup base or condiment), tamari (a fermented soy sauce product), bean sprouts, and sea vegetables such as kombu, kelp, nori, spirulina, and other forms of algae, as well as products labeled as nutritional yeast in natural foods stores, are reliable sources of B12. They aren't, so be careful. These foods contain mostly inactive analogs of vitamin B12, rather than the cyanocobalamin that the human body needs. Red Star Vegetarian Support Formula is the form of nutritional yeast that contains the form of vitamin B12 that your body can use.

Considering vitamin supplements

For vegetarians who need a reliable source of vitamin B12, using a supplement is a foolproof way to go. You can use a supplement instead of or in addition to eating fortified foods.

Buy the lowest dose of vitamin B12 supplement that you can find. You need only 2.4 micrograms per day, but you'll probably find that most of the supplements supply much more. Your body adapts its level of absorption according to its needs and the amount in your diet. If you flood yourself with vitamin B12, your body is going to absorb only what it needs and you'll waste the rest. So there's no point in taking mega-doses of the vitamin.

Omega-3s and Your Health

The idea that vegetarians need to pay attention to their intake of *omega-3 fatty acids* or *n-3 fatty acids* is relatively new. Research suggests that vegetarians get lots of *n-6 fatty acids* in their diets, such as linoleic acid from corn, cottonseed, sunflower, and soybean oils. Linoleic acid is an *essential fatty acid,* a substance essential for health that your body can't produce and that you have to get from food.

Vegetarians get fewer n-3s, including α-linolenic acid, another essential fatty acid that's found in flaxseeds, canola oil, cooked soybeans, tofu, and other foods. For some vegetarians, lower intakes of n-3s may create an imbalance that causes them to have insufficient amounts of eicosapentaenoic acid (EPA) and docosahexaenoic acid (DHA), n-3 fatty acids that are physiologically active and essential for humans. N-3s such as α-linolenic acids get converted into EPA and DHA in your body.

If you eat fish or eggs, you have a direct source of EPA and DHA in your diet, so no conversion is necessary. If you don't, you have to eat enough n-3 fatty acids to enable your body to convert them to EPA and DHA.

Spotting the best vegetarian sources of omega-3s

Sea vegetables such as nori, kelp, kombu, dulse, and others are direct sources of EPA and DHA. You can find these vegetables at Asian markets and natural foods stores. Many vegetarians enjoy cooking with these foods.

Supplements made from microalgae rich in DHA are also available now in natural foods stores. They're made with non-gelatin capsules, so they're suitable for vegans to use.

Other good sources of n-3 fatty acids for vegetarians include ground flaxseeds, flaxseed oil, canola and walnut oils, and whole or chopped walnuts.

Getting the omega-3s you need

Don't worry about micromanaging your n-3 fatty acid intake. You don't have to count the grams of fat. Just stay aware of the need, follow the eating guide in Chapter 7, and think about using some of the good food sources of n-3 fatty acids more often in your diet.

If you aren't confident that you're getting enough n-3s, consider taking a DHA supplement of 200–300 micrograms per day. Consult your healthcare provider for individualized advice. DHA supplements made from *microalgae* (microscopic forms of algae) are well-absorbed and can boost blood levels of DHA and EPA. Some brands of soymilk, such as Silk, are also available fortified with DHA.

Other Vitamins and Minerals

A long list of vitamins, minerals, and other substances are necessary for your health. Those I mention in this section — vitamin D, zinc, and vitamin B2 (also known as riboflavin) — merit discussion because, like other nutrients I cover in this chapter, they're generally associated with animal products. You may wonder if it's difficult for vegetarians to get enough of these.

Don't worry — just read on.

The sunshine vitamin: Vitamin D

Vitamin D is actually a hormone that regulates your body's calcium balance for proper bone mineralization. Your body is designed to make vitamin D when your skin is exposed to sunlight. Theoretically, you don't have to get any from foods.

If your skin is light, you need from 5 to 15 minutes of summer sun on your hands and face each day to make enough vitamin D for your body to store over the winter months, when you're exposed to less sunlight. People with darker skin need more sun exposure.

Of course, that's how it works in the ideal world.

In reality, many people are at risk of vitamin D deficiency because they live in large, smog-filled northern cities where they don't have as much exposure to sunlight as they may need. Others at risk include people with very dark skin (especially those living at northern latitudes), people who are housebound and rarely see the light of day, and people who don't get regular exposure to sunlight because they cover their entire bodies with clothing or sunscreen every time they go outdoors. As you age, your ability to produce vitamin D also diminishes, so older people may also be at greater risk of deficiency.

Food sources of vitamin D

Few foods are naturally good sources of vitamin D, and most vegetarians don't eat the ones that are. Eggs are a source, and so is liver, but liver isn't recommended for anybody because of its high cholesterol content and because an animal's liver is a primary depository for environmental contaminants.

In the United States, milk and other dairy products have been fortified with vitamin D for many years as a public health measure to protect people who may have inadequate sunlight exposure. If you drink milk, you get a vitamin D supplement in effect.

Vitamin D is added to fluid milk, but the amount present has been determined to be unreliable. Some samples of milk contain very little vitamin D, and other samples contain many times the amount permitted by the government. The problem has to do with difficulties the milk industry has in dispersing the vitamin D evenly throughout large vats or tanks of milk.

Vegetarians and vegans have several potential sources of vitamin D in addition to sunshine. Some nondairy milks — soymilk and rice milk, for example — are fortified with vitamin D, as are some brands of margarine and breakfast cereals. Read the food labels to be sure. (Whether vitamin D is dispersed more evenly in nondairy milk than it is in cow's milk is unknown.)

Foods can be fortified with different forms of vitamin D, including vitamin D2 or vitamin D3. Vitamin D2, called *ergocalciferol,* is made by irradiating provitamin D (from plants or yeast) with ultraviolet light. Vitamin D3, called *cholecalciferol,* may not be vegetarian and is definitely not vegan. It's made from fish liver oils or *lanolin,* which is sheep wool fat. Usually, the food label states the form of vitamin D used. If it doesn't, you may have to check with the manufacturer to determine the source.

Vitamin D supplements

If you get little exposure to sunlight and you aren't sure whether you're getting enough vitamin D from fortified foods, check in with your healthcare provider or a registered dietitian for advice. It may be necessary to take a vitamin D supplement to ensure that you're getting enough. I include more advice about the pros and cons of taking supplements in Chapter 4.

For your health: Zinc

Your body needs zinc for proper growth and development. Zinc plays a role in various chemical reactions that take place in your body, as well as in the manufacturing of new proteins and blood, and it helps to protect your body's immune system.

Some vegetarians have trouble meeting recommendations for zinc intake, and the phytates in some plant foods may bind with zinc and prevent your body from absorbing it. Even so, vegetarians in North America and other Western lands haven't been found to have zinc deficiencies.

Meat is high in zinc, but vegetarians can find plant sources, including dried beans and peas, mushrooms, nuts, seeds, wheat germ, whole-wheat bread, and fortified cereals. Dairy products and eggs are also good sources of zinc for vegetarians who eat those foods.

Some methods of cooking and preparing foods may increase the availability of zinc. For example, sprouting seeds, cooking or serving foods in combination with acidic ingredients such as tomato sauce or lemon juice, and soaking and cooking legumes may make zinc more available to the body.

The other B: Riboflavin

Another name for vitamin B2 is *riboflavin.* Riboflavin has multiple functions in the body, primarily related to its role in enabling enzymes to trigger various chemical reactions.

Riboflavin is present in small amounts in many foods, and many of the foods that are the most concentrated in riboflavin come from animals. Among the richest sources are milk, cheddar cheese, and cottage cheese. Other animal sources include organ meats, other meats, and eggs.

So vegetarians who eat dairy products have options, but what about those poor vegans? Isn't it a challenge for vegans to get enough riboflavin?

Nope. You'll find lots of good plant sources of riboflavin, including almonds, asparagus, bananas, beans, broccoli, figs, fortified soymilk and breakfast cereals, kale, lentils, mushrooms, peas, seeds, sesame tahini, sweet potatoes, tempeh, tofu, and wheat germ.

Vegans haven't been found to have symptoms of riboflavin deficiency, even though some studies have shown that they have lower intakes of riboflavin than nonvegetarians.

You don't have to eat mountains of broccoli to get enough riboflavin. Riboflavin is spread widely throughout the plant world, and you have many choices. As always, one of the chief strategies for ensuring that your diet is adequate is to limit the junk foods and to pack your meals with as many nutrient-dense foods as you can.

Chapter 4

Supplement Savvy

In This Chapter

- Taking a critical look at supplements
- Making sure you use supplements safely
- Understanding the role of herbs and probiotics
- Finding information about supplements

If your household is like most, you have a stash of vitamin and mineral supplements in your pantry or medicine cabinet. In fact, the supplement industry estimates that American consumers spend several billion dollars every year on nutrient-packed pills, gel tabs, and capsules. Considering the popularity of these products, what I tell you in this chapter may come as a surprise.

If you take vitamin and mineral *supplements* — concentrated doses of individual nutrients — you may think of them as little daily doses of nutritional insurance. In one sense, that idea may contain a tidbit of truth. If your diet is heavy on French fries and soft drinks, and your idea of a vegetable is a dollop of ketchup, you're probably missing out on a wide range of essential nutrients. In that case, it may not hurt — and it may even help, at least a little — to fill in some of the gaps with a tablet. Something is better than nothing, right?

Perhaps, but it's not ideal.

The truth is, the only foolproof way to get the complete array of vitamins, minerals, and other essential substances you need is to get them the way they're packaged naturally — in whole foods. No single food has it all — not bananas, milk, a sandwich, an energy bar, or a smoothie. Different foods provide different constellations of the nutrients found in nature. Take them together, though, in a diet that has enough calories to meet your energy needs and a reasonable variety of wholesome foods, and the nutritional puzzle pieces typically fall into place to give you the complete complement of what you need.

Unfortunately, your diet may not be picture-perfect all the time. That's why it's important to understand how best to focus your efforts on improving it and what role, if any, supplements should play.

Some people have the mistaken notion that vegetarians are more prone to nutritional deficiencies and, therefore, that they have a greater-than-average need for supplements. Not so, and if you're considering becoming a vegetarian, this chapter will help disabuse you of that worry. If you're already following a vegetarian eating style and relying on outdated information about vitamin and mineral supplements, this chapter may help you save some money and improve your overall health.

In this chapter, I help you simplify your vitamin and mineral regime. I give you the latest on which supplements science says are useful and not useful, and I help you understand when and if a particular supplement makes sense for you. Most important, I help you stay safe if you do decide to use supplements. I also talk about dietary supplements that are relatively new to store shelves — including herbs and probiotics — and explain what they are, how to use them, and where to find reputable sources of ongoing information about all these products.

Examining What the Science Says

It may seem like a no-brainer: "If a little bit of [fill in the name of your favorite vitamin or mineral] is good for me, more must be even better." If only it were that simple.

As it turns out, some of the vitamins and minerals you may have been taking for years may not be doing you any good, and some may even be causing you harm. A substantial body of research conducted in recent years has called into question the use of some of the most common supplements, including vitamins C and E and even beta carotene. On the other hand, science has also shed light on the real benefits of a few select supplements — like folic acid — for some people under certain circumstances.

Sound complicated? It is and it isn't.

The science of nutrition is complex, and as the years go by, scientists learn more and more about how our amazing bodies use the nutrients we need for health. Thirty years ago, that nutrient list included a few dozen vitamins and minerals. Today, scientists know that foods contain hundreds of substances — and maybe more — that support your health. Many of these substances work in synergy with one another when they're all present together in whole foods, but they act differently in your body when you take them separately in concentrated doses.

It can be frustrating sometimes when recommendations change, but that's the nature of science. Dietary recommendations are based on the best evidence available at the time, and they get better — and more accurate — as more research findings become available.

The good news is that I translate all this information into straightforward, practical advice that can help you make the most of your diet — whether you're a part-time or full-time vegetarian — taking only the supplements you actually need to support your health. Your wallet may benefit, too!

In this section, I answer a few common questions you may have about the science of supplements.

Do supplements work?

If you're wondering whether supplements work, the answer, for the most part, is no.

That's the short answer. In general, vitamin and mineral supplements give most people a false sense of security. People greatly overestimate the likely benefits of supplements, and they don't recognize the potential pitfalls.

Although supplements are a huge business — packing the shelves at supermarkets, drugstores, natural foods markets, and specialty shops at the mall — little, if any, evidence has shown that they're effective in lowering the risks for chronic disease for healthy adults. Despite a couple of exceptions — folic acid for some women of childbearing age, and vitamin B12 for older adults, for example — the bulk of the research on vitamin and mineral supplements generally doesn't support the exaggerated benefits claimed by the supplement industry.

A paper published in February 2009 in the Archives of Internal Medicine, for example, found that in a study of 160,000 older adult women, those who took vitamin and mineral supplements had no significant differences in the risk of death, heart disease, and certain forms of cancer than those who didn't take supplements. And the Institute of Medicine, part of the prestigious National Academy of Sciences, has concluded that the available evidence doesn't support the health claims for some of the most popular supplements in recent years, including vitamins C and E, selenium, and beta carotene.

This news can be hard to accept. After all, you, like me, may have grown up taking children's chewable vitamins in the morning and feeling virtuous as an adult for dutifully downing your once-a-day nutritional insurance pill. Old ideas about health can be hard to shake, especially those that have been as entrenched as the supplement routine. Even if you can understand the issue intellectually, it can be difficult to let go emotionally. You may be familiar with the concept of insurance, but when you apply that concept to nutritional supplements, it falls short.

Can they hurt me?

It may seem hard to believe that something you need for good health can have a downside. After all, the vitamin C in a fresh orange or a plate of lightly steamed broccoli florets is an important ingredient in many of your body's physiological processes. If you don't get enough of this vital nutrient, you'll develop the deficiency disease scurvy.

The same is true of other vitamins, minerals, and nutrients you get from foods. If you don't get enough of these important substances, your health can suffer. That's why over the years, nutritionists have chiefly focused on problems associated with the lack of — and not the abundance of — these nutrients. After deficiency diseases were identified, nutrition scientists pushed for everyone to get enough calcium, iron, vitamins A, B, C, D, and E, and more.

And supplements came to be. Extracted from foods or produced in laboratories, supplements became the easy way to replace nutrients that were in short supply in people's diets. This seemed to make sense — most people figured that supplements couldn't hurt, and maybe they could help.

More recently, though, that way of thinking has been called into question. Science suggests that nutrients behave differently when they're not a part of the total package of a whole food. In other words, vitamin E from a gel cap may not have the same benefits as the vitamin E found in whole seeds, nuts, or leafy greens.

Supplements may, in fact, cause problems under certain circumstances. Taking high doses of individual nutrients, for example, can cause nutrient imbalances that didn't exist before. If you take large amounts of supplemental zinc, for instance, you can deplete your body's copper stores. That's because the zinc and copper interact with each other. If they were present together in a whole food instead, the effects of the other nutrients in that food may provide some balance and counteract the effects of the individual minerals. So supplements lack the natural checks and balances found in foods that contain the entire spectrum of nutrients. Even multivitamin and mineral tablets aren't complete. They contain a long list of vitamins and minerals, but they don't include phytochemicals and other substances important to health, some of which probably haven't been identified yet.

Worse, some supplements once thought to provide special protection against heart disease and some forms of cancer have more recently been linked with increased health risks instead. For example, research has found that people at high risk for lung cancer — especially smokers — can actually increase their risk for lung cancer by taking beta carotene, a supplement that was previously thought to prevent cancer. Even vitamin E, taken in high doses, has been found in some studies to increase the risk of death.

What can I do instead?

Looking for a great way to save money? Skip the supplements. Switch to nutrient-dense, whole foods instead.

Just a single serving of many of these nutritional powerhouses can provide you with a hefty dose of the vitamins, minerals, and other substances you need in greater quantities to support health. Good choices include:

- **Dark green, leafy vegetables:** A healthy serving provides you with a big dose of dietary fiber, folic acid, iron, vitamins A and C, potassium, and sometimes calcium. Add a fistful of baby spinach leaves to a pita pocket sandwich, or roll the fresh greens into a bean burrito. Ribbons of cooked kale add a nutritious punch to a plate of black beans and rice, and you can stir cooked, chopped collard greens or spinach into a pot of lentil soup.
- **Deep orange fruits and vegetables:** Think peaches, apricots, sweet potatoes, and orange, yellow, or red bell peppers. These foods are rich in vitamins A and C, as well as dietary fiber. Add a thick slice of a juicy, red tomato to a sandwich, or add a generous handful of grated carrots to a hummus wrap sandwich. Stir some chopped, dried apricots into a bowl of hot oatmeal.
- **Whole, unprocessed grains:** In addition to being the B vitamin storehouses, grains are a rich source of dietary fiber and trace minerals. You get a nice dose of all these when you reach for a bowl of shredded wheat or bran flakes cereal, a sandwich on whole-wheat bread, or a steaming bowl of seven-grain cereal.
- **Dried beans and other legumes:** Think of these as the multivitamin and mineral supplements that they are, jam-packed full of dietary fiber, protein, folic acid, manganese, iron, and a wide range of other minerals. They're versatile, too — enjoy bean or lentil soup, bean burritos and nachos, black bean dip, or a bowl of chili.
- **Seeds and nuts:** Sprinkle them here, there, and everywhere. Examples include almonds, sunflower seeds, walnuts, pecans, and pumpkin seeds. Add them to salads, hot cereal, and casseroles. I like a handful of chopped walnuts on a bowl of rice pudding. You can't beat seeds and nuts as a source of health-supporting essential fatty acids, and they're a good source of protein, dietary fiber, and many vitamins and minerals, too.

Start thinking differently about what it means to supplement your diet. Fresh, wholesome foods like these are an enjoyable — and low-cost — way to ensure that you get the nutrients you need to be well.

Recognizing When a Supplement Makes Sense

Supplements may not be the best route to good nutrition for vegetarians, but they can be a reasonable crutch to support you in times when you don't have a better option. Many vegans, in fact, need to take vitamin B12 supplements, and others may benefit from supplements of vitamin D, calcium, and possibly

DHA unless they eat foods fortified with these nutrients. In this section, I help you understand how to identify the times when you may need a supplement, and I explain how to be selective in using vitamin and mineral supplements. In most cases, that means your supplement routine will be inexpensive and include just a short list of products.

Keep in mind that supplements are only meant to fill in the nutritional gaps in your diet. If somebody tells you that you need a long list of high-cost supplements, a warning bell should go off in your head, triggering a critical examination of the recommendation and, perhaps, a second opinion from your healthcare provider.

Special situations that call for a supplement

Supplements aren't ideal, but they may be worth considering if your circumstances are such that the benefits of taking a supplement outweigh the risks.

I cover those cases in Chapter 3, where I discuss the nutritional nuts and bolts of living vegetarian. Though you can easily get enough iron, calcium, omega-3 fatty acids, B vitamins, vitamin D, and zinc from plant foods, in some instances, a supplemental boost of one or more of these may be merited if you aren't getting enough from the foods you choose.

If you're healthy, your body should generally be able to adapt to varying dietary conditions, absorbing more of a particular nutrient when you need more and absorbing less when you need less. If you get enough calories to meet your energy needs and you eat a reasonable variety of wholesome foods but your diet alone isn't enough to keep nutrients at normal levels in your body, a more complicated problem is likely the culprit.

Another time when a nutritional supplement may be recommended is during pregnancy. I discuss the use of prenatal supplements in Chapter 19. Vegetarians aren't more likely than nonvegetarians to need prenatal supplements, but some people mistakenly think that women who don't eat meat when they're pregnant miss out on vital nutrients. This isn't so, but if you're pregnant or planning to get pregnant, your healthcare provider may still recommend a prenatal vitamin and mineral supplement.

The bottom line is that, with the exception of vitamin B12 for vegans, vegetarians aren't more prone than meat eaters to develop nutritional deficiencies. Vegetarians can generally get the nutrition they need from foods. They don't need supplements unless they have reason to believe that their intake isn't sufficient or that an underlying health problem is causing a nutrient imbalance.

Not all health professionals are equally familiar with vegetarianism and qualified to counsel vegetarians about their diets. If you do need help sorting out the pros and cons of vitamin and mineral supplementation, seek out a registered dietitian who's knowledgeable about vegetarian diets. A *registered dietitian* is a food and nutrition expert who has met specific criteria for education and experience as set forth by the Commission on Dietetic Registration, has passed a national registration exam, and has completed continuing education requirements. Search online for a registered dietitian who practices in your geographic area and has expertise in vegetarian nutrition by going to the American Dietetic Association Web site: `www.eatright.org/cps/rde/xchg/ada/hs.xsl/home_fanp_consumer_ENU_HTML.htm`.

In addition, the National Academy of Sciences recommends that certain groups take supplements, whether they're vegetarian or not. Specifically, the academy recommends that all adults over the age of 50 take a vitamin B12 supplement to ensure that they're getting enough. The academy also recommends that women in their childbearing years take a folic acid supplement. Research suggests that folic acid supplementation may help prevent neural tube defects, and many women have difficulty reaching the recommended level of 400 micrograms of folic acid per day. Vegetarian women can take a bow, though: Their folic acid intakes are typically much higher than those of nonvegetarians.

Nutritional insurance

Do you need to take out a nutritional insurance policy by popping a daily multivitamin and mineral supplement? The consensus among nutritionists is that filling dietary gaps with supplements isn't ideal and that whole foods are best.

Still, nutritionists aren't in complete agreement about what role, if any, once-a-day supplements should play in your diet, whether you're a vegetarian or not. Some suggest taking them every so often — once or twice a week, for example — just for good measure.

No evidence exists to suggest that taking low, periodic doses of multivitamin and mineral supplements is helpful, but no evidence exists to indicate that it's a problem, either. The choice is really a personal judgment call. Some nutritionists do advocate taking a supplement during times when you feel you aren't eating particularly well.

Vitamins and minerals are necessary ingredients in your diet to help ensure that various processes in your body function normally. They don't give you energy by themselves, as some people mistakenly think. Only *macronutrients* can do that. Macronutrients are components of your diet that your body needs in relatively large quantities, including carbohydrates, proteins, and fats.

As for me, I don't go out of my way to take any supplements at all. Of course, you and I do get some unintentional supplements in some of the foods we eat. Many ready-to-eat breakfast cereals, including the ones I buy, are *fortified* with a smattering of vitamins and minerals, and the soymilk I drink is fortified with vitamin B12.

Fortification is the practice of boosting the nutritional value of a product by adding vitamins and minerals not already present in that food.

So nutritionists don't exactly know what the value of a supplement is if you take it in small doses or infrequently. If it makes you feel better because you're taking it, then go for it — the risk of harm is small.

Using Supplements Safely

The benefits of nutritional supplements are largely unproven, and the side effects can be undesirable. If you're considering using supplements, think carefully about what you plan to take, how much the tablet or capsule contains, and how often you plan to take it.

When it comes to supplements, it's important to get individualized advice. The recommendations in this book are meant to apply to a general adult audience. Your individual needs may be somewhat different, especially if you're being treated for a disease or condition. In that case, be sure to follow the advice of your doctor or other healthcare provider.

Daily supplements versus high-potency formulas

Most common, once-a-day multivitamin and mineral supplements supply about 100 percent of the amounts recommended on a daily basis. Some even provide less than 100 percent of some nutrients. That's okay — these are supplements, not replacements for the nutrients in foods.

Many people believe that it's a good thing when a supplement contains five or ten times the daily recommended dosage. (After all, more is better, right? Wrong!) If you want to take a multivitamin and mineral supplement, choose one that supplies *no more* than 100 percent of the amounts recommended on a daily basis. Check the bottle's label to see what the supplement includes and how much.

High-potency formulas that contain multiple times the recommended daily amounts of nutrients are sometimes marketed as special formulations for people under stress (who isn't?) or for people with other needs requiring super-high dosages. Don't fall for these marketing ploys. They play on those hard-to-shake beliefs that more must be better. If you have a problem requiring unusually high doses of certain nutrients, leave it to your healthcare provider to prescribe the supplement.

Multivitamins versus individual vitamins and minerals

Ordinary, once-a-day multivitamin and mineral supplements are probably benign when they provide up to 100 percent of the recommended daily amounts of nutrients. Taken now and then — or even daily — they probably won't harm you. If you buy generic or store brand varieties, they may not cause much damage to your pocketbook, either.

Similarly, taking low doses of individual vitamins and minerals — those sold as isolated, separate nutrients — once in a while is probably harmless. However, many individual nutrients come in capsules or tablets that contain multiple times the recommended daily amount.

You increase the chances of creating problems for yourself when you take individual vitamins and minerals instead of taking them as part of a larger grouping of the nutrients that occur in foods naturally. That's especially true when you take dosages that are many times more than what you need in a day.

Remember that the best sources of nutrients are the whole foods that contain the array of nutrients in amounts and combinations found in nature. The next best option, though not ideal, is the multivitamin and mineral solution. Your last choice should be individual nutrients, unless your doctor or other healthcare provider specifically recommends them.

When supplements act like drugs: Being aware of interactions

Vitamins, minerals, and other nutrients can have drug-like effects in your body when you take them in high doses. A good example is niacin, one of the B vitamins. High doses of niacin change the production of blood fats in your liver. People at risk for coronary artery disease can use a special prescription form of niacin to lower their triglyceride and LDL ("bad cholesterol") levels and raise their HDL ("good cholesterol") levels. That's a pretty powerful use of a vitamin.

Other nutrients can have similarly potent effects when you take them in large doses. Like niacin, they can cause physiological changes in your body. High doses of vitamin C, for example, can cause your body to absorb less copper. Nutrients can also interact with one another, changing the natural balance of nutrients. High doses of zinc, for example, can deplete your copper stores. Certain vitamins and minerals in high doses can also interfere with the action of prescription drugs you may be taking and can either enhance or diminish the effects of those medications. For example, folic acid supplements may lessen the effects of phenytoin (Dilantin). (Herbs may have an effect as well. I cover those in the next section.)

If you already take supplements, including herbs or other alternatives, tell your healthcare provider what you're taking. Better yet, provide a written list. It will help ensure that you don't forget any individual item, and it will give your healthcare provider an accurate and up-to-date picture of everything you're taking. The more information your healthcare provider has about you, the more likely he or she will be able to warn you of potential side effects.

Considering Herbs and Probiotics

People who follow vegetarian diets tend to be more aware than the average eater of food and nutrition matters. As a vegetarian, you're more likely to hear messages — accurate or not — about your vitamin and mineral needs and about alternative dietary supplements.

Of course, you can supplement your diet with all sorts of things. In discussions about vegetarian diets, most people focus on vitamin and mineral supplements. But some people — vegetarian or not — add bran to their diets for a natural laxative, and others add flaxseed oil for the essential fatty acids. In the broadest sense, anything you add to your diet can be considered a supplement, including a peanut butter cracker.

In 1994, the U.S. Congress passed into law a similar but slightly narrower description of dietary supplements in the Dietary Supplement Health and Education Act (DSHEA). That law defines a dietary supplement as any product meant to supplement the diet with one or more dietary ingredients, taken by mouth in tablets, capsules, powders, liquids, and in other forms, and labeled as a dietary supplement.

In this section, I zero in on two specific kinds of supplements — herbs and probiotics — that have been the subject of increasing interest among vegetarians and others interested in diet and health. In both cases, the science is evolving, and the federal government is putting research dollars into studying the benefits and possible risks of each.

Understanding what they are

Herbs and probiotics are completely different things, but both are forms of dietary supplements used by many vegetarians and others interested in their potential health benefits. If you want to be a well-informed vegetarian, you should have at least a rudimentary knowledge of these two dietary embellishments.

Herbs

Herbs are plants used in small quantities for various purposes. Herbs can be grouped into two categories:

- **Culinary herbs,** such as rosemary and sage, are used for the flavor and aroma they add to foods. These plants are considered herbs and not vegetables, simply because they're used in such tiny amounts. Vegetarians love to use herbs and spices in cooking. *Spices* are usually made from the roots, bark, and other, non-leafy parts of the plants.
- **Medicinal herbs** are generally thought to confer health benefits, and many people use them for their potential therapeutic effects. (Some people refer to these as *botanicals,* another name for herbs.) Just as some people believe that vitamin and mineral supplements protect health by boosting immunity or preventing heart disease and cancer, some people believe that certain herbs may have similar effects.

Some herbs are used for both medicinal and culinary purposes. Ginger is safe and effective during pregnancy, for example, in small quantities for short periods of time to relieve nausea and vomiting, according to the National Center for Complementary and Alternative Medicine, a unit of the National Institutes of Health. Ginger also happens to taste good in a variety of vegetarian dishes, including soups and stir-fry.

When I talk about herbal supplements, though, I'm generally referring to medicinal herbs taken for their therapeutic benefits, rather than culinary herbs. Herbal supplements can be composed of a single herb or a combination of two or more herbs.

Examples of popular medicinal herbs include

- Black cohosh
- Echinacea
- Feverfew
- Gingko
- Kava

- Red clover
- Saint-John's-wort
- Valerian

A comprehensive list of popular herbs, including their common names, uses, and possible side effects, is available online from the National Center for Complementary and Alternative Medicine at `http://nccam.nih.gov/health/herbsataglance.htm`.

Probiotics

Enough about plants, though. Bugs are important, too.

In addition to herbal supplements, vegetarians often read about *probiotics,* or friendly bacteria. These are the good kind of bugs to have in your food.

The idea is that if you take these live microorganisms as supplements in capsule or powder form, or if you eat them in whole foods, they can replace or increase your body's natural supply of good bacteria.

If you take an antibiotic to fight an infection, for example, the medicine kills off the bad bacteria as well as the friendly bacteria that are naturally present in your intestines. These friendly bacteria are necessary for the normal functioning of your intestines.

You may be able to use probiotics to restore the natural supply of friendly bacteria in your intestines, though. Probiotics include yeast and a range of other microorganisms. Supplements, or yogurt with *active cultures,* such as Lactobacillus bulgaricus or Streptococcus thermophilus, may help to repopulate your intestines with these good guys.

Some scientific support also exists for the benefits of probiotics in the treatment of diarrhea and irritable bowel syndrome, the prevention and treatment of urinary tract infections, and the prevention and treatment of eczema in children. It's too early to make concrete recommendations, though. In fact, it's even possible that some benefits may be caused by a *placebo effect.* In other words, the benefits may all be in your head.

You can buy probiotics as dietary supplements in natural foods stores and pharmacies, or you can get them in whole foods. Common vegetarian foods that contain probiotics include some brands of yogurt, soymilk and other soy products such as miso and tempeh, kefir (a drinkable form of yogurt), and fermented milk (such as buttermilk). Fermented vegetables such as sauerkraut and kimchi — a traditional Korean dish made from pickled vegetables and seasonings — also contain probiotics.

Many brands of yogurt contain active cultures. Look for those words — "active cultures" — on the package to be sure that you're getting a dose of friendly bacteria when you eat that product. Other products, such as kefir, may use the word "probiotic" on the label to show that the product contains active cultures. Still others, such as buttermilk, may simply list "bacterial cultures" on the product's ingredient list.

Knowing how to use them safely

You may be surprised to discover that federal guidelines regulating the marketing and sales of nutritional supplements are quite loose. The products aren't held to the same standards as prescription drugs or over-the-counter medications. In fact, companies can start marketing nutritional supplements without having to prove that their products are safe or effective at all.

That alone should cause you to hesitate — and to do a little research of your own — before you decide to use a dietary supplement.

Herbal supplements, in particular, merit special attention. Just as with vitamin and mineral supplements, if you use herbal supplements, be sure to let your healthcare provider know. He or she understands your total care plan and may recognize when a supplement has the potential to cause a problem.

Talking to your healthcare provider is especially important if you're on medications or contemplating surgery. Like high levels of vitamin or mineral supplements, some herbal supplements have the potential to interfere with medications you may be taking. Others may affect your risk for bleeding during surgery or may alter the way you react under anesthesia.

If you're pregnant or breastfeeding, it's also best to let your healthcare provider know if you're taking an herbal supplement. Before giving an herbal supplement to a child, check with your child's healthcare provider, too. Though you may have nothing to worry about, it's better to be safe. Most research on herbal supplements has been conducted on adults, and little if any has been conducted on pregnant women, nursing mothers, or children.

Be aware that discrepancies are sometimes found between what's in the supplement bottle and what the label says the product contains. The plant species used or the amount of the active ingredient listed can be different from what you may think you're getting. Products may also be contaminated with pesticides, other herbs, and other substances that may not be safe to ingest.

In contrast to herbal supplements, probiotics don't seem to pose much risk, though more research is needed on their use by young people, older adults, and people with depressed immune systems. For healthy adults, the most common side effects are gas and bloating — uncomfortable, perhaps, but not serious.

Locating Reliable Sources of More Information

A great deal of research is being conducted on dietary supplements, so recommendations can change in relatively little time. That's why it's important that you know where to go for accurate, up-to-date information about their use and safety.

One first-line source of information is your healthcare provider. Physicians, registered dietitians, pharmacists, physician assistants, and other providers don't always have all the answers. They may not know, for example, where the science stands on the use of evening primrose oil or ginseng.

They do know where to go to get the information, though, and they usually have access to medical libraries or researchers and other experts who can provide reliable feedback.

I also highly recommend the following resources:

- **The National Center for Complementary and Alternative Medicine:** Access this clearinghouse online at `http://nccam.nih.gov/` for updates on research about herbal supplements and probiotics. Fact sheets and other materials are available in English and Spanish.
- **The Office of Dietary Supplements at the National Institutes of Health:** Find it online at `http://ods.od.nih.gov/`. Resources include comprehensive fact sheets with information about specific vitamins, minerals, and herbal supplements; advice for supplement users and tips for older adults; and background information about dietary supplements.
- **MedlinePlus, a service of the U.S. Library of Medicine and the National Institutes of Health:** Find it online at `http://medlineplus.gov/`. This site lets you search by the first letter of herbs and other supplements for detailed information compiled from a variety of authoritative sources. It also provides a link to the Dietary Supplements Labels Database, which you can use to compare the ingredients in more than 3,000 brands of dietary supplements. If you prefer, you can go to the database directly at `http://dietarysupplements.nlm.nih.gov/dietary/`.

Chapter 5

Making the Transition to Meat-Free

In This Chapter

- Converting to vegetarianism quickly or gradually
- Planning the details of your dietary switch
- Steering clear of animal ingredients
- Getting started on your lifestyle change

"Where are you going?" the cat asked.

"I don't know."

"Well, either road will get you there."

—Lewis Carroll, *Alice's Adventures in Wonderland*

Don't be like Alice, who doesn't know which route to take to get where she wants to go. If you want to change your eating style, *how* you do it isn't as important as having a plan for getting there. Without a plan, you're less likely to achieve your goal.

We all have different styles of doing things. What's important is that you find a method that works for you. There's no right or wrong way of making a lifestyle change, as long as you're successful. So do what's comfortable for you. In this chapter, I cover key points to think about in formulating your plan of action as you make the switch to a vegetarian eating style.

Finding the Right Approach

In this section, I present two approaches for you to consider as you plan your transition to a vegetarian lifestyle. I also cover pros and cons to consider as you weigh the options for making the transition.

Going cold tofu: Instant vegetarian

One day he's chewing on a 10-ounce rib-eye at a local steakhouse, and the next day he's ordering the tempeh burger at his neighborhood natural foods cafe. It's not a likely scenario, but some people do make an overnight decision to go vegetarian. They see a video or hear a lecture that inspires them, or they read a book or story about the horrors of the slaughterhouse. Worse: They visit a slaughterhouse in person.

Instant vegetarian

The overnight transition to vegetarianism usually isn't flawless, but those who do it are motivated to make the change as soon as possible. Some people just prefer to make big changes quickly as opposed to dragging them out over months or years. That's fine.

Benefits of the quick switch

Making the switch to a vegetarian diet in one fell swoop has some benefits:

- People who make big changes right away tend to notice the benefits quickly and enjoy the benefits sooner. That may be especially appealing if you have health concerns. For example, if you're overweight, you may notice weight loss sooner, or if you're a diabetic, your blood sugar level may drop.
- You get immediate gratification. Your personality may be such that you need the satisfaction that comes with taking immediate action and reaching your goal as soon as possible.
- You know you'll get there in this lifetime. In comparison to people who take the gradual approach (see the later section "Taking your time: The gradual approach" for details), people who make the change all at once jump one big hurdle to arrive at their goal. They don't run the risk that the others do of getting stuck in a rut along the way.

The overnight approach works best for people who

- Have done some homework: Even a little bit of reading or talking with another vegetarian about basic nutrition questions and ideas for quick and easy meals can help immensely.
- Are surrounded by support: The overnight approach is easiest for people who live or work with other vegetarians and have someone they can emulate or question about basic nutrition and meal-planning issues.
- Are relatively free of other distractions: Making a big change overnight is easier if you don't have a new baby, a new job, or an 80-hour workweek.

If you switch to a vegetarian diet and find yourself tired and hungry or irritable, you may not be eating enough. Some people who switch overnight haven't had time to figure out what they can eat. They end up eating only a few different types of foods, and they often don't get enough calories.

Don't fall prey to the "iceberg lettuce syndrome" — the situation where the only thing new vegetarians know they can eat is lettuce. You can eat all kinds of foods — see Part III for ideas.

If you're a diabetic and you switch to a vegetarian diet, visit your healthcare provider. Diabetics who adopt a vegetarian diet frequently need less insulin or oral medication and need to have their dosages adjusted. In some cases, diabetics' blood sugar levels decrease enough to allow them to discontinue their medication altogether. Don't attempt to change your medications without checking with your doctor or other healthcare provider first, however.

Drawbacks of instant vegetarianism

If you take the instant route to gratification, you'll soon find out that you have no time to develop necessary new skills and to put supports in place. It takes time to soak up the background information about nutrition and meal planning, as well as to learn how to deal with all sorts of practical issues such as eating out and handling questions from friends and family. These are some of the things that are necessary for a successful transition, and if they're not taken care of before you dive in, your entry can be a bit sloppy, or you may even do a big belly flop.

If you opt for the overnight approach, use your wits to do the best you can at the outset, but come up with a plan for smoothing the transition as soon as possible. Get yourself educated, get a plan, and get some support. I include plenty of tips in Parts II, IV, V, and VI.

Taking your time: The gradual approach

When it comes to making a major lifestyle change, most people fare best by taking the gradual approach. You probably will, too.

Let your diet evolve at your own pace as you develop new skills and educate yourself about the new eating style. As you master each new skill, you'll become more secure and comfortable with meal planning and handling a variety of food-related situations. In Parts II and IV of this book, I provide you with lots of practical information to help make the transition to a vegetarian lifestyle as easy as possible.

Benefits of the gradual approach

Advantages to making a gradual switch to vegetarianism include:

- Your new eating habits are more likely to stick. Making changes gradually allows you to collect the information and support you need to make it work and helps you build a strong foundation.
- The gradual approach may be less disruptive to your routine. If you have more time to adapt to each change every step of the way, you may have a better chance of sticking with it.

Drawbacks of easing in

The gradual approach has disadvantages, too. Keep these in mind and try to avoid them:

- Getting stuck in a rut along the way. You can get stuck anywhere along the line and never make the transition to a full-fledged vegetarian diet. Some people cut out red meat as an initial step, but they never make it any further. They're forever stuck in the chicken-and-fish rut, or they get as far as substituting cheese and eggs for meat. Their blood cholesterol levels soar because they're living on cheese omelets, macaroni and cheese, and grilled cheese sandwiches.

  Don't let this happen to you. Have a plan to keep moving.
- Procrastinating and dragging the change out too long. If you take too long to make the change, it may never happen. Don't delude yourself. If you want to adopt a vegetarian diet but it's been a year since you started, it's time to put pen to paper and develop a more structured plan with dated goals for getting there.

Easing the Way

Whether you choose to make the switch to a vegetarian diet overnight or gradually, you can tackle the change in a variety of ways. It's just a matter of personal preference. You may be especially interested in nutrition and devour books on the subject, or you may love to cook and find yourself experimenting with recipes long before you get around to thinking about vegetarian meals while traveling. You may like to read, or you may prefer talking to other vegetarians or attending lectures instead.

How you choose to proceed is up to you. The following guide is only one suggestion for planning a reasonably paced transition to a vegetarian eating style.

Defining some simple steps

Have you ever had to write a term paper or thesis? Cringe! Any large project can seem overwhelming when you look at it in its entirety.

To tackle a big project, break it up into smaller pieces, focusing on only one piece at a time. Writing down the steps you plan to take and checking them off as you master each one can give you a sense of accomplishment and help keep you motivated along the way.

Draw up a weekly or monthly plan of action outlining the things on which you want to focus each step of the way. For instance, you may want to concentrate on reading several books and other written materials for the first two months. From there, you may pick up a few vegetarian cookbooks and experiment with recipes, adding a few meatless meals per week to your schedule for the next month. Step it up to five meatless days per week for the next few months, and begin attending local vegetarian society meetings.

Your plan should reflect your lifestyle, your personality and preferences, and any constraints that you may have to factor in, such as cooking for a family or traveling frequently.

It helps to set time goals in your plan for transitioning to a vegetarian diet. If you can keep these deadlines in your head, fine. If not, or if you find yourself dragging your feet getting to the next step, you may need more structure. Write your plan on paper, and break it down week by week. Give yourself a reasonable amount of time to complete each task, and stick to your plan.

Educating yourself

Education is a good first step in any plan for lifestyle change. If you like to read, you're in luck, because you can find many excellent resources on vegetarian diets. It's a great idea to spend several weeks to months reading everything you can get your hands pertaining to vegetarianism as you make the transition.

You'll find materials on all aspects of vegetarianism. Some books, magazines, and online materials may seem to cover the same subjects, but it's worth reading them all, because each author presents the information in his or her own style, and the repetition of the subject matter can help you learn and build a solid foundation for your transition.

If you don't particularly care to read, seek out lectures about vegetarian diets and related topics on college campuses, at natural foods stores, and in community centers. Or drop in on meetings of local vegetarian societies where guest speakers are often featured.

There is one caveat, however. Although many excellent resources are available, some materials contain inaccuracies or misinformation. I include references to many reliable resources in this book, including some listed in the later section "Tracking down resources for up-to-date advice."

Reducing your meat intake

While you're doing some background reading about vegetarian diets, start reducing your meat intake in some easy ways:

- Add two or three meatless main meals to your diet each week. Begin with some easy and familiar entrées, such as spaghetti with tomato sauce, vegetarian pizza, bean burritos, vegetable lasagna, and pasta primavera.
- Try some vegetarian convenience foods, such as veggie burger patties, veggie breakfast meats, frozen vegetarian dinners, and veggie hot dogs. They're quick and convenient. Most supermarkets carry these products, and you can also find a large selection at natural foods stores. Substitute them for their meat counterparts.
- When you do eat meat, make it a minor part of the meal rather than the focal point of the plate. Use meat as more of a condiment or side dish. Keep portions small, and extend the meat by mixing it with rice, vegetables, pasta, or other plant products. For instance, rather than eat a chicken breast as an entrée, cut it up and mix it into a big vegetable stir-fry that feeds four to six people.

From there, just keep going. Add a few more meatless meals to your weekly schedule and stick with it for a month or so. Set a date beyond which you'll be eating only vegetarian meals. Mark that date with a big star on your calendar, and then cross off each day that you're meat-free after that.

Be aware that interruptions in your routine, such as a vacation, holiday, or sickness, can trigger a lapse in your eating plan. In times of stress or a break in routine, it's common to fall back into familiar patterns, including old ways of eating. Give yourself a break, and then start fresh again. With time, you'll discover how to handle breaks in routine and you won't be derailed by them.

Setting your goals

After you start the switch to vegetarianism, it's all about practicing and continuing to expand your knowledge base and experience. Expect some bumps in the road, but keep going.

Experiment with recipes, new foods, and vegetarian entrées at restaurants. Invite friends and family members over to your place for meals, read some more about nutrition and meal planning, and allow yourself more time to get comfortable with your new lifestyle.

All this is easier to accomplish if you set specific goals and a timeline for achieving them. Think about steps that seem realistic for you to achieve, and estimate the length of time you think it will take to master each step. Sketch out a plan in a notebook or set up a spreadsheet or table on your computer.

Table 5-1 shows a sample schedule for making the transition to a vegetarian lifestyle. Your own transition may take more or less time. Adjust the timeline — and the specific steps — to suit your own needs and preferences.

Table 5-1 Twelve Months to a Vegetarian Eating Style

Time Frame	*What to Do*
Months 1 and 2	*Read about basic nutrition, meal planning, dealing with social situations, and other aspects of vegetarian diets.* Borrow books from the library, buy a few good resources at a bookstore, visit vegetarian organizations online for materials, attend lectures, and subscribe to vegetarian magazines. Absorb information.
Month 3	*Begin to reduce your meat intake.* Add two or three meatless meals each week, experiment with new products, make a list of all the vegetarian foods you already enjoy, and when you do eat meat, make it a minor part of the meal, rather than the focal point of the plate.
Months 4 through 6	*Cut back even more on your meat intake.* Plan five meatless days each week. Limit meat to use as a condiment, a side dish, or a minor ingredient in a dish. Plan a cutoff date after which you'll eat only vegetarian meals. Mark that date on your calendar. A week or two before that date, stop buying meat and products containing meat, such as soup with ham or bacon and baked beans with pork. Keep a diary or log of everything you eat for several days. You'll refer to it later to gauge your progress. Be specific about the ingredients. For example, if you eat a sandwich, note the kind of bread and filling you chose.
Months 7 and 8	*Look into joining a local vegetarian society or attending a national vegetarian conference.* Seek out occasions when you can be surrounded by like-minded individuals. Continue reading and absorbing information. Continue experimenting with new recipes. Go out to eat and order vegetarian entrées at restaurants.
Months 9 through 12	*Practice.* Continue to seek new information. Socialize and invite friends and family members to your home for vegetarian meals. Look back over the past year and evaluate how you've handled holidays and special occasions, vacations, and breaks in your routine. Do some situations need attention, such as eating away from home or finding quick and easy meal ideas? Keep a food diary for several days and compare it to your first one (which I suggest starting in months 4 through 6).

Monitoring your progress

After you complete the initial steps in the transition to a vegetarian diet, keep a food diary from time to time for several days or weeks — whatever you can manage. Spot-checking your diet now and then helps you stay aware of what you're eating. Recording not only what, where, and when you eat but also how you feel and who you're with may help you become aware of behavior patterns you may want to target for change.

Keep a low-cost food diary with a pen and pad of paper, or try one of the several online food diaries available. Some charge a monthly fee. An added bonus of maintaining your diary online is that some, such as `www.myfooddiary.com`, will tally your daily intake of calories and nutrients and also let you know how many calories you're burning with exercise.

Making Sure It's Meat-Free

Avoiding obvious sources of meat, fish, and poultry is relatively easy. You probably recognize a chicken leg or steak when you see one. Living vegetarian gets trickier when animal ingredients and their byproducts are added to products in small quantities that may be hard to see or taste.

In this section, I help you find those hidden ingredients.

Being wary of hidden animal ingredients

Many foods contain hidden animal ingredients. Some are present in very small amounts. Casein and whey — both derived from dairy products — are acceptable for vegetarians but not for vegans. Others, such as rennet (which comes from the stomach lining of calves and other baby animals), are typically unacceptable to all vegetarians. I include a list of hidden animal ingredients in Table 5-2. Decide for yourself what you are and aren't willing to eat.

In some cases, an ingredient may originate from either an animal or plant source, and you can't determine which from the label. You may need to contact the manufacturer and ask (see the later section "Communicating with food companies" for more info).

Table 5-2 Hidden Animal Ingredients

Ingredient	*What It Is*	*Where You Find It*
Albumin	The protein component of egg whites	As a thickener or texture additive in processed foods, such as creamy fillings and sauces, ice cream, pudding, and salad dressing
Anchovies	Small, silver-colored fish	Worcestershire sauce, Caesar salad dressing, pizza topping, Greek salads
Animal shortening	Butter, suet, lard	Packaged cookies and crackers, refried beans, flour tortillas, ready-made pie crusts
Carmine (carmine cochineal or carminic acid)	Red coloring made from a ground-up insect	Bottled juices, colored pasta, some candies, frozen pops, "natural" cosmetics
Casein (caseinate)	A milk protein	As an additive in dairy products such as cheese, cream cheese, cottage cheese, and sour cream; also added to some soy cheeses, so read the label
Gelatin	Protein from bones, cartilage, tendons, and skin of animals	Marshmallows, yogurt, frosted cereals, gelatin-containing desserts
Glucose (dextrose)	Animal tissues and fluids (some glucose can come from fruits)	Baked goods, soft drinks, candies, frosting
Glycerides (mono-, di-, and triglycerides)	Glycerol from animal fats or plants	Processed foods
Isinglass	Gelatin from the air bladder of sturgeon and other freshwater fish	As a clarifying agent in alcoholic beverages, some jellied desserts

(continued)

Table 5-2 *(continued)*

Ingredient	*What It Is*	*Where You Find It*
Lactic acid	An acid formed by bacteria acting on the milk sugar lactose	Cheese, yogurt, pickles, olives, sauerkraut, candy, frozen desserts, chewing gum, fruit preserves
Lactose (saccharum lactin, D-lactose)	Milk sugar	As a culture medium for souring milk and in processed foods such as baby formulas and sweets
Lactylic stearate	Salt of *stearic acid* (tallow, other animal fats and oils)	As a conditioner in bread dough
Lanolin	Waxy fat from sheep's wool	Chewing gum
Lard	Fat from the abdomens of pigs	Baked goods, refried beans
Lecithin	Phospholipids from animal tissues, plants, and egg yolks	Breakfast cereal, candy, chocolate, baked goods, margarine, vegetable oil sprays
Natural flavorings, unspecified	Can be from meat or other animal products	Processed and packaged foods
Oleic acid (oleinic acid)	Animal *tallow* (solid fat of sheep and cattle separated from the membranous tissues)	Synthetic butter, cheese, vegetable fats and oils; spice flavoring for baked goods, candy, ice cream, beverages, condiments
Pepsin	Enzyme from pigs' stomachs	With rennet to make cheese
Propolis	Resinous cement collected by bees	Sold in chewing gum and as a traditional medicine in natural foods stores
Stearic acid (octadecanoic acid)	Tallow, other animal fats and oils	Vanilla flavoring, chewing gum, baked goods, beverages, candy
Suet	Hard white fat around kidneys and loins of animals	Margarine, mincemeat, pastries
Tallow	Solid fat of sheep and cattle separated from the membranous tissues	Margarine

Ingredient	What It Is	Where You Find It
Vitamin A (A1, retinol)	Vitamin obtained from vegetables, egg yolks, or fish liver oil	Vitamin supplements, fortification of foods such as breakfast cereals, low-fat and skim milk, and margarine
Vitamin B12	Vitamin produced by microorganisms and found in all animal products; synthetic form (cyanocobalamin or cobalamin on labels) is vegan	Supplements, fortified foods such as breakfast cereals, soymilk, and rice milk
Vitamin D (D1, D2, D3)	D1 is produced by humans upon exposure to sunlight; D2 (ergocalciferol) is made from plants or yeast; D3 (cholecalciferol) comes from fish liver oils or lanolin	Supplements, fortified foods such as milk, margarine, yogurt, and orange juice
Whey	Watery liquid that separates from the solids in cheese-making	Crackers, breads, cakes, processed foods

Communicating with food companies

If you aren't sure whether a product is meat-free, check company Web sites for ingredient information. If you don't find the answers you need, check the Web site or package for a phone number you can call to speak with a customer service representative.

Some ingredient questions may be beyond the ability of a customer service representative to answer. If you sense the individual you're speaking with doesn't know the answer, ask to speak with a supervisor. Another alternative: Put your question in writing and send a more formal inquiry to the company. Food companies take customer inquires about diet matters very seriously — lawsuits have resulted from false or misleading information about ingredients being given to consumers.

It may take some digging and persistence, but you can usually get answers to your questions about the sources of ingredients in food products.

Nonfood products that contain animal ingredients

Animal ingredients may turn up in some surprising places, including cosmetics, lotions, perfume, and household supplies such as glue and ink. Many vegans avoid such products, although it can be difficult for anyone to avoid them completely, considering the extent of their use.

Glycerides, lanolin, lecithin, oleic acid, stearic acid, tallow, and vitamin A, for example, are often used to make cosmetics. Lanolin is used in ointments and lotions, and many soaps contain oleic acid, stearic acid, or tallow. Read product labels carefully if you want to avoid these ingredients, and check with the manufacturer if you're in doubt about the source.

Tracking down resources for up-to-date advice

The Vegetarian Resource Group publishes the "Guide to Food Ingredients," listing the uses, sources, and definitions of more than 200 food ingredients, as well as whether ingredients are vegan, vegetarian, or nonvegetarian. You can order the guide online for a nominal cost at `www.vrg.org/catalog/fing.htm`.

The North American Vegetarian Society also provides information about hidden animal ingredients. Visit the organization online at `www.navs-online.org/`.

Vegans can find information about animal-free nonfood products by visiting People for the Ethical Treatment of Animals (PETA) online at `www.peta.org/`.

Applying More Advice for Getting Started

You're on your way! This section offers more tips to make the journey a little easier. Some people are their own worst critics. Although you don't want to delude yourself into thinking that you're making progress when you're not, you also don't want to be too harsh with yourself if you have a setback now and then. Change takes time and patience.

Occasional slips are normal. If you have a lapse, pick up where you left off and start again. No one is keeping score but you.

Scouting out supermarkets

Living vegetarian doesn't require you to buy special foods or ingredients. However, after you make the switch to a meatless diet, you'll find a big world of new products you'll probably want to try, as well as new ways to put familiar old ingredients together.

Now's the time to encourage creativity. Break with your usual shopping routine and visit some new stores. Walk up and down all the aisles, including the frozen foods case. You'll find dozens of vegetarian foods. Buy a few on this trip, and then try a few more next time.

Whole foods — products that are as close to their natural state as possible — are the healthiest. However, prepared foods and vegetarian specialty products — such as frozen entrées, veggie burger patties, and boxed mixes — can be an acceptable convenience. They're nice to have on hand for occasions when you don't have time to cook and need something quick and easy.

TIP

When you sample new products, expect to find a few duds. Next time, try another brand of the same type of product, because different brands of the same product can vary. Veggie burger patties and soymilk are two examples of products that vary greatly in flavor and texture, depending on the brand.

After you start experimenting with new foods, you'll discover a long list of great products that will become regulars in your kitchen. I cover much more about ingredients and shopping in Chapters 6 and 7.

Scanning cookbooks and magazines

Go to the library and check out all the vegetarian cookbooks and magazines you can find. If you have friends who are vegetarians, borrow their cookbooks. Then begin to page through each one.

Read the names of the recipes. If you see one that grabs you, look at the list of ingredients. If the recipe inspires you, make the dish. When you find a cookbook with lots of recipes that appeal to you, go to a bookstore and buy it, or order it online. Natural foods stores also carry a good selection of popular vegetarian cookbooks.

When you orient yourself to the options on a vegetarian diet, you may be surprised at the variety. You may also notice lots of ethnic foods from cultures that have vegetarian traditions. A vegetarian diet has more variety than a meat-centered diet, and reviewing cookbooks will help you see that.

Listing vegetarian foods you already like

If you're new to a vegetarian diet, it's helpful to make a list of the vegetarian foods you already enjoy. Fix them often. You probably eat a number of meatless foods and haven't thought of them as being vegetarian until now. Here are some examples:

- Baked potato topped with broccoli and cheese
- Bean burritos and tacos
- Grilled cheese sandwiches
- Lentil soup
- Macaroni and cheese
- Pancakes
- Pasta primavera
- Spaghetti with tomato sauce
- Vegetable stir-fry over rice
- Vegetarian lasagna

If your favorite vegetarian foods happen to contain dairy products such as cheese or milk, gradually reduce the amount and switch to nonfat varieties. You can probably modify some of your other favorite foods to make them vegetarian. Some examples:

- **Bean soups:** Leave out the ham or bacon.
- **Breakfast meats:** Use veggie versions of link sausage or sausage patties and bacon. They taste great and can replace meat at breakfast or in a BLT.
- **Burgers:** Buy veggie burger patties instead of meat.
- **Chili:** Make it with beans, tempeh, or textured vegetable protein instead of meat.
- **Hot dogs and luncheon meats:** Buy veggie versions. All natural foods stores carry them, and so do some supermarkets. They look and taste like the real thing but are far better for you.
- **Meatloaf:** Even this all-American food can be made with lentils, chopped nuts, grated vegetables, and other ingredients. Many vegetarian cookbooks include a number of delicious variations. Try the recipe for Everybody's Favorite Cheese and Nut Loaf in Chapter 15.

- **Pasta sauces:** Why add meat or meat flavorings when you have basil, oregano, mushrooms, red peppers, sun-dried tomatoes, pimentos, black olives, and scores of other delicious ingredients to add flavor to sauces? Great on cannelloni, stuffed shells, manicotti, ravioli, fettuccine, and steamed or roasted vegetables.
- **Sandwiches:** Load up on lettuce, tomatoes, mustard, chopped or shredded vegetables, and a little cheese if you like, but leave out the meat. Use pita pockets, hard rolls, and whole-grain breads to give sandwiches more character.
- **Stuffed cabbage:** Instead of using pork or beef, fill cabbage leaves with a mixture of rice and garbanzo beans and seasonings.

Part II
Planning and Preparing Your Vegetarian Kitchen

"Let's see—maybe there's a better place for your vegetarian cookbook than on the butcher block between the steak knives and the A1 sauce?"

In this part . . .

Consider your kitchen Command Central. It's the heart of your vegetarian food operation — where you prepare delicious and nutritious meals for you, your family, and guests.

By preparing and eating more of your meals at home, you'll not only save money but also have more control over the composition of your diet. That's why it's so important that you have what you need at your fingertips.

In this part, I coach you on how to set up your kitchen for maximum efficiency. You want to have on hand the most common and versatile ingredients for vegetarian meals. I help you understand where to shop to get what you need, and I give you strategies for getting the best values. I cover the practical equipment you should have and the basic cooking techniques you should be familiar with to get the best results from your kitchen adventures.

Chapter 6

Getting Familiar with Common Vegetarian Ingredients

In This Chapter

- Identifying the staples of the vegetarian diet
- Adding specialty products to your diet
- Understanding natural and organic labels

Vegetarian meals don't have to be exotic, made with ingredients you can't pronounce or buy at your local supermarket. In fact, they can be as all-American as corn on the cob, baked beans, and veggie burgers.

But if you're interested, vegetarian cooking can be an introduction to a whole new world of flavors, textures, and aromas based on culinary traditions from around the world. Many vegetarian cookbooks draw on these traditions, which contribute greatly to the diversity of foods that many vegetarians enjoy.

In this chapter, I describe useful ingredients in vegetarian cooking. You may already use some, but others may be unfamiliar to you. All are healthful, multipurpose ingredients worth knowing about and keeping on hand. I also shed some light on the often-confusing topic of natural and organic products to help you make well-informed choices when you shop.

Building the Foundation of the Vegetarian Diet

It's important to keep on hand ingredients in a variety of forms. Fresh is best, of course, but fresh isn't always the most convenient option. So plan to keep a range of fresh, frozen, canned, and packaged foods in your kitchen.

Bringing home the beans

Beans are highly nutritious and extremely versatile. You can mix several types of beans (such as garbanzos, pintos, red kidney beans, and white kidney beans) to make many-bean chili. You can mash them for dips, soups, and spreads. You can combine them with rice, pasta, or other grains to make a variety of interesting vegetarian entrées. Among others, try the following:

- Black beans
- Black-eyed peas
- Cannellini beans (white kidney beans)
- Garbanzo beans
- Kidney beans
- Navy beans
- Pinto beans

Forms of beans

Beans come in several forms, all of which are useful in vegetarian cooking. Here's a rundown on the types you'll see at the store:

- **Canned beans:** Canned beans are one of the most nutritious and versatile convenience foods you can buy. Conventional supermarket varieties are fine, although natural foods stores carry a wide variety of organic canned beans.

  Rinse canned beans in a colander before using them. Rinsing removes excess salt and the fibrous outer skins of the beans that otherwise tend to flake off and look unsightly in finished dishes.

  If you have leftover canned beans, add them to a can of soup or stir them into pasta sauce to use them up. You can also spoon them on top of a green salad or mash them with a little garlic and add them to a casserole.

- **Dried beans:** Beans were sold dried in bags long before canned beans were available. Some people still prefer to buy and prepare their beans the old-fashioned way. Preparing dried beans is more time-consuming, but some people prefer the flavor and texture that result.

  Dried beans are often a little firmer than canned beans. In fact, they can be too hard if you don't soak them long enough before cooking them. Also, dried beans don't usually contain added salt; you have to season them yourself or they're bland. (The exception is bags of beans packaged with their own spice mixtures, which are often very salty.)

- **Frozen beans:** You can find a few types of beans, such as lima beans (also called butter beans) and black-eyed peas, in the freezer section of some supermarkets. They taste the same as fresh and are just as nutritious. Compared to canned beans, frozen beans have less sodium and may be slightly more nutritious, because nutrients can leach out of canned beans into the water in the can.
- **Bean flakes:** This form of dried beans is convenient and fun to use. You just shake some into a bowl, add boiling water, stir, and cover. In five minutes, you have a smooth, creamy bean paste that you can use to make nachos, burrito or taco filling, bean dip, or bean soup. The more water you add, the thinner the beans. The bean flakes themselves keep in the cupboard for months, just like dried and canned beans. After you mix the bean flakes with water, leftovers keep in the refrigerator for up to a week or in the freezer for up to six months.

  You're more likely to find bean flakes at a natural foods store than in a supermarket. Fantastic Foods and Taste Adventure are two brands that make both dehydrated black bean flakes and pinto bean flakes.

Whether frozen, canned, or dried, beans have approximately the same nutritional value. Convenience and flavor are the major differences.

Common dishes from around the world

Beans are used in many different cultures to make a wide range of traditional foods and dishes. Examples include:

- American navy bean soup
- Chinese stir-fried tofu and vegetables
- Ethiopian puréed lentil and bean dishes
- Greek lentil soup
- Indian curries and *dal* (Indian-style lentil soup)
- Indonesian tempeh
- Italian *pasta e fagioli* (pasta with beans)
- Mexican burritos, tacos, and nachos
- Middle Eastern hummus and *falafel* (small, round, fried cakes made from mashed garbanzo beans and spices)
- Spanish black bean soup

You can substitute many types of beans for others in recipes. For example, you can easily substitute black beans for pinto beans in tacos, burritos, nachos, and bean dip. The dark color often makes a striking contrast to other components of the meal, such as chopped tomatoes, salsa, or greens.

Making vegetarian-style beans and franks

If you loved beans and franks as a nonvegetarian, you'll love this vegetarian version. Add two sliced meatless vegetarian hot dogs, a handful of minced onion, and a couple of tablespoons of molasses to a 15-ounce can of vegetarian baked beans. If you like, you can also add a couple of tablespoons of ketchup and dark brown sugar. Stir, pour into a 1½-quart casserole dish, and bake, uncovered, at 350 degrees for 40 minutes, or until hot and bubbly.

Eating more vegetables and fruits

The best vegetables and fruits are locally grown, in season, and fresh. Most of them are packed with vitamins, minerals, and phytochemicals. Frozen foods run a close second in terms of nutrition, and they're a perfectly acceptable alternative to fresh.

However, canned vegetables and fruits are sometimes more convenient or less expensive than fresh or frozen. Because heat destroys some vitamins, and time on the shelf can allow some nutrients to leach into the water used in packing, canned vegetables and fruits tend to be less nutritious than fresh or frozen. However, that doesn't mean that they're worthless! On the contrary, canned produce can make a sizable nutritional contribution.

Varying your vegetables

Vegetarian cooking incorporates many kinds of vegetables, but some of the more popular recipe ingredients include white potatoes, sweet potatoes, onions, bell peppers, tomatoes, greens, broccoli, and many varieties of squash. Ethnic dishes make use of such less common vegetables as bok choy (in stir-fries), kohlrabi (in stews and casseroles), and arugula (steamed or in salads). I include recipes incorporating some of these in Part III.

When you use canned vegetables, you can save vitamins and minerals that may have leached out by incorporating the liquid from the can into the dish you're making. For example, if you're making soup or sautéed vegetables to serve over rice or pasta, dump the entire contents of the can into the pot. The packing liquid may contain salt, but you can compensate by leaving out any additional salt that's called for in the recipe.

Some vegetables are worth extra attention because of their superior nutritional value. Deep yellow and orange vegetables, for instance, are particularly rich in beta carotene, which your body converts into vitamin A. Sweet potatoes, acorn squash, tomatoes, red peppers, and pumpkin are examples. Other vegetables are potent sources of vitamin C, including white potatoes, green bell peppers, broccoli, cauliflower, cabbage, and tomatoes, which are high in both beta carotene *and* vitamin C.

Cruciferous vegetables are another example of *super vegetables.* They contain cancer-fighting substances as well as copious amounts of vitamins and minerals. Following are a few popular examples from this group:

- **Arugula** (pronounced a-*roo*-guh-la) is a leafy green vegetable that's distinctive for its sharp, peppery flavor. It grows wild throughout southern Europe and is native to the Mediterranean. You eat arugula raw, and you can mix it with other salad greens or let it stand alone. I include a recipe for arugula salad in Chapter 11.
- **Daikon** (or Chinese radish) is a large, white, Asian radish. In contrast to the small red radishes that we know in North America, daikon is usually eaten cooked. However, it's sometimes used raw in salads.
- **Kale, Swiss chard, and bok choy** (also known as Chinese cabbage) are thick, green, leafy vegetables that are best when cut or torn into strips and steamed, sautéed, or stir-fried. Many vegetarians add a little olive oil and minced garlic to these cooked greens. You can eat them plain or mix them into casseroles, soups, and other dishes.
- **Kohlrabi** is a white, turnip-shaped vegetable commonly eaten in Eastern Europe. Like other cruciferous vegetables, it's full of potentially cancer-preventing phytochemicals and is very nutritious and flavorful. Kohlrabi is usually diced, boiled, and eaten cooked or added to stews.

Picking fruits

You can use fruits in vegetarian meals in many creative ways — to make cold soups, warm cobblers, pies, and quick breads, for instance. I like to add bits of fruit — mandarin oranges, strawberries, and chunks of apple and pear — to green salads. Many of the dessert recipes in Chapter 14 call for fruit, which cuts the calories and adds important nutrients to sweet treats.

Strawberries and grapes often contain substantial amounts of pesticide, and sometimes people don't wash them well before eating them. Wash all fruits under running water, rubbing with your hand or a soft cloth, or brush the fruit to remove pesticide residues, especially if you're going to eat the skin. In addition, although you may not eat the peel, be sure to wash the outsides of such fruits as melons, grapefruit, and oranges before setting them on your cutting board to slice them. If you don't, you may contaminate your cutting board surface. You may also drag any bacteria or contaminants present on the fruit's surface into the edible inner portion.

Choose plenty of deeply colored fruits to get the benefit of their rich nutritional content. Apricots, peaches, papayas, and mangoes, for example, are densely packed with many important vitamins and minerals. Watermelon, cantaloupe, oranges, and grapefruits are also super healthful.

Choosing breads and cereals

Grains are versatile — you can eat them by themselves or combine them with other ingredients to make a seemingly endless collection of dishes. Grains can be an accompaniment to a meal, such as a slice of bread or a couscous salad, or they can be the foundation of an entrée, such as rice pilaf or a plate of spaghetti. Around the world, grains are familiar ingredients in soups, salads, baked goods, main dishes, side dishes, desserts, and other foods. Examples include:

- Chinese vegetable stir-fry with steamed white rice
- Ethiopian *injera* (flatbread made from teff flour)
- Indian saffron rice with vegetables
- Indonesian *tempeh* (made with whole soybeans and a grain such as rice)
- Irish porridge (oatmeal)
- Italian *polenta* (typically made with cornmeal, baked or pan-fried, and often served with a sauce on top)
- Mexican enchiladas made with corn tortillas
- Middle Eastern couscous with cooked vegetables
- Middle Eastern *tabbouleh* (wheat salad)
- Russian or Eastern European *kasha* (toasted buckwheat) with bowtie pasta and cooked mushrooms
- Spanish rice
- Thai jasmine rice with curried vegetables

Grains come in two forms: whole and processed. Both forms are indispensable in vegetarian cooking.

Whole grains are grains that have had only the outer hull removed, leaving a small, round kernel, often called a *berry.* The major advantage to using whole grains in cooking is that they're slightly more nutritious than processed grains, *and* they taste better. After you get used to the flavor of whole-wheat bread, for example, you'll probably prefer it to soft, white bread.

The major drawback to whole grains is that they take more time to cook than processed grains. Whole-grain hot cereal, for example, may take 10 minutes to cook, compared to the 1 or 2 minutes it takes to fix instant hot cereals.

You can find many types of grains at most supermarkets and natural foods stores. Some may be new to you. Here are some examples of how they're used:

- **Amaranth:** An ancient grain eaten for centuries in Central and South America. You can cook it and eat it as a side dish or hot cereal, or use it to make casseroles, baked goods, crackers, pancakes, and pasta. Use it as you'd use rice in recipes. Amaranth flour is also available.
- **Barley:** Barley comes in two forms: hulled and pearl.
  - Hulled barley is the more natural, more nutritious, unrefined form, having only its outermost husks removed. Not surprisingly, hulled barley is brown in color. You can use it in soups, casseroles, and stews.
  - Pearl barley is more processed than hulled barley. In addition to having the inedible outer hull removed, pearl barley also has the outer bran layer removed. Pearl barley is less nutritious than hulled barley, but it cooks faster.
- **Corn:** Corn kernels are actually the seeds of a cereal grass. You can use corn in cornbread, corn tortillas, and polenta.
- **Kamut:** An ancient type of wheat used in Europe for centuries, usually in baked goods.
- **Millet:** Tiny, beadlike seeds that come from a cereal grass. You use millet in baked goods and as a cooked cereal.
- **Oats:** Oats come from cereal grasses. Use oats as a hot cereal and in baked goods, veggie burger patties, and loaves.
- **Quinoa** (pronounced *keen*-wah): An ancient grain eaten for centuries in Central and South America. You can cook quinoa and eat it as a side dish or hot cereal, or you can use it to make casseroles, salads, a side dish with cooked vegetables, or pilaf. Use it as you'd use rice in recipes.

  Quinoa has to be processed to remove the seed coat, which contains saponin and is toxic. Before you cook quinoa, rinse it several times with water to remove any remaining saponin. The saponin is soapy; you'll know that you've rinsed the quinoa well enough when you no longer see suds. Packaged quinoa is often pre-rinsed and needs only minimal rinsing. Quinoa purchased from bulk bins often is not pre-rinsed.
- **Rice:** Rice has been grown for centuries in warm climates, where the seeds are harvested from cereal grasses that grow in wet areas called *rice paddies.* If you're a diehard white rice eater, that's fine. However, brown rice has a slightly better nutritional profile, and many people like its hearty flavor. Several quick-cooking brown rice varieties are on the market; you can find them at regular supermarkets and natural foods stores.

Supermarkets also carry many specialty rice varieties, such as aromatic jasmine rice, basmati rice, and *arborio rice* (a medium-grain rice used to make the Italian dish risotto). Jasmine and basmati rice are available in white or brown form; the brown form is slightly less refined. You can use these various forms of rice in casseroles, baked goods, puddings, and side dishes.

Rice is one of the least allergenic and most easily digestible foods. That's why rice cereal is one of the first solid foods introduced to infants. People who are allergic to wheat and other grains can use rice flour to make baked goods such as breads and cookies.

- **Rye:** A form of cereal grass grown for its seeds, which are ground to make flour. You use rye in baked goods, such as bread and rolls.
- **Spelt:** An ancient type of wheat that has been used in Europe in baked goods for centuries.
- **Teff:** One of the oldest cultivated grains in the world, teff is used in Ethiopia to make *injera,* a flat, spongy bread that's a staple food there. You can also use it in baked goods, soups, and stews.
- **Wheat:** Wheat is grown around the world as a cereal grass and used in baked goods and pasta.
- **Wheat berries:** Whole-wheat grains that have had only the outer hull removed. Use wheat berries in baked goods and as a hot cereal.

When you make a pot of rice (or any grain), make more than you need for one meal. You can store the leftovers in an airtight container in the refrigerator for up to two weeks. You can reheat leftover rice and serve it with steamed vegetables, bean burritos, or tacos; top it with vegetarian chili; add it to a filling for cabbage rolls; or use it to make rice pudding.

Selecting seeds and nuts

Seeds and nuts are excellent additions to many vegetarian dishes because they're rich in protein and other nutrients, including health-supporting phytochemicals. A handful of chopped walnuts or almonds or a sprinkling of toasted sunflower seeds makes a nice addition to a casserole, salad, or bowl of hot cereal.

Here are a couple of products made from seeds and nuts that are common in vegetarian cooking:

- **Tahini:** *Tahini* is a smooth, rich paste made from sesame seeds. It's a traditional Middle Eastern ingredient in *hummus* — a spread made with garbanzo beans, garlic, lemon juice, and olive oil — and you can also use it to flavor salad dressings, sauces, dips, and other foods. Tahini is sold in cans or bottles. A layer of sesame oil usually floats at the top of the container; you have to stir it in each time you use the product.

- **Nut butters:** Don't stop with peanut butter. Cashew butter and almond butter are equally, if not more, nutritious, and they're just as delicious. You can use them in all the same ways that you use peanut butter — for sandwiches, cookies, and in cooking.

If you buy a jar of natural peanut butter, you'll notice a layer of oil at the top. Don't make the mistake of pouring the oil off, thinking that you'll reduce the fat content. If you do, you'll need a pickax to get the dry nut butter out. Storing the jar upside down to keep the oil dispersed is also risky — the oil may ooze out onto the shelf. Instead, just stir the oil back into the nut butter when you're ready to use it.

Fitting in Specialty Foods and Products

Living vegetarian can be extraordinarily simple. No fancy equipment or special food products are necessary — unless you want them.

Several specialized products do exist that many vegetarians find especially convenient. All these products are nutritious, and many are much better for your health than their animal-product counterparts. You can find most of the products I mention in this section in your neighborhood supermarket; you can find them all in a natural foods store.

Introducing soy foods and their variations

Soy foods come in dozens of forms and can take the place of meat, cheese, eggs, milk, and other animal products in recipes. Nutritionally, soy products are far superior to their meat counterparts because they're cholesterol-free, low in saturated fat, and usually lower in sodium. They're also free of *nitrates* and *nitrites,* preservatives that are commonly found in processed meat products and that may cause cancer.

Table 6-1 lists several types of soy products that are available and ways you can use them.

Table 6-1 Soy Foods and Their Uses

Product	*Description and Uses*
Meat substitutes	Meat substitutes are made from soy protein, tofu, and vegetables. Some are made to resemble traditional cold cuts, hot dogs, sausage, and bacon in appearance, texture, flavor, and aroma. You can use them in the same ways as their meat counterparts.
Miso	A rich, salty, East Asian condiment used to flavor soups, sauces, entrées, salad dressings, marinades, and other foods. This savory paste is made from soybeans, a grain (usually rice), and salt. It's combined with a mold culture and aged for at least one year. It doesn't have much nutritional value and is high in sodium, so, like most condiments, you should use it sparingly.
Soy cheese	A substitute for dairy cheese that comes in many different forms, such as mozzarella style, jack style, American style, and cream cheese style. You can use it to make sandwiches, pizza, casseroles, and spreads.
Soy margarine	Find it in natural foods stores. It's free of casein and other dairy byproducts that other brands of margarine usually contain, so it's vegan-friendly.
Soy mayonnaise	Several brands are available at natural foods stores. Use soy mayonnaise in all the ways you use regular mayonnaise.
Soymilk	A beverage made from soaked, ground, and strained soybeans. You can use it cup for cup in place of cow's milk in recipes, or drink it as a beverage.
Soy sauce	A rich, salty, dark brown liquid condiment made from fermented soybeans. Use it to add punch to stir-fried vegetables and as a dip for dumplings and egg rolls.
Soy yogurt	Made from soymilk, it often contains active cultures and is available in many flavors. Use it as a substitute for dairy yogurt.
Tamari	A type of soy sauce and byproduct of the production of miso. It has a rich flavor.
Tempeh	A traditional Indonesian soy food made from whole soybeans mixed with a grain and a mold culture, fermented, and pressed into a block or cake. You can grill it and serve it as an entrée or use it in sandwich and burrito fillings, casseroles, chili, and other foods.

Product	Description and Uses
Textured soy protein (TSP) or textured vegetable protein (TVP)	A product made from textured soy flour that's usually sold in chunks or granules. It takes on a chewy, meatlike texture when rehydrated and is used in such foods as vegetarian chili, vegetarian sloppy Joes, veggie burgers, and other meat substitutes.
Tofu	Soybean curd that's made in a process similar to cheese making, using soymilk and a coagulant. It's bland-tasting and is available in a variety of textures and densities. You can use it in numerous ways: cubed and added to a stir-fry, marinated and baked, as a substitute for eggs and cheese, as a sandwich filling, and as an ingredient in dips, sauces, desserts, cream soups, and many other foods.

You can use different styles of tofu in different recipes. For instance, soft tofu works well for making dips and sauces because it's easy to blend. Firm tofu works well in baked goods and as an egg replacer in many recipes. Extra-firm tofu is the style of choice for making stir-fries because the cubes can hold up when they're jostled around in the pan. Unlike cheese, tofu is never coagulated with rennet, an enzyme taken from the stomach linings of baby animals. It's always safe for consumption by vegetarians and vegans alike.

Like cheese, eggs, meat, and other high-protein foods, tofu and tempeh can spoil if left unrefrigerated. Don't leave out foods made with tofu and tempeh at room temperature for more than two hours. After a meal, cover the leftovers and put them back in the refrigerator as soon as possible.

Making the most of milk substitutes

In addition to the soymilk, soy cheese, and soy yogurt listed in Table 6-1, you may also be interested in rice, almond, and oat milks packaged in *aseptic* (shelf-stable) packages. People who are allergic to soy often use these alternatives.

You can buy potato milk in natural foods stores in powdered form. Some soymilk is still sold in powdered form, too. The liquid soymilk is more palatable and much more convenient, though. Also try soy and rice frozen desserts and soy-based nondairy coffee creamer.

Soymilk is generally more nutritious than any of the other milk alternatives, such as rice milk and almond milk. Fortified soymilk has extra calcium and vitamins A, D, and B12. If you use substantial amounts of a milk alternative, fortified soymilk is usually the best choice in terms of nutrition.

If you try soymilk for the first time and don't care for it, give it a second chance and try another brand. Soymilk varies in flavor considerably from one brand to the next. You may have to taste three or four before you find your favorite. Most of the soymilk sold in the United States and Canada is made from whole soybeans, so it has a bit of a beany aftertaste, which some people find pleasant but others don't like. If you object to the mild bean flavor, you may prefer a flavored soymilk to plain. Soymilk does grow on you — if you use it for a while, you'll probably grow to love it.

If you do include dairy products such as cheese, milk, and ice cream in your diet, buy the nonfat varieties. Even the so-called low-fat items are too high in fat. The problem with dairy fat is that two-thirds of it is saturated, so the fat in dairy products is a real artery-clogger.

Incorporating egg replacers

Powdered vegetarian egg replacer is an egg substitute made from a mixture of vegetable starches. You just mix a teaspoon and a half of the powder with two tablespoons of water to replace one egg in virtually any recipe. It keeps on a shelf in your cupboard almost indefinitely and is always on hand when you need it. It's free of saturated fat and cholesterol and doesn't pose a salmonella risk. Ener-G Egg Replacer is a common brand that you can find in most natural foods stores.

Considering meatless burgers, dogs, sausages, and cold cuts

Some vegetarian burger patties are made from soy (see Table 6-1), and some are grain- or vegetable-based. A wide variety is available, and they all taste different, so you need to experiment to find your favorites. Some are meant to look and taste like meat, while others don't look or taste anything like meat. Vegetarian hot dogs are typically made of soy, and they taste and look like the real thing. You can cook meat substitutes on a grill and serve them the same way that their meat counterparts are served.

One of the advantages of these products is that you can take them to picnics and cookouts and join in on the food fun with all the nonvegetarians. One note: Vegetarians will want to reserve one side of the grill — or an entire grill — to themselves so that their veggie burgers and dogs aren't cooked on the same surfaces as meat. Call it a "meat-free" zone!

Meatless sausages and cold cuts are also versatile. Let these fakes stand in for their meat counterparts in sausage biscuits, sandwiches, on top of pizza, and in other creative ways. The nutritional advantage of these foods is that they're low in saturated fat, they're cholesterol-free, and many even contain a good dose of dietary fiber. They're quick, convenient, and good tasting, and they work well as transition foods for people making the switch to a vegetarian diet.

Including other vegetarian convenience foods

Seitan (pronounced *say*-tan) is a chewy food made from wheat *gluten,* or wheat protein. You can buy mixes for seitan at natural foods stores, but most people find it far more convenient to buy it ready-made. Look for it in the refrigerated sections of natural foods stores.

One of the best ways to sample seitan is at a Chinese or Thai restaurant, where you can try it prepared in different ways. It's often served in chunks or strips in a stir-fry.

Seitan dishes are absolutely scrumptious. Don't be surprised if you get hooked on the stuff. But because seitan is relatively unfamiliar to most Westerners, you aren't likely to find it in small-town or even medium-sized-city restaurants. You'll usually see it in big-city Asian restaurants where the menu is fairly extensive.

After you've been living vegetarian for a while, you'll discover many more interesting and delicious vegetarian specialty products.

Exploring Natural and Organic Alternatives

Living vegetarian doesn't necessarily mean that you have to buy and eat natural or organic foods. If you *are* a vegetarian, though, or if you're just interested in eating healthfully, you probably want to know something about these distinctions. In this section, I help you understand what these terms mean and how to assess their importance.

Going au naturel

No legal definition exists for the term *natural foods,* but the term is generally understood to mean that a food has been minimally processed and is free of artificial flavors, colors, preservatives, and any other additives that don't occur naturally in the food.

Assessing the benefits of natural foods

Natural foods are frequently better choices than their mainstream counterparts because they lack the undesirable ingredients of many commercially processed foods and they contain more nutritious ingredients.

For instance, natural breakfast cereals, baking mixes, and bread products contain no partially hydrogenated fats, so they're trans fat-free. They're usually made with whole grains, so they contain more dietary fiber, vitamins, and minerals than products made with refined grains.

Natural foods are usually minimally sweetened, and when they are sweetened, less sweetener is used. Fruit juice is frequently used as a sweetener in lieu of refined sugar. Natural foods also typically contain less sodium than other foods, and they contain no monosodium glutamate or nitrites. The ingredients used in natural foods products are often organically grown, too.

Knowing when natural isn't enough

Just because a food is natural doesn't mean it's nutritious. Believe it or not, natural junk foods *do* exist. Many natural sweets, candies, and snack foods have little nutritional value, just like their conventional counterparts. Even though they're natural, you should limit your consumption of them. Don't let these foods displace more nutritious foods in your diet.

Finding natural foods

Even within a particular category of food products, you usually find more healthful choices in a natural foods store than in a regular supermarket. For instance, in a regular supermarket you may find several types of whole-grain pasta. In a natural foods store, you're likely to find an entire shelf of them. The same is often true of breakfast cereals and many other product categories.

Opting for organic

The word *organic* applies to foods that are grown without persistent or toxic fertilizers and pesticides, using farming methods that are ecologically sound and earth-friendly. Organic foods must be produced and processed without the use of antibiotics, synthetic hormones, genetic engineering, sewage sludge, or irradiation.

The widest variety of organic products is available at natural foods stores, but most conventional grocery stores now also stock organic produce and organic processed and packaged foods. Natural foods stores in particular often carry seasonal organic produce from local farmers. Buying locally supports small farmers in the community and helps ensure the freshness of the products.

If you don't have access to a big natural foods store with a good supply of organic produce, seek out local farmers at roadside stands or at farmer's markets where you can buy locally grown fruits and vegetables in season. Ask the farmer about his or her use of synthetic fertilizers and pesticides. Some small farmers don't go to the hassle and expense of applying for organic certification of their crops, even though they use sustainable farming methods and their products may, in fact, be organic. Get to know your farmer, and ask.

Organically grown fruits and vegetables generally have the same nutritional value as conventionally grown produce, but they're often not as pretty as those grown with chemical fertilizers and pesticides. They may have small blemishes, and they won't be glossy like their nonorganic counterparts, which often have a waxy coating added. Don't let the appearance of organically grown foods deter you from buying them.

Deciphering organic labels

For decades, the natural foods industry self-regulated and adhered to its own high standards for organic farming and food processing practices. In recent years, though, the market for organic foods has grown dramatically. As more commercial interests entered the organic foods market, longstanding organic standards became threatened.

In conjunction with the 1990 Farm Bill, the U.S. Congress passed the Organic Foods Production Act. Years of public discussion and debate followed, including considerable debate and disagreement between longstanding organic foods advocates and commercial newcomers who wanted to loosen the standards. In 2002, the U.S. Department of Agriculture (USDA) implemented a final rule establishing national organic standards. The rule maintains the high standards originally established by the old-guard natural product industry.

Organic growers and food manufacturers can have their products certified as "USDA Organic" through third-party state and private agencies accredited by the USDA. Labeling regulations permit products to be labeled "100 Percent Organic" if all ingredients meet the organic standards. Both these products and those containing at least 95 percent organic ingredients (the remaining 5 percent must also be approved for use in organic products) may display the "USDA Organic" label.

Products that contain at least 70 percent organic ingredients may be labeled as such on package ingredient lists.

Weighing the choice: Organic versus conventional produce

I'll convey the most important point first: Eat more fruits and vegetables, whether they're organic or conventional. Most of us need to increase substantially our intake of produce, and that's easier to do for vegetarians than meat-eaters.

After you've done that, you can think about whether you want to eat organic or conventional produce.

Very few studies have looked at the potential health effects of long-term exposure to low doses of pesticides, herbicides, and other contaminants, but it makes sense to avoid unnecessary exposure when possible. You may never know whether your diet ultimately delayed or prevented illness, but it seems prudent to err on the side of safety.

On the other hand, organic foods have disadvantages, too: cost and access. Not everyone can afford to pay extra for organic. And despite the growth in sales of organic foods, not everyone lives in neighborhoods where supermarkets stock a big supply of them.

If you don't have access to organic produce or can't afford it, cheer up. If you eat a diet that's high in fruits, vegetables, and whole grains, the fiber content of your diet will help remove environmental contaminants from your body. Roughage binds with contaminants and helps move waste through your system faster, leaving contaminants less time to be in contact with the lining of your intestines.

You can also take a practical approach to organics and avoid the dirty dozen — the 12 fruits and vegetables that are most likely to be contaminated, according to the Environmental Working Group. Avoid the conventional versions of these items if you can and instead buy organic: apples, bell peppers, carrots, celery, cherries, imported grapes, kale, lettuce, nectarines, peaches, pears, and strawberries. A pocket guide or iPhone application is available online from the EWG at `www.foodnews.org/index.php`.

Chapter 7

Shopping and Stocking Your Vegetarian Pantry

In This Chapter

- Buying the food items you need
- Deciding where to shop
- Using shopping strategies
- Saving money when shopping

Much of the challenge of eating well on a vegetarian diet is having the staple ingredients on hand when you need them. Shopping for these supplies can be a rewarding adventure, partly because living vegetarian can open you to culinary options you may never have considered.

This chapter is devoted to helping you understand how to go about the task of getting the right supplies into your house — at the right price. I encourage you to go beyond your neighborhood supermarket to explore places you may never have set foot in to search for good foods. I also cover some important shopping strategies, including ways to save money without sacrificing quality or nutritional value.

Figuring Out What You Need

Before you head to the store, you should have a concrete plan for what you want to buy. You'll be more likely to end up having the range and types of foods you need to build healthy meals at home, and you'll be less likely to come home with impulsive, less-than-ideal purchases, or to forget to buy a key ingredient.

The first step in creating a plan is to look at your overall nutritional goals. From there, you draft a shopping list.

Sketching out your meal plans

In Chapter 3, I cover the basics of vegetarian nutrition and provide background information about meeting your needs for vitamins, minerals, protein, and other components of a health-supporting diet.

This nutrition information is shown in the quick-reference guide in Figure 7-1. Use this information to help you think about the foods that belong on your plate and on your shopping list.

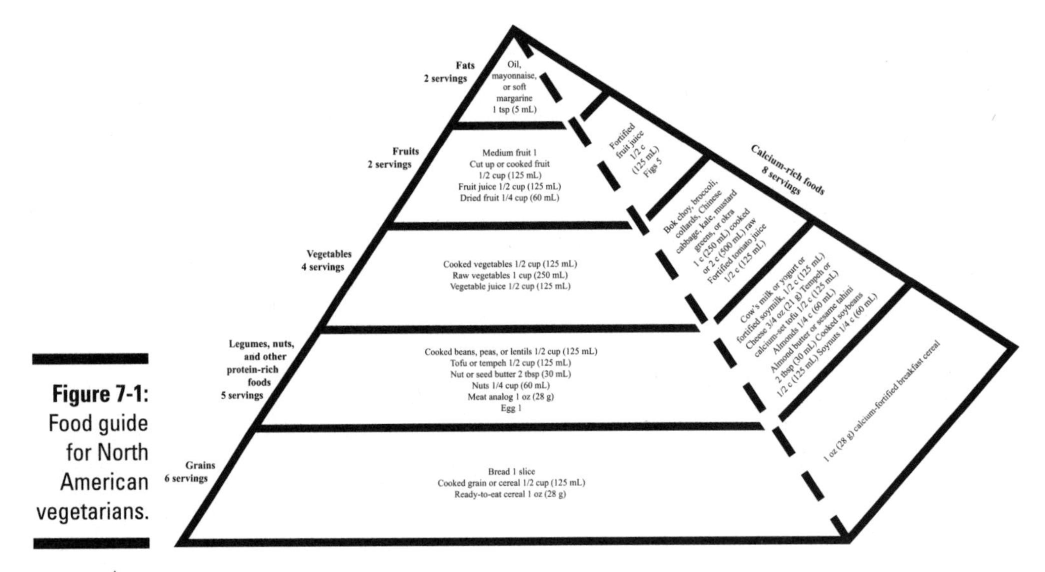

Figure 7-1: Food guide for North American vegetarians.

Use the following tips in conjunction with Figure 7-1 to help you sketch out your meal plans:

- Eat a variety of foods, and get enough servings to meet your energy needs.
- Eat at least 8 servings of calcium-rich foods from the food guide in Figure 7-1. These foods can pull double-duty and count as servings from other food groups at the same time. For example, a serving of calcium-fortified orange juice also counts as a serving from the fruit group.

- Every day, include 2 servings of foods that contain n-3 fats. Find these foods in the legumes/nuts group and in the fats group. Examples of one serving of these foods include the following:
  - 1 teaspoon (5 milliliters) of flaxseed oil
  - 3 teaspoons (15 milliliters) of canola or soybean oil
  - 1 tablespoon (15 milliliters) of ground flaxseed
  - ¼ cup (60 milliliters) of walnuts

  Use primarily canola oil and olive oil for cooking to help ensure an optimal balance of fatty acids in your diet.
- You can count nuts and seeds as servings from the fat group.
- Be sure to get enough vitamin D through sun exposure, eating fortified foods, or taking a supplement.
- Get at least three reliable food sources of vitamin B12 in your diet every day. One serving is equal to the following:
  - 1 tablespoon (15 milliliters) of Red Star Vegetarian Support Formula nutritional yeast
  - 1 cup (250 milliliters) fortified soymilk
  - ½ cup (125 milliliters) cow's milk
  - ¾ cup (185 milliliters) yogurt
  - 1 ounce (28 grams) of fortified breakfast cereal

  If you don't consistently get enough vitamin B12 from food sources, take a B12 supplement. Aim for 5 to 10 micrograms each day, or take a weekly supplement of 2,000 micrograms.
- Keep your intake of sweets and alcohol to a minimum to help ensure that you have enough room in your diet for the nutritious foods listed in the food guide.

Keeping a grocery list

People have different shopping styles, just as they have different personalities. Some people like to plan their week's meals, draw up a corresponding shopping list, and buy only what they need to make the meals they've planned. Other people prefer the casual approach — walk up and down the aisles, and put whatever strikes their fancy into their basket. At home, they decide what they're having for dinner based on what they feel like making and the supplies they have on hand.

Do what works for you. However, be aware that maintaining a list does have its advantages, especially when you've made a change in your eating habits and are developing new eating skills. Planning ahead gives you more control over your meals.

The grocery lists that follow are only suggestions of what you may want to buy to stock your vegetarian kitchen. You can adapt them to suit your individual preferences. If you like them just the way they are, you may want to photocopy them and make enough copies to last a few months. Tape a fresh list to the door of your refrigerator after each shopping trip. Check off items that you need and add any that aren't listed. I provide separate lists for items that you buy weekly and items that you buy less often.

Shopping by the week

You have to purchase the foods on this list fairly frequently because they're perishable and will keep in the refrigerator for only a week or two before spoiling. Use the list to help you decide which foods you want to buy this week. Vary your choices depending on seasonal availability of fruits and vegetables or just to help ensure you get a good mix of nutrients in your diet.

Fresh Fruits (especially locally grown)

Apples

Apricots

Bananas

Blueberries

Cantaloupe

Cranberries

Grapefruit

Grapes

Honeydew

Kiwi

Lemons

Limes

Mangoes

Nectarines

Oranges

Peaches

Pears

Pineapples

Plums

Strawberries

Watermelon

Other _____

Prepared Fresh Fruits

Bottled mango or papaya slices

Cut fruits

Fresh juices — orange, grapefruit, apple cider

Packaged fruit salad

Other _____

Fresh Vegetables (especially locally grown)

Asparagus

Bean sprouts

Beets

Bell peppers

Bok choy

Broccoli

Brussels sprouts

Cabbage

Carrots

Celery

Collard greens

Corn

Cucumbers

Kale

Leeks

Mustard greens

Onions

Potatoes

Spinach

Sweet potatoes or yams

Tomatoes

Other _____

Prepared Fresh Vegetables

Fresh herbs — basil, dill, mint, rosemary, others

Fresh juices — beet, carrot, carrot/spinach

Packaged cut vegetables and mixed greens

Other _____

Fresh Deli Items

Four-bean salad

Fresh marinara (seasoned tomato) sauce

Fresh pizza (with veggie toppings; try cheeseless)

Fresh salsa

Hummus

Other _____

Dairy/Dairy Replacers, Eggs/Egg Replacers, and Tofu

Eggs (or vegetarian egg replacer)

Nonfat cheese or soy cheese

Nonfat yogurt or soy yogurt

Skim milk or soymilk

Tofu (fresh or aseptically packaged)

Other _____

Breads and Other Grain Products (especially whole grain)

Bagels (eggless for vegans and lacto vegetarians)

Bread loaves

Breadsticks or rolls

English muffins

Pita pockets

Tortillas

Other _____

Meat Substitutes

Meatless deli slices

Meatless hot dogs

Veggie burger patties

Other _____

Shopping by the month

Some items keep for a long time in a cupboard, refrigerator, or freezer, so you can shop for them fairly infrequently. The following list suggests items you may want to buy on a monthly basis. You may be able to buy many of these nonperishables in large quantities to save money.

Cupboard staples

Canned Goods

Applesauce

Artichoke hearts

Beans — vegetarian baked beans, garbanzo, black, pinto, kidney, navy, split

Bean salad

Fruits — any fruits, cranberry sauce, fruit cocktail

Pasta sauce

Peas — green peas, lentils, black-eyed peas

Pumpkin and fruit pie fillings

Soups — lentil, tomato, vegetarian split pea

Tomato sauce and paste

Vegetables — asparagus, carrots, corn, green beans, peas, tomatoes

Vegetarian refried beans

Other _____

Snacks and Treats

Baked potato chips and baked tortilla chips

Bean dip

Flatbreads (including matzo) and breadsticks

Granola bars

Popcorn (bag kernels or microwave)

Rice cakes and popcorn cakes

Whole-grain cookies

Whole-grain crackers

Other _____

Herbs and Spices

Basil

Bay leaves

Black pepper

Cinnamon

Cumin

Curry powder

Dill

Garlic

Ginger

Oregano

Paprika

Salt

Vegetable bouillon

Other _____

Beverages

Bottled fruit juice

Club soda

Juices — beet, carrot, or tomato

Plain or flavored mineral water or seltzer water

Soymilk (aseptically packaged)

Sparkling cider or grape juice

Other _____

Dry Goods

Beans and bean flakes

Coffee and tea

Cold cereal (whole-grain) — raisin bran, shredded wheat, bran flakes, others

Couscous (whole-grain, if available)

Grains — barley, millet, bulgur wheat, kasha, amaranth, spelt, teff, quinoa, kamut, others

Hot cereal (whole-grain) — oatmeal, whole-wheat, mixed grain

Pasta (eggless for vegans and lacto vegetarians)

Rice — basmati, jasmine, brown, wild, arborio, others

Soup mixes or cups

Textured vegetable protein (TVP)

Tofu (aseptically packaged)

Vegetable oil spray

Vegetarian egg replacer

Whole-grain bread, pancake, and all-purpose mixes

Whole-wheat flour, other flours

Other _____

Condiments

Chutney

Fruit spreads, jams, and jellies

Honey (for non-vegans)

Horseradish, marinades, BBQ sauce

Ketchup

Mustard

Olives

Pickles and pickle relish

Salad dressing

Salsa

Soy mayonnaise

Stir-fry sauce

Sun-dried tomatoes

Syrups and molasses

Vinegar — balsamic, herbed, fruited, malt, rice, other

Other _____

Dried Fruits

Apples

Apricots

Blueberries

Cherries

Currants

Dates

Figs

Mixed fruits

Prunes

Raisins

Other _____

Freezer staples

- Frozen bagels (eggless for vegans and lacto vegetarians)
- Frozen entrées
- Frozen fruit and juice bars
- Frozen fruit juice
- Frozen pasta (eggless for vegans and lacto vegetarians)
- Frozen vegetables
- Frozen waffles and pancakes
- Italian ices and popsicles
- Meatless bacon
- Meatless sausage links and patties
- Muffins and dinner rolls
- Nondairy ice cream
- Nonfat frozen yogurt
- Sorbet
- Vegetarian burger patties
- Vegetarian hot dogs
- Veggie burger crumbles
- Other _____

Let's Go Shopping! Considering the Options

Many people — especially vegetarians — find that they purchase certain items at one store and others at another store. For instance, you may shop at a natural foods store to buy frozen vegetarian entrées, specialty baking mixes, whole-grain pastas and breakfast cereals, and your favorite brand of soymilk by the case. Then perhaps you shop at a conventional supermarket for such items as toilet paper, fresh produce, and toothpaste.

It all depends on your individual preferences. This section includes a quick rundown of your options, listed generally in order of most to least common.

Your neighborhood supermarket

Conventional supermarkets have undergone a transformation during the past ten years because of the rapid expansion of the natural foods and organic foods markets. In years past, your neighborhood supermarket may have had a special aisle for "health foods" where the soymilk and vegetarian specialty items were shelved. Today, these foods are more likely to be integrated with all the other foods in the store.

Your neighborhood supermarket also probably carries a much wider variety of natural products, organic foods, and vegetarian specialty items than it did in the past. Depending on the store, the selection may be comprehensive enough that you can do most of your shopping in one stop.

Warehouse stores

Buying in volume can be economical, but it isn't always the best approach. Think carefully before you buy a giant jar of pickles, a huge bottle of ketchup, or a 20-pound bag of rice! Not only can storage space be a challenge, but some foods may spoil before you can reasonably use them up.

An extra large bottle of vegetable oil, for example, may go rancid before you can use it all. Buy too many croutons, and they may become stale before you get to the bag's bottom.

On the other hand, warehouse stores can be a good place to find excellent values on such vegetarian staples as canned beans, tubs of hummus, veggie burger patties, and other meatless specialty items.

Natural foods stores

If you've never set foot in a natural foods store, now's the time to do it. You may not shop there for all your groceries and supplies, but you should take a look at the many vegetarian specialty items and other great choices available. Natural foods stores include large national chains such as Whole Foods, regional chains such as Earth Fare in the Southeast, and local mom-and-pop stores.

Some people buy everything they need at natural foods stores, including nonfood items, because natural foods stores often carry a wider selection of environmentally-friendly cleaning supplies and paper goods than other stores. Big natural foods stores are common in large, metropolitan areas, and prices may be better and the variety greater than in cities that have only small stores.

Farmer's markets and CSA farms

Many communities have farmer's markets where small farmers sell locally grown, seasonal fruits and vegetables. You can often find locally produced honey, jams and jellies, and baked goods at the farmer's market, too.

Not all sellers who grow and produce their foods organically pay to have their foods certified as organic, so ask your local farmers for details about their approach. That's an advantage to buying from local farmers — you can establish a relationship with these people over time and grow to know and trust their food.

Be careful, though. Some farmer's markets permit large wholesalers to sell out-of-state products. Ask questions about the source of the produce you see to make sure you're getting locally grown food, if that's important to you.

An alternative or supplement to seasonal produce from the farmer's market is food from a *CSA* farm. CSA stands for *community-supported agriculture.* Residents of a community pay a local farmer a predetermined amount of money upfront. In return, they get a portion of the harvest throughout the growing season. If you subscribe to a CSA farm, expect to pick up your food at a centrally agreed-upon site. In some cases, home delivery may also be available.

The advantages to getting your food from farmer's markets and CSA farms is that foods are more likely to be fresh and at their peak of nutritional value, and you support your local farmers by giving them your business.

Ethnic food markets

Shopping at ethnic supermarkets can be an entertaining and enjoyable experience. Go with an open mind and be ready to experiment with a few foods you've never tried before.

You're more likely to find these markets in larger metropolitan areas and in communities with large concentrations of immigrants from different countries. In some cities, you'll find many different types of ethnic grocery stores — Indian, Asian, Mexican, Middle Eastern, kosher, and many more. Each one carries foods and ingredients that are commonly eaten in those cultures but are frequently difficult to find in mainstream U.S. stores.

In Indian stores, for instance, you'll be amazed to see the many types of lentils that you can buy — red, orange, yellow, brown, and so on. You may find small, skillet-sized pressure cookers used in India to quick-cook dried beans and lentils. You'll also find unusual spices and a variety of Indian breads.

You may see the same products — Indian heat-and-serve, single-serving entrées, for example — at conventional supermarkets at substantially higher prices. Go with a list, especially if you have to drive a distance to get to the store.

In Asian stores, you'll find vegetables such as some Chinese greens that are shipped to the store from overseas but that aren't typically found in U.S. stores. You'll find unusual condiments as well.

In all these stores, you'll see a wide range of traditional foods, many of which are vegetarian or just different and which you can incorporate into your vegetarian menus at home. More than anything, a trip to one of these stores may inspire you to sample foods of another culture and to expand your own culinary repertoire.

Food cooperatives

Many communities also have food cooperatives, or co-ops. A *co-op* is a group that purchases foods in volume for distribution to its members, usually at a reduced cost. The costs are reduced because of the power of the group to buy in volume, so larger groups may be able to get better prices than smaller groups. Co-ops frequently emphasize natural foods products, and the arrangement can be especially helpful in small towns and rural areas, where people don't have access to large natural foods stores with good variety and competitive prices.

Membership rules vary from one co-op to another. Some co-ops ask their members to contribute a certain number of hours each month to help unload trucks and bag or distribute groceries. Others don't require that you work, but they may give an additional discount to those who donate their time.

Gourmet stores

Upscale or gourmet foods stores are a fun place to look for interesting specialty items such as jams and preserves, other condiments, canned soups, pasta sauce, and a wide variety of packaged foods, often from boutique producers or international sources. Many of these products are appropriate for vegetarians. The downside: They're often expensive. Gourmet stores can be a good place to shop, though, if you need a special gift for the vegetarian on your list.

Web sites and catalogs

Another alternative for people who don't have time to shop or can't find certain food items in their local stores is to shop via mail-order catalog or online stores such as Pangea (`www.veganstore.com`) or the Mail Order Catalog for Healthy Eating (`www.healthy-eating.com`). Be aware, though, that shipping and handling charges can be substantial.

Other places to try

Bargain stores such as Grocery Outlet are good places to check for deeply discounted vegetarian specialty products. Inventories change frequently, so you may not be able to find a specific item if you need it. If you're open to surprises, though, you may find some good buys.

Similarly, specialty stores such as Trader Joe's often carry a sizable selection of vegetarian foods at competitive prices. Look for name brands here, as well as products such as soymilk and rice milk sold under the store's private label.

Making the Most of Your Shopping Adventure

Regardless of where you shop, a few more tips and suggestions may help ensure that you have a good supply of nutritious ingredients on hand at home. The ideas I include in this section have worked for me and for many other vegetarian shoppers.

Slowing down to see what's new

Every now and then, spend a little more time than usual roaming up and down each aisle of your favorite store to find out what's new. It's a good way to notice products you may like but didn't know were available.

That's especially true in the produce department. Many people make a beeline for the iceberg lettuce and tomatoes and don't notice much else. Take an extra ten minutes to peruse the piles of colorful fruits and vegetables, and pick up something you've never tried before.

Experimenting with new foods and products

Speaking of trying foods that are new to you . . .

Expect to find a few duds. Along the way, though, you're also likely to find some new favorite foods. Unless you take a risk and try something new, you may never know what you're missing!

That's the way I discovered — and grew to love — mangoes and papayas. I didn't like the flavor of mango the first time I tasted it, but I tried it again. By the second or third try, I was hooked.

If you haven't tried soymilk, soy yogurt, meatless hot dogs, or meatless burger crumbles (for tacos, burritos, and spaghetti sauce), I highly recommend them. If you've never eaten an arugula salad or fresh kale, let the recipes in Chapters 11 and 12 be your introductions.

Don't see what you need? Ask the manager

If you want to buy a specific product but you don't see it in your favorite store, ask to speak with the manager. Chances are good that she can place a special order for you.

Store managers can often arrange for you to purchase certain foods such as soymilk and canned or bottled foods by the case at a discount. If the demand is high enough, the store may even begin to stock that product regularly.

Keeping Your Costs Under Control

Living vegetarian can be a very cost-effective lifestyle. Diets that include little or no meat, eggs, cheese, or other dairy products tend to be built around foods that are relatively inexpensive, such as beans, pasta, rice, and vegetables.

The economy of a vegetarian diet can be lost, though, if you find yourself depending on too many specialty products and ready-made foods. To hold down the cost of your meals, keep the suggestions in this section in mind.

Collecting the building blocks to keep on hand

It's important to keep a good supply of basic ingredients on hand so that you can whip up a meal in a relatively short amount of time. The most practical ingredients are shelf-stable — you can leave them in your pantry for several months before using them. Good examples include pasta, canned soups, several varieties of canned beans, pasta sauce, rice, dry cereal, hot cereal, and condiments. Most of these items are inexpensive.

Good freezer basics include several types of frozen vegetables, veggie burger patties, whole-grain rolls and bread, and frozen, mixed berries (for a quick dessert, see the Berry Cobbler recipe in Chapter 14).

Buying in volume — or not

Buying in volume isn't always the most economical way to shop. If food spoils before you can use it, you may end up spending more than you otherwise would have spent on a smaller quantity of food. If you shop at a warehouse store, you may also want to factor in the membership cost in addition to the cost of the food itself when evaluating whether buying in large quantities is economical.

One remedy: Consider saving by buying in volume and splitting the item with a friend. For example, you may buy a case of oranges or a large container of pre-washed salad greens and each take half.

Perusing private labels and store brands

Compare prices among brands. Private label and store brand products — including breakfast cereals, canned goods, soymilk, peanut butter, and many other foods — can cost significantly less than similar brand-name products. The quality is often indistinguishable from one brand to the next, too. You won't know until you try!

Scaling back on specialty items

Expensive specialty items — gourmet condiments, frozen entrées licensed by famous chefs, and many boutique-brand packaged foods — can take a big bite out of your food budget. Give them as gifts now and then, and enjoy the champagne raspberry preserves when a jar comes your way. In general, though, these foods are too costly to be staples for most of us. Scale back on these items if you're working on keeping your costs under control.

Similarly, many specialty products in natural foods stores can be pricey. One way to save money at natural foods stores is to buy staples — beans, rice, lentils, granola, seeds, and nuts — from bulk food bins. You won't pay for unnecessary packaging, and you can take only the amount you need.

Getting the best value — nutritiously

Sometimes, the best value isn't the least expensive option.

For example, a standard bag of carrots from your supermarket may remain in your refrigerator so long that the veggies turn to rubber. You can pay more for a bag of prepeeled carrots (or any prewashed, preprepared vegetable), and if the added convenience helps ensure that you eat your veggies, the extra cost may be worth it.

On the other hand, junk foods — such as many packaged cookies, cakes, and other desserts — are also relatively expensive. If you have them in your house, you're very likely to eat them. By keeping them out of the house in the first place, you not only save money but also protect the nutritional quality of your diet, too.

Cooking meals at home

The typical American spends about 40 percent of his food budget on meals eaten away from home. That means that eating more meals at home can potentially save you a lot of money.

Establish the habit of making simple meals at home. The information I cover in Parts II and III in this book — as well as in Chapter 16 — can help you get started.

Consider growing some of your own ingredients, too. Tomatoes, bell peppers, lettuce, and herbs such as basil and parsley are easy to grow, even in small spaces. You'll save money, and fresh ingredients for your homecooked meals will be right outside your back door.

Chapter 8

Cooking Tools and Techniques

In This Chapter

- Assessing your equipment needs
- Knowing the fundamentals of vegetarian cooking
- Giving new life to old recipes
- Getting familiar with alternative ingredients

Adopting a vegetarian diet is a good reason to brush up on basic cooking skills. If you understand how to use simple tools and techniques in the kitchen, you'll be able to enjoy more healthful, good-tasting vegetarian meals in your own home.

That's important.

The more meals you eat at home instead of away, the more money you save, and the more control you have over the ingredients in the food. Preparing your own meals gives you the flexibility to adapt traditional recipes by removing the meat and other animal ingredients. It's also a good way to improve the overall quality of your diet, because you can cut back on added sodium and sugar at the same time that you limit artery-clogging saturated fat and cholesterol from meat, eggs, cheese, and other dairy products.

Cooking your own vegetarian meals at home from scratch or by using some convenient shortcuts also gives you an opportunity to be creative. After you develop confidence in your cooking skills, you'll be able to experiment with new and delicious combinations of basic ingredients — vegetables, rice, pasta, beans, and fruit. You'll be less likely to get into a rut and more likely to be the mealtime destination of choice among your family and friends.

In this chapter, I introduce the tools you need to fix simple vegetarian meals. I cover the ins and outs of cooking with some common vegetarian ingredients and convenience products, as well as how to modify your favorite nonvegetarian family recipes to make them veg-compatible. The emphasis is on simple and practical ways to be your own vegetarian chef.

Tools You Really Need

The world is full of kitchen gadgets and appliances that make the work easier and save you time. You can do without them, but having them makes cooking more enjoyable.

Then there are the true necessities, the tools without which it would be very difficult to cook from scratch. In this section, I outline the equipment you need to outfit a basic vegetarian kitchen.

Use the lists I provide to do a quick inventory of your own kitchen. Consider what's missing and what you need to buy. Yard sales and kitchen supply outlets are great places to pick up inexpensive but serviceable pieces.

Spend the time to make sure you have the equipment you need for a fully functional kitchen. If you don't have what you need, you may find it frustrating — and even unsafe — to perform simple cooking tasks. Using a serrated knife instead of a paring knife to peel an apple, for example, is a good way to do a poor job and cut your finger while you're at it.

Pots and pans

You don't need a large collection and they don't have to be top-of-the-line, but you do need several good-quality pots and pans in various sizes. Invest now and you'll use these to make a lifetime's worth of health-supporting vegetarian meals.

Stainless steel is clean, safe to use, and a good choice for most pots and pans. Choose four to six of the following to handle most of your cooking needs, from pasta sauce and potatoes to chili and soup:

- 1-quart saucepan
- 2-quart saucepan
- 3-quart saucepan
- 4-quart saucepan
- 4½-quart pot
- 6-quart pot or larger (such as a stock pot or stew pot), if you cook for a crowd

A skillet is also invaluable. You'll use it for everything from frying pancakes and grilling sandwiches to sautéing onions and mushrooms. Choose a heavy-duty, 10-inch, cast-iron skillet from a company with a reputation for making such high-quality products, like Lodge (`www.lodgemfg.com`). Stainless steel is fine, too. I also recommend getting one or more of the following:

- 7-inch skillet
- 9-inch skillet
- 12-inch skillet

Try to get at least one skillet lid — it comes in handy for steaming greens and other times when you need to hold the heat in with what you're cooking. If you buy the same brand of pots, pans, and skillets, it's likely that one or more of the lids that come with some of the pieces will fit other pieces as well.

Knives

You may have noticed the large wooden knife blocks that some people keep in their kitchens, filled with an assortment of scary-looking blades that have figured prominently in several bad movies.

Vegetarians have no need for all those flesh-piercing utensils.

All you need is four different kinds of knives, and only one looks like it could play a role in an action flick (Figure 8-1 shows you what some of these knives look like):

- 6-inch, serrated knife. Use it for cutting fruits and vegetables, especially tomatoes. You'll use this knife more than any other.
- Paring knife. This is for cutting certain fruits like apples and pears, peeling potatoes, and similar tasks. The blade is short — 3 or 4 inches — so it's easy to maneuver.
- 10-inch, serrated bread knife. Use this for cutting cakes and similar foods.
- 10-inch chef's knife. This one has a long blade that's widest near the handle, gradually tapering to a point at the end. Holding the point on the cutting board, you rock the blade up and down from the handle to chop salad greens, fresh vegetables, fruits, and other ingredients.

Figure 8-1: The basic knives you need for vegetarian cooking.

Assorted extras

Most vegetarian cooks need some baking supplies, including pans, sheets, measuring cups, and utensils. At a minimum, you need:

- Two cookie sheets
- Muffin tin for baking cupcakes and muffins
- 8-inch square baking pan
- 9-x-13-inch baking pan
- Two loaf pans
- Two pie tins (cake pans aren't as useful, so they're optional)
- Set of two or three stainless steel mixing bowls
- At least one set of dry measuring cups
- At least one set of measuring spoons
- 2-cup liquid measuring cup

You may be able to whip a yogurt dressing with a fork and spread hummus into a serving bowl with the flat side of a place knife, but having the proper utensils for certain uses in the kitchen gives you better results. The following items are so practical that I deem them necessary for a complete vegetarian kitchen:

- Three or four wooden spoons in various shapes and sizes
- Two or three silicon spatulas
- Stainless steel spatula for flipping pancakes and veggie burgers or removing cookies from a cookie sheet
- Soup ladle
- Stainless steel colander
- Pizza cutter
- Pastry brush

- Pastry blender
- Pasta fork
- Two stainless steel whisks
- Hand-held bean or potato masher
- Large, stainless steel, slotted spoon

A well-equipped kitchen also includes attractive and functional serving bowls and plates, as well as at least a few clean, attractive tablecloths, placemats, cloth napkins, and decorative hot pads or trivets to place under hot dishes on the table. These features are important, because the way in which you present food can influence how well others receive the food. If it looks appealing, people are more likely to expect it to taste good.

Handy Appliances You May Actually Use

If you're like many people, you have an appliance graveyard lurking in the back of a kitchen cupboard or pantry. Those bread machines and waffle irons that seemed like a good idea at the time are collecting cobwebs in dark corners around the house.

One key to positioning yourself to cook your own vegetarian food at home is to clear some space. If your countertop and drawers are jammed with appliances and gadgets that you seldom use, get rid of them. Donate them or pack them away, but remove them from your kitchen. Make room so that you have easy access to the few appliances you may actually use to make daily meal prep easier and faster.

Which tools you need depends on the kinds of foods you like best and how you feel about performing certain kitchen tasks. For example, I like using an old-fashioned, crank-style can opener that I keep in a kitchen drawer. An advantage is that I don't have an electric can opener taking up space on my countertop.

And not everybody needs a large, countertop mixer. A hand-held mixer may work just fine for your needs, or maybe all you need is a simple whisk. Think about your equipment needs and remove what you don't need to free up precious kitchen space.

In the following sections, I tell you about several appliances that many vegetarians find particularly handy.

High-speed blenders

Blenders are useful for whipping up fresh fruit smoothies and fresh vegetable blends. An advantage of a high-speed blender over a juicer — another popular appliance — is that it preserves the dietary fiber in fruits and vegetables and may be easier to clean.

Ordinary kitchen blenders are inexpensive, but their motors tend to overheat easily when the blender is used to process thick mixtures such as hummus (see the recipe in Chapter 10) and some smoothies (Chapter 9 includes two smoothie recipes). A heavy-duty, restaurant-quality blender such as the Vita-Mix (`www.vita-mix.com`) holds an advantage over ordinary blenders because it has a larger capacity and stronger motor. The trade-off is that the Vita-Mix is much more expensive.

Make cold melon soup by placing fresh cantaloupe or honeydew chunks, nonfat plain yogurt, and a few teaspoons of honey in a blender and processing until creamy. For thinner soup, add orange or apple juice by the tablespoon until the soup reaches the right consistency. Serve in a rimmed soup bowl with a dash of cinnamon or nutmeg and a sprig of fresh mint on top.

Food processors

Mini food processors — those with only a one- or two-cup capacity — can be a great help in grating, mincing, and chopping small amounts quickly. I use mine to grate carrots or a small chunk of red cabbage for color in a salad, or when I don't feel like chopping an onion by hand.

Larger, full-capacity food processors are a must-have in many vegetarian households. They make quick work of larger chopping jobs, and the results are often more consistent than when you chop by hand.

It's amazingly simple and quick to make fresh *gazpacho* — a cold, traditional Spanish soup made with fresh vegetables — using a food processor. You'll find many variations, but most include one or more of the following ingredients, blended until smooth: fresh, uncooked tomatoes, onions, cucumbers, bell peppers, celery, olive oil, chives, minced garlic, red wine vinegar, and lemon juice. I include a recipe for Classic Gazpacho in Chapter 11.

Rice cookers

One of the pitfalls of cooking rice the traditional way — in a pot on the stovetop — is that if you don't watch it carefully, the rice can quickly boil over and make a mess. Enter the rice lover's best friend: the rice cooker.

Some people swear by their rice cooker for that reason: It lets you make perfectly steamed rice while you're out of the kitchen doing something other than watching the stove. Rice is also a mainstay of many ethnic vegetarian meals — as an accompaniment to Indian and Asian entrées as well as with a variety of bean dishes. You can find several recipes in Chapter 12.

A great way to store leftover rice is to portion it into airtight plastic bags and place them in the freezer. The rice will keep for months that way. When you're ready to use it, transfer the frozen rice onto a plate or into a bowl and reheat it in a microwave oven. Eat it with leftover chili, stew, stir-fry, or Indian entrées.

Pressure cookers

Name a gadget that can cook a two-hour meal in ten minutes. Nope, it's not a magic wand — it's a *pressure cooker.* A pressure cooker enables you to cook a vegetarian casserole, stew, or bean dish in the time it takes to set the table.

The way a pressure cooker works is simple. You place foods such as beans, rice, lentils, or root vegetables in the cooker with water or another liquid. You tightly seal the pot and set it over high heat on the stove. As the steam inside the pot builds, the pressure inside the pot increases, raising the temperature to 250 degrees — far above the boiling point of 212 degrees. Under these conditions, foods get soft and cook faster than they would by other means.

Some people remember their mother's — or grandmother's — old jiggle-top pressure cooker from 30 years ago or more. Those old-fashioned pressure cookers hissed, rattled, and clanged so violently that at times they seemed as if they were about to blow a hole through the ceiling.

Newer jiggle-top pressure cookers from companies such as Mirro and Presto are much easier and safer to use and include features that prevent too much pressure from building up.

European companies such as Kuhn Rikon, Magefesa, and Zepter make pressure cookers with stationary pressure regulators instead of the old jiggle-top variety. Stationary regulators don't clog as easily, making it safer to leave the kitchen while the cooker is in use. These newer pressure cookers are quieter and may require less water to reach high pressure. After the food is done, you can release the pressure on the stovetop, in contrast to old models that had to be set in the sink and doused in cold water to bring the pressure down.

You can find a variety of pressure cookers in specialty stores, as well as some good buys online. You may even find some sold in Indian food markets, including short, squatty, pressure frying pans. Pressure cookers are a standard tool in Indian cooking because they're so useful in shortening the cooking time for lentils, rice, and many of the dried beans used in Indian dishes.

Pressure cookers come in several shapes and sizes. If you want only one, a 6- or 8-quart cooker is the most practical size to buy. It may look big, but when you use a pressure cooker, you fill it from only one-half to three-quarters full. The remainder of the space is needed for the steam that builds up inside the pot.

Note that European models designate the volume of the pot in liters rather than quarts. A 6-liter pressure cooker is roughly equivalent to a 6.5-quart cooker; a 7-liter cooker is about the same size as a 7.5-quart cooker. Prices vary widely by make and model.

If you want to try your hand at using a pressure cooker, a good resource is *Great Vegetarian Cooking Under Pressure* by pressure cooker guru Lorna J. Sass (William Morrow Cookbooks). Another good resource is registered dietitian Jill Nussinow, "The Veggie Queen." Visit her Web site at `www.pressurecookingonline.com`.

Slow cookers

A '70s-era kitchen aid that's enjoying a comeback is the *slow cooker,* a removable stoneware crock with a lid set into a heating element. Slow cookers allow you to dump ingredients into a pot, turn on the heat, and walk away. At the end of the day, you come home to the smell of dinner, ready to serve, wafting through your home.

Newer models include timer and temperature controls, whereas the original model had only a few settings. Slow cookers come in various sizes, from very small — good for singles and couples — to family-sized models. Others are made with locking lids, carrying handles, and travel bags for people who want to cook and carry a dish to get-togethers with family and friends.

Slow cookers work well for making vegetarian-style meals with combinations of vegetables, beans, lentils, and grains such as rice, oatmeal, bulgur wheat, and others. For example, you can use a slow cooker to make the recipes for Fruited Oatmeal (see Chapter 9), Lentil Soup (see Chapter 11), Cashew Chili (see Chapter 11), Cuban Black Beans and Rice (see Chapter 12), and Cajun Red Beans and Rice (see Chapter 12).

Electric teakettles

Electric teakettles are relatively inexpensive — under $40, or about the cost of a common coffee maker — and they can save you time. Just add hot water and flip the switch. You get boiling water in less time than it takes to heat water on the stovetop.

It's a quick and very convenient way to have hot water in a hurry when you want to make a single serving of oatmeal, a soup cup, or a hot drink. It's also useful for rehydrating bean flakes and instant couscous used in some vegetarian dishes, and for giving a jump-start to a pot of hot water for cooking pasta or rice.

Vegetarian Cooking Basics

You don't have to be a professional chef to fix good vegetarian meals. You don't even have to impress your friends. All you have to do is satisfy yourself (and your family — maybe).

If you're already an accomplished cook, you probably need no reassurance that you'll be adept at fixing vegetarian masterpieces after you figure out what you want to make. That won't take long, because you're already skilled and creative.

On the other hand, if you're like most people, you're no pro, but you somehow manage to put together reasonably appealing meals most of the time. Usually, that means whipping dinner together in a half hour or less, with the emphasis on less.

Vegetarian cooking can be as simple or as complicated as you want it to be. In this section, I cover some of the basic skills that any beginning vegetarian needs.

Mastering simple cooking skills

Becoming comfortable and confident in the kitchen takes time and practice. Like driving a car or knitting a sweater, you get better at it over time.

Each of the cooking techniques I describe requires hands-on experience. If you're a beginning cook, consider asking an experienced friend or family member to guide you at the start, or look into taking a cooking class at your local natural foods store or community college.

Baking

You'll use your oven to make many vegetarian foods, including casseroles, baked potatoes, breads and rolls, pies, winter squash, and more. A few key points are important to follow when you bake:

- **Preheat the oven.** Set the oven to the temperature called for in the recipe. Give the oven at least 10 or 15 minutes to reach the correct temperature before putting the dish in.
- **Use the timer.** Don't rely on your nose to tell you the food is ready. Set the oven timer, or use a free-standing timer. Make sure that you're in a place in the house where you'll hear the timer going off when the food is ready.
- **Measure precisely.** Think of dry ingredients in cakes, breads, rolls, and other baked goods like a science experiment. You have to add the flour, baking soda, and baking powder in the right proportions for the ingredients to react together properly. Use measuring spoons and cups to measure amounts accurately.
- **Check for doneness.** Poke a toothpick or knife into the center of the baked good to see whether it's ready to come out. If the toothpick or knife comes out clean, the food is probably done. If not, check again in another few minutes.

Boiling

You need to know how to boil water on the stovetop for cooking grains, dried beans, pasta, rice, and a number of other foods. In most cases, this entails bringing a pot or saucepan of water to a boil over high heat, adding the food, and then turning the heat down to let the food simmer for a period.

When cooking some foods, such as a pot of rice, you need a lid to keep the steam inside the pot. You can cook other foods, such as pasta, without a lid to help ensure the water doesn't boil over the sides.

Setting a timer is helpful so that you know when a food should be done cooking. Be careful not to overcook some foods, such as pasta, which can become mushy if it's left in the water too long.

Steaming

Steaming vegetables such as broccoli, cauliflower, and cabbage by cooking them for a short time in a small amount of water is a great way to preserve their nutrients. It takes only minutes to steam most vegetables so that they're tender enough to eat while retaining their color and crispness.

The microwave oven is the quickest and easiest way to steam foods. Set the food in a glass container and add a few tablespoons of water. Cover with a paper towel or plastic wrap and heat. A medium-sized baking potato needs six to eight minutes to cook, while a small serving of broccoli may need only a minute or two.

To steam foods on the stovetop, add several tablespoons of water to a pot and heat it to boiling. Add the food, cover with a lid, and let the food steam for as long as necessary. When the food is done, remove the pot from the stove and drain the excess water into the kitchen sink.

Be careful when you drain the water after steaming foods, whether you cook the food in a microwave oven or on the stovetop. When you remove the lid or cover, hold the pot or bowl away from your face so that escaping steam doesn't burn you.

If you're looking for kitchen gadgets specifically for steaming, consider trying one of the following:

- **Bamboo or stainless steel steamer baskets:** You set them into a pot of boiling water and they hold the food just above the bottom of the pan. The advantage in using these baskets is that they help fragile foods such as dumplings or tiny vegetables stay intact better and prevent nutrients from being leached into the cooking water.
- **A free-standing, electric steamer:** These small appliances are inexpensive and simple to operate and clean. You can find them in most discount and home stores.

Sautéing

Many vegetarian recipes begin by asking you to sauté onions, garlic, celery, or bell peppers before adding other ingredients. To *sauté* is to fry foods in a small amount of fat — usually vegetable oil — over high heat.

Begin by adding a few tablespoons of olive oil or another vegetable oil to a skillet to keep food from sticking and to help distribute the flavors. When the skillet is hot, add the food and stir frequently to prevent sticking. Onions are done when they're translucent. Garlic may brown a bit, and celery and peppers become soft when done.

Caring for cast-iron cookware

Follow the manufacturer's recommendations for cleaning and seasoning cast-iron pots, pans, griddles, and other cookware. Instructions may vary from one brand to the next, depending on how the cookware is made. In general, though, you should clean cast-iron cookware by using nothing but a stiff, nylon brush or a sponge and hot water. After removing food residue, use a clean towel to dry your cookware. Air-drying can cause cookware to rust. Use a paper towel to rub the cooking surface with a light coat of vegetable oil, and then store the cookware in a cool, dry place. Place a piece of newspaper or a paper towel between pots and their lids to promote air circulation and prevent rust formation.

Stewing

Soups, stews, and chili are examples of foods that are often left to *stew* or simmer on low heat on the stovetop for an hour or more before serving. They have to be stirred every so often with a long-handled spoon to keep food from sticking to the bottom of the pot.

A deep stew pot or stockpot is ideal for making large batches of foods like these, which tend to splatter if the heat is too high or the pot is too full. A lid — tipped to let steam escape — can help keep your stovetop and burners clean.

Prepping fruits and vegetables

The first step in preparing most fruits and vegetables is to wash them thoroughly, even if you're using organic produce. Harmful bacteria can be present on any produce, and you need to take precautions to be safe.

Using your hands or a soft brush, rub fruits and vegetables as you hold them under fast-running water for several seconds, removing all visible traces of dirt and debris. Soap isn't necessary, and research has found little or no advantage to commercial fruit and vegetable rinses.

For fruits and vegetables such as broccoli and strawberries that have lots of crevices, soak them in a pan or sink full of water, swishing to help dislodge tiny bits of debris, and then rinse them well.

Even if you don't plan to eat the peel, it's important to wash the outsides of foods like watermelon, cantaloupe, and oranges before setting them on a cutting board or slicing them. If you don't, bacteria on the outside of the food could contaminate the cutting board's surface and anything you set on it afterward. Cutting unwashed foods can also permit bacteria to get into the part of the food that you're going to eat.

The right knives can make peeling, chopping, and cutting much easier. Use a paring knife to peel and cut apples, pears, peaches, cucumbers, bell peppers, and radishes. A chef's knife is useful for chopping romaine lettuce, herbs, greens, and pineapple. Use a serrated knife to slice soft fruits and vegetables such as tomatoes and kiwi fruit. (For more information and an illustration of these knives, refer to Figure 8-1 and the earlier section on knives.)

Save time fixing meals by pre-preparing fruits and vegetables. Take a half hour when you return from shopping to wash and cut up produce. Store precut fruits and vegetables in airtight containers in the refrigerator so that they're ready when you start fixing a meal.

Look for ways to be efficient when you prepare fresh fruits and vegetables. Chop several bell peppers or onions at a time, for example, and store the extra in 1⁄4-cup portions in airtight plastic bags in the freezer. Do the same thing with fruit. You can cut up peaches and strawberries, freeze them, and use them later to make smoothies or to serve for dessert with ice cream.

Cooking extra now for later

Batch cooking is a great way to save you time and effort in the kitchen. It means fixing a larger amount of whatever recipe you're making and setting part of it aside for later. For little additional effort, you can have enough food for two or more meals in the time it takes you to prepare one.

Batch cooking works best for foods that you can store in the refrigerator for at least a few days or in the freezer for several weeks. For example, you can serve a big pot of chili for dinner one day and store the remainder in the refrigerator and use it for lunches for the next couple of days. Or you can freeze part of a large lasagna and reheat it for a quick meal another day.

Some salads even store well for days in the refrigerator. Pasta salad, bean salad, and other marinated vegetable salads, for example, can keep in the refrigerator for up to several days.

Discovering a few tricks for cooking with tofu and tempeh

People often talk about tofu and tempeh as if they're bookends and go together like a matching pair. The fact is, they look and taste very different, and their functions in recipes vary, too. Another common misconception is that if you're a vegetarian, you probably eat a lot of tofu. (Most nonvegetarians have never heard of tempeh.) It's not necessary to eat tofu and tempeh to be a full-fledged vegetarian, but you can certainly make some good-tasting dishes with these nutritious soy foods.

If you do care to experiment with tofu and tempeh, here are a few tips to make the most of these ingredients:

- When you use tofu as a meat substitute, you usually cut it up into cubes and stir-fry it (extra-firm tofu works best for this), or you marinate it and cook it in slabs or chunks. If it's frozen first and then thawed, tofu develops a chewy texture that resembles that of meat.

- Try using tofu to make cream soup without milk or cream. Just purée some soft tofu with vegetable broth and mix it into the soup.
- You can use tempeh in ways that meats are more traditionally used. For instance, you can grill, barbecue, bake, or broil strips or blocks of tempeh. You can use chunks of tempeh with vegetable pieces for shish kebabs or to make stews, casseroles, and other combination dishes.

Adapting Traditional Recipes

When I think about adapting traditional recipes to make them suitable for vegetarians, I can't help but remember a 1970s TV show that left a lifelong impression on me. That show — *Gilligan's Island* — was all about a group of castaways from a tourist boat, stranded on a deserted tropical island.

Living off the land, this group somehow managed to bake the most luscious-looking coconut cream and banana cream pies. As a kid with a mother who baked, I was mesmerized — and confounded — by the idea that they could do this without milk, eggs, or butter.

"Well, they had coconut milk," I thought. But no eggs or butter. And those pies looked so good.

The show was fiction, but the idea that plant-based ingredients can perform just as well in recipes as animal ingredients is the truth. You'll be surprised to discover how easy it is to cook the vegetarian way, and you may be equally as pleased to know that you don't have to throw away your old family favorites!

When you substitute plant-based ingredients for animal ingredients in recipes, anticipate that the characteristics of the finished products may be at least somewhat different from the qualities of foods made with animal products.

I encourage you to try substituting plant-based ingredients for animal products in your favorite nonvegetarian recipes. If you do, however, one caution is in order: Expect to experiment before you get it right. If you don't use a vegetarian cookbook in which the recipes have already been tested, you'll have to fiddle around with traditional, nonvegetarian recipes until you hit on the ingredient substitutions and proper amounts that work best.

In the following sections, I show you how to make many of your favorite foods without using any animal products whatsoever. That way, you'll be prepared to adapt recipes for any kind of vegetarian diet. Of course, if you're a vegetarian who eats dairy products and eggs, you may not care to work those ingredients out of your recipes. However, if you're a vegan or you just want to lower your intake of saturated fat and cholesterol, the information is very useful.

In many cases, you can use any of several different foods as a substitute for an animal product in a recipe. Experiment to find the one that gives you the best result. Figure 8-2 shows some common substitutions.

When you modify a recipe, use a pencil to note on the original recipe the substitutions you made and how much of each ingredient you used. When the product is finished, jot down suggestions for improvements or any minor adjustments that you want to make next time. Whenever you make an adjustment, erase and update your notes on the recipe.

Replacing eggs

How many times have you wanted to bake a batch of cookies, only to find that you were missing a vital ingredient, such as eggs?

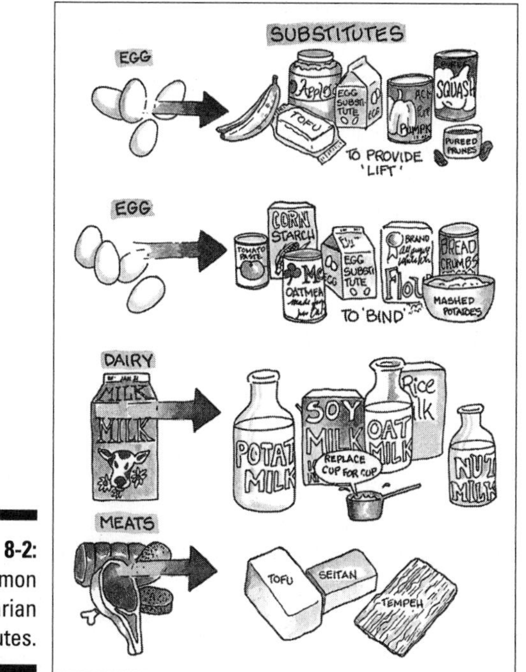

Figure 8-2: Common vegetarian substitutes.

Bet you didn't know that those eggs aren't so vital. If you had known that you could substitute any of several other foods for the eggs in your recipes, you wouldn't have had to waste a minute running to the store.

Eggs perform a number of functions in recipes, including binding ingredients, leavening, affecting texture, and affecting color (such as in a spongecake or French toast). So your choice of a substitute depends on how well it can perform the necessary function. In some cases, the egg's effect in the recipe is so slight that you can leave it out altogether, not replace it with anything else, and not even notice that it's missing.

In this section, I look at some of the foods that you can use to replace eggs in recipes.

Baking without eggs

In baked goods, you usually use eggs for *leavening,* or lightness, and to act as a binder. In some recipes, eggs are beaten or whipped, incorporating air into the product and decreasing its density. The type of baked good determines whether you can leave out the egg entirely, or whether you need to replace it with another ingredient to perform the egg's function in the original recipe.

For example, in baked goods that are relatively flat and don't need a lot of leavening, such as cookies and pancakes, you can often get away with leaving out the egg and not replacing it. That's particularly true when the original recipe calls for only one or two eggs. In recipes that call for more eggs, the eggs probably play a much greater role in leavening or binding, and you'll find that the recipe fails if you don't replace them.

If you omit the eggs in a recipe and don't replace them, add a tablespoon or two of additional liquid — soymilk, fruit juice, water, and so on — for each egg you omit to help the product retain its original moisture content.

In baked goods that are light and have a fluffy texture, you want to replace eggs with an ingredient that provides some lift. Try any of the following to replace one whole egg in a recipe:

- Half of a ripe mashed banana. This works well in recipes in which you wouldn't mind a banana flavor, including muffins, cookies, pancakes, and quick breads.
- ¼ cup of any kind of tofu, blended with the liquid ingredients in the recipe — soft or silken tofu works especially well.
- ½ teaspoon commercial vegetarian egg replacer, such as EnerG Egg Replacer, mixed with 2 tablespoons of water. This product is a combination of vegetable starches and works wonderfully in virtually any recipe that calls for eggs. Natural foods stores usually sell this in a 1-pound box.

- ¼ cup applesauce, canned pumpkin, canned squash, or puréed prunes. These fruit and vegetable purées may add a hint of flavor to foods. If you want a lighter product, also add an extra ½ teaspoon of baking powder to the recipe, because using fruit purées to replace eggs can make the finished product somewhat denser than the original recipe.
- 1 heaping tablespoon soy flour or bean flour mixed with 1 tablespoon water. This combination thickens and adds lift.
- 2 tablespoons cornstarch beaten with 2 tablespoons water. This mix also thickens and adds some lift.
- 1 tablespoon finely ground flaxseeds whipped with ¼ cup water. This blend also thickens and adds some lift.

Holding food together without using eggs

In foods where the ingredients need to stick together, such as vegetable and grain casseroles, lentil loaves, and vegetarian burger patties, you need an ingredient that acts as a binder. Eggs traditionally serve that function in non-vegetarian foods like meatballs, meatloaf, hamburgers, and many casseroles, but you can find plenty of alternatives if you want to omit the eggs.

REMEMBER

Moist foods such as casseroles or some vegetarian loaves may not require any additional moisture when you're removing or replacing eggs. In these cases, you'll need to experiment to determine whether the finished dish is moist enough without extra liquid to compensate for the eggs you've left out.

TIP

With the following substitutions, you'll probably find that you have to experiment a bit to determine just the right amount of an ingredient to serve the purpose in a specific recipe. A good starting point with most recipes is 2 or 3 tablespoons of any of these ingredients, or a combination of them, to replace one whole egg. If the original recipe calls for two eggs, start with 4 to 6 tablespoons of egg substitute.

Try any of the following to replace eggs in a recipe where eggs are used to bind food:

- Arrowroot starch
- Cornstarch
- Finely crushed breadcrumbs, cracker meal, or matzo meal
- Mashed potatoes, mashed sweet potatoes, or instant potato flakes
- Potato starch
- Quick-cooking rolled oats or cooked oatmeal
- Tomato paste
- Whole-wheat, unbleached, oat, or bean flour

EnerG Egg Replacer is one of my favorite egg substitutes, and only 1½ teaspoons of the powder mixed with 2 tablespoons of water replaces one whole egg in most recipes.

When working with dry ingredients such as arrowroot or cornstarch, some recipes may work best if you mix the dry ingredients with water, vegetable broth, or another liquid (about 1½ teaspoons of dry ingredient to 2 tablespoons of liquid) first, and then add the mixture to the recipe in place of the egg.

Some of the egg substitutes may affect the flavor of the finished product, too, so you should consider that when you decide which ingredients to use. For instance, if you add sweet potato to a burger patty, you may be able to taste it in the finished product, whereas if you mix some sweet potato into a casserole, its flavor may be more disguised by the other ingredients. On the other hand, the extra flavor that some of these ingredients add may be a pleasant surprise.

If you're looking for another substitution, try replacing one whole egg with ¼ cup of any kind of tofu blended with 1 tablespoon of flour. (See the nearby sidebar "Imitating egg with tofu" for more information on tofu as an egg substitute.)

Cooking with dairy substitutes

Replacing dairy products in recipes is incredibly easy. The dairy products you're most likely to find in recipes are milk, yogurt, sour cream, butter, and cheese. You can easily substitute good nondairy alternatives for any of these products.

Getting the cow's milk out

You can replace cow's milk in recipes with soymilk, rice milk, potato milk, nut milk, or oat milk. Just substitute any of these alternatives cup for cup for cow's milk. Though oat milk has a neutral flavor without sweetness, some nut milks, such as almond, are too sweet for savory dishes. Use them in desserts and smoothies.

Plain and vanilla flavors are the most versatile varieties of soymilk because the mild flavors blend in with just about any recipe. You can try plain soymilk in savory recipes such as some main dish sauces and soups. Use vanilla soymilk in sweeter dishes, such as puddings and custards, and on cereal, in baking, and for smoothies. Carob or chocolate soymilk is also delicious in some smoothies and puddings.

Make your own version of nondairy buttermilk by adding 2 teaspoons of lemon juice or vinegar to 1 cup of soymilk or any other milk substitute.

Imitating egg with tofu

Tofu can stand in for eggs in all sorts of recipes. Usually it's invisible as an ingredient, but sometimes it works as an egg imposter. Here are some examples:

- Use chopped firm tofu or extra-firm tofu in place of egg whites in recipes for egg salad sandwich filling. Just make your favorite egg salad recipe, but use chopped tofu instead of hard-boiled eggs. You can even use soy mayonnaise instead of regular mayonnaise for a vegan version. See the recipe for Tofu Salad in Chapter 11.
- Add chopped firm tofu to mixed green salads or spinach salad in place of chopped hard-boiled eggs. You can also add chopped or minced tofu to bowls of Chinese hot and sour soup.
- Make scrambled tofu instead of scrambled eggs. Natural foods stores stock *tofu scrambler* spice packets, and you may also see them in the produce section of your regular supermarket, next to the tofu. Vegetarian cookbooks also give recipes for making scrambled tofu. The recipes usually include turmeric to give the tofu a yellow color, similar to that of scrambled eggs. Try the recipe for Scrambled Tofu in Chapter 9. Use scrambled tofu to fill pita pockets or as a sandwich filling on hoagie rolls.

Removing sour cream and yogurt

You can use soy yogurt and soy sour cream in most of the same ways that you use the dairy versions, including baking, making sauces and dips, and eating as is. Because these substitutes sometimes separate when they're heated on the stove, they may or may not work in certain sauce recipes. Most of the time, however, you'll find that you have no problems substituting them for dairy yogurt and sour cream.

Choosing cheesy alternatives

Nondairy cheese alternatives may not melt as well as regular, full-fat dairy cheeses, but they generally melt better than nonfat dairy cheeses. That's partly because they tend to be high in fat — albeit vegetable fat.

Experiment with cheese replacers to find the brands you like the best and the varieties that work best in your recipes. Most cheese substitutes do well as an ingredient in a mixed dish, such as a casserole, in which the cheese doesn't stand alone but is mixed throughout the dish. Cheese substitutes, including nonfat dairy cheeses, melt better this way, too.

Make your own nondairy substitute for ricotta cheese or cottage cheese by mashing a block of tofu with a fork and mixing in a few teaspoons of lemon juice. You can use this "tofu cheese" to replace ricotta cheese or cottage cheese in lasagna, stuffed shells, manicotti, Danish pastries, cheese blintzes, and many other recipes. Nutritional yeast works well as a substitute for Parmesan cheese on casseroles, salads, baked potatoes, popcorn, and pasta. It has a savory, cheesy flavor. You'll find it in natural foods stores.

Making better butter choices

You can use stick-style soy margarine in recipes in place of regular margarine or butter, but it's not ideal. That's because any kind of regular, stick-style margarine is made by using a process that creates artery-clogging trans fat.

Instead, try replacing stick margarine and butter with soft, tub-style, trans fat-free margarine available in natural foods stores. Examples include Earth Balance and Canoleo. Read the labels and look for those containing zero trans fat and less than a gram of saturated fat per serving. Experiment with these in recipes, because the soft nature of the product may cause it to perform differently in different recipes.

You can also use liquid vegetable oil to replace butter in some recipes. Use about ⅞ cup of vegetable oil to replace 1 cup of butter. This substitution may not work as well in recipes for baked goods as it does in other recipes, so experiment to find the amount that works best.

Using meat substitutes

Some vegetarian dishes are originals — they've been meatless from the start, and they don't cry out for a meat-like ingredient. Examples include falafel (deep-fried Middle Eastern chickpea balls), spinach pie, ratatouille (a spicy vegetable dish), and pasta primavera.

Other dishes were created with a meat-like ingredient in mind. Without meat — or a suitable substitute — they lack something. Examples include burgers (a burger without the burger is just a bun), sloppy Joes, and recipes that call for chunks of meat, such as stews and stir-fries.

Some meat products are stand-alone traditions — hot dogs, sausage, cold cuts, and bacon, for instance. Believe it or not, you can find some very good imposters to replace even these stand-alones. Whatever the recipe, you have lots of choices when it comes to replacing the meat. Consider the following, which I've generally listed in order of their popularity:

- **Meatless strips, nuggets and patties:** Soy- and vegetable-based substitutes that resemble chicken and beef strips, chicken nuggets, burger and chicken patties, and similar products are sold in the refrigerator and freezer cases in supermarkets and natural foods stores. Use them in stir-fry recipes, over rice, in casseroles, and in other creative ways.

- **Tofu:** Tofu is a smooth, creamy food that has very little flavor or odor. It picks up the flavor of whatever it's cooked with. It can be seasoned with herbs, spices, and sauces and can be baked, fried, or sautéed and used like meat as a main dish or as an ingredient in other dishes, such as the stir-fry recipe I include in Part III.
- **Tempeh:** Tempeh can be crumbled and used to make such foods as tempeh sloppy Joes, tempeh mock chicken salad, and tempeh chili.
- **Textured vegetable protein (TVP):** This product can replace the ground meat in taco and burrito fillings, sloppy Joe filling, and spaghetti sauce. If you toss a handful into a pot of chili, people won't likely be able to tell the difference between the TVP and ground meat. You can find dry TVP in natural foods stores and through mail order catalogs, though it's largely being replaced by newer products, such as the meatless ground "beef" products sold in the refrigerator and freezer cases in supermarkets and natural foods stores.
- **Bulgur wheat:** Bulgur wheat has a nutty flavor, and you can use it in some of the same ways that TVP is used. For example, you can toss a handful of bulgur wheat into a pot of chili and get much the same effect as you would if you used TVP. It absorbs the liquid in whatever it's cooked with, and it has the appearance of ground meat and a chewy texture.

Factoring in other replacements for animal ingredients

Don't let any animal ingredient stand in your way of enjoying a favorite recipe.

Replace beef and chicken broths in soup with vegetable broth, available in a variety of flavors and forms. For convenience and flavor, brands that are packaged in aseptic, shelf-stable boxes and sold in many supermarkets and natural foods stores work well. Plain vegetable broth works well in most soup recipes, but consider tomato and red pepper, ginger carrot, and other variations, too. You can also use bouillon cubes, powders, and canned vegetable broths.

And if you think being vegetarian means an end to molded gelatin salads, consider this: Vegetarian sources of plain and fruit-flavored gelatin are available. They're made from sea vegetables and are sold in natural foods stores. Agar is one form, and it's made from red algae. You can use these products in the same ways that nonvegetarian gelatin is used.

If you use vegetarian gelatin in recipes in which regular gelatin is usually liquefied and then added to cold ingredients, you may need to change the process a bit. Vegetarian gelatin may begin to set immediately, so you need to add the rest of the recipe's ingredients right away. If a recipe doesn't turn out well when you use the original preparation method, try blending the vegetarian gelatin into cold liquid first, and then bringing the liquid to a boil.

Selecting Vegetarian Cookbooks

Though you can modify traditional nonvegetarian recipes to remove the animal ingredients, it can help to have a few vegetarian cookbooks on hand as well.

Vegetarian cookbooks are helpful for a couple of reasons. They provide explicit instructions for making recipes that may be new to you. You can even use a recipe as a starting point and modify it to make the dish uniquely your own.

And some people like to peruse cookbooks just for the ideas they can inspire. If you find yourself in a rut, thumbing through a favorite cookbook may remind you of something you haven't made in a while.

Dozens and dozens of excellent vegetarian cookbooks are available, ranging from vegan to lacto ovo vegetarian and including cuisines from many different cultures. I'm not going to attempt to list them here. However, you should realize that it's common to find one or two favorite cookbooks that you return to over and over again. Other cookbooks may contain only one or two recipes that you make now and then.

As you shop for vegetarian cookbooks, be especially attuned to the complexity of the recipes and the number and type of ingredients used. In my experience, the most practical cookbooks for home cooks are those with simple recipes requiring only basic cooking skills and equipment. Ingredient lists should be short and be relatively free of exotic ingredients that you aren't likely to use often or that may be expensive or hard to find.

In fact, I'm a strong advocate of learning to cook using no recipes at all. After you develop confidence in cooking, you should be able to put together foods that taste good — like the ingredients in a veggie burrito or a pot of vegetarian chili — without the need to measure. That's freedom in cooking!

You do have to use precise measurements in baking, which requires that certain ingredients be present in exact proportions. For most other simple recipes, though — fruit salad, a batch of hummus, or veggie lasagna, for instance — you can use a little more or less of this or that and everything will turn out just fine.

Part III

Meals Made Easy: Recipes for Everyone

"Do I like arugula? I love arugula!! Some of the best beaches in the world are there."

In this part . . .

Try your hand at some of the starter recipes I include in this part of the book. The recipes cover the range of food categories. I hope many of these become regular additions to your table.

Notice the short ingredient lists and simple preparation instructions. Preparing vegetarian meals can be quick and easy. None of the recipes I include in this book require more than basic cooking skills.

Most of these recipes can be adapted to suit your own needs and tastes. Some are vegan, and others can be easily modified to add or subtract animal ingredients (like butter, milk, and eggs) using information in Chapter 8 and throughout the book.

As you test these recipes, use a pencil to jot notes in the margins about what you liked, what you didn't like, and ideas for changes you'd like to try the next time you fix each recipe. If you make a modification, write a note in the book to help you remember what you changed and how much of the ingredient you added or took away.

Enjoy!

Chapter 9

Beyond Cereal and Toast: Whipping Up Breakfast Basics

In This Chapter

- Blending breakfast to go
- Replacing eggs with tofu
- Updating old favorites
- Waking up to a miso morning

Recipes in This Chapter

- Very Berry Smoothie
- Orange Juice Smoothie
- Scrambled Tofu
- Mushroom Quiche with Caramelized Onions
- Mom's Healthy Pancakes
- Fruited Oatmeal
- Favorite Cinnamon Rolls
- Vegan French Toast
- Morning Miso Soup

Although breakfast for many people is a bagel or glass of juice as they run out the door, eating a nutritious meal in the morning doesn't take a lot of planning or preparation. That's fortunate, because breakfast is important. It gives you an energy lift as you start your day and helps you stay alert and perform your best.

Some good vegetarian and fuss-free options include a bowl of dry cereal with milk, a piece of fruit, a cup of yogurt, or toast and jam. When you do have more time to prepare a meal in the morning, making something a little more substantial can be fun. You can make most breakfast classics, including breakfast breads and egg dishes, in a variety of ways to include as many or as few animal ingredients as you care to eat.

I include several examples in this chapter to get you started.

Getting Off to a Smoothie Start

Unlike their cousin the milkshake, smoothies made with fresh fruit and either nonfat milk or nondairy alternatives are low in calories and saturated fat and contain no cholesterol. They take only minutes to make in a blender and make a refreshing and convenient take-and-go breakfast alternative. I include two examples here — use your imagination to create other variations.

Vegans can replace frozen yogurt with nondairy alternatives available in natural foods stores. Ice cubes, crushed in blending, improve the mouth feel or texture of smoothies, making them frostier.

Very Berry Smoothie

Rich purple jewel tones dotted with tiny black flecks create a light, attractive refreshment. Serve this smoothie in a clear glass to show off the pretty colors. This one's a crowd-pleaser, so if you have company, plan to multiply the recipe.

Preparation time: *5 minutes*

Yield: *One large (16-ounce) serving or two smaller servings*

1 cup vanilla soymilk

½ ripe banana

1 cup frozen mixed berries (strawberries, blackberries, blueberries, and raspberries)

2 tablespoons pure maple syrup

1 Place all the ingredients in a blender.

2 Blend on high speed for about 1 minute or until smooth, stopping every 15 seconds to scrape the sides of the blender with a spatula and to push the solid ingredients down to the bottom of the blender.

3 Pour into a tall (16-ounce) tumbler or two smaller (8-ounce) glasses and serve immediately with an iced tea spoon and a straw.

Per serving: *Calories 190 (From Fat 27); Fat 3g (Saturated 0g); Cholesterol 0mg; Sodium 63mg; Carbohydrate 39g (Dietary Fiber 3g); Protein 4g.*

Orange Juice Smoothie

This delicious drink is reminiscent of the orange Creamsicles some of us enjoyed in summers past. Add a sprig of fresh mint as a garnish.

Preparation time: *10 minutes*

Yield: *Two 12-ounce servings*

1 cup orange juice

2 cups frozen nonfat vanilla yogurt

½ cup fresh orange sections (remove every bit of peel and white membrane — see Figure 9-1 for instructions)

1 teaspoon pure vanilla extract

5 or 6 ice cubes

1 Place all the ingredients in a blender.

2 Blend on high speed for about 1 minute or until smooth, stopping every 15 seconds to scrape the sides of the blender with a spatula and to push the solid ingredients down to the bottom of the blender. Thin the mixture as needed with a little more orange juice until it reaches the desired consistency.

3 Pour into two tall (16-ounce) tumblers and serve immediately with iced tea spoons and straws.

Vary It! *The fresh orange tastes great in this recipe, but you can leave it out if it's too much work or if you happen to be out of oranges.*

Per serving: *Calories 304 (From Fat 0); Fat 0g (Saturated 0g); Cholesterol 0mg; Sodium 132mg; Carbohydrate 65g (Dietary Fiber 1g); Protein 9g.*

Figure 9-1: The easy way to remove the white membranes of an orange.

Sectioning an Orange to Eliminate Membranes

1. cut cut

2. the dividing membrane

Using Tofu to Take the Place of Eggs

A big surprise to many people who are new to vegetarianism is that tofu works wonderfully as a stand-in for eggs in certain recipes. Although it can't take the place of an egg fried sunny side up, tofu can hold its own as an egg substitute in many other ways.

You can mix tofu with spices and fry it in a skillet with sliced onions and bell peppers to make a tasty variation of traditional scrambled eggs. With toast and hash browns, it makes a hearty and healthful breakfast alternative. Tofu also works wonderfully as an egg substitute in quiche.

Scrambled Tofu

Tofu picks up the flavors with which it's cooked, so this dish tastes very much like its egg counterpart — a savory blend of onions, bell peppers, and spices. Turmeric gives the dish its yellow hue and helps the tofu stand in for eggs. Serve this dish hot with Seasoned Home Fries (see Chapter 11), whole-grain toast, and juice. You can also serve it hot or cold as a sandwich filling in a pita pocket or on a kaiser roll.

Preparation time: *10 minutes*

Cooking time: *10 minutes*

Yield: *4 servings*

2 tablespoons olive oil

1 medium onion, chopped

2 teaspoons minced garlic

2 cups bell pepper strips (green, red, yellow, or mixed)

Two 12-ounce bricks of firm tofu

½ teaspoon black pepper

1 teaspoon turmeric

1 tablespoon soy sauce

Salt to taste

1 Heat the olive oil in a large skillet. Over medium heat, cook the onion, garlic, and bell peppers, stirring occasionally, until the onions are translucent and the peppers are soft — about 8 minutes.

2 Crumble the tofu into the onion-bell peppers mixture. Add the black pepper, turmeric, and soy sauce, and mix everything together with a wooden spoon or spatula. Heat thoroughly for about 2 minutes, mixing and scrambling the ingredients continuously.

Per serving: *Calories 149 (From Fat 90); Fat 10g (Saturated 1g); Cholesterol 0mg; Sodium 547mg; Carbohydrate 11g (Dietary Fiber 3g); Protein 8g.*

Mushroom Quiche with Caramelized Onions

The base of this quiche is tofu, and the texture and consistency are similar to — but not exactly like — those of a traditional quiche made with eggs. Popular among many vegetarians as a hot breakfast item, this dish may take some getting used to for those accustomed to the flavor and texture of egg-based quiche. The dish is delicious served with home fried potatoes (see the recipe for Seasoned Home Fries in Chapter 11) or a muffin and a seasonal fresh fruit salad. You might even add a side of soy-based link sausages or sausage patties. For brunch or lunch, consider serving this quiche with potatoes and steamed greens, such as kale or spinach. Cheese-eaters may want to mix in a handful of low-fat shredded cheddar cheese in Step 4.

Preparation time: *45 minutes*

Cooking time: *60 to 70 minutes*

Yield: *6 servings*

Two 12-ounce bricks of firm tofu

3 tablespoons olive oil

1 medium sweet onion, chopped

½ cup canned mushrooms, drained, or 2 cups sliced fresh mushrooms

2 tablespoons soy sauce

1 teaspoon dry mustard

½ teaspoon salt

¼ teaspoon black pepper

2 teaspoons minced garlic

2 tablespoons flour

2 tablespoons lemon juice

9-inch pie shell, unbaked (if you buy yours from a store, read the nutrient facts label and make sure the pie shell is lard- and trans fat-free)

Paprika

1 Slice the tofu and place it between two clean towels. Set a heavy cutting board or similar weight on top and press the tofu for 30 minutes.

2 Preheat the oven to 350 degrees. In a small skillet, heat the olive oil. Add the onion and slowly cook over medium heat until the onion is caramel-colored, stirring often. This will take 30 to 40 minutes. If you're using fresh mushrooms, sauté them with the onion.

3 In a medium mixing bowl, combine the tofu, soy sauce, dry mustard, salt, black pepper, garlic, flour, and lemon juice. Mash with a pastry blender or fork, and mix the ingredients well.

4 Add the onion and mushrooms to the tofu mixture. Stir until the ingredients are well blended.

5 Pour the mixture into the unbaked pie shell and spread the filling evenly. Sprinkle the top lightly with paprika.

6 Bake for 60 to 70 minutes, or until the crust is lightly browned and the quiche is set and looks firm in the middle when you jiggle the pan. Serve immediately.

Per serving: *Calories 377 (From Fat 225); Fat 25g (Saturated 4g); Cholesterol 0mg; Sodium 625mg; Carbohydrate 20g (Dietary Fiber 3g); Protein 21g.*

Putting a Vegetarian Spin on Breakfast Favorites

Pancakes, French toast, hot cereal, fresh cinnamon rolls, and other breakfast staples have always been vegetarian, but you can give these everyday foods a twist that most nonvegetarians don't. Health-conscious vegetarians often prefer to use whole grains, add such extras as dried fruit and nuts, and substitute nondairy alternatives for fatty dairy products to cut down on saturated fat, cholesterol, and animal protein. Taste for yourself.

Mom's Healthy Pancakes

These pancakes are fluffy yet hearty. As kids, my siblings and I simply called them "Healthies," and we'd beg my mother to make a batch on weekend mornings. Make these pancakes as soon as possible after mixing the batter, because the leavening action of the baking powder begins as soon as it mixes with the liquid ingredients. Batter left too long will begin to lose its leavening power and can result in flat, dense pancakes. Leftover pancakes keep well in the refrigerator or freezer, and you can reheat them in the microwave oven.

Preparation time: *10 minutes*

Cooking time: *5 minutes*

Yield: *6 servings (about twelve 5-inch pancakes)*

1 cup whole-wheat flour

½ cup white flour

⅓ cup wheat germ

1 teaspoon baking soda

1 teaspoon cinnamon

1 tablespoon baking powder

½ teaspoon salt

1¾ cups skim milk or soymilk

¼ cup vegetable oil

1 egg

2 egg whites, beaten stiff (see note)

1 Measure the dry ingredients into a medium-sized bowl.

2 Add the milk or soymilk, oil, and whole egg, and stir well using a whisk. Break up any remaining chunks of flour by using the back of a spoon.

3 Fold in the beaten egg whites with a wooden spoon or rubber spatula. The batter will be thick but light and somewhat foamy.

4 Pour the batter by ⅓-cup measures onto a hot, oiled skillet. When the pancakes are bubbly all over and the edges are browned, turn them over and cook for about 30 seconds, or until the undersides are browned. Serve immediately.

Tip: *Egg whites are beaten stiff when they've thickened enough to form a peak when you pull the beaters out of the bowl. If you keep beating egg whites past this point, they can collapse and become thin again. The key is to stop beating when you notice that the whites are forming soft peaks that can stand up on their own.*

Per serving: *Calories 253 (From Fat 99); Fat 11g (Saturated 2g); Cholesterol 37mg; Sodium 461mg; Carbohydrate 30g (Dietary Fiber 4g); Protein 10g.*

Fruited Oatmeal

This recipe is very simple, but the tartness of the Granny Smith apples makes it special. Expect an oatmeal-like texture with a bit of crunch from the nuts and some sweetness added by the apples and brown sugar. You can eat this oatmeal as is, but it's delicious served with vanilla soymilk on top.

Preparation time: *5 minutes*

Cooking time: *5 minutes*

Yield: *Two 1-cup servings*

1¾ cup water

1 cup quick-cooking rolled oats

½ Granny Smith apple, peeled and finely diced

2 tablespoons chopped walnuts or pecans

1 teaspoon cinnamon, plus extra for dusting

2 tablespoons brown sugar

1. Bring the water to a boil in a medium-sized saucepan.
2. Add the oats and boil for 1 minute, stirring constantly.
3. Add the apple, nuts, and cinnamon, and cook on low heat for an additional 2 to 3 minutes, or until heated through. The mixture will be thick and creamy.
4. Remove from the heat and ladle into serving bowls. Top each serving with a dusting of cinnamon and 1 tablespoon of brown sugar.

Vary It! *For softer apples, add the apples with the water in Step 1. Then add the oats and the remaining ingredients and cook as directed.*

Per serving: *Calories 273 (From Fat 72); Fat 8g (Saturated 1g); Cholesterol 0mg; Sodium 13mg; Carbohydrate 48g (Dietary Fiber 6g); Protein 6g.*

Favorite Cinnamon Rolls

These rolls are soft, chewy, and sweet and will fill your home with the welcoming fragrance of cinnamon and freshly baked bread. Egg whites replace whole eggs in this version to reduce saturated fat and cholesterol without sacrificing flavor.

Preparation time: *45 minutes, plus an hour and 45 minutes for dough to rise*

Cooking time: *20 minutes*

Yield: *24 rolls*

Rolls:

1 cup whole-wheat flour

3 cups all-purpose flour

1 package (¼ ounce) active dry yeast

1 cup plain soymilk or milk

¾ cup sugar

¼ cup vegetable oil

1 teaspoon salt

4 egg whites

¼ cup trans fat-free margarine

2 teaspoons cinnamon

Glaze:

1 cup powdered sugar

½ teaspoon pure vanilla extract

2 tablespoons soymilk or milk

1. In a large mixing bowl, combine the whole-wheat flour, 1 cup of the all-purpose flour, and the yeast. Set aside.
2. Combine the soymilk, ¼ cup of the sugar, the oil, and the salt in a saucepan and heat on low until warm (no more than 110 degrees, or just warm to the touch). Stir to blend, and then add to the flour and yeast mixture. Whisk in the egg whites.
3. Beat on high speed for about 4 minutes, stopping occasionally to scrape down the sides.
4. Stir in enough (most) of the remaining all-purpose flour to make a stiff dough.
5. Remove the dough from the mixing bowl and set it on a floured surface. Knead the dough for about 10 minutes, adding more flour by the tablespoon as needed to prevent sticking. When you're finished kneading, the dough should be smooth and elastic.
6. Set the dough in an oiled bowl, and then turn it to oil the top of the dough ball. Cover with a towel or waxed paper and let it rise in a warm place until doubled in size — about 1 hour.
7. After the dough has doubled, punch it down and divide it into two pieces. Roll each piece of dough into a rectangle about ¼ inch thick.
8. Melt the margarine and brush it onto each rectangle of dough. In a small cup, mix the cinnamon and the remaining ½ cup of sugar, and sprinkle the mixture evenly over both rectangles of dough.

9 Roll up the rectangles, starting at the widest ends. Pinch the ends shut with your fingers and press the seams into the dough (see Figure 9-2).

10 Cut the rolls into 1-inch pieces, and place the pieces cut side down in two oiled baking dishes or 9-inch round nonstick pans. Cover each dish with a towel or waxed paper, and let the rolls rise in a warm place until doubled in size — about 45 minutes. Preheat the oven to 375 degrees.

11 Bake for 20 minutes, or until the rolls are slightly browned. Don't overcook.

12 While the rolls are baking, prepare the glaze by mixing the confectioner's sugar, vanilla, and 1 tablespoon of the soymilk. Add additional soymilk in increments of 1 teaspoon until the glaze is thick but pourable. Drizzle the glaze over the warm rolls and serve.

Vary It! *Add ½ cup chopped walnuts, raisins, or currants to the filling in Step 8.*

Per serving: *Calories 164 (From Fat 41); Fat 5g (Saturated 1g); Cholesterol 0mg; Sodium 131mg; Carbohydrate 28g (Dietary Fiber 1g); Protein 3g.*

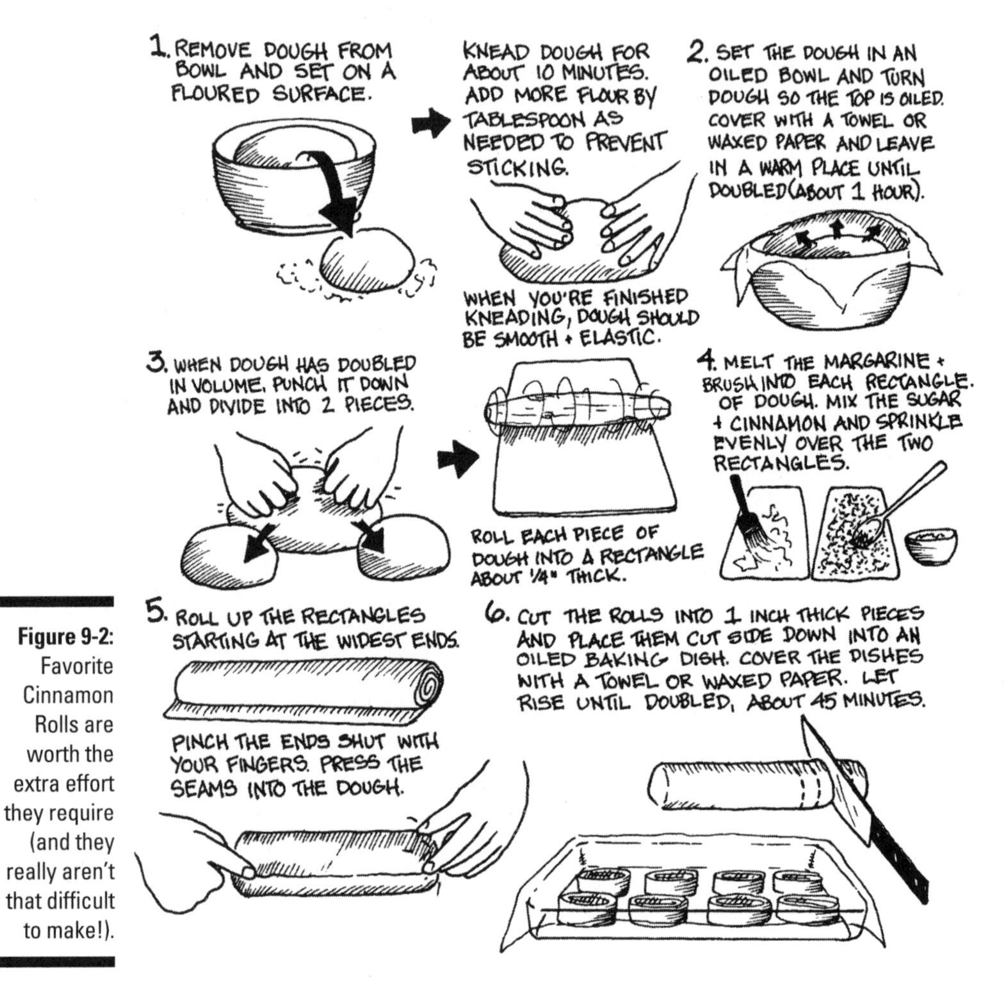

Figure 9-2: Favorite Cinnamon Rolls are worth the extra effort they require (and they really aren't that difficult to make!).

Vegan French Toast

Bananas and soymilk take the place of eggs in this version of French toast. Don't expect this to look or taste exactly like traditional French toast, but it's an interesting adaptation that's flavorful and nutritious.

Preparation time: *10 minutes*

Cooking time: *5 minutes*

Yield: *4 servings*

2 large ripe bananas

1 cup plain or vanilla soymilk

¼ teaspoon nutmeg

½ teaspoon pure vanilla extract (go ahead and use it even if you're using vanilla soymilk)

1 or 2 tablespoons of vegetable oil to grease the skillet

8 slices multigrain or whole-wheat bread

Powdered sugar

Maple syrup

Sliced kiwi fruit and strawberry halves for garnish (optional)

1 In a blender or food processor, purée the bananas, soymilk, nutmeg, and vanilla. Pour the mixture into a shallow pan such as a pie tin, cake pan, or 8-x-8-inch baking pan.

2 Generously oil a griddle and heat it until a drop of water spatters when flicked onto the pan.

3 Dip both sides of each slice of bread into the soymilk mixture, and transfer each slice to the griddle. The first side will take about 2 minutes to brown. Turn gently and carefully, because the bread tends to stick to the griddle. Cook the second side for about 3 minutes.

4 Carefully remove the French toast from the griddle and turn it so that the brown side is face up. Dust each slice with powdered sugar, and serve with a pitcher of warm maple syrup. Garnish with thin slices of kiwi fruit and strawberry halves (if desired).

Per serving: *Calories 245 (From Fat 27); Fat 3g (Saturated 0g); Cholesterol 0mg; Sodium 229mg; Carbohydrate 48g (Dietary Fiber 5g); Protein 7g.*

Starting Your Day the Miso Way

Miso is a fermented soy condiment — a rich, salty, savory paste — that's a key ingredient in many East Asian dishes, including soups, sauces, gravies, and salad dressings. You can find it in natural foods stores and Asian markets. Although you may not think of using miso at breakfast time, it's a common sight on breakfast tables in Japan, where miso soup is the traditional way to start the day. I provide more information about miso in Chapter 6.

Morning Miso Soup

This simple recipe for miso soup takes only minutes to make — quick enough for even the most harried of mornings. A big mugful is comforting at breakfast, but you can eat this soup as a snack or as part of a meal anytime.

Preparation time: *Less than 10 minutes*

Cooking time: *5 minutes*

Yield: *4 servings*

2 cups vegetable broth

2 cups hot water

4 tablespoons miso

1 teaspoon fresh ginger root, grated (optional)

½ cup thinly sliced mushrooms (see Figure 9-3)

½ cup diced firm tofu

3 tablespoons thinly sliced scallion greens

1 Pour the vegetable broth and 1 cup of the hot water into a medium saucepan.

2 In a bowl, dissolve the miso in the remaining 1 cup of hot water. Mix well, and then add to the contents of the saucepan.

3 Add the ginger root (if desired) and mushrooms and heat until simmering — about 5 minutes.

4 Remove from the heat and stir in the tofu and scallion greens. Serve in a mug or bowl.

Per serving: *Calories 76 (From Fat 27); Fat 3g (Saturated 0g); Cholesterol 0mg; Sodium 1217mg; Carbohydrate 9g (Dietary Fiber 2g); Protein 6g.*

How to Trim and Slice Mushrooms

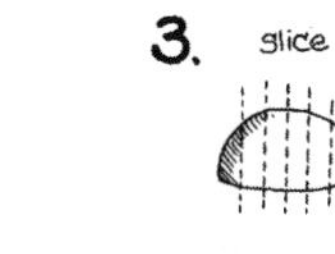

Figure 9-3: Sliced mushrooms are a key ingredient in Morning Miso Soup.

Chapter 10

Serving Simple Starters

In This Chapter

- Enjoying delicious dips and spreads
- Including flavors from around the world
- Fixing finger foods

Appetizers are versatile. You can eat dips and spreads now or as sandwich fillings tomorrow. You can pack leftover hors d'oeuvres in a bag lunch, or reheat them and serve them with a salad for a quick lunch.

You can also assemble several appetizers on a platter for a sophisticated — and often lighter — alternative to an entrée for dinner. The samples I include in this chapter are tasty, easy to make, and good for you, too.

Making Dips and Spreads

Recipes in this section borrow flavors from around the world. The first, hummus, is perhaps the most popular vegetarian dip. You'll find it on the menu at every Middle Eastern restaurant, and most vegetarian restaurants as well. Many variations exist, but the basic dip is made with garbanzo beans and olive oil, and it's very easy to make.

Beans, a popular ingredient in many cultures, are a staple in most vegetarian diets. They make great dips because you can puree them to a creamy texture and flavor them with a variety of ingredients. For more information about beans as ingredients in vegetarian cuisine, see Chapter 6.

The other recipes in this section are also tasty and versatile. From the cooling cucumber and dill yogurt dip and guacamole to zippy mango salsa and flavorful roasted garlic spread, these dips and spreads have wide appeal among vegetarians and meat-eaters alike. Use them to add color, flavor, and pleasing texture to your meals.

Basic Hummus with Toasted Pita Points

Hummus is a smooth, creamy, garlicky dip or spread that's made primarily from garbanzo beans, or chickpeas. This healthful dish is popular throughout the Middle East and is typically served in a shallow bowl, drizzled with a few teaspoons of olive oil and fresh lemon juice, with warm wedges of pita bread for dipping. You can also serve hummus as a dip for raw vegetables and as a pita pocket filling.

Preparation/cooking time: *15 minutes*

Yield: *Eight ¼-cup servings*

2 pieces of whole-wheat pita bread, each cut into 12 wedges

About 1 tablespoon olive oil for brushing onto pita wedges

One 15-ounce can garbanzo beans, rinsed (about 1¾ cups)

¼ cup water

1 large clove garlic, minced

¼ cup tahini

¼ cup lemon juice

¼ teaspoon cumin

2 teaspoons olive oil

½ fresh lemon

Paprika

1 Preheat the oven to 350 degrees. Brush olive oil onto the pita wedges using a pastry brush. Arrange them on a baking sheet and place into the oven.

2 While the pita wedges are browning in the oven, place the garbanzo beans, water, garlic, tahini, lemon juice, and cumin in a blender or food processor and process until smooth and creamy. Scoop into a shallow bowl.

3 Drizzle the olive oil over the hummus, followed by a squeeze of fresh lemon juice and a dusting of paprika.

4 Remove the pita wedges from the oven after about 15 minutes, or when the bread is lightly browned and toasted. (If you prefer warm and chewy pita wedges, start checking the bread after 7 to 10 minutes and remove it sooner.) Arrange the pita wedges around the bowl of hummus for dipping, or serve separately in a small basket or bowl.

Vary It! *Add 1 roasted red bell pepper or 1 tablespoon chopped fresh dill in Step 1. Increase garlic by 1 or 2 cloves for a greater garlicky punch.*

Per serving: *Calories 146 (From Fat 69); Fat 8g (Saturated 1g); Cholesterol 0mg; Sodium 155mg; Carbohydrate 17g (Dietary Fiber 3g); Protein 5g.*

Spicy Black Bean Dip

This dip has a mild flavor and a smooth, creamy texture. With the salsa added, it works well as a dip for tortilla chips and for raw vegetable pieces such as broccoli and cauliflower florets, baby carrot sticks, and bell pepper strips. Depending on what you have on hand, you can garnish the dip with parsley sprigs, minced green onions or tomatoes, grated cheddar or Jack cheese, a dollop of sour cream or mashed avocado, or any combination of these ingredients. If the dip sets for about 30 minutes, it thickens enough to be used as a filling for burritos or tacos. If you plan to use this as a filling or spread, the salsa is optional.

Preparation time: *10 minutes*

Cooking time: *10 minutes*

Yield: *Eight ¼-cup servings*

One 15-ounce can black beans, rinsed

½ cup warm water

½ small onion, minced

2 teaspoons minced garlic

¼ cup mild salsa (optional)

1 Combine the beans and water in a 2-quart saucepan. Cook over medium heat for 2 to 3 minutes, or until the beans are hot.

2 Remove from the heat and mash the beans well with a potato masher or fork. Add the onion and garlic and stir well.

3 Return the mixture to the stovetop and heat on low for 5 minutes, stirring constantly.

4 Stir in the salsa (if desired) and heat until the beans are hot and bubbly. Add more water by the tablespoon, if necessary, until the dip reaches the desired consistency.

5 Remove from the heat and serve.

Vary It! *In this recipe, the onions remain crunchy. If you prefer the onions cooked, sauté them for a few minutes in a teaspoon of olive oil before adding them to the beans in Step 2.*

Per serving: *Calories 55 (From Fat 9); Fat 1g (Saturated 0g); Cholesterol 0mg; Sodium 218mg; Carbohydrate 10g (Dietary Fiber 4g); Protein 4g.*

Tahini

Tahini is a paste made from ground sesame seeds. It has a mild sesame flavor and is used as an ingredient in some dips and salad dressings. You can find tahini at natural foods stores, Middle Eastern and specialty stores, and some supermarkets. It's often sold in a can with a plastic lid or in a jar, like natural peanut butter, with a layer of oil floating on top. Stir in the oil before scooping out the tahini.

Guacamole

Enjoy this rich, mildly flavored dip with tortilla chips, spread a layer on a sandwich, or fold some into a burrito. Avocado has a buttery consistency whether you mix the dip by hand or make it smooth in a blender or food processor. Note that you should use this recipe soon after preparation; the dip will turn dark on the surface. You can delay the discoloration by covering the surface of the dip with plastic wrap and keeping it chilled.

Preparation time: *10 minutes*

Yield: *Four ¼-cup servings*

1 medium avocado

¼ cup finely diced sweet onion (such as Vidalia)

1 teaspoon minced garlic

2 tablespoons lime juice

2 tablespoons salsa (optional)

1 Cut the avocado in half, remove the seed and peel, and place the meat in a small bowl (see Figure 10-1 for peeling and seeding instructions). Mash the avocado with a potato masher or fork until it's fairly smooth.

2 Add the remaining ingredients and mix well by hand. For a much smoother result, you can puree everything in a blender or food processor. If you do so, however, stir in the salsa by hand after processing.

Per serving: *Calories 94 (From Fat 63); Fat 7g (Saturated 1g); Cholesterol 0mg; Sodium 66mg; Carbohydrate 9g (Dietary Fiber 4g); Protein 2g.*

Figure 10-1: Extracting the meat of an avocado for blending into Guacamole.

Mango Salsa

You may be in the habit of buying bottled salsa from the store, but salsa is quick, easy, and inexpensive to make at home, and you can vary the flavors depending on the ingredients you have on hand. Try this version, made with fresh mango. Serve with black bean burritos, as a relish on veggie burgers, or with your favorite tortilla chips.

Preparation time: *15 minutes (plus chilling time)*

Yield: *Six ½-cup servings (may vary depending on size of produce used)*

1 large, ripe mango, pitted, peeled, and diced

½ small cucumber, peeled and finely diced

1 large ripe tomato, seeded and diced

¼ green bell pepper, seeded and chopped

¼ red bell pepper, seeded and chopped

¼ red onion, minced

1 tablespoon finely chopped jalapeño pepper

Juice of 1 large lime

3 tablespoons olive oil

2 tablespoons chopped cilantro (optional)

Salt and pepper to taste

1 Mix all the ingredients in a small bowl. Season to taste with salt and pepper.

2 Cover and chill for at least an hour, and stir before serving.

Vary It! *Add sliced avocado or fresh pineapple chunks. Adjust the amount of jalapeño peppers to accommodate preferences for mild, medium, or hot salsa.*

Per serving: *Calories 97 (From Fat 63); Fat 7g (Saturated 1g); Cholesterol 0mg; Sodium 101mg; Carbohydrate 9g (Dietary Fiber 1g); Protein 1g.*

Cucumber and Dill Yogurt Dip

You serve this dip cold with whole-grain crackers, toasted pita bread points, or fresh vegetable sticks. It's similar to Greek tzatziki sauce or Indian raita — both yogurt-based dips — so it makes a good accompaniment to spinach pie (Greek spanakopita) and spicy Indian entrees. You can also use it to top a bean burrito!

Preparation time: *10 minutes (plus chilling time)*

Yield: *Six ¼-cup servings*

½ medium cucumber, peeled and finely chopped

1 large clove garlic, minced

1 tablespoon fresh dill, chopped

One 8-ounce carton of nonfat, plain yogurt

1. Peel the cucumber, and then chop it into fine pieces (see Figure 10-2). Mince the garlic and chop the fresh dill.
2. Scoop the yogurt into a small bowl. Add the cucumbers, garlic, and dill, and stir to mix the ingredients.
3. Cover and chill for at least an hour before serving.

Vary It! *Get creative. Instead of dill, try using fresh cilantro and a dash of cumin, or omit the garlic and substitute fresh mint for dill. Adding salt and pepper to taste or a squeeze of fresh lemon are also good options.*

Per serving: *Calories 24 (From Fat 1); Fat 0g (Saturated 0g); Cholesterol 1mg; Sodium 30mg; Carbohydrate 4g (Dietary Fiber 0g); Protein 2g.*

CHOPPING A CUCUMBER

1. WASH AND PEEL THE CUCUMBER USING A VEGETABLE PEELER (OPTIONAL).

2. SLICE OFF ABOUT ½" OF THE ENDS OF THE CUCUMBER

3. FOR CUCUMBER ROUNDS, MAKE EVEN, ⅛" SLICES WITH A FRENCH KNIFE.

4. TO CHOP CUCUMBER INTO SMALLER PIECES, FIRST CUT LENGTHWISE. SLICE EACH HALF INTO SEVERAL LONG STRIPS, THEN SLICE THE STRIPS INTO PIECES.

Figure 10-2: Chop the cucumber into small pieces.

Roasted Garlic Spread

This dish is as simple as can be to make, and it makes the house smell so good. Pop a garlic bulb into the oven for a nutritious spread on French bread rounds or homemade rolls.

Preparation time: *1 minute*

Cooking time: *1 hour*

Yield: *4 servings*

1 large garlic bulb

2 teaspoons olive oil

1. Preheat the oven to 350 degrees.
2. Pull the outer, loose layers of tissue off the garlic bulb, leaving enough so that the cloves remain intact.
3. With a sharp knife, cut off the top of the bulb (about ¼ inch of the top — see Figure 10-3) to expose the tops of the cloves inside. Drizzle the tops of the cloves with the olive oil.
4. Wrap the bulb loosely in aluminum foil, seal the foil completely, and place in the hot oven.
5. Bake for about 1 hour, or until the cloves are soft. Remove from the oven.
6. Place the garlic bulb on a small serving dish. Lift the cloves out of the bulb with a dinner knife and smear the softened garlic on bread. If the cloves are hard to spread, squeeze the bulb to loosen them.

Per serving: *Calories 20 (From Fat 9); Fat 1g (Saturated 0g); Cholesterol 0mg; Sodium 1mg; Carbohydrate 2g (Dietary Fiber 0g); Protein 0g.*

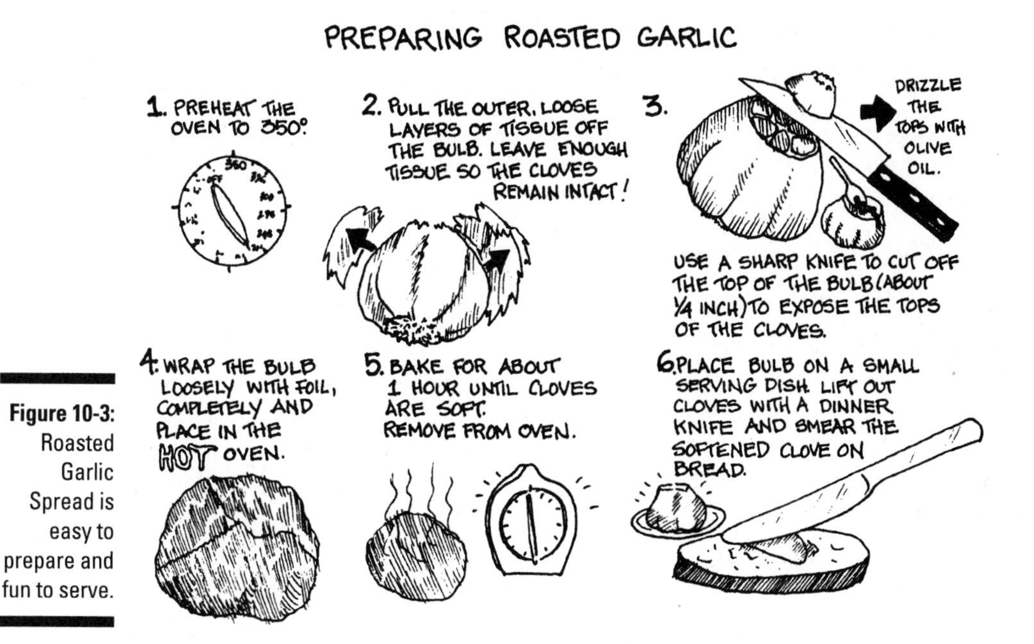

Figure 10-3: Roasted Garlic Spread is easy to prepare and fun to serve.

Creating Other Easy Appetizers

Bite-sized portions of many foods — pizza, sandwiches, and spinach pie, for example — lend themselves to serving as hors d'oeuvres or simple starters before a meal. Following are three more examples of meatless appetizers with widespread appeal.

Cheesy Pesto French Bread Rounds

Fill your kitchen with the intoxicating aroma of basil and warm bread. Serve these as an appetizer or accompanied by a green salad or bowl of soup for a light meal.

Preparation time: *10 minutes*

Cooking time: *20 minutes*

Yield: *6 servings (3 rounds each)*

1 French baguette sliced into eighteen ¼-inch-thick rounds (about 2 inches in diameter)

¾ cup pesto (use the pesto recipe from Pesto Pasta Primavera in Chapter 12, or use ready-made pesto)

1 heaping cup grated, part-skim mozzarella cheese

1. Preheat the oven to 350 degrees. Spray or rub a cookie sheet or pizza pan with vegetable oil. Alternatively, line the cookie sheet with foil, and then rub the foil with a tablespoon of olive oil to keep cheese from baking onto the cookie sheet.
2. Spread each round with 2 teaspoons of pesto.
3. Top each round with about 1 tablespoon of grated cheese.
4. Bake in the oven for about 20 minutes, or until cheese is lightly browned and bubbly. Remove from the cooking sheet and serve immediately.

Vary It! *Add 1 teaspoon of black olive tapenade or sun-dried tomatoes to tops of rounds.*

Per serving: *Calories 444 (From Fat 208); Fat 23g (Saturated 7g); Cholesterol 21mg; Sodium 665mg; Carbohydrate 45g (Dietary Fiber 4g); Protein 15g.*

Grilled Vegetable Quesadilla

Cut quesadilla into wedges and serve as an appetizer with Mango Salsa (see the recipe earlier in this chapter). These also make a quick meal that kids love. Serve half of a large quesadilla with a small green salad for a light lunch or supper.

Preparation time: *15 minutes*

Cooking time: *15 minutes*

Yield: *1 large quesadilla (serves four as an appetizer or two as a light meal)*

½ cup chopped broccoli florets

¼ cup grated carrots

¼ cup chopped onion

½ cup chopped zucchini squash

¼ cup red or yellow bell peppers, seeded and chopped

Salt and pepper to taste (optional)

4 tablespoons olive oil

2 large whole-wheat flour tortillas

¾ cup grated, reduced-fat Jack cheese or cheddar and Jack cheese mixture

Salsa; plain, nonfat yogurt; or reduced-fat sour cream and sliced black olives for garnish (optional)

1 Prepare broccoli, carrots, onions, zucchini, and bell peppers and place them into a glass bowl. Add 2 teaspoons of water, cover, and heat in a microwave oven for 3 to 5 minutes, until tender. Drain and set aside. Salt and pepper to taste (if desired).

2 Smear 2 tablespoons of olive oil on the bottom of a large skillet. Place a flour tortilla into the skillet.

3 Sprinkle half the cheese evenly on top of the flour tortilla, and then distribute steamed vegetables evenly on top of the cheese. Top the vegetables with the remaining cheese.

4 Top with second flour tortilla. Brush the top of the second tortilla with remaining olive oil.

5 Cook on medium heat. After several minutes, lift the edge of the bottom quesadilla with a spatula to check for doneness. When the bottom tortilla appears lightly browned and the cheese is melted (about 5 to 7 minutes), carefully turn the quesadilla over and cook for an additional 2 or 3 minutes, until the other side is lightly browned.

6 Remove from the heat and cut into wedges for serving. Top with a scoop of salsa, a dollop of plain, nonfat yogurt or reduced-fat sour cream, and sliced black olives.

Vary It! *Substitute ¾ cup cooked, drained, chopped spinach for the vegetables to make a spinach quesadilla.*

Per serving: *Calories 263 (From Fat 176); Fat 20g (Saturated 5g); Cholesterol 15mg; Sodium 91mg; Carbohydrate 15g (Dietary Fiber 2g); Protein 8g. Analyzed for 4.*

Stuffed Mushrooms

This appetizer goes well with Cheesy Pesto French Bread Rounds (see the recipe earlier in this chapter) and a selection of olives. A mini-chopper or small food processor can make preparation easier and faster.

Preparation time: *15 minutes*

Cooking time: *25 minutes*

Yield: *6 servings (3 mushrooms each)*

18 medium white mushrooms (about 1 pound)

2 tablespoons olive oil

1 small onion, minced (about ½ cup)

2 cloves garlic, minced

½ cup breadcrumbs

1 teaspoon dried oregano

¼ cup chopped fresh parsley

Salt and ground black pepper to taste

¼ cup grated Parmesan cheese (optional)

Fresh rosemary or basil sprigs for garnish

1. Preheat the oven to 375 degrees.
2. Rinse and pat the mushrooms to dry them. Trim off the bottoms of stems. Separate the stems from the mushroom caps to create a small crater in the bottom of each cap. Chop the stems finely and set aside.
3. In a medium skillet, add olive oil and sauté the mushroom stems, onions, and garlic on medium heat until the onions are translucent, about 5 minutes. Stir in the breadcrumbs, oregano, parsley, salt and pepper, and cheese.
4. Distribute the filling equally among mushroom caps, pressing lightly to set filling into each cap.
5. Place mushroom caps on an oiled baking sheet. Bake for about 20 minutes, or until the mushrooms are hot. Remove from the baking sheet and place on a serving dish. Garnish with fresh rosemary or basil sprigs.

Vary It! *Experiment with additions to filling mixture. Add a few tablespoons of chopped walnuts or pecans, basil, or sundried tomato pesto.*

Per serving: *Calories 97 (From Fat 47); Fat 5g (Saturated 1g); Cholesterol 0mg; Sodium 179mg; Carbohydrate 11g (Dietary Fiber 1g); Protein 3g.*

Chapter 11

Enjoying Easy Soups, Salads, and Sides

In This Chapter

- Making savory soups with beans and lentils
- Using salads as light meals, sides, and snacks
- Rounding out meals with side dishes

Recipes in This Chapter

- Lentil Soup
- Classic Gazpacho
- Cashew Chili
- Arugula Salad with Pickled Beets and Candied Pecans
- Easy Four-Bean Salad
- Italian Chopped Salad
- Tofu Salad
- Tabbouleh
- Seasoned Home Fries

Soups, salads, and sides complete a meal by complementing the entrée and by adding flavor, texture, and color to the plate. They're versatile, too. Alone, a hearty bowl of soup, generous salad, or side — or any combination of two or more — can make the perfect lunch or dinner. Soup is even on the menu in Chapter 9 — for breakfast!

I love the convenience of soups, salads, and sides, so I keep plenty of the ingredients in my pantry and refrigerator. Use them to add variety to meals and as quick, light meals and snacks by themselves.

Serving Soups for All Seasons

Beans and lentils are the foundations of many filling, nutritious vegetarian soups. High in protein and fiber, they give soups a stick-to-your ribs quality that makes them well suited for the main course at meals. Use canned beans or dried, or try your hand at using a pressure cooker to soften dried beans more quickly (I include more information about pressure cookers in Chapter 8). Recipes in this section include traditional lentil soup; a classic, cold vegetable soup (gazpacho); and chili with a twist, made with raisins and cashews.

Lentil Soup

This soup has a rich, savory flavor, and the lentils cook quickly because of their small size and flat shape. If you have leftover cooked spinach, stir it into this soup to add color and a healthful nutrient boost.

Preparation time: *10 minutes*

Cooking time: *About 1 hour*

Yield: *8 servings*

1¼ cups dried lentils (½ pound)

5 cups water

1 medium onion, chopped

1 teaspoon minced garlic

½ teaspoon salt

½ teaspoon black pepper

2 tablespoons olive oil

One 16-ounce can stewed or crushed tomatoes

1 bay leaf

1 Rinse the lentils in a colander or strainer.

2 Combine the lentils, water, onion, garlic, salt, pepper, and olive oil in a large saucepan. Cover and cook on medium-high heat until boiling — about 14 minutes.

3 Stir in the tomatoes and bay leaf.

4 Reduce the heat to low, cover, and let simmer for 45 minutes, or until the lentils are tender.

Vary It! *In place of regular crushed tomatoes, try fire-roasted crushed tomatoes if you can find them at your grocery store.*

Per serving: *Calories 144 (From Fat 36); Fat 4g (Saturated 1g); Cholesterol 0mg; Sodium 325mg; Carbohydrate 22g (Dietary Fiber 3g); Protein 6g.*

Classic Gazpacho

This spicy Spanish soup is served cold. Many variations exist, from very smooth to chunky. Adjust the heat according to your taste by adding more or less cayenne pepper and hot pepper sauce (this version is relatively mild). The soup's color and taste may vary slightly, depending on the tomatoes you use.

Preparation time: *15 minutes, plus time for bread to marinate*

Yield: *5 servings*

1 slice stale white bread

2 tablespoons red wine vinegar

1 tablespoon olive oil

1 teaspoon minced garlic

One 28-ounce can whole peeled tomatoes

½ cup finely chopped onion

1 green bell pepper, chopped fine (see Figure 11-1)

1 cucumber, peeled and sliced

¼ teaspoon cayenne pepper

1 teaspoon hot pepper sauce

Juice of 1 fresh lemon (about 2 tablespoons)

1 Break the bread into small pieces and place it in a small cup or bowl. Pour the vinegar and oil on top and add the garlic. Mash with a fork, and then set aside for at least 30 minutes.

2 Place the bread mixture in a blender and add 1 cup of the liquid from the canned tomatoes, plus the onion, green bell pepper, and cucumber (save a few tablespoons of onion and green bell pepper for garnish). Blend at low speed for 1 minute.

3 Add the remaining tomatoes and juice, cayenne pepper, hot pepper sauce, and lemon juice. Give the mixture a whirl in the blender to break up the whole tomatoes. Chill.

4 Garnish with a sprinkling of chopped onions, green bell peppers, or croutons and serve.

Per serving: *Calories 99 (From Fat 27); Fat 3g (Saturated 0g); Cholesterol 0mg; Sodium 409mg; Carbohydrate 14g (Dietary Fiber 3g); Protein 3g.*

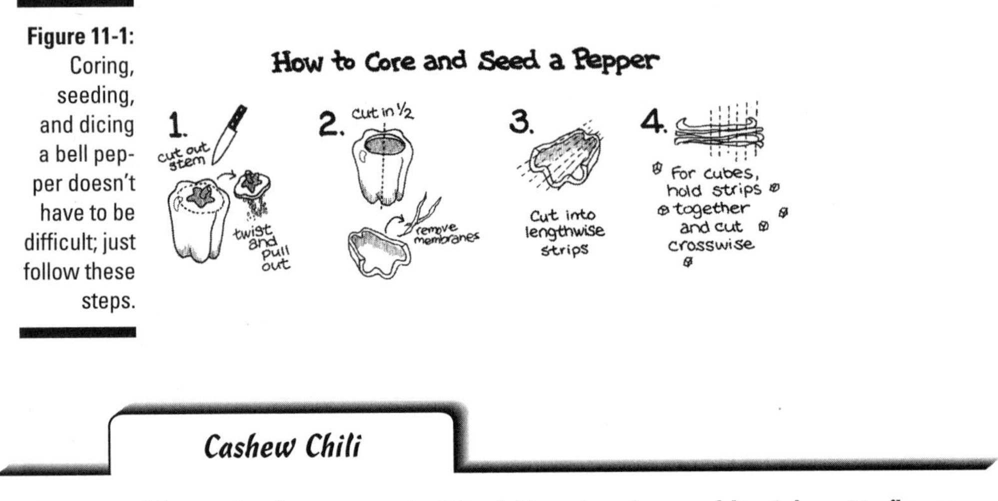

Figure 11-1: Coring, seeding, and dicing a bell pepper doesn't have to be difficult; just follow these steps.

Cashew Chili

Raisins add a touch of sweetness to this chili, and cashews add a rich, nutty flavor. Serve it over steamed rice with cornbread and a green salad on the side. This chili is thick, and it thickens considerably more if left overnight.

Preparation time: *20 minutes*

Cooking time: *About 55 minutes*

Yield: *6 large servings*

2 tablespoons olive oil

4 medium onions, chopped

2 large green bell peppers, seeded and chopped

2 stalks celery, minced .

4 teaspoons minced garlic

1 teaspoon dried basil

1 teaspoon oregano

½ teaspoon chili powder

1 teaspoon ground cumin

1 teaspoon black pepper

One 15-ounce can tomato sauce

One 16-ounce can stewed tomatoes or whole peeled tomatoes

2 tablespoons red wine vinegar

1 bay leaf

1 cup cashew pieces

⅓ cup raisins

3 cups cooked dark red kidney beans (two 15-ounce cans)

1 In a large pot, heat the olive oil. Add the onions, bell peppers, and celery and cook over medium heat until the onions are translucent — about 10 minutes.

2 Stir in the garlic, basil, oregano, chili powder, cumin, and pepper.

3 Add the tomato sauce, tomatoes (with their juice), vinegar, and bay leaf. Reduce the heat to low and continue cooking for 2 to 3 minutes.

4 Stir in the cashews and raisins and cook over low heat for another 16 to 17 minutes.

5 Add the beans and cook for an additional 25 minutes, stirring frequently. The chili is done when all the ingredients are well blended and soft and the chili is thick and bubbly.

Vary It! *In place of dark red kidney beans, try using a mixture of garbanzo beans, pinto beans, and red kidney beans for a change of pace. A handful of fresh or frozen corn kernels adds color.*

Per serving: *Calories 401 (From Fat 144); Fat 16g (Saturated 3g); Cholesterol 0mg; Sodium 1,067mg; Carbohydrate 57g (Dietary Fiber 11g); Protein 14g.*

Going Beyond Iceberg Lettuce

Chilled salads made with fresh, seasonal ingredients are palate-pleasing refreshments in summer months and add contrast to the warmth and texture of cooked foods in cold-weather months. These dishes are delicious and very easy to make.

Arugula Salad with Pickled Beets and Candied Pecans

This salad is a delicious and nutritious alternative to mixed green salads. Bottled, pickled beets from your pantry and candied pecans available in many supermarkets make this special salad a snap to assemble.

Preparation time: *15 minutes*

Yield: *6 servings*

6 ounces goat cheese or chèvre

2 tablespoons freshly ground black pepper for dipping

6 heaping cups arugula, washed, with heavy stems removed

1 cup sliced, pickled beets

1 cup candied pecan halves

1. Divide the goat cheese or chèvre into 12 portions. Let the cheese stand at room temperature for 10 to 15 minutes, and then roll it into balls and dip one side of each ball into the pepper.
2. Arrange the arugula on 6 plates. Cut the beet slices in half, and then distribute the beets and pecans evenly over the arugula. Put 2 balls of cheese on each plate.
3. Drizzle the salad with your favorite vinaigrette dressing and serve immediately.

Vary It! *Vegans omit the goat cheese. Also good: Sprinkle the salad with dried cherries or dried cranberries.*

Per serving: *Calories 351 (From Fat 291); Fat 32g (Saturated 9g); Cholesterol 22mg; Sodium 404mg; Carbohydrate 22g (Dietary Fiber 1g); Protein 9g.*

Arugula

Arugula, also called *rocket salad* or *roquette,* is a dark, leafy green known for its distinctive sharp, spicy, peppery flavor, similar to that of mustard greens. Arugula is native to the Mediterranean and grows wild throughout southern Europe. You can serve it cooked, but it's most popular as a salad green. Arugula is becoming very popular in the U.S., and you can find it in many supermarkets.

Easy Four-Bean Salad

My advice: Double this recipe. It's delicious, and the leftovers taste even better after a day or two in the refrigerator. If you prefer a less sweet salad, reduce the amount of sugar to your taste.

Preparation time: *10 minutes, plus chilling time*

Yield: *8 to 10 servings*

One 15½-ounce can cut green beans, drained and rinsed

One 15½-ounce can cut wax beans, drained and rinsed

One 15½-ounce can dark red kidney beans, drained and rinsed

One 15½-ounce can garbanzo beans, drained and rinsed

½ cup chopped green bell pepper

½ red onion, chopped

½ cup sugar (or less, to taste)

⅔ cup red wine vinegar

⅓ cup vegetable oil (avoid olive oil, as it solidifies in the refrigerator)

½ teaspoon salt

2 teaspoons black pepper

1. Place the green beans, wax beans, kidney beans, garbanzo beans, bell pepper, and onion in a large bowl.
2. Combine the sugar, vinegar, and oil in a small bowl or cup, stir well, and pour over the bean mixture. Add the salt and pepper and toss to coat.
3. Chill overnight. Toss again to coat the beans before serving.

Per serving: *Calories 218 (From Fat 88); Fat 10g (Saturated 1g); Cholesterol 0mg; Sodium 399mg; Carbohydrate 29g (Dietary Fiber 5g); Protein 5g.*

Italian Chopped Salad

Everyone remembers his or her first Italian chopped salad. Mine was in the beach town of Lahaina on the western coast of Maui. Exact proportions or measurements aren't necessary. The key is that you chop everything into tiny pieces. That's what makes this salad so addictive — the texture. You'll eat heaping bowls full of this salad — a good thing, because it's as nutritious as it is delicious.

Preparation time: *45 minutes*

Yield: *6 servings*

1 large head of romaine lettuce, chopped into ½-inch squares

1 cup canned garbanzo beans, rinsed, drained, and chopped

½ cucumber, peeled and minced

½ green bell pepper, seeded and finely chopped

½ yellow or orange bell pepper, seeded and finely chopped

2 or 3 radishes, minced

1 stalk celery, finely chopped

½ medium red onion, minced

1 large tomato, seeded and chopped into ¼-inch pieces

2 or 3 tablespoons of finely chopped fresh parsley

⅓ cup chopped black olives

½ cup grated Parmesan cheese or finely chopped, reduced-fat Swiss cheese (optional)

Vinaigrette dressing to taste

Freshly ground black pepper

1 Mix all the vegetables and cheese (if desired) in a salad bowl. Add your favorite Italian or vinaigrette dressing, and toss well.

2 Divide the salad into six bowls and add pepper to taste. Serve immediately.

Vary It! *Use up whatever salad ingredients you have on hand. Finely chopped yellow summer squash or zucchini, banana peppers, grated carrot, and jicama are good in this salad.*

Per serving: *Calories 205 (From Fat 128); Fat 14g (Saturated 3g); Cholesterol 5mg; Sodium 519mg; Carbohydrate 14g (Dietary Fiber 4g); Protein 7g.*

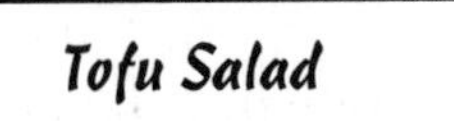

Tofu Salad

If you've never tried tofu salad, you're in for a nice surprise. Many variations exist. This recipe is similar to traditional egg salad, but tofu takes the place of egg whites, and turmeric and mustard provide the color. The result is a mock egg salad that tastes and looks very much like the real thing. Serve it as a sandwich filling on bread, crackers, or rolls, or in a pita pocket with grated carrots. This salad also works well served on a bed of salad greens or as stuffing for a fresh, ripe tomato.

Preparation time: *10 minutes*

Yield: *6 servings*

1 pound firm tofu

½ teaspoon salt

½ cup soy mayonnaise

2 teaspoons yellow mustard

½ teaspoon garlic powder

½ teaspoon black pepper

¼ teaspoon turmeric

¼ cup finely chopped celery

2 green onions, finely chopped

1 Mash the tofu with a fork or potato masher until crumbly.

2 Add the salt, mayonnaise, mustard, garlic powder, pepper, and turmeric and mix well.

3 Add the celery and green onions and mix well. Chill before serving.

Per serving: *Calories 110 (From Fat 63); Fat 7g (Saturated 1g); Cholesterol 0mg; Sodium 132mg; Carbohydrate 6g (Dietary Fiber 1g); Protein 6g.*

On the Side

You can serve any of the soups or salads I include in this chapter as simple side dishes at almost any meal. A small cup of soup on the side of the plate, a scoop of tofu salad, or a handful of green salad serve as nutritious garnishes — a touch of color or cool crunch — at the same time that they round out your meals.

I wrap up this chapter with two popular side dishes that you can serve with a variety of foods.

Tabbouleh

Lemon juice gives this Middle Eastern dish its tangy flavor, and mint provides its characteristic fragrance. You can serve tabbouleh as a side dish with sandwiches and burgers, or add it to hummus in a pita pocket.

Preparation time: *20 minutes, including time to cook bulgur wheat and let it cool*

Cooking time: *20 minutes*

Yield: *6 servings*

2 cups water

1 cup bulgur wheat

¼ cup olive oil

Juice of 1 large lemon (about ¼ cup)

½ teaspoon salt

3 green onions, chopped (white and green parts)

¼ cup chopped fresh mint leaves

½ cup chopped fresh parsley

½ cup canned garbanzo beans, drained and rinsed

2 medium tomatoes, finely chopped

1 Bring the water to a boil in a saucepan. Add the bulgur wheat, reduce the heat to low, cover, and simmer until the bulgur absorbs all the water — about 8 to 10 minutes.

2 Remove the bulgur wheat from the stovetop and set it aside to cool for about 15 minutes.

3 In a medium bowl, combine the cooled bulgur wheat and the remaining ingredients (except the tomatoes) and toss well.

4 Chill the tabbouleh for at least 2 hours before serving. It's best when you chill it overnight. Add tomatoes and toss again just before serving.

Vary It! *Reduce the mint (or leave it out entirely) if you're not a big fan.*

Per serving: *Calories 238 (From Fat 90); Fat 10g (Saturated 1g); Cholesterol 0mg; Sodium 206mg; Carbohydrate 34g (Dietary Fiber 7g); Protein 5g.*

Seasoned Home Fries

This is the ultimate easy recipe — you don't have to measure anything, and it comes out great every time. These potatoes are hugely popular with everyone I serve them to, and the leftovers are good reheated. The cayenne pepper and paprika give the potatoes a nice coppery color, but you can leave out or reduce the amount of the cayenne pepper if you like your food less spicy. You can pair this dish with Scrambled Tofu or Mushroom Quiche with Caramelized Onions from Chapter 9, or with any sandwich and many entrées.

Preparation time: *10 minutes*

Cooking time: *40 minutes*

Yield: *About 8 servings*

Vegetable oil spray

6 medium white potatoes

Several shakes each of garlic powder, oregano, cayenne pepper, and paprika

1 Preheat the oven to 350 degrees.

2 Coat a baking sheet with vegetable oil spray or a thin layer of olive oil.

3 Wash the potatoes. Leaving the peels on, cut the potatoes into wedges and place them in a mixing bowl. (You can quarter small potatoes. Cut fist-sized potatoes into 6 or 8 wedges each.)

4 Sprinkle the potato wedges with several shakes of each of the spices, more or less to your taste. Toss to coat.

5 Spread the potatoes in a single layer on the baking sheet. Spray the tops with a thin film of vegetable oil spray, or brush with olive oil.

6 Bake for about 40 minutes, or until the potatoes are soft. Serve plain, with ketchup or salsa, or with malt vinegar for dipping.

Per serving: *Calories 117 (From Fat 9); Fat 1g (Saturated 0g); Cholesterol 0mg; Sodium 6mg; Carbohydrate 26g (Dietary Fiber 2g); Protein 2g.*

Chapter 12

Making Meatless Main Dishes

In This Chapter

- Preparing bean-based entrées
- Using pasta in creative combinations
- Enjoying old favorites with a twist
- Adding an Asian influence

Recipes in This Chapter

- Cajun Red Beans and Rice
- Basic Bean Burrito
- Cuban Black Beans
- Pesto Pasta Primavera
- Rotini with Chopped Tomatoes and Fresh Basil
- Vegetarian Lasagna
- Roasted Vegetable Pizza
- Tempeh Sloppy Joes
- Vegetable Stir-Fry
- Soy-Ginger Kale with Tempeh

If you're new to a vegetarian diet, visualizing meatless meals may be hard at first. Western culture tends to define meals by the meat that's served. Ask some nonvegetarian friends what they're having for dinner tonight, and they're likely to mention the chicken or fish — not the salad or the peas and carrots.

In fact, vegetarian meals often have no main dish or entrée. Instead, vegetarians can make a meal of any number of side dishes. Beans, rice, vegetables, and other foods come together and fill the plate in colorful combinations of healthful, tasty foods.

Who says you need a focal point?

When there is a focal point, vegetarian entrées often resemble meat-based entrées — minus the meat. Lasagna, burritos, and pizza, for example, can all be made meat-free.

Other vegetarian main dishes reflect cuisines from cultures around the world. Delicious stir-fries, bean and rice dishes, and soy foods are some examples I include in this chapter.

Beans: Versatility in a Can

Beans are a staple ingredient that nearly every culture uses. Inexpensive and one of the most nutritious sources of protein, they should be a mainstay on your vegetarian menus. I include three favorite recipes here.

Cajun Red Beans and Rice

This dish is a tradition in deep-South coastal states such as Louisiana, where seasoned beans and rice are a favorite among people of mixed French, Spanish, and African descent. If you make the rice ahead of time, this dish takes very little time to prepare. Toss a salad and steam some broccoli or greens, and dinner's ready.

Preparation time: *10 minutes*

Cooking time: *15 minutes*

Yield: *8 servings*

3 tablespoons olive oil

1 large onion, chopped

3 cloves garlic, minced

½ cup chopped green bell pepper

1 stalk celery, including green leaves, chopped

½ teaspoon salt

1 teaspoon cumin

1 tablespoon chili powder

½ teaspoon thyme

Two 15-ounce cans dark red kidney beans, rinsed and drained (about 3 cups)

4½ cups cooked rice

Fresh parsley for garnish

1 In a large skillet, heat the olive oil. Cook the onions, garlic, bell pepper, and celery in the oil over medium heat until the onions are translucent — about 7 minutes. Add the salt, cumin, chili powder, and thyme, and stir to combine.

2 Add the beans and mix well. Reduce the heat to low and continue cooking for several minutes until the beans are hot, stirring frequently to prevent sticking.

3 Add the rice to the bean mixture and mix all the ingredients well. Cook for about 5 minutes to heat the rice through before serving. Garnish with sprigs of parsley.

Per serving: *Calories 266 (From Fat 54); Fat 6g (Saturated 1g); Cholesterol 0mg; Sodium 164mg; Carbohydrate 45g (Dietary Fiber 7g); Protein 9g.*

Basic Bean Burrito

You can doctor up this burrito anyway you like, but here's a bare-bones version that virtually everyone loves. If you use whole beans, mash them with a potato masher or fork and mix them until they're smooth in consistency. Cheese eaters can add a couple tablespoons of grated, low-fat cheddar cheese or cheese alternative to the topping if desired.

Preparation time: *5 minutes*

Cooking time: *7 minutes*

Yield: *4 burritos*

2 cups canned vegetarian refried beans, plain mashed pinto beans, or black beans

Four 10-inch whole-wheat flour tortillas

Chopped tomato, romaine lettuce or spinach, and green onions (any combination, about 1 cup total)

¼ cup plain nonfat yogurt or low-fat sour cream (vegans can use soy yogurt or nondairy sour cream)

1 cup salsa

1 In a small saucepan, warm the beans over medium heat until heated through — about 7 minutes.

2 For each burrito, follow these instructions:

Lay a tortilla flat on a dinner plate. (Warm it first, if you like, by heating it on each side for a minute or two in a hot skillet). Spoon about ½ cup of the beans onto the center of the tortilla.

Fold one end of the tortilla toward the middle, and then fold the sides toward the middle. Leave the burrito on the plate with the end of the fold tucked underneath so that the burrito doesn't unroll (see Figure 12-1).

3 Top with salsa, chopped tomato, greens, and green onion. Add a dollop of plain yogurt and serve immediately.

Vary It! *Express yourself by adding banana or jalapeño peppers, sliced black olives, sliced avocado, mashed sweet potatoes, strips of cooked tempeh . . . or whatever suits your fancy.*

Per serving: *Calories 275 (From Fat 41); Fat 5g (Saturated 0g); Cholesterol 0mg; Sodium 197mg; Carbohydrate 45g (Dietary Fiber 6g); Protein 14g.*

Cuban Black Beans

Serve these beans over a plate of your favorite steamed rice, or thin the beans with vegetable broth and serve them as black bean soup. This flavorful, hearty dish goes well with a refreshing green salad and a chunk of crusty bread.

Preparation time: *15 minutes*

Cooking time: *50 minutes*

Yield: *8 servings*

¼ cup olive oil

1 large onion, chopped (see Figure 12-2)

1 green bell pepper, chopped

2 stalks celery, including green leaves, chopped

4 garlic cloves, minced

4 cups (two 20-ounce cans) black beans, drained and rinsed (or 2 cups dried black beans, soaked or cooked in a pressure cooker)

2 teaspoons salt, or to taste

1 bay leaf

2 teaspoons ground cumin

½ teaspoon oregano

2 tablespoons lemon juice

1 In a large skillet, heat the olive oil. Cook the onions, bell pepper, celery, and garlic in the oil over medium heat until the onions are translucent (about 10 minutes).

2 Add the beans, salt, bay leaf, cumin, oregano, and lemon juice, and stir well to combine.

3 Cover and simmer for another 35 or 40 minutes, stirring occasionally to prevent sticking. Remove the bay leaf and serve over rice.

Per serving: *Calories 195 (From Fat 63); Fat 7g (Saturated 1g); Cholesterol 0mg; Sodium 593mg; Carbohydrate 25g (Dietary Fiber 9g); Protein 8g.*

FOLDING A BURRITO

1\. FOLD ONE END OF THE TORTILLA TOWARD THE MIDDLE,

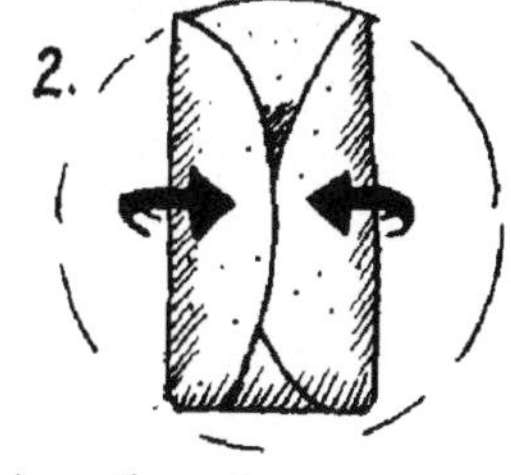

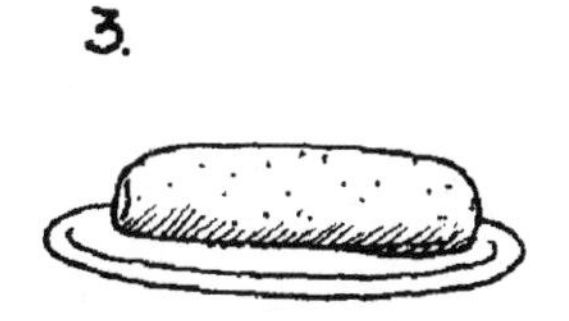

Figure 12-1: How to fold a burrito so that it doesn't unroll on the plate.

CHOPPING ONION

1\. CUT ONION IN HALF, STEM TO BOTTOM. SLICE OFF TIP

2\. PLACE ONE HALF, CUT SIDE DOWN, ON A CUTTING BOARD.

3\. SLICE IN 1/8" PARALLEL SLICES, STOPPING ABOUT A HALF AN INCH FROM THE ROOT END.

STOP HERE

4\. TURN THE ONION HALF AND SLICE ACROSS THE PARALLELS SO THAT THE SMALL ONION PIECES FALL AWAY.

Figure 12-2: Follow these steps to chop an onion.

Pasta-Mania

Pasta is a fun food. Multi-colored, multi-shaped varieties served with different types of sauce and extras make pasta almost endlessly versatile. There's a variation to suit nearly every palate. Leftover pasta keeps in the refrigerator for several days or in the freezer for several months.

Pesto Pasta Primavera

Pesto is a sauce that's traditionally made with basil, pine nuts, Parmesan cheese, garlic, and olive oil. This dish is best in the summertime, when basil is fresh and you can make a batch of pesto from scratch. Pesto is a snap to make, and it fills your home with the wonderful aroma of fresh basil. This recipe makes about double the amount of pesto that you use for a pound of pasta. Store the leftover pesto in an airtight container for one week in the refrigerator, or freeze it in a small jar or airtight plastic bag for several months. When you can't make your own pesto, substitute ready-made pesto from the supermarket and have dinner ready in minutes.

Primavera means "spring" in Italian. In culinary circles, primavera dishes are made with a variety of fresh vegetables and herbs.

Preparation time: *15 minutes*

Cooking time: *About 20 minutes*

Yield: *6 servings*

Pesto:

3 cups chopped fresh basil leaves, lightly packed

4 cloves garlic, peeled

4 tablespoons pine nuts

6 tablespoons Parmesan cheese or a combination of Parmesan and Romano cheeses

¼ teaspoon black pepper

⅓ cup olive oil

Pasta:

1 pound whole-wheat fettuccine or other pasta

2 tablespoons olive oil

1 medium onion, minced

1 clove garlic, minced

1 pound asparagus, trimmed and sliced diagonally in ¼-inch pieces

1 medium zucchini, sliced

1 medium carrot, sliced very thin

½ pound fresh mushrooms, sliced

1 cup frozen peas, thawed

2 green onions, chopped

½ teaspoon black pepper

½ cup grated Parmesan cheese

½ cup pine nuts

6 cherry tomato halves and several sprigs of parsley for garnish

1. Place the basil, garlic, pine nuts, cheese, and pepper in a blender or food processor. Process until well blended and smooth, and then dribble in the olive oil. Continue processing until the mixture is the consistency of a smooth paste. Set aside.
2. In a large pot, cook the pasta according to package instructions. Drain and set aside. (Follow Steps 2 through 5 while the pasta is cooking.)
3. In a small skillet, heat 1 tablespoon of olive oil. Add the pine nuts and cook on medium heat, stirring frequently, until the pine nuts begin to sizzle. Continue cooking until they become golden brown (about 3 minutes), taking care not to let them burn. Remove them from the heat and set them aside.
4. In a large skillet, heat the remaining olive oil. Cook the onion and garlic in the oil over medium heat until the onions are translucent — about 7 minutes.
5. Add the asparagus, zucchini, carrot, and mushrooms, and cook over medium heat for 5 minutes.
6. Add the peas and green onions. Heat for 2 minutes.
7. Add the pepper, and then add the cooked pasta, cheese, pine nuts, and pesto, tossing until the ingredients are well mixed. Serve immediately. Garnish with cherry tomato halves and parsley.

Vary It! *For a vegan version, omit the Parmesan cheese, or substitute nutritional yeast for all or part of it. Alternatively, leave out the pesto sauce and instead toss the pasta with herbs, spices, vegetables, and an additional 2 to 3 tablespoons of olive oil. You can also substitute chopped walnuts for toasted pine nuts.*

Per serving: *Calories 627 (From Fat 271); Fat 30g (Saturated 6g); Cholesterol 9mg; Sodium 264mg; Carbohydrate 73g (Dietary Fiber 15g); Protein 25g.*

Rotini with Chopped Tomatoes and Fresh Basil

You can make this light, simple dish any time of the year, but I like to make it in the summertime, when tomatoes and basil are in season and fresh. Vegans can substitute soy cheese for regular cheese, or leave out the cheese entirely. You can also substitute nutritional yeast for all or part of the Parmesan cheese. Leftovers are delicious the next day, after the pasta has absorbed the flavors of the sauce.

Preparation time: *15 minutes*

Cooking time: *10 minutes or less for pasta*

Yield: *6 servings*

3 cups diced ripe roma tomatoes (scoop out and discard most of the seeds)

½ cup chopped fresh basil

8 ounces shredded part-skim mozzarella cheese

1 cup grated Parmesan cheese

½ cup chopped walnuts

1 pound whole-wheat rotini (spiral-shaped) pasta

4 tablespoons olive oil

Freshly ground black pepper to taste

Black olives and parsley for garnish

1. In a medium bowl, toss together the tomatoes, basil, mozzarella cheese, Parmesan cheese, and walnuts.
2. In a large pot, cook the rotini according to package instructions and drain.
3. In the pot or in a large bowl, toss the pasta with the olive oil. Next, add the tomato mixture and toss until well combined.
4. Serve on individual plates and add freshly ground black pepper to taste. Garnish each plate with a black olive and a sprig of parsley.

Per serving: *Calories 585 (From Fat 242); Fat 27g (Saturated 9g); Cholesterol 33mg; Sodium 439mg; Carbohydrate 64g (Dietary Fiber 8g); Protein 28g.*

Vegetarian Lasagna

Bottled pasta sauce and no-boil lasagna noodles make the preparation for this dish quick and easy. This dish appeals to vegetarians and nonvegetarians alike. Serve it with a simple green salad and crusty bread.

Preparation time: *20 minutes*

Cooking time: *1 hour*

Yield: *10 servings*

1 tablespoon olive oil

1 large onion, chopped

1 large head or 2 small heads broccoli

2 medium zucchini and/or yellow (summer) squash

1 cup carrots, grated

3 cups reduced-fat ricotta cheese

4 egg whites

1 teaspoon oregano

1 teaspoon thyme

2 tablespoons dried parsley

½ teaspoon black pepper

¾ cup grated Parmesan cheese

1 jar (about 24 ounces) prepared pasta sauce

8 ounces no-boil whole-wheat lasagna noodles

12 ounces shredded part-skim mozzarella cheese (3 cups)

1 Preheat the oven to 350 degrees.

2 Chop the broccoli and zucchini or squash into small pieces. In a medium skillet, heat the olive oil. Cook the onions in the oil over medium heat until they're translucent — about 7 minutes. Add the broccoli, zucchini, and carrots, and steam for 2 to 3 minutes to partially cook. Remove from the heat.

3 In a medium bowl, mix the ricotta cheese, egg whites, oregano, thyme, parsley, pepper, and half the Parmesan cheese.

4 In a lightly oiled, 13-x-17-inch pan, layer the lasagna as follows:

Put a generous layer of pasta sauce on the bottom of the pan.

Lay 4 uncooked lasagna noodles on the sauce (or the number of noodles needed to fit the pan in a single layer; break the noodles if necessary to fit).

Add a layer of the ricotta cheese mixture, and then half the vegetable mixture, followed by half the mozzarella cheese.

Repeat the layers, and then top it off with another layer of noodles, more pasta sauce, and the remaining Parmesan cheese.

5 Bake for 1 hour, or until brown and bubbly.

Vary It! *If you have sun-dried tomato pesto or basil pesto on hand, spread some of that on top of the lasagna along with the pasta sauce (or mixed in with the pasta sauce) in Step 4.*

Per serving: *Calories 354 (From Fat 126); Fat 14g (Saturated 7g); Cholesterol 43mg; Sodium 666mg; Carbohydrate 32g (Dietary Fiber 7g); Protein 25g.*

All-Time Favorites

Who says you can't enjoy a slice of pizza or a sloppy Joe on a vegetarian diet? These days, just about any old favorite recipe has a vegetarian counterpart. I include two popular examples in this section.

Roasted Vegetable Pizza

Pizza in its simplest form — with fresh tomato sauce and mozzarella cheese — is vegetarian. In this recipe, I show you how to take it one step further, adding roasted vegetables for rich flavor. But don't stop there. With a homemade base of pizza dough, you can be creative and try many different topping combinations. Another option: Leave off the cheese. You'll be surprised how good a vegan, cheeseless pizza tastes. And without all that high-fat dairy cheese, pizza can be a guiltless indulgence.

This recipe makes two pizzas. Make both pizzas at the same time, or save half the dough for another day. Pizza dough keeps in the freezer for up to three months.

Preparation time: *About 50 minutes (including time for dough to rise)*

Cooking time: *65 to 70 minutes*

Yield: *2 large pizzas (4 servings, or 8 slices per pizza)*

Pizza Dough:

1½ teaspoons yeast

1¾ cups warm water (110 degrees)

1 tablespoon honey or sugar

1 teaspoon olive oil

½ teaspoon salt

2 cups whole-wheat flour

2 cups all-purpose flour

Extra oil to grease the pizza pan

Pizza Toppings:

2 cups combined of your choice of washed, chopped vegetables (whatever you have on hand): broccoli, mushrooms, green bell peppers, onions, black olives

2 tablespoons olive oil

2 tablespoons balsamic vinegar

¾ cup bottled pizza sauce (more or less to suit your taste)

2 cups shredded reduced-fat mozzarella cheese

1 Preheat the oven to 375 degrees.

2 In a large mixing bowl, dissolve the yeast in warm water (about 110 degrees) and add the honey or sugar. Add the oil and salt and stir.

3 Gradually add the flour, alternating between whole-wheat and white. Mix well after each addition using a wooden spoon or your hands to make a soft dough.

4 Turn out the dough onto a floured board and knead for 5 minutes, adding more flour if needed.

5 Lightly oil the sides of the mixing bowl. Put the dough in the bowl, turn the dough over once, cover with a towel or waxed paper, and let set in a warm place for about an hour.

6 While the dough rises, wash and chop the vegetables for the pizza toppings.

7 Pour olive oil and balsamic vinegar onto the bottom of a 13-x-9-inch baking dish. Add the vegetables and toss to coat them with oil and vinegar. Spread evenly on the bottom of the baking dish and roast in the oven for 35 to 40 minutes, stirring occasionally, until the vegetables are cooked and browned. Set aside.

8 Divide the dough in half. Reserve one half for later if desired (store in the refrigerator or freezer).

9 Give the dough a quick knead, and then pull and stretch it out onto an oiled, 14-inch pizza pan. You can also roll the dough out first on a floured surface, or simply press the ball of dough onto the pizza pan and distribute it evenly by using your hands. If it springs back too much, let it rest for 5 minutes, and then stretch it again.

10 Spoon on the pizza sauce and spread it to within ½ inch of the edge of the dough.

11 Sprinkle the shredded cheese evenly over the pizza, and then add the vegetable toppings.

12 Place the pizza in the oven and bake for 30 minutes, until the crust and/or cheese begins to brown lightly. Do not overcook.

13 Remove from the oven and let set for 5 minutes.

14 Cut into slices and serve.

Per serving: *Calories 362 (From Fat 85); Fat 9g (Saturated 3g); Cholesterol 10mg; Sodium 479mg; Carbohydrate 54g (Dietary Fiber 6g); Protein 17g.*

Tempeh Sloppy Joes

This is many people's favorite alternative to burgers and hot dogs. Serve this sandwich filling on whole-grain burger buns. It's also good served on whole-grain toast. Sloppy Joe filling made with tempeh tastes similar to — but not as greasy as — the kind made with meat.

Preparation time: *15 minutes*

Cooking time: *About 20 minutes*

Yield: *4 servings*

1 tablespoon olive oil

1 clove garlic, minced

1 medium onion, chopped

½ green bell pepper, chopped

One 8-ounce package tempeh (any variety), crumbled into small pieces

2 tablespoons reduced-sodium soy sauce

½ cup spaghetti sauce

1 teaspoon brown mustard (or other mustard)

2 tablespoons cider vinegar

2 teaspoons sugar

4 whole-grain burger buns

1 In a medium saucepan, heat the olive oil. Add the garlic, onion, and bell pepper, and cook over medium heat until the onions are translucent — about 10 minutes. Add the tempeh and soy sauce and stir well. Cook for an additional 2 minutes.

2 Add the spaghetti sauce, mustard, vinegar, and sugar. Mix well and simmer for 10 minutes. Serve on whole-grain buns.

Per serving: *Calories 301 (From Fat 99); Fat 11g (Saturated 2g); Cholesterol 0mg; Sodium 634mg; Carbohydrate 39g (Dietary Fiber 3g); Protein 17g.*

Asian Alternatives

The two recipes in this section are good examples of easy, tasty, Asian-style vegetable dishes that incorporate soy foods as rich, healthful protein sources. Enjoy these often.

Vegetable Stir-Fry

Chinese stir-fries traditionally use peanut oil to cook and flavor the vegetables. More convenient and just as effective for flavor is peanut butter, which this recipe uses. Feel free to vary the amounts or varieties of vegetables according to your preferences and what you have on hand. This meal is highly nutritious — full of vitamins, minerals, and fiber. Leftovers aren't a problem — this dish tastes great reheated.

Preparation time: *20 minutes*

Cooking time: *About 15 minutes*

Yield: *4 servings*

4 tablespoons reduced-sodium soy sauce

1 teaspoon sugar

¼ teaspoon ground ginger

2 cloves garlic, minced

2 tablespoons vegetable oil

1 pound firm or extra-firm tofu, sliced into rectangles (approximately ¼ inch x 1½ inch)

1 large onion, chopped

3 stalks celery, thinly sliced

3 carrots, thinly sliced

1 cup broccoli florets

½ cup sliced canned water chestnuts

1 cup sliced fresh mushrooms

3 cups sliced bok choy (Chinese cabbage)

2 cups mung bean sprouts

1 tablespoon creamy peanut butter

1 tablespoon cornstarch

¼ cup water

1 In a small bowl, combine the soy sauce, sugar, ginger, and garlic. Set aside.

2 In a large skillet, heat the vegetable oil over medium heat. Add the tofu and fry it on both sides until browned. Remove, place on a plate lined with paper towels, and set aside.

3 Add the onion, celery, and carrots to the skillet, and cook over medium heat for 3 to 4 minutes, stirring frequently.

4 Add the remaining vegetables, peanut butter, and soy sauce mixture, and continue to cook for an additional 2 minutes. Stir.

5 Add the cornstarch and water and continue to cook until the liquid thickens and all the ingredients are steaming hot. Serve immediately over rice.

Per serving: *Calories 238 (From Fat 117); Fat 13g (Saturated 1g); Cholesterol 0mg; Sodium 144mg; Carbohydrate 22g (Dietary Fiber 7g); Protein 13g.*

Soy-Ginger Kale with Tempeh

This is a staple in my home from spring through fall, when the CSA (community-supported agriculture) farm to which I subscribe delivers organic kale nearly every week. (See Chapter 7 for additional information on CSA farms.) The soy-ginger sauce makes this dish slightly sweet and delicious. I serve it with steamed brown rice or seasoned couscous.

Preparation time: *10 minutes*

Cooking time: *30 minutes*

Yield: *6 servings*

1 bunch kale (about 1 pound)

2 tablespoons olive oil

2 cloves garlic, minced

½ cup chopped onion (about half of a medium-sized onion)

⅓ cup plus ¼ cup of your favorite bottled soy-ginger sauce

One 8-ounce block of tempeh (any variety), chopped into ½-inch cubes

A few tablespoons of water

1 Rinse the kale leaves well, removing and discarding the thick stems. Cut the rinsed leaves into ½-inch-wide strips (do not pat dry). Set aside.

2 In a large skillet over low heat, cook the olive oil, garlic, and onions for about 5 minutes, stirring to prevent sticking.

3 Add the kale, and then pour ⅓ cup of the soy-ginger sauce over the mixture. Cover and cook over medium heat for about 10 minutes, or until tender, stirring occasionally to make sure the kale doesn't stick (add a few extra tablespoons of soy-ginger sauce, if necessary).

4 In a small skillet over medium heat, add ¼ cup of soy-ginger sauce, followed by the tempeh cubes. When the sauce and tempeh begin to sizzle (about 5 minutes), turn the tempeh using a spatula. Reduce heat, add a few tablespoons of water, and cook for an additional 5 minutes, or until the tempeh is browned on both sides. Be careful not to let the sauce burn. Remove from heat.

5 Remove the lid from the kale. Lift and stir the mixture with a spatula, and then transfer it to a shallow serving bowl. Top with the tempeh cubes and serve.

Vary It! *Replace kale with other greens, such as Swiss chard or collards.*

Per serving: *Calories 167 (From Fat 89); Fat 10g (Saturated 2g); Cholesterol 0mg; Sodium 3839mg; Carbohydrate 13g (Dietary Fiber 4g); Protein 9g.*

Chapter 13

Baking Easy Breads and Rolls

In This Chapter

- Incorporating whole-grain goodness
- Keeping home-baked goodies on hand

Recipes in This Chapter

- Whole-Wheat Crescent Rolls
- Banana Chip Muffins
- Easy Cornbread
- Cinnamon Applesauce Muffins
- Zucchini Bread
- Pumpkin Biscuits

Bread — the "staff of life" — is a dietary staple around the world. It's hard to think of anything more inviting in a home than the fragrance of freshly baked bread.

There's room for bread on the table at breakfast, lunch, or supper. Muffins, quick breads, and homemade biscuits make good snacks anytime, too. I discuss in more detail in Chapter 6 the different types of cereal grains used to make breads and rolls, and in Chapter 3 I cover the nutritional merits of grains (plenty!). The recipes I include in this chapter are a sampling of the many types of breads you can enjoy on a vegetarian diet.

If you eat out, be aware that meats such as bacon and minced ham are sometimes added to biscuits, loaf breads, and rolls. In the American South, cracklin' cornbread is made by stirring small, crunchy pieces of fried pork fat into the batter.

Most important, I want to convey the notion that breads are easy to bake and enjoy at home and that they complement and round out many vegetarian meals. Enjoy a slice or share a batch!

Whole-Wheat Crescent Rolls

These rolls fill the house with the delicious aroma of yeast bread and make any meal special. Their texture is soft and slightly chewy. They stay fresh for two or three days, and they also freeze well. The addition of whole-wheat flour makes these rolls more nutritious and flavorful than rolls made only with refined white flour. It also gives them a nice light brown color.

Preparation time: *15 minutes, plus 2½ hours for the dough to rise*

Cooking time: *20 minutes*

Yield: *32 rolls*

1 package (¼ ounce) active dry yeast

¼ cup warm water

¾ cup warm milk

2 tablespoons sugar

2 tablespoons honey

1 teaspoon salt

2 egg whites

¼ cup vegetable oil

1½ cups whole-wheat flour

2 cups all-purpose flour

Olive oil

1 In a large mixing bowl, dissolve the yeast completely in the water.

2 Add the milk, sugar, honey, salt, egg whites, oil, and whole-wheat flour. Stir until all the ingredients are well combined and the dough is smooth.

3 Add the all-purpose flour and mix until the dough forms a ball. If the dough is too sticky to handle, add several more tablespoons of flour as necessary.

4 Place the dough on a floured surface and knead for several minutes until it's smooth and elastic. (See Figure 13-1 for the proper technique for kneading bread dough.)

5 Oil a large bowl with olive oil. Place the dough ball in the bowl and turn it over once so that you lightly coat the top of the dough ball with oil. Cover with a towel or waxed paper and let the dough rise in a warm place until it doubles in size — about 2 hours.

6 Punch down the dough ball with your fist; then divide the ball into two pieces.

7 Place the dough balls one at a time on a floured surface. Roll each piece of dough into a circle approximately 12 inches in diameter. Lightly brush each circle with vegetable oil. With a sharp knife, cut the circle in half, and then in quarters, continuing until you have 16 wedges of dough (see Figure 13-2).

8 Preheat the oven to 350 degrees. Beginning at the wide end of each wedge, roll the dough into a crescent shape, ending by pressing the tip onto the roll. Set each roll on a lightly oiled baking sheet and bend the ends slightly to give the roll a crescent shape.

9 Cover and let the rolls rise for about 30 minutes before baking. Bake for 20 minutes, or until the rolls are lightly browned. Do not overcook.

Per serving: *Calories 74 (From Fat 18); Fat 2g (Saturated 0g); Cholesterol 0mg; Sodium 80mg; Carbohydrate 12g (Dietary Fiber 1g); Protein 2g.*

Figure 13-1: Use these steps to knead bread dough.

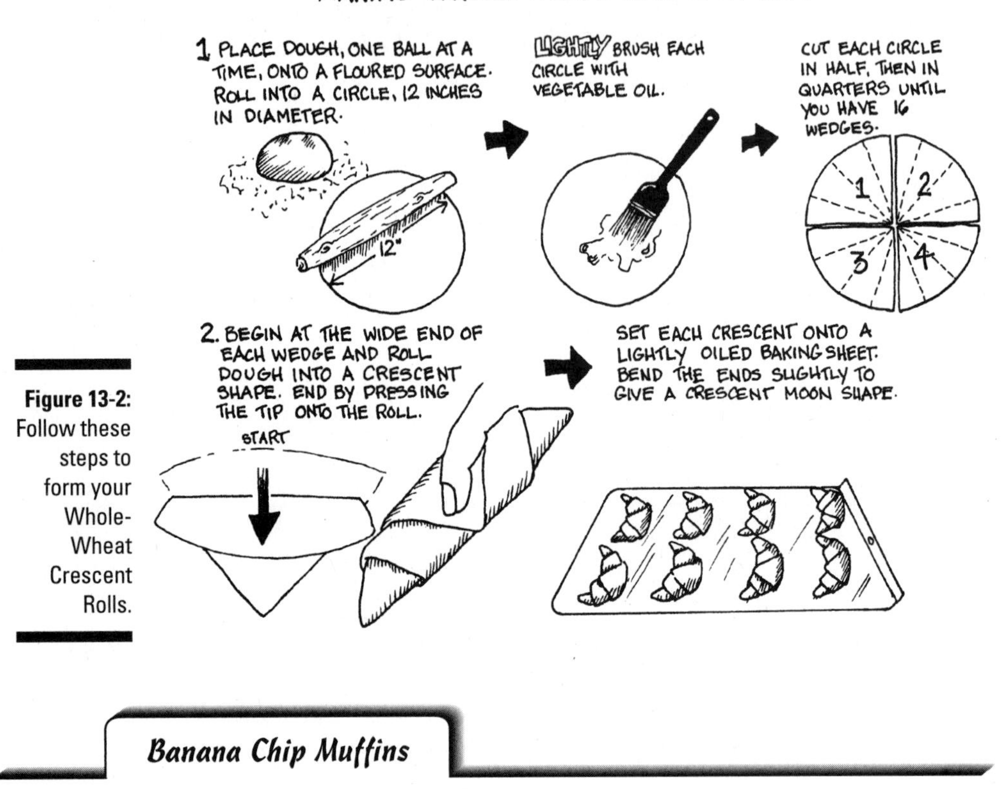

Figure 13-2: Follow these steps to form your Whole-Wheat Crescent Rolls.

Banana Chip Muffins

Everyone loves these muffins, and they're a great way to use up bananas that are becoming too ripe. They freeze well, so make a batch whenever you have extra bananas and you'll have muffins on hand when you want them.

Preparation time: *15 minutes*

Cooking time: *20 minutes*

Yield: *12 muffins*

½ cup vegetable oil

1 cup packed brown sugar

*1 tablespoon powdered vegetarian egg replacer (EnerG Egg Replacer) blended with 4 tablespoons water**

2 teaspoons pure vanilla extract

2 ripe bananas, mashed

1½ cups all-purpose flour

½ cup whole-wheat flour

¼ cup wheat germ (optional)

1 teaspoon baking powder

1 teaspoon baking soda

¼ teaspoon salt

6 ounces chocolate chips (about 1 cup)

** or use 4 egg whites instead of egg replacer and water*

1 Preheat the oven to 350 degrees.

2 In a medium bowl, whisk together the oil, sugar, egg replacer/water, and vanilla. Add the bananas and blend well with a whisk or an electric mixer.

3 Add the flour, wheat germ (if desired), baking powder, baking soda, and salt. Fold the dry ingredients into the wet ingredients with quick strokes — be careful not to overmix.

4 Stir in the chocolate chips.

5 Spoon the batter into lined or oiled muffin cups and bake for 20 minutes, or until muffins are golden brown. When cool enough to handle, remove muffins from cups.

Vary It! *Substitute carob chips or butterscotch chips for the chocolate chips. Add a handful of chopped walnuts or pecans.*

Per serving: *Calories 309 (From Fat 120); Fat 13g (Saturated 3g); Cholesterol 0mg; Sodium 200mg; Carbohydrate 47g (Dietary Fiber 3g); Protein 4g.*

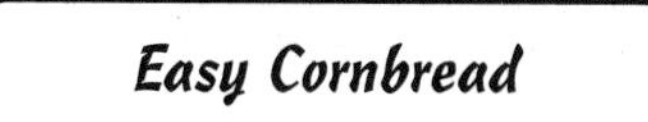

Easy Cornbread

This cornbread is flavorful and moist. Serve it warm alongside Cashew Chili (see Chapter 11), with vegetable stews and casseroles, or with a bowl of soup.

Preparation time: *10 minutes*

Cooking time: *30 minutes*

Yield: *4 servings*

2 tablespoons olive oil
¾ cup cornmeal
2 egg whites
1½ cups plain nonfat yogurt
½ teaspoon baking soda
½ teaspoon salt
½ cup chopped onion
¼ teaspoon black pepper

1 Preheat the oven to 425 degrees. Oil a 1-quart casserole dish or skillet.

2 Combine all the ingredients in a medium-sized bowl and stir until completely mixed.

3 Pour the batter into the oiled casserole dish or skillet and bake for 30 minutes, or until set. Do not overcook.

Vary It! *Add ¼ cup chopped sun-dried tomatoes.*

Per serving: *Calories 210 (From Fat 72); Fat 8g (Saturated 1g); Cholesterol 2mg; Sodium 554mg; Carbohydrate 27g (Dietary Fiber 2g); Protein 9g.*

Cinnamon Applesauce Muffins

These simple muffins are good for breakfast and in bag lunches. The whole-wheat flour adds dietary fiber. They taste even better the second day, and they freeze well.

Preparation time: *15 minutes*

Cooking time: *20 minutes*

Yield: *12 muffins*

⅓ cup vegetable oil

⅓ cup packed brown sugar

*1 tablespoon powdered vegetarian egg replacer (EnerG Egg Replacer) blended with 4 tablespoons water**

1½ cups unsweetened applesauce

2 cups whole-wheat flour

2 teaspoons baking powder

½ teaspoon baking soda

¼ teaspoon salt

1 teaspoon cinnamon

½ teaspoon nutmeg

** or use 4 egg whites instead of egg replacer and water*

1 Preheat the oven to 375 degrees.

2 In a medium bowl, whisk together the oil, sugar, and egg replacer/water. Add the applesauce and blend well with a whisk or an electric mixer.

3 Add the flour, baking powder, baking soda, salt, cinnamon, and nutmeg. Fold the dry ingredients into the wet ingredients with quick strokes — be careful not to overmix.

4 Spoon the batter into lined or oiled muffin cups and bake for 20 minutes, or until muffins are golden brown. When cool enough to handle, remove muffins from cups.

Vary It! *Add a handful of chopped walnuts or pecans, chopped dates, dried cranberries, cherries, currants, or raisins.*

Per serving: *Calories 163 (From Fat 60); Fat 7g (Saturated 1g); Cholesterol 5mg; Sodium 175mg; Carbohydrate 24g (Dietary Fiber 3g); Protein 3g.*

Zucchini Bread

This is another good breakfast or snack bread in the summer, when zucchini is in season and abundant. The recipe makes two loaves — serve one and freeze the other.

Preparation time: *30 minutes*

Cooking time: *1 hour*

Yield: *2 loaves (24 slices)*

2 cups sugar

1 cup vegetable oil

*4½ teaspoons powdered vegetarian egg replacer (EnerG Egg Replacer) blended with 6 tablespoons water**

3 teaspoons pure vanilla extract

2 cups coarsely grated, unpeeled zucchini, packed

2 cups all-purpose flour

¼ teaspoon baking powder

2 teaspoons baking soda

3 teaspoons cinnamon

1 teaspoon salt

1 cup chopped walnuts

** or use 6 egg whites instead of egg replacer and water*

1 Preheat the oven to 350 degrees.

2 Combine the sugar, oil, egg replacer/water, and vanilla in a large bowl. Add the zucchini and mix well.

3 In a separate bowl, combine the flour, baking powder, baking soda, cinnamon, and salt. Stir to blend the dry ingredients well.

4 Add the dry ingredients to the zucchini mixture, mix well, and then stir in the nuts.

5 Pour into two greased and floured 9-x-5-x-3-inch loaf pans (make sure that the pans are well greased!). Bake for 1 hour, or until the tops of the loaves are golden brown.

Vary It! *If you prefer a lighter, less oily bread, reduce the amount of vegetable oil to ¾ cup.*

Per serving: *Calories 215 (From Fat 108); Fat 12g (Saturated 1g); Cholesterol 0mg; Sodium 207mg; Carbohydrate 25g (Dietary Fiber 1g); Protein 2g.*

Pumpkin Biscuits

Serve these with honey for breakfast in the fall, or with the Cashew Chili in Chapter 11 in lieu of cornbread for a change of pace.

Preparation time: *15 minutes*

Cooking time: *20 minutes*

Yield: *12 biscuits*

1 cup unbleached, all-purpose flour

½ cup whole-wheat flour

3 teaspoons baking powder

1 teaspoon salt

¼ cup sugar

1 teaspoon cinnamon

¼ teaspoon nutmeg

¼ teaspoon allspice

⅓ cup Earth Balance or other trans fat-free margarine (make sure margarine is cold and hard)

¾ cup canned pumpkin or mashed sweet potatoes

¾ cup plain soymilk or nonfat cow's milk

1. Preheat the oven to 450 degrees.
2. In a medium bowl, sift together the flour, baking powder, and salt. Stir in the sugar, cinnamon, nutmeg, and allspice.
3. Using a pastry blender, cut the margarine into the dry ingredients until crumbly.
4. Fold in the pumpkin, and then gradually add milk, stirring to make dough. Use your clean hands if necessary to shape the dough into a ball.
5. Roll the dough out onto a cool, floured surface to about ½ inch thick. Using a juice glass or a small, round cookie cutter, cut the dough into rounds and place them onto an oiled baking sheet. Push together and roll the dough scraps to cut out the remaining biscuits.
6. Bake for about 20 minutes, or until lightly browned. Be careful not to overcook.

Vary It! *Add a handful of chopped pecans or dried cranberries.*

Per serving: *182 Calories (From Fat 104); Fat 12g (Saturated 5g); Cholesterol 0mg; Sodium 412mg; Carbohydrate 17g (Dietary Fiber 2g); Protein 2g.*

Chapter 14

Dishing Out Delicious Desserts

In This Chapter

- Checking out some chocolate alternatives
- Making greater use of fruits
- Creating new versions of old classics

Recipes in This Chapter

- Cocoa Pink Cupcakes
- Vegan Chocolate Chip Cookies
- Rich Chocolate Tofu Cheesecake
- Easy Vegan Chocolate Cake
- Baked Apples
- Pear Cranberry Crisp
- Mixed Berry Cobbler
- Rice Pudding
- Cherry Oatmeal Cookies

Most people think of dessert as a guilt-inducing indulgence that's not very nutritious and is loaded with calories. Where most commercial cakes, cookies, and pies are concerned, this is usually true!

Though most desserts contain no meat, many do contain oodles of artery-clogging saturated fat and calories from rich dairy ingredients and eggs. Desserts made with gelatin — a vegetarian no-no — are usually junky, with lots of added sugar and artificial colors and flavors, too.

That's where the vegetarian alternative can help. Make no mistake — just because a dessert is vegetarian doesn't mean it's guaranteed to be better for your health. In general, though, desserts made from scratch at home using wholesome ingredients and fewer animal products tend to be better for you.

The recipes in this chapter will convince you to extend the vegetarian way to the desserts you bake every day. Many of these recipes incorporate fruit to boost the nutritional value and lower the calorie content, and, where possible, whole-wheat flour replaces white.

Chocolate Desserts

Sometimes, nothing but chocolate will do. The desserts in this section will satisfy the chocolate lovers in your life and add a delicious finishing touch to any meal.

The first two recipes are kid-pleasers and lunchbox staples — cupcakes and cookies. They freeze and travel well. The remaining two recipes include a decadent — but much more healthful — tofu version of a classic cheesecake and a foolproof, make-it-in-minutes chocolate cake that I guarantee everyone will love.

Cocoa Pink Cupcakes

I adapted this recipe from one that has been a favorite in my family since I was a little girl. These cupcakes have a not-too-sweet, mild cocoa flavor. The cocoa gives them a rosy brown color on the inside.

Preparation time: *30 minutes*

Cooking time: *25 minutes*

Yield: *About 22 cupcakes*

½ cup whole-wheat flour

2½ cups all-purpose flour

1 tablespoon cocoa

1 teaspoon salt

¾ cup vegetable oil

1¼ cup sugar

*1 tablespoon powdered vegetarian egg replacer (EnerG Egg Replacer) blended with 4 tablespoons water**

1 teaspoon pure vanilla extract

1 teaspoon baking soda

1 cup cold water

6 ounces (1 cup) semi-sweet chocolate chips

½ cup chopped walnuts

Confectioners' sugar

** or use 4 egg whites instead of egg replacer and water*

1 Preheat the oven to 375 degrees.

2 In a small bowl, combine both types of flour with the cocoa and salt, stir to blend, and then set aside.

3 In a large mixing bowl, cream the oil and sugar together, and then add the egg replacer and vanilla.

4 In a cup, combine the baking soda and water and stir to dissolve.

5 Add the baking soda mixture alternately with the dry ingredients to the creamed mixture, beginning and ending with the dry ingredients and blending well on low speed after each addition.

6 Stir half the chocolate chips and nuts into the batter.

7 Line muffin tins with paper liners and fill each a little more than half full with batter.

8 Sprinkle the remaining chocolate chips and nuts over the tops of the cupcakes.

9 Bake for about 25 minutes. The cupcakes should appear done but no more than very lightly browned. Be careful not to overcook them, or they'll become dry.

10 After the cupcakes cool, sift confectioners' sugar over the tops.

Per cupcake: *Calories 234 (From Fat 108); Fat 12g (Saturated 3g); Cholesterol 0mg; Sodium 165mg; Carbohydrate 29g (Dietary Fiber 1g); Protein 3g.*

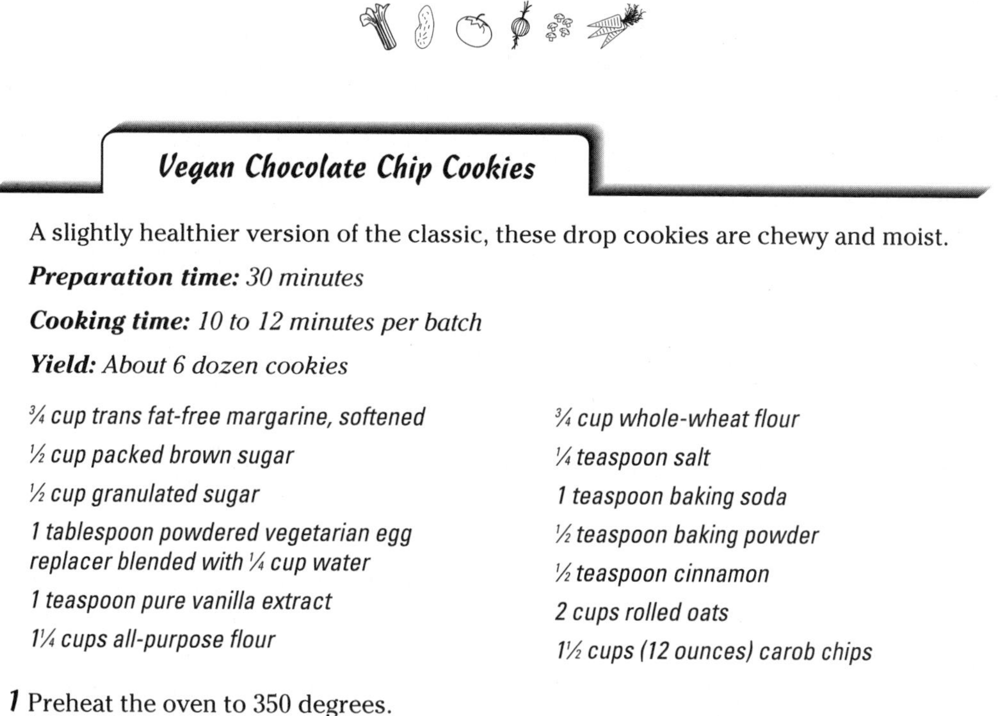

Vegan Chocolate Chip Cookies

A slightly healthier version of the classic, these drop cookies are chewy and moist.

Preparation time: *30 minutes*

Cooking time: *10 to 12 minutes per batch*

Yield: *About 6 dozen cookies*

¾ cup trans fat-free margarine, softened

½ cup packed brown sugar

½ cup granulated sugar

1 tablespoon powdered vegetarian egg replacer blended with ¼ cup water

1 teaspoon pure vanilla extract

1¼ cups all-purpose flour

¾ cup whole-wheat flour

¼ teaspoon salt

1 teaspoon baking soda

½ teaspoon baking powder

½ teaspoon cinnamon

2 cups rolled oats

1½ cups (12 ounces) carob chips

1 Preheat the oven to 350 degrees.

2 In a mixing bowl, cream the margarine, sugars, egg replacer, and vanilla until smooth.

3 In a separate bowl, mix both types of flour with the salt, baking soda, baking powder, and cinnamon. Add to the butter mixture and mix well.

4 Stir in the oats. The batter will be very stiff.

5 Add the carob chips and mix thoroughly using clean hands.

6 Drop the dough by rounded teaspoonfuls about 3 inches apart on an ungreased cookie sheet. Bake each batch for 10 to 12 minutes, or until the cookies are lightly browned.

Per serving: *Calories 71 (From Fat 26); Fat 3g (Saturated 2g); Cholesterol 0mg; Sodium 52mg; Carbohydrate 10g (Dietary Fiber 1g); Protein 1g.*

Rich Chocolate Tofu Cheesecake

Tofu makes a fabulous replacement for cream cheese in cheesecake recipes. This cheesecake has a rich chocolate flavor and a smooth, creamy texture. I adapted it from a recipe in *Tofu Cookery* (25th Anniversary Edition) by Louise Hagler (Book Publishing Company). You can also use a ready-made 9-inch graham cracker crust (which holds the same volume as the 8-inch springform pan used in this recipe).

Preparation time: *30 minutes*

Cooking time: *40 minutes (plus 2 hours to chill)*

Yield: *8 servings (1 pie)*

Filling:

1¼ pounds firm tofu

1½ cups sugar

3 ounces (3 squares) semi-sweet baking chocolate

1 teaspoon pure vanilla extract

½ teaspoon pure almond extract

⅛ teaspoon salt

Crust:

1¼ cups graham cracker crumbs

1 tablespoon sugar

⅓ cup trans fat-free margarine

1 Remove tofu from packaging and place it in a clean sink. Cover the tofu with waxed paper and a heavy weight, and let it drain for 20 minutes. While the tofu drains, make the crust.

2 For the crust, combine the graham cracker crumbs and sugar in a medium bowl. Melt the margarine in a saucepan over low heat, and then add it to the crumb mixture. Stir until blended, and then press the crumbs into the bottom and sides of an 8-inch, non-stick springform pan. Set aside.

3 In a blender or food processor, blend the drained tofu, ¼ pound at a time, with 1 cup of the sugar, adding ¼ cup sugar with each addition of tofu (set aside the remaining ½ cup sugar for later). Process until the ingredients are well blended.

4 Pour the tofu-sugar mixture into a bowl. Preheat the oven to 350 degrees.

5 In a double boiler or saucepan, melt the chocolate. While the chocolate is melting, place the graham cracker crust in the oven and cook for about 8 minutes; then remove, set aside, and let cool.

6 After the chocolate has melted, add it to the tofu mixture. Mix in the vanilla, almond extract, salt, and remaining ½ cup sugar. Blend well.

7 Pour the mixture into the graham cracker crust and bake for about 40 minutes. When the cheesecake is done, the edges will rise slightly, with small cracks on the surface. The middle won't rise, but it will be springy to the touch and have a dry, firm appearance. Chill the cheesecake for at least 2 hours after baking.

Vary It! *Top the cheesecake with fresh raspberries or a thawed bag of frozen raspberries mixed with a few tablespoons of sugar. You can also top the cheesecake with canned cherries.*

***Per serving:** Calories 380 (From Fat 144); Fat 16g (Saturated 6g); Cholesterol 0mg; Sodium 201mg; Carbohydrate 56g (Dietary Fiber 1g); Protein 7g.*

Easy Vegan Chocolate Cake

This is the recipe I turn to when there's nothing in the house for dessert and I want to make something delicious — fast. This family favorite is a no-fail, quick and easy recipe using common ingredients you probably have in your pantry. Add your own frosting if you'd like, but I just dust with confectioners' sugar before serving. I adapted this recipe from the Moosewood Collective's *Moosewood Restaurant Book of Desserts* (Three Rivers Press).

Preparation time: *6 minutes*

Cooking time: *30 minutes*

Yield: *8 squares*

1½ cups unbleached white flour
⅓ cup unsweetened cocoa powder
1 teaspoon baking soda
½ teaspoon salt
1 cup sugar
½ cup vegetable oil
1 cup cold water or cold coffee
2 teaspoons pure vanilla extract
2 tablespoons cider vinegar
Confectioners' sugar, for serving

1. Preheat the oven to 375 degrees.
2. Oil an 8-inch square cake pan. Sift together the flour, cocoa, baking soda, salt, and sugar directly into the cake pan.
3. In a 2-cup measuring cup, mix the oil, cold water or coffee, and vanilla.
4. Pour the liquid ingredients into the cake pan and mix the batter with a fork or a small whisk. When the batter is smooth, add the vinegar and stir quickly. The batter will have pale swirls as the baking soda and vinegar react. Stir just until the vinegar is evenly distributed throughout the batter.
5. Bake for about 30 minutes and set aside to cool. Using a mesh strainer, shake confectioners' sugar on top of the cake immediately before serving.

Vary It! *Serve with nonfat or nondairy ice cream or sliced fruit.*

***Per serving:** Calories 317 (From Fat 132); Fat 15g (Saturated 1g); Cholesterol 0mg; Sodium 304mg; Carbohydrate 45g (Dietary Fiber 2g); Protein 3g.*

Fruit Desserts

Fruit desserts are a good choice because they add fiber and other health-supporting nutrients while still satisfying a sweet tooth. They also tend to be lower in calories than desserts made with lots of refined flours, sugar, and fat.

The desserts that follow are wholesome enough that you can even enjoy them for breakfast or as a snack.

Baked Apples

Baked apples are quick and convenient to make when you already have the oven heated to bake a casserole or loaf of bread. They keep in the refrigerator for up to three days, and they're good any time of the day or night — warm with ice cream, cold for breakfast, or as a snack.

Preparation time: *10 minutes*

Cooking time: *1 hour*

Yield: *4 servings*

4 large, tart apples

½ cup packed brown sugar

1 teaspoon cinnamon

1 tablespoon trans fat-free margarine

½ cup apple juice (optional)

1 Preheat the oven to 350 degrees.

2 Wash and core the apples, stopping the coring just before the bottom of the apple (so that the hole doesn't go all the way through). Using a paring knife, cut away the peel from the top to about one-third of the way down each apple (see Figure 14-1).

3 In a small dish or cup, combine the brown sugar and cinnamon. Spoon one-quarter of the mixture into the center of each apple.

4 Divide the margarine into quarters and put one chunk in the center of each apple.

5 Set the apples in an 8-x-8-inch baking dish or a 1-quart casserole dish and add water or apple juice (if desired) to a depth of about ½ inch.

6 Bake uncovered for 1 hour, or until the apples are soft. Serve warm or chilled.

Per serving: *Calories 255 (From Fat 32); Fat 4g (Saturated 1g); Cholesterol 0mg; Sodium 41mg; Carbohydrate 60g (Dietary Fiber 6g); Protein 0g.*

Figure 14-1: Filling Baked Apples.

Pear Cranberry Crisp

This beautiful dish is sweet, tart, and totally satisfying. The cranberries add a festive color and flavor, so this dish works well as a simple but lovely dessert for holidays and other gatherings.

Preparation time: *20 minutes*

Cooking time: *40 minutes*

Yield: *12 servings*

Filling:

6 large, soft, ripe pears, peeled, cored, and sliced

1½ cups fresh or frozen cranberries

¾ cup sugar

2 tablespoons all-purpose, unbleached flour

Topping:

1 cup rolled oats

½ cup packed brown sugar

⅓ cup all-purpose, unbleached flour (or mix half and half with whole-wheat flour)

¼ cup trans fat-free margarine

½ cup chopped pecans

1. Preheat the oven to 375 degrees. Oil a 9-x-13-inch baking dish.
2. In a large bowl, combine the pears, cranberries, sugar, and flour. Toss to coat the fruit. Spread the fruit over the bottom of the baking dish.
3. In a small bowl, combine the oats, brown sugar, flour, and margarine. Using a fork or pastry blender, combine the ingredients until the mixture is crumbly and the margarine is well incorporated. Stir in the pecans.
4. Sprinkle the topping evenly over the fruit and pat it down with your fingers.
5. Bake for 40 minutes, or until the fruit is bubbly and the topping is browned. Serve warm or chilled.

Vary It! *Use tart apples in place of the pears and walnuts in place of the pecans.*

Per serving: *Calories 222 (From Fat 70); Fat 8g (Saturated 2g); Cholesterol 0mg; Sodium 46mg; Carbohydrate 38g (Dietary Fiber 3g); Protein 2g.*

Mixed Berry Cobbler

I adapted this recipe from one of my favorite vegetarian cookbooks, *The Peaceful Palate* by Jennifer Raymond (Heart & Soul Publications). It's delicious, easy to make, and much lower in fat than a berry pie. It's one of the staple desserts at my house. You can use whatever type of berries you like, or a mixture of berries.

Preparation time: *15 minutes*

Cooking time: *25 minutes*

Yield: *9 servings*

5 to 6 cups fresh or frozen berries (boysenberries, blackberries, blueberries, strawberries, raspberries, or a mixture)

3 tablespoons all-purpose, unbleached flour

½ cup plus 2 tablespoons sugar

1 cup all-purpose, unbleached flour (or mix half and half with whole-wheat flour)

1½ teaspoons baking powder

¼ teaspoon salt

2 tablespoons vegetable oil

½ cup soymilk, rice milk, or nonfat cow's milk

1. Preheat the oven to 400 degrees.
2. Spread the berries in a 9-x-9-inch baking dish. Mix in the flour and ½ cup of the sugar.
3. In a separate bowl, mix the flour and the remaining 2 tablespoons of sugar with the baking powder and salt.
4. Add the oil to the flour mixture and mix it with a fork or your fingers until the mixture resembles coarse cornmeal.
5. Add the soymilk, rice milk, or cow's milk and stir to combine.
6. Spread this mixture over the berries (don't worry if they're not completely covered) and bake for about 25 minutes, or until golden brown.

Vary It! *For a real treat, top the hot cobbler with a spoonful of nondairy or reduced-fat ice cream.*

Per serving: *Calories 304 (From Fat 36); Fat 4g (Saturated 0g); Cholesterol 0mg; Sodium 141mg; Carbohydrate 68g (Dietary Fiber 5g); Protein 3g.*

Other Classic Comforts

Rice pudding and oatmeal cookies are two of my favorite comfort foods. The recipes in this section are variations of two traditional recipes. These versions are healthful enough to eat every day.

Notice that the rice pudding recipe that follows calls for soymilk in place of the cow's milk traditionally used in puddings and custards. Any nondairy milk alternative, including rice milk and almond milk — plain or vanilla — would work, though. Use whatever you have on hand. I like vanilla soymilk in this recipe because of its creaminess and the extra vanilla flavor it adds.

Rice Pudding

This rice pudding requires no eggs because it thickens in the pan as the rice absorbs the liquid in the recipe. The result is a rich, creamy rice pudding that's simple to make and requires minimal time supervising the stovetop.

Preparation time: *5 minutes (plus time to chill, if desired)*

Cooking time: *50 minutes*

Yield: *6 servings*

3 cups vanilla soymilk

½ cup long-grain white rice

¼ teaspoon salt

1 tablespoon trans fat-free margarine

¼ teaspoon ground cinnamon

¼ teaspoon ground nutmeg

¼ cup sugar

1 In a medium saucepan over high heat, bring the soymilk to a boil, stirring constantly (about 5 minutes).

2 Add the rice, salt, margarine, cinnamon, nutmeg, and sugar, and stir to combine. Reduce the heat to low and cover.

3 Cook, covered, for about 45 minutes, or until all the liquid has been absorbed. Lift the lid and stir every 15 minutes, covering again tightly each time you remove the lid.

4 After all the liquid has been absorbed, remove the pudding from the heat. Cool to warm, and then serve or place in the refrigerator to chill for at least 2 hours before serving.

Per serving: *Calories 186 (From Fat 32); Fat 4g (Saturated 1g); Cholesterol 0mg; Sodium 163mg; Carbohydrate 33g (Dietary Fiber 0g); Protein 4g.*

Cherry Oatmeal Cookies

The cherries in these thick, chewy cookies are a nice surprise. The cookies freeze well and are a healthful and delicious treat to bring to family and friends when you visit.

Preparation time: *20 minutes*

Cooking time: *12 minutes per batch*

Yield: *36 large cookies*

1 cup trans fat-free margarine

1 cup packed dark brown sugar

½ cup sugar

2 large eggs

1 teaspoon pure vanilla extract

¾ cup whole-wheat flour

¾ cup all-purpose flour

½ teaspoon salt

½ teaspoon baking soda

½ teaspoon baking powder

1 teaspoon cinnamon

2 cups rolled oats

1 cup dried cherries

½ cup chopped walnuts or pecans

1 Preheat the oven to 350 degrees. In a large mixing bowl, cream the margarine and sugars. Add the eggs and vanilla and beat until light and fluffy.

2 In a separate bowl, combine both types of flour with the salt, baking soda, baking powder, and cinnamon. Add the dry mixture to the margarine mixture and beat until combined.

3 Stir in the oats, cherries, and nuts and blend well.

4 Drop the cookie dough by the tablespoonful onto lightly greased cookie sheets. Bake for 12 minutes, or until lightly browned.

Per serving: *Calories 142 (From Fat 60); Fat 7g (Saturated 2g); Cholesterol 12mg; Sodium 110mg; Carbohydrate 19g (Dietary Fiber 1g); Protein 2g.*

Chapter 15

Celebrating the Holidays, Vegetarian-Style

In This Chapter

- Creating festive vegetarian meals
- Adapting old traditions
- Sampling special holiday recipes

Recipes in This Chapter

- Everybody's Favorite Cheese and Nut Loaf
- Stuffed Squash
- Fluffy Mashed Potatoes with Mushroom Gravy
- Roasted Roots
- Candied Sweet Potatoes
- Maple Pecan Pumpkin Pie

Holidays are made special by the family and friends we share them with and by the traditions that accompany those special days. Chief among those traditions is the food!

We associate certain foods with specific holidays. What's the centerpiece of most Thanksgiving tables? The turkey, of course. That is, unless you're a vegetarian, in which case you can make use of a variety of plants to round out your holiday meals.

The selection of recipes I include in this chapter underscores the diversity of the vegetable world. Meatless holiday meals are characterized by an abundance of colorful, flavorful, and festive foods.

Adopting New Traditions and Adapting the Old

Food traditions vary, and some even change over time. Christmas dinner may mean antipasto and ravioli to one family and a tofu turkey with all the trimmings to another. It's all a matter of what has become familiar to you and your family or friends over the years.

Many of the recipes in this book are well suited to special-occasion meals. They're colorful and will fill your home with delicious aromas.

Some make use of seasonal foods that evoke a particular mood or time of year. Pear Cranberry Crisp, for instance (see Chapter 14), may bring late autumn and the winter holidays to mind.

If you're new to a vegetarian diet and your traditions still center on foods of animal origin, don't worry. Choose the aspects of special-occasion meals that still fit and find replacements for those that don't.

For example, most of a traditional Thanksgiving dinner is already vegetarian. Mashed potatoes, candied sweet potatoes, green peas, cranberry sauce . . . these foods can stay on the menu. You can easily swap the turkey for Everybody's Favorite Cheese and Nut Loaf, or Stuffed Squash (recipes for both are in this chapter), or any of a number of other main dishes.

To supplement your menu, add Whole-Wheat Crescent Rolls (Chapter 13), Arugula Salad with Pickled Beets and Candied Pecans (Chapter 11), and Stuffed Mushrooms or Roasted Garlic Spread (Chapter 10). Use your imagination and think about the foods that you enjoy most. Over time, these will become your new traditions, as dear to you as the old ones once were.

Tips for Entertaining for Special Occasions

When you peruse the recipes that follow, think about what makes holiday meals special for you. Do you eat in the dining room rather than the kitchen? Do you place candles or a centerpiece on the table? Do you cover the table with your best linen tablecloth and fine china? Maybe you play soft music in the background.

The atmosphere you create in the room where you eat and the manner with which you present the food set the tone for the meal and can set a meal apart from the everyday routine. You can serve the recipes in this chapter anytime, but if you reserve them for special occasions and serve them with flair, they say, "This meal is special!"

Holiday Recipes to Savor

Many vegetarians have traditions of their own that are linked with specific holidays. In my family of longtime vegetarians, we make Everybody's Favorite Cheese and Nut Loaf at Thanksgiving, for Christmas Eve, and for some special events. The dish is so good that it's become legendary among my extended circle of family and friends. Some of the nonvegetarians who sometimes join us for holiday meals have adopted the tradition for themselves.

Everybody's Favorite Cheese and Nut Loaf

This savory dish is reminiscent of meatloaf in texture and has a nutty flavor. It's festive if it's presented on a platter garnished with parsley and cherry tomatoes. My family now makes two loaves to ensure we have enough for second helpings — leftovers are wonderful reheated or eaten cold in a sandwich.

Preparation time: *20 minutes*

Cooking time: *30 minutes*

Yield: *8 servings*

2 tablespoons olive oil

1 large onion, chopped fine

½ cup water

1½ cups whole-wheat breadcrumbs, plus extra for topping

2 cups grated, reduced-fat cheddar cheese (vegans may substitute nondairy cheese)

1 cup chopped walnuts

Juice of 1 fresh lemon (about 2 tablespoons)

6 egg whites (or vegetarian egg replacer equal to 3 eggs)

Parsley, bell pepper rings, and cherry tomato halves for garnish

1 Preheat the oven to 400 degrees.

2 In a large skillet, heat the olive oil. Over medium heat, cook the onions in the olive oil until they're translucent.

3 Add the water and breadcrumbs and mix well. Remove from the heat.

4 Add the cheese, walnuts, lemon juice, and egg whites and mix well.

5 Scoop the mixture into a greased, 9-x-5-x-3-inch loaf pan or casserole dish. Top with a sprinkling of breadcrumbs.

6 Bake for about 30 minutes, or until golden brown.

7 Turn out onto a platter and garnish with parsley sprigs and cherry tomatoes, or slices of red and yellow bell peppers.

Vary It! *Experiment with this recipe by replacing part of the walnuts with an equal amount of chopped water chestnuts, or a portion of the cheese with an equal portion of grated carrots. You can also serve this dish with ½ cup ketchup mixed with ½ cup salsa.*

Per serving: *Calories 260 (From Fat 180); Fat 20g (Saturated 3g); Cholesterol 0mg; Sodium 509mg; Carbohydrate 11g (Dietary Fiber 1g); Protein 11g.*

Stuffed Squash

Stuffed squash makes a festive main course for holiday meals and is a convenient and colorful centerpiece for the table. Any kind of squash will do, but acorn and butternut squash are the easiest to find in supermarkets. This recipe shows you how to make one large stuffed squash, enough for four ample servings (¼ squash each). Double or triple the recipe as needed to make enough servings to feed the number of guests you expect. You can also use the filling to make several small (3 to 4 inch), individual stuffed pumpkins if you prefer.

Preparation time: *1 hour (including time to prebake squash)*

Cooking time: *20 minutes*

Yield: *4 servings*

1 butternut squash, cut in half, seeds and strings removed (or use 2 small acorn squashes for 4 individual servings — see Figure 15-1 for an illustration of both types of squash)

2 tablespoons olive oil

1 medium onion, chopped

½ cup sliced fresh mushrooms

1 clove garlic, minced

¼ cup finely minced celery (leaves and stem)

¼ teaspoon black pepper

2 tablespoons minced fresh parsley

½ teaspoon sage

½ teaspoon thyme

Juice of 1 lemon (about 2 tablespoons)

¼ cup diced, peeled Granny Smith apple

¼ cup chopped walnuts

¼ cup golden raisins

3 slices whole-wheat bread, coarsely crumbled

¾ cup grated, reduced-fat cheddar cheese (use nonfat cheese or a mix to lower the saturated fat content further; vegans may substitute nondairy cheese)

1 Preheat the oven to 350 degrees.

2 Cut the squash in half. Scoop out and discard the seeds and strings.

3 Place the squash halves in a small, oiled baking dish and fill the dish ½ inch high with water. Place the dish in the oven and bake, covered loosely with foil, for 30 minutes, or until the squash is tender but still firm (don't let it cook so much that it falls apart when you move it later).

4 While the squash is baking, heat the olive oil in a small skillet. Cook the onions, mushrooms, garlic, and celery in the oil until the onions are translucent. Stir in the black pepper, parsley, sage, thyme, and lemon juice and remove from the heat.

5 Transfer the onion mixture to a mixing bowl. Add the apples, walnuts, raisins, bread, and cheese. Mix until well combined.

6 If desired, remove the squash from the oven and transfer it, cut side up, to a decorative baking or casserole dish. Otherwise, carefully pour out the water in the pan. Fill each squash half with half the stuffing mixture and press down lightly.

7 Cover tightly with foil and bake the squash for another 20 minutes, or until the cheese is melted and the stuffing is browned.

Per serving: *Calories 412 (From Fat 162); Fat 18g (Saturated 5g); Cholesterol 15mg; Sodium 320mg; Carbohydrate 56g (Dietary Fiber 11g); Protein 12g.*

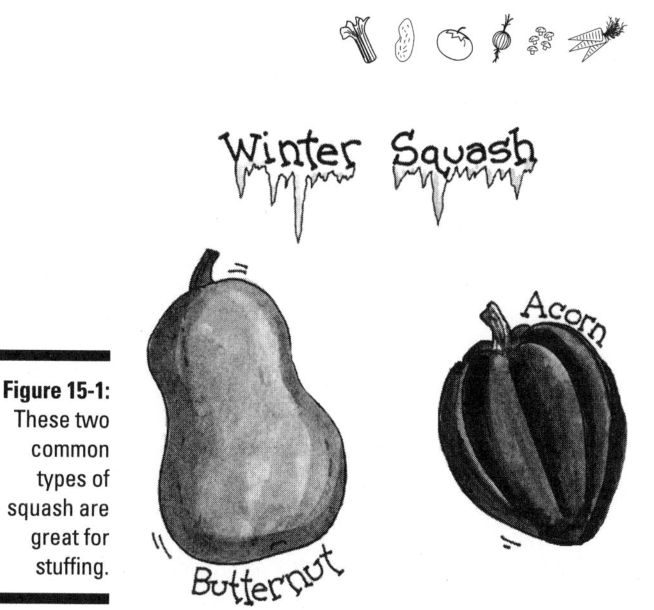

Figure 15-1: These two common types of squash are great for stuffing.

Fluffy Mashed Potatoes with Mushroom Gravy

Yukon Gold potatoes lend a rich color to this dish. The gravy goes well on the potatoes, as well as over Everybody's Favorite Cheese and Nut Loaf (see the recipe earlier in this chapter).

Preparation time: *15 minutes*

Cooking time: *30 minutes*

Yield: *8 servings*

Potatoes:

5 pounds Yukon Gold potatoes, peeled and cut into 2-inch chunks (or halved if they're small)

¼ cup melted trans fat-free margarine

1 cup plain soymilk or skim milk

¼ teaspoon salt

¼ teaspoon black pepper

Mushroom Gravy:

1 tablespoon olive oil

1 pound mushrooms, thinly sliced

1 medium onion, chopped

2 tablespoons flour

1 vegetable bouillon cube

¼ teaspoon salt (if using a sodium-free bouillon cube; otherwise omit)

¼ teaspoon black pepper

1 cup plain soymilk or skim milk

1 Place the potatoes in a pot and fill with cold water to cover them. Cover and cook over medium-high heat until boiling; then reduce the heat, tilt the cover to allow steam to escape, and simmer for 30 minutes, or until the potatoes are tender.

2 While the potatoes are cooking, prepare the Mushroom Gravy. Heat the olive oil in a medium skillet. Over medium heat, cook the mushrooms and onions in the oil until the onions are translucent — about 5 minutes.

3 Add the flour, and crumble the bouillon cube into the skillet. Add the salt (if needed) and pepper and stir.

4 Add the soymilk. Cook and stir for 2 to 3 minutes, until the gravy thickens and the ingredients are well blended. Pour into a small pitcher or serving dish and set aside until the potatoes are ready. If you don't serve it immediately, the gravy may need to be reheated in a microwave oven, or you can hold it on the stovetop, warming, until ready to eat.

5 When the potatoes are done, drain them. In a large bowl, combine the potatoes, margarine, milk, salt, and pepper. Mash with a potato masher until smooth and well blended. Use a wooden spoon to help blend the ingredients if necessary. Transfer to a serving dish and serve with Mushroom Gravy.

Vary It! *Add 1 tablespoon chopped parsley or chives in Step 5.*

Per serving: *Calories 346 (From Fat 77); Fat 9g (Saturated 3g); Cholesterol 0mg; Sodium 282mg; Carbohydrate 56g (Dietary Fiber 5g); Protein 10g.*

Roasted Roots

Herbs and the roasting process bring out the delicious flavors of the root vegetables in this dish. Peel the vegetables before cutting them. The beets stain the other vegetables a rich red color.

Preparation time: *20 minutes*

Cooking time: *1 hour, 5 minutes*

Yield: *8 servings*

2 medium carrots, sliced into 1-inch pieces

1 medium turnip, sliced into 1-inch wedges

3 to 4 beets, quartered

1 parsnip, cut lengthwise and then sliced

1 medium onion, quartered

2 small leeks, split and washed

1 medium white potato, cut into eighths

1 medium sweet potato, cut into eighths

1 bulb fennel, quartered

⅓ cup olive oil

4 cloves garlic (chopped or whole cloves)

Three 2-inch sprigs fresh rosemary, chopped

½ teaspoon salt

½ teaspoon black pepper

5 tablespoons balsamic vinegar

1 Preheat the oven to 350 degrees.

2 Toss the vegetables with the olive oil, garlic, rosemary, salt, and pepper. Set the fennel and leeks aside and arrange the remaining vegetables in one layer in an oiled roasting pan or on a greased cookie sheet. Do not cover.

3 Roast the vegetables for 15 minutes, and then add the fennel and leeks. Roast all the vegetables for an additional 50 minutes, or until tender. Pay particular attention to the beets and make sure that they're soft before removing them from the oven.

4 Toss the vegetables with the balsamic vinegar and serve.

Per serving: *Calories 198 (From Fat 81); Fat 9g (Saturated 1g); Cholesterol 0mg; Sodium 216mg; Carbohydrate 27g (Dietary Fiber 5g); Protein 3g.*

Candied Sweet Potatoes

You can make this simple holiday classic healthier by substituting trans fat-free margarine for butter.

Preparation time: *5 minutes*

Cooking time: *12 to 14 minutes*

Yield: *6 servings*

8 medium sweet potatoes (about 3 pounds)

¾ cup packed brown sugar

2 tablespoons trans fat-free margarine

Juice of a fresh lemon (optional)

1 Wash the potatoes and cut them into ½-inch segments. Place them in a medium saucepan, cover with water, and bring the water to a boil. Let the potatoes simmer until tender — about 12 to 14 minutes. Cook longer if necessary, but don't let the potatoes get mushy.

2 Remove the potatoes from the heat and drain. Add sugar and margarine, and stir gently with a large wooden spoon to coat the potatoes.

3 Turn the potatoes into a serving bowl. Squeeze the juice of one lemon over the tops of the potatoes before serving (if desired).

Vary It! *Substitute ⅓ cup maple syrup for the brown sugar. Add ¼ teaspoon ground ginger in Step 2. Another option: Add a handful of chopped pecans and/or ⅓ cup crushed pineapple in Step 2.*

Per serving: *Calories 349 (From Fat 40); Fat 5g (Saturated 1g); Cholesterol 0mg; Sodium 81mg; Carbohydrate 76g (Dietary Fiber 4g); Protein 3g.*

Maple Pecan Pumpkin Pie

Tofu replaces the eggs and milk in this pumpkin pie recipe. The filling is thick, rich, and smooth.

Preparation time: *15 minutes*

Cooking time: *40 minutes*

Yield: *One 9-inch pie (8 slices)*

One 15-ounce can pumpkin (about 2 cups)

8 ounces silken firm tofu

½ cup pure maple syrup

½ teaspoon ground ginger

1 teaspoon cinnamon

¼ teaspoon nutmeg

1 tablespoon flour

One 9-inch pie shell

¼ cup chopped pecans

1. Preheat the oven to 350 degrees.
2. Place the pumpkin, tofu, maple syrup, ginger, cinnamon, nutmeg, and flour in a blender or food processor and process until smooth.
3. Pour the pumpkin mixture into the pie shell.
4. Sprinkle the chopped nuts evenly over the top of the pie.
5. Bake for about 40 minutes, or until set. Cool and then serve.

Per serving: *Calories 236 (From Fat 99); Fat 11g (Saturated 2g); Cholesterol 0mg; Sodium 137mg; Carbohydrate 31g (Dietary Fiber 3g); Protein 5g.*

Part IV

Living — and Loving — the Vegetarian Lifestyle

"Easy—there it is. The wild rutabaga, and I think you've got a clear shot."

In this part . . .

You have the nuts and bolts behind you — an understanding of the various approaches to vegetarianism, sound reasons to adopt a vegetarian lifestyle, and the nutrition basics. You know how to begin, and you've gotten your kitchen ready. You may have tried fixing a few of the recipes in this book.

Now it's time for some advanced advice for those of you who want to dig more deeply into making the transition to a vegetarian lifestyle complete.

In this part, I share strategies for keeping the peace in households where you may be the only vegetarian at the table, and I discuss how to deal with the nonvegetarians you encounter every day outside your home. I also offer advice for maintaining your vegetarian diet when you're a guest at someone else's home or when you eat out.

Chapter 16

Getting Along When You're the Only Vegetarian in the House

In This Chapter

- Working with multiple food preferences
- Using strategies to gain support from nonvegetarians
- Accommodating nonvegetarian visitors

Some food fights are easy to settle. He likes chunky and she likes creamy? Keep two jars of peanut butter in the cupboard. Butter versus margarine, or mayo against Miracle Whip? Not a problem — just keep both on hand.

If only all differences in food preferences were so easy to handle.

If you're a vegetarian, chances are good that you're the only vegetarian in your household. Like May–December romances, partnerships between a steak-and-potatoes type and a veggie lover can present challenges. If you're going vegetarian alone at home, you, your partner, and other immediate family members have to address several issues. Some will be unique to your living situation, but other problems are common in many people's experiences. An awareness of the issues and the manner in which you approach them can have a significant effect on your relationships.

Managing Family Meals

Should you fix a standard meat-two-vegetables-a-starch-and-a-salad meal, dodge the meat, and fill up on the rest? Or do you fix a separate vegetarian entrée for yourself?

Some families feel burdened by this problem. Nobody wants to get stuck cooking two meals for the same table every night. Should all meals be vegetarian, so that everyone can eat them? After all, most nonvegetarians enjoy many meatless dishes.

I don't have all the answers. The situation requires discussion and negotiation among all parties concerned. Unless you face it head on, though, it will bite you from behind.

Thinking through some of the options may help.

Fixing both meat and vegetarian foods

Serving both a meat entrée and a vegetarian alternative is the most obvious solution to the problem of preparing meals that everyone can eat. It's also the most work. For vegetarians who happen to be the family cook, it may also be out of the question if they don't want to handle and cook meat.

Deciding who prepares what

If you're the head cook in your household and you're a vegetarian, should you be expected to fix meat for the nonvegetarians? Maybe it's time for a nonvegetarian family member to learn how to cook meat. Another option: Prepare meals together. You can be in charge of everything but the meat.

On the other hand, maybe you don't want to subject yourself to the smell of cooking meat, or you don't want meat to touch your kitchen counter or pots and pans. I can relate to that point of view, but you'll have to hash this one out with your family. It may help to know that you're not alone — lots of people like you are out there.

If your household has meat-eaters, sit down and try to work out compromises that everyone can live with. Some options include the following:

- You may agree that any meat brought into the house is prepared on waxed paper so that it doesn't touch the countertops, and that it's cooked in a pan dedicated solely to meat.
- You may decide to cook meats outdoors on a grill to keep offending odors and drippings out of the house.
- You may choose to buy only precooked meats such as deli or luncheon meats, which can be kept wrapped in the refrigerator and need no cooking or further preparation in the house.

If you're having difficulty negotiating meal issues with family members, it may help to talk to other vegetarians who've dealt with the same challenge. Consider joining a vegetarian chat group on the Internet, participating in a blog, or joining a local vegetarian society. The Vegetarian Resource Group maintains a list of these and other resources online at `www.vrg.org/links/`.

Respecting others' food preferences

What you eat is a highly personal decision. I've found over the years that it's best to let friends and family members make their own food choices without pressure from me — a strategy you may want to consider as well.

For a while, you may have to live with two variations at every meal. However, fixing both vegetarian and nonvegetarian options for family meals has its advantages. As any experienced vegetarian knows, the vegetarian option tends to look really tasty. Try setting a vegetarian pizza on the table with other pizzas at a party or any gathering. If you're a vegetarian, claim a slice quickly, because the nonvegetarians often want the vegetarian choice. If you're not quick enough, you may find your vegetarian option is gone.

Likewise, put a bowl of pasta topped with marinara sauce on the table side by side with pasta and meat sauce. Conduct a bit of research, and see which bowl empties fastest. I'll put my money on the marinara — I've seen it happen too many times.

So the idea is that if you set out vegetarian entrées, nonvegetarians are going to want them, too. Eventually, you may be making nothing but vegetarian meals, and the other people in your household may not even notice the transition. And all that arguing and teeth gnashing will have been for nothing.

Making do when you need to

Instead of fixing two meals, another option is for the vegetarian to eat large servings of whatever meatless foods are on the table. Mashed potatoes, steamed broccoli, salad, and a dinner roll make a great meal. You don't have to have a hole on your plate where the meat should be — you can cover up that bare spot with extra servings of the foods you like.

Many vegetarians use this same approach when they eat out at the homes of friends or others who don't serve vegetarian meals. They eat what they can of the available meatless foods. They don't draw attention to themselves (some vegetarians prefer it that way), and they don't give others the impression that it's hard to be a vegetarian by whining about not having an entrée.

This approach is passive and fairly nonconfrontational. Problems are most likely to arise if another family member is threatened by the fact that you're behaving outside the norm of the family. For some, it may appear that you're rejecting family values and traditions. You're not, of course — you're just changing them.

Finding the vegetarian least common denominator

Another sensible strategy is to consider vegetarian foods to be the least common denominator. In other words, everyone — vegetarians and nonvegetarians alike — can eat vegetarian foods, so when you plan meals, start with vegetarian choices. Choose foods that are vegetarian as prepared but to which a nonvegetarian can add meat if he so chooses.

For example, make a stir-fry using assorted vegetables, and serve it on steamed rice. A nonvegetarian can cook chicken, shrimp, pork, or beef separately and add it to her own dish. Rather than actually making two different entrées, you're modifying an entrée for a meat-eater. That should be considerably easier than fixing two distinct dishes.

The following foods are vegetarian, and meat-eaters can add small bits of meat if they so desire:

- Bean burritos, nachos, or tacos
- Italian stuffed shells, manicotti, cannelloni, or spaghetti with marinara sauce
- Mixed green salad (large dinner-style)
- Pasta primavera
- Pesto pasta
- Vegetable jambalaya
- Vegetable lasagna with marinara sauce
- Vegetable soup or stew
- Vegetable stir-fry
- Vegetarian chili
- Vegetarian pizza

Gaining Support from Nonvegetarians

When it comes to handling family meals, several options that have worked for others may work for you, too. Because every family is different, though, you may have to adapt these ideas to fit your own circumstances.

For starters, take some simple steps (which I outline in this section) to make the other members of your household more receptive to vegetarian meals. Remember that you're asking them to trade in some of their favorite foods.

What would it take for you to give up something that you like and are comfortable with? You'd probably want a substitute that tastes as good as or better than the food you gave up. You wouldn't want to be inconvenienced. And you wouldn't want the change to leave you feeling dissatisfied.

Keep these points in mind when you ask for your family's support.

Employing strategies for compromise

If your family is agreeable, you may be able to get away with fixing nothing but vegetarian meals, provided that everyone likes the choices. Whether you go vegetarian all the time or part of the time, it's a good idea to go out of your way to make foods that everyone likes.

Begin by making a list of family members' favorite foods that just happen to be vegetarian. For instance, your list may include veggie-topped pizza, bean soup and chili, stuffed baked potatoes, macaroni and cheese, and pasta primavera.

To that list, add some transition foods — foods that look and taste like familiar meat-based foods — that appeal to most people. These alternatives taste good, and you can use them in the same ways as their meat counterparts. Let them serve as crutches or training wheels until people become more comfortable planning meals without meat. Examples include vegetarian burger patties and hot dogs, vegetarian cold cuts for sandwiches, vegetarian bacon and sausage links and patties, and frozen, crumbled, textured vegetable protein for recipes that call for ground beef.

Let family members have input into planning, and incorporate their preferences into the menu. People who are involved in any aspect of meal planning are more likely to be interested in eating the food. Before long, you'll have a list of family favorites a mile long.

Setting a positive example

Don't push. It's that simple. When you preach to people and push them to do something they aren't ready to do — like change the way they eat — you're likely to get the opposite of the result you want.

Instead, teach by your own example. If your family or partner isn't ready to make a switch to a vegetarian diet, quietly go about it by yourself. Your actions will make an impression, and chances are good that, given time, the others will notice, get interested, and make some dietary changes of their own.

I know this from experience. I grew up in Michigan in a typical, traditional, meat-and-potatoes household. One day, just as my family was sitting down to dinner, my mother briefly announced that from that day forward she would be a vegetarian. She didn't say another word about it. No explanation except, "This is what I've decided is right for me." She sat down, and we ate our dinner.

My mother didn't discuss her vegetarianism with anyone. With the exception of her family and close friends, most people didn't even realize that she didn't eat meat. At family meals, she continued to prepare meat for my father and siblings, but she would make a cheese omelet or toasted cheese sandwich for herself (a Wisconsin native, she was stuck in the cheese-and-eggs rut!) and eat the same salad and vegetables that the rest of us ate.

Time went by. Then, one by one, my sisters and brother and I quietly went vegetarian ourselves. Nobody offered more of an explanation than to say to our mother, "You don't have to fix meat for me anymore." Eventually, the only person my mother was fixing meat for was my father. Many years later, meat faded out of his diet, too. Decades later, we're still vegetarians.

This process repeated itself when I met my husband, a barbecue-eating, native North Carolinian. Within a couple of years and without a word from me, he began to eat less meat. Pretty soon, he began referring to himself as a vegetarian. Now, after ten years on a meat-free diet, he's a pro. He's 6 feet 3 inches tall and weighs more than 200 pounds, and people are surprised to learn that such a big guy doesn't eat meat. But the greens and beans he takes to work for lunch draw comments from colleagues ("You win the prize for the best-smelling food"), and he's healthier and trimmer than he has been in years.

The only reasonable alternative if you want others to change is to let them decide for themselves.

Negotiating the Menu When Guests Come

In addition to working out a family meals plan that promotes harmony, you may want to discuss how to handle such events as guests staying at your house and friends dropping in for dinner.

For some people, this kind of situation can cause friction, particularly when the lifestyle of the vegetarian in the household is seen as an intrusion or unwelcome deviation from the norm. Having outsiders come in can cause built-up tensions to flare. The meat-eater's point of view may be that if eating meat is the norm of the guests, the preferences of the majority should rule.

In other circles, vegetarianism may be no big deal. In fact, some of your guests may well be vegetarians.

If you live with someone who doesn't support your diet choice, it's a good idea to get the issue on the table and resolved before your guests arrive. You may decide to eat out, or you may opt to bring takeout food home. However you solve the dilemma, the important thing is to anticipate the problem and try to find a solution ahead of time.

Then enjoy the visit.

Giving your guests options

Many people are comfortable having some vegetarian dishes and some meat dishes available, with everyone free to choose what she pleases. This works well when the household already has a custom of serving two entrées — one meat-based and one meatless.

But nonvegetarians will often have a helping of the meatless choice. You want the nonvegetarians to have the option of tasting that delicious-looking vegetarian entrée if they care to, but you also want to ensure that enough is left for the vegetarians. So when you invite guests to come to your home, ask whether anyone has special dietary needs. Do your best to estimate the number of people who may be interested in the vegetarian option — and then double that amount.

I once spent several days on a small boat on the Great Barrier Reef on a scuba diving trip. I was one of two vegetarians in a group of about 15 divers and crew. For our first meal — spaghetti — the cook set aside a small amount of pasta with meatless sauce for me and the other vegetarian on board. She labeled it as meatless and set it out on the serving table alongside the larger bowl of pasta with meat sauce.

I made the mistake of being near the end of the serving line. By the time I got to the head of the line, the vegetarian spaghetti was gone. From that day forward, the two vegetarians were permitted to go through the serving line first.

And I made an important realization — the vegetarian option often appeals to everyone, vegetarian or not.

Serving meals with mainstream appeal

One easy way to handle meals for guests is to rely on old standbys — such as vegetable lasagna or pasta primavera — that are familiar to both vegetarians and nonvegetarians alike. Serve either with a loaf of good bread and a beautiful mixed green salad or a bowl of minestrone soup, and finish the meal with coffee and a delectable dessert.

Nobody will notice the meat was missing.

Minimizing the focus on meat-free

Another strategy to redirect guests' attention away from the fact that they're eating a meatless meal is to knock their socks off with a delicious, sophisticated, gourmet vegetarian meal, or an ethnic entrée such as an Indian curry dish or a Middle Eastern spinach pie. Many guests would be impressed and also enjoy the change of pace.

Don't forget about the importance of setting the mood, too. Dim the lights and put on some soft music. Set an attractive table. Use a tablecloth or placemats. Place a vase with a sprig of foliage or flowers from the yard on the table. Arrange foods in serving bowls or plates using parsley and sliced fruit as garnishes.

Emphasize the experience of enjoying good food with good company. The fact that there's no meat on the table will be irrelevant.

Chapter 17

Vegetarian Etiquette in a Nonvegetarian World

In This Chapter

- Making a good impression
- Being a gracious guest
- Arranging meals for special events
- Dating meat-eaters
- Managing your diet on the job

Maintaining a vegetarian lifestyle at home is one thing. Doing it away from home with style and grace is quite another thing.

You owe no one an explanation of your eating preferences. Keep in mind, though, that despite the fact that vegetarianism has become more mainstream in recent years, the vast majority of people still eat meat. You'll likely stand out, especially if you eat a vegan diet. For that reason, you're the one who has to do most of the adapting in social situations.

This chapter focuses on common situations for which you need to develop skills to relate with nonvegetarians. You'll become more comfortable managing these situations as you gain experience.

Mastering the Art of Diplomacy

You may be quite content to live a vegetarian lifestyle without explaining your choice to people or attempting to convert them. If so, you're probably the type who doesn't rub people the wrong way. You're an undercover vegetarian, not a walking-and-talking advocate, and that's fine. Even in your relative silence, you're making a statement.

On the other hand, if you frequently find yourself engaged in debates or earnest discussions about the merits of a vegetarian diet — whether you initiate the conversations or not — you should give some thought to your approach in handling these interactions.

Diplomacy is key, whether you prefer to do your own thing quietly or hope to inspire others to go vegetarian. You're going to need skills in dealing with social situations in which your vegetarian lifestyle converges with the nonvegetarian world. That takes time and experience. To help you get started, consider some of the ideas I share in this section.

Watching how you present yourself

Maybe you've seen the features on fashion do's and don'ts in women's magazines — photos of people whose looks are perfectly and tastefully coordinated next to photos of people with visible panty lines and other fashion gaffes. Your physical appearance can give people the impression that you have it all together or that you're a frazzled wreck.

Similarly, the manner in which you conduct yourself in the presence of others affects the way they perceive you. If you're belligerent, bossy, or brash, you come off as being uncouth. Even if your point of view is valid, others are likely to discount your message at best, or be totally repulsed and reject it outright at worst.

Bumper stickers, buttons, and shirts with vegetarian slogans can express your politics loudly and clearly. One makes a statement. Two emphasize the point. But when your car and body are blanketed with these items, you're shouting. Some people may perceive you to be hysterical, depending on the degree of coverage of your car and clothing. Consider that you may reach a point of diminishing returns in terms of the extent to which people will listen to you when you express your views this way.

Activism done right

Some people feel that the in-your-face approach has a place. Opinions are neither right nor wrong, of course. The bra burners of the 1960s drew attention to women's rights issues with their very visible demonstrations of activism, and other women advocated for change in subtler ways. Likewise, a peace activist may wage a hunger strike to draw attention to the cause, while others promote peace behind the scenes. Ultimately, it's up to you whether — and how — you advocate for vegetarianism. Attacking others, however, is a negative approach that's likely to alienate people and make them less receptive to your message.

Responding to questions about your vegetarianism

> "Why did you become a vegetarian?"

Most vegetarians have heard that question at least 25 times.

> "If you don't eat meat, how do you get enough protein?"

They've heard that one just as many times.

> "What do you eat for dinner?"
>
> "Don't you just want a good steak once in a while?"
>
> "If you're a vegetarian, why do you wear leather shoes?"

When someone inquires about your vegetarianism, the best response is a simple, straightforward answer. You don't have to expound on the basics — just the facts will do. This is generally also not the time for the hard sell, nor is it the time to push someone else's buttons by criticizing his lifestyle.

You need to use your own judgment, but most of the time when people ask questions such as these, they're just exploring. They may not know much about vegetarianism, but you've piqued their interest. They may also be testing to determine what you're all about. Answer their questions concisely, but leave 'em wanting more. They'll ask when they want to know — and they may start asking you how they can become vegetarians themselves.

Being an effective role model

The world would be a better place if more people were vegetarians. Those of us who've already made the switch know how nice it can be to live a vegetarian lifestyle. A vegetarian lifestyle is good for you, good for the animals, and good for the environment and planet.

As much as possible, try to be positive in the way you interact with others where food is concerned. Your attitude affects how people feel about you, and people will associate that feeling with how they feel about the concept of vegetarianism.

The truth is, being a vegetarian isn't particularly difficult, and vegetarianism has many benefits. Why not advertise that fact? Some vegetarians present themselves in a way that makes it appear that living a vegetarian lifestyle is difficult and problematic. Instead, aim to show people how easy and pleasurable a vegetarian lifestyle can be. Take the positive tack if you want other people to give vegetarianism serious consideration.

Filling your plate to show vegetarian variety

Show people that eating a vegetarian diet doesn't mean you have to go hungry when you eat meals away from home. Fill your plate with food. Let people see how colorful and appealing a meal without meat can be. Don't leave a hole on your plate or allow it to look empty, as if you're deprived.

Demonstrating flexibility in difficult situations

All vegetarians get stuck at a truck stop or a fast food restaurant occasionally, where the menu has nothing for them to eat. Those are the times when you kick into survival mode and take a can of juice, a side order of slaw, and a bag of chips and just hang on until the next opportunity to eat some real food.

Most of the time, though, vegetarians have plenty of food choices, or they can make do with what there is. Whether you're a guest at someone's home, eating out on business, traveling, or out with friends, try to be flexible when it comes to your meals. By being flexible, you show others that you can be a vegetarian in a nonvegetarian society without your lifestyle creating too much conflict. Send a positive message.

Flexibility and planning are especially important for vegans. Vegans do find it more difficult than other types of vegetarians to eat away from home, especially when they're in the company of nonvegetarians or nonvegans and don't have total control over where a group goes to eat.

I was once with a group of people, including two vegans, at a conference in Kansas City. When it came time to find a place to have dinner the first night, we literally roamed the streets for over an hour, wearily walking into one restaurant after another, reading the menus, finding few vegan options, and walking out again. Eventually, after everyone was frustrated and overly hungry, we settled on a Chinese restaurant. We ended up eating there every night thereafter for the rest of that week. The impression I was left with was that it was no fun at all to be a vegan, and that it was tiresome and difficult to find something to eat away from home.

In a mixed group of vegetarians and nonvegetarians, or vegans and nonvegans, try to anticipate the limitations and preempt them. Steer your party to a restaurant at which you'll be most likely to have numerous choices. If you can't do that, do your best to make do wherever you land.

Some vegetarians eat a vegan diet at home but occasionally make exceptions and eat foods containing dairy and/or eggs when they're away and have limited choices. The reason? They prefer not to isolate themselves socially from nonvegans. They feel that by being 90 percent of the way there, they're achieving most of the benefits of a vegan lifestyle without sending others a negative message.

When you're setting an example for others, inspiring lots of people to drastically reduce their intake of animal products may be more important than inspiring one or two to go vegan.

Reconciling your approach and withholding judgment

Just as your decision to live a vegetarian lifestyle is a personal one, the manner in which you interact with people is also your call. You can be tolerant or intolerant, compassionate or confrontational.

Your approach may be different from someone else's. A concession on the part of one near-vegan to eat a food containing dairy or eggs when away from home may be completely unacceptable to another vegan, just as someone's decision to eat a piece of turkey on Thanksgiving may be an abhorrent choice to other vegetarians.

In understanding others' choices, remember how complicated and deeply personal the subject of vegetarianism can be for everyone. Reconcile your approach as you see fit. Then work at projecting a positive image. Show others that you're confident, content, and at ease with your lifestyle.

Handling Dinner Invitations

Being invited to a nonvegetarian's home for dinner is one of the most common and most stressful events that many vegetarians encounter. How you approach the situation depends on the degree of formality of the occasion and how well you know your host.

If you're a dinner guest at a nonvegetarian's home, carrying on about the evils of meat is a major faux pas. Even if another guest eggs you on and attempts to engage you in an earnest discussion or debate about the merits of vegetarianism, resist the temptation to participate. To do otherwise is just plain uncouth.

Letting your host know about your diet

You're a vegetarian. Out with it.

You need to find the right moment to tell your host that you're a vegetarian, but do say something right away. If the invitation comes by phone or in person, the best time to say something is when you get the invitation.

Say something as simple as, "Oh, thanks, I'd love to come. By the way, I'd better tell you that I'm a vegetarian. You don't have to do anything special for me — I'm sure I'll find plenty to eat — but I just didn't want you to go to the trouble of fixing meat or fish for me."

The chances are good that your host will ask whether you eat spaghetti, lasagna, or some other dish that she's familiar with serving and thinks will suit your preferences. If your host suggests a shrimp stir-fry, you can also nip that one in the bud by letting her know that you don't eat seafood (if you don't). In that case, you may want to toss out a few more suggestions of foods you do eat, or you could reassure your host that you'll be fine with extra servings of whatever vegetables, rice, potatoes, salad, and other side dishes she plans to serve.

If you get an invitation to dinner by mail, through a spouse or partner, or in some other indirect way, you have two choices:

- Show up for dinner and eat what you can.
- Call your host, thank her for the invitation, and mention that you're a vegetarian.

If you're invited as the guest of the invitee (your spouse, partner, date, and so on), you may want that person to call the host on your behalf. Examine the situation and do what feels most appropriate and comfortable.

Initiating a conversation about your vegetarianism may feel a bit awkward, but telling your host upfront that you're a vegetarian may be less uncomfortable than showing up for dinner and finding nothing you can eat, or discovering that the host made something special for you that you have to decline. In either of these situations, not only do you run the risk of not getting enough to eat, but you also risk disappointing your host, or causing her to puzzle over why you aren't eating what she served.

Some people may get a little nervous when you say that you're a vegetarian. Most people who invite guests to dinner want them to enjoy the food and have a good time, and they may not be very familiar with what a vegetarian will and won't eat.

Reassure your host that you'll be fine. Most people appreciate knowing ahead of time that you won't be eating the meat or fish, as it saves them from preparing something that won't be eaten. It also saves everyone the embarrassment of scurrying to find something for you to eat.

It also saves you from being stuck in a situation where the rice is cooked with chicken stock, the salad is sprinkled with bacon, the appetizer is shrimp cocktail, and steak is the main course. You can be a bit conspicuous if all you have on your plate is a dinner roll.

If you think you may have trouble finding something you can eat at your host's house, have a snack before you leave home. That way, you won't be starving if you discover that there isn't much to eat.

Offering to bring a dish

Another way to handle dinner invitations is to offer to bring a dish that you and everyone else can share. This works best in casual situations when you know your host well or the setting is relatively informal. You may not feel as comfortable suggesting this, for instance, if you're invited to a formal company function at the home of someone with whom you don't interact regularly. In that case, bringing your own food could look tacky.

Offer, but don't push. If you tell your host that you're a vegetarian and he sounds worried or unsure about how to handle it, you might say something like this:

> "If you'd like, I'd be happy to make my famous vegetable paella and saffron rice for everyone to try. It often complements other entrées, and a salad and bread would go well with it."

If your host bites, great. If he brushes the suggestion aside, just consider your job done and leave it at that. Reassure him that you think you'll be fine eating the salad and side dishes that don't contain meat or seafood and that you're looking forward to coming.

Graciously declining nonvegetarian foods

Sometimes you tell your host you're a vegetarian, and she tries to fix a special vegetarian meal for you but "misses." For example, she may prepare a gelatin mold salad, use eggs in the baked goods, put bacon on the salad, or serve the salad with Caesar dressing (it contains anchovies). What should you do?

Be the best actor that you can be and just eat whatever you can of the other parts of the meal. Make your plate look as full as you can so your host doesn't feel as though you don't have enough to eat. Assure her that you're fine. Enjoy the company, and eat when you get home.

If you can't finesse a refusal without your host detecting it, go ahead and fess up. Explain as gently and simply as possible that you prefer to pass up the dish and why. You might acknowledge that many people are surprised to discover that vegetarians don't eat gelatin (or lard, or beef broth, for example). Be sure to thank your host for her efforts and reassure her that you have plenty to eat.

Being a stealth vegetarian: What to do if your host doesn't know

If you decide for whatever reason not to tell your host in advance that you're vegetarian, downplay the fact during the meal or stay undercover. Pointing out that you can't eat the food will make everyone uncomfortable and may put a damper on the gathering. Your host may feel badly and may even ask why you didn't say something sooner.

Eat what you can. If you have to ask about ingredients in the food or can't eat something because it's not vegetarian, be sure to play down any inconvenience. Reassure your host that you'll be just fine with what there is.

Managing Invitations to Parties and Other Special Events

It's fairly easy to manage invitations to weddings, banquets, and other functions with lots of people where the food is served from a buffet line. Even if a sit-down meal is served, a hotel or restaurant usually caters the food. You may find it much easier to speak to hotel or restaurant staff about getting a vegetarian meal than to ask the host directly.

Handling parties at private homes

If the function happens to be held at a private residence, follow the advice I give earlier in this chapter for dinner invitations. If the occasion involves a large number of guests, lots of food will likely be served and your host won't have to do anything special for you.

Making your way through public venues

If the event is at a hotel, country club, community center, or restaurant, call ahead and ask to speak with the person in charge of food service. You don't have to say anything to your host. It's perfectly fine to speak to the chef or caterer and request a vegetarian or vegan entrée as a substitute for whatever the other guests are served. Most places will provide an alternate dish at no extra charge, but if you're worried, you can ask to be sure.

Be clear about what you can and can't eat. For example, if you eat no meat, fish, or poultry, say so. Otherwise, you may end up with fish or chicken, because *vegetarian* only means "no red meat" to some people. You might also ask about how the chef will prepare other items. For instance, if Caesar salad is on the menu, see whether a different type of dressing or salad can be set aside for you (and remind the kitchen that you don't eat bacon on your salad).

When talking with a member of the food service staff, be sure to mention whether you eat dairy products or eggs. If you aren't explicit and you're vegan, you may end up with egg pasta, an omelet or a quiche, grated cheese on your salad, or cheese lasagna.

If you call the restaurant or hotel kitchen to request a vegetarian meal for an event, feel free to suggest a dish that you know you can eat and that the kitchen is likely to be able to make. For example, most kitchens have the ingredients on hand to make pasta primavera or a large baked potato topped with steamed vegetables. Those foods are simple to fix, and they can be preferable to that ubiquitous steamed vegetable plate.

As service begins and the waiter arrives at your table, quietly mention that you've ordered a vegetarian meal. If a plate containing meat lands at your place before you can say a word, let the food sit there until you can flag the waiter down to request your vegetarian meal.

If the meal is a buffet, you may not even need to mention to your host that you're a vegetarian, unless you think the menu may be limited. If the buffet is at a restaurant or hotel, it will likely include a wide variety of dishes and plenty of foods you can eat.

If you're vegan, however, it may still be wise to phone ahead and ask about the menu, just to be certain that it will have dairy- and egg-free choices. You may have to request that some rice, pasta, or vegetables are set aside for you if everything on the buffet line is doused with butter or cream sauce. The chef may even be able to make you a separate, suitable dish.

Dating Nonvegetarians

If your date is also a vegetarian, you're off to a good start. All you have to worry about (food-wise, anyway) is whether you're in the mood for Mexican or Thai or Ethiopian food. Even if you're a lacto ovo vegetarian and he's vegan, at least you're simpatico — you'll work it out.

On the other hand, if your date isn't a vegetarian, broach the subject early on. You don't have to make a big deal of the fact that you don't eat meat, but it's probably best to mention that fact sooner rather than later, and probably before you get together for a meal.

If you think the relationship has the potential to become serious, ask yourself the following:

- Do I expect this person to stop eating meat eventually?
- What would the ground rules be if we decided to live together or get married?
- Would I be willing to have meat in the house?
- Would we cook two different meals or fend for ourselves?
- What would we do if nonvegetarian guests came over for dinner — serve them meat or not? How would we resolve the conflict if we had different views about what to do?
- Can I live and let live, or would I want my partner to see things as I do and maintain a vegetarian lifestyle?

This last question is especially important if your vegetarianism is rooted in ethical beliefs or attitudes. Compromising is difficult to do when your personal philosophy is on the line. You have to decide whether your life has room for someone who lives differently.

Some vegetarians find it harder to connect with a partner who doesn't share the same sensitivities. If it doesn't matter now, it may later. You don't have to arrive at any hard-and-fast conclusions about these issues today, but if you've never faced them before, it's a good idea to begin thinking about them.

If you do decide to get serious with a nonvegetarian, be sure to think about how to make the relationship work over the long run. If you want children, talk about whether you'd raise them as vegetarians.

Some vegetarians won't even date a nonvegetarian. (Kiss someone with greasy cheeseburger lips? No way!) Would you? If you restrict yourself to dating only fellow vegetarians, you may spend many a Saturday night alone in front of the TV. After all, vegetarians make up only a few percent of the population, and only a fraction of that number is in the market for dating.

Vegetarian singles groups are active in some cities. Check with your local vegetarian society to see whether such a group exists in your area. These groups plan restaurant outings, outdoor activities, and other events with the needs and interests of singles in mind.

Working It Out: Vegetarianism on the Job

In many professions, your success at the office is partly dependent on how well you blend into the organization's culture. If your job involves entertaining clients or meeting with colleagues over lunches and dinners, your eating habits will be on display for all to observe.

If you're a vegetarian, the manner in which you conduct yourself can lead others to mark you as either sensible and health-conscious or eccentric and difficult to get along with. Your vegetarianism may be an asset or a liability, depending on how you play it.

To tell or not to tell?

Your dietary preferences may matter more in some work settings than in others. An advertising agency or a marketing department within a company, for instance, is likely to be comfortable with creative people who do things differently.

On the other hand, if you're in a more conservative line of work such as law or banking — especially in rural areas where vegetarianism may be less prevalent — it may be more important for you to appear compatible with your peers. Attitudes about vegetarianism have changed considerably in recent years, but some people are still likely to be less receptive to your lifestyle than others.

Why should you care about what your co-workers think of your eating habits?

You should care, because in some settings, the perception that you're a social outlier may negatively affect your chances of getting a promotion or a choice job opportunity. If you play your cards right, however, you can turn the difference in your lifestyle into an asset that will reflect positively on you and enhance the company's and clients' impressions of you.

To increase your likelihood of success, don't preach about the benefits of being a vegetarian. With your actions as a model, you can communicate an understated message to your co-workers that they'll get loud and clear. Show them that you care about your health, that you're socially and environmentally conscious, and that you have the self-discipline and determination to persevere with healthy lifestyle habits.

Handling meals during job interviews

Job interviews often involve lunching with your prospective employer and colleagues. If that happens, you don't need to hide the fact that you're a vegetarian. At the same time, you don't need to make a big deal of it. Just approach the subject in a confident, matter-of-fact way. Remember the following:

- First impressions count. In job interviews, you want your prospective employer and colleagues to focus on your skills and ideas, not on the fact that you don't eat meat. When they reflect back on their meeting with you, you want them to think about how well you'll fit into their environment and how your unique set of abilities will be an asset to their organization.
- When you're on a job interview, the people interviewing you often choose the restaurant where you'll eat. As a result, you may not have an opportunity to steer them to a restaurant with lots of vegetarian choices.

  If that happens, size up your menu options as quickly as possible, and downplay inconveniences due to your diet. It's fine to ask the waiter whether the soup is made with chicken stock or a beef base, or if the beans contain lard. But if you do, make it as low-key as you can. If the answer is affirmative, have an alternative choice in mind so that you don't hold up the table's order while you struggle to find another dish.
- Have an off-menu choice in mind, in case you have trouble finding something on the menu. In a restaurant that serves pasta, for example, it's usually easy for the kitchen to whip up pasta tossed with olive oil, garlic, and vegetables, or pasta served with a marinara sauce. If you find yourself at a steakhouse, go for a baked potato, salad, and vegetable side dishes.

You may find yourself at a job interview being served a catered luncheon in a conference room. If you're served a preplated ham sandwich, eat what you can from your plate and leave the rest. No explanation is necessary unless someone asks why you've left the food. In that case, all you have to say is, "I don't eat ham." Reassure your host that you're okay. The key is not to make a big deal out of the fact that they served you meat and you aren't eating it.

Leaving a positive impression

The best way to handle your dietary difference is to present yourself with confidence and in a matter-of-fact way. Let your co-workers view you as being sensible and health conscious, virtues that would reflect upon you positively in most business settings.

Just keep it low-key. After all, if one of your co-workers had values that didn't suit you, would you want to hear about it every day?

Whatever you do, don't allow yourself to be sucked into silly conversations or debates about the political or ethical issues relating to vegetarianism. These subjects have great merit, but you'll be a loser if you bring them up in a business setting.

Chapter 18

Dining in Restaurants and Other Venues

In This Chapter

- Taking a positive approach
- Considering the restaurant options
- Combining vegetarian menu items
- Staying on course when you travel

Despite the 16-ounce steaks and entire chicken halves that are typically on the menu, finding vegetarian entrées when you go out to eat is much easier nowadays than it used to be. Even so, you may find it handy to have a few tricks up your sleeve to ensure that you get what you want when you order meals away from home.

This chapter covers some of the strategies and know-how that seasoned vegetarians rely on when they go out to eat in the nonvegetarian world, whether at a neighborhood restaurant or farther away from home.

Adopting the Right Attitude

Eating out is one of life's great pleasures. Your vegetarian diet shouldn't get in the way of enjoying good food in the company of other people.

In fact, you may find just the opposite is true: Living vegetarian can open up a wide range of new food options at restaurants that serve foods you don't make or can't easily make at home.

Eating out on a vegetarian diet can be fun. It's all a matter of your perspective and of understanding some ways to expand your choices wherever you go.

Staying flexible

One key to enjoying vegetarian meals away from home is your ability to adapt to the environment you find yourself in. At certain times and in certain circumstances, you may find that you don't have much control over where you eat out.

Keep an open mind-set. If you find yourself at a restaurant where the closest thing to a vegetarian entrée is the macaroni and cheese side dish, get creative. Order a few side dishes and piece together your own meal. Eat less for dinner but enjoy a favorite dessert. Be ready to adapt to your setting as needed.

Savoring the atmosphere and the companionship

Enjoy your surroundings and the people you're with when you eat out. The setting and company are as much a part of the experience as the food.

Let's face it: Most of us aren't at risk of withering away from lack of food. If you find yourself in a situation where you have few good meatless menu options, order what you can. Then turn your attention to the beautiful view, the soothing music, and the stimulating conversation. Have a snack later if you get hungry before your next meal.

Choosing the Restaurant

The good news for vegetarians is that the restaurant industry has gotten the word that a sizable percentage of the population wants meatless foods when they eat out. Not all these people are full-fledged vegetarians, but they see vegetarian meals as being healthful and tasty alternatives.

Even some of the most traditional restaurants have responded by adding at least one or two vegetarian options to their menus. Many have added veggie burgers because burgers are so familiar and popular.

Wherever you go, you have options now. Look for vegetarian restaurants, and try cafes nestled into natural foods supermarkets, where high-quality, delicious, meatless entrées are nearly always available. Mix it up by eating at ethnic restaurants serving Middle Eastern, Chinese, Indian, and other cultural cuisines known for their many good meatless dishes. (For more on different cultural cuisines, see the "Ethnic options" section later in the chapter.)

Chains versus fine dining

Finer restaurants and family-run eateries are often in a good position to accommodate special requests. They tend to prepare their menu items at the time you order, so you usually have an opportunity to make substitutions or other modifications, such as ordering a salad without bacon or a plate of pasta without the meat.

In contrast, chain restaurants are less likely to be able to help you out, because many of their menu items are premade. They may not be able, for instance, to cook the rice without chicken stock — the rice may have been made in a large batch the day before — or to fix the pasta without meat in the sauce.

Ask questions about how the food is prepared *before* you order. If you prefer to avoid cream, butter, grated cheese, anchovies, and other animal products, check to be sure that the restaurant doesn't add these ingredients to the food, rather than sending the food back after it comes to the table with the offending ingredient.

Be aware of cooking terms that can be clues that a menu item contains an animal product. For example, *au gratin* usually means that the food contains cheese, *scalloped* means that the dish contains cream, *sautéed* can mean that the food is cooked with oil or butter, and *creamy* usually means that the item is made with cream or eggs.

If you have plans to go to a restaurant that's unfamiliar to you, try to look at the menu ahead of time. Look online to see whether the restaurant has it posted. If you don't see an obvious vegetarian option, call the restaurant and ask for recommendations on what may be possible for you to order. Given enough notice, finer restaurants are usually happy to prepare a special meal for you.

Be reasonable about special requests, especially if you can see that the restaurant is very busy. If it's a Friday or Saturday night and the place is packed, make your requests as simple as possible.

Vegetarian choices on the run

Fast-food restaurants usually have only limited vegetarian options, but some have several. Examples include pancakes with syrup, muffins, mixed green salads, bean burritos (you can order them without the cheese), bean soft tacos or regular tacos, bean tostadas, beans-and-cheese, veggie burgers (or ask for a cheeseburger, hold the meat), baked potatoes with toppings from the salad bar, and veggie wrap sandwiches. Note that yogurt served at some fast-food restaurants may contain gelatin.

Vegetarian restaurants and natural foods cafes

Many natural foods stores have a cafe tucked away in a corner where you can get a quick bite to eat. Some carry cold salads and ready-made sandwiches, and others have a more extensive menu, including a salad bar, hot foods, and baked goods. These cafes are popular places for many people to pick up a takeout meal for lunch during the workweek or for dinner on the way home from work.

Vegetarian restaurants, on the other hand, can run the gamut from homey little vegetarian hideaways to large metropolitan restaurants that stay packed for lunch and dinner. Some are raw foods specialists or macrobiotic eateries, others concentrate on vegan cuisine, and many serve foods with an ethnic bent, such as Indian or Middle Eastern, or a blend of various ethnic cuisines.

Some vegetarian restaurants have an atmosphere that harks back to the counterculture feel of the 1960s and 1970s, with an earthy environment that's usually quite casual. You're likely to find brown rice rather than white rice on the menu, and whole-grain breads rather than white, refined dinner rolls. Even desserts such as cookies and pie crusts may be made from whole grains.

Many vegetarian restaurants flag menu items that are vegan. You'll find packets of raw sugar on the table, and the menu will offer natural soft drinks, herbal teas, dishes made with tempeh and tofu, and foods made with minimally processed ingredients. You may find soymilk on the menu, too. All these features — as well as a rustic, homey atmosphere — give many vegetarian restaurants a distinctive personality that sets them apart from other eateries.

Ethnic options

Many countries, such as India and China, have vegetarian traditions. Not everyone in these countries is vegetarian, but many are, and everyone — vegetarian or not — is familiar with the concept of vegetarianism and vegetarian foods.

Other countries — including those along the Mediterranean Sea, in the Middle East, and in parts of Latin America — have some traditional foods that happen to be vegetarian, despite the fact that the diets in these cultures often rely heavily on meats and other animal products. For example, people in the Middle East eat lamb in large amounts, but falafel, spanakopita, and hummus are also traditional favorites that happen to be vegetarian.

What you order when you eat at an ethnic restaurant depends on what type of vegetarian diet you follow and how that restaurant prepares specific foods.

In one Italian restaurant, for example, the pasta with marinara sauce may be topped with cheese, and in another restaurant, it may not. If you're vegan, you probably won't eat the spanakopita at a Greek restaurant, because it's usually made with cheese. On the other hand, you may be perfectly happy with a Greek salad (minus the feta cheese), some pita bread, and an order of boiled potatoes with Greek seasonings.

Wherever you eat, ask the wait staff for specifics about how the restaurant prepares the food, especially if you're vegan. One restaurant's baked ziti in marinara sauce may be free of all dairy products, while another's may come to the table smothered in melted mozzarella. Some restaurants prepare marinara sauce using beef broth, or use pasta made with eggs.

The following are common vegetarian foods served in a variety of ethnic restaurants:

At Italian restaurants, try:

- Fresh vegetable appetizers (sometimes called antipasto) with or without mozzarella cheese
- Mixed green salads
- Minestrone soup, lentil soup, or pasta e fagioli (pasta with beans)
- Focaccia and Italian bread with olive oil or flavored oil for dipping
- Vegetable-topped pizzas (with or without cheese)
- Pasta primavera

- Spaghetti with marinara sauce
- Pasta with olive oil and garlic
- Italian green beans with potatoes
- Cappuccino or espresso
- Italian ices and fresh fruit desserts

At Mexican restaurants, try:

- Tortilla chips with salsa and guacamole
- Bean nachos
- Mixed green salads
- Gazpacho
- Bean soft tacos, tostadas, and burritos
- Spinach burritos or enchiladas
- Cheese enchiladas
- Bean chalupas (fried tortillas layered with beans, lettuce, tomato, and guacamole)
- Chiles rellenos (cheese-stuffed green pepper, usually batter-dipped and fried, topped with tomato sauce)
- Flan (custard dessert made with milk and eggs)

Vegetarians should make it a habit to ask whether beans are cooked with lard, a common practice at Mexican restaurants. Restaurant menus don't always note whether the beans are prepared vegetarian-style.

At Chinese restaurants, try:

- Vegetable soup or hot and sour soup
- Vegetarian spring rolls or egg rolls
- Sweet and sour cabbage (cold salad)
- Vegetable dumplings (fried or steamed)
- Minced vegetables in lettuce wrap
- Sesame noodles (cold noodle appetizer)
- Sautéed greens
- Chinese mixed vegetables with steamed rice

- Broccoli with garlic sauce
- Vegetable lo mein
- Vegetable fried rice
- Szechwan-style green beans or eggplant (spicy)
- Family-style tofu or Buddha's Delight
- Other tofu and seitan (gluten) dishes

At Indian restaurants, try:

- Dal (lentil soup)
- Cucumber and yogurt salad
- Chapati, pappadam, naan, and roti (Indian breads)
- Samosas and pakoras (vegetable-filled appetizers)
- Mutter paneer (tomato-based dish made with cubes of cheese and peas)
- Vegetable curry with steamed rice
- Palak paneer (spinach-based dish with soy cheese)
- Lentil, chickpea, and vegetable entrées
- Rice pudding

At Ethiopian restaurants, try:

- Fresh vegetable salads
- Injera (large, round, flat, spongy bread — tear off small pieces and use them to pinch bites of food from a communal tray)
- Bean, lentil, and vegetable-based dishes served directly on a sheet of injera on a tray or platter

At Middle Eastern restaurants, try:

- Hummus
- Spinach salad
- Dolmades (stuffed grape leaves)
- Baba ghanoush (a blended eggplant dip or appetizer)
- Fatoush (minced fresh green salad)
- Tabbouleh salad (wheat salad)

- Spanakopita (spinach pie)
- Lentil soup
- Falafel plate or sandwich (chickpea patties)
- Vegetarian stuffed cabbage rolls (filled with rice, chickpeas, and raisins)
- Halvah (sesame dessert)
- Muhallabia (ground rice pudding; contains milk but no egg)
- Ramadan (cooked, dried fruit with nuts, often served with cream and nutmeg)

Working with Menu Choices

You may occasionally encounter a menu with virtually nothing for a vegetarian. Some fast-food restaurants and truck stops are good examples — the baked beans contain pork, the biscuits are made with lard, the vegetables are cooked with bacon, and the staff just chuckles and looks incredulous when you ask whether the menu has any meatless items.

You're stuck.

In most cases, though, choices abound — you just have to get creative. With a little practice, getting a vegetarian meal at a nonvegetarian restaurant is a relatively simple matter.

Asking about appetizers

Combine several appetizers to create a vegetarian plate. For example, you might order a mixed green salad, a bowl of gazpacho, an order of stuffed mushrooms, and an appetizer portion of grilled spinach quesadilla.

This approach is a great way to taste a sampling of several foods. It's also a good way to control the amount of food you eat. Restaurant portions — especially for entrées — are often excessively large. By ordering a few appetizers, you can help hold your meal to a reasonable amount of food.

Surveying the sides

Just as you may create a meal from the appetizer menu, you can also piece together a full plate from a sampling of sides.

Pay attention to the descriptions of entrées on the menu, including the sides listed with meat dishes. Choose three or four and check with your server to find out how they're prepared. With a salad and some good bread, your meal may be the envy of the others at your table.

Coming up with creative combinations

When a restaurant isn't vegetarian-friendly, scan the menu to get an idea of the ingredients the restaurant has on hand. In many cases, you can ask for a special order using ingredients that the restaurant uses to make other menu items.

For example, if the restaurant serves spaghetti with meat sauce, you know it has pasta. The tomato sauce may be already mixed with meat, but the restaurant may have vegetables on hand for side dishes. Ask for pasta primavera made with olive oil and garlic and whatever vegetables the chef can add.

If all else fails, ask for a baked potato and a salad, or a salad and a tomato sandwich on whole-wheat toast. Be creative. If the restaurant serves baked potatoes, ask for one as your entrée. You can top it with items from a salad bar, if the restaurant has one. Try adding broccoli florets, salsa, black olives, sunflower seeds, or whatever looks good to you.

Making sensible substitutions

Take a good look at meat-containing entrées and determine whether the restaurant can prepare them as vegetarian dishes instead. For instance, a pasta dish mixed with vegetables and shrimp could easily be made without the shrimp. A club sandwich could be fixed with grilled portobello mushrooms, avocado, cheese, or tomato slices instead of the meat. This is most likely to work at restaurants where the food is made to order and not premade.

Working with Restaurant Staff

Your attitude and level of confidence can make a difference in how well you do when you order a vegetarian meal at nonvegetarian restaurants. Your relationship with the wait staff is key.

Try these tips to make the most of your dining experience:

- **Be assertive:** When you speak with restaurant staff about your order, be clear about what you want. For instance, instead of just saying, "I'm a vegetarian," explain specifically what you do and don't eat. Let the wait staff know whether dairy and eggs are okay, or if you don't want your foods flavored with chicken broth or beef broth.

  Don't be afraid to press to get the information you need to order effectively. Ask your server whether your special request will increase or decrease the cost of the meal, too.

- **Smile and say thank you:** As your mother taught you, "please" and "thank you" go a long way. If you need special attention from restaurant staff to get a meal you can eat, ask for it. But be as pleasant as possible. Ask the wait staff for suggestions if you can't figure out what to order.

Traveling Vegetarian

When you travel, you have less control over meals, you may be in unfamiliar surroundings, and you have to contend with the challenges that a change in your normal routine can bring. That can put you at risk of eating poorly, particularly if you can't find the meatless options you need.

For vegetarians, avoiding meat and possibly other animal products away from home requires a set of skills acquired with time and experience. In this section, I help you get started by explaining ways you can improve your chances of getting what you need when you're on the road, in the air, or on the sea.

Tips for trippin' by car, bus, or train

If you get hungry when you're traveling by car or bus, you're usually limited to whatever is near the exit off the highway. Your choices are likely to be a fast-food restaurant, a family chain eatery, or a truck stop with limited choices for vegetarians.

Traveling by train, your food choices may vary, depending on the type of train you're riding. Commuter trains may offer no food service whatsoever. Trains with longer routes may have meal cars serving pre-packaged snacks and drinks, or table service with limited menus.

In all these cases, consider carrying your own food. Pack a cooler or bag of food to take along. On car trips, it'll save you time not having to stop for meals. You'll save money and eat more healthfully, too.

If you do pack your own food, portable, nutritious choices include:

- Bagels
- Bottles of mineral water or flavored seltzer water
- Deli salads
- Fresh fruit
- Homemade, whole-grain quick breads, muffins, and cookies
- Hummus sandwiches
- Individual aseptic (shelf-stable) packages of soymilk
- Individual boxes of dry breakfast cereal
- Instant soup cups or instant hot cereal cups that you can mix with hot water from the coffee maker at a gas station food mart
- Peanut butter or almond butter sandwiches on coarse, whole-grain bread
- Peeled baby carrots
- Small cans or aseptic packages of fruit juice
- Snack-sized cans of fruit or applesauce
- Tofu salad sandwiches

If you like to take trips by bicycle, or if you hike, you need to pack foods that are light and portable and don't require refrigeration. Dried fruit and nut mixtures, small containers of soymilk or fruit juice, crackers and peanut butter, and fresh fruit are good choices.

Food for fliers

Airlines have cut back on meal service, especially on shorter flights. So don't be surprised if the next time you fly, you aren't served anything more than peanuts or snack mix and a beverage. Be prepared. Many travelers now routinely carry on their own food from home.

Most airlines still offer full meal service on most longer and overseas flights, and they'll usually accommodate requests for vegetarian meals if you give them enough notice. Call the airline's reservation desk at least 24 hours before your flight, or make the request when you make your reservations. You may or may not be able to post the request online.

Meal service options are in a state of flux because of continuing cutbacks in service at most airlines. In the past, most airlines offered several options for vegetarians, including lacto ovo, lacto vegetarian meals, and vegan meals. Some even offered fruit plates, Asian vegetarian meals, or Indian vegetarian meals. Check with your airline to find out which options are available.

On most flights, the crew has a list of passengers who've ordered special meals. They may identify you as you're taking your seat, or they may ask you to ring your flight attendant call button just before meal service begins. On other flights, you get served a regular meal unless you speak up and tell the flight attendant that you've ordered a vegetarian meal.

If you're seated near the back of the plane, you may want to ring your call button early during meal service to let attendants know that you're expecting a vegetarian meal. That ensures that they don't give it away before they get to you, which can happen when someone seated in front of you asks for a vegetarian meal despite not having ordered one, or when the airline hasn't loaded enough of a particular special request onto the aircraft.

Meatless at sea

Many cruise ships are like spas or small communities on the water, with a wide selection of restaurants and foods, including healthful, vegetarian options. Some have separate menus for health-conscious people, or they flag specific entrées and menu items as being "healthy." In many cases, that means meatless.

If meat is included in the dish, it can often be left out of meals that are cooked to order. For example, a large salad topped with chicken strips can be fixed without the chicken, and a stir-fried vegetable and meat dish can be made with all the same ingredients, minus the meat.

If you find that the menu choices on a particular day don't include enough meatless options, handle the situation the way you would at a fine restaurant. Explain your needs to your server or the chef and ask for a special order. If possible, let the kitchen know what you want the day before, especially if the ship is going to seat a large number of people (for the captain's dinner, for instance).

If you have concerns about whether a cruise line can accommodate your food preferences, have your travel agent request information about meals and get sample menus to examine. You can also call the cruise line's customer service office yourself and ask for more details, or check online.

In addition to elaborate, sit-down meals, cruise ships are known for their expansive buffets. The sheer volume and variety of foods at these buffets makes it easy for most vegetarians to find enough to eat. You'll have to sidestep the gelatin salads, and vegans may have to bypass items made with cream and cheese. Fill up on good-for-you fresh tropical fruit salads, rice, pasta, vegetables, and breads.

Coming up with alternatives: When plans go awry

Be prepared to improvise when you travel. You may not be able to anticipate every complication that affects your meals, but if you're flexible and creative, you can deal effectively with most of them.

Most travel snafus involving meals happen when you travel by plane. As any frequent flyer will tell you, ordering a vegetarian meal doesn't ensure you'll get it. This isn't always the airline's fault — sometimes a last-minute change of aircraft means that the meals meant for your flight aren't on that plane.

If you miss a flight and have to take an alternate flight, you won't get the meal that you specially ordered. If you upgrade your ticket to first class just before you board your flight, your vegetarian meal may be back in coach, and you may have another (maybe better) menu from which to choose. Your flight attendant can usually retrieve your meal from coach, though, if you still want it.

If you have a particularly long travel day and want to be extra sure that your vegetarian meal has been ordered, phone the airline a day or two before you leave, just to confirm that the request has been noted. This is especially important if you've made a schedule change, because the agent may not have carried your meal request over to your new reservation.

If you do find yourself stuck with a ham sandwich instead of your vegetarian meal, here are a few things you can do:

- Eat what you can of the meal you've been served. Picking the meat off a sandwich and eating the rest isn't an option for many vegetarians, nor is pushing the sausage away from the stack of pancakes that it's been leaning against. But your only other option may be peanuts and tomato juice, so if you're really hungry, eat the salad and dessert.
- Ask your flight attendant for cookies, crackers, nuts, pretzels, or juice. A flight attendant who knows you haven't received your special order will often help you out with some extra snacks or beverages.
- Pull out the reserves (fresh fruit, a bagel, a sandwich) that you took along in your carry-on bag, just in case.

Whatever your mode of travel, packing your own provisions is good insurance for times when best-laid meal plans fall through. Some good ideas include small containers of aseptically packaged soymilk or fruit juice, packaged crackers, pretzels, nuts, dried fruit, or trail mix. You can sprinkle trail mix over a green salad during in-flight meals or at airport food courts to turn a salad into a substantial meal or snack.

Take opportunities to restock while you're traveling. Many hotels have bowls of fruit at the checkout desk. Before you leave your hotel, pick up a piece and stow it in your carry-on bag or luggage. If you don't find fruit at your hotel, buy a piece or two in the airport terminal or at a supermarket or cafeteria. Fruit is good to have on hand in case you get hungry while you're traveling, and it's a good source of dietary fiber and other nutrients that people tend to neglect when they travel.

Part V
Living Vegetarian for a Lifetime

"Relax – another helping of vermicelli, and I'll be done with your precious shredder."

In this part . . .

I tell you what you need to know when the urge for pickles and soy ice cream strikes. If your baby is already here, I explain how to handle first foods, food fights, and other challenges to raising young children on a vegetarian diet.

Feeding older vegetarian children and teens raises other issues, including understanding growth rates, balancing nutrition and treats, and getting kids interested in good foods whether they're at home or away. I finish up with a discussion about the changing needs of adult vegetarians and special health advantages of eating green as you age. Aging healthfully includes remaining active the vegetarian way for as long as you want to, and I include information for athletes of all skill levels, too.

Chapter 19

When You're Expecting: Managing Your Vegetarian Pregnancy

In This Chapter

- Preparing for a healthy pregnancy
- Minding your nutritional needs while pregnant
- Quelling the queasies

There's no better time to be a vegetarian than when you're expecting. A healthful vegetarian — and even vegan — diet can provide a woman and her baby with all the nutrients needed for a healthy start in life.

But don't be surprised if your family and friends don't know it.

Tell people you're pregnant and that you're a vegetarian, and all sorts of alarms go off. Your family and friends may not have personal experience with a vegetarian diet, and they may not have a current understanding of the science. They're anxious, and anxieties heighten when a baby's involved.

Of course, vegetarians all over the world have been having healthy babies for centuries. Here at home, though, the idea of meat-free diets for babies and children still sounds like a radical idea.

That's slowly changing. But until more people have a better understanding of vegetarian diets, concern over your vegetarian pregnancy may seem like an irritating intrusion.

Take comfort in knowing that people care. The information in this chapter will help put to rest some of the most common worries.

Before Baby: Ensuring a Healthy Start

Not every pregnancy is planned, but if you think you may get pregnant soon, it's a perfect time to be proactive and get yourself into great nutritional shape. The longer you have to eat well before you become pregnant, the better for you and your baby.

Prepregnancy advice for vegetarians is similar to advice for nonvegetarians:

- Follow your healthcare provider's advice about taking a regular-dose, daily multivitamin and mineral supplement for several months before you get pregnant. In particular, it's important to have adequate folic acid before pregnancy.

  Folic acid may help prevent a *neural tube defect* such as *spina bifida* in your baby. Spinal bifida is a type of birth defect that involves an incomplete closure of the spinal cord.
- Limit sweets and junk foods. They displace more nutritious foods from your diet.
- If you drink coffee, limit it to two cups a day. Avoid alcohol and tobacco.
- Get regular physical activity and drink plenty of water.

Vegetarians often go into pregnancy with an edge over nonvegetarians, because vegetarians are more likely than other women to have close-to-ideal body weights. Vegetarians are also more likely to have been eating plenty of folic-acid-rich foods, which lower the risk of neural tube defects.

Paying careful attention to your diet and fitness level can go a long way toward ensuring a healthy pregnancy and baby. Drinking plenty of fluids, especially water, and getting plenty of rest round out a healthful prepregnancy lifestyle. In this section, I tell you what you need to know about nutrition and exercise when you're thinking about becoming pregnant.

Maximizing nutrition before you get pregnant

The longer you can eat well before you become pregnant, the better off you and your baby will be. People who are well nourished have strong immune systems, and they're less likely to succumb to common illnesses such as colds and the flu.

By limiting your intake of junk foods and stocking up on plenty of nutritious vegetables, whole grains, and fresh fruits, you can substantially increase your intake of folic acid. Having high folic acid intakes *before* pregnancy increases the chances that your baby is protected from neural tube defects, because the period of risk occurs in the very earliest stage of pregnancy, before many women even realize that they're pregnant.

Ensuring that your iron stores are high before you get pregnant is also important. Many North American women go into pregnancy with low iron stores and put themselves at risk for iron deficiency while they're pregnant. Maternal blood volume increases by about 50 percent during pregnancy, and women who go into pregnancy with low iron stores risk becoming anemic. See Chapter 3 for more about iron.

If you follow a strict vegetarian or vegan diet, you need to have a reliable source of vitamin B12 before, during, and after pregnancy. See Chapter 3 for details on getting vitamin B12.

Staying physically fit

The beginning of a pregnancy isn't the best time to start a new or vigorous exercise program. If you establish an exercise routine prior to pregnancy, though, it's likely you can continue that level of activity throughout your pregnancy. Get individualized advice from your healthcare provider about exercising while you're pregnant.

Staying physically active helps you maintain muscle tone and strength and promotes normal stools during a time when many women experience problems with constipation and hemorrhoids. The fiber in your vegetarian diet also helps where that's concerned!

Eating Well for Two

Okay, the test strip turned blue, and you're on your way to becoming a mommy. The questions are beginning to trickle in:

- Now that you're eating for two, how are you going to get enough protein?
- Are you getting enough calcium, iron, and other nutrients?

The questions you're being asked now that you're pregnant are probably the same ones you fielded when you first became a vegetarian. Understanding the ways that pregnancy changes your nutritional needs and how you can meet those needs on a vegetarian diet gives you the confidence to enjoy your vegetarian pregnancy with fewer worries. Figure 19-1 shows some vegetarian foods that are especially good for you during pregnancy.

In this section, I explain how to maximize the nutritional value of your diet to help ensure a healthy pregnancy for you and your baby.

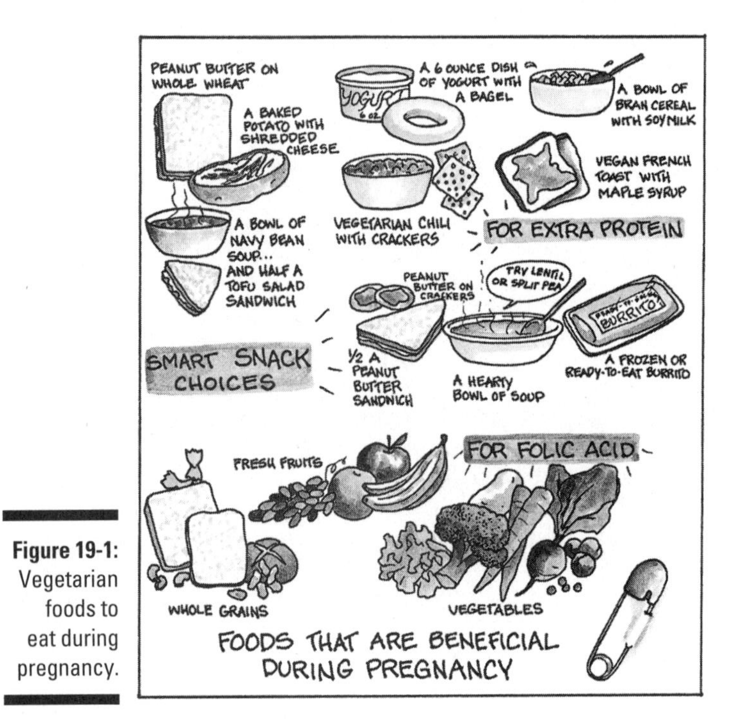

Figure 19-1: Vegetarian foods to eat during pregnancy.

Watching your weight gain

Vegetarians are more likely than nonvegetarians to go into pregnancy at body weights that are close to ideal. Women who become pregnant when they're close to their ideal weight can expect to gain from 25 to 35 pounds. If you're overweight when you become pregnant, you may gain less — between 15 and 25 pounds. If you're thin when you become pregnant, it's healthy to gain more weight — between 28 and 40 pounds. Weight gain varies with individuals, however, so it's important to get prenatal guidance from a healthcare provider, whether that's your medical doctor, nurse practitioner, or midwife.

In the first three months of pregnancy, it's common to gain very little weight — a few pounds at most. Weight gain picks up in the second and third trimesters, and a weight gain of about one pound a week is typical. When you begin gaining weight after the third month of pregnancy, you need about 300 more calories per day than what you needed before you became pregnant. Women who need to gain more weight need slightly more calories, and women who need to gain less weight need fewer calories to maintain a healthy pregnancy weight.

Most vegetarian women have weight gains that follow patterns similar to those of nonvegetarian women. However, vegan women are more likely to be slender going into pregnancy. They may be more likely than other women to have low calorie intakes because their diets tend to be bulky — high in fiber from lots of fruits and vegetables.

If you're having trouble gaining weight, consult your healthcare provider for individualized advice. The following tips may also be helpful:

- Snack between meals. Foods that are light and easy to fix are good choices. Examples include a bowl of cereal with soymilk, half a peanut butter sandwich or peanut butter on crackers, a bowl of hearty soup such as lentil or split pea soup, or a frozen, ready-to-heat bean burrito.
- Substitute starchy, calorie-dense foods for bulkier, low-calorie foods. For example, instead of filling up on a lettuce and tomato salad, try a thick soup made with vegetables and barley. Choose starchy vegetables such as potatoes, sweet potatoes, peas, and corn more often than low-calorie choices such as green beans and cucumbers. (Don't give up the folic-acid-rich greens, though!)
- Drink extra calories. Shakes and smoothies are an easy way to get the calories you need. If you use dairy products, make shakes or smoothies using ice milk or frozen yogurt mixed with fresh fruit. Vegans can use tofu, soymilk or other milk substitutes, and nondairy ice cream substitutes.

If you have the opposite problem and are beginning your pregnancy overweight, now is not the time to actively reduce. At most, you want to control your weight gain by limiting sweets and fatty or greasy foods that are calorie-rich but nutrient-poor. Plan to lose weight gradually with diet and exercise after the baby is born.

Putting nutritional concerns in perspective

Friends and family mean well, but constant badgering about your vegetarian diet can make even the most confident woman worry that her diet and pregnancy don't mix. Not true. In this section, I give you some information you can use to reassure others that you're just fine.

During pregnancy, it's important to keep the junk foods in your diet to a minimum to ensure that your diet has enough room for the good stuff.

Protein

Protein is the first thing people ask vegetarians about, but it's the nutrient about which most vegetarians have the least need to be concerned. Even if you're pregnant, protein still should be the least of your concerns.

The recommended level of protein intake during pregnancy is 60 grams per day for most women. That's about 10 grams more protein than a woman needs when she's not pregnant. Most women, including vegans, already exceed that level of protein intake before they become pregnant. They typically get even more during pregnancy because their calorie intakes increase by 300 calories a day, and protein makes up part of those 300 calories. So you can see that getting enough protein during pregnancy is nothing to worry about.

Each of these food choices adds an extra 10 grams of protein or more to your diet:

- A peanut butter sandwich on whole-wheat toast
- A baked potato topped with 1 ounce of shredded cheese (regular or soy)
- A 6-ounce dish of flavored yogurt (regular or soy) with a bagel
- A bowl of vegetarian chili and a couple of crackers
- A bowl of bran cereal with soymilk
- Two pieces of vegan French toast with maple syrup
- A bowl of navy bean soup with half a tofu salad sandwich

Calcium

Counting the milligrams of calcium in your diet wouldn't be much fun, and fortunately, it's not necessary. Getting enough calcium isn't a problem, even if you're a vegetarian.

In general, pregnant women need about the same amount of calcium as women who aren't pregnant. Current recommendations call for 1,000 milligrams of calcium per day during pregnancy and while a woman is breastfeeding her baby. Because vegans typically get less calcium than other vegetarians and nonvegetarians, it may be more challenging for them to meet the recommended level of calcium intake.

On the other hand, evidence suggests that the body becomes more efficient at absorbing and retaining calcium during pregnancy, and that may offset lower calcium intakes. Nevertheless, vegans and other pregnant women should aim for the 1,000 milligram target. The best way to do this is to try hard to get three or four servings of calcium-rich foods each day. See Chapter 3 for a list of good calcium sources.

If you need help getting enough calcium in your diet when you're pregnant, try these ideas:

- Go for big portion sizes of calcium-rich foods. Instead of a wimpy half-cup serving of cooked kale, go for the gusto and take a 1-cup helping.
- It may be easier to drink your calcium than to chew it. Have some calcium-fortified orange juice, or make smoothies with calcium-fortified soymilk, fresh or frozen fruit, and tofu processed with calcium.

Vitamin D

Vitamin D goes hand in hand with calcium to ensure that your baby's bones and teeth develop normally, so be sure that your vitamin D intake is adequate. Vegetarian women and nonvegetarian women alike need vitamin D, but it's easy to come by. You can get vitamin D from exposure to sunlight or from fortified foods. Either way, just make sure you get it. For more on vitamin D, see Chapter 3.

Iron

If you pumped up your iron stores before you got pregnant (see the earlier section "Maximizing nutrition before you get pregnant" for details), it's more likely you began your pregnancy with adequate iron stores. This is important, because going into pregnancy with low iron stores is associated with iron deficiency anemia in the later stages of pregnancy.

All pregnant women need additional iron during the second and third trimesters of pregnancy when maternal blood volume skyrockets and iron levels plummet. It's common for healthcare professionals to advise women to take an iron supplement of 30 milligrams per day during this time.

Vitamin B12

A pregnant vegetarian woman needs no more vitamin B12 than any other pregnant woman. However, if you're a vegan, you have to make a special effort to get it. I can't stress this point enough. Vitamin B12 is found in eggs and dairy products, which vegans don't eat. Not only does your baby need vitamin B12 from you while she's developing, she also needs it while breast-feeding. So vegan women need a reliable source of vitamin B12 before, during, and after pregnancy. See Chapter 3 for more information about vitamin B12.

Docosahexaenoic acid (DHA)

Getting enough DHA may be important for brain and eye development for developing fetuses and babies. Some research suggests that vegetarian and vegan women may have lower DHA levels in their breast milk than nonvegetarians.

To be on the safe side, vegetarians should make sure they get a regular dietary supply of *alpha-linolenic acid,* a fatty acid that gets converted into DHA in the body.

Eggs are a direct source of DHA, but vegetarians who don't regularly eat eggs can get alpha-linolenic acid from the following sources:

- Canola, flaxseed, soybean, and walnut oils
- Cooked soybeans
- Ground flaxseeds
- Tofu
- A vegetarian DHA supplement made from microalgae
- Walnuts

Pregnant vegetarian women should limit foods containing *linoleic acid,* another type of fatty acid found in corn, safflower, and sunflower oils. They should also avoid trans fats found in foods made with partially hydrogenated vegetable oils. These fats can interfere with your body's ability to convert linolenic acid into DHA.

Caffeine and pregnancy

Whether you're a vegetarian or not, it's best to limit or eliminate your use of caffeine when you're pregnant. Studies have demonstrated problems in pregnancies resulting from the equivalent of five cups of coffee or more per day. Until otherwise demonstrated to be benign, it's best to err on the side of caution and eliminate caffeine during pregnancy, or limit your use to one or two cups of caffeine-containing beverages per day, including coffee, tea, cola, and other soft drinks that contain caffeine.

Keeping mealtime simple

What most pregnant women need is someone to make their meals for them. Now *that's* meal planning made easy! Okay, so you don't see that happening within the next nine months. Here's what you do: Take some shortcuts for a while. Here's how:

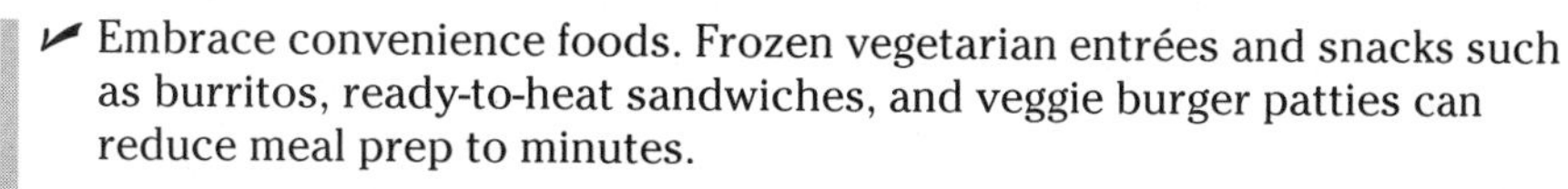

- Embrace convenience foods. Frozen vegetarian entrées and snacks such as burritos, ready-to-heat sandwiches, and veggie burger patties can reduce meal prep to minutes.
- Order out when you're too tired to cook. Chinese vegetable stir-fry with steamed rice, a falafel sandwich and a side of tabbouleh, or a vegetarian pizza may be just what you need.
- Open a can or carton. Eat a can of lentil soup with whole-wheat toast, or have a bowl of cereal with soymilk for dinner. Canned beans, canned soups, breakfast cereals, frozen waffles, burritos, and microwave popcorn are nutritious and quick.
- Become a weekend cook. Make a big batch of vegetarian chili or lasagna and freeze part of it. You can take it out and reheat it when you don't feel like cooking. Muffins and quick breads freeze well, too. You can also fix a big, fresh fruit salad or mixed green salad one day, and eat it over the next three days.

When you're pregnant, you don't really have to eat different foods than when you're not pregnant — you just need about 300 *more* calories per day. Give yourself permission to take the easy way out. Forget about fancy meals until a time when you have more energy. For now, eat according to your appetite, and pay attention to the overall quality of your food choices.

Handling Queasies and Cravings

Pregnancy is a time for strange food aversions and even stranger cravings. They're usually harmless and pass by the end of the first three months of pregnancy. Not always, but usually. The same is true of nausea — better known as morning sickness, and known by some as morning, noon, and night-time sickness!

These aspects of pregnancy are no different for vegetarians than they are for nonvegetarian women. In this section, I help you understand how to cope with these issues.

Dealing with morning sickness any time of day

Morning sickness can be one of the most uncomfortable symptoms of pregnancy, whether you're a vegetarian or not. For many women, overcoming nausea is a matter of waiting it out. In the meantime, here are some things you can do to minimize its effects:

- Eat small, frequent meals or snacks. Don't give yourself a chance to get hungry between meals, because hunger can sometimes accentuate feelings of nausea.
- Eat foods that are easy to digest, such as fruits, toast, cereal, bagels, and other starchy foods. Foods high in carbohydrates such as these take less time to digest. In contrast, fatty or greasy foods such as chips, pastries, cheese, heavy entrees, and rich desserts take longer to digest and are more likely to give you trouble.
- If nausea or vomiting keep you from eating or drinking for more than one day and night, check with your healthcare provider for guidance. It's important to get help to ensure you don't become dehydrated.
- If vitamin or mineral supplements make you nauseous, try varying the time of day you take them, or try taking them with a substantial meal. If the smell of the supplement is offensive, try coating it with peanut butter before swallowing.

Managing the munchies

If you're having cravings for something outlandish — like, say, tofu cheesecake with chocolate sauce, or a hummus and green tomato sandwich — the best thing to do is . . . go for it! This phase isn't going to last forever, and, let's face it, pregnancy is a time when all sorts of hormonal changes are taking place and things can be topsy-turvy for a while. You can't do much about it, it's not necessary to control it, and it's probably not going to hurt you. As long as you aren't chewing on radial tires or the clay in your backyard, you should be just fine.

What should you do if you crave meat or another animal product that you've banned from your diet? It's your call. But if you want to stay meat-free and you find yourself daydreaming about bacon, try a look-alike. Soy- and vegetable-based substitutes for cold cuts, hot dogs, burgers, and bacon are available in supermarkets. I talk about these in more detail in Chapter 6.

Chapter 20

Raising Your Vegetarian Baby

In This Chapter

- Beginning with breast milk or baby formula
- Selecting solids for your infant's first year
- Taking care of your toddler's nutritional needs

Enter a world in which the little people wear mashed potatoes in their hair and toss more of their food onto the floor than into their mouths. Hey, when was the last time you smeared smashed peas up and down your arms and hid cracker bits behind your ears? You were probably 2 feet tall and had three chins. Me, too.

This scenario is pretty much the same whether your child is a vegetarian or not. In this chapter, I help guide you in the progression from breast or bottle to baby food and beyond, and I address special considerations for vegetarians.

Many healthcare providers are neither familiar with nor comfortable with the idea of vegetarian diets for children. Be assured that vegetarian diets are perfectly safe and adequate for children and that they hold numerous advantages over nonvegetarian diets. Although attitudes are changing, vegetarian diets are still largely outside the North American cultural mainstream, and that's the primary reason you may meet with resistance from healthcare providers. You may also find unnecessary cautions and a lack of support for vegetarian diets in baby and childcare books. Be assured that these opinions aren't consistent with scientific knowledge. The American Dietetic Association and the American Academy of Pediatrics support the use of well-planned vegetarian diets for children.

Taking Vegetarian Baby Steps

We all begin our lives as vegetarians. Think about it: How many newborns eat hamburgers or chicken nuggets? Milk is our first food. Humans make milk for human babies. The alternative, infant formula, is as close a replica as can be made in a laboratory. Babies should be breast-fed or bottle-fed exclusively for the first four to six months of life. They need no other food during this time.

In fact, if you start solid foods too soon, your child is more likely to become overweight and to develop allergies. Resist the temptation to start solids too early.

In this section, I walk you through the basics of a vegetarian diet during the first year of life.

First foods: Breast and bottle basics

Breast milk is the ideal first food for babies, because it's tailor-made for them. At no other time in our lives do most of us have a diet so well-suited to our needs. From birth through at least the first six months — and longer, if possible — breast-feeding your baby is the best choice, bar none, for several reasons:

- The composition of breast milk makes it the perfect food for human babies. Even baby formulas may not compare. It's quite possible that breast milk contains substances necessary for good health that haven't yet been identified and, therefore, aren't available in synthetic formulas. Just as researchers are discovering new phytochemicals and other substances in whole foods that aren't found in synthetic vitamin and mineral supplements, our knowledge about the nutritional value of breast milk continues to evolve.
- Breast milk contains protective substances that give your baby added immunity, or protection against certain illnesses. Breast-fed babies are also less likely than others to have problems with allergies later in life.
- Breast-fed babies are more likely than others to maintain an ideal weight throughout life.
- Breast milk is convenient and sterile.
- Breast-feeding is good for Mom because it helps the uterus return to its former size more quickly and aids in taking off excess baby fat.

Babies of vegetarian women are at an advantage over babies of nonvegetarians, because vegetarian women have fewer environmental contaminants in their breast milk. The diets of vegetarian women contain only a fraction of the amount of pesticide residues and other contaminants that nonvegetarian women unknowingly consume. These contaminants are concentrated in animal tissues and fat, and women who eat the animal products store the contaminants in their own tissues and fat. Consequently, when they produce breast milk, they pass the contaminants on to their babies through their milk.

Getting the calories and nutrients you need

You may be surprised to discover that you need even more calories when you're breast-feeding — about 200 more calories per day — than you do when you're pregnant. Therefore, most breast-feeding women need a total of 500 more calories a day than they needed before they were pregnant. No wonder many women find that during the time they're breast-feeding, they begin losing some of the extra weight they gained while they were pregnant.

You also need a few more nutrients when you're breast-feeding compared to when you're pregnant. For instance, you need about 5 grams more protein daily. You can easily get the extra nutrients in the additional calories that you eat. Be sure to include plenty of fluids, too; water is always the best choice. And remember that vegans need to be sure to get a reliable source of vitamin B12. (See Chapter 3 for more about getting enough B12.)

Women who breast-feed their babies get lots of applause, but women who can't breast-feed shouldn't be chastised. Some women can't or don't want to breast-feed, for a variety of reasons. They need to feed their babies a synthetic baby formula instead. Just as with breast-fed babies, formula-fed babies need their formula — and nothing but their formula — for at least the first four to six months, if not longer.

Several brands of baby formula are on the market, and your healthcare provider may recommend a few to you. Most are based on cow's milk, altered for easy digestibility and to more closely resemble human milk. Others may contain animal fat or other animal byproducts. Vegans don't use these formulas, but some formulas contain no animal products and are soy-based. These formulas, including such brands as Isomil, Prosobee, and Soyalac, are acceptable for use by vegans. Read product labels or talk with your doctor or a registered dietitian if you're unclear about the source of ingredients in baby formula. (I include a list of hidden animal ingredients — ingredients you may not realize come from animals — in Chapter 5.)

Commercial soymilks aren't the same thing as infant soy-based formulas, and they aren't appropriate for infants. If you don't breast-feed your baby, be sure that you feed her a commercial infant formula, *not* the commercial soymilks that are meant for general use (such as Edensoy and Silk). When your child is older, these are fine, but not in infancy and toddlerhood.

If you bottle-feed your baby, don't put anything in the bottle except breast milk, formula, or water for the first six months. Sugar water drinks, soft drinks, and iced tea are inappropriate for babies and small children, and you shouldn't introduce fruit juices and diluted baby cereals until after the six-month point. Nothing is more nutritious or beneficial to your baby for the first six months than breast milk or infant formula.

Solids second

When your baby is about four to six months of age, you want to begin gradually introducing solid foods when any of the following occur:

- Your baby reaches about 13 pounds in weight, or doubles his birth weight.
- Your baby wants to breast-feed eight times or more during a 24-hour period.
- Your baby takes a quart of formula or more in a 24-hour period and acts as if she's still hungry and wants more.

You won't find any hard-and-fast rules about how to introduce solid foods to babies — just some general guidelines. For instance, all babies — whether their families are vegetarians or not — start out eating foods that are the easiest to digest and least likely to cause problems, such as allergic reactions or choking. These foods are cooked cereals and mashed or pureed fruits and vegetables and their juices.

Baby rice cereal is the most common first solid food for babies because almost every baby can tolerate it. It's best to give your baby iron-fortified baby rice cereal until he's at least 18 months old.

Rather than abruptly discontinuing breast or bottle feedings, just supplement them by gradually introducing small amounts of solid foods. Begin by mixing baby rice cereal with breast milk or infant formula and offering a few tablespoons. Work up to two feedings a day, totaling about a half cup. From there, gradually add other foods, one at a time, a little at a time, and your baby will increase the amounts at her own pace. Introducing foods one at a time helps you pinpoint the culprit if your baby has a sensitivity to a particular food.

You should wait until a bit later to introduce protein-rich foods and foods that are high in fat. After your baby becomes accustomed to cooked cereals and mashed and pureed fruits and vegetables, he can move on to table foods.

Don't give cow's milk to infants under the age of one year. Cow's milk can cause bleeding in the gastrointestinal tract of human babies and lead to anemia. Studies have also linked cow's milk given to infants with an increased risk for insulin-dependent diabetes.

Adding foods throughout the first year

Table 20-1 suggests schedules for feeding vegan babies from 4 through 12 months of age. Notice that the guide excludes all animal products. If you prefer a lacto ovo vegetarian approach, you can substitute milk-based baby formula for soy formula, eggs for tofu, and low-fat cheese and yogurt for soy varieties.

Table 20-1 Feeding Vegan Babies Ages 4 to 12 Months

	4–7 Months	*6–8 Months*	*7–10 Months*	*10–12 Months*
Milk	Breast milk or soy formula	Breast milk or soy formula	Breast milk or soy formula	Breast milk or soy formula (24–32 ounces)
Cereal & Bread	Begin iron-fortified baby cereal mixed with breast milk, soy formula or soymilk	Continue baby cereal and begin other breads and cereals	Continue baby cereal and other breads and cereals	Continue baby cereal until 18 months; total of 4 servings per day (1 svg. = ¼ slice bread or 2–4 tbsp. cereal)
Fruits & Vegetables	None	Begin non-citrus juice from cup (2–4 oz. vitamin C-fortified varieties such as apple, grape, and pear) and begin mashed vegetables and fruits	4 oz. of juice and pieces of soft/cooked fruits and vegetables	Table-food diet: Allow 4 servings per day (1 svg. = 2–4 tbsp. fruit and vegetable, 4 oz. juice)
Legumes & Butters	None	None	Gradually introduce tofu, and begin casseroles, pureed legumes, soy cheese, and soy yogurt	2 servings daily, each about ½ oz; don't start nut butters before 1 year, and don't introduce peanut butter before the second birthday

Note: Overlap of age ranges occurs due to varying rates of development.

Adapted from Simply Vegan, Fourth Edition, by Debra Wasserman and Reed Mangels, PhD, RD, 2006. Reprinted with permission from The Vegetarian Resource Group, P.O. Box 1463, Baltimore, MD 21203; phone 410-366-8343; Web site `www.vrg.org`.

And while I'm on the topic . . .

You should be aware of a few more issues concerning your baby's nutritional needs during the first 12 months:

- Remember: If you're a vegan mom, it's critical that you take a vitamin B12 supplement while you're breast-feeding to ensure that your baby has a source of vitamin B12, too.
- Babies over the age of three months who have limited exposure to sunlight need a vitamin D supplement of not more than 400 IU per day. If you suspect your baby may not be getting enough sunlight to make adequate amounts of vitamin D, consult your healthcare provider for advice about whether a supplement is needed.
- After the first four to six months of age, mothers usually start their breast-fed babies on iron supplements, whether the babies will be vegetarians or not. After solid foods are added to the diet, iron-fortified baby cereal and other foods provide additional iron in your baby's diet. Your healthcare provider can give you individualized advice about the need for your baby to have an iron supplement.

Tracking Your Toddler

Vegetarian diets are associated with numerous health advantages, so children who start their lives as vegetarians are more likely to develop good eating habits that will follow them into adulthood. Even so, after young vegetarian children from 1 to 3 years of age begin eating table foods, you need to be aware of a few issues.

Planning meals

No real trick exists for feeding most children a healthful vegetarian diet. The key is to offer a variety of foods and encourage children to explore new tastes. Table 20-2 presents a daily feeding guide for vegetarian toddlers and preschoolers ages 1 through 3 years.

If you need assistance adapting the food guide for your individual child, or you need more help with menu planning, contact a registered dietitian who's familiar with vegetarian diets and accustomed to working with families.

Table 20-2 Meal Planning Guide for Vegetarian Toddlers and Preschoolers Ages 1 to 3

Food Group	*Number of Servings*	*Example of One Serving*
Grains	6 or more	½ to 1 slice bread; or ¼ to ½ cup cooked cereal, grain, or pasta; or ½ to ¾ cup ready-to-eat cereal
Legumes, nuts, and seeds	2 or more	¼ to ½ cup cooked beans, tofu, tempeh, or textured vegetable protein; or 1½ to 3 ounces *meat analog*; or 1 to 2 tablespoons nuts, seeds, or nut or seed butter (meat analogs are soy-based substitutes such as veggie hot dogs, burger patties, and cold cuts; note that hot dogs must be cut up to prevent choking risk)
Fortified soymilk or the like	3	1 cup fortified soymilk, infant formula, breast milk, or low-fat cow's milk
Vegetables	2 or more	¼ to ½ cup cooked or ½ to 1 cup raw vegetables
Fruits	3 or more	¼ to ½ cup canned fruit; or ½ cup juice; or 1 medium fruit
Fats	3	1 teaspoon trans fat-free margarine or oil; use 1/2 teaspoon flaxseed oil or 2 teaspoons canola oil daily to supply omega-3 fatty acids

Adapted from Simply Vegan, Fourth Edition, by Debra Wasserman and Reed Mangels, PhD, RD, 2006. Reprinted with permission from The Vegetarian Resource Group, P.O. Box 1463, Baltimore, MD 21203; phone 410-366-8343; Web site www.vrg.org.

Adjusting to food jags

Young children are notorious for wanting to eat nothing but mashed potatoes one week and shunning them the next. Fixations on certain foods — and aversions to others — are typical at this age, and they sometimes last as long as weeks, or even months.

Toddlers grow out of food jags, and these fickle food preferences seldom do any nutritional harm. Your best strategy is not to overreact. I cover ways to encourage your child to eat vegetables and other healthful foods in Chapter 21.

Getting enough calories

Vegetarian diets — especially low-fat or vegan diets — can be bulky. Many plant foods are high in fiber and relatively low in calories. Because young children have small stomachs, they may become full before they've had a chance to take in enough calories to meet their energy needs. For this reason, it's important to include plenty of calorie-dense foods in the diets of young vegetarian children.

Some adults try to keep their fat intakes to a minimum, especially to control their weight, but this very aspect of plant-based diets that helps adults control their weight can backfire for young children. If fat intakes are overly restricted in young vegetarian children, those children may have trouble getting enough calories to meet their energy needs.

Liberally using plant sources of fat can help provide young children with the extra calories they need during a period of their lives in which they're growing and developing rapidly. So, for instance, adding a slice of avocado (nearly all fat) to a sandwich is fine, or using nut and seed butters on sandwiches and vegetable sticks is also a good idea.

Vegan or vegetarian? Determining what's appropriate for young children

After your baby makes the transition to a toddler diet and solid foods, you have to decide the extent to which you'll include animal products in her diet.

Young children can thrive on vegetarian diets that include some animal products or none at all. Table 20-2 shows you what to expect toddlers and preschoolers to eat.

The protein needs of vegan children are a bit higher than those of nonvegan children because of differences in the digestibility and composition of plant and animal proteins. Assuming vegan children get enough calories to meet their energy needs, though, and get a reasonable variety of plant foods, they should get plenty of protein in their diet.

An added caution is that vegan children need a reliable source of vitamin B12 in their diet. I list examples in Chapter 3. If a vegan child has limited exposure to sunshine, consult your healthcare provider to determine whether you should add a vitamin D supplement or fortified foods to his diet.

Serving sensible snacks

Another way to ensure that vegetarian kids get enough calories is to offer them between-meals snacks.

Be sure you don't give toddlers foods they can choke on. For instance, grapes and whole tofu hot dogs are dangerous because they can easily get stuck in the esophagus. If you want to offer these foods to a small child, be sure to slice them in half or into quarters. Also be careful with chips, nuts, and other small snacks that can lodge in a small child's throat. Be sure to supervise young children while they're eating.

A few nutritious snack ideas for vegetarian toddlers include

- Cooked or dry cereal with soymilk
- A dab of smooth peanut or almond butter on a cracker
- Graham crackers
- Single-serving, shelf-stable boxes of 100 percent fruit juice or soymilk
- Small pieces of fresh fruit
- Soy yogurt
- Tofu processed with calcium, served as cubes or made into smoothies or pudding
- Unsweetened applesauce
- Whole grain cereal O's

Chapter 21

Meatless Meals for Children and Teens

In This Chapter

- Dispelling concerns about growth rates
- Reviewing nutritional needs for youngsters
- Involving kids in meal planning and cooking
- Eating vegetarian at school
- Managing weight concerns
- Understanding food allergies

In polls conducted for the nonprofit Vegetarian Resource Group over a period of more than ten years, 2 to 3 percent of U.S. children ages 8 to 18 consistently report never eating meat, poultry, fish, or seafood. And 11 percent of 13- to 15-year-old females say they're consistently vegetarian. That's a substantial number of young vegetarians!

Many of these kids are the only vegetarians in their household. You've got to admire their conviction, but when Mom and Dad aren't vegetarians as well, the situation may cause parents to worry about the nutritional adequacy of the diet and how best to manage family meals without meat. Even in homes where vegetarianism is the norm, people have questions about vegetarian nutrition for kids.

Feeding any child — vegetarian or not — takes time, patience, and care. A haphazard diet that's heavy on chips and soft drinks and light on fruits and vegetables — or otherwise poorly planned — isn't likely to meet the needs of any growing child. On the other hand, a well-planned vegetarian diet offers health advantages over nonvegetarian diets for kids, and it helps put into motion healthful eating habits that can become a pattern for a lifetime.

In this chapter, I discuss the most common questions and concerns about vegetarian diets for children and teenagers.

Watching Your Kids Grow

Some people worry that children who don't eat meat won't grow as well as they should. Reports in the scientific literature over the years have addressed poor growth rates in vegetarian children, but a closer look at those studies should help you put concerns into perspective.

According to these studies, growth problems in vegetarian children occurred when:

- The children lived in poverty in developing countries and didn't have enough to eat.
- The children did not live in poverty but had bizarre, inadequate diets that were severely limited in variety and calories.

Malnutrition, not vegetarianism, causes growth retardation. Any child, vegetarian or not, who doesn't have enough to eat will suffer from nutritional deficiencies and may have difficulties developing properly.

I advocate vegetarian diets that contain adequate calories and a range of foods to ensure that children meet their nutritional needs. Reasonable vegetarian diets not only meet nutritional needs but are associated with health advantages as well.

Understanding issues about growth rates

Talking about growth rates in vegetarian children raises the question of what a normal rate of growth is for a child.

Your pediatrician has growth charts that she uses to plot your child's height and weight at regular intervals, comparing them to population norms. You may even be doing this yourself at home. Growth rates are usually reported in percentiles. For instance, one child may be growing at the 50th percentile for height and weight, while another child of the same age may be growing at the 90th percentile. Still another same-aged child may be growing at the 25th percentile. Is one child healthier than the others? Not necessarily.

Within any group of people, you should expect to see 50 percent growing at the 50th percentile, 25 percent growing at the 25th percentile, and so on. That's called *normal distribution*. In other words, variation is normal within population groups. A child growing at the 25th percentile isn't necessarily healthier or less healthy than a child growing at the 90th percentile. What's important is for a child growing at a particular rate to continue to grow at that same rate.

However, you want to investigate a decline in your child's growth rate. So a child growing at the 35th percentile who continues to grow at the 35th percentile is probably fine, but if that child's rate of growth falls to the 25th percentile, the parent or healthcare provider should look into possible causes.

Putting size into perspective

Parents expect kids who eat a meat-and-potatoes diet to go through growth spurts at certain ages. Most people think of that as a good thing. It's important for your child to grow up to be big and strong, right? Many people hope for football player-sized kids and worry that the playground bully is going to pick on their child if he's too small.

The good news is that the growth rates of lacto and lacto ovo vegetarian children are similar to the growth rates of nonvegetarian children. (I give details about what different kinds of vegetarians eat in Chapter 1.) However, you find very little information about growth rates of vegan children in the U.S., although a peek at the growth rates of children in China eating a near-vegan diet has given scientists some idea of what you may expect to see.

In a population study called the China Project, which began in 1983, scientists found that children eating a near-vegan diet grew more slowly than U.S. children eating a standard, Western-style diet containing meat and milk. The Chinese children attained full adult stature eventually, but they took longer to get there. They grew over a period of about 21 years, as compared to American children, who stopped growing at about the age of 18.

Chinese girls reached *menarche* (had their first menstrual cycle) at an average age of 17, compared to 12 for U.S. girls. This later age of menarche was associated with lower rates of breast cancer in Chinese women, theoretically because they were exposed to high levels of circulating estrogen hormones for a shorter period.

So it's possible that U.S. vegan children may grow more slowly than other children. Nobody knows, though, if that's good, bad, or indifferent.

The most important thing for you to know if you have a vegetarian child is that your child's growth rate should be constant or increase. If growth takes a nose dive, that's the time to investigate and intervene. In the meantime, if your child is growing and is otherwise healthy, you have no need to worry.

Feeding Fundamentals

People generally associate vegetarian diets with health advantages for everyone, but a few of the characteristics that make vegetarian diets good for you can be pitfalls for children if you aren't aware of them and don't take precautions. The primary issue is the bulkiness of a vegetarian diet and the fact that some children can fill up before they've taken in enough calories.

In this section, I discuss this issue and a few others to be aware of.

Making sure kids get enough calories

It's important to be aware of the potential bulkiness of a vegetarian diet and to give your child plenty of calorie-dense foods. A child whose diet is mostly low-calorie, filling foods such as large lettuce salads and raw vegetables could run into problems getting enough calories. It's okay to include those foods in your child's diet, but don't let them displace too many higher-calorie foods.

Feel free to use vegetable sources of fat in your child's diet, such as seed and nut butters, olive oil, and avocado slices. These fats are a concentrated source of calories. Though some adults may want to limit fatty foods in their own diet to control their weight, kids need the extra calories for growth.

If you and your family have recently switched to a vegan diet and your child loses weight or doesn't seem to be growing as quickly, include more sources of concentrated calories. Add vegetable fats, and serve more starchy foods — breads, cereals, peas, potatoes, and corn — and fewer low-calorie, bulky foods such as lettuce and raw vegetables. Like other vegetarian diets, vegan diets can be healthful for children.

Reviewing the ABCs of nutrition for kids

When it comes to designing a vegetarian diet for older children and teens, a few key nutrients — protein, calcium, iron, and vitamins B12 and D — deserve special attention. (If you need more information about any of these nutrients, I cover them in detail in Chapter 3.)

I could list numerous other nutrients and discuss their roles in the growth and development of children, as well as the importance of including good food sources in the diet. When it comes right down to it, though, the major points you need to remember are to make sure kids get enough calories, a reasonable variety of foods, a reliable source of vitamin B12 for vegan and near-vegan children, and adequate vitamin D. Kids should limit junk foods, too, so they don't displace more nutritious foods.

Supplement with snacks

Give your kids nutritious snacks between meals. In Chapter 20, I include a list of good snacks for younger children. For older children and teens, you can expand that list to include

- Bagels
- Bean burritos and tacos
- Cereal and soymilk or skim cow's milk
- Dried fruit
- Fresh fruit
- Fresh vegetable sticks with hummus or black bean dip
- Frozen fruit bars
- Frozen waffles with maple syrup or jam or jelly
- Individual frozen vegetarian pizzas
- Muffins
- Popcorn
- Sandwiches
- Smoothies made with soymilk and fresh fruit, ice cream substitute, or nonfat dairy products
- Veggie burgers
- Whole-grain cookies
- Whole-grain crackers
- Whole-grain toast with jam or jelly

Protein power

Protein and calorie malnutrition go hand in hand. When a child's diet is too low in calories, her body will burn protein for energy. When her diet contains sufficient calories, her body can use protein for building new tissues instead of burning it for energy.

The exact amount of protein your child needs each day depends on several factors, including her size, her activity level, and the number of calories she needs each day. The Food and Nutrition Board of the Institute of Medicine in the National Academies estimates that 5 to 20 percent of calories should come from protein for children and teens ages 4 through 18.

The most practical way to ensure that your child has enough protein is to be sure she's getting adequate calories from a reasonable variety of foods, including vegetables, grains, legumes, nuts, and seeds (fruits contain no protein). Particularly good sources of protein that are also likely to be hits with children include

- Bean burritos and tacos
- Hummus or other bean dips with vegetable sticks or tortilla chips
- Nonfat or soy cheese on crackers
- Nonfat or soy yogurt
- Peanut butter on apple chunks or celery sticks

- Peanut butter sandwiches made with whole-grain bread or crackers
- Soymilk and fruit smoothies made with milk
- Tempeh sloppy Joes
- Tofu or nonfat ricotta cheese and vegetable lasagna
- Tofu salad sandwiches
- Vegetarian pizza
- Veggie burgers
- Veggie hot dogs

Some scientific evidence suggests that young vegan children's bodies may use protein better if they space meals close together. Eating snacks between main meals, for example, may be helpful.

Keeping up with calcium

Because children and teens grow rapidly, they need plenty of calcium to accommodate the development of teeth and bones. The Food and Nutrition Board of the Institute of Medicine in the National Academies recommends that children and teens ages 9 through 18 get 1,300 milligrams of calcium each day. Several factors, though — including the presence of vitamin D and the absorption and retention of calcium — are as important to maintaining a healthy body as having adequate amounts of calcium in the diet.

Still, it's a good idea to encourage children and teens to eat three servings of calcium-rich foods each day. (Counting servings of calcium-rich foods, rather than counting milligrams of calcium, is a practical way to ensure your children get what they need.)

Calcium-rich foods include dark green, leafy vegetables (such as kale, collard greens, turnip greens, and bok choy), broccoli, legumes, almonds, sesame seeds, calcium-fortified orange juice and soymilk, and nonfat dairy products such as milk and yogurt. I list more foods in Chapter 3.

Aim for big servings — at least a cup at a time — and include one or two servings of a fortified food such as calcium-fortified soymilk or juice. If your kids won't eat their vegetables, keep reading — I cover that later in this chapter.

Iron, too

Many high-calcium foods also happen to be high in iron, so serving these foods to your family gives you a double benefit. Good plant sources of iron include dark green, leafy vegetables, soybeans and other legumes, bran flakes, and blackstrap molasses.

The Food and Nutrition Board of the Institute of Medicine in the National Academies recommends that boys and girls ages 9 to 13 get 8 milligrams of iron in their diets each day. Boys 14 to 18 need a little more — about 11 milligrams of iron each day — and girls in that age range need about 15 milligrams of iron each day.

Don't forget that it's important for vegetarians to have good food sources of vitamin C present at meals to increase the body's absorption of the iron in those meals. So give your kids plenty of fruits and vegetables to help ensure they get enough calcium, iron, and vitamin C.

Vitamin B12

Everyone, including children and teens, needs a reliable source of vitamin B12. The Food and Nutrition Board of the Institute of Medicine in the National Academies recommends that boys and girls ages 9 to 13 get 1.8 micrograms of vitamin B12 in their diets each day. Boys and girls 14 to 18 need a little more — 2.4 micrograms each day.

If your kids are eating a vegan or near-vegan diet, they should be eating vitamin B12-fortified foods regularly, or taking a vitamin B12 supplement. If you have any doubt about whether fortified foods are providing enough vitamin B12, the safest bet is to have your kids take a supplement.

Children who regularly include animal products in their diets — milk, eggs, cheese, and yogurt, for example — should have no problem getting necessary amounts of vitamin B12. If you aren't sure if your child is getting enough, ask a registered dietitian for individualized guidance.

Vitamin D

The important thing to remember about vitamin D and children is that vitamin D, in concert with calcium, is critical for the normal growth and development of bones and teeth. Our primary source of vitamin D isn't food at all — our bodies produce vitamin D after our skin is exposed to sunlight. Vitamin D-fortified cow's milk and fortified soymilk, rice milk, and almond milk are also good sources.

The Food and Nutrition Board of the Institute of Medicine in the National Academies recommends that boys and girls ages 9 to 18 get 5 micrograms (200 IUs) of cholecalciferol, a form of vitamin D, in their diets each day in the absence of adequate sunlight exposure. I discuss vitamin D in more detail in Chapter 3.

If you have any doubts about whether your child is at risk of not getting enough vitamin D, ask a registered dietitian or your healthcare provider for an assessment and recommendations.

Planning healthy meals

In Table 21-1, I provide a vegetarian meal-planning guide for school-aged children up to 13 years of age.

Table 21-1 Meal-Planning Guide for Vegetarian Children Ages 4–13

Food Group	*Number of Servings for 4- to 8-year-olds*	*Number of Servings for 9- to 13-year-olds*	*Example of One Serving*
Grains	8 or more	10 or more	1 slice of bread; or ½ cup cooked cereal, grain, or pasta; or ¾ cup ready-to-eat cereal
Protein Foods	5 or more	6 or more	½ cup cooked beans, tofu*, tempeh, or textured vegetable protein; 1 cup fortified soymilk*; 1 ounce of meat analog; ¼ cup nuts or seeds*; 2 tablespoons nut or seed butter*
Vegetables	4 or more	4 or more	½ cup cooked or 1 cup raw vegetables*
Fruits	2 or more	2 or more	½ cup canned fruit; or ½ cup juice; or 1 medium fruit
Fats	2 or more	3 or more	1 teaspoon trans fat-free margarine or oil
Omega-3 Fats	1 per day	1 per day	1 teaspoon flaxseed oil; 1 tablespoon canola or soybean oil; 1 tablespoon ground flaxseed or ¼ cup walnuts

Starred food items: 6 or more for 4- to 8-year-olds; 10 or more for 9- to 13-year-olds (1 serving = 1⁄2 cup calcium-set tofu; 1 cup calcium-fortified soymilk, orange juice, or soy yogurt; 1⁄4 cup almonds; 2 tablespoons tahini or almond butter; 1 cup cooked or 2 cups raw broccoli, bok choy, collards, kale, or mustard greens).

Notes: You can increase the calorie content of the diet by adding greater amounts of nut butters, dried fruits, soy products, and other high-calorie foods. Also, use a regular source of vitamin B12 like Vegetarian Support Formula nutritional yeast, vitamin B12-fortified soymilk, vitamin B12-fortified breakfast cereal, vitamin B12-fortified meat analogs, or vitamin B12 supplements. I also recommend adequate exposure to sunlight — 20 to 30 minutes of summer sun on the hands and face two to three times a week — to promote vitamin D synthesis. If sunlight exposure is limited, use supplemental vitamin D.

Adapted from Simply Vegan, Fourth Edition, by Debra Wasserman and Reed Mangels, PhD, RD, 2006. Reprinted with permission from The Vegetarian Resource Group, P.O. Box 1463, Baltimore, MD 21203; phone 410-366-8343; Web site www.vrg.org.

Feeding furry family members

Dogs are naturally omnivorous and can fare well on a diet that excludes meat. Cats, on the other hand, are carnivorous and need the nutrients found in meat. Specifically, cats must have a source of the amino acid taurine in their diets, and taurine doesn't exist in the plant world. If you don't feed your cat meat, you must provide a taurine supplement. Your veterinarian may or may not be receptive to the idea of a vegetarian diet for cats and dogs, just as many human healthcare providers have no training in the fundamentals of vegetarian diets for people.

Teaching Your Children to Love Good Foods

As a kid, what did you do with the vegetables you didn't want to eat? I lined my peas up along the edge of my plate, hidden along the underside of my knife. As you probably know by now, trying to force people to do things they don't want to do is usually a losing battle, and kids are no exception.

You can employ some strategies, however, that may increase the likelihood your child will eat his vegetables. More important, you can do some things — which I outline in this section — to increase the likelihood that your child will grow up enjoying healthful foods and will make them a part of his adult lifestyle.

Modeling healthy choices

Model the behavior you want your children to adopt. If you'd like your children to eat broccoli and sweet potatoes, for example, let them see you enjoying these foods yourself. But don't pretend to like something you don't, because children can spot a fake. If you don't care for a food, fix it for the others in your household, and don't make a big show out of the fact that you don't have any on your plate.

Present foods with a positive attitude — an air that says you expect others to like this food. If your child expresses dislike for a food that you want to see her eat, play it low key. She may come around in time. If not, you have no need to fret — you can choose from among hundreds of different vegetables, fruits, and grains. If your child doesn't like one, plenty of other options are out there to take its place.

Giving kids the freedom to choose

Children, like everyone, prefer a measure of freedom. If your child turns up his nose at a particular food, offer one or two other healthful choices. For instance, if your child says no to cooked carrots, offer a few raw carrots with dip instead. If he refuses these, let it go. The next meal will bring new choices, and not eating any vegetables at one meal won't make a difference in your child's health.

Of course, your child may also have special requests. It's okay to entertain them, within reason. In the next section, I talk about the advantages of empowering your children to make good food choices by getting them involved in meal planning.

Getting kids involved in meal planning

Children are more likely to eat what they've had a hand in planning and preparing. Give your child frequent opportunities to get involved in meals.

You can start by taking her shopping. If you buy apples or pears, let her pick out two or three. Let older children have even more responsibility. Send your teen to the opposite side of the produce department to pick out a head of cauliflower. Who cares if it's the best one? It's more important that your kids become involved.

When it comes to cooking, supervise young children and let them help with simple tasks like retrieving canned goods from the pantry or dumping prepared ingredients into a pot. Older kids can help wash and peel fruits and vegetables for salads and can assemble other ingredients for casseroles and stir-fries.

By planting a garden and tending it together, you can teach children how their foods grow to help them gain an interest in and appreciation for fresh foods. Short on space or time? Try a window-sill herb garden, or grow a pot of tomatoes on your back porch or apartment balcony.

Troubleshooting Common Challenges

Older vegetarian children and teens share diet problems similar to those faced by their nonvegetarian peers. In this section, I offer guidance on the meat-free approach to addressing these issues.

Making the most of school meals

It isn't easy to find a healthy school lunch, and it's tougher to find a healthy vegetarian school lunch. School systems have poured lots of time, energy, and money into improving their meals programs in recent years to try to bring them into compliance with contemporary dietary recommendations.

Most aren't there yet, but changes in federal regulations have made more health-supporting vegetarian options feasible. For instance, schools can now serve yogurt as a meat replacement, and they can serve nondairy cheeses in lieu of dairy cheeses if they're nutritionally similar.

But schools still lack many vegetarian options, and kids who want vegan options have even fewer choices. What to do? Here are some ideas:

- Take a bag lunch. Figure 21-1 shows some ideas to get you started. I also include a list of ten good-tasting and practical ideas for bag lunches in Chapter 25.
- Get a copy of the cafeteria menu. You and your child can peruse the menu for the best choices each day.
- Talk to school food service personnel. The circumstances are different for each student and each school, so try sitting down with food service personnel and discussing practical solutions that both your child and the school can live with.

Following are two options that may help your child make the most of school meals:

- **The National Farm to School Program:** This program is one example of efforts to bring more fresh, locally grown produce to schools. The program has great potential to improve the nutritional quality of school meals and increase options for vegetarians.

  The program began with pilot projects in California and Florida in the late 1990s as a movement to connect community farmers and local schools with the goals of supporting local agriculture, improving meals in school cafeterias, and educating students about nutrition.

  By 2004, Congress recognized the program in the 2004 Child Nutrition Reauthorization — legislation renewed every five years authorizing federal support for child nutrition programs, such as the National School Lunch Program.

  Today, more than 2,000 Farm to School programs operate in 39 states. The National Farm to School Program is managed jointly by the Center for Food & Justice, a division of the Urban and Environmental Policy Institute at Occidental College, and the Community Food Security Coalition. For more information about Farm to School, go to `www.farmtoschool.org/`.

- **Nutrient-based menu planning system:** Schools also have the option of using this system, which enables them to evaluate meals based on overall nutritional composition rather than on whether a meal consists of a specified number of servings from various food groups.

  Theoretically, a nutrient-based system could make it easier to serve meatless menus, because servings from the meat group wouldn't be mandatory. In reality, few schools in the U.S. are using this system, because changing from the old food groups system would take more resources than many schools can afford.

Figure 21-1: Popular vegetarian choices for school lunchboxes.

Supporting a healthy weight

Obesity is a major public health problem for children in North America and around the world. In fact, health professionals worry that today's young people may face more health problems earlier than their parents and grandparents did because of excess weight. The problem is that diets are too high in calories and kids get too little physical activity.

The good news is that vegetarian children are more likely than other children to be at an ideal body weight. Their diets also contain substantially more fiber and less total fat, saturated fat, and cholesterol than diets that include meat.

If your vegetarian child happens to be overweight, give him a junk check (more about that in the next section, "Dejunking your child's diet"). If the diet is already up to snuff, the problem is likely due, at least in part, to lack of exercise.

Pull your child away from the computer or video games and require her to get moving. For teens, aerobics classes, school sports, and weight lifting are excellent ways of burning calories and increasing cardiovascular fitness and overall strength. Encourage biking, hiking, or canoeing. Switch activities depending on the season: Snow ski in the winter and swim in the summer. Mix it up to avoid getting into a rut.

Older children and teens are also very body-conscious. Teen boys are more likely than teen girls to feel they're too skinny. Nature will probably take its course, and today's string bean will be tomorrow's 40 regular. It just happens sooner for some than for others.

So if your teen wants to gain weight, the way to do it is simply to eat more of the good stuff. Increase serving sizes at meals and add healthful snacks between meals. Smoothies and juice blends add easy, quick calories. Increasing weight-bearing exercise — within limits, of course — also stimulates the growth of more muscle tissue.

Your teen should steer clear of special — and often expensive — protein powders, shakes, bars, or other supplements. Flooding his body with extra protein from supplements won't build muscle. In fact, it could be harmful. Byproducts of protein breakdown have to be filtered from the bloodstream, increasing the workload for the kidneys. Muscle is made from hard work and food — nothing fancier than the nutrients found in an ordinary diet that contains a mix of health-supporting foods.

Eating disorders are more common in teens — especially girls — than in adults. No cause-and-effect relationship exists between vegetarian diets and eating disorders, such as *anorexia nervosa* (self-starvation) and *bulimia* (bingeing and purging), but research suggests that some girls adopt vegetarian diets to mask eating disorders. The effects of anorexia can also cause a loss of the taste for meat, leading some anorexics to stop eating meat. Eating disorders have psychological origins, and people with eating disorders need psychiatric or psychological intervention.

Dejunking your child's diet

Older children and teens have more freedom than younger children to make their own food choices. Too often, they reach for chips, soft drinks, fast foods, and other junky foods that provide little nutrition in exchange for the calories.

Encourage your child to clean up her diet. Keep fresh fruits on hand at home, and serve bigger helpings of vegetables, whole grains, and legumes. Give your child a big apple to take to school in her backpack. Most important, model a junk-free diet yourself.

Having some appealing and acceptable — though not nutritionally perfect — alternatives on hand may also be helpful. Diet drinks and noncaloric, flavored seltzer water, for example, are preferable to caloric soft drinks and other sweetened beverages. (All caloric drinks, including fruit juices, contribute to excessive calorie intakes for many children, teens, and adults and may contribute to obesity.) Popping plain corn kernels or steaming a bowl of *edamame* (fresh or frozen soybeans in the pod) and adding a dash of salt may satisfy the craving for a salty snack with only a fraction of the sodium contained in store-bought, processed snacks.

Dealing with food allergies

Food allergies are an extreme response by the body's immune system to proteins in certain foods. Symptoms include hives, rashes, nausea, congestion, diarrhea, and swelling in the mouth and throat. Severe reactions can cause shock and death.

So for some vegetarians, a stray peanut can be life threatening. Other vegetarian foods that are among the most common food allergens include eggs, milk, soy, tree nuts, and wheat.

About 2 percent of adults and 8 percent of children have true food allergies, according to the American Academy of Allergy, Asthma, and Immunology. Food allergies are different from food intolerances, which seldom cause symptoms so severe. One example is lactose intolerance — the inability to digest the milk sugar lactose. (I discuss lactose intolerance in Chapter 3.)

Food companies are now required by law to use easily recognizable terms on food labels to describe common food allergens. Prior to this law, food companies listed terms such as *albumin* and *casein* on food packages without explaining that they were byproducts of eggs and milk. The law also requires companies to disclose potential allergens if they're present in natural flavorings, natural colorings, and spices added to packaged foods.

Food allergies have no cure. Children may outgrow them, but allergies to peanuts and tree nuts usually last for life. For more help with food allergies, I recommend:

- Food Allergy & Anaphylaxis Network (FAAN), online at `www.foodallergy.org/`.
- American Academy of Allergy, Asthma, and Immunology, online at `www.aaaai.org/`.
- *Food Allergies For Dummies,* by Robert A. Wood, MD, with Joe Kraynak (Wiley).
- *Food Allergy Survival Guide,* by Vesanto Melina, Jo Stepaniak, and Dina Aronson (Healthy Living Publications).

Chapter 22

Aging Healthfully: Vegetarian Lifestyles for Adults of All Ages

In This Chapter

- Understanding how age changes nutrient needs
- Maintaining your health as you get older
- Eating properly for an active lifestyle

Many people are healthy and active into old age these days, though age-related health problems do increase over the years. Within the field of *gerontology,* the study of normal aging, scientists are gaining insights into aspects of aging that most of us take for granted as part of growing old.

Everyone experiences age-related changes, but these changes occur at different rates for different individuals. Some people never develop certain diseases and conditions. Why? Genetic differences among people may be one reason, but lifestyle factors probably play a role, too.

Diet is one lifestyle factor that makes an undeniable difference in the way people age. Physical activity makes a substantial difference, too. Eating a nutritious diet and maintaining a lifelong habit of regular — preferably daily — physical activity may not stop the clock, but it can delay evidence of aging that some people experience later than others.

For example, you may consider constipation, hemorrhoids, and weight gain to be normal parts of the aging process, or you may figure that diabetes and high blood pressure are inevitable at some point.

But vegetarians, as a group, have a different experience.

Vegetarians have lower rates of many chronic diseases and conditions. They generally live longer than nonvegetarians, too. Health and longevity differences may be due, in part, to lifestyle factors typical of vegetarians, including a higher level of physical activity, not smoking, and eating a health-supporting diet.

In this chapter, I explain how you can leverage an active vegetarian lifestyle to help delay or prevent some age-related changes in your health. I discuss nutritional issues that affect adults as they age, and I cover the basics of eating for optimal athletic performance — whether you're a neighborhood walker or a contender for the Tour de France.

Monitoring Changing Nutrient Needs

Older adults — roughly defined as people age 50 and above — have been at the back of the line when it comes to research on the body's nutritional needs throughout the life cycle. Very little is known about how the aging process affects the body's ability to digest, absorb, and retain nutrients.

For now, recommended intakes for most nutrients for older adults are simply extrapolated from the recommendations for younger people. Scientists do know a bit about how metabolism and the body's need for certain nutrients changes with age, though. Read on to discover the smartest way to get the nutrients you need as you age.

Getting more for less

Yes, what you've always heard is true: Your metabolism declines as you age. Unfair! Unfair! But the sad fact is that you need fewer calories the older you get, assuming that your physical activity level stays the same. In fact, if your activity level decreases, then your calorie needs decline even further. Oh, woe!

It gets worse. If you consume fewer calories, your intakes of protein, vitamins, minerals, and other nutrients also decrease. Unfortunately, as far as anyone knows right now, your nutritional needs don't diminish. Your needs for certain nutrients may actually rise.

That means that you have to be extra careful to eat well. You have to get the same amount of nutrition that you got when you were younger (and eating more calories), but you have to get it in less food. That means that you have to eat fewer *empty-calorie foods* — less junk. Fewer sweets, snack chips, cakes, cookies, candy, soft drinks, and alcohol. Empty-calorie foods are nutritional freeloaders — they displace nutrient-dense foods and provide little nutrition in exchange for the calories.

Paying special attention to specific nutrients

More research is needed on how nutritional needs change for people age 50 and beyond, but scientists are sure that needs do change. That may be partly because the body's ability to absorb certain nutrients declines gradually over time.

For example, older people produce less stomach acid, which is vital in helping the body absorb vitamin B12. Scientists estimate that 10 to 30 percent of adults over age 50 may have difficulty absorbing enough vitamin B12 from their food.

That's why older adults should eat foods that are fortified with vitamin B12 or take a B12 supplement. This is especially noteworthy for older vegans, who need to be careful to have a reliable source of vitamin B12 in their diets.

For older vegetarians and nonvegetarians alike, recommendations for calcium, vitamin D, and vitamin B6 are higher than for younger people. If you're an older vegetarian who doesn't drink milk or eat other dairy products and your exposure to sunlight is limited, use vitamin D-fortified foods such as some brands of soymilk, or get your vitamin D from a supplement.

The bottom line is that it's important for older folks to be frugal with their calories and save them for nutritious foods, rather than filling up on sweets and junk.

Celebrating the Vegetarian Advantage

You knew you were getting old when your eyebrows started turning gray.

Besides the obvious outward signs of aging — wrinkles, lines, and gray hair — older people have other common complaints. Most have to do in some way with the digestive tract. People start getting constipated, or they have more trouble with heartburn and indigestion. Some of these problems are a result of a decrease in the production of the stomach secretions that aid digestion, or they're in some other way a result of the body not functioning as efficiently as it once did.

But many problems are the result of a lifelong poor diet, or a lack of regular exercise, or any of a host of other destructive habits, such as smoking or abusing alcohol. For instance, if you've been exercising, eating plenty of fiber, and drinking enough water for the past 20 years, you're much less likely than other people to have hemorrhoids or varicose veins.

The general dietary recommendations for older adults are similar to those for younger people: Get enough calories to meet your energy needs and maintain an ideal weight, and eat a variety of wholesome foods, including fruits, vegetables, whole grains, legumes, and a limited number of seeds and nuts. Drink plenty of fluids, and limit the sweets and junk.

Eating a vegetarian diet can help prevent or delay many of the common problems associated with getting older, and a vegetarian diet can also help alleviate some of the problems after they've developed.

That's something to celebrate — and take advantage of. In this section, I explain how.

Being fiber-full and constipation-free

Constipation is nearly always caused by diet. Regardless of your age, you need plenty of fiber in your diet from fruits, whole grains, vegetables, and legumes. Enough fluids and regular exercise are important factors, too.

Older adults develop problems with constipation when their calorie intakes dip too low. When you eat less, you take in less fiber. If you eat too many desserts and junk foods, you may be getting even less fiber. Older people are also notorious for being physically inactive. Both of these factors — low fiber intake and low activity level — can cause you to become constipated.

Constipation can be made worse if you're taking certain medications, including antacids made with aluminum hydroxide or calcium carbonate, or if you're a habitual laxative user.

If you have problems with constipation, take these steps to get moving again:

- **Add fiber to your diet.** Eat plenty of fresh fruit, vegetables, whole-grain breads and cereal products, and legumes.
- **Drink fluids frequently.** Water is best. You don't have to count eight glasses per day, but keep a pitcher of water on your desk and a bottle of water in your car, and drink them regularly. Stopping for a sip every time you pass a water fountain is a good idea.
- **Eat prunes and drink prune juice.** Prunes have a laxative effect for many people.
- **Keep fatty foods and junk foods to a minimum.** These foods are usually low in fiber and will displace other foods that may contribute fiber to your diet.
- **Hold your fat intake down.** If you do, you won't need as many antacids. Fat takes longer to digest than other nutrients, so it stays in the stomach longer and can promote indigestion and heartburn.

- **Get regular physical activity.** It keeps your muscles (including those in your abdomen) toned and helps prevent constipation.

Vegetarian diets tend to be lower in fat and higher in fiber than nonvegetarian diets, so it's not surprising that older vegetarians are less likely than older nonvegetarians to have problems with constipation.

Heading off heartburn

Because vegetarian diets tend to be lower in fat than meat-based diets, they can be useful in helping to minimize problems with heartburn. If you're a vegetarian and have trouble with heartburn, examine your diet. You may be eating too many high-fat dairy products or greasy junk foods such as chips, donuts, and French fries.

In addition to cutting the fat in your diet, you can help prevent heartburn by avoiding reclining immediately after a meal. If you do lie down for a nap after lunch, put a couple of pillows under your back so that you're elevated at least 30 degrees and aren't lying flat. Avoid overeating, and try eating smaller, more frequent meals.

Getting a grip on gas

You may prefer to call it "flatulence." By any name, it's intestinal gas, which can be caused by a number of things, including the higher fiber content of a vegetarian diet.

More than just a social problem, gas can cause discomfort in your abdomen, and it can cause you to belch or feel bloated. Before you incriminate the beans and cabbage, though, a few other causes of gas may be exacerbating your problem, including drinking carbonated beverages, swallowing too much air when you're eating, and taking certain medications.

If you do think your diet is the problem, you have a few options. Consider these gas-busters:

- Single out the culprits. Beans? Cabbage? Onions? Eliminate one at a time until you reduce your gas production to a level you can live with. The foods that cause gas in one individual don't necessarily cause gas in everyone else, and unfortunately, the foods that do cause gas are among the most nutritious.
- Try using a product such as Beano. It uses enzymes to break down some of the carbohydrate that causes gas. Products such as these come in liquid form. Squeeze a few drops on the food before you eat it. Its effectiveness varies from person to person.

- Get active. People who exercise regularly have fewer problems with gas.
- Give it time. If you're new to a vegetarian diet, your body will adjust to the increased fiber load over several weeks, and your problem with gas should subside.
- Avoid carbonated beverages.
- Eat slowly and chew your food thoroughly to minimize the amount of air that you take in with each bite.

Living vegetarian is good for what ails you

You're diabetic and you follow a special diet? Maybe you're on a special diet for high blood pressure or heart disease? It doesn't matter. A vegetarian diet is compatible with restrictions for any diet, and in many cases, a vegetarian diet is ideal for people with medical conditions.

For instance, if you have diabetes, you may be able to reduce the amount of medication or insulin you take if you switch to a vegetarian diet. (Be sure to talk to your healthcare provider, though, before making any changes in your medications!) The fiber content of vegetarian diets helps to control blood sugar levels.

Anyone taking medications for high blood cholesterol levels should know that the fiber in vegetarian diets helps to lower blood cholesterol levels, too.

Vegetarians have weights that are closer to ideal than nonvegetarians. Weight control is an important component of the dietary management of diabetes, heart disease, high blood pressure, arthritis, and many other conditions.

There's no reason that you can't eat a vegetarian diet, whatever your ailment. If you need help adapting a vegetarian diet to your special needs, contact a registered dietitian with expertise in vegetarian nutrition. The American Dietetic Association's online referral service can help you find a dietitian in your area. Go to `www.eatright.org`.

Staying Active the Vegetarian Way

Regular exercise is an important component of a healthy lifestyle, and it's just as important for older adults as it is for younger people. When you're regularly and vigorously physically active, you burn more calories and are more likely to keep your weight at an ideal level.

Inspired by Murray Rose

I was 16 years old and a competitive swimmer when a book by Murray Rose caught my eye and inspired me to change my lifestyle forever. Murray Rose was an Australian swimmer who attributed his athletic endurance to his vegetarian diet.

Rose was a three-time Olympic gold medalist and world record holder in swimming in Melbourne in 1956. He won a gold medal, a silver medal, and another world record at the 1960 Olympic Games in Rome. I'm not sure the switch to a vegetarian diet improved my own athletic performance, but it didn't hurt, and the diet stuck.

Rose was an athlete. I was, too — then. I practiced hard for two hours every day, after which I would swim two lengths of the high school pool underwater. If I tried that today, I'd need the Emergency Medical Service.

You can also eat more! The more food you consume, the more likely it is that you'll get the nutrients you need. You're likely to preserve more bone and muscle tissues, too, when you exercise regularly, especially when the activity is weight-bearing exercise, such as walking or using weight sets.

In this section, I explain how adults of all ages can eat for optimal athletic performance, whether you're a recreational athlete or a pro.

Nourishing the weekend warrior

In this chapter, I use the term *athlete* to describe a person who is vigorously physically active most days of the week for extended periods of time. Swimmers or runners, or triathletes who are training hard, are considered athletes.

They are in the minority of people who are active.

If you go to the gym three times a week to work out on the stair climber and lift weights, it's great, but you're not an athlete. It's also not enough activity to make any appreciable difference in your nutritional needs. Ditto for golfers and weekend warriors.

Unless your activity level rises to the level of an elite athlete, it's unlikely that you need to take any special dietary measures on account of your activity.

Giving elite athletes the edge

Like everyone, an athlete's diet should consist primarily of carbohydrates, with adequate amounts of protein and fat. Can athletes do well on a vegetarian diet? You bet. Just ask vegetarian baseball player Prince Fielder (Milwaukee Brewers); basketball players Salim Stoudamire (Milwaukee Bucks) and Raja Bell (Charlotte Bobcats); football player Ricky Williams (Miami Dolphins); legendary track star Carl Lewis; and mixed martial arts fighter Mac Danzig.

Experts have different opinions about how much protein athletes need. Some question whether physical activity level affects the body's need for protein at all. Others feel that increased physical activity level does necessitate higher levels of protein, depending on the kind of activity. The extra protein isn't needed primarily for muscle development, though, as you may think. Instead, it's needed to compensate for the protein that athletes burn up as fuel.

Some athletes need a tremendous number of calories to meet their energy needs, and if they don't have enough fuel from carbohydrates and fats, their bodies turn to protein for energy. When that happens, they need extra protein in their diets so that enough will be available for building and repairing tissues.

Because athletes need more calories than people who are less active, they tend to consume more protein in the extra food they eat. If you're an athlete, just getting enough calories to meet your energy needs and eating a reasonable variety of foods is probably enough in itself to ensure that you get the protein you need.

This is especially true for endurance athletes, such as swimmers, cyclists, runners, and triathletes. If you're into strength training (weight lifting, wrestling, or football, for example), you need to be more aware of getting enough protein-rich foods in your diet, especially if your calorie intake is low.

Any athlete restricting calories to lose weight while training should also be more careful to get enough protein. The lower your calorie intake, the harder it is to get what you need and the less room there is for junk.

A little protein is a good thing, but too much isn't. In times past, it was a tradition for coaches to serve their football players big steaks at the training table before games. A baked potato with butter and sour cream and a tossed salad with gobs of high-fat dressing rounded out the meal. High-protein, high-fat meals like that actually make players sluggish and are generally bad for your health. Instead of a steak, athletes are better off with a big plate of spaghetti with tomato sauce, a heaping helping of stir-fry with vegetables and tofu, or bean burritos with rice and vegetables.

Crunching the carbs

A diet that consists primarily of carbohydrates — vegetables, pasta, rice and other grains and grain products, fruits, and legumes — promotes an athlete's stamina and results in better performance. Athletes who restrict their carbohydrate intake show poorer performance levels. It isn't surprising, then, that vegetarian diets have advantages for athletes, because vegetarian diets tend to consist primarily of carbohydrate-rich foods.

Athletes used to "carb load" before an event. They would restrict their carbohydrate intake and load up on fat and protein for a few days, and then they would gorge on carbohydrate-rich foods for a couple days just before the event. The idea was that this method would maximize their muscles' storage of fuel and result in better performance. Now we know that it's more effective just to eat a high-carbohydrate diet all the time. Most vegetarians are, in effect, in a constant state of carbohydrate loading.

When you eat a diet that's high in carbohydrates, your body stores some of it in your muscles and liver in a form of sugar called *glycogen.* Your body taps into its stockpile of glycogen for energy during athletic events. Whether you engage in endurance events such as swimming and cycling or in shorter, high-intensity activities such as running a sprint or the high hurdles, your muscle and liver glycogen stores are a vital energy supply and a critical determinant of your ability to perform your best.

Getting enough calories and carbohydrate in your diet helps to ensure that the protein in your diet is available for the growth and repair of tissues and that it doesn't have to be sacrificed and burned for fuel. But not just any carbohydrates will do.

Soft drinks, candy, snack cakes, and other junk foods consist mainly of carbohydrates, but these are empty-calorie foods. Anyone who depends on their diet to help them feel and perform their best needs to take pains not to let the junk displace nutrient-dense foods from their diet. Junk-food forms of carbohydrates do provide calories, so they can help keep your body from needing to burn protein for fuel. But you need the nutrients in the more wholesome foods, too. You've got to look at the big picture, not just a little piece of it.

Good choices for carbohydrates include

- All fruits
- All vegetables
- Beans
- Breads
- Cold and hot cereals
- Lentils
- Pasta

- Potatoes
- Rice
- Soymilk

Some athletes have tremendously high calorie needs, and many carbohydrate-rich foods are bulky. They can be so filling that some athletes can get full at meals before taking in enough calories. If this is your experience, go ahead and include some low-fiber, refined foods in your diet, despite the fact that most other people need more fiber. For instance, you may choose to eat refined breakfast cereals or white bread instead of whole-grain products.

Meeting vitamin and mineral needs

Generally speaking, a well-planned vegetarian diet that emphasizes adequate calories and variety and limits the junk should provide athletes with all the nutrients they need. Under certain circumstances, though, a few nutrients may deserve some special attention.

- **Calcium:** Athletes have the same needs for calcium as nonathletes, but some female athletes who train intensely may be at risk if their level of training causes *amenorrhea,* the cessation of regular menstrual cycles.

  Amenorrhea isn't caused by vegetarian diets; any female athlete may stop having periods when training too intensely. Like postmenopausal women, amenorrheal women have reduced levels of estrogen, and that can lead to accelerated loss of calcium from the bones.

  Several factors affect bone health, and the amount of calcium you absorb and retain from your diet is more significant than how much calcium your diet contains in the first place. Nevertheless, if you're a female athlete who has stopped having periods or who skips periods, the recommendations for calcium intake are higher for you. That means you need to push the calcium-rich foods and make less room in your diet for junk. (See Chapter 3 for more information about calcium.)

- **Iron:** All athletes — vegetarian or not — are at an increased risk of iron deficiency due to iron losses in the body that occur with prolonged, vigorous activity. Female endurance athletes are at the greatest risk, as are athletes who have low iron stores. (See Chapter 3 for a refresher on iron.) You don't need to take a supplement unless blood tests show that you need one. Men, in particular, should avoid taking unnecessary iron supplements because of the connection between high intakes of iron and coronary artery disease.

- **Other vitamins and minerals:** Vegans, in particular, need to have a reliable source of vitamin B12 in their diets. Some studies also show that exercise raises all athletes' needs for riboflavin and zinc. The most practical advice for any athlete is simply to do your best to eat well to help ensure you get what you need.

Meal planning for peak performance

Whether you're in training or getting ready for an athletic event, what you eat and when you eat it can make a difference in your level of performance.

Here are a few pointers for vegetarian athletes, or any athlete wanting to maximize athletic performance:

- If your calorie needs are high, add snacks between meals. Dried fruit mixtures, bagels, fresh fruit, soup and crackers, hot or cold cereal with soymilk, a half sandwich, soy yogurt or a smoothie, a bean burrito, or baked beans with toast are all excellent choices.
- If you have trouble getting enough calories with meals and snacks, reduce your intake of the bulkiest foods, such as salad greens and low-calorie vegetables, and eat more starchy vegetables such as potatoes, sweet potatoes, peas, beans, and lentils. Liquids can be an especially efficient way to add extra calories — try fruit or soy yogurt shakes and smoothies. Make your own blends, using wheat germ, soy yogurt and soymilk, and frozen or fresh fruit. Remember, too, that some added vegetable fats can be fine for athletes who need a compact source of extra calories.
- Take a portable snack in your gym bag, backpack, or bike pack for immediately after your workout to replace carbohydrates and protein and provide calories. Fresh fruit, a sports bar, a bagel, or a package of crackers and bottle of fruit juice are good choices.
- Keep fluids with you when you work out — fruit juice or bottled water are best.
- Don't work out when you're hungry. Your session will suffer. Take a break and have a light snack first.
- If you get the pre-event jitters, eat only foods that are easy to digest and low in fiber. Some athletes who get too nervous to eat any solid food before an event may find that it's possible to drink a smoothie or eat some nonfat or soy yogurt.

The following sections give you specific diet information to follow before, during, and after your athletic events.

Before an athletic event

If you generally eat a high-carbohydrate vegetarian diet, you're ahead of the game already. When it gets closer to the time of the event in which you'll be competing, it's time to pull a few more tricks out of your sleeve.

In the hours before the event, you want to eat foods that are easy to digest and will keep your energy level up. High-carbohydrate foods are good choices, but now's the time to minimize your fat and protein intakes. Fat, in particular, takes longer to digest than other nutrients. You want your stomach to empty quickly to give your food time to get to the intestines, where it can be absorbed before you burst into action. That's the reason to keep your fat intake low at this point. Avoiding foods that are concentrated in protein is also a good idea immediately before an event because protein also takes a bit longer to digest than carbohydrate. That leaves fruits, vegetables, and grains as the best choices in the hours before an event.

It's also a good idea to avoid foods that are excessively high in sodium or salt because these foods can make you retain fluids, which may impair your performance or make you feel less than your best. Foods that are especially high in fiber are probably best saved for after the event, too, because you'll probably want your large intestine to be as empty as possible during the event. High-fiber foods can cause some people to have diarrhea and others to become constipated before athletic events, especially when they're anxious.

The best rule is to give yourself one hour before the event for every 200 calories of food you eat, up to about 800 calories. In other words, if you eat a meal that contains 400 calories, it's best to eat it two hours before the event.

Here are some good pre-event light meal and snack ideas:

- A bagel with jam
- A banana and several graham crackers
- A bowl of cereal with soymilk
- Cooked vegetables over steamed rice
- Pancakes or waffles with syrup
- Pasta tossed with cooked vegetables or topped with marinara sauce
- A soymilk and fruit smoothie
- Toast or English muffins and 1 cup of soy yogurt
- A tomato sandwich and a glass of fruit juice

During an athletic event

Have you ever played a set of tennis in the blazing sun on a hot summer day, or paddled a canoe or kayak for several hours on a river in the middle of August? If so, you may have needed as much as 2 cups of water every 15 minutes to replace the fluids your body lost during heavy exercise in the extreme heat. Many athletes don't pay enough attention to fluid replacement, yet it's critical to your health and optimal performance.

It's important to drink 1⁄2 to 1 cup of water every 10 to 20 minutes while you're exercising and, when possible, when you're competing. If you're working out in a gym, make it a point to take frequent breaks to visit the water fountain. One good gulp or several sips can equal 1⁄2 to 1 cup of water. In other settings, keep a water bottle with you on a nearby bench, in your boat, or strapped to your bike.

Water is the best choice for exercise sessions or athletic events that last up to 90 minutes. After that, there's a benefit to getting some carbohydrates in addition to the water to help boost your blood sugar and prolong the period of time before your muscles tire out. Eating or drinking carbohydrates sooner may also be of some benefit when the activity is of very high intensity, such as racquetball or weight training.

In these situations, it's a good idea to aim for about 30 to 80 grams of carbohydrates per hour. Sports drinks are fine for this purpose and may be more convenient than eating solid food. An added advantage is that they provide fluid as well as carbohydrates. For most brands of commercial sports drinks, that means aiming for ½ to 1 cup every 15 minutes, or twice as much every ½ hour. In lieu of commercial sports drinks, you may prefer to drink fruit juice diluted 1:1 with water. For example, mix 2 cups of apple juice or cranberry juice with 2 cups of water, and drink it over a 1-hour period.

Fruit is also a good choice for a carbohydrate boost. A large banana contains at least 30 grams of carbohydrates, and so do two small oranges.

After an athletic event

After the event or exercise session, protein can come back to your table. You need protein for the repair of any damaged muscles. Your body also needs to replenish its stores of muscle and liver glycogen, amino acids, and fluids. So calories, protein, carbohydrate, and fluids are all very important immediately after an athletic event. The sooner you begin to replace these nutrients, the better. In fact, studies have shown that your body is more efficient at socking away glycogen in the minutes and hours immediately following the event than it is if you wait several hours before eating.

Part VI
The Part of Tens

"Of course you're better off eating grains and vegetables, but for St. Valentine's Day, we've never been very successful with, 'Say it with Legumes.'"

In this part . . .

Like all *For Dummies* books, this one ends with tidy tidbits of information presented in handy lists to make it easy for you to grab at a glance.

These final chapters succinctly summarize why it makes sense to go vegetarian. I give you practical advice about how to make it happen, including simple ingredient substitutions and easy lunchbox ideas.

Chapter 23

Ten Sound Reasons for Going Vegetarian

In This Chapter

- Eating for planetary and individual health
- Making the compassionate (and less expensive) choice
- Being a model for your children's health

If you need any more persuading to become a vegetarian, this chapter lists ten sound reasons for making the switch.

Talk to veteran vegetarians and you'll probably discover that they were first compelled to kick the meat habit by one of these reasons. After they started on the vegetarian path, other reasons gradually became apparent, adding conviction to the original decision.

Which of these reasons speaks to you?

Vegetarian Diets Are Low in Saturated Fat and Cholesterol

Generally speaking, vegetarian diets tend to be lower in total fat, saturated fat, and cholesterol than nonvegetarian diets. The fewer animal products in the diet, the less saturated fat and cholesterol the diet usually contains. This lower intake of saturated fat and cholesterol probably contributes to the decreased rates of coronary artery disease in vegetarians compared to nonvegetarians. Vegan diets are generally lower in saturated fat than lacto or lacto ovo vegetarian diets, and vegan diets contain zero cholesterol. (Chapter 1 tells you more about the differences between these varying types of vegetarian diets.)

Vegetarian Diets Are Rich in Fiber, Phytochemicals, and Health-Supporting Nutrients

Unless you're a junk food vegetarian, you're likely to get far more dietary fiber on a vegetarian diet than a diet that includes meat.

Fiber-rich vegetarian foods provide the bulk that helps prevent constipation, hemorrhoids, varicose veins, and diverticulosis. Bean burritos, lentil soup, pasta primavera, vegetable stir-fry, and four-bean salad are a few examples. Find easy recipes for all these and more in Part III.

Fruit-, grain-, and vegetable-rich vegetarian diets are also high in beneficial vitamins, minerals, and phytochemicals such as vitamins E and C, selenium, beta carotene, lycopene, and isoflavones, which promote and protect your health when you get them naturally from whole foods. These substances are associated with many of the health advantages of vegetarian diets, including a decreased risk for coronary artery disease and cancer.

Vegetarians Are Skinnier

Because vegetarian diets tend to contain lots of bulky fiber foods, the diet helps you fill up before you fill out. In other words, you stop eating before you take in too many calories, because you feel full sooner. Therefore, vegetarians are usually leaner than nonvegetarians. Who needs fancy weight-loss diets when you can be a vegetarian and stay slim the easiest, and healthiest, way of all?

Vegetarians Are Healthier

Not surprisingly, vegetarian diets are associated with a long list of health advantages. Not only are vegetarians slimmer and less prone to coronary artery disease, but they also have lower low-density lipoprotein cholesterol levels, lower blood pressure, lower overall cancer rates, and lower rates of type 2 diabetes.

Science suggests that several characteristics of vegetarian diets are responsible for these health advantages, including lower intakes of saturated fat and cholesterol and higher intakes of fruits, vegetables, whole grains, nuts, soy foods, and beneficial substances such as the aforementioned dietary fiber and phytochemicals.

Vegetarian Diets Are Good for the Environment

Livestock grazing causes *desertification* of the land by causing erosion of the topsoil and drying out of the land. Topsoil is being destroyed faster than it can be created because of people's appetite for meat. All over the world, irreplaceable trees and forests are being lost to make way for cattle grazing. By eating a vegetarian diet, you can help to minimize this devastation.

But that's not all. Animal agriculture is one of the greatest threats to the world's supply of fresh water. Factory farms suck up tremendous quantities of water from aquifers deep beneath the earth's surface to irrigate grazing lands for livestock. Animal agriculture also pollutes rivers and streams by contaminating water supplies with pesticides, herbicides, and fertilizers used to grow food for the animals. Nitrogenous fecal waste from the animals themselves compounds the problem. By choosing a vegetarian lifestyle, you can do your part *not* to support this contamination.

The production of meat, eggs, and dairy products also makes intensive and extensive use of fossil fuels (like oil) to transport animal feed and animals and also to run the machinery on the factory farms where animals are raised. By eating fewer animal products, you can do your part to conserve fossil fuels by not supporting the production and distribution of animal-derived goods or food.

Vegetarian Diets Are Less Expensive

Sure, if you opt to buy all your groceries at gourmet stores or the ready-made food section of your local deli, your vegetarian diet can soon become more expensive than a diet that includes meat.

But assuming you eat most of your meals at home and take reasonable care to shop for value, eating vegetarian is an inherently economical way to go. Wholesome, simple vegetarian meals made from fresh, frozen, or canned vegetables and fruits, whole-grain breads and cereals, dried or canned beans and peas, and a smattering of seeds and nuts are all you need. It costs less to build a meal from these ingredients than it does to make meat the focal point of the plate.

Vegetarian Diets Are More Efficient

Vegetarian diets can sustain more people than diets that center on meat and other animal products. That's because it takes far more food energy for an animal to produce a pound of meat or a cup of milk than humans get in return when they eat those foods.

Eating plants directly — rather than eating them after they've passed through animals first — is a more sustainable and efficient way to nourish the world.

Vegetarian Diets Are the Compassionate Choice

The philosopher and Nobel Peace Prize-winner Albert Schweitzer said, "Until he extends his circle of compassion to include all living things, man will not himself find peace."

And the great artist Leonardo da Vinci said, "I have from an early age abjured the use of meat, and the time will come when men such as I will look on the murder of animals as they now look on the murder of men."

Enough said?

Vegetarian Foods Are Diverse and Delicious

Vegetarians often hear, "If you don't eat meat, what *do* you eat?" If meat eaters only knew. . . .

The variety in food is in the colorful, flavorful, healthful plant kingdom. Cultures around the world serve satisfying and nutritious dishes such as curried vegetables, garlicky pasta, savory spinach pies, thick soups, and spicy couscous mixtures made with a mouth-watering assortment of fresh, meatless ingredients.

Endless combinations of these foods prepared in creative ways make vegetarian meals exciting. See the recipes in Part III for some suggestions to get you started cooking the vegetarian way.

Vegetarian Diets Set a Good Example for Children

Your children notice what you eat. They learn through the models they experience at home and at school. The dietary practices they adopt when they're young carry forward into their adult lives. Who doesn't want the best for their children?

Children everywhere are at high risk for obesity and obesity-related diseases such as type 2 diabetes, high blood pressure, and coronary artery disease. Consider, too, that they'll inherit this earth someday. Encouraging your children to move toward a vegetarian diet is one way to support their health and the health of their world.

Chapter 24

Ten Simple Substitutes for Vegetarian Dishes

In This Chapter

- Making recipes dairy-free
- Using convenient meat-free ingredients
- Substituting stocks in soups

Sometimes, all that stands between a traditional meat eater's meal and a vegetarian version is one small ingredient. With the creative use of alternatives, you can modernize an old family favorite to accommodate everyone's food preferences, with delicious results!

You'll be surprised at the ways you can accomplish these substitutions. In most cases, even the most discerning foodie won't mind the difference. In fact, kitchen wizards may be stunned to discover the tricks I describe in this chapter. For example, how many people know that you can swap mashed banana for an egg in quick breads and cookies? Or that you can fool your friends by using tofu instead of egg whites in your favorite recipe for an egg salad sandwich?

You may take for granted that you need eggs, milk, and butter to make baked goods, or that spaghetti sauce and sloppy Joes require ground beef, or that vegetarian foods look, smell, or taste "different" from the foods to which you're accustomed.

It doesn't have to be that way. Use the easy techniques I show you in this chapter to assist you in your menu makeovers. I list them roughly in the order in which I think you'll use them — from those you'll use most often to those you may use less often.

In some examples, I show you easy ways to work the meat out. In others, I show how simple it is to work out other animal ingredients such as dairy products and eggs, depending on the extent to which you want to avoid animal ingredients.

Replace Eggs with Mashed Bananas

Replacing eggs with mashed bananas is a no-brainer, as well as a great way to use up those bananas that have gone from just right to too brown.

In recipes that call for an egg, you can use half of a ripe, mashed banana in place of one whole egg. This trick works best in recipes for foods such as pancakes and muffins in which you wouldn't mind a mild banana flavor. Add 1 to 2 tablespoons of a liquid — soymilk, fruit juice, or water, for example — for each egg omitted to restore the recipe to its original moisture content. Because banana tastes best in foods that are slightly sweet — pancakes, cookies, and muffins — adding sweet liquids such as vanilla soymilk or fruit juice generally works fine. If in doubt, though, use an unsweetened liquid such as any unflavored milk or water.

I don't recommend this substitution for recipes that rely on eggs to provide lift, as many cakes or a soufflé require. Flat foods such as pancakes, waffles, and cookies, and even dense baked goods such as quick breads and muffins, are perfect candidates for the banana trick, though.

Substitute Soymilk or Rice Milk for Cow's Milk in Any Recipe

You can use soymilk or rice milk (or any other kind of nondairy milk) cup for cup instead of cow's milk in most recipes. Nobody will know the difference.

Use plain or vanilla flavor in sweet dishes such as rice pudding and smoothies. Use plain soymilk or rice milk in savory recipes like mashed potatoes and cream soups.

Replacing cow's milk in recipes is an easy way to cut the saturated fat content and make foods such as puddings and soups appropriate for vegans and people who are lactose intolerant.

Use Vegetable Broth in Place of Chicken Stock and Beef Broth

A savory broth is the foundation for many soup recipes. But even vegetable soups — such as minestrone, barley, potato and leek, navy bean, and others — are often made with chicken or beef stock, rendering the recipe out of the question for vegetarians. But it doesn't have to be that way. Replacing chicken or beef broth with vegetable broth is one of the easiest ways to convert a nonvegetarian soup to a version that works for everyone.

Vegetable broths are widely available in a variety of flavors and forms. My favorites, because of their convenience and flavor, are those packaged in aseptic, shelf-stable boxes sold in many supermarkets and natural foods stores. Bouillon cubes, powders, and canned vegetable broths are also available.

Of course, you can also save money by making your own from scratch at home. You don't need a recipe, because there's no need for precision. First, save vegetable scraps — carrots, onions, celery, bell peppers, and others — in an airtight container in the freezer. When you're ready, simmer them in a large pot of water for at least an hour. Salt and pepper to taste, and drain. After the broth cools down to room temperature, you can freeze it in ice cube trays and save the little blocks of broth to use later as needed.

Plain vegetable broth works well in most soup recipes, adding just the right amount of richness and flavor. For a change of pace, try using other mixtures available in stores, including tomato and red pepper, ginger carrot, and other tasty variations.

Stir in Soy Crumbles Instead of Ground Meat

In some recipes, your guests won't know the difference between ground beef and soy crumbles. Not that your bean burritos or meatless spaghetti sauce need the extra ingredient, but if you're aiming for the flavor, look, and feel of fillings and sauces made with ground beef, meatless soy crumbles and similar products are an excellent choice. Use them in the same way you used ground beef or ground turkey in the past. You can brown the crumbles in a skillet or just toss them into a pot of sauce while it's heating.

Find soy crumbles in the freezer section of most supermarkets and natural foods stores. Boca and Morningstar Farms are two popular brands.

Make a Nondairy Version of Ricotta or Cottage Cheese

Italian favorites like lasagna, manicotti, and stuffed shell pasta are familiar crowd-pleasers. If you make a good-tasting and versatile substitute for ricotta or cottage cheese, your traditional recipes can be suitable for anyone who prefers a nondairy alternative.

Making the substitution for ricotta or cottage cheese is simple. Just mash a block of tofu — any firmness will do — and add a few teaspoons of lemon juice. Mix well. You can mix with your clean hands if you find it helps distribute the lemon juice more evenly throughout the tofu.

Then, simply proceed with your recipe as usual, incorporating your substitution.

Add fresh, spicy marinara sauce, a crisp green salad, and crusty bread to round out your revamped recipe and make a complete meal.

Take Advantage of Soy "Bacon" and "Sausage"

Soy "bacon" and soy "sausage" have come a long way since food companies first tried to replicate the texture and flavor of the real things. Now they're better than their meat counterparts, because they not only taste great but also are free of nasty nitrates and contain less sodium than the original, with little or no saturated fat and cholesterol. Many of these products are seasoned to taste just like their meat counterparts. Others may taste slightly different or have a different texture. Experiment to see which brands you like best.

Use soy-based products such as Lightlife Fakin' Bacon and Smart Bacon, or Morningstar Farms Veggie Bacon Strips, to make a new-fashioned BLT sandwich. Crumble them over a spinach salad and stir them into German-style potato salad. The same companies make veggie sausage patties and links that you can serve with pancakes and waffles. Use sausage crumbles on pizza, or make a healthier sausage biscuit by serving a patty on a toasted English muffin.

There's no going back. Trust me.

Top a Tofu Hot Dog with Vegetarian Chili

Veggie dogs are here to stay — they're that good. You can grill them, boil them, or heat them in the microwave oven. Serve them in a whole-wheat bun or sliced into a pot of baked beans. Top one with some meatless bean chili (you can find the recipe for my Cashew Chili in Chapter 11) and a scoop of freshly minced onions or slaw.

Try Lightlife Tofu Pups or Smart Dogs, Morningstar Farms America's Original Veggie Dogs, and Yves hot dogs and brats.

Why not enjoy your first health-supporting chili dog? Have two.

Create a Nondairy Substitute for Buttermilk

Buttermilk adds a tangy flavor to pancakes, biscuits, and creamy salad dressings. You can get the same effect without the saturated fat by making your own using plain soymilk or rice milk.

Just add 2 teaspoons of lemon juice or white vinegar to 1 cup of plain soymilk, rice milk, or other nondairy milk. Stir, and use this mixture cup for cup in place of regular buttermilk made from cow's milk. The acid in the lemon juice or vinegar lends the tangy flavor of buttermilk. No need to wait for the milk to clabber or curdle the way traditional buttermilk is made, though the milk may thicken slightly if left for a few days before using.

Add Flaxseeds Instead of Eggs

If you want to replace eggs with flaxseeds, here's what to do: Using a small whisk, whip 1 tablespoon of finely ground flaxseeds with 1⁄4 cup of water. This mixture replaces one whole egg. Double it if the recipe calls for two eggs.

That's all there is to it! The flaxseeds gel and bind with the other ingredients in the recipe, sticking everything together, just like an egg would do. It's also a nice way to add some heart-healthy omega-3s to your diet.

This substitution works well in a variety of recipes whether sweet or savory, including quick breads, muffins, casseroles, and loaves. Don't use it in recipes for foods that rely on eggs for lift, though, including many cakes and soufflés.

Swap Tofu for Hard-Boiled Eggs

Love an old-fashioned egg salad sandwich but want to avoid eggs? No problem.

In traditional recipes, hard-boiled eggs are mashed and mixed with mayo and a little mustard. Some recipes call for minced celery or a teaspoon of pickle relish.

Follow your favorite recipe, but replace the egg with an equal amount of firm or extra-firm silken tofu. You can even use eggless mayonnaise in place of regular mayonnaise to make this filling suitable for vegans. I include a sample Tofu Salad recipe in Chapter 11.

Chapter 25

Ten Vegetarian Lunchbox Ideas

In This Chapter

- Packing portable meals and snacks
- Taking advantage of leftovers
- Thinking outside the (lunch) box

Packing your own lunch can save you money and is likely to be healthier than a meal you buy in a restaurant or from a vending machine. But coming up with fresh ideas for tasty foods to take to work or school can be a challenge, whether you're a vegetarian or not.

Free your mind to be creative; creativity is key to packing an appealing bag lunch. Who says you must have a main course, or that lunch has to include a sandwich, an apple, and a bag of chips?

Try some of these delicious ideas and look forward to lunch again.

Almond Butter Sandwich with Granny Smith Apple Slices on Whole-Wheat Bread

Get creative — you can find alternatives to peanut butter. Almond butter is one example.

To make an almond butter sandwich with Granny Smith apple slices, start with two slices of whole-wheat bread. Top one slice with a sticky smear of almond butter. Add thinly sliced apple, finish with the second slice of bread, and cut diagonally into quarters for four little tea sandwiches. These sandwiches are perfect with a serving of leftover pasta salad and a cup of hot tea.

For a change of pace, soy butter, cashew butter, and pumpkin butter make good alternatives to peanut butter. Other ingredients you can substitute or add to the mix include pear, banana, fruit compote, or preserves.

Bean Burrito

To make a tasty bean burrito, start with a flour tortilla and add a scoop of black beans or refried pinto beans. Add chopped lettuce and tomato or a handful of tossed salad from last night's dinner. What's nice is that you can vary the burrito at the same time that you use up leftovers.

You don't have to stop there. Add leftover cooked vegetables or the odd serving of cooked rice. If you like salsa or black olives, include those, too.

Fold up one end of the tortilla, and then roll the whole thing into a neat package. Wrap tightly in aluminum foil or waxed paper. When you're ready to eat, heat and serve. If the idea of warm lettuce bothers you, leave it out or add it later.

My favorite burrito recipe includes mashed sweet potatoes as an option. Find the recipe in Chapter 12.

Easy Wraps

Like burritos and pita pocket sandwiches, wraps are a handy sandwich style because they envelope loose ingredients and help to keep them from falling out.

Use a large flour tortilla or a Middle Eastern, lahvash-style flatbread to make a hummus wrap. (See Chapter 10 for a recipe to make your own hummus.) I like to use hummus as the base because you don't need to warm it before eating, and its stickiness helps bind it with other ingredients, such as leftover salad, chopped tomatoes, grated carrots, black olives, sprouts, and cooked rice.

Sliced low-fat cheese or tempeh strips also make good bases for a wrap sandwich. Add your favorite ingredients, leaving a few inches free of filling on the far end of the bread. Begin rolling from the end closest to you. Extra filling will be squeezed onto the bare end that you strategically left for that purpose. A tightly wrapped sandwich holds together well, but rolling it in waxed paper or aluminum foil helps it survive intact in a backpack or lunchbox until mealtime.

Fresh Fruit Salad with Nonfat Vanilla Yogurt

In contrast to the granola parfait (see the next section for details), this light, one-dish meal is mostly fresh fruit.

Use whatever's in season — melon balls, blueberries, peach slices, strawberries, or cut-up apples with raisins and cinnamon. Toss fruit with a simple dressing made by mixing nonfat, plain, vanilla, or lemon yogurt with a few tablespoons of orange juice and 1 to 2 teaspoons of honey. This is a no-measure recipe. Use enough yogurt to make a creamy dressing to coat the fruit, and add just enough juice and honey to thin and sweeten the dressing to your liking.

A big bowl of fresh fruit may be all you need during a day spent sitting at your computer. Fruit is a nutritious, filling, and hydrating choice.

Granola Parfait

Make a granola parfait by using plain or flavored nonfat yogurt and your favorite granola. Simply alternate layers of yogurt and cereal in a narrow glass or plastic cup with a snug lid. This works especially well with plain, vanilla, or lemon yogurt, and you can substitute soy yogurt for yogurt made with cow's milk, too.

For extra pizzazz, add sliced strawberries, bananas, peaches, melon, or blueberries. Add a shot of nutrients by sprinkling a tablespoon of wheat germ on top.

This works well for a portable breakfast on the run, too.

Leftovers from Last Night's Dinner

Yesterday's vegetable curry or pasta primavera makes a great lunch tomorrow. Round out the meal with a chunky slice of good bread and a piece of fresh fruit. Other tasty vegetarian dinners that reheat and transport well for tomorrow's lunch include

- Beans and rice
- Macaroni and cheese casserole
- Spinach pie (Greek spanakopita)
- Vegetable lasagna
- Vegetable pizza
- Vegetable stir-fry
- Vegetarian chili

Pita Pocket Sandwich

The nice thing about pita pockets is the way they keep crumbly or messy ingredients all together and off your lap. Buy whole-wheat pita bread instead of white: It tastes better and is more nutritious.

Fill a pita pocket with tofu salad and baby spinach leaves, hummus, and shredded carrots; baked beans and coleslaw; or tossed salad with a drizzling of vinaigrette dressing. You can find the recipe for hummus in Chapter 10 and one for tofu salad in Chapter 11.

Use waxed paper to wrap your pockets tightly and keep the ingredients intact until you're ready to eat.

Soup Cup

Single-serving soup cups are light and easy to carry. All they require when you're ready to eat is a source of hot water and a spoon. Peel back the top, add piping hot water, stir, and enjoy.

Complement a steamy soup cup with a few whole-grain crackers and a cooling side of fresh fruit salad.

Another variation on the theme: Try a hot cereal cup when you don't have time to eat before you leave home in the morning. Tote it to the office and enjoy it as you ease into your day. Hot oatmeal and multigrain hot cereals are appropriate and healthful alternatives for any meal.

Vegetarian Chili

Like hot soup or cooked cereal, vegetarian chili is hearty and portable. Carry it in a squatty Thermos or in a glass container that you can place in a microwave oven for reheating.

Whole-grain crackers make good scoops for hot chili in lieu of a spoon. Cucumber slices with ranch dressing dip round out this easy meal.

I include my favorite vegetarian chili recipe — Cashew Chili — in Chapter 11.

Veggie Burger on a Bun

Assemble your burger before you leave home. Heat the patty and place it in a whole-grain bun. Pack lettuce and a tomato slice separately and add them when you're ready to eat. Packing the burger fully assembled risks a soggy sandwich.

Add mustard and ketchup, or vary it by topping your burger with salsa or a scoop of corn relish or mango chutney. Pack a side of crispy carrots and bell pepper slices.

Veggie burgers on a bun even taste good cold. Many brands and varieties are available to choose from to suit most anyone's taste. Experiment with several to find those you like the best.

Index

• Numerics •

• A •

• D •

• E •

• F •

• G •

• H •

• I •

• M •

• N •

• S •

• T •

• U •

• V •

• W •

• X •

• Y •

• Z •

Notes